RECENT TRENDS IN COMMUNICATION AND ELECTRONICS

PROCEEDINGS OF THE INTERNATIONAL CONFERENCE ON RECENT TRENDS IN COMMUNICATION AND ELECTRONICS (ICCE-2020), GHAZIABAD, INDIA, 28-29 NOVEMBER, 2020

Recent Trends in Communication and Electronics

Editors

Sanjay Sharma

KIET Group of Institutions, Delhi-NCR, Ghaziabad, U.P., India

Astik Biswas

Stellenbosch University, South Africa

Brajesh Kumar Kaushik

Indian Institute of Technology, Roorkee, India

Vibhav Sachan

KIET Group of Institutions, Delhi-NCR, Ghaziabad, U.P., India

CRC Press
Taylor & Francis Group
Boca Raton London New York

CRC Press is an imprint of the
Taylor & Francis Group, an **informa** business

A BALKEMA BOOK

CRC Press/Balkema is an imprint of the Taylor & Francis Group, an informa business

© 2021 Taylor & Francis Group, LLC

Typeset by Integra Software Services Pvt. Ltd., Pondicherry, India

Library of Congress Cataloging-in-Publication Data

A catalog record has been requested for this book

Published by: CRC Press/Balkema
            Schipholweg 107C, 2316XC Leiden, The Netherlands
            e-mail: enquiries@taylorandfrancis.com
            www.routledge.com – www.taylorandfrancis.com

ISBN: 978-1-032-04572-6 (Hbk)
ISBN: 978-1-032-04573-3 (Pbk)
ISBN: 978-1-003-19383-8 (eBook)
DOI: 10.1201/9781003193838

Table of contents

Preface

Dear Delegates,

The AICTE Sponsored 4th International Conference on Recent Trends in Communication & Electronics (ICCE-2020) was the 4th Conference of its series organized by Elcctronics and Communication Engineering Department of KIET Group of Institutions, Delhi- NCR, Ghaziabad, India. It was considered as a forum to bring together researchers, scientists, academicians, graduate students and Electronics and Communication Engineers around the globe to explore and discuss new science, technology and engineering ideas and achievements of Electronics & Communication Engineering and allied areas.

The motive of this conference was of great importance. The conference attracted many participants working in various fields of engineering: IOT, Machine Learning and Artificial Intelligence, Opto-electronics, etc. The success of the conference inspired the organizers to turn the conference into series of event. The conference comprised of various parallel sessions including keynote sessions. Each session was addressed by outstanding experts who highlighted the recent advances in various facets of Electronics and Communication Engineering. It also offered the budding researchers an opportunity to present their work before eminent experts of their fields.

The organizing committee is extremely gratified by the tremendous response to the call for papers. The conference is approved by CRC Press (Taylor & Francis Group) and Sponsored by Govt. of India, AICTE New Delhi. The proceedings will be published online by CRC Press (Taylor & Francis Group). The conference received more than 250 papers. The acceptance rate of the conference is about 70%. Speakers and participants across the globe participated in the conference. Keynote speakers from more than nine countries delivered their expert lectures on latest technologies.

Furthermore, I express my heartfelt thanks to the Staff and Management of the Institute for their cooperation and support, and especially Dr. A Garg, Honorable Director, KIET Group of Institutions, Delhi-NCR, Ghaziabad, Dr. Manoj Goel, Honorable Joint Director, KIET Group of Institutions, Delhi-NCR, Ghaziabad, and all members of the Program Committee and the Organizing Committee for their hard work in preparing and organizing the conference. Last but not least, I thank Taylor & Francis Group, CRC Press/Balkema for its professional assistance and particularly Dr. Janjaap Blom, Senior Publisher, Taylor & Francis Group, CRC Press/Balkema who supported this publication.

Chief Editor

Technical Chair Committee

- Dr. Balaanand Muthu, *V.R.S College of Engineering & Technology, Tamilnadu, India*
- Dr. Madhuri Yadav, *Govt College Rohtak, Haryana, India*
- Dr. Vijay Anant Athavale, *PIET, Panipat, Haryana, India*
- Dr. M S Choudhary, *DTU, New Delhi, India*
- Dr. Rajmohan Pardeshi, *Karnatak Arts Science and Commerce College Bidar, Karnatak, India*
- Dr. Purnima K Sharma, *Sri Vasavi Engineering College, Andhra Pradesh, India*
- Dr. Sandeep Dahiya, *BPS Women University, Khanpur-Kalan, Sonipat, Haryana, India*
- Dr. Laxman Singh, NIET, *Gr. Noida, Uttar Pradesh, India*
- Mr. Sanjeev Kumar, *J.C. Bose University of Science And Technology YMCA, Faridabad, Haryana, India*
- Mr. Nitin Arora, *Indian Institute of Technology, Roorkee, India*
- Dr. Amrita Rai, *G L Bajaj Institute of Technology & Management, Greater Noida, Uttar Pradesh, India*
- Mr. Kuldeep Narayan Tripathi, *IIT Roorkee, India*
- Dr. Amit Sehgal, *G L Bajaj Institute of Technology & Management, Greater Noida, Uttar Pradesh, India*
- Dr. Akansha Singh, *Amity University, Noida, Uttar Pradesh, India*
- Dr. Pushpa, *IEC College of Engg. & Technology, India*
- Prof. Poonam Sharma, *Amity University, Noida, Uttar Pradesh, India*
- Dr. Amit Kumar Manocha, *Maharaja Ranjit Singh Punjab Technical University, Bathinda, India*
- Dr. Rajendra Kumar Sharma, *Faculty of Engineering And Technology, Agra, India*
- Dr. Anuradha Dhull, *The Northcap University, Gurugram, Haryana, India*
- Dr. Prasanna Kumar Singh, *Noida Institute of Engineering & Technology, Greater Noida, India*
- Dr. Pallavi Gupta, *Sharda University, Greater Noida, India*
- Dr. Mohammad Farukh Hashmi, *NIT Warangal, Telangana, India*
- Dr. Himanshu Shekhar Pradhan, *NIT Warangal, Telangana, India*
- Dr. Amarjit Kumar, *NIT Warangal, Telangana, India*
- Dr. Gopi Ram, *NIT Warangal, Telangana, India Dr. Koushik Guha, NIT Silchar, Assam, India*
- Dr. Banani Basu, *National Institute of Technology, Silchar, India*
- Dr. Arnab Nandi, *National Institute of Technology, Silchar, India*
- Dr. Krishnamoorthy K, *National Institute of Technology Karnataka, Surathkal, India*
- Dr. Aniruddha Chandra, *NIT Durgapur, India*
- Prof. V. Rama, *National Institute of Technology, Warangal, India*
- Dr. Ekta Goel, *National Institute of Technology, Warangal, India*
- Dr. R Boopathi Rani, *National Institute of Technology Puducherry, India Dr. Prakash Kodali, National Institute of Technology Warangal, India*
- Dr. Gaurav Singh Baghel, *NIT Silchar, India*
- Dr. Harpal Thethi, *LPU, Phagwara, Punjab, India*
- Dr. Ajay Yadav, *Bennett University, Greater Noida, India*
- Dr. Kapil Tyagi, *JJIT Noida, India*
- Prof. Rajesh Kr. Tyagi, *AMITY University, Gurgaon*
- Dr. Bajrang Bansal, *JIET, Noida*

Organizing Committee

Chief Patron
Shri Sarish Aggarwal
Honorable Chairman, *KIET Group of Institutions, Delhi-NCR, Ghaziabad, U.P.*

Patron
Dr. A Garg
Director, *KIET Group of Institutions, Delhi-NCR, Ghaziabad, U.P.*

Co - Patron
Dr. Manoj Goel
Joint Director, *KIET Group of Institutions, Delhi-NCR, Ghaziabad, U.P.*

Dr. Anil Ahlawat
Dean Academics, *KIET Group of Institutions, Delhi-NCR, Ghaziabad, U.P.*

Editors
Dr. Sanjay Sharma, *Dean (R&D) & HoD (ECE), KIET Group of Institutions, Delhi-NCR, Ghaziabad, U.P.*
Dr. Astik Biswas, *Dept. Of Electrical and Electronics Engineering, Stellenbosch University, South Africa*
Dr. B K Kaushik, *Department of Electronics and Communication Engineering, IIT Roorkee*
Dr. Vibhav Kumar Sachan, *Additional HoD, ECE, KIET Group of Institutions, Delhi-NCR, Ghaziabad, U.P.*

Convener
Dr. Parvin Kumar Kaushik
Dr. Ruchita Gautam
Dr. K K Singh
Technical Program Committee Chair
Dr. Sachin Kumar
Prof. KYUNGPOOK, *National University Republic of Korea*
Dr. Ashok Prajpati
Editor in Chief, *INT. Journal, Michigan, USA*
Dr. Shelly Garg
Prof. & Head GNOIT, *Greater Noida*

Publication Chair
Dr. Parvin Kumar Kaushik
Dr. Abhishek Sharma
Dr. Vaibhav Bhushan Tyagi
Mr. Shubham Shukla

Organizing Chair

Dr. Dharmendra Kumar

Dr. Pravesh Singh

Dr. Richa Srivastava

Mr. Sachin Tyagi

Mr. Mohit Tyagi

Sponsored & Publicity Chair

Dr. Himanshu Sharma

Mr. Vipin Kumar

Ms. Vaishali Kikan

Ms. Diksha Singh

Ms. Nikita Goel

Ms. Pooja Tyagi

Acknowledgement

This database contains the proceedings for the 4th International Conference on Recent Trends in Communication & Electronics (ICCE-2020). The database contains 114 papers from various regions across India. The proceedings were edited by Dr. Sanjay Sharma (Dean (R&D) & HoD (ECE), KIET Group of Institutions, Delhi-NCR, Ghaziabad, U.P.), Dr. Astik Biswas (Dept. of Electrical and Electronics Engineering, Stellenbosch University, South Africa), and Dr. B. K. Kaushik (Department of Electronics and Communication Engineering, IIT Roorkee).

ICCE 2020 was held at the Electronics and Communication Engineering Department, KIET Group of Institutions, Delhi-NCR, Ghaziabad, U.P. during November 28 and 29, 2020. The main theme of the ICCE series was Recent Trends in Communication & Electronics.

This database is initiated by the CRC Press (Taylor & Francis Group) and Sponsored by AICTE, Govt. of India, New Delhi.

Recent Trends in Communication and Electronics – Sharma et al. (Eds)
© 2021 Taylor & Francis Group, LLC, ISBN 978-1-032-04572-6

Wavelet transform based enhancement of medical images

Renu Sharma
Research Scholar Jaypee Institute of Information Technology, Noida, India

Madhu Jain
Associate Professor Jaypee Institute of Information Technology, Noida, India

ABSTRACT: Now-a-days Radiographic image enhancement is a vital necessity. This paper proposed an algorithm which improves visual quality of X-Ray and MRI image. Dual Tree Complex Wavelet Transform (DTCWT) is used for decomposition of Image. Since edge information is present in High Sub-band Image, so it requires further processing for quality improvement. Quantitative analysis is done by computing Peak Signal to Noise Ratio (PSNR), Mean Square Error (MSE), Structural Similarity Index Measure (SSIM) and NIQE index

1 INTRODUCTION

Improving the visual aspect of medical images is required for effective diagnosis of patient. In this field, medical image enhancement plays vital role. For providing effective treatment to the patient, resolution and contrast of image must be effective. Enhancement can be done disease specific or it can be for general quality improvement. Radiographic images such as X-Ray play an essential role in diagnosing the patient. This paper proposed an algorithm which improves the resolution along with contrast of the image.

Earlier, contrast enhancement was performed by spatial domain based methods which includes Histogram Equalization (Fu et al. 2000), Adaptive Histogram Equalization (Fu et al. 2000), (Li et al. 2010) and Contrast Limited Adaptive Histogram Equalization (CLAHE) (Li et al. 2010). On the other hand, transform based methods are curvelet transform, shearlet transform, Fourier transform etc. Dual Tree Complex Wavelet Transform (DTCWT) (Zafar et al. 2014), (Sharma et al. 2015) based resolution and contrast enhancement is proposed in this paper. Wavelet Transform is preferred as compared to Fourier Transform because former provides time and frequency localization.

2 PROPOSED WORK

Proposed algorithm is divided into four stages. Firstly, pre-processing stage in which input image is converted to grayscale image. Further even number of rows and columns are checked, for applying DTCWT. Second stage is for enhancing contrast of the pre-processed image. Singular value decomposition is used for this purpose. Third stage is for resolution enhancement. Lanczos interpolation (Zafar et al. 2014) is used for decomposition of image. Decomposition factor taken for this algorithm is 2. Lastly, visual quality of the output image is improved using Gaussian and median filter.

Figure 1 demonstrates the block diagram of the proposed work. Singular Value decomposition is used for contrast enhancement. It is a tool which decomposes the image into three

DOI: 10.1201/9781003193838-1

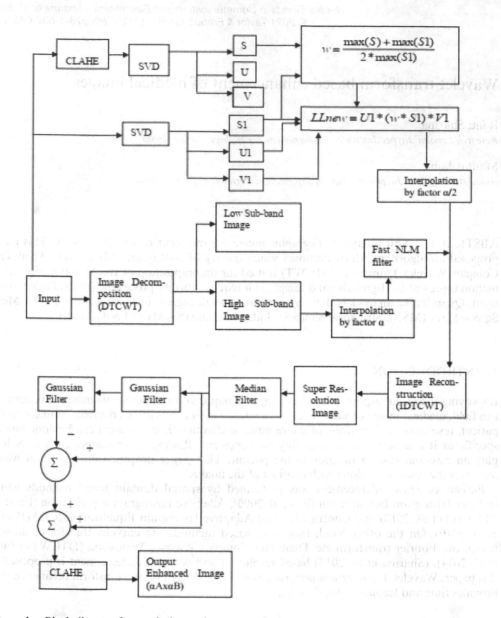

Figure 1. Block diagram for resolution and contrast enhancement.

equal size images such as S, U and V. A weighting function is calculated. To remove the artifacts, high sub-band images are interpolated using Lanczos interpolation.

3 RESULTS & DISCUSSION

Radiographic images such as X-Ray and MRI image are considered. Brain MRI image is taken from MR technology information portal. Figure 2(a) shows the 512 x 512 original image. Figure 2(b) shows the 128 x 128 down-sampled input image. Figure 2(c) shows the DTCWT-NLM-SVD-RE method based image. Figure 2(d) DTCWT-Fast NLM-CLAHE method based image. As seen from the images, visual quality of Figure 2(d) is much better than other three images.

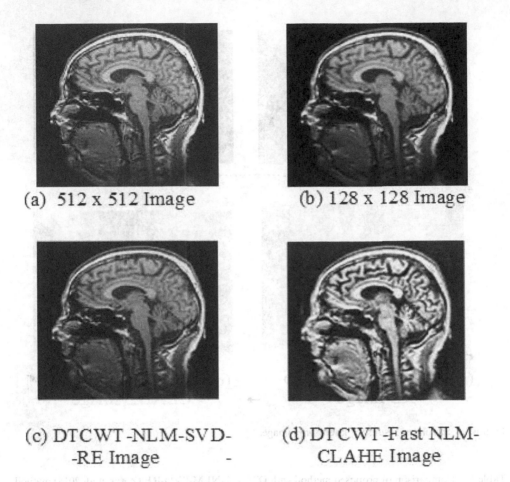

(a) 512 x 512 Image   (b) 128 x 128 Image

(c) DTCWT-NLM-SVD-
-RE Image

(d) DTCWT-Fast NLM-
CLAHE Image

Figure 2.   Simulated results for brain MRI image.

Figure 3 shows the simulated result for knee X-Ray image, Figure 3(a) presents original image having size 1954 x 2256, Figure 3(b) presents the input image having size 977 x 1128, Figure 3(c) presents the DTCWT-NLM-SVD-RE method based image and Figure 3(d) shows the DTCWT-Fast NLM-CLAHE method based image. Proposed method showed better visual quality as compared to other existing methods.

3.1 *Parameter calculation*

For quantitative analysis of the proposed work, several parameters are calculated. Parameters are as follows:

$$PSNR = 10 * \log\left(\frac{MAX_I^2}{MSE}\right) \qquad (1)$$

where MAX$_I$ denotes the highest intensity available in the image. If we are considering gray scale image, then its value is 255. MSE in Eqn. (1) is mean square error.

Table I shows the comparison of proposed algorithm i.e., DTCWT-Fast NLM-CLAHE meth- od with DTCWT-NLM-SVD-RE method. This table shows the simulated result for various parameters such as PSNR, MSE, SSIM (Zhou et al. 2002) and NIQE index. It can be viewed that proposed meth- od DTCWT-Fast NLM-CLAHE has better result as compared to DTCWT-NLM-SVD-RE (Zafar et al. 2014) method.

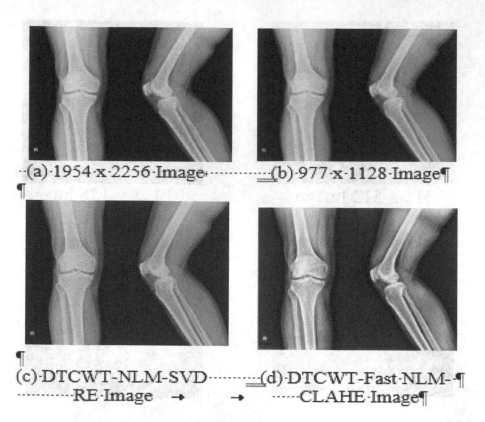

(a) 1954 x 2256 Image   (b) 977 x 1128 Image

(c) DTCWT-NLM-SVD-   (d) DTCWT-Fast-NLM-
RE Image   →   →   CLAHE Image

Figure 3. Simulated results for knee X-ray image.

Table 1. Comparison of proposed method with DTCWT-NLM-SVD-RE (Zafar et al. 2014) method.

IMAGES	METHODS PARAME TERS	DTCWT-NLM-SVD-RE [1]	DTCWT-Fast NLM-CLAHE (Proposed)
BRAIN MRI	PSNR	26.5500	**30.0080**
	MSE	0.00200	**0.00090**
	SSIM-Index (Zhou et al. 2002)	0.83000	**0.85100**
	NIQE	3.66700	**3.98900**
KNEE X-RAY	PSNR	26.7534	**32.5136**
	MSE	0.00210	**0.00050**
	SSIM-Index (Zhou et al. 2002)	0.95250	**0.89970**
	NIQE INDEX	4.48270	**3.74530**
ELBOW X-RAY	PSNR	26.2240	**37.4470**
	MSE	0.00230	**0.00018**
	SSIM-Index (Zhou et al. 2002)	0.95150	**0.87430**
	NIQE INDEX	5.06860	**3.95900**

4 CONCLUSION

For providing effective treatment to the patient, visual quality of the radiographic image must be good. Proposed algorithm considered Brain MRI, Knee and Elbow X-ray images. For quantitative analysis of the proposed work different parameters are calculated. Simulated results showed that proposed DTCWT-Fast NLM-CLAHE method is better. For qualitative analysis, simulated result images are shown in Figure 2 and Figure 3. Visual appearance of proposed simulated image is good as compared to other image. By using Fast non local mean (NLM) filter, computation time of the algorithm is reduced compared to previously existed algorithms.

5 FUTURE SCOPE

Image enhancement can be used for other practical applications. Proposed work can be used for detecting exact shape and size of tumor, for detecting fractured bone etc. In Radiographic Image different types of noise are also present which can be reduced by using other filters.

REFERENCES

Zafar I. M., Abdul G., Masood S.A., Mohsin R. M. and Umar K. 2014. Dual-tree complex wavelet trans form and SVD based medical image resolution enhancement, Elsevier, Signal processing.10: 430–437.

Zhou W. and Alan C. B. 2002. A universal image quality index, IEEE Signal Processing Letters. 9: 81–84.

J.C. Fu, H.C. Lien and S. T. C. Wong. 2000. Wavelet-based histogram equalization enhancement of gastric sonogram, Elsevier, Computerized Medical Imaging and Graphics. 25: 59–68.

Li Y., Yanmei L and Hailun F. 2010. Study on the methods of image enhancement for liver CT images, Elsevier, Optik. 121: 1752–1755.

Sharma R, Chopra P. K. 2015. Enhancing X-Ray Images Based on Dual Tree Complex Wavelet Transform, International Journal of Science, Engineering and Technology Research (IJSETR). 4: 2179–2184.

Recent Trends in Communication and Electronics – Sharma et al. (Eds)
© 2021 Taylor & Francis Group, LLC, ISBN 978-1-032-04572-6

Crop disease prediction and prevention system using genetic analysis

V. Kakulapati & M. Dhanaraju
Sreenidhi Institute of Science and Technology, Yamnampet, Ghatkesar, Hyderabad, Telangana

ABSTRACT: Farming is the art and science of flourishing the plant and other crops for rising economic genetic algorithm. However, the diseases which influence the crops have made the impact of agriculture production. This disease commonly gets infected by pathogens such as bacteria, fungus and virus. In order to address the above issue, a novel recommendation system for earlier detection of crop diseases. The crop disease data is gathered from disparate resources including social web pages and evaluated through the Hadoop system. In this work, we came up with a framework which gives a better investigation of various crop diseases with help of classification algorithms like genetic algorithm which bring about solutions to the development problem. In this work, we utilized genetic algorithm is implement for finding out the optimal solution and practice by solving problems in disease detection. The conclusion of the proposed work is providing a better recommendation solution to avert the disease in the initial stages and yield the better agricultural production.

Keywords: agriculture, production, prevent, disease, state, optimization

1 INTRODUCTION

Generally, India is known as a cultivated country; about 70% of the people depend on farming which has the most compelling place in India in points of GDP, exports, food safety, livelihood and the comprehensive economic growth. Threats to agriculture are multiple as well as complex, results in very serious disaster overtime. Studies on agriculture critical situation have brought out that desolation of the land, water and other resources. Various diseases of the plant cause by of micro-organisms (Samiksha et.al. *2017*) or plant pathogen resulting the growth of plants. Cure of these diseases are difficult due to damage to plant prior to utilizing control measure. To overcome this detecting the diseases in the early stage and take immediate measures should be done. Before giving recommendation for various diseases, accurate detection of disease is essential.

Pathogen which is an agent causing disease engenders this disease (Patil et.al. 2011). Very often, pests or disease is identified on the plant leaves and stems. The vital aspect of the lucrative cultivation of crops is the identification of the pest or disease in the plants, leaves, stems, the pest percentage or disease occurrence, pest symptoms or the hit of the disease. It also encompasses a survey on divergent the disease categorization technique. This survey gives plant leaf disease data which is an influential and cardinal tool for the detection of the disease in plant leaves done by adapting a genetic algorithm.

GA (genetic algorithm) (Singh et al. 2017) is enforced over the problem field where the outcome is very arbitrary and the process of the outcome encompasses complex interrelated modules. Also, a genetic algorithm is very fitting for such class of the problem where the problem specification is very difficult to put together.

DOI: 10.1201/9781003193838-2

In accordance with the prediction result (Luo et al. 2009), precise and accurate preventive methods, standards and appropriate treatment measures are made by decision-makers and users in the pursuit of attaining the utmost profitable benefits with the investment minimum capital.

Farmers are facing many obstacles in the cultivation of crops. These crops are being affected by diverse diseases. We have to collect the data of different diseases causing pathogens from the different websites. And have to perform analysis of the data for the better recommendation of pesticides for crops.

At present, no scenario of implementing the proper measures and precautions in order to bar crop diseases and advise proper measures to the farmers and so there is a need of facilitating such scenario digitally as manual proposals overtaken by the digital world by providing everything in detail on online platforms itself. There is a situation in front of us to provide a methodology to the farmers in implementing some proper measures towards their crops so that by some dependent attributes.

2   RELATED WORKS

There are numerous amounts of problems encountered by Indian agriculture to bloat productivity (Subbarao 2002). Many researchers implemented new techniques in agriculture, cultivars and cultivation of the crop, and control techniques of pest. There is no expected result in the farming community in terms of obtaining the upper bound yield (Reddy et.al. 2005). Prototypes in which the entire paddy sample (Radhiah 2012) will pass during the RGB computation prior to carry on to the binary translation. This process of disease identification will detect Diseases like Paddy Blast, Brown Spot, and Narrow Brown Spot.

Automatic leaf ailment (Lalitha et.al. 2015) recognition and categorization for noticing the disease and apprising the cultivator the category of disease is made feasible by this prototype which will also facilitate immediate actions to safeguard and to protect the healthy plants and to come up antidotes to save plants which are infected.

The GNB (Gaussian Naïve Bayes) and RARM (Rapid Association Rule mining) are utilized by a concurrent decision-making system (Tripathy et al. 2011), which can forecast pests through a different procedure. Based on the components used in the system, they held some work to make predictions of when pests will attack and infect the plant depending on different kinds of information.

Recently using the Principal Component Analysis and Partial Least Square, YRFHB (yellow rust Fusarium head blight) which are also called as fungous ailments of wheat and barley (Lu et.al. 2018) were identified and detected. Applying PCA (Whetton et.al. 2018) on healthy and contaminated yields at dissimilar development levels and examined chronological model and sequential automatic association. The outcomes are recommended by utilizing PLS for accurate prediction on every level of growth. They utilized PLS (Whetton et.al.2018) regression with disappear-one-out cross-validation for ailments. It was concluded with outcomes exhibits that the same regression model extended for fungal diseases in wheat berry can be useful to forecast like ailments in barley.

Diverse categorization methods can be employed for the classification of plant, leaf disease. For example, KNN is considered to be the most appropriate in addition to austere of all classification prediction algorithms used. One of the drawbacks of SVM is that it is quite hard and difficult to determine and estimate the optimal parameters in SVM due to the training data is not linearly distinguishable. (Luo et al. 2009).

To make leaf ailment recognition (Sivasangari et al. 2017) and retrieval recommendations utilizing image processing techniques which explores the features are specified concerning leaf ailment and its precautionary measures. The genetic algorithm used three operators like selection, cross over and mutation. The feature selection is carried out using Genetic algorithm (GA). The extracted features are inserted into the selection process.

3 FRAMEWORK

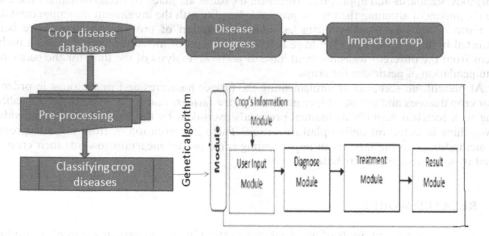

Figure 1.   The prosed frame work of crop disease analysis.

Figure 1 shows that system framework provides us how the system is designed and gives us the brief details about the entire architecture of the proposed work.

3.1   *The crop disease database*

For predicting the crop-disease, taken the dataset of the plant leaf images. In this work, analyze above 54 thousand plant leaf images. These images are classified with 38 class labels and every class label belongs to a crop- disease pair.

3.2   *Multi-linear regression*

It is a statistical tool which investigative on the relation of multiple autonomous variables to a needy variable. When multiple variables relate to dependent variable is recognized, taking all the autonomous variables information and utilizing it to formulate a more genuine and accurate forecast.

3.3   *Genetic algorithm*

These applications usually incorporate the target of the decision, cataloguing, optimization, and reproduction of solid problems. The excellence of a GA is measured and estimated with regard to the speed, accuracy, and domain of applicability. The advantage of GA in solving difficult predicaments in the domain of farming systems is recognized, analyzed and presented here."

Genetic Algorithm works with a set of individuals, representing possible a solution to the above task (problem) is as follows. In view of the expecting and desire outcome, an evaluation for the individual, the selection principle is applying.

The steps involved in the genetic algorithm are as follows:

- Initialization of the population of random solutions
- Determining the best solution in the population
- Starting the Loop
- Selection of two parents from the population
- Making two children from parents
- Placing children into the population
- Making and placing an immigrant into the population

- Checking if a new best solution exists
- Ending the loop
- Returning the best solution found

4 METHODOLOGY

We are experimenting two modules in our program. One of them is a crop module which consists of crop details and another one is a recommendation module which consists of crop diseases and their recommendations.

Pre-processing of the data is removing the undesirable things from data as Duplicate fields and empty fields, etc. After facilities, the input files, take the required fields needed and remove the unwanted fields. Here we are using linear regression to predict the outcome based on the effect of percentage. Then we apply multi linear regression algorithm based on more than one field i.e. crop diseases and the effect of percentage. We apply a genetic algorithm to optimize the predicted variable which we got in the regression model. We implemented step by step procedure of the genetic algorithm which includes Initialization, fitness function, selection, cross-over, mutation and termination. The fitness function tells how to fit an individual is and the probability that an individual will be selected for optimizing the outcome.

Again, we are applying linear regression and multiple regressions for the prediction whether our outcome is optimized or not.

5 EXPERIMENTAL EVALUATION

Table 1. A sample of crop disease dataset.

S.No.	Crop name	Disease	Percentage	outcome
1	1	1001	28.23	1
2	1	1002	50.11	1
3	1	1003	32.54	0
4	1	1004	16.12	1
5	1	1005	19.19	0
6	1	1006	20.01	0
7	2	2001	25.98	1
8	2	2002	36.79	1
9	2	2003	76.98	1
10	2	2004	78.82	0
11	2	2005	45.01	1
12	2	2006	88.87	0
13	2	2007	75.54	1
14	2	2008	55.45	0
15	2	2009	21.12	0

Chilli-1 ->
Phytophthore root rot-1001
Basai stem rot -1002
Bacterial wilt-1003
Anthracnose-1004
Viral disease-1005
Root knot nematode-1006

Tomato-2->
Bacterial wilt-2001

Basai stem rot -2002
Root knot nematode-2003
Late blight-2004
Classification   0   1
                47   42
Optcutoff : 0.6755667
Mis classification error [1]   0.3871

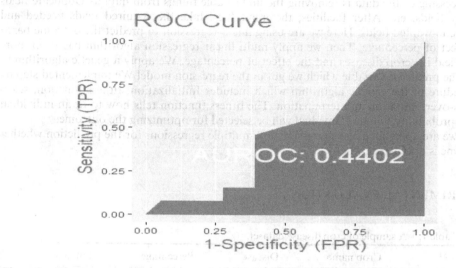

Figure 2.   The Roc curve of sensitivity and specificity.

$ Concordance
[1] 0.4316239
$discordance
0.5683761
$Tied
[1] 0
$ pairs
[1] 234
Confusion Matrix
            0   1
      0 18   12
      1   0   1
Coefficients
(Intercept)   out
42.959   3.084

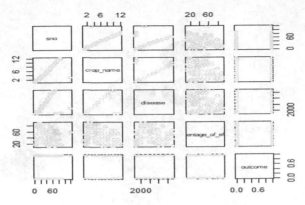

Figure 3. Multi regression analysis of crop disease.

Name of the crop   Infection   effect percentage
1.158947   1.023455   1.156009
predict (model, new data=data. frame (effect per =15.33, name of crop=1,infection=1004))
1   2
42.95851 46.04238

GA settings:
Category = real-valued
Size of population = 50
Quantity of generations = 100
Superiority = 2
Probability of Crossover = 0.8
Probability of Mutation = 0.1
Search domain = x1
Minimum 10
Maximum 150
GA results:
Iterations = 100
Value of Fitness function = 149.8353
Solution = x1
[1,] 149.8353
    Residuals:
    Minimum   1Quartile   Median   3Quartile   Maximum
    -0.5531   -0.4663   -0.4079   0.5283   0.6081
Coefficients:

	Estimate	Std. Error	t value	Pr(>|t|)
(Intercept)	4.208e-01	1.950e-01	2.158	0.0338 *
Name of crop	3.835e-04	1.684e-02	0.023	0.9819
Infection	-5.819e-06	2.055e-05	-0.283	0.7777
Effect				
percentage	1.611e-03	2.595e-03	0.621	0.5364

Codes of Significance: 0 '***' 0.001 '**' 0.01 '*' 0.05 '.' 0.1'' 1
Standard error Residual: 0.5094 on 85 degrees of freedom
Multiple R-squared: 0.005676,   Adjusted R-squared: -0.02942
F-statistic: 0.1617 on 3 and 85 DF, p-value: 0.9218

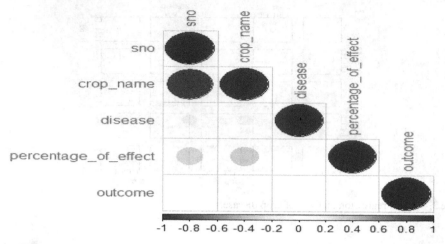

Figure 4.　The percentage of effect of crop disease.

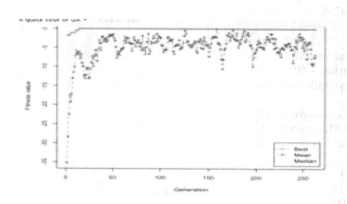

Figure 5.　The mean, median and standard deviation analysis of crop disease data.

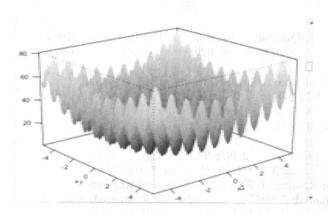

Figure 6.　Optimization technique by using GA.

6 CONCLUSION

This work provides the best digital e-care facility for the field of agriculture in order to provide proper measures to the farmers of various crop fields based on the crop's specification and the extent of crop effect which provides the best recommendation system for pests and also gives the information to what extent the crop may be healthier with proper prediction techniques. Digitizing such crop related diseases and some helpful instructions will benefit the farmers as well as crop production for a healthy and abundant improvement in agriculture especially for nations such as India.

7 FUTURE SCOPE

By enhancing the present system with additional features and good interface it can form a base station for the field of agriculture which is the backbone of our country. It can also provide proper preventive measures during drought conditions for the formers. In addition to this, it may suggest proper pesticide instructions, suggestions and several other measurements with a friendly interface by means of some private and government undertakings.

REFERENCES

Samiksha Bhor, Shubha Kotian., Aishwarya Shetty, and Prashant Sawant *Feb*, 2017 "Developing an agricultural web portal for crop disease prediction using data mining techniques", *International Journal of Recent Scientific Research, Vol. 8, Issue, 2*, pp. 15507-15509, ISSN: 0976-3031.

Patil J.K, Kumar R, June, 2011 "Advances in Image Processing for Detection of Plant Diseases", Journal of Advanced Bio Informatics Applications and Research ISSN 0976-260,4Vol 2, Issue 2, pp 135–141.

Vijai Singh, Misra A.K, 2017, Detection of plant leaf diseases using image segmentation and soft computing techniques, Information Processing in Agriculture, Volume 4, Issue 1.

Luo, J., Huang, W., Wang, J. and Wei, C., 2009, in IFIP International Federation for Information Processing, Volume 294, Computer and Computing Technologies in Agriculture II, Volume 2, eds. D. Li, Z. Chunjiang, (Boston: Springer), pp. 937–945.

Subbarao, I. V., Dec, 2002, Indian agriculture – Past laurels and future challenges. In *Indian Agriculture: Current Status, Prospects and Challenges*, Commemorative Volume, 27th Convention of Indian Agricultural Universities Association, pp. 58–77.

Reddy P K and Ankaiah, R., June 2005, A framework if information technology-based agriculture information dissemination system to improve crop productivity, Current Science, vol. 88, Num.12, pp. 1905–1913.

Radhiah Binti Zainon, Dec 2012" Paddy Disease Detection System Using Image Processing "

K. Lalitha, K. Muthulakshmi, A. Vinothini, 2015 "Proficient acquaintance-based system for citrus leaf disease recognition and categorization", International Journal of Computer Science and Information Technologies, Vol. 6 (3), 2519–2524.

Tripathy A K, Adinarayana J, Sudharsan Merchant D, SN, Desai UB,. 2011 "Data Mining and Wireless Sensor Network for Agriculture Pest/Disease Predictions", World Congress on Information and Communication Technologies, pp.1229–1234, 2011.

Lu J, Ehsani R, Shi Y, de Castro AI, Wang S, 2018, Detection of multi-tomato leaf diseases (late blight, target and bacterialspots) in different stages by using a spectral-based sensor. Sci Rep; 8:2793.

Whetton RL, Hassall KL, Waine TW, Mouazen AM, 2018, Hyperspectral measurements of yellow rust and Fusariumhead blight in cereal crops: Part 1: a laboratory study. BiosystEng; 166:101–15.

Whetton RL, Waine TW, Mouazen AM, 2018, Hyperspectral measurements of yellow rust and fusarium head blight in cereal crops: Part 2: on-line field measurement. BiosystEng; 167:144–58.

Sivasangari A, Priya K, Indira K, 2017, "Cotton Leaf Disease Detection and Recovery Using Genetic Algorithm", International Journal of Pure and Applied Mathematics, Volume 117 No. 22, 119–123 ISSN: 1311-8080 (printed version); ISSN: 1314-3395 (on-line version).

Soil moisture sensor based automatic irrigation water pump controlling system with GSM technology

Upasana Sharma
Assistant Professor, ECE Department, ABES Engineering College, Ghaziabad, Uttar Pradesh, India

Akash Gupta, Swati Khantwal, Vipin Kumar & Vipin Singh
Student, ECE Department, ABES Engineering College, Ghaziabad, Uttar Pradesh, India

ABSTRACT: In this paper, GSM based water system framework with control of dampness level and water level is depicted. It is another remote system standard. The utilization of GSM based android gadget in water system framework is made. According to writing overview parcel of pre assumptions like, client have ability in horticulture and control gadgets has been worked according to client's desire as opposed to sensor information to determine challenges utilizing this innovation. The proposed framework assesses the water necessity for crops based on soil dampness information assembled by a few sensor hubs sent in the field. Mechanization demonstrated time and save in a way which is of incredible help to the business and just as for business visionary in all the measurements. Observing strategies actualized utilizing the Microcontroller (ATMEGA 328), Soil Moisture sensor and GSM.

Keywords: AVR Microcontroller (ATMEGA328), GSM, enlistment engine, water system, soil dampness sensor, Agriculture

1 INTRODUCTION

Water system may be a logical strategy for artificially giving water to the land or soil that would be refined. For the foremost part in dry locals neither having nor little precipitation water must tend for fields whichever over channels and indicator taps (Hebbarand & Vara Prasad 2017) tube wells. Unsurprising water system techniques had extreme issues, for example, ascend in remaining task at hand of farm work and regularly it lead to issue, for example, over water system or under-water system, and substance of soil. Remote innovations have been developing quickly as of late. These innovations are grouped relying upon the scope of correspondence between sensor hubs. For shorter separations Infrared (IR) sensors hooked in to light, point to multi point correspondence Wireless Personal Area Network (WPAN), significant range correspondence GSM/GPRS/DTMF and Bluetooth, Zigbee are often utilized. That remote sensing-based framework screens a water necessity to yield constantly and directs an input for the basic framework that governs a water stream (Topp et al. 1980).

This framework is sufficiently smart to impart the signs on portable of the far metro control the water system time and water stream. Sufficient water stream or water system augment the proficiency and creation also as recoveries the time and work of ranchers.

GSM and android based water system arrangements have been believed to be structured with present day information and segments that fulfil many building plan restrictions for example, affordable or vitality replaceable. In this paper, engineering and usage of a robotized water system framework dependent on GSM is introduced. Remotely observed frameworks

DOI: 10.1201/9781003193838-3

for water system reason have built up another prerequisite for ranchers to spare vitality, time and cash. Likewise (Jury & Vaux 2007) ultrasonic separation estimation gadget is utilized to gauge water level present in store. This will help in turning fundamental siphon ON or OFF in auto mode as indicated by level of water present in the capacity. Again it will forestall dry run state of fundamental siphon, if water isn't accessible in source during auto alongside manual mode activity of the principle siphon.

II EXISTING PROBLEM

This robotized water frameworks are served for aiding the ascending of horticultural harvests, upkeep of scenes or revival of vegetation of upset soil which have low water content hence the duration is lacking precipitation. This is an unnatural utilization of moisture content in the soil. By this a mechanized water-flexibly framework is affected by ignorance in technique for developing plants, absence of utilization in innovation, and so forth. As it may, some programmed frameworks (Purnima & Reddy 2005) exist in advertise, yet it is costlier and for the most part not appropriate for little gardens. In this manner, the new framework are required for pouring water naturally so that it is utilized at everyday life and can be produced simply. An upside of that thought incorporates minimal effort, simple to deal with low support and low force utilization. It is very well known that excess water in nursery would influence development in the plant.

It could be overwhelmed via computerized framework utilizing dampness sensor that holds proportion of the moisture percentage in the dirt. That is used to run a siphon just when moisture percentage in the dirt is beneath an ideal position (Hellín et al. 2016).

In the ordinary framework, that is essential that we required a somebody for turning ON-OFF, a siphon with a goal that the farming turns out to be troublesome procedure in everyday life. In current accessible framework peruses just the moisture sensor esteem. Some of the time these qualities are right according to the plan, however according to the barometrical condition manual revisions additionally compulsory. During circumstances, physical tasks notably starting and ending the present accessible frameworks likewise these activities control at the field. This includes extra labour. By utilizing GSM innovation, client could start ON and OFF siphon from wherever as required by sending suitable order through the versatile phone.

One progressively significant element passed up a great opportunity on existing frameworks. That is dry run assurance for the engines. In summer season, at some point well will dry on the off chance that the engine runs persistently without siphoning the water, at that point engine will wear out. Furthermore, checking the line voltages before turn over the engine is additionally significant. If the line voltage is not in the predetermined furthest reaches of the engine, at that point the engine will harm. Various plants require diverse dampness levels. It can't be changed the edge estimation of right now accessible frameworks without any problem.

III LITERATURE REVIEW

In water method range soil dampness sensor temperature sensor are put in base of plant and microcontroller addresses the sensor data and transmit detail.

A straightforward strategy to collection control issue utilizing counterfeit neural system controller the anticipated framework is connected with ON/OFF of controller and we found that ON/OFF controller based framework flops wretchedly in light of its constraints. On the other hand ANN based framework can spare heaps of assets (vitality and water) and can distribute improved result to sort of agribusiness territories (Idso et al. 1981).

IV PROPOSED SYSTEM

The proposed framework comprises of microcontroller unit, analog to digital converter, induction engine, soil dampness sensor.

A Microcontroller Unit

Microcontroller may be a gadget utilized rather than processors wherever a need is constrained or knowledge rate required may not be large matter. ATmega328 is 8-piece elite microcontroller of Atmel's mega AVR family with little force utilization. ATmega328 depends on RISC design with 131 directions. The Microcontroller consist of a 32 KB ISP, streak memory, EEPROM (1 KB), SRAM (2 KB)

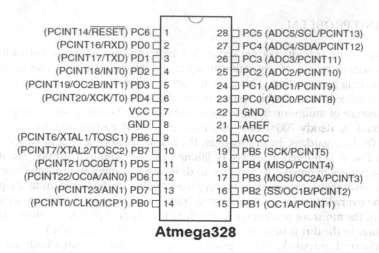

Figure 1. Pin configuration of atmega328.

B A to D converter

Analog to Digital converter may be a gadget which is employed to vary over the yield of the sensor which is straightforward in nature as microcontroller comprehends advanced information.

C Induction motor

Acceptance engine is favored over dc engine since it is ease, roughness, littler size and weight. Aside from the essential plan it can add dusty and dangerous condition in light of the very fact that the brushes aren't being utilized.

D Soil moisture sensor

Soil dampness sensor may be the gadget that quantifies a substance of water for the dirt .Further it is that the lesser water amount in soil means that the water stream prerequisite. The sensors that are having a capacitance are built up as lately which are otherwise called recurrence field sensors. Soil dampness sensors compute the volumetric water content straight by utilizing another asset of the dirt, for instance, electrical opponent, dielectric static or cooperation with neutrons.

E GSM

The Global System for portable (GSM) correspondence is the second age of versatile innovation. Despite the fact that the world is moving toward third and fourth era yet GSM has been the best and across the board innovation in the correspondence. Here in our venture we are interfacing GSM module with MC miniaturized scale controller. The message will be send to a specific GSM versatile number utilizing AT orders with the help of MC.

V BLOCK DIAGRAM

The circuit is done that gives constant DC power gracefully for the microcontroller in restraint to another electronic gadget. It utilized within the whole circuit, 220 volt AC is modified over to 24 volt AC by a step down transformer. Soil dampness sensor sum the water content in the dirt and its opposition different likewise and henceforth yield voltage as well.

A diode connect rectifier is utilized to change over 24 volt AC to 24 volt DC and afterward the voltage controller (L7805) cuts down the voltage to 5 volt dc. The yield of the dirt dampness sensor is associated with a microcontroller pin (Bhore et al. 2014). In barren express a obstruction of soil stickiness sensor is around 1.7 million ohms having wet status i.e. when dampness is available the opposition drop down 20.7 million on the voltage over the sensor in the dry state (0% gauged water content) is 1.9 volt and in the weight it is 0.1 volt (half volumetric water content).

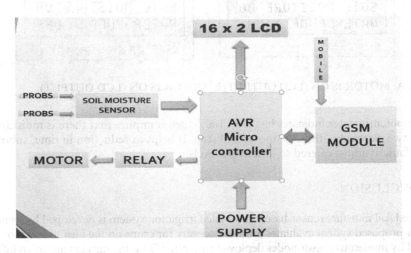

Figure 2. Block Diagram.

VI RESULTS

In the wake of structuring and executing the proposed framework, the accompanying outcomes were watched. At whatever point dampness amount of the dirt falls, it is detected by the dirt dampness sensor and therefore the span of engine activity is monitored counting on the dampness substance of the dirt. In this manner it is conceivable to possess a particular figuring of the heaviness of water from the heaviness of damp soil and therefore the heaviness of barren soil. Additionally, the examination for the mass level of water assists with having a sufficient water guideline for crops and defeats both the antagonistic states of abundance and lack of water at various areas during a land. Moderate Flood-type approaches expend

Figure 3 . A: Output on LCD When MOTOR ON b: Output on LCD When Motor Off.

17

a serious measure of water, yet the zone between crop lines ruins dehydrated and get damp-ness just from the accidental precipitation, (Wang et al. 2010) hand siphons while this water system procedure gradually exerts a touch quantity of water to the plant through cannels.

The framework rises the generate adequacy and abatement rancher's outstanding task at hand. There is productive use of water. The time spent is a smaller amount there by giving more flows. Controls the event of weeds (Yuan et al. 2004) decrease the compost. Consump-tion of soil might be halted absolutely by utilizing this type of a frame work.Urges advance-ment of a price dynamic water system control framework, secures electrical vitality. When the moisture content is negligible in soil and we can say the potentiometer is 100 percent, the potentiometer indicates infinite impedance. It will occur between the two electrodes .Therefore the motor will start as ON condition.

Figure 4.   A: MOTOR IS ON (LCD OUTPUT) b: MOTOR IS ON (LCD OUTPUT).

When the potentiometer indicates less than 100% then it implies that there is moisture content in mid of two electrodes, thus the motor will stop. It helps in reduction in time, subtraction of operator fault in adjust offered soil moisture levels.

VII   CONCLUSION

The induced soil moisture sensor based automated irrigation system is developed by using wireless sensors. A proposed system evaluates a water necessity for crops on the idea of soil moisture data assembled by numerous sensor nodes deployed within the field, the farmers are often informed by linking the system with transceiver and hence place a turn to the farmer's mobile number.

REFERENCES

Hebbarand, Santhosh & Vara Prasad, Golla. 2017. Automatic Water Supply System for Plants by using Wireless Sensor Network. International conference on I-SMAC (IoT in Social, Mobile, Analytics and Cloud) IEEE.
Purnima & Reddy, S.R.N. 2005. Configuration of remote monitoring furthermore, control System with Automatic Irrigation System utilizing GSM-Bluetooth.
Topp, G.C., Davis, J.L. & Annan, A.P. 1980. Electro-magnetic determination of soil water contented measurements in axial transmission lines, Water Resources Research 16, 574–582.
Navarro Hellín, H., Martínez del-Rincon, J., Domingo Miguel, R., Soto Valles, F. & Torres Sánchez, R.. 2016. A decision support system for managing irrigation in agriculture. Computers and Electronics in Agriculture.
Bhore, Prof. Arti, V.V., Mane, Poul, Mr. S.B, Patil, Mr. V.G. & Patil Mr. S.S. 2014. Automated Survey Analysis System" IJIRCCE Vol. 2, 1.
Jury W. A. & Vaux H. J. 2007.The emerging global water crisis, Managing scarcity and conflict between water users"Adv. Agronomy, vol. 95, pp. 1–76.
Wang, X., Yang, W., Wheaton A., Cooley, N & Moran, B. 2010.Efficient registration of optical and IR images for automatic plant water stress assessment.Comput. Electron. Agricult., vol. 74, no. 2, pp. 230–237.
Yuan, G., Luo, Y, Sun, X & Tang, D. 2004. Evaluation of a crop water stress index for detecting water stress in winter wheat in the North ChinaPlain. Agricult. Water Manag., vol. 64, no. 1, pp. 29–40.
Idso, S. B., Jackson, R.D., Pinter, P. J., Reginato, R. J. & Hatfield, J. L. 1981.Normalizing the stress-degree-day parameter for environ-mental variability.Agricult. Meteorol., vol. 24, pp. 45–55.

Recent Trends in Communication and Electronics – Sharma et al. (Eds)
© 2021 Taylor & Francis Group, LLC, ISBN 978-1-032-04572-6

IOT-enabled air pollution meter with digital dashboard on smartphone

Upasana Sharma
Assistant Professor, ECE Department, ABES Engineering College, Ghaziabad, Uttar Pradesh, India

Shruti Parashar, Pratham Jadoun, Piyush Katiyar & Rishabh Varshney
Student, ECE Department, ABES Engineering College, Ghaziabad, Uttar Pradesh, India

ABSTRACT: From direct observation, the conservative air pollution monitoring system is having anaccuracy that is high, however it is high bulk-wise and cost-wise, single datum class-make it non-feasible for large-scale installation. It aims to introducethe internet of things into the area of environmental safety.Our research focuses on an IOT enabled system which checks pollution and gives output accessible by a smartphone through an application. Core electronics and a bit of mobile application management are used in this project. Pollution is perpetual problem since many years and this device will help to provide data of the current pollution levels and the concerned people can act accordingly.

Keywords: monitoring system, air quality monitoring, GSM, internet of things, air pollution forecast, infrared sensor, android

1   INTRODUCTION

In the case of pollutionmonitoringthree majorproblems are seen, including environment, pollution detection and pollution source localization (Chen et al. 2015) At environmental pollution monitoring, firstly the environmental data is being sampled and then transmitted in real time to the data center. Contaminations monitoring are of great importance in the field of environment pollution monitoring. Pollution is mostly caused by elements called as pollutants that contaminate the airthenliquid. If the pollution source can be located when a pollution event is detected, it is helpful for solving the contamination accident timely. Thus, in pollution monitoring, we studied the three difficulties, that are environment status observation, pollution detection and pollution source localization. At environment pollution monitoring, an environmental data is being transmitted to data center in real time. With the help of sampling analysis the problem of pollution detectionis to decide. A system tells temperature, CO2 concentration and humidity. Networks having sensors have a lot ofbenefits such as comparatively dense circulation, high monitoring range also no limitations of geographical locations as a result of which they are often adopted in pollution monitoring. Anentityof research is the pollution which is due to the causes in air and water. It is havingcomparable properties in air as well as water (Sowe et al. 2014). Thus In these environmentswe can use different general monitoring frames of network. Ifcircumstances inchanged environments are taken into consideration .An Internet of thing enabled with digital dashboard air pollution meter at Smartphone is a system to determine and analyze the air quality at a given area. It can be used in an indoor or outdoor setup alike. (Bellavista et al. 2013) This system will detect particulate matter, harmful gases as well as determine the temperature and humidity of the environment that is placed in. It is hard

DOI: 10.1201/9781003193838-4

for people to afford various commercial air pollution sensors so this is an Endeavour to provide them with a compact, low-cost system with the same perks (Lehmann et al. 2010).

2 PREVIOUS ANALYSIS

As known to us anindustrial growth isextremely increasing, environmental pollution associatedmatters come into existence.In this research we established a system known as IOT, in which sensing devices are connected with a wireless embedded system. Earlier that was completed with GSM that willprovide alerts buta history is saved. Hence, we move to Internet of things that previous system is having disadvantage thatthe past value monitoring is impossible. We willoverwhelm toa systemplanned here.

3 SUGGESTED STRUCTURE

We see carbon factory in front of our college every day. Not one day goes without complaining about the black smog and pollution it creates. There are many such places which add immensely to pollution but are unaware or complacent about the fact(Stankovic 2014). There is a need for a system to determine if the pollution is under control or if the area is inhabitable or not. That's when the air pollution meter comes into play. It would be convenient in deciding the pollution levels and inhabitability of the place.

BLOCK DIAGRAM

The circuit is designed and connections made using the ESP8266 Node MCU, sensors and jumper wires.

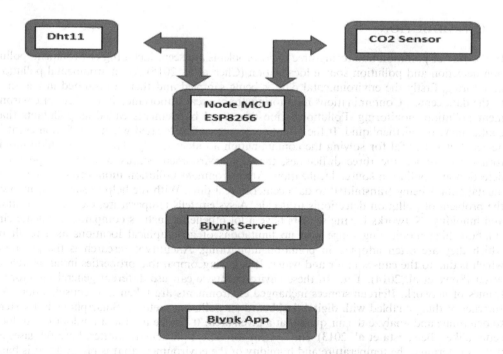

Figure 1. Block diagram of air pollution meter.

The sensors were chosen for the project. The project uses two sensors one the Infrared CO2 sensor module and the other is the DHT11 temperature and humidity sensor. These sensors have to be connected to the ESP8266 Node MCU which is programmed in the Arduino IDE. The code is run and checked for any modifications and changes required. Soldering is done to integrate the components into the circuit. The data generated by the sensors is sent to the Blynk app server.

4 COMPONENTS

4.1 *ESP 8266 node MCU*

This is a low cost open-source platform for the Internet of Things.Wi-Fi is integrated in this module making it a choice firmware for prototyping. The MCU is a microcontroller unit for carrying out various commands. It is programmed in Lua scripting language.

4.2 *IR CO2 gas sensor module*

For detecting the presence of CO2,We use MH-Z14 NDIR Infrared gas module, it is a common type and sensor having small size which uses infrared principle which is non-dispersive for detecting a presence of CO2 in the air. It has a great selectivity, nil oxygen dependence and durability. In-built temperature sensor, temperature compensation having digital output along with analog voltage output. That common infrared gas sensor is made by an addition of infrared absorbing gas detection technology, superior circuit design and precision optical circuit design.

4.3 *Sensor DHT11 for temperature and humidity*

This DHT11 as compound sensor which has a standardized digital signal output for humidity as well as temperature. It has a measurement component for humidity that is resistive-type also a NTC temperature measurement device. Its output pin is associated. This sensor is comparatively low-cost and cheapfor their presentation.

4.4 *Blink app*

Blynk application can control hardware remotely. It is capable of displaying the sensor's data, store it and then visualize it. This application permits the user for creating edges for research with the help of different widgets.

4.5 *Blynk server*

This isaccountable for entirely the communication taking place between the android smart phones withthe hardware. Anyone canuse either Blynk Cloud or private Blynk server. This is an open source system which is used to handle up to thousands of devices. One can also launch it on Raspberry Pi.

4.6 *Blynk libraries*

We use libraries for hardware platforms. They help to enabling communication with the server; also process all the incoming commands and the outgoing commands.When one presses a button in Blynk app on the Smartphone, the signal moves to the Blynk Cloud, where then it finds its way to the hardware. It works the same way vice-versa and everything happens in a blink of an eye.

4.7 *Smartphone and digital dashboard*

Android phone here, is using by way ofsmart meter. Blynk app is very well-designed builder of interface. This works well on IOS as well as android, alsodelivers easy-to-use widgets for the LCD, push buttons, on/off buttons, LED indication, RTC and others (Chen et al. 2010). Using Blynk digital dashboard features in a smart phone helps in reducing the cost of the project.

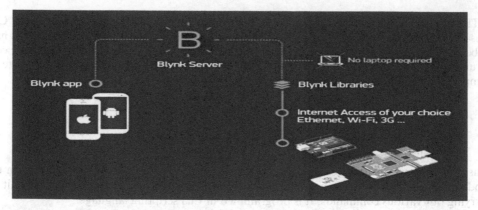

Figure 2. Blynk Server, Blynk gets connected with embedded hardware.

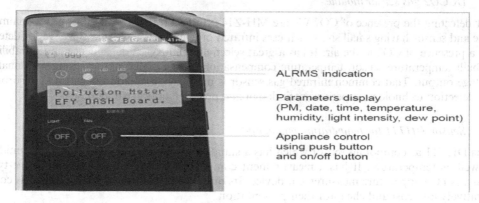

Figure 3. Digital dashboard on android phone.

5 LITERATURE SURVEY

Big data has proved to be a pioneer that connects point between things and objects on the internet. In this cyber-physical space, various types of sensors connect and interact over wireless networks, collect the data and deliver services varying from environmental pollution monitoring, disaster management and recovery to improving the lifestyle in homes and smart cities.Though, there are two sides of everything (Fang et al. 2014). The IOT has its own advantages and disadvantages. A few drawbacks include how to design suitable infrastructure for collecting and storing large quantity of heterogeneous sensor records, searching hands-onusage for a presentsensor data and manages IOT communities significantly that operators can endlessly search, find, and use the sensor data.For addressing these issues, this paper describes an IOT-based system combined with cloud computing. The sensors collect data of temperature, humidity and carbon dioxide concentration and with the help of a microcontroller it is sent to the cloud server. It can all be accessed by the help of an android Smartphone (Welbourne et al.2009). Newly, the WHOproclaimed that socialfitness is affected due to harmful effects of air pollution. Pollution is proving to be a threat to human health across the world. According to a July 2014 Public Facilities indoor air quality measurement,organizations were found to be in substandard position of indoor air pollution standards.

6 RESULT

This IOT-based system is used for monitoring an air quality with the help of ESP8266 Node MCU will help us take measures to improve air around us. IOT helps to enhancemonitoring process of various aspects of environment (Oh et al. 2015). Use of DHT11 and CO2 gas sensor tells us the various types of dangerous gases present. ESP8266 microcontroller is a main part forproject that helps controlling the overall system. Wi-Fi module attaches an entire process to internet and Smartphone is used for the visual output. An IOT-enabled air pollution meter with digital dashboard on Smartphone is a system for analyzing the air quality at a given area (Rao et al. 2016). It can be used indoors or outdoors alike. All components are assembled together and integrated into a working model. It can be used in heritage sites too as a precautionary measure in pollution sensitive areas like heritage sites. The caves microclimate can be determine and protected once the harmful gases are detected and it will stop decaying in homes, hospitals, schools and industries.

7 CONCLUSION

It is concluded that the characteristics of the air pollution monitoring application are large scale long lived with fixed sensor nodes, static physical topology, cost driven and no delay in controlof an air pollution problem. Collected data is treated as indicative rather than absolute and Error rate decreases.

This system for monitoring the air quality using ESP8266 Node MCU, IOT Technology is proposed to help take measures for improvement in quality of air. The usage of internet of thing technology improves the monitoring of severalfeatures of environment such as air quality monitoring including temperature, carbon dioxide concentration and humidity issue. Now using the DHT11 as well as CO2 gas sensor tells an idea of various hazardous gases also ESP8266 works as a star of research paper that controls the overall thing. Wi-Fi module helps system connect to internet and Smartphone gives visual output. An IOTenabled air pollution meter with digital dashboard on Smartphone is a system to determine and analyze the air quality at a given area. It can be used in an indoor or outdoor setup alike.

REFERENCES

Sowe, Sulayman K., Kimata,Takashi., Dong, Mianxiong., Zettsu, Koji. 2014. Managing Heterogeneous Sensor Data on a Big Data Platform. IoT Services for Data-Intensive Science.IEEE.

Chen, Xiaojun.,Xianpeng, Liu., & Xu, Peng. 2015. IEEE. IOT-based air pollution monitoring and forecasting system. IEEE. 978-1-4799-1819-5/15.

Oh, C. S., Seo, M. S., Lee, J. H., Kim, S. H., Kim, Y. D., & Park, H. J. 2015. Indoor air quality monitoring systems in the IoT environment. The Journal of Korean Institute of Communications and Information Sciences, 40(5), 886–891.

Welbourne, E., Battle, L., Cole, G., Gould, K., Rector, K., Raymer, S. 2009. Building the internet of things using RFID: The RFID experience, IEEE internet comput., vol. 13, no. 3, pp.48–55.

Fang, Shifeng., Xu, Li Da., Zhu, Yunqiang., Ahati, Jiaerheng., Pei Huan, Jianwu. Liu, Yan Zhihui. 2014. An integrated system for regional environmental monitoring and management based on internet of things, IEEE Transactions on Industrial Informatics, vol.10, no. 2, pp.1596–1605.

Stankovic, J. A. 2014. Research directions for the Internet ofThings.IEEE Internet Things, vol. 1, no. 1, pp. 39.

Chen, Shanzhi, Liu, Hui Xu, Hu, Dake, Wang, Bo, Hucheng, Atzori, L., Iera, A. & Morabito, G. 2010. The internet of things: A survey, Comput. Netw., vol. 54, no. 15, pp. 27872805.IEEE

Bellavista, P., Cardone, G., Corradi, A., & Foschini, L. 2013. Convergence of MANET and WSN in IoT urban scenarios.IEEE, vol. 13, no. 10, pp. 35583567.

Rao, BulipeSrinivas., Rao,Srinivas., Ome, Prof. Dr. K., & Ome. Mr. N. 2016. Internet of Things (IOT) Based Weather Monitoring system, IJARCCE Journal, vol. 5, no. 9.

Lehmann, Grzegorz, Rieger, Andreas, Blumendorf, Marco, DAI, Sahin Albayrak. 2010. A 3-Layer Architecture for Smart Environment Models/Amodel-based approach/LaborTechnische University Berlin, Germany. IEEE.978-1-4244-5328-3/10.

Recent Trends in Communication and Electronics – Sharma et al. (Eds)
© 2021 Taylor & Francis Group, LLC, ISBN 978-1-032-04572-6

An automated chilli yield estimation approach based on image processing

Chanki Pandey, Jaya Vishwakarma & M.R. Khan
Department of Electronics and Telecommunication Engineering, GEC Jagdalpur, CG, India

Sharad Chandra Rajpoot
Department of Electrical Engineering, GEC Jagdalpur, CG, India

Prabira Kumar Sethy & Bibhuti Bhusan Nayak
Department of Electronics, Sambalpur University, Burla, India

Santi Kumari Behera
Department of Computer Science and Engineering, VSSUT Burla, India

Preesat Biswas
Department of Electronics and Telecommunication Engineering, CVRU Bilaspur, CG, India

ABSTRACT: India is the largest consumer and exporter of chilli. Exports of chillies sum up to around one lakh tons, which makes 33% of the total spices exported from the country. The contribution of chilli powder to the spice market depends on the production of ripen chilli. So, yield estimation of ripen chilli is necessary to make the spice export more viable. The manual yield estimation based on human visual information is tedious, labour-intensive, prone to error and time-consuming. This manuscript suggests an automated chilli yield estimation method based on image processing. The recommended method able to fit the bounding box in ripen chilli and also count the number of chillies on the natural environment. It achieved 99.64% of accuracy on 25 number of on-tree image samples.

Keywords: Chilli yield estimation, Image processing, Object detection, Binarization

1 INTRODUCTION

Chilli peppers have been a part of the human diet since at least 7500 BC. India is also one of the prominent exporters of chilli to the world, as it exports one- fourth of the total quantity. Andhra Pradesh produces a maximum share of about 49% of chilli in India. Chillies fruits are of various shapes and sizes which depend upon the commercial variety of the drug. In general, it is oblong, conical, 10-20mm long and 4-7 mm wide. "It is cultivated in an area of 2,298,076 acres and produces 800,100 tons of yield" (Subbiah and Jeyakumar 2012). Most commonly, it is known by hot pepper. There is a wide number of varieties available across India such as Jwala, Byadgi, Wonder hot, Sannam, and LC 334 and used for various purposes, i.e. for spicing, oil, medicine, and colour, etc. The fruit grouping and quality evaluation by visual investigation cause blunder because of outer impacts, for example, weariness, retribution, and predisposition. Despite expert administrators, the characterization of fruits in the fruit industry drives irregularities due

DOI: 10.1201/9781003193838-5

to varieties in visual observation. Thus, a mechanized framework is required to dissect the products to provide reliable information. Hence, quick, canny, and non-damaging methods are necessary for this application area (Behera *et al.* 2020).

An automated chilli yield estimation will generate an overall report for the big chilli farm, i.e. about the production and quality of the chilli. So that the farmer ready to collect the red chilli and perform the post-harvest process like storage and export. Mostly red chilli is used for commercial purposes as green chilli cannot store for a long period. The main motive of the research article is automatic red chilli yield estimation. "For counting green and red pepper fruits, a methodology is proposed based on Bag of word model with the average accuracy of 74.2% by correlating the manual and machine counting of fruits" (Song *et al.* 2014). "Another framework is presented based on the image process for yield estimation of chilli crop with an accuracy of 92.7% along with the average error of 7.3%" (Bhookya *et al.* 2020). The measurement of chilli pepper plant size was carried out by (Gao *et al.* 2011), based on the mathematical morphology, which is quite more reasonable and accurate than the artificial measure.

In the last decade, a vast number of research works have been reported on different aspects related to fruit industry for yield estimation like grading (Al Ohali 2011), fruit quality assessment (Patel *et al.* 2012, Bhargava and Bansal 2018) prediction of volume and mass (Omid *et al.* 2010) and defect detection (Devi 2013). In the last few years, much work has been done for on-tree detection (Hemming *et al.* 2014, Meng and Wang 2016, Cheng *et al.* 2017), but very limited research is going on for the red chilli yield detection.

2   MATERIAL AND PROPOSED

The exploration study was actualized utilizing the MATLAB 2019a. All the applications were run on a laptop, i.e. HP Pavilion Core i5 5th Generation with basic NVIDIA GEFORCE. The photographs were captured using a smartphone, i.e. Redmi Note 8 Pro 64 MP quad camera. The flowchart of suggested method for chilli yield estimation is depicted in Figure 1.

2.1   *Image acquisition*

Images are captured using a smartphone that has a 64-megapixel resolution in natural light of the day, between 9 A.M. to 4 P.M. Total 25 sample images of red chilli plants, laden with fruits have been captured and all the images have been captured in stationary mode with each image having 1000 x 1800 pixel resolution.

2.2   *Red component extraction*

The captured image is in RGB form. To extract the red component from the image, we first convert the RGB form image to gray form and subtract the gray image from the original image. Then, the red component extraction is given by equation 1.

$$\text{Red Component Extraction} = \text{RGB Image} - \text{Gray Image} \qquad (1)$$

2.3   *Filtering out noise*

After extracting the red component of the image, we filter out the noise, i.e. light shadow, speckle noise, and impulsive noise by using a median filter concerning the window size of 3 x 3. Along with filtering out the noise from the image median filter sharpen contrast, highlight contours and detect edges.

25

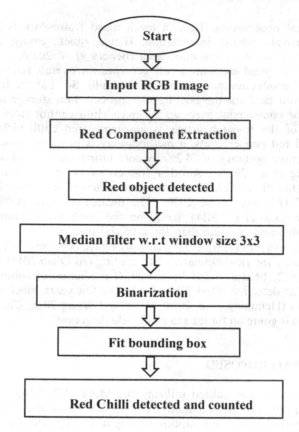

Figure 1. Flow chart of proposed methodology.

2.4 *Binarization*

After filtering out the noise from the image, we convert the image into a binary image to detect the red chilli. For this purpose, we took a constant red threshold value = 0.25.

2.5 *Red chilli region detected*

As the region is detected and at last bounding box is added for getting the exact region of red chilli. Hence, red chilli detected and counted successfully.

3 RESULTS AND DISCUSSION

In this study, we proposed an approach based on image analysis for an automated chilli yield estimation. The total 25 images are used and about 99.64% of accuracy achieved and hence, we detected and counted the red chilli successfully. The sample of input colour images of red chilli plants, laden with fruits is captured using a smartphone, i.e. Redmi Note 8 Pro 64 MP quad-camera as shown in Figure 2a. The binary image with the red objects as white shown in Figure 2b. Hence, red chilli detected and counted successfully with deploying bounding box as shown in Figure 2c.

The accuracy of the proposed approach is calculated, as shown in equation 2.

(a)	(b)	(c)

Figure 2.   Yield estimation process of Chilli (a) input RGB image (b) binary Image (c) Ripen chilli enclosed inbounding box.

$$Accuracy(\%) = 100 - AverageError(\%) \tag{2}$$

Where Average Error (%) is the average of total error as given by equation 3.

$$Average\ Error\ (\%) = \sum_{a=1}^{n} \frac{E_a}{n} \tag{3}$$

The Average Error (%) = 0.360. So, the accuracy of the proposed approach is given by, Accuracy (%) = 100-0.360 = 99.64 %. The comparison between manual counting and the proposed approach is shown in Figure 3.

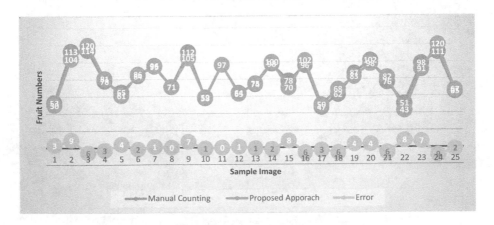

Figure 3.   Comparison of manual counting Vs Automated count.

4 CONCLUSION

The manuscript mainly focused on an automated chilli yield estimation for this, we proposed a novel approach based on image analysis. The total 25 images are used and about 99.64% of accuracy achieved and hence, we detected and counted the red chilli successfully along with the average error of 0.36%. This research may be extended by reducing false negative, and false positives. This work may be further improved by machine learning and deep learning approach.

REFERENCES

Behera, S.K., Rath, A.K., and Sethy, P.K., 2020. Maturity status classification of papaya fruits based on machine learning and transfer learning approach. *Information Processing in Agriculture*.

Bhargava, A. and Bansal, A., 2018. Fruits and vegetables quality evaluation using computer vision: A review. *Journal of King Saud University - Computer and Information Sciences*.

Bhookya, N.N., Malmathanraj, R., and Palanisamy, P., 2020. Yield Estimation of Chilli Crop using Image Processing Techniques. *In*: *6th International Conference on Advanced Computing and Communication Systems, ICACCS 2020*. Institute of Electrical and Electronics Engineers Inc., 200–204.

Cheng, H., Damerow, L., Sun, Y., and Blanke, M., 2017. Early Yield Prediction Using Image Analysis of Apple Fruit and Tree Canopy Features with Neural Networks. *Journal of Imaging*, 3 (1), 6.

Devi, P.L., 2013. Defect Fruit Image Analysis using Advanced Bacterial Foraging Optimizing Algorithm. *IOSR Journal of Computer Engineering*, 14 (1), 22–26.

Gao, Y., Li, X., Qi, K., and Chen, H., 2011. Measurement of chili pepper plants size based on mathematical morphology. *In*: *IFIP Advances in Information and Communication Technology*. Springer New York LLC, 61–70.

Hemming, J., Ruizendaal, J., Willem Hofstee, J., and van Henten, E.J., 2014. Fruit detectability analysis for different camera positions in sweet-pepper. *Sensors (Switzerland)*, 14 (4), 6032–6044.

Meng, J. and Wang, S., 2016. The Recognition of Overlapping Apple Fruits Based on Boundary Curvature Estimation. *In*: *6th International Conference on Intelligent Systems Design and Engineering Applications, ISDEA 2015*. Institute of Electrical and Electronics Engineers Inc., 874–877.

Al Ohali, Y., 2011. Computer vision based date fruit grading system: Design and implementation. *Journal of King Saud University - Computer and Information Sciences*, 23 (1), 29–36.

Omid, M., Khojastehnazhand, M., and Tabatabaeefar, A., 2010. Estimating volume and mass of citrus fruits by image processing technique. *Journal of Food Engineering*, 100 (2), 315–321.

Patel, K.K., Kar, A., Jha, S.N., and Khan, M.A., 2012. Machine vision system: A tool for quality inspection of food and agricultural products. *Journal of Food Science and Technology*.

Song, Y., Glasbey, C.A., Horgan, G.W., Polder, G., Dieleman, J.A., and van der Heijden, G.W.A.M., 2014. Automatic fruit recognition and counting from multiple images. *Biosystems Engineering*, 118 (1), 203–215.

Subbiah, A. and Jeyakumar, S., 2012. *Production and Marketing of Chillies*.

Recent Trends in Communication and Electronics – Sharma et al. (Eds)
© *2021 Taylor & Francis Group, LLC, ISBN 978-1-032-04572-6*

Deep learning based intelligent violence detection surveillance system

Aaditya Kumar, Abhijeet Anand, Aniket Tomar & Piyush Yadav
Department of Electronics and Communication Engineering,
G.L. Bajaj Institute of Technology and Management, Greater Noida, UP, India

Krishna Kant Singh
Department of Electronics and Communication Engineering,
KIET Group of Institutions, Delhi-NCR, Ghaziabad, India

ABSTRACT: Rising incidents of violence among people in public places is increasing concern in cities and mostly in highly dense cities of the country. The purpose of this research study is to design and develop an intelligent surveillance system based on deep learning technology which helps to detect violence and trigger an alert system. Using deep learning algorithms VGGNET, this study analyze the training and designing of model with violence dataset of hockey fights, movie fight scenes etc. which is able to detect violence in any live video stream. In this research the training data played a very big role in making the model accurate and configurable with various video monitoring systems. This study discussed the detailed design and implementation of embedded deep learning algorithms in Raspberry PI architecture to make smart surveillance system. The Results shows that the Algorithm proposed in this paper provides higher accuracy of 92.5%, which is higher in comparison to other existing algorithm.

Keywords: Surveillance, Violence, Deep learning, Neural network

1 INTRODUCTION

Deep Learning has become the buzz word of this decade and the rapid growth in research have made this possible that it is being used in multiple disciplines from medical science to household management system (Dubey, S. et al., 2020). It can also be used to detect the violence happening in real-time data. A number of violent fights (Baba, M. et al., 2019), stone pelting, vehicle burning, political agenda clashes etc. keep happening in public spaces where innocent civilians have to suffer loss of life and money.

Nowadays broad variety of surveillance systems have been developed by the several scientists and designers that are based on some of the application like remote checking, alarming, notification provider, traffic monitoring (Juan, I. E. et al., 2017), violence detection and actions recognition (Mabrouk, A.B. et al., 2018). The sole aim is to design and implement the whole system which are accessible, feasible and economical as well. D. Jeevanand used hardware as Raspberry Pi for the implementation of video capture through distributed networked system. The system was intended to active in constant circumstances and working proceed as first capture the video and as per the requirement of customer notify the authorized person by providing SMS alert (Bottou, L. et al. 2018).

DOI: 10.1201/9781003193838-6

M. Baba et. el. depicted a system which uses hardware as Raspberry Pi in (Baba, M. et al., 2019) with camera that is IP enabled for the surveillance purpose. The purpose of the configured system is to shows the captured images instantaneously in browser via TCP/IP and along that they also proposed a face detection algorithm which empowers to detect human faces even on spot video monitoring. U. Kumar et. el. proposed to design a wireless monitoring system by utilizing the hardware as Raspberry Pi (Kumar, U. et al., 2014) and customary wireless CCTV cameras (Kumar, U. et al., 2014). Along with his group member he executed a system in that the image obtained must be moved to the drop box by utilizing a 3G web dongle. This was effectively actualized by the use of Raspberry Pi and 3G dongle (Kumar, U. et al., 2014).We have proposed a new network architecture to increase the accuracy as compared to older methods. Our new method uses Darknet19 + VGG Neural Net + LSTM to detect violence from videos in real-time.

2 PROPOSED METHOD

The formulation of proposed network architecture as shown in Figure 1. The Darknet layer of the network is pre-trained by using ImageNet and are preserved during the period of training while the rest of the layers of model are trained on the video dataset (Juan, I.E. et al., 2017). Initially all the video frames which are provided as input are put in for pre-processing. Furthermore, the process of feature extraction is proceed where first layer is of VGG-16 which is used to extract spatial features and then second layer is of LSTM (Karpathy, A. et al.,2014) which is used for global feature extraction. In the end the feature extracted from LSTM are fed to fully connected layer for classifying as violent or non-violent. Along the feature extraction layer of LSTM which is intended for the global temporal feature from data, the feature which is obtained from the optical flow i.e local temporal is also major and trained them to reduce the error obtained in classification pocess as an alternative of shrinking E.P.E (end point error) which drastically improves recognition performance and accuracy. A fully connected layer is designed in such a way that its output becomes binary and gives output as two different value representing two classes i.e. fight and non-fight class.

3 DATA ANALYSIS

Test dataset videos are converted into shape of (N, VF,...., 3) here N represents total videos in dataset, VF represents total frames in a video. Pre- processing is done on these videos in these steps:

- Every input frame of every video is converted to the size i.e. 224 x 224 x 3 totaling 150528 pixels parameters. Techniques such as zooming, flipping and cropping etc. are also used as an approach to increase the model accuracy using data augmentation.
- Afterward, the hockey fight and movie (Varol, G. et al., 2018) datasets are randomly rearranged individually and the dataset is split into ratio of 80 and 20 as training and validation percentage respectively.
- After analysis output shape of this VGG layer is N x VF x 224 x 224 x 3.

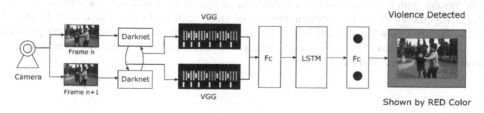

Figure 1.   Basic architecture of the model to be implemented.

3.1 Identification and use of features

Mainly there are two features which we try to identify which helps us to determine the violence in the image i.e. spatial features (Varol, G. et al., 2018) (features related to 3D space) and temporal features(Varol, G. et al., 2018) (features related to reference of time) by using convolutional layers of VGGNET which is pre-trained on ImageNet dataset and the layers of LSTM which is efficient in identifying temporal features (Nievas, E.B. et al., 2011). The main reason of the use of LSTM model here is to process the sequence of images because any action has to be a sequence of frames of still images. Output of this VGG layer is of dimension N*VF* F, here F is total features acquired from each frame. The activation function used is tanh (works better for hidden layers) and bias equals 0.

3.2 Classification

In the third step, completely connected layer sequence is used to do the classification which are placed in a stack after the stack of convolutional layers, in which the 1st layer contains 2048 channels, the 2nd layer contains 1024 neurons and the end (or last) layer which consists of a softmax function contains just 2 neurons each for violence & non-violence class. RELU activation function is used in 1st and 2nd layer, whereas soft-max activation function is used in the end (or last) layer in the suggested model. Here uniform initializer is used for initializing the neurons because it generally works better for layer which use RELU as activation function.

Normalization is a technique which is used to transform different given parameters of the image to obtain more convenient values. Using this technique the intensity of an image can be transformed from the normal range of 0-255 to between 0 and 1 and that will be very easy to process.

For reducing the effect of overfitting, we used two methods. In 1st method, we use "L2 normalization" as regularization parameter (λ) in all of the FC layers (with *alpha (λ)* = 0.2). In the 2nd approach, after first connected layer, we apply drop out only with 20% probability. Dropout regularization technique is also used to reduce overfitting of the model by restricting the learning of model.

3.3 Datasets

The proposed model is given training using two of the clean datasets i.e. hockey fight and movie datasets. In addition, we used self-recorded video for the purpose of testing as well as finetuning of the model. After the training of the model, it is then used to detect violence and non-violence in real-life by analyzing the videos streamed through the surveillance camera. The dataset consists of a total of 1200 videos where the data split is done as 480 videos for Testing, 600 videos for Training and 120 videos for Validation set. This data splitting is based on careful analysis of mean absolute errors corresponding to different splitting ratios. We have taken 50% of total data for training, 40% of total data for testing and 10% of total data is used for validation set.

- The Hockey Dataset contains 1002 videos which have been separated into 501 violent videos and 501 non-violent videos. It has been gathered from hockey matchesin which every video contains 55 frames and each frame has a size of 360 x 288. The videos have the similar background. In this, to increase the probability of detecting violence, 20 frames were fed as input from every video to the suggested model.
- Movie Dataset contains 202 videos which have been separated into 101 violent and 101 non-violent videos. The violence dataset has been gathered from scenes of movies, in which each video contains 50 frames with the size 720 x 576. The movie dataset contains dissimilar backgrounds. In this data training, 16 frames are inserted to the suggested model as input beginning from the first frame.

4 RESULTS AND DISCUSSIONS

After training, performance analysis of proposed model in real-life with the help of camera stream is performed and the corresponding result will be discussed.

4.1 Platform requirement

The model suggested is developed using python library i.e. Keras with TensorFlow and other helping libraries. We have used Stochastic Gradient Descent optimizer with alpha (learning rate) being set to 0.2 . The training is done with batch size set to 4 and 40 frames in a video. The training of the model is done on Google Colaboratory with TPU RAM as 35GB and Disk Size as 107GB. The model is also tested remotely using Raspberry Pi 4 Model-B having 2 GB RAM and Quad-core processor with live video using web camera connected to the microcontroller device.

4.2 Performance analysis

Our two colleagues, as shown in Figure 2 a), are fighting which is streamed through the camera and then fed as input to our trained model. Similarly, as shown in Figure 2 b), our two colleagues are handshaking which is again streamed through the camera and then fed as input toour trained model.

The trained model takes the streamed video as input and then determines whether there is violence or not in the video. As we can see in Figure 2 a), the model detected correctly that the video in which the two colleagues are fighting is a violent video and marked it with RED Color to show that the video is violent. Similarly, the trained model takes the second streamed video as input and then determines whether there is violence or not in the video. As we can see in Figure 2 b), the model detected correctly that the video in which the two colleagues are handshaking is a non-violent video and marked it with GREEN Color to show that the video is non-violent.

4.3 Result analysis

In the Table 1 it is shown that the 3-diamentional convolution network is giving less accuracy in comparison to latter one i.e. Dartnet19 + VGG + LSTM because the latter one consists of spatial as well as temporal features. The results are represented in Table 1 below.

Table 2 depicts various performance parameters like Loss, precision, recall, f1 score and their values.

(a) Violence Detected   **(b)** Non-Violence Detected

Figure 2. Successfully detecting violence and Non-violence.

Table 1. Model performance.

Neural Model	Accuracy
Conv3D (Previous Method)	91.00%
Darknet19 + VGG + LSTM	92.25%

Table 2. Model parameters.

Loss	Frame Accuracy	Frame threshold	AccuracyPrecision		Recall	F1 Score
0.3208	0.8981	2	0.9225	0.9122	0.9350	0.9235

5 CONCLUSION

Surveillance is very beneficial for the governments and law enforcement body to maintain social control, acknowledge and monitor threats, and prevent/investigate criminal activity. There are a lot of violence cases which are not even tracked because of delay in the reporting of the event to the police. Video surveillance cameras are used in shopping centers, public places, banking institutions, companies and ATM machines. In this research, the proposed model contains VGG-16 model pre-trained on ImageNet datasct which extracts spatial features, and then LSTM extracts the temporal features. Last layer consists of a chain of 3 FC layers for doing the classification task. Two open datasets are used in the modelthat includes Hockey Fights and Movie Fights. The dataset that has been proposed contains a wide variety of rough conditions. The non-violence category includes normal movement of human activities. A process of fine tuning has been used to avoid over-fitting and for the improvement of test accuracy of the model. The model is not able to perform well in low light conditions but it can be improved using better resolution video camera having better aperture. This model with a modification can also detect theft happening in the area under surveillance.

REFERENCES

Baba, M.; Gui, V. & Cernazanu, C. 2019. A Sensor Network Approach for Violence Detection in Smart Cities Using Deep Learning", Sensors, 19, 1676;pg 1-17.
Bottou, L. 2018. Optimization methods for large-scale machine learning," Siam Review, vol. 60, no. 2, pp. 223–311.
Dubey, S.; Singh, P.; Yadav, P. & Singh, K.K. 2020. Household Waste Management System Using IoT and Machine learning. Procedia Computer Science, Vol.167, pp1950–1959.
Juan, I.E.; Juan, M. & Barco, R. 2017. A low-complexity vision-based system for real-time traffic monitoring. IEEE Trans. Intell. Transp., 18, 1279–1288.
Karpathy, A.; Toderici, G.; Shetty, S.; Leung, T.; Sukthankar, R. & Li, F. 2014, Large-scale video classificationwith convolutional neural networks. In Proceedings of the Conference on Computer Vision and Pattern Recognition, Columbus, OH, USA, 23–28; pp. 1725–1732.
Kumar, U.; Manda, R.; Sai, S. & Pammi, A. 2014 "Implementation Of Low Cost Wireless Image Acquisition And Transfer To Web Client Using Raspberry Pi For Remote Monitoring. International Journal of Computer Networking, Wireless and Mobile Communications (IJCNWMC).," vol. No. 4, no. 3, pp. 17–20.
Mabrouk, A.B. & Ezzeddine, Z. 2018. Abnormal behavior recognition for intelligent video surveillance systems: A review. Expert Syst. Appl., 91, 480–491.
Nievas, E.B.; Suarez, O. D.; Bueno, G. & Sukthankar, R. 2011. Violence detection in video using computer vision techniques. In Proceedings of the International Conference on Computer Analysis of Images and Patterns, 29–31; pp. 332–339; Seville:Spain.
Varol, G.; Laptev, I.; Schmid, C., 2018. Long-term temporal convolutions for action recognition. IEEE Trans. Pattern Anal. Mach. Intell., 40, 1510–1517.

A review on content-based image retrieval: Relating low level features to high level semantics

Priyesh Tiwari & Kulwant Singh
Department of Electronics & Communication Engineering, Manipal University Jaipur, Jaipur, India

Shivendra Nath Sharan
NIIT University Nimrana, Neemrana, India

Shiv Narain Gupta
Department of Electronics & Communication Engineering, Greater Noida Institute of Technology (GNIOT), Greater Noida, U.P., India

ABSTRACT: The technique through which a system can find similar images to a query image among an image dataset is known as Content Based Image Retrieval (CBIR). This technique works by highlighting a few features from the image in question to the images in the dataset to provide the best matching outcome. This review deals with economical and correct retrieval in CBIR.

1 INTRODUCTION

As there was a sudden spike in the use of image acquisition, the digital image information also up-graded itself. The need of economical retrieving system was the need of hour, so researchers came up with Text-based and Content-Based retrieval techniques. In Text based image retrieval (TBIR), the images are annotated with a text description and the result is displayed after matching that annotation of the image with text query of user. In Content based image retrieval (CBIR), user gives an image as an input and the image is then matched by system in its database to find the exact image on basis of the description of image and other features such as colour, texture and shape of images to provide an output. Figure 1 shows framework of CBIR:

(a) Image-Database: Digital device can be used to store acquire data in database.
(b) Image Pre-processing: Enhancement of image before the retrieval process.
(c) Image Acquisition: A feature vector database is created with features like texture, colour or shape of the images. It can be features divided into two parts i.e. High level and low level features.
(d) Similarity Matching: It tells us the percentage about how much an image is same.
(e) Output/Retrieved Images: Final outcome after the process.
(f) User Interaction/Feedback: User give their feedback on the result being relevant or not.

CBIR system allows the user to get their desired output by matching the content of image. Content at this point refers to colour, spatial location, texture or shape (Liu et al., 2007; Vassilieva et al., 2009).

The foremost intention is to reduce semantic-gap. The paper is alienated into 5 sub-parts. Semantic gaps deals in Section 2. Low level semantics cover in Section 3 and high level semantics covers in section 4. The performance measure is cover in Section 5 and paper is concluded in last segment.

DOI: 10.1201/9781003193838-7

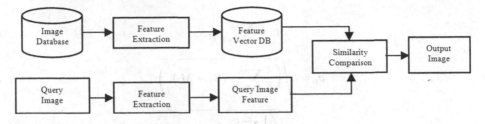

Figure 1.   Basic architecture of CBIR.

2 SEMANTICS

The CBIR nevertheless fails to linkage the space between HL semantics and LL features even after tremendous research. Human thought is handiest successful to recognize HL features and system understands LL features (Zhou et al., 2000; Mojsilovic et al., 2001. In Ref (Song et al., 2001), Semantic may be categorized as:

2.1 *Local semantic level*

It retrieves either single object or combination of objects as output.

2.2 *Thematic or Global semantic level*

It correlates each item within the photo with the person question to provide relatable output. In Ref (Eakins et al., 1999), the CBIR query is classified majority into three district levels:

(a) *Level 1:* Retrieval of the image on the basis of LL features present.
(b) *Level 2:* Retrieve image having a certain object on it.
(c) *Level 3:* Retrieve image having a certain object which might be taken as query.

   The breathing space between level 1 and level 2 is known as semantics gap (Song et al., 2001). Queries beneath Level 2 and Level 3 are less difficult for human understanding. CBIR ought to attempt to ensure the gap among LL capabilities and HL semantics are minimized (Zhou et al., 2000).

3 LOW LEVEL FEATURES

The framework of CBIR focuses extra on characteristic extraction. A characteristic database is made to save the characteristic vectors of photograph (Liu et al., 2007). Local or global is the categories of features descriptors. Global descriptors often used to take image for entire and local descriptors, which take features of items to constitute the entire image (Liu et al., 2007).

3.1 *Colour*

The colour feature is extensively utilized in CBIR. Colour feature cannot be changed with reference to the orientation and size of the object. Colour descriptor is used to extract the precise share of shade from the photograph in question. A technique used to signify the colour explanation of an image is:

3.1.1 *Colour moments (Duanmu, 2010; Huang et al., 2010)*
Colour Moment (CM) tells approximately the allocation of the colours in picture. CM is rotation & scaling invariant. The first, second and third order CM is representing mean, standard deviation and colour skewness are shown by equation (1), (2) and (3) respectively.

$$M_{ij} = \sum_{j=1}^{K} \frac{1}{K} S_{ij} \qquad (1)$$

$$\sigma_{ij} = \sqrt{\left(\frac{1}{K}\sum_{j=1}^{K}\left(S_{ij} - M_i\right)^3\right)} \qquad (2)$$

$$SK_i = \sqrt[3]{\left(\frac{1}{K}\sum_{j=1}^{K}\left(S_{ij} - M_i\right)^3\right)} \qquad (3)$$

3.1.2 *Colour Correlogram (CC) (Huang et al., 1997a, b)*
Colour Correlogram (CC) is obtained by combining spatial correlations of pairs of colours in the image and pixel's colour distribution.

3.1.3 *Colour histograms (Huu et al., 2012)*
Colour Histograms depicts a diverse form of colourations found in the picture after which shape shade organizations and the amount of pixels found in each colour groups. It is similarly labeled right into a nearby global colour histogram and colour histogram. The global histrogram does not take the spatial place of pixels. inattention and take the entire picture as one vicinity whereas; nearby histogram divides the picture into m quantity of blocks.

3.1.4 *Colour coherence vector (Wen, 2008; Hongli et al., 2004)*
It works with the aid of using combining each colour histrogram and spatial positioning information. These vectors similarly classify every pixel into both coherent and in-coherent.

3.2 *Texture (Tamura et al., 1978)*

The texture is the frequency of visual pattern in an image which has uniform properties. The structure and specification properties of texture.

3.2.1 *Structure of texture patterns*
The texture may be imagined as a fixed of macro-scopes regions and its shape is a re-taking place sample with inside the photo that is labeled by 'placement rule'. The texture is similarly categorized into 'Structural' and 'Statistical'.

3.2.2 *Specification of the feel styles*
In order to obtain accurate result there are five visual texture properties:

(a) Coarsness: It is used to investigate the quantity and length of texture styles. The texture pattern classifies the pattern in coarse or fine.
(b) Contrast: The values acquired right here is the distinction among the depth values of pixels that are adjoining in a photo.
(c) Directionality: It is worldwide asset of the area of the photo which tells us approximately the inclination of the free styles.
(d) Line Likeness: Line likeness offers with texture primitives
(e) Regularity: Regularity deals with texture styles when it is uniformly organized.

3.3 *Shape (Zhang et al., 2004)*

The shape is one of the most significant features which provide information of facial appearance. For accurate output, shape descriptors desire to locate equivalent shapes from the database of images. The shape descriptors are categorized into two components based on contour or region methods. Figure 2 shows the shape representation classification and description technique.

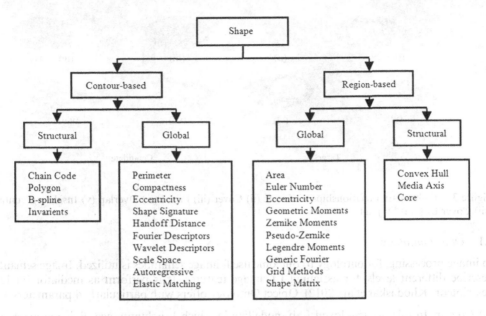

Figure 2.   The categorization of shape representation & its description technique.

3.3.1 *Contour based methods*

This approach describes and provides information about shape of boundary. It is again sub-divided into classes that Discrete and Continuous. In the discrete, the image is again sub-divided into smaller components. While in continuous, image is not sub-divided into a variety of sub-parts and create an attribute vector from original shape.

3.3.2 *Region based methods*

In this method image are categories on the basics of region present along with boundary. It is further divided into two categories, structural and global.

3.4   *Spatial location*

It provides us with the information on how objects in an image are related spatially to each other. Spatial Context Modeling (Egenhofer et al., 2004) is based on the relationship between objects in image. Eight Spatial relationships are denoted in Figure 3.

4   HIGH LEVEL SEMANTICS

In CBIR as discussed earlier, LL features are extract; in which it does not go with the user point of view. This difference is known as semantic gap and to overcome it, researchers came up with HL semantics retrieval method. The proposed methods to reduce semantic gap:

(a) By using Object-ontology; HL concepts is defined (Stanchev et al., 2003; Mezaris et al., 2003).
(b) Using ML tools to connect LL features with query concepts (Feng et al., 2003; Shi et al., 2004).
(c) To understand user needs, use of Relevance feedback (RF) into the retrival loop (Mezaris et al., 2003; Zhang et al., 2001).
(d) Semantic template is generated to support, HL image retrieval (Smith, 1998; Zhuang et al., 1999).
(e) Using both textual and visual information obtained from World Wide Web (www) image retrieval (Zhuang, 1999; Cai et al., 2004; Feng et al., 2004).

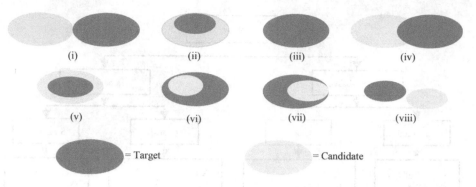

Figure 3.   Eight spatial relationships (i) Touch (ii) Cover (iii) Equal (iv) Overlap (v) Inside (vi) contains (vii) Cover-by (viii) Disjoint.

4.1   *Object ontology*

In image processing, for ontology to outcome itself, image semantics is utilized. Image semantics describe different levels, for assigning LL image features. It is perform as mediator for level descripteor (Khodaskar et al., 2012). Object Ontology offers with particularly 4 parameters:

(a) Colour: In this, we employed Lab modelling in which L is luminance, R is red-green and b is blue-yellow. Whenever neither of their two colours is present, luminance classified it as none, otherwise its classified it into very low, low, medium, excessive.
(b) Position: It may be analyzed with reference to horizontal and vertical axis.
(c) Size: Dimension of the vicinity as large, medium or small.
(d) Shape: Defines the form of the vicinity into oblong, medium oblong or very oblong.

4.2   *Machine learning*

There are primarily two types of Machine learning that is Supervised and Unsupervised learning. In supervised learning, image is taken and a classifier like Decision tree or Bayesian Classifier is used to determine the semantic category label. But in unsupervised learning, image is classified in a groups and a group label assign to each of individual group. The predominant thought is to restrict the inter group similarity and augment intra institution similarity along these lines, that during resultant image from analogous group are acquired.

4.3   *Importance of feedback*

This is to make the wholly process more interactive. User opinions are provided to system for better output.

4.4   *Semantic Templates (ST)*

A semantic template (Zhuang et al., 1999) is a set of mappings among HL semantic values and LL visual capacities. On the other hand it also enhances the performance of image retrievals.

4.5   *WWW image retrieval*

Web image has URL which can basically inform the framework concerning the classification of images. A few web image retrieval frameworks are utilizing text based methodology. A great deal of examination is likewise continuing for recovering images from web by blend of text and visual methodologies (Cai et al., 2004).

5 PERFORMANCE MEASURES

We measured the performance of system through Recall and Precision. The standard definitions of these two methods are given in Table 1 (Goel et al., 2012):

Table 1. Contingency Table for Retrieval Results.

	Relevant	Non-Relevant	
Retrieved	TP	FP	TP+FP
Non Retrieved	FN	TN	FN+TN
	TP+FN	TN+FP	(TN+FP) + (TP+FN)

Recall (Sensitivity) is the ratio of accurately predicted positive observations in actual class:

$$Recall = TP/(TP + FN)$$

Precision is the ratio of accurately positive outcomes to the total predicted positive outcomes:

$$Precision = TP/(TP + FP)$$

6 CONCLUSION

This paper gives us a writing audit on content based image retrieval (CBIR) is likewise done in paper. CBIR is a technique which needs more work and research to get better accurate results. CBIR can be the next big thing in terms of obtaining the results on web search (Alsmadi, 2020). The performance measures are there to know how well our CBIR model is performing to give us the results. Now there is no sequential technique available at the moment which can reduce the semantic gap completely, future research area is recommended.

REFERENCESS

Alsmadi, M.K. 2020. Content-Based Image Retrieval Using Color, Shape and Texture Descriptors and Features. *Arabian Journal for Science and Engineering*: 1.

Cai, D., He, X., Li, Z., Ma, W.Y. & Wen, J.R. 2004. Hierarchical clustering of WWW image search results using visual, textual and link information. In *Proceedings of the 12th annual ACM international conference on Multimedia, USA, October 2004*. Association for Computing Machinery.

Duanmu, X. 2010. Image retrieval using color moment invariant. In *2010 Seventh International Conference on Information Technology: New Generations, USA, 7–8 April 2008*. IEEE.

Eakins, J.P. & Graham, M.E. 1999. Content-based image retrieval, a report to the jisc technology applications programme.

Egenhofer, M.J. & Franzosa, R.D. 2006. Point-Set Topological Spatial Relations. *Classics from IJGIS: 12th years of the International Journal of Geographical Information Science & Systems* 5(2): 141.

Feng, H. & Chua, T.S. 2003. A bootstrapping approach to annotating large image collection. In *Proceedings of the 5th ACM SIGMM international workshop on Multimedia information retrieval, USA, November 2003*. IEEE.

Feng, H., Shi, R. & Chua, T.S. 2004. A bootstrapping framework for annotating and retrieving WWW images. In *Proceedings of the 12th annual ACM international conference on Multimedia, USA, October 2004*. Association for Computing Machinery.

Goel, N. & Sehgal, P. 2012. A refined hybrid image retrieval system using text and color. *International Journal of Computer Science Issues (IJCSI)* 9(4): 48.

Hongli, X., De, X. & Yong, G. 2004. Region-based image retrieval using color coherence region vectors. In *Proceedings IEEE 7th International Conference on Signal Processing, 2004, China, 31 Aug.-4 Sept. 2004*. IEEE.

Huang, J., Kumar, S.R. & Mitra, M. 1997a. Combining supervised learning with color correlograms for content-based image retrieval. In *Proceedings of the fifth ACM international conference on Multimedia, Seattle, WA, USA, November 1997*. Association for Computing Machinery, United States.

Huang, J., Kumar, S.R., Mitra, M., Zhu, W.J. & Zabih, R. 1997b. Image indexing using color correlograms. In *Proceedings of IEEE computer society conference on Computer Vision and Pattern Recognition, USA, 17–19 June 1997*. IEEE.

Huang, Z.C., Chan, P.P., Ng, W.W. & Yeung, D.S. 2010. Content-based image retrieval using color moment and Gabor texture feature. In *2010 International Conference on Machine Learning and Cybernetics, China. 11–14 July 2010*. IEEE.

Huu, Q.N. & Thu, H.N.T. 2012. Content based image retrieval with bin of color histogram. In *2012 International Conference on Audio, Language and Image Processing, China, 16–18 July 2012*. IEEE.

Khodaskar, A. & Ladke, S.A. 2012. Content Based Image Retrieval with Semantic Features using Object Ontology. *International Journal of Engineering Research & Technology*.

Liu, Y., Zhang, D., Lu, G. & Ma, W.Y. 2007. A survey of content-based image retrieval with high-level semantics. *Pattern recognition* 40(1): 262–282.

Mezaris, V., Kompatsiaris, I. & Strintzis, M.G. 2003. An ontology approach to object-based image retrieval. In *IEEE Proceedings 2003 International Conference on Image Processing, Japan, 14–17 Sept. 2003*. IEEE.

Mojsilovic, A. & Rogowitz, B. 2001. Capturing image semantics with low-level descriptors. In *Proceedings IEEE 2001 International Conference on Image Processing, Greece*.

Shi, R., Feng, H., Chua, T.S. & Lee, C.H. 2004. An adaptive image content representation and segmentation approach to automatic image annotation. *In International conference on image and video retrieval, Ireland, 21–23 July 2004*. Springer, Berlin, Heidelberg.

Smith, J.R. 1998. Decoding image semantics using composite region templates. In *Proceedings. IEEE Workshop on Content-Based Access of Image and Video Libraries, USA, 12 June 2000*. IEEE.

Song, Y., Wang, W. & Zhang, A. 2003. Automatic annotation and retrieval of images. *World Wide Web* 6(2): 209–231.

Stanchev, P., Green Jr, D. & Dimitrov, B. 2003. High level color similarity retrieval.

Tamura, H., Mori, S. & Yamawaki, T. 1978. Textural features corresponding to visual perception. *IEEE Transactions on Systems, man, and cybernetics* 8(6): 460–473.

Vassilieva, N.S. 2009. Content-based image retrieval methods. *Programming and Computer Software* 35 (3): 158–180.

Wen, L. & Tan, G. 2008. Image retrieval using spatial multi-color coherence vectors mixing location information. In *2008 ISECS International Colloquium on Computing, Communication, Control, and Management, China, 3–4 August 2008*. IEEE.

Zhang, L., Lin, F & Zhang, B. 2001. Support vector machine learning for image retrieval. In *Processing 2001 International Conference on Image Processing, Greece, 7–10 Oct. 2001*. IEEE.

Zhang, D. & Lu, G. 2004. Review of shape representation and description techniques. *Pattern recognition* 37(1): 1–19.

Zhou, X.S. & Huang, T.S. 2000. CBIR: from low-level features to high-level semantics. *In Image and Video Communications and Processing 2000* 3974(1): 426–431.

Zhuang, Y., Liu, X. & Pan, Y. 1999. Apply semantic template to support content-based image retrieval. In *Storage and Retrieval for Media Databases2000*. International Society for Optics and Photonics.

Recent Trends in Communication and Electronics – Sharma et al. (Eds)
© 2021 Taylor & Francis Group, LLC, ISBN 978-1-032-04572-6

Mishap detection and alerting system using GPS and GSM

Anjana Bhardwaj, Ankit Chaudhary, Ayush Giri, Akanksha Bhadauria & Ashutosh Tiwari

ECE Department, ABES Engineering College, Ghaziabad, Uttar Pradesh, India

ABSTRACT: A big cause of fatality in today's date is the accidents caused by vehicles. In this paper, the Accident Alert System requires an approach by which accident detection can be done on an early basis to save human life. So first of all the GPS will record the incoming information from satellite and will also store positional data. If it is required to track a particular vehicle, it is required to send an SMS to owner of the vehicle, so that the accident alert system will be activated. The accident alert system may also be activated by activation of the pressure sensor. Once the message is send to GSM, it will get activated and will check the last gotten scope and longitude esteems from the stored values and GSM will also be used to send the accidents occurred information to the emergency server, that can be the hospital, police station and the relative of the user, this information is predefined in the program which is written in Arduino Uno. To design this Accident Detection and Alerting system we are required to use various components like Arduino, Vibration Sensors, and GPS & GSM Modules.

Keywords: GSM, GPS, TPMS, ABS

1 INTRODUCTION

With the high number of vehicles out and about, the number of vehicle related accidents have also increased. So to prevent the number of casualties from increasing a programmed ready framework for mishaps is canvassed in this manuscript. This framework can recognize mishaps in very small time and sends the required information to the emergency services within (this information will have geographical coordinates, the time of vehicle accident).

Figure 1 shows the concept to be used for accidents detection. This system provides a perfect solution for the streets mishaps in the most doable manner.

2 LITERATURE REVIEW

Cloud-Based Cyber-Physical Robotic Mobile Fulfilment Systems: A Case Study of Collision Avoidance has analysed that why the mishaps are rising in number day by day (Keung et al. 2019). Design and application of Internet of Things-based warehouse management system for smart logistics titled has a discussion about IoT to manage a ware house (Lee et al. 2018). Effective Coverage Control for Mobile Sensor Networks with Guaranteed Collision Avoidance in his research paper discussed various technologies to avoid vehicle collision and accident; and this was very useful for my study and research. (Hussein et al. 2007). Using Smart phones to Detect Car Accidents and Provide Situational Awareness to Emergency Responders with GPS and GSM technology for mishap detection and in rescue of human life by emer-

DOI: 10.1201/9781003193838-8

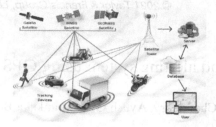

Figure 1.   Overview of vehicle tracking.

gency responding technology was a effective approach (Thompson et al. 2005). In Traffic-incident detection-algorithm based on nonparametric regression author used nano-technology for detection of traffic (Tang et al. 2005) (Deep et al. 2014). A Wireless Vehicular Accident Detection and Reporting System were introduced (Megalingam, et al. 2010).

3 METHODOLOGY

The system is very efficient to be implemented. This system can be implemented in vehicles and to inform about any incident. The system is executed simply by use of GSM & GPS technologies.

In this system, controlling is given to the Arduino. Figure 2 shows the block diagram for mishap detection and alerting system, here it contains Arduino Uno; power supply to operate the system, 16x2 LCD to display the notification message, LPG gas detector is used to find if any type of gas is releasing and can be harmful, vibration sensor, GPS receiver and GSM Module. The force gracefully supplies electrical vitality to the load. Here the entire component in the system works at 12V dc supply (Sun et al. 2019). The Vibration Sensor can gauge and break down straight speed, removal or quickening. This is to the spring and can identify vibration occurred on any point. Various kinds of GSM modules are accessible in the market. SIMCOM created various frequencies modules (Sadeghi et al. 2015). We have selected the SIM900A for our project. Global Position System (GPS) is constellation of 24 satellites & multiple ground stations. It uses satellite as a reference point to calculate accurate position.

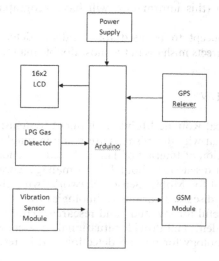

Figure 2.   Block Diagram.

42

4 ALGORITHM

Speed is the most significant and hazardous while driving. It affects the rising of accidents and furthermore builds danger of being associated with an accident. Individuals need some preparing chance to choose to respond and afterward to take action. At high speeds it takes time to stop vehicle after pulling the brakes. The time takes to stop vehicle after brakes is proportional to the square of speed [8]. Therefore, it is difficult to avoid collision at high speed. As we know that any of the moving body contains a kinetic energy as:

$$KE = \frac{1}{2}mv^2$$

Where, $KE = Kinetic\ Energy,\ m = mass\ of\ object\ \&\ v = speed\ of\ vehicle$

As the accident take place this Energy is transferred to a devastating force (Thompson et al. 2005), an injury may cause because of this devastating force. As the brakes are applied in the vehicle, two forces are generated in the vehicle, which work diminishes speed of vehicle. One of these is gravitational power (g) and second is brake power (f). As the coefficient of friction at the plain surface is 0.8 and gravitational force is 9.8 m/s^2. So the time taken by vehicle to reduce the speed from u to v can be calculated as:

$$t = \frac{v - u}{a}$$

Where,
$v = final\ speed,\ u = initial\ speed\ \&\ a = acceleration\ or\ deceleration$

5 HARDWARE DESCRIPTION

LCD MODULE: LCD screen is used to display the message to the user. Here we are use a 16*2 LCD, which means the LCD is having 16 Columns & 2 Rows, so we will be able to write 32 characters at the screen at one time. Figure (3) is showing an LCD 16*2. Each of these characters is made by 5*8 Pixel Dots. We can also customize the words by providing binary values in 5*8 matrixes, '1' will show the character is present and '0' will the character is absent. The LCD used to operate at 4.7-5.3V voltage and consumes a current of 1mA. It can be available in green and blue background. It can display the alphanumeric, mean Alphabets as well as numeric letter.

ARDUINO UNO: The ARDUINO UNO is electronics board; it supports both analog and digital input/output that may be easily interfaced with other boards and various circuits. Arduino can also be used for ADC and DAC. The operating voltage for Arduino board is 5 Volts. It can also be used for serial communication using UART, I2C & SPPI (Zhang et al. 2018).

UBLOX NEO 6M GPS RECEIVER MODULE: UBLOX NEO 6M is the main GPS recipient which is intended for low force utilization and minimal effort. UBLOX NEO 6M GPS collector module can withstand the natural conditions and electrical testing which are performed for street vehicles.

GLOBAL POSITION SYSTEM (GPS): It is a constellation of 24 satellites & multiple ground stations. It uses satellite as a reference point to calculate accurate position. It gives the location and time data of in every climate conditions.

SIM800A GSM: In Arduino community SIM 800A is the most used GSM module. SIM800A has a property which helps its interaction with Arduino easier to help the beginners.

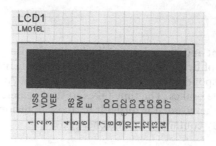

Figure 3.   LCD 16*2.

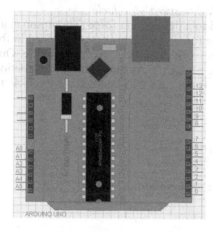

Figure 4.   Arduino Uno.

Figure 5.   GSM SIM800A Module.

Figure (4) is showing an Arduino Uno and Figure (5) shows the GSM SIM800A Module used in this project.

6 SIMULATION

Proteus 8 is used for simulation and testing before going for hardware, it helps in checking if any error is occurred in the program. Figure 6 shows the simulation diagram of accident detection system. Here we have used Arduino Uno, vibration Sensor, LCD, GSM & GPS Module and all these components are already described.

7 RESULT

The system was able to detect the accidents of the vehicle is occurred and then sends the message to the monitoring centre through GSM module. At the monitoring centre it can identify the exact location of accident on Google Map through GPS Module. Then the monitoring centre will inform any ambulance which is in the same area the accident is occurred, so that the rescue of the accidental person can be done with a small time gap.

In this way this Mishap Detection and Alerting System Using GPS and GSM shown in Figure 7, the prototype shown in Figure 7 can identify the occurrence and location of accident and human life can be saved.

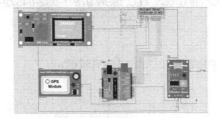

Figure 6.   Simulation circuit.

Figure 7.   Hardware of project.

REFERENCES

Keung, Lee & kam 2019. Cloud-Based Cyber-Physical Robotic Mobile Fulfillment Systems: A Case Study of Collision Avoidance. Volume 8, pp. 89318–89336.

Lee, Lv, Ng & Choy 2018. Design and application of Internet of Things-based warehouse management system for smart logistics. Int. J. Prod. Res. vol. 56, pp. 2753–2768.

Sadeghi, C. 2015. Security and privacy challenges in industrial Internet of Things. Proc. 52nd Annu. Design Autom. Conf. (DAC). pp. 1–6.

Sun & Tian 2019. A Model Predictive Controller with Switched Tracking Error for Autonomous Vehicle Path Tracking. Volume-7, pp 1–12.

Zhang, Zhu & Lv. 2018. CPS-based smart control model for shop or material handling. IEEE Trans. Ind. Informant., vol. 14, no. 4, pp. 1764–1775.

Hussein, & M. 2007. Effective Coverage Control for Mobile Sensor Networks With Guaranteed Collision Avoidance. IEEE Transactions on Control, Volume: 15 Issue: 4, pp 642–657.

Deep, & K. 2014. Wireless Reporting System for Accident Detection at Higher Speeds. in IJERA. Vol. 4, Issue 9, pp 17–20.

Megalingam, Nair & Prakhya 2010. Wireless Vehicular Accident Detection and Reporting System. in International Conference on Mechanical and Electrical Technology. pp. 636–640.

Thompson, White & D. 2005. Using Smart phones to Detect Car Accidents and Provide Situational Awareness to Emergency Responders. in Third International ICST Conference on Mobile Wireless Middleware, Operating Systems, and Applications.

Tang & Gao 2005. Traffic-incident detection-algorithm based on nonparametric regression. In IEEE Transactions on Intelligent Transportation Systems, vol. 6, pp. 38–42.

Virtanen, Schirokoff & J. 2005. Impacts of an automatic emergency call system on accident consequences. In eighteen proceeding of ICTCT. pp. 1–6.

Recent Trends in Communication and Electronics – Sharma et al. (Eds)
© 2021 Taylor & Francis Group, LLC, ISBN 978-1-032-04572-6

Review paper on power generation by piezoelectric footstep technique

Anjana Bhardwaj, Krishna Gupta, Pranshul Sharma, Tushar Kansal & Lovekesh Singh
ECE Department, ABES Engineering College, Ghaziabad, India

ABSTRACT: Electricity is a basic need of human civilization. To a large extent, our energy needs are fulfilled by non-renewable sources of energy which are depleting quickly. They also cause a lot of other problems such as global warming, ozone depletion and extreme weather. In our current model, the human weight is used to apply pressure (strain) to the piezo crystal to produce electricity. With this piezo-electric device or other techniques used to converting mechanical energy into electrical energy, we can create an alternative that uses human power for its sustenance. Over years research has been done extensively in the fields of renewable form of energy (Nechibvute et al. 2013).

Keywords: Piezoelectric current, inverters, sensors

1 INTRODUCTION

Electricity is a fundamental requirement for the modern human and this demand for electricity has always increased with its inception. These sources are also limited in number and take millennia to replenish naturally. In this paper we shall see at relatively new form of energy. Piezo-electric footstep power generation system is new system which works on the simple principle that "Energy can neither be created nor destroyed" (Li et al. 2014). Foot step power generation mechanism is to converting mechanical energy into electrical energy (Boby et al. 2016).

In this type of system, a human being simply walks on the piezo surface and this energy is stored within the crystal as t contracts due to pressure and when the crystal relaxes, the mechanical energy is simply converted into electrical with the help of a transducer. Figure 1 show the Piezoelectric Power Producing Mechanism, here we can see the piezoelectric sensor is placed between two metallic planes. As a person press the sensor by his or her feet, the compression is used to produce electricity.

2 LITERATURE SURVEY

A simple idea to produce electricity with the help of ocean currents was given. An Oscillating Water Column (OWC) uses piezoelectric concept to produce electricity by converting and storing energy from ocean waves (Takamura et al. 2018). An equivalent idea of resistance control for maximum power transfer method using piezoelectric substance for power generation was undertaken (Varshney et al. 2017). Evaluation on mechanical impact parameters in piezoelectric power generation was carried out (Basari et al. 2015). Electrical power generation using foot step for urban area energy applications explained that how a power can be generated in urban areas (Ghosh et al. 2013) (Koyama et al. 2008).

DOI: 10.1201/9781003193838-9

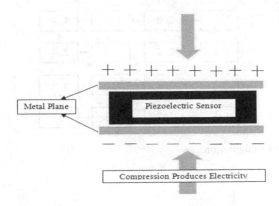

Figure 1.   Piezoelectric Power Producing Mechanism.

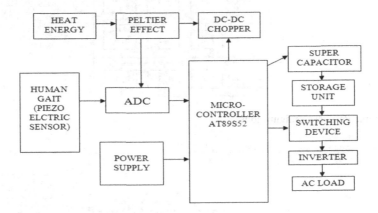

Figure 2.   Block diagram.

3   VARIOUS TECHNOLOGIES FOR POWER GENRATION

As the flywheel is attached with gear wheel of the system that is further connected to rotate the shaft and shaft is coupled with the P.M.D.C. generator to generate DC current which is stored in the batteries (Taghavi et al. 2016).

TYPES OF CONFIGURATION

A. *Simple Series Configuration:* In this configuration, a high voltage is produced for a very low current. Here the final voltage is calculated by adding the voltages across each block of piezoelectric crystal. Figure 3 shows the simple series configuration; here all the blocks are connected in series, i.e., one after other.

B. *Simple Parallel Configuration*: In this configuration the voltage of the system is same as the voltage at each terminal of piezoelectric crystal, and on the other hand the final current is produces by adding all the currents across each terminal. Figure 4 shows a simple parallel configuration by which we can obtain a high current.

C. *Parallel-Series Configuration*: The drawbacks of previous two configurations are overcome by this configuration. As here both series and parallel connections of crystals are available so this configuration provides a good value of voltage and current as well. So finally, it also gives a power produced as compared to previous one. Figure 5 shows a simple parallel configuration by which we can obtain a high current.

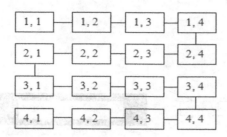

Figure 3.   Simple Series Configuration.

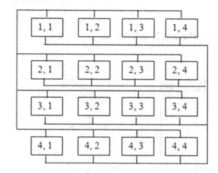

Figure 4.   Simple Parallel Configurations.

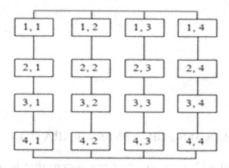

Figure 5.   Parallel-Series Configuration.

D. *Total cross tied (TCT) Configuration:* In this configuration all interconnections are done at column of the junction of parallel-series configuration through ties and the TCT configuration is shown in Figure 6, it is new approaching piezoelectric system. We can see that the TCT configuration is the best overall configuration.

4 COMPARISON

As the different innovations are talked about before that how we can design or piezoelectric sensor to create power. Presently we will think about all the advancements based on current, voltage and force age. As we can see in Figure 7 that the simple parallel configuration produces maximum current and simple series configuration produces the minimum. The TCT configuration produces maximum voltage and simple parallel configuration produces the minimum, the comparison can be seen in Figure 8.

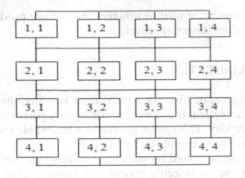

1, 1	1, 2	1, 3	1, 4
2, 1	2, 2	2, 3	2, 4
3, 1	3, 2	3, 3	3, 4
4, 1	4, 2	4, 3	4, 4

Figure 6.   Total cross tied Configuration.

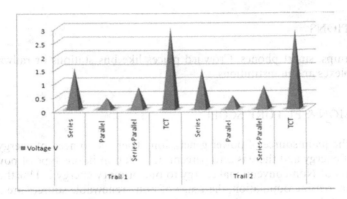

Figure 7.   Voltage by all configurations.

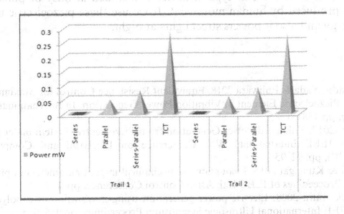

Figure 8.   Power Generation by all configurations.

So TCT configuration provides a higher efficiency as compared to the entire configuration. So, this technology should be used for further work.

5   RESULT AND DISCUSSION

The V-I characteristics of both piezoelectric materials discussed were studied in order to understand the effect associated with the various pressures and the complexity applied to them. Ammeter and voltmeter are used to measure the voltage produced across piezoelectric objects and their current flow value respectively.

6 ADVANTAGES & LIMITATION

The power can be easily generated by either walking or running; this technology is not using any type of fuel. The piezo system is ordinary to use by any person. A battery will be used to collect the generated electricity. This system generates a renewable energy. The main feature is that is not harming environment by pollution or smoke.

The piezoelectric transducer is used for dynamic measurement only and has high temperature sensitivity. Some of the piezoelectric crystals are water soluble and get dissolve in high humid environment.

7 APPLICATIONS

Power street lamps, smart phones. Crowded places like bus stations or railway stations and shopping complexes and in institutions.

8 CONCLUSION & FUTURE SCOPE

Footsteps are the main source of power generation. There is no need of energy from conventional source of energy and there is zero percent of pollution in this type of power generation. The contribution of Non-conventional energy to our primary energy is 11% that is a common fact. With the increasing population, the usage of non –renewable sources are also increasing. Due to overuse of that energy one day may extinct from our planet. Piezoelectricity is an alternative source of energy that can we used as a renewable energy nowadays at very low cost and almost one- time investment. This type of device is installed at busy or public place so that more energy is produced by human movement. If we use these piezoelectric in roads, it produces more energy and also it powers street lights at night.

REFERENCES

Takamura, Yamada, Yada & Fujiwara 2018. Equivalent Resistance Control for Maximum Power Transfer Method of Piezoelectric Element in Vibration Power Generation. In International Power Electronics Conference, pp 1381–1385.

Varshney & Shivi 2017. Enhanced Power Generation from Piezoelectric System under Partial Vibration Condition. In IEEE International WIE Conference on Electrical and Computer Engineering (WIECON-ECE), pp. 92–95.

Basari, A., Awaji & Kumagai 2015. Evaluation on mechanical impact parameters in piezoelectric power generation. In Proceedings of IEEE 10th Asian Control Conference, pp. 1–7.

Koyama and Nakamura 2008. Electric power generation using a vibration of a polyuria piezoelectric thin film. In IEEE International Ultrasonic Symposium Proceedings, pp 935–945.

Taghavi, Majid, Stinchcombe & Ieropoulos 2016. Self-sufficient wireless transmitter powered by foot-pumped operating wearable MFC. In Bio-inspiration & bio-mimetics 11, no. 2, pp. 016101.

Ghosh, Joydev, Sen, & Basak 2013. Electrical power generation using foot step for urban area energy applications. In Proceedings of IEEE International Conference on Advances in Computing, Communications and Informatics (ICACCI 2013), pp. 1366–1370.

Boby 2016. Footstep Power Generation Using Piezoelectric Transducers. In International Journal of Engineering and Innovative Technology (IJEIT pp.1–5.

Li & Strezov 2014. Modeling piezoelectric energy harvesting potential in an educational building. In Energy Convers. Manag., vol. 84, pp. 432–445.

Nechibvute, Chawanda, & Luhanga 2013. Piezoelectric Power Harvesting Devices: An Alternative Energy Source for Wireless Sensors. In Hindawi Publishing Corporation Smart Materials Research, pp.1–4.

Recent Trends in Communication and Electronics – Sharma et al. (Eds)
© 2021 Taylor & Francis Group, LLC, ISBN 978-1-032-04572-6

Design and development of interactive digital colorimeter

Rakhi Kumari, Himani, Shilpa Srivastava & Sanjay Kumar Singh
ABES Engineering College, Ghaziabad, Uttar Pradesh, India

ABSTRACT: Digital Colorimeter is an instrument used to measure the absorbance of wavelength of light at a particular frequency (color) by a sample available. The instrument available so far requires lots of manual operations. The novel Colorimeter meter developed in this paper is efficiently used to overcome that problem. With the help of microcontroller based system it can calculate the absorbance value of solution of different concentration more effectively. A touch screen display has been used for providing an interactive system. This paper provides designing of an interactive, portable and cost effective Digital Colorimeter that is capable of measuring absorbance values more effectively with the provision of storage also.

1 INTRODUCTION

Traditionally in different fields color is used as optical characteristics to control object properties. The interaction of light of different colors with the object is used for analyzing the properties of the object (Snizhko et al. 2017) (Mukesh et al. 2013). Colorimeter is a device that is used for absorbed color measurements. This device has high sensitivity that it can respond for even small change in color which can't be detected by the naked eyes. The sample under test develops a particular color when comes under contact of chemical agents and thus shows the concentration of solute in a specific solution. Colorimeter measures the absorbance of particular wavelength of light by a specific solution. It comprises of two main components an illuminant and a photo detector. It is commonly used to determine the concentration of a known solute in a given solution by the application of the Beer– Lambert law (Mukesh et al. 2013) (Anzalone et al. 2013). As per the Beer–Lambert law the light absorbance capacity or optical density of any solution is directly proportional to concentration of solution in moles per litre and length of the light path passes through the solution.

$$A \propto cl \tag{1}$$

Where,
A = Absorbance/Optical density of solution
c = Concentration of solution in moles/L
l = Path length.
Absorbance A is given by

$$A = \log P_0/P \tag{2}$$

Where, P0 is the input power and P is the output power of the light after passing through the sub-stance. With this in mind, we can begin designing our device.

This paper discusses the design and development of an Interactive Digital Colorimeter which provides an efficient approach to measure the readings and the micro-controller ensures

DOI: 10.1201/9781003193838-10

that the calculations are done on standard measurements (Firdaus et al. 2019). Colorimeter can detect the absorbance value of different concentration of solvent as this will be able to help in various clinical operations as for example in pathological lab for detecting various illness and disease, concentration of glucose or any other solvent in blood is to be checked and absorbance value which we are calculating here can easily tell us the concentration value as they are directly proportional.

2   THE PROPOSED SYSTEM

The system is digital and is equipped with data accusation system. The proposed colorimeter comprises of ATMEGA 328P, Nextion Human Machine Interface (HMI), Real Time Clock (RTC), data logger and a photo sensor as show in Figure 1.

Arduino uno an open source electronics prototyping platform is used as the microcontroller unit. A 3.2 inch interactive touch screen is used for efficient display as well as easy and effective control of the equipment. The data logger helps in saving the measured values which could produce graphical results for analysis and maintenance of record for future use. Battery operation enhances the portability of the equipment which improves the usage of the equipment. The designed system can be operated and controlled by a mobile using IoT, thereby increasing the reach out for the acquired data. As the IoT devices are embedded with sensors and provides internet connectivity it allows the systems communication and control via the web.

3   HARDWARE SELECTION OF THE SYSTEM

3.1   *Arduino Uno*

The Microcontroller board used in the proposed project is an Arduino Uno on the ATmega328P. This microcontroller board consists of input/output pins of total number 14 of which 6 pins can be used as a PWM outputs, 6 analog inputs, a ceramic resonator of resonating frequency16 MHz, a USB connection, a power jack, an ICSP header and a reset button. The calculations regarding incoming voltage in photodiode and the corresponding absorbance values of each solution at different frequencies corresponding to different colors of the filter is done by the Arduino.

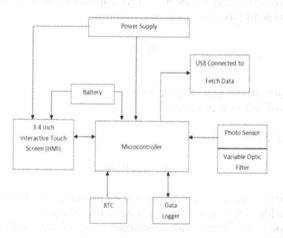

Figure 1.   Block Diagram of the proposed Colorimeter.

3.2 Human Machine Interface (HMI)

The system is operated by Nextion NX4024T032 3.2 Inch HMI Intelligent Smart USART UART Serial Touch TFT LCD Screen Module. It is Human Machine inter-face that provides us with the complete control and helps us to visualize the process or application etc. which we are undergoing within machine. It is a very good alternative for previously used display devices like LCD and LED Nixie tube. With the Nextion Editor software, we as a user can able to create our own imagination in the form of design and interface that to the display of Nextion. The HMI is designed to automatically be interfaced with Aurdino Microcontroller.

3.3 Data logger

A data logger has find its application in various applications where there is a need to store data or information for future use. So, it is a device which records the information in the running time for future reference. Sometimes it is inbuilt inside the system or sometimes it is externally added with the system.

3.4 RTC

Real time clocks (RTC), is a Clock Module that displays time, date etc related to real time. The real time clock IC DS1307 is an 8 pin IC using an I2C inter-face. This IC also uses a low power for its operation, it also consist 56 bytes SRAM for battery backup. The clock/calendar in this IC provides seconds, minutes, hours, day, date, month and year qualified data.

3.5 Software design

For the development of Interactive digital colorimeter the programming of the microcontroller is in C language with the use of μ-Vision IDE-Kiel. Programming module is needed for the calculating the absorbance value in addition to having communication with HMI for interactive control of the device and display of.

4 STEPS FOLLOWED TO MEASURE ABSORBANCE

In the developed interactive digital colorimeter shown in Figure 2, following steps are followed for the measurement of absorbance of solutions of different concentration.

Step 1: Colorimeter is termed ON, a welcome screen Figure 4 with a Logo appears on the touch screen and prompts the user to GET STARTED touch button. At the same time in the circuit as shown in the block diagram of Figure 3 a transmitter comprising of an LED is turned ON.

Step 2: After pressing the GET STARTED button on HMI a calibration screen as shown in Figure 5 appears where calibration steps are given. First calibration step is to set the absorbance value to zero for this practically a distilled solution is added in a small tub and held in a square cross section, sealed at one end, called a cuvette shown in Figure 2. The filter is adjusted to a particular wavelength and the LED light passes through this sample solution and reaches the photodiode. The photodiode converts the light signal in analog voltage which is further processed by signal processing unit to be accessible by microcontroller Figure 3.

Step 3: After calibration the system is ready to measure the absorbance of unknown solution. On pressing the arrow next page of SELECT FILTER appears on the screen of HMI Figure 6. There are options of filter selection from 400-680 nm. By pressing the touch buttons any filter can be selected.

Step 4: Operating Wavelength and the absorbance value of the solvent under test appears on the screen.

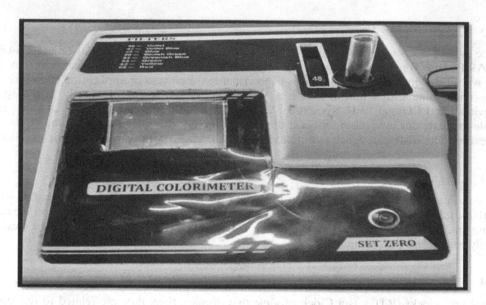

Figure 2.   Developed Colorimeter.

Figure 3.   Block diagram for measurement of absorbance value.

Figure 4.   Welcome Screen on GUI.

Step 5: Keep on adding solvent to the distilled water by keeping the selection of filter same and we will able to note down the absorbance value for different concentration.

Step 6: Again by pressing or selecting the next filter step 4 & Step 5 is repeated again.

Step 7: Keep on changing filter and checking the absorbance value for different concentration of glucose in the distilled water.

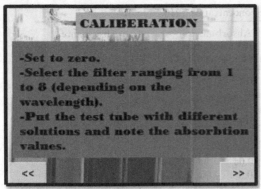

Figure 5.   Steps for calibration.

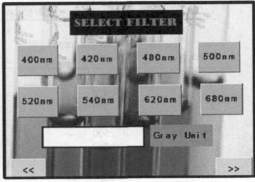

Figure 6.   Filer selection.

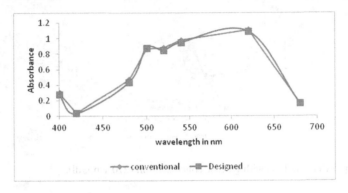

Figure 7.   Comparative analysis with glucose concentration (59.9mg/dl).

5   RESULTS

The designed digital colorimeter was tested using different concentration of glucose in the distilled water and results were compared with the conventional colorimeter. The comparison of these results is presented in Table 1 and Table 2. The designed digital colorimeter shows similar values when compared to the conventional colorimeter.

Table 1.   Comparative readings with glucose concentration (59.9 mg/dl).

Filter No.	Wave-length (nm)	Absorbance (Conventional Colorimeter)	Absorbance (Digital Colorimeter)
1	680	0.12	0.123
2	620	0.01	0.01
3	540	0.92	0.924
4	520	0.11	0.11
5	500	0.58	0.56
6	480	0.05	0.05
7	420	-0.17	-0.17
8	400	0.55	0.54

Table 2. Comparative readings with glucose concentration (210.1 mg/dl).

Filter No.	Wave-length (nm)	Absorb-ance (Con-ventional Colorimeter)	Absorbance (Designed Digital Colorime-ter)
1	680	0.17	0.168
2	620	1.11	1.09
3	540	0.98	0.97
4	520	0.88	0.86
5	500	0.87	0.89
6	480	0.47	0.45
7	420	-0.04	-0.039
8	400	0.3	0.31

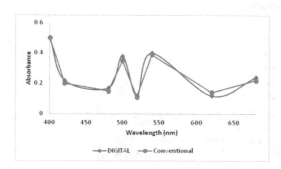

Figure 8. Comparative analysis with glucose concentration (210.1 mg/dl).

6 CONCLUSION

The testing of the developed colorimeter was done with different solutions at different wave-lengths. The performance of the digital colorimeter designed with this methodology was successfully demonstrated. The designed system is user friendly, accurate and much cost-effective.

REFRENCES

Snizhko, D. V., Sushko, O. A., Reshetnyak, E. A., Shtofel, D. H., Zyska, T., Mussabekov, N., & Kalizhanova, A. 2017. Colorimeter based on color sensor.

Mukesh, M. Z. J., & Shinde, M. A. 2013. Absorbance Measurement by Colorimeter. *International Journal*, *3*(10).

Anzalone, G. C., Glover, A. G., & Pearce, J. M. (2013). Open-source colorimeter. *Sensors*, *13*(4), 5338–5346.

Firdaus, M. L., Aprian, A., Meileza, N., Hitsmi, M., Elvia, R., Rahmidar, L., & Khaydarov, R. (2019). Smartphone coupled with a paper-based colorimetric device for sensitive and portable mercury ion sensing. *Chemosensors*, *7*(2), 25.

Recent Trends in Communication and Electronics – Sharma et al. (Eds)
© 2021 Taylor & Francis Group, LLC, ISBN 978-1-032-04572-6

Designing and development of a low cost electronic braille pad using electromagnetic actuation

Tuhin Rana & Ankit Mitra
Department of Electrical Engineering, National Institute of Technology Durgapur (NITDgp), WB, India

Masum Billah
Department of Electronics and Communication Engineering, National Institute of Technology Durgapur (NITDgp), WB, India

Nirupam Mondal
Department of Electrical Engineering, National Institute of Technology Durgapur (NITDgp), WB, India

Tushar Kanti Bera
Department of Electrical Engineering, National Institute of Technology Durgapur (NITDgp), WB, India
Center for Biomedical Engineering and Assistive Technology (BEAT), National Institute of Technology Durgapur (NITDgp), WB, India

ABSTRACT: Visual and hearing disabilities put the people in difficulties during communication with others. Fast communication is still found very difficult for disabled people in real world as the braille and other assistive technologies are generally designed for some specific purposes such as education or they are costly. In this paper, a low cost electromagnetic actuation based Electronic Braille Pad (EBP) is developed for electronic braille based communication technology. Each character cell of a brail letter is developed with a push-type solenoid actuator consisting of a copper wire and mild steel core. A brail pad unit is developed with six actuators all of which are connected with an electronic controller. The controller has been developed with an Arduino UNO board interfaced with a custom made electronic keypad. The basic English letters are typed through the electronic keypad as per the Braille technology and the actuation of the braille pad is studied.

1   INTRODUCTION

Visual and hearing disabilities (Gordon et al. 2017, Crow, 2008) in people put a big barrier in their day to day life especially during the communication with their community. For educational and other purposes the braille technology is being used for communication and understanding information among the visual and hearing disabled persons. In braille medium, the braille alphabets (Jiménez et al. 2009) are made with a protruded points or projected dots in a braille cells (Figure 1a) containing six points or in 3×2 matrix arrangement (Figure 1a). All the English alphabets are represented by distinct braille alphabets which all are the different combinations/patterns of projected dots (Figure 1b). Though some other assistive technologies have been developed for the disabled people during their education and working in home, but one to one communication is still found very difficult in real world. Electronic braille communicators (Fewell, 1980, Shubhom et al. 2017) developed with piezoelectric materials (Tichý et al. 2010, Cho et al. 2006, Smithmaitrie et al. 2008) are not only expansive but also suffer from the wear and tear problem with time.

DOI: 10.1201/9781003193838-11

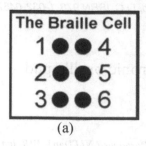

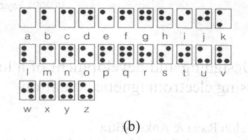

Figure 1. (a) A braille cell, (b) English alphabets and their braille-equivalent representations.

In this paper, an electromagnetic actuation based Electronic Braille Pad (EBP) (Supriya & Senthilkumar, 2009) is developed for electronic braille based communication technology (Supriya & Senthilkumar, 2009). The electromagnetic actuation system has been developed as a low cost electronic braille based communication technology. Each character cells of a brail pad is developed with a push-type solenoid actuator which consists of a copper wire and mild steel core. Six identical electromagnetic actuators all are assembled together and connected with the electronic controller to develop an electronic brail pad unit. The braille pad controller has been developed with a push button electronic switch based signal generator board which is basically a custom made electronic switch assembly or switch board. The electronic switch assembly works as an electronic keypad powered by a DC power supply and is used to select- ively provide the power to the electromagnetic actuators to work as the braille cells. The elec- tronic input to type the Basic English letters are sent through the keypad as per the Braille character protocol and the actuation of the braille pad is studied.

2 MATERIALS AND METHOD

The proposed electromagnetic braille pad is developed with an electromagnetic actuator panel powered and controlled by a Braille Pad Controller which is nothing but an elec- tronic switch board controlled by an Arduino UNO. Each of the electromagnetic actu- ator assembly has been developed with six push type electromagnetic actuators each of which is controlled by the Braille Pad Controller. Each electromagnetic actuation system works to move a piston or core which corresponds to a dot of the six-dot based braille cell. To control all the character cells of a brail pad DC signal provided to the copper coil wound over a mild steel core.

2.1 *Electromagnetic Braille Pad (EBP)*

The Electromagnetic Braille Pad (EBP) has been developed using custom made solenoid actu- ators which works with the electromagnetic principle. Each of these actuators consists of a ferromagnetic core and a ferromagnetic body (Figure 2a). The schematic of the proposed actuator with its ferromagnetic core, ferromagnetic body, axial arm and plunger has been show in the Figure 2a. Figure 2b shows a complete electro-magnetic actuator with copper coils wound over the core. The ferromagnetic core and ferro-magnetic body help to establish the necessary magnetic fields to pull the ferromagnetic actuator axis which is attached to a non-ferromagnetic arm that moves axially through the ferromagnetic core. Six such solenoid actuators (length: 4 cm) were put together to make the electromagnetic actuator assembly (Figure 2c). The entire electromagnetic actuator assembly is put in a wooden box containing a flat top cover plate which has six holes to allow the movement of six actuator cores or axes. The actuator axes passes through the holes of the perforated flat plate and the surface profile of the cover plate produces a required pattern corresponding to a braille alphabet (Figure 2d). The actuation of each of the actuators was controlled individually by an Arduino UNO board

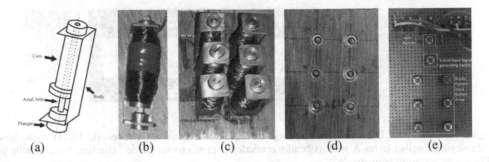

(a) (b) (c) (d) (e)

Figure 2. (a) Schematic of solenoid actuator unit; (b) Physical model of designed solenoid actuator; (c) assembly of six solenoid actuators; (d) Face of the Braille display cell, (e) Push button switch based electronic key pad for controlling the braille keyboard.

through switching transistors. To create a particular pad-plate surface profile, an input command is given by the array of 6 pushbutton switches (Figure 2e) correspond to the six dots of a standard Braille cell pattern. Each of the pushbuttons generates a HIGH signal (5V) when pressed, rest of the time it produces a LOW signal (0V).

2.2 Solenoid driver circuit

The solenoids are driven by Arduino board through transistor switches. The solenoids are connected in diode clamped switching circuits to protect the semiconductor switches (Transistors) from the inductive energy stored in the solenoid coils as shown in schematic of Figure 3a. The on state losses cause in heat generation in the transistors, so, the transistors should have less ON state resistance and it must be able to withstand the heat generated in it. Taking this into account the solenoid driver circuit is designed by using D882 n-p-n power transistors as in the Figure 3b. The driver panel contains six individual solenoid driver circuits with a common ground.

2.3 Arduino UNO based controller board

The braille pad controller has been developed with a push button electronic switch based signal generator board (Figure 2e) which is an assembly of six push button switches which are arranged in a 3×2 matrix. To type the Basic English letters in braille format, the electronic inputs are sent to the actuator assembly through the keypad. Electronic inputs for the different English alphabets are applied as per the braille character protocol and the actuation of the

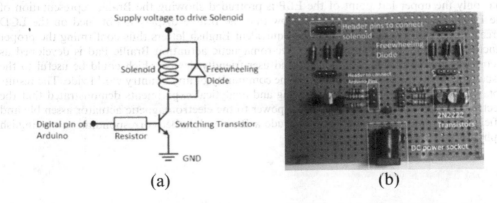

(a) (b)

Figure 3. (a) Solenoid driving circuit, (b) PCB of the receiving end of the solenoid driver circuit.

| (a) | (b) | (c) |

Figure 4. (a) The LCD board for displaying the English and Braille Alphabets (b) LCD display board is displaying English letter A and its braille equivalent, (c) electromagnetic actuation based braille pad (The upper left point is protruded).

braille pad is studied and the Braille-pad performance is assessed. The input given to the keypad has been used to display the input on a liquid-crystal display (LCD) display. The inter-face between the keypad, the LCD display, and the EBP is performed and controlled by an ArduinoUNO board.

2.4 Liquid-crystal display

The input given to the keypad has been used to display the input on an LCD display. The inter-face between the keypad and the LCD display is performed and controlled by an Arduino UNO based controller board. A liquid-crystal display (LCD) is a flat-panel display which uses the light-modulating properties of liquid crystals combined with polarisers. Figure 4a illustrates the LCD along with its associated circuit where the LCD is connected with an Arduino using potentiometer and resistors. It is used to display the alphabets which the person presses on the keypad input device through the push buttons. The corresponding Braille equivalent is simultaneously displayed on the line below. This can be used as a reference to verify that the sender and the receiving end gets the same signal in the form of alphabet as well as whether the EBP displays the correct Braille pattern.

3 RESULTS

The electromagnetically actuated Braille-pad device is tested by displaying the Braille symbols for one of the English alphabets at a time. The push button based controller is used to type a Braille character and the Braille character is displayed on a LCD display panel. A particular test case for the English alphabet "A" is displayed in Figure 4b. Figure 4b shows the LCD for the test case whereas Figure 4c shows the EBP surface for that particular test case. It is noticed that only the upper left point of the EBP is protruded showing the Braille representation of the English alphabet "A". Test runs show that the Braille display is obtained on the LCD screen along with the same output and equivalent English letters thus confirming the proper function of the device. The proposed electromagnetic actuation Braille Pad is developed as a cost effective, portable, light weight and user friendly device which could be useful to the visually impaired persons from low income community of any country worldwide. The results obtained from the instrumentation testing and analytical experiments demonstrated that the electronic key board successfully fed the power to the electromagnetic actuator assembly and efficiently activate the assembly to protrude a Braille alphabet corresponding to an English alphabet.

4 CONCLUSIONS

In this paper, an electromagnetic actuation based Electronic Braille Pad (EBP) is developed for electronic braille based communication technology. Each character cells of a brail pad is developed with a push-type solenoid actuator which consists of a copper wire and mild steel core. Six identical electromagnetic actuators are assembled together and connected with the electronic controller to develop an electronic brail pad unit. The braille pad controller has been developed with a push button electronic switch based signal generator board which is interfaced with an Arduino UNO based signal controller. The electronic input to type the Basic English letters are sent through the keypad as per the Braille character protocol and the actuation of the braille pad is studied. The instrumentation testing and analytical results demonstrated that the electronic key board successfully provides the electric power to the solenoids and efficiently activate the electromagnetic actuator assembly to protrude a braille alphabet corresponds to an English alphabet. It is found that, the proposed electromagnetic actuation Braille Pad is developed as a cost effective, portable, light weight and user friendly device which could be useful to the visually impaired persons from low income community of any country worldwide.

REFERENCES

Sanchez-Gordon, S., Mejía, M., & Luján-Mora, S. (2017, April). *Model for adjusting workplaces for employees with visual and hearing disabilities*. In 2017 Fourth International Conference on eDemocracy & eGovernment (ICEDEG) (pp. 240–244).

Crow, K. L. (2008). *Four types of disabilities: Their impact on online learning*. TechTrends, 52(1), 51.

Jiménez, J., Olea, J., Torres, J., Alonso, I., Harder, D., & Fischer, K. (2009). *Biography of louis braille and invention of the braille alphabet*. Survey of ophthalmology, 54(1), 142–149.

Fewell, W. B. (1980). *U.S. Patent No. 4,215,490*. Washington, DC: U.S. Patent and Trademark Office.

Shubhom, V. T., Keerthan, S., Swathi, S., Abhiram, G., & Shashidhar, R. (2017, October). *BRAPTER: Compact braille transput communicator*. In 2017 IEEE International Conference on Consumer Electronics-Asia (ICCE-Asia) (pp. 164–168).

Tichý, J., Erhart, J., Kittinger, E., & Privratska, J. (2010). *Fundamentals of piezoelectric sensorics: mechanical, dielectric, and thermodynamical properties of piezoelectric materials*. Springer Science & Business Media.

Cho, H. C., Kim, B. S., Park, J. J., & Song, J. B. (2006, October). *Development of a Braille display using piezoelectric linear motors*. In 2006 SICE-ICASE International Joint Conference (pp. 1917–1921).

Smithmaitrie, P., Kanjantoe, J., & Tandayya, P. (2008). *Touching force response of the piezoelectric Braille cell*. Disability and Rehabilitation: Assistive Technology, 3(6), 360–365.

Supriya, S., & Senthilkumar, A. (2009, June). *Electronic Braille pad*. In 2009 International Conference on Control, Automation, Communication and Energy Conservation (pp. 1–5).

Recent Trends in Communication and Electronics – Sharma et al. (Eds)
© 2021 Taylor & Francis Group, LLC, ISBN 978-1-032-04572-6

Analysis of segmentation techniques using morphological operators on brain images

Akanksha Kulshreshtha & Arpita Nagpal
G.D.Goenka University, Gurgaon, India

ABSTRACT: Image segmentation is an important step in medical image diagnosis system. There are various segmentation methods to partition an image into different subregions on the basis of edge detection, area based or clustering based methods. This paper provides a thorough analysis of different segmentation techniques with morphological operators for brain tumor detection. The different segmentation techniques are k-mean, Entropy filtering, area-based segmentation. After segmenting the image, morphological operators are used to mask the segments. Manual segmentation is used to construct the gold standard for comparing the segmented image. Comparison is performed using performance parameters such as F1 score, Sensitivity, Specificity, Loss Function, precision and Jaccard Coefficient. The elapsed time is also been observed for the segmentation process. The experimental results show that combining segmentation techniques with morphological operation of erosion gives an improved performance over dilation.

1 INTRODUCTION

Medical image segmentation used to handle different medical modalities like Medical resonance imaging (MRI), computed tomography, ultrasound, multimodal, digital mammography, X-ray, Chest radiography etc. The common challenge in brain MRI images is to extract the white matter, gray matter and cerebrospinal fluid (Bauer et al.2012). The brain MRI image segmentation includes the partitioning of basic three regions i.e. white matter, grey matter and cerebrospinal fluid. These anatomical features help the radiologist to detect the exact shape, size and appearance of tumor (Christe et al. 2010 & Vijay and Subhashini 2013). Brain MRI image segmentation comprised of few important steps like pre-processing of image, feature extraction, edge detection and tumor detection. The pre-processing of the image is basically the filtering stage, in which the different types of filters are employed to remove the noise present in the image (Hooda et al. 2014 & Joseph et al. 2014). Technically Image Segmentation can be defined as the differentiating the digital image into sub-regions.

This paper includes the analysis of performance parameters of different segmentation methods for brain MRI images. The remainder of the paper contains related work, segmentation techniques, observations & results and conclusion.

2 RELATED WORK

Researchers gave commendable contribution in the field of image segmentation. In 1997, a feature extraction algorithm based on density gradient was proposed (Comaniciu, Meer 1997). Colored image of 512X512 dimension was analyzed by considering mean shift algorithm for feature space processing. Cooper combined variance and no. of regions to reduce

DOI: 10.1201/9781003193838-12

the NP problem (Cooper 1998). He analyzed the reconstruction of 3D from 2D image by relaxing the no range information condition. Schimdt et.al demonstrated the segmentation using alignment features. They analysed the algorithm quantitatively by considering 4 different types of alignment- based feature encoding for supervised pixel classification (Schmidt, et al. 2005). In 2003, the boundary predictions were estimated by using graph -based estimation. This method preserves the low variability image regions and ignore the details (Felzenszwalb et al. 2004).

Patil et.al discussed about the various image segmentation techniques like region based, threshold based, segmentation based on clustering and region merging. The detailed classification of these segmentation techniques was discussed and tabulated for the purpose of research (Patil & Deore 2013). The Local entropy segmentation of the image can be determined by gray level peaks of its histogram. The gray level co-occurrence matrix was analysed to observe the Haralick texture feature for image segmentation (Chaudhary & Kulkarni 2013). The clustering techniques proved a good solution for efficient image segmentation. K-mean and fuzzy C mean clustering algorithms can be used to find out the regions of similarity. In 2015, the hybrid clustering technique was proposed for image segmentation in which K-mean and Fuzzy C mean clustering algorithms were integrated (Abdel et al. 2015).

In 2015, the morphological operation along with the fuzzy C mean clustering-based segmentation was implemented to improve the segmentation result (Chudasama et al. 2015). One of the parameters is dice similarity coefficient was discussed for probabilistic segmentation (Shamir et al. 2019). In 2018, the manual segmentation for brain MRI images was discussed (Tiahyaningtijas, 2018). By using performance parameters, different segmentation algorithms are analyzed for various gaussian noise levels (Zhang, et al. 2019).

3 IMPLEMENTATION OF SEGMENTATION METHODS

Thresholding is suitable for images having light objects on dark backgrounds. It is used to figure out the edges or regions by reducing the noise impact in the images. There are two types of thresholding methods for medical images, Global and local thresholding. (Ratha et al. 2019). The region-based segmentation is based on detection and division of areas (Tiahyaningtijas 2018). Figure (1) is showing the output of different segmentation methods on brain MRI image. The area is separated according to the variation in intensities at the borders. Watershed segmentation can be demonstrated by considering the holes in each local minimum, which immerses in water and will rises until local maximums are not achieved. Clustering algorithms can also be used for segmentation of images with multiple intensity variations. Clustering is mainly based on forming the classes of pixels of similar types. The quality of result depends on the measurement of similarity. K- mean, Fuzzy clustering are the most commonly used clustering techniques used for brain medical image segmentation (Alam, et al. 2019). Entropy

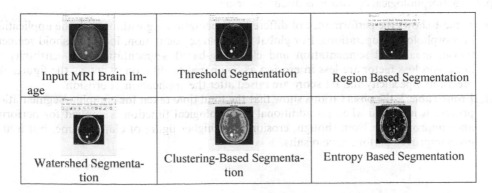

Input MRI Brain Image	Threshold Segmentation	Region Based Segmentation
Watershed Segmentation	Clustering-Based Segmentation	Entropy Based Segmentation

Figure 1. Output of different segmentation methods on brain MRI image.

thresholding is based on the information present in gray level histogram. The entropy of the histogram is used to extract the expected amount of information needed to specify the state of the system.

4 PROPOSED METHOD

The analysis of few segmentation algorithms is based on the observed values of their respective performance parameters before and after applying the morphological operations

   Step 1: Input a brain MRI image.
   Step 2: Pre-processing of image by noise filtering.
   Step 3: Apply Segmentation technique with morphological operations
   Step 4: Create the gold standard of the same image by manual segmentation
   Step 5: Observe the values of performance parameters by comparing the region of interest in both images

The effectiveness of the proposed algorithm is validated by using some of the performance parameters. The performance parameters used are F1 score, sensitivity, precision, Specificity, loss function and Jaccard coefficient. The overview of these parameters is as follows:

SENSITIVITY = TP/(TP+FN),   PRECISION = TP/(TP+FP)

SPECIFICITY = TN/(TN+FP) ,   JACCARD COEFFICIENT = TP/(TP+FP+FN)

LOSS FUNCTION= 1- ((TP + TN)/(TP+TN+FP+FN))

F1 SCORE= 2(PRECISION*SENSITIVITY)/(PRECISION+RECALL)

ELAPSED TIME: It is the time taken by algorithm to segment the image

TP- Tumor exists & is detected correctly TN- Tumor does not exist & is not detected

FP- Tumor does not exist but is detected FN- Tumor exists but is not detected

5 OBSERVATIONS & RESULTS

The structuring element in morphological operations can be of square shape, diamond shape or cross shaped, depending upon the requirement. The basic morphological operations are erosion, dilation, opening and closing, that can be expressed by logical notations. In the two basic operations of erosion and dilation the pixels are removed and added from the boundaries, depending on the shape and nature of the structuring element.

   Another parameter for analysis is elapsed time, which can be defined as the time taken from starting of the segmentation process to its termination. The variation in elapsed time after applying morphological operators is discussed below:

a.)   Table 1 shows the performance of different segmentation algorithms with the application of morphological operations. For global threshold segmentation, local threshold segmentation, area-based segmentation and clustering-based segmentation the sensitivity to noise and loss factor are less in erosion as compared to dilation, whereas the figures for precision, Specificity and F1 score are raised after the application of erosion.

b.)   From Table 2, the observations show that the total time taken for complete segmentation process is increased when an additional morphological function is applied for performance improvement. Even though, erosion gives higher figure of elapsed time, but it also gives improved performance results.

Table 1. Observed values of performance parameters of segmentation methods.

Segmentation Algorithm	Performance Parameter	Value	After erosion	After Dilation
Global Threshold Segmentation	Sensitivity	0.7277	0.7675	0.933
	Precision	0.1049	0.5255	0.0528
	Specificity	0.042	0.2512	0.004
	F1 Score	0.18326	0.6239	0.1
	Loss	0.8166	0.3761	0.9
	Jaccard coefficient	0.0996	0.4534	0.0526
Local Threshold Segmentation	Sensitivity	0.711	0.901	0.81
	Precision	0.15	0.074	0.16
	Specificity	0.1	0.12	0.11
	F1 Score	0.255	0.148	0.2601
	Loss	0.9812	0.9520	0.977
	Jaccard coefficient	0.02	0.074	0.0116
Region Based Segmentation	Sensitivity	0.8143	0.2312	0.8338
	Precision	0.2723	0.9	0.074
	Specificity	0.0786	1	0.0157
	F1 Score	0.4081	0.3756	0.1359
	Loss	0.5918	0.5244	0.8641
	Jaccard coefficient	0.2564	0.2312	0.0729
K-Mean clstering based segmentation	Sensitivity	0.8771	0.6929	0.6687
	Precision	0.1726	0.4263	0.1393
	Specificity	0.0284	0.2477	0.0742
	F1 Score	0.2884	0.5278	0.2306
	Loss	0.7116	0.4722	0.7694
	Jaccard coefficient	0.1685	0.3585	0.1303

Table 2. Observed values of elapsed time for segmentation techniques.

Sno.	Algorithm	Elapsed Time (seconds)	After erosion	After Dilation
1	Area/Region Based Segmentation	12.74322	16.8675	12.71465
2	K-Mean Clustering For Segmentation	2.867885	10.964	7.931
3	Global Threshold segmentation	8.77396	12.5174	15.1915
4	local Threshold Segmentation	1.50479	9.6235	11.21843

6 CONCLUSION

In this paper, the different segmentation techniques for brain MRI image are discussed for tumor detection. The impact of morphological operations: dilation and erosion are observed by evaluating the performance parameters. Manual segmentation is used to create a reference image (gold standard) for comparing the segmented image. The performance is evaluated on the basis of observed values of F1 score, Sensitivity, Specificity, Precision and Loss Function and Jaccard Coefficient. From observations, it can be concluded that application of morphological operation improves the performance of segmentation methods, whereas the elapsed time is observed to be raised with an improved performance. The erosion function gives better results than dilation in terms of F1 score, sensitivity, loss function, Specificity, Precision and Jaccard Coefficient in the discussed segmentation algorithms.

REFERENCES

Bauer, S., Fejes, T., Slotboom, J., Wiest, R., Nolte, L.P. and Reyes, M., 2012, October. Segmentation of brain tumor images based on integrated hierarchical classification and regularization. In MICCAI BraTS Workshop. Nice: Miccai Society (p. 11).

Christe, S.A., Malathy, K. and Kandaswamy, A., 2010. Improved hybrid segmentation of brain MRI tissue and tumor using statistical features. ICTACT J Image Video Process, 1(1), pp.34–49.

Vijay, J. and Subhashini, J., 2013, April. An efficient brain tumor detection methodology using K-means clustering algoriftnn. In 2013 International Conference on Communication and Signal Processing (pp. 653-657). IEEE.

Hooda, H., Verma, O.P. and Singhal, T., 2014, May. Brain tumor segmentation: A performance analysis using K-Means, Fuzzy C-Means and Region growing algorithm. In 2014 IEEE International Conference on Advanced Communications, Control and Computing Technologies (pp. 1621-1626). IEEE.

Joseph, R.P., Singh, C.S. and Manikandan, M., 2014. Brain tumor MRI image segmentation and detection in image processing. International Journal of Research in Engineering and Technology, 3(1), pp.1–5.

Comaniciu, D. and Meer, P., 1997, June. Robust analysis of feature spaces: color image segmentation. In Proceedings of IEEE computer society conference on computer vision and pattern recognition (pp. 750-755). IEEE.

Cooper, M.C., 1998. The tractability of segmentation and scene analysis. International Journal of Computer Vision, 30(1), pp.27–42.

Schmidt, M., Levner, I., Greiner, R., Murtha, A. and Bistritz, A., 2005, December. Segmenting brain tumors using alignment-based features. In Fourth International Conference on Machine Learning and Applications (ICMLA'05) (pp. 6-pp). IEEE.

Felzenszwalb, P.F. and Huttenlocher, D.P., 2004. Efficient graph-based image segmentation. International journal of computer vision, 59(2), pp.167–181.

Patil, D.D. and Deore, S.G., 2013. Medical image segmentation: a review. International Journal of Computer Science and Mobile Computing, 2(1), pp.22–27.

Chaudhari, A.K. and Kulkarni, J.V., 2013, February. Local entropy- based brain MR image segmentation. In 2013 3rd IEEE International Advance Computing Conference (IACC) (pp. 1229-1233). IEEE.

Abdel-Maksoud, E., Elmogy, M. and Al-Awadi, R., 2015. Brain tumor segmentation based on a hybrid clustering technique. Egyptian Informatics Journal, 16(1), pp.71–81.

Chudasama, D., Patel, T., Joshi, S. and Prajapati, G.I., 2015. Image segmentation using morphological operations. International Journal of Computer Applications, 117(18).

Shamir, R.R., Duchin, Y., Kim, J., Sapiro, G. and Harel, N., 2019. Continuous dice coefficient: a method for evaluating probabilistic segmentations. arXiv preprint arXiv:1906.11031.

Tjahyaningtijas, H.P.A., 2018, April. Brain tumor image segmentation in MRI image. In Proc. IOP Conf. Ser., Mater. Sci. Eng. (Vol. 336, p. 012012).

Zhang, C., Shen, X., Cheng, H. and Qian, Q., 2019. Brain tumor segmentation based on hybrid clustering and morphological operations. International Journal of Biomedical Imaging, 2019.

Ratha, P. and Mukunthan, B. 2019, Brain Tumor Detection and Segmentation using Histogram and Optimization Algorithm. International Journal of Innovative Technology and Exploring Engineering Special Issue, vol. 8, no. 10S, pp. 125–129

Alam, M.S., Rahman, M.M., Hossain, M.A., Islam, M.K., Ahmed, K.M., Ahmed, K.T., Singh, B.C. and Miah, M.S., 2019. Automatic Human Brain Tumor Detection in MRI Image Using Template-Based K Means and Improved Fuzzy C Means Clustering Algorithm. Big Data and Cognitive Computing, 3(2), p.27.

Recent Trends in Communication and Electronics – Sharma et al. (Eds)
© 2021 Taylor & Francis Group, LLC, ISBN 978-1-032-04572-6

Application of machine learning and artificial intelligence in civil engineering: Review

Uzair Khan & Rajat Verma
Department of Civil Engineering, ABES Engineering College, Ghaziabad, Uttar Pradesh, India

Rizwan Ahmad Khan
Department of Civil Engineering, Z.H College of Engineering & Technology, Aligarh Muslim University, Aligarh

S Anbu Kumar
Department of Civil Engineering, Delhi Technological University, Delhi

Harshit Varsheny
Department of Civil Engineering, Z.H College of Engineering & Technology, Aligarh Muslim University, Aligarh

ABSTRACT: The emergence of modern technologies has brought advancement in the multi-disciplinary fields. This advancement has come up with the concept of machine learning and artificial intelligence. These modern technologies ensure that revolutionary changes are taking place in the field of civil engineering. The prime focus basically deals about the proper management and maintenance of structures and the optimization of the design work. This provides an idea to have practical implementation of these techniques in the field of civil engineering.

Keywords: Machine Learning (ML), Artificial Intelligence (AI), Multi-disciplinary field, civil engineering

1 INTRODUCTION

Nowadays two major technologies Machine Learning and Artificial Intelligence have large number of applications. These applications include from medicals, education, sports, research, agriculture, engineering, etc. The traditional methods available for modelling and having optimizing complex structures need huge resources but these two techniques provide solutions to deal with the problems.

The concept of Artificial Intelligence came in 1956, when at Dartmouth College; during meeting it was first used. This innovative branch of department of computer science is concerned with interaction of various disciplinary fields to do research and design work to perform our computer work with intelligence. It is believed that civil engineering problems including the construction management, design selection and making decisions were also influenced by the experience of experts. This tool ensures the use of imitate experts to handle the complex problems at different experts' level *(Lu et al. 2012)*. The goal of this technique basically deals with the imitation and execution of intelligent operation of human brain. While the machine learning is considered to be the subset branch of this artificial intelligence. It provides

DOI: 10.1201/9781003193838-13

computer a capability to perform without having the explicit programming. This machine learning basically focuses on doing sorting and making predictions on given problems. It works on the concept of modelling tool related to inform and performs usually data analysis. The machine language techniques are considered as non-parametric and approximating functions. It is viewed as a tool of knowledge generation (*Reich 1996*). Both Artificial Intelligence and Machine Learning plays major role in the practical works of civil engineering. A detailed progress on the structure having active tensegrity uses AI which revolves around the work of doing self-oriented diagnosis, controlling the multi objective shapes and studying the processes of reinforcement (*Adam & Smith 2008*). Machine learning involves detection of occupancy in case of buildings and accessing the monitoring of traffic system with the help of case studies. Both the techniques have provided its user number of advantages. It includes to do working with efficiency, coordinating among the team members to share their experiences and solving complex problems with better understanding and careful analysis. It also helps in doing proper evaluation and making critical reviews. These tools will be considered as one of the most successful tools in the field of civil engineering.

2 LITERATURE REVIEW

2.1 *Artificial intelligence: Applications in civil engineering*

Artificial Intelligence has become the most famous evolutionary techniques in the multi-disciplinary fields. It is a science that deals with the research and application of all the activities and its laws related to human intelligence. It basically involves to acquiring the required data, its manipulation, testing and interpretation of results. This includes intimation and execution of intelligent role of the human brain. This will result in the development of new technology products (*Lu et al. 2012*). The algorithms discussed in Figure 1 is used by them has become the major source to deal with the practical applications of civil engineering.

Among these, Genetic Algorithm is an algorithm which uses the concept of stimulation of the natural process of evolution of biological entity. It is a branch of evolutionary algorithm. The basic procedure of genetic algorithm includes choosing the initial set of values, evaluating their present status and repetition to be continued till termination. Further selection of best individual and evaluation take place. The least fit is replaced by upcoming new dataset. The shift from one generation to the other involves four basic parts (*Yadav & Prajapati 2012*):-

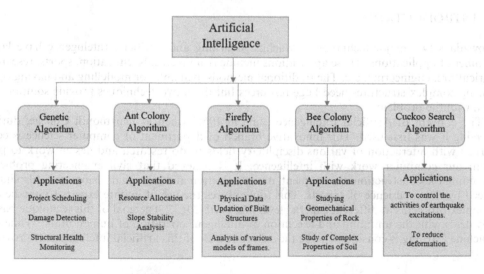

Figure 1. Common adopted algorithms of Artificial Intelligence and their applications.

1. Selection: - To select new initials as per data set.
2. Crossover: - Mixing of knowledge of two datasets.
3. Mutation: - A random string deformation having certain probability.
4. Sampling: - A process required domain of new dataset from older ones.

Genetic algorithm includes that the representation of problem is done through chromosomes. It is based on the principle of evolution given by Darwin. It is useful in monitoring the structural health of structures. It helps in detection of damages with its own potential capability. The physical health of bridges with cables is determined by estimation based on genetic algorithm and particle swam optimization (PSO). A multi objective-based optimization model was developed to prepare a schedule for construction projects. It helps the managers of construction to help in the generation and evaluation of optimum schedule plans for project. This algorithm is based on the concept of survival of the fittest. It is considered to be the search technique on the basis of natural selection.

Another algorithm, Ant Colony Algorithm is developed by Dorigo with the concept of finding the shortest route from its home to food source *(Dorigo et al. 1991)*. This will help to determine the problem in case of analysis of stability of the slope. This can be done by identification of critical slip having non-circular shape. This algorithm will provide the location of critical slip in case of slope *(Gao 2016)*. Based on this technique a combination is made with critical path method to ensure proper allocation of resources in the management of construction project *(Wang et al. 2016)*.

The Firefly algorithm is also based on the artificial intelligence. It is brought by Yang in 2008. The theme of this algorithm was based on the nature of fireflies of doing flickering while travelling *(Yang 2008)*. The algorithm finds its best use in updating data of structures of experimental model using numerical models. It helps in further reducing the distinction between the data of analytical and experimental models of structures of cantilever beam and frames *(Kubair & Mohan 2018)*. The method is considered to be most accurate and have optimized speed of performing task. The algorithm provides enough knowledge in validating both analytical and experimental data.

The new technique naming Bee Colony Algorithm is developed by Karaboga in 2005. It is based on the concept of inverse analysis. The simple concept involves that bees while doing collection of honey do the task with the method of division of the labour and exchange of the information. It considered fitness function as the base for evolution *(Karaboga 2005)*. This algorithm finds application in knowing the parameters of geo-mechanical properties which helps in studying the properties of the rock. Specific boundary condition is also mentioned to find optimum parameters *(Zhao & Yin 2016)*. To deal with the problems related to complex nature of soil properties, bee colony algorithm is used to have response surface. It will also help in approximating the limit function *(Kang & Li 2016)*.

One of the algorithms namely, Cuckoo Search is based on the concept of nesting the parasitic behaviour the cuckoos *(Yang & Deb 2009)*. It is useful in controlling the activities of earthquake-excitations and reducing the deformation occurred due to isolation system with no increase in the acceleration of superstructure *(Zamani et al. 2017)*. A detailed study by Bilal et al (2016) was done on construction waste analysis to have high performance computation and large-scale data storage. Big data architecture was prepared for this analysis for the analysis and management of unprecedented data. These tools have benefit of time saving and cost beneficial in civil engineering applications.

2.2 *Machine learning: Application in civil engineering*

Machine Learning is a tool used to solve the complex problems. These problems are solved through various dimensions. These dimensions include: representation of input data and knowledge gain through it, the procedure and its functional operations, its learning mode, supplementary knowledge required to proceed while learning, ability to manage missing data, complexity level of algorithms and the interaction between user and its learner. Solutions to different problems through machine learning involves engineering model to solve complex problems. In case of civil engineering, machine learning is used in the design work of architecture, monitoring the treatment plants of water and analysing the problems related to structures. Machine learning involve the method of problem analysis, data collection and their representation, selection of optimum solution and at last doing evaluation, interpretation and deployment of solution. Machine learning basically goes into two stages. This includes shallow and deep learning *(Reich 1996)*. Machine learning finds its application in occupancy detection. Occupancy Detection plays significant role in improving the efficiency of energy and security. The method uses vibration sensor equipped with the concept of machine learning in identification of occupancy and localization of buildings. Motion sensors, cameras and heating sensors are present. The sensing method is responsible to identify the people entering the area where this equipment is present.

When computer vision is used with machine learning, it will play an important role in detection of vehicles moving of road. This will help in the estimation of traffic flow on road and also help in doing planning and maintenance work of road This work involves foreground detection, segmentation of image, tracking, counting and classification *(Kalman 1960)*. Machine learning comes out to be a significant tool in monitoring and controlling the structural components of both large scale and movable bridges. The tool is based on tracking error in structural components through statistical and regression analysis applications. Vibrations are monitored through accelerometer. It also follows the procedure of processing large data. The prediction of item bid prices during construction of highway is also done through machine learning. It helps in the improvement of estimation of cost. A proper bid is prepared. It includes the location of contract, total value added, time period, all cost relevant to project, etc. This is especially done in items like guardrails by determining its unit price. The compressive strength of concrete of its mix design is predicted through machine learning. Its helps in doing manipulation of algorithms in a way to get god results as per prediction *(Chou et al. 2014)*. Also there have been studies on the different civil engineering applications. For knowledge extraction, Arciszewski et al 1987 studied the feasibility of wind bracing to understand the methodology of acquisition of design knowledge. Machine learning provides advantages of better understanding and careful analysis of complex problems. It generally uses supervised learning tools. Machine learning basically works on three attributes. It includes Supervised, Unsupervised and Reinforcement Learning attributes. Supervised tools work on the concept of acquisition of knowledge and its labelling. Unsupervised tools work where dataset is unknown to the user. A plotting of cluster map is done through unsupervised learning. Reinforcement Learning involves works on enabling an agent to gain knowledge through environment interaction. Applications of civil engineering follow supervised attribute. *(Mitra 2017)*.

3 CONCLUSIONS

The study on application of artificial intelligence and machine learning provide a vast field of study and research. Both these tools will help the inexperienced users to solve the engineering problems and with better efficiency. The careful analysis of task and their interactive formulations will reduce complexity level of problems through various algorithms. This will provide a better approach to understand the problem and their applications will help to bring new developments in different fields. The problems related to structural maintenance of structures and detection of cracks can easily be done through these techniques. These techniques provide enough flexibility in handling the task. These tools will help researchers from civil engineering field to deliver more sustainable and practical based solution. The trend of using these techniques in the fields of civil engineering will increase day by day to develop more optimum solutions. The future trend of civil engineering shows that the research and development play an important role to test the reliability of solutions.

REFERENCES

Adam, B. & Smith, I.F.C. 2008. Active tensegrity: a control framework for an adaptive civil- engineering structure. Computers & Structures, 86(23-24), 2215–2223.

Arciszewski, T., Mustafa, M. & Ziarko, W. 1987. A methodology of design knowledge acquisition for use in learning expert systems. International Journal of ManMachine Studies, 23–32.

Bilal, M., Oyedele, L. O., Akinade, O. O. & Ajayi, S. O. 2016. Big data architecture for construction waste analytics (CWA): a conceptual framework. Journal of Building Engineering, 6, 144–156.

Chou, J. S., Tsai, C. F., Anh-Duc, P., Lu, Y.H. 2014. Machine learning in concrete strength simulations: Multi-nation data analytics. Construction and Building materials, 73, 771–780.

Dorigo, M., Maniezzo, V. & Colorni, A. 1991. Positive feedback as a search strategy. Technical Report, 91–106.

Gao, W. 2016. Determination of non-circular critical slip surface in slope stability analysis by meeting ant colony optimization. Journal of Computing in Civil Engineering, 30(2).

Kang, F. & Li, J. J. 2016. Artificial bee colony algorithm optimized support vector regression for system reliability analysis of slopes. Journal of Computing in Civil Engineering, 30(3).

Kalam, R. 1960. A new Approach to linear filtering and prediction problems, Transactions of the ASME-Journal of Basic Engineering, 82(1), 35–45.

Karaboga, D. 2005. An Idea Based on Honey Bee Swarm for Numerical Optimization, Erciyes University, Turkey.

Kubair, K. S. & Mohan, S.C. 2018. Numerical model updating technique for structures using firefly algorithm. International Conference on Recent Advanced in Materials, Mechanical and Civil Engineering, 330.

Lu, P., Chen, S. & Zheng, Y. 2012. Review article artificial intelligence in civil engineering. Mathematical Problems in Engineering, 1–23.

Mitra, S. 2017. Applications of machine learning and computer vision for smart infrastructure management in civil engineering. Master's Theses and Capstones, 1138.

Reich, Y. 1996. Machine learning techniques for civil engineering problems. Computer-Aided Civil and Infrastructure Engineering, 12(4), 1–27.

Wang, C., Abdul-Rahman, H. & Chow, P.S. 2016. Development and test run of civil engineering schedule acceleration model through ant colony optimization. Journal of Civil Engineering and Management, 22(8), 1009–1020.

Yadav, P.K. & Prajapati, N.L. 2012. An Overview of Genetic Algorithm and Modeling, International Journal of Scientific and Research Publications, 2(9), 1–4.

Yang, X S. 2008. Nature- inspired metaheuristic Algorithms. Luniver Press, UK.

Yang, X. S. & Deb, S. 2009. Cuckoo search via Levy flights. Proceedings of World Congress on Nature and Biological Inspired Computing, 210–214.

Zamani, A. A., Tavakoli, S. & Etedali, S. 2017. Fractional order PID control design for semi-active control of smart base isolated structures: a multi-objective cuckoo search approach. ISA Transactions, 67, 222–232.

Zhao, H. B. & Yin, S. D. 2016. Inverse analysis of geomechanical parameters by the artificial bee colony algorithm and multi-output support vector machine. Inverse Problems in Science and Engineering, 24(7), 1266–1281.

Recent Trends in Communication and Electronics – Sharma et al. (Eds)
© 2021 Taylor & Francis Group, LLC, ISBN 978-1-032-04572-6

Sequencing web link using randomized algorithm

Santosh Kumar & Chirag Anand
ABES Engineering College, Ghaziabad, India

ABSTRACT: Internet has become the essential source of information for everyone holding the smart device in their hands. For the retrieval of any kind of information each and every one rely on the web through internet. In today's scheme it is sole duty of the data provider to provide the most important and relevant information for users every query posed. Since there is high increment in the amount of data to its highest unit as peta bytes has made it difficult for the user to access the most relevant information for them. In this paper, a Randomized Simulated annealing-based approach for selecting top-ranked web links has been proposed. This approach is using content, usage and structured data to select top quality web links. Further proposed algorithm WDPSA, has been experimentally compared with Iterative improvement-based approach and it has been found that SA based approach is better in selecting top-T web links.

Keywords: web mining, web page ranking, simulated annealing, iterative improvement

1 INTRODUCTION

From the early stages of WWW (World Wide Web), various methods has been developed to rank the pages by search engines. In the current scene, the instances present in a search phase is one of the presiding factor in ranking algorithms applied by search engines. The frequency of a tag to be searched can be ranked by keyword density or by its frequency inside page or document by HTML tags. Link popularity concept was developed for the superiority search results and to resist automatically generated web pages basically based on content specific ranking criteria. Depending on this concept, general importance is measured by the number of inbound links. Hence, the importance and priority of a web page is high, if other web pages interlink to it. Good ranking for pages are often escaped by the concept of link popularity, which are only created to deceive search engines and having no implication within the web, but numerous web provide manipulate just by creating multitude of incoming links for doorway pages from just as meaningless other web pages.

2 RELATED WORK

Author has proposed combination of strategies for the link analysis algorithm. It is proposed for getting the pages which are more relevant to the user's domain. For the prototype in this paper structured data is used for betterment of the results. The experimental results show that there is minute increment in the rate of harvest of the results. Rating the links is major part on which crawler depends which in return enhance the mechanism of discovery (Chahal et al. 2014). Author proposed a model to find importance of any web page using back link location. Web page link and content is annotated by its location and this has been used to find geographic rank of the web page. Accurate geo-ranking results has been shown using location based queries. Re-ranking of the search engine results are also based on the Location-based

DOI: 10.1201/9781003193838-14

query (Koundal. 2014). Local graph is always build for each query as in the HITS (Hyperlink induced Topic Search) algorithm and finds out the new geographic rank from the graph based upon the locations annotated with the back links. As HITS and PageRank requires initial rank-score, this model do not require rank-score as for root data set. Model offers more accurate rank of results model for locale web pages. Searching based on the location is more popular because of the availability of web services. As in the general search globally important webpages are being shown same is with geographic search as it shows important local results for particular location. Some modifications are being made in PageRank algorithm power & spread formulas and HITS algorithm is being used for constructing local graph and for geographic-score calculation for the webpages. Topically and geographic ranking search results are more accurate by using Geo-ranking algorithms (Koundal. 2014). Author used text and visual data as a basis for prediction of second hand items (Bar-Ilan et al. 2015). Author has proposed the model by interpretation of the structure of the hyperlink for web-page ranking and information retrieval. Page importance is being measured by hyperlink structure analysis by calculating the weight of the pages based on the links. This method involve normalizing the links on a page not just by counting the links from the pages. K-elements can be defined as the weight of the keywords which is being computed from the anchors, elements &keywords (Asadi et al. 2009). For ranking the web pages a proposed methodology is the linear combination of the weights of keywords and the hyperlink structure. Page weight is measured iteratively by hyperlink analysis that is connecting to the number of links to those pages. Page weight is computed in terms of various keywords with the help of the K-elements by the extension of the Dublin core elements to anchors (Asadi et al. 2009). from the perspective view of Big data that how it is being used with cloud computing, here are some approaches: (i) Since, huge amount of data is being processed and stored on cloud, which will be managed using by big data technologies. So data should be encrypted using keys, so that hackers cannot steal the data.(ii) Whenever nodes joins the clusters it must be authenticated. If it turned out to be a malicious cluster then such types of nodes must be authenticated (Greco et al. 2001). Author has discussed about two types of link analysis ranking methods. 1st is Site Rank, i.e. it is basically an modification of the page rank to the web site granularity and 2nd is popularity rank, i.e. based on the number of times of the user clicks on the out links in the webpage which is being tracked by the navigation session of the users. The most importantly, the distributions of the site rank and popularity rank are very close each other (Fathalla et al. 2020). Therefore the reasonable first order approximation is provided by the site rank and popularity rank is provides the aggregate behavior with in the web site. User is the random surfer for the site rank and user navigation is according to the previous navigation frequencies for different sessions (Fathalla et al. 2020). Author has proposed the idea for solving the problems which are critical and has strong relationship b/w massive biomedical text and heterogeneous typed entities. Working with this type of data is also very critical task of the type of data which is very much noisy in nature and also unstructured. A graph is also build showing the relationship and this graph has been build from the text data (Ganeshiya et al. 2014). The approach used in this paper by author is EntityRel. If entities are closely coupled with each other than these entities are heterogeneous types. Though author is mainly focusing on the medical data but also has said that this algorithm is efficient enough that it can also work on different type of field data. Author has also given the idea about the future scope of the work and also show the interest on the Meta path weight learning (Ganeshiya et al. 2014). Author has given the document based on the semantic web ranking, nor based on the keywords but for the conceptual instances between them also (Lai Jun at al. 2006). Which will directly result the relevant page on the top the result page. The author has also calculated fraction of the relations between the keywords of intention of the users (Lai Jun et al. 2006). While using the semantic web as it provides the relevant results and it is further improved by page ranking algorithms. The probability is basically based on the ontology, webpage and query for ranking the web pages which has been retrieved. The scheme proposed by the author measures the page relevance by calculating the combined probability by web page and the ontology and then adding that result with the Probability (webpage) with the query. The future progress would be design for more meaning full and exhaustive results, so that it can analyses the result more precisely (Lai Jun et al. 2006). Researcher explains the different approach in which first structure of the base set is analyzed for computing the authoritative page (Danasingh et al. 2016). A statistical approach is being applied i.e. co-citation matrix is for most co-cited pages and it is combined with a link analysis approach for the evaluation of the content page. Also a new approach has been also presented in this paper such as Topic distillation on the web, in which structure of the base set is being analyzed for the computation of the authoritative page. In this paper thus technique is shown more efficient which is based on the mutual reinforcement method. The proposed approach has many interesting properties:It can be used for the unconnected graphs (Danasingh et al. 2016). No need to make matrix inversion and has no convergence problem in the approximation form. By using the closed form not ill-conditioned matrix can be implemented (Danasingh et al. 2016). Author has presented the authentic ranking of results based on variant of h-index for directed information system such as web. H-index is used for assessing the importance for the webpage (José et al. 2006). As in page rank the computation is based on the integral web graph but in the h-index it is based in the local computation i.e. only neighbors of the neighbors are taken into consideration. Early results show a strong correlation between the PageRank and Hw-index though hw-index is simpler and less complex in terms of computation than PageRank (José et al. 2006). Author has given an algorithm which is to improve the relevancy of the result, based on the keywords (Ji M et al. 2015). The result of the proposed algorithm is being compared with the other algorithms such as PageRank along with topic distillation with page characteristics result and query dependent link. An algorithm is discussed based on the keyword ratio which is calculated as (ratio of no. of query dependent unique keywords) to total unique keywords extracted from the webpage (Ji M et al. 2015). Authors has proposed algorithms for ranking the web documents (Kumar et al. 2020). They have discussed the relevant points of the algorithms and how efficient they are in nature. The discussions is also made on the rapid increase in the data on the web since the internet was introduced the data has been increased from MB's to PB's (petabytes) (Kumar et al. 2020). Many factors have been also compared such as (Input, approaches, input parameters, usefulness, limitations and result's quality). Also conclude that it will be challenge to get out the relevant results with accordance to users query and that's why optimization is needed with the help of optimization techniques (Kumar et al. 2020).

3 SIMULATED ANNEALING

It's a comprehensive optimization method and belongs to the stochastic optimization algorithms. It is conceptual inspired by the process of annealing in metallurgy. Process of annealing, the metal is heated to a very high temperature so that the molecules of the metal can move openly and then it is frigid under predefined conditions, so molecules can take thin path till all of them reach their lowest temperature collectively. i.e.in simple words annealing is the somatic process that occurs when a certain metal's temp is increased and then slowly frigid. By this similarity, each move of the simulated annealing try to replace the prevailing solution by random solution till we get the desired output. At every step SA decides to stay in current state or move the system to new state. System takes this type of decision probabilistically.

3.1 *Representation of candidate solution*

Sample solution has been taken consisting set of URL's with unique Id's. Sample solution is represented having Unique Id's in Figure 1.

URL Unique Id's

65	87	98	12	34	54	66	21	89	23

Figure 1.   Candidate solution - Top-10 URLs.

3.2 Fitness evaluation

The presence of the similarities between the given solution and that of ideal solution for problem is described as the fitness function. Its parameters are given in Table 1.

Table 1. Cost function parameters.

Cost Function Parameters	Description
Access Frequency	The no. of times a webpage is accessed by a user or redirected by some other page is known as access frequency.
Number of Distinct Visitors	Webpage accessed in a fixed time frame by the particular visitors is known as Unique Visitors. This also tell the importance of webpage.
Time Duration	An important measure for relevant web page and provides an average time spent on that website by a single user.
Hubs	Defined as collection of pages containing the redirection links to different other inter related pages. Cost can be calculated as: - No. of URL pointed by kth hub on Tth web page=$\sum ni=1\sum HUij=1N$.
Authorities	Web has ample amount of pages that has list of links pointing towards many pages on a relevant topic.Let, wehave n number of authorities andrepresented as $AT_1, AT_2, AT_3, \ldots .ATn$. Cost of authority= $\sum_{i=1}^{n} AT_i$

3.3 Web links cost function

$$Total\_MAX\_ \quad COST \quad = MAX\_ACCESS \quad \_FREQ + MAX\_VISITO \quad R + MAX\_DURATI \quad ON$$
$$+ MAX\_HUBS \quad + MAX\_AUTHOR \quad ITIES$$

$$Fitness\_Va \quad lue = Cost = c1. \sum_{i}^{topT} AF_i + c2. \sum_{i}^{topT} DUR_i + c3. \sum_{i}^{topT} UNQV_i + c4. \sum_{i}^{topT} HUB_i + c5. \sum_{i}^{topT} AUTH_i$$

$$c1 = \frac{Total\_MAX\_COST}{MAX\_ACCESS\_FREQ \times topT} \quad c2 = \frac{Total\_MAX\_COST}{MAX\_DURATION \times topT} \quad c3 = \frac{Total\_MAX\_COST}{MAX\_VISITOR \times topT}$$

$$c4 = \frac{Total\_MAX\_COST}{MAX\_HUB \times topT} \quad c5 = \frac{Total\_MAX\_COST}{MAX\_AUTHORITIE S \times topT}$$

Algorithm Initialization

1. Initial state generated randomly, S; set of topT web links=$\{WD_1, WD_2, WD_3, \ldots, WD_{topT}\}$
2. Initial temperature, $T=T_0=2*WD\_COST(S)$
3. Optimal state, maxS=S
    1. findNeighbors(S,n) //generates N neighbors of state S
    2. randomN(neighborS[]) //returns a random neighbor form set of neighbors
    3. WD_COST(S) //Returns total view evaluation Cost of any state S
    4. randomV() //generates a random number between 0 and 1

Method:

1. *while T>1 or (maxS unchanged for 5 stages)*
2. *for Generation=1 to 100*
3. *neighborS[ ]=generateNeighbors(S,n)*
4. *S'=randomN(neighborS[ ])*
5. *deltac=WD_COST(S')-WD_COST(S)*
6. *if deltac<=0 then*
7. *S=S'*
8. *k⁻e deltac/T*
9. *if deltac>0 then*

10. rand_var=randomV( )
*11. if rand_var<k **then***
12. S=S'
*13. if WD_COST(S) < WD_COST(maxS) **then***
14. maxS=S
15. T=0.90 T // reduce T by some quantity*
*16. **return** maxS*

4 EXPERIMENTATION AND RESULT

The algorithms web links prioritization using Simulated Annealing and web links prioritization using Iterative Improvement were implemented in JDK1.8, Intel based windows 10 environment. The following reading were taken to verify that for which value to reduce the temperature is providing better cost and It has been observed from Table 2, that for 0.90 value, better cost is obtained. So, this value is considered for further experiments.

Further experiments have been carried out to find the generation value for which, algorithm is conversing and it has been observed from Fig. that after 100 generation it is conversing. So program is run for these number of generations. Then both the algorithms were compared by considering the above values, and It can be seen from Figure 2 and Figure 3 that Simulated Annealing based approach is able to select better quality top-T web links.

In Figure 2, the proposed algorithm, WDPSA has been compared for different top-T values. The set of top-10 documents selected by the proposed algorithm have better quality web links as compared to others as contribution of cost in evaluation parameter is more in case of Top-10. The Figure 2 shows the comparison of both the algorithms for different top-T

Table 2. Top-T web links and their cost.

Top-T Web Docs	Cost SA T=0.75*T	Cost SA T=0.80*T	Cost SA T=0.85*T	Cost SA T=0.90*T	Cost SA T=0.95*T
5	26099	26144	26170	26341	26081
6	25970	25970	26023	25849	26076
7	25852	25924	25825	25828	25736
8	25812	25818	25755	25793	25690
9	25575	25803	25603	25768	25687
10	25572	25646	25593	25763	25588
Average	25813.33	25884.17	25828.17	25890.33	25809.67

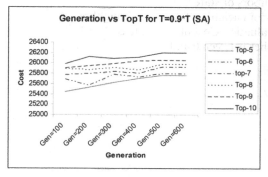

Figure 2. Generation vs Top-T web links.

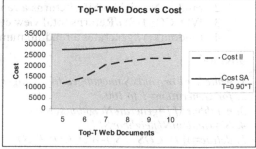

Figure 3. Cost vs Top-T web links (WDPSA and WDPII).

76

values and it is clearly visible from the graph that WDPSA is able to select better quality Top-T web links for T={5, 6, 7, 8, 9, 10}.

5 CONCLUSION

In this paper, a simulated annealing based approach for selecting top-T web links has been proposed. This approach is using content, usage and structured data to select top quality web links. The parameters, access frequency, access count, time spent on web page, hubs and authorities have been considered to evaluate web links. Further proposed algorithm WDPSA, has been experimentally compared with Iterative improvement based approach and it has been found that SA based approach is better in selecting top-T web links.

REFERENCES

Asadi, S., Zhou, X. & Yang, G. 2009.Using Local Popularity of Web Resources for Geo-Ranking of Search Engine Results. World Wide Web 12, 149–170.

Bar-Ilan, J. & Levene, M. 2015. Scientometrics 102: 2247. Doi:10.1007/s11192-014-1477-2.

Chahal, P., Singh, M. & Kumar, S. 2014. An Efficient Web Page Ranking for Semantic Web. J. Inst. Eng. India Ser. B 95, 15–21. doi:10.1007/s40031-014-0070-7.

Danasingh., A, Antony & Tamizhpoonguil., B. & EPIPHANY, J. L. 2016. A Survey on Big Data and Cloud Computing. IJRITCC. 4. 273–277.

Fathalla, A., Salah, A., Li, K. 2020. Deep end-to-end learning for price prediction of second-hand items. KnowlInfSyst . https://doi.org/10.1007/s10115-020-01495-8.

Ganeshiya, D. & Sharma, D. 2014. Keyword ratio oriented webpage rank algorithm. 9th International Conference on Industrial and Information Systems, ICIIS.

Greco, G., Greco, S. & Zumpano, E. 2001. World Wide Web 4: 189.

Ji, M., He, Q. & Han, J. 2015. Mining strong relevance between heterogeneous entities from unstructured biomedical data. Data Min Knowl Disc 29, 976–998.

José, B. & Mark, L. 2006. Ranking Pages by Topology and Popularity within Web Sites. World Wide Web. 9. 301–316. 10.1007/s11280-006-8558-y.

Koundal, D. 2014. Prioritizing the ordering of URL queue in focused crawler. Journal of AI and Data Mining. 2. 10.22044/jadm.2014.146.

Kumar, S. & Anand, C. 2020. Comparative Study of Web Page Ranking Algorithms. International Journal of Advanced Science and Technology, 29(5s), 322–331.

Lai, Jun., Soh, B. & Fei, Chai. 2006. A Web Page Ranking Method by Analyzing Hyperlink Structure and K-Elements. 3983. 179–186.10.1007/11751632_19.

Advancement of 5G Technology

Ajay Suri & Rashi Sharma
ABES Engineering College, Ghaziabad, India

ABSTRACT: Commercial 5G networks are live throughout the world. One factor you might notice will be that when you listen to about 5G, you notice a lot of chat about speed. Obviously, regarding newbies like kids, the particular speed of 5G has been the easiest thing to know. But 5G is a lot, a lot more...... The world's on-line needs are changing. International mobile data traffic will be anticipated to multiply by five prior to the end of 2024. Particularly in dense downtown areas, the latest 4G sites simply will not able to be able to keep up. 5G smartphones will always be available in beginning regarding 2019. The significance of 5G is definitely the opportunity it offers for people, business along with the world at large: sectors, regions, towns and urban centers which might be more connected, better and much more sustainable. It's enabling industries to reinvent on their own.

1 INTRODUCTION

Today the planet economy depends upon many things and something of the primary foundations of the actual major economic corporations usually continue steadily to use, including it in neuro-scientific interaction, is utilizing the latest inventions. Software applications for information interaction has gained a significant place in business and the overall economy. Wireless cellular revenue and marketing and sales communications using a selection of techniques and equipment contain transcended classic communication devices and provided the energy to improve this global economic condition. With the development of World Wide Web document, visitors got resulted in the capability prerequisites of contemporary 4G and 3G radio stations solutions.

Strenuous analysis on cordless interaction systems inside the 5th century will certainly create in lots of grounds. Hopefully that 5G technology will still be used until 2020. This document removes a number of the 5G activities linked to European research, contemporary literature, along with the 5G white paper of this major wireless technology players. The target is to know very well what 5G can be and just how many 5G initiatives have already been developed to accomplish it.

What 5G methods will be normalized around 2020? It's prematurely for me to say it confidently. However, it really is acknowledged that 4G sites are smaller sized than 5G systems. A 5G community must have countless potential equipment, 10-fold unrealistic proficiency, critical performance and 10-fold degree of data (eg Gbit/s for reduced mobility and optimum data rates of speed of just 1GB/s for excessive freedom) and a variety of 25 (Mitra et al. 2015). In several occasions, the niche is building cable connections based on smooth and ubiquitous conversation between everyone (individual to individual), everything (equipment to device), anyplace and in line with the device/provider or electronic system you need. Which means that some special conditions that aren't backed by 4G and 5G systems must discover a way to control the communication technique (for instance, for high-speed consumers). High-speed systems can reach rates of speed of 350 to 500 kilometers/h, while checking with 4G sites only

DOI: 10.1201/9781003193838-15

supports conversation scenarios around 250 kilometers/h. The label 5G identifies the continuing future of this telecommunications (Andrews et al. 2014). This can be a huge step over cellular technology forward that is usually an excellent exemplary case of innovative changes to broadcast interfaces.

2 LITERATURE SURVEY

A large number of human publications are usually related to 5G, with more articles appearing almost every month. Therefore, the literary edition selected in this section corresponds to the latest articles at the level of popular magazines and selected newspapers.

Literature asks us to reconsider the relationship between persistence and spectral competence (EE vs. SE). The combined design must be an integral part of the 5G study. There is no more network. This is another claim that 5G recommends for switching from cell-oriented minds to soft client and C-RAN models (Bhushan et al. 2014, Bangerter et al. 2014). The next step is to signal and handle components for different types of site visitors. The fourth aspect (Wunder et al. 2014) usually introduces the concept of an invisible earth station. This prevents correct placement of a large GUITTO in the form of an unusual antenna array, where the antenna elements can be integrated into the environment (so that the following stations are practically invisible). Finally, Full De Dos Niveles Stereo is available as part of a useful 5G technology. Like the usual previous content list of five devastating 5G ideas at (Queseth et al. 2014, Jungnickel et al. 2014). The typical design of base stations (uplink/downlink, control/ data program) actually offers proposals for a highly flexible device-oriented architecture where various types of visitors and network devices can be handled better.

However, with global networks there are so many that a 5G system is unusable if there is any doubt. Therefore, the Internet, which will be consumer-focused in the future, will benefit from scientific advances in content delivery, control (response time), and monitoring (persistence) (Nam et al. 2014). Subsequent thematic 5G surveillance was discovered by the METIS project in terms of methodology, prerequisites and conditions for potential cellular availability systems in the future (Zhou et al. 2019). Previous METIS cases and horizontal drivers have been identified in Zone II. Another article (Liang et al. 2019, Gupta et al. 2015) proposes a change in the pathway to 5G. Usually the last generation would have succeeded in growing micronutrients. Closely coordinated macroeconomic coexistence can, however, be a more useful way in the future.

3 TECHNOLOGIES

3.1 *Millimeter waves*

These wireless sites have encountered an issue: More people and machines are usually eating more information than once you want before, nonetheless it continues to be crammed on the very same bands using the radio-frequency array that mobile service providers possess always utilized. This implies considerably less bandwidth for anybody, creating slower support and additional lowered connections (Moysen et al. 2018).

They might be known as millimeter waves given that they vary within duration from 1 to be able to 10 mm, in comparison with stereo system waves that offer these smartphones, which gauge tens of centimeters in proportions. But utilizing millimeter waves to add mobile consumers with some type of nearby base train station is generally a totally new strategy.

3.2 *Small cells*

Type Small cells can be portable miniature basic channels that want minimal work capability and can get easily positioned through the town every 300 meters or even more. To prevent indicators being deleted, sellers can install a large number of these channels in the town to create a dense community that functions as a relay staff, receives signs from various other lower channels, and sends info to users just about everywhere. These dimensions difference makes it easier to connect fabric to lighting columns apart from the building over. This completely different network structure must be sure the usage of more targeted and much more efficient ribbons.

3.3 *Massive MIMO*

4G base stations currently have twelve antenna slot machine games that take care of all cellular site visitors: 8-10 for transmitters and many for receivers. On the other hand, the 5G BTS can help a hundred ports, meaning extra antennas can match exactly the same assortment usually. This can imply that the bottom station can send and also receive signals from more users at exactly the same time, increasing the ability to accommodate cellular networks by way of a factor of 22 or higher (Shao-Yu et al. 2019). This technology is recognized as massive MIMO. Just about all be determined by MIMO, which frequently signifies numerous inputs numerous inputs. MIMO describes a radio device that uses several transmitters and receivers to receive and send more data simultaneously.

3.4 *Beamforming*

The beam transformation could be a cell base stop traffic signaling program that determines probably the most efficient method of transmit info to certain customers and decreases disruption to shut users in this technique. With regards to the situation and know-how, there are many methods that 5G sites may use to put into action it (Lama et al. 2020). Beamforming permits large MIMO arrays to utilize frequencies around them better. The biggest task for that MIMO giant would be to decrease interference and at exactly the same time send more info from additional antennas at the same time.

3.5 *Full Duplex*

BTS and cellphones today accept transceivers that has to move if they receive and send data on a single frequency, or take care of very different frequencies when customers want to receive and send files at an equal period. With 5G, transceivers are prepared to receive and send understanding at exactly the same time and rate of recurrence. To accomplish complete duplex on personalized devices, researchers have to develop circuits that may route inbound and outgoing signs in order that they usually do not collide once the antenna is transmitting and receiving understanding at the same time.

4 PERFORMANCE COMPARISON

When you are usually connected to the Internet, the speed usually depends on the specific signal strength, which is displayed in letters such as 2G, 3G, 4-G etc. in the next line to display the signal bar on your home screen. Each generation is identified as a set of criteria for telephone networks detailing the technical implementation of a particular mobile system.

The purpose of wireless communication would be to provide top quality, reliable communication exactly like wired communication (optical fiber) and each brand-new generation of solutions represents a large step (a leap relatively) for the reason that direction. This advancement journey was were only available in 1979 from 1G which is still carrying on to 5G. Each one of the Generations has benchmarks that must definitely be met to formally utilize the G terminology. 1G - This is the first era of mobile technology. The first generation of professional cellular networks began in the late 1970s with fully implemented standards maintained in the 1980s. Australia was introduced by Telecom (now known as Telstra) in 1987 and gained its first cell phone community using the 1G analog method (Rabia et al. 2020). 1G can be an

Table 1. Comparison Between 1G, 2G, 3G, 4G & 5G.

Parameter	1G	2G	3G	4G	5G
Bandwidth	2Kbps	14-64 Kbps	2 Mbps	200mbps	>1gbps
Technology	Analog Cellular	Digital Cellular	CDMA/IP Technology	Unified IP And Seamless Combination of LAN/PAN/WAN	4G+WWWW
Service	Mobile Telephony	Digital Voice/Short Messaging	Integrated High Quality Audio/Video	Dynamic Information Access	Dynamic Information Access With AI Capabilities
Multiplexing	FDMA	TDMA/ CDMA	CDMA	CDMA	CDMA
Switching	Circuit	Circuit	Packet Except For Air Interface	All Packet	All Packet
Core Network	Pstn	Pstn	Packet Network	Internet	Internet
Handoff	Horizontal	Horizontal	Horizontal	Horizontal & Vertical	Horizontal & Vertical

analog technology, and cellphones in general have poor battery life and good sound quality looks great without much security, and sometimes cell phone calls are dropped. 2G - phones have received the first major increase when switching from 1G to 2G. The main difference between your two cellphone technologies (1G and 2G) is that the radio pulses used by 1G networks are usually analogous, whereas 2G locations are digital. The original purpose of this technology was originally to provide a reliable and safe connection route. He is aware of the idea of CDMA and GSM. Lack of support for information such as SMS and MMS. The 2G function is achieved by allowing multiple users on the same route to multiplex. During 2G, cellphones are used for information along with voice. 3G - This technology sets the standard for most wireless technologies that we are familiar with and enjoy. The third technology creates internet browsing, email, video downloads, pictures from several others and technology to discuss smartphones. 3G regularly uses a new technology called UMTS as a central network structure - Universal Mobile or portable telecommunications method. This network combines the 2G system area with completely new technology and new methods to achieve much faster data loading. 3G supports multimedia services and the current trends are better known. 4G - 4G is a totally different technology from 3G and basically has been made possible almost exclusively because of technological advancements in the past decade. The main technology that allows this can be MIMO (many multi-input services) and OFDM (multiplexing orthogonal consistency). The two main benchmarks for 4G are usually WiMAX (running away) and LTE (divided). LTE (Long-term Evolution) is a series of improvements to existing UMTS technology and converted from the previous Telstra 1800 MHz belt. 5G - 5G is an era in development to improve 4G. 5G claims, among other things, faster price info, greater connection thickness, lower latency. A variety of 5G programs include communication between devices, much better battery usage, and increased overall wireless coverage. The highest speed of 5G definitely aims to stay because the speed is at 35.46 Gbps, which is more than 35 times faster than 4G.

5 CONCLUSION

Despite the fact that the 5G concept is still being developed, evaluations show emerging similarities. Improved performance is largely expected through a mixture of network compression (e.g. small cells, D2D), increased spectrum (increased provider aggregation, frequency division, frequencies above 6 GHz) and improved wireless communication technology. (e.g. large GUITTO, new waveform, RAT with virtual zero latency). This type of machine in terms of communication will have an increased share of community connections and data traffic. The combination of highly reliable cellular and communication networks really requires new alternatives due to stringent special requirements under difficult distribution conditions. Working with cases, scenarios and the distribution of diversity generally leads to large variability, so when using the overall concept of the 5G system, the greatest flexibility, scalability, and reconfiguration are required.

REFERENCES

Mitra, R.N.& Agrawal, D.P. 2015. 5-G Mobile Technology: A Survey. *ICT Express, Science Direct*; 1(3): 132–137.

Andrews, J.G., Buzzi, S., Choi, W., Hanly, S.V., Lozano A., Soong, A.C.K.& Zhang, J.C. 2014. What will 5G be?. *IEEE J. Sel. Areas Commun*; 32:1065–1082.

Bhushan, N, Li, J., Malladi, D, Gilmore, R., Brenner, D, Damnjanovic, A., Sukhavasi, R.T., Patel, C.& Geirhofer, S. 2014. Network densification: theominant theme for wireless evolution into 5G. *IEEE Commun. Mag;* 52:82–89.

Bangerter, B., Talwar, S., Arefi, R. & Stewart, K. 2014. Networks and devices for the 5G era. *IEEE Commun. Mag.*; 52: 90–96.

Wunder, G., Jung, P.,Kasparick, M., Wild, T., Chen, Y., Brink, S.,Gaspar, M., Michailow, M., Festag, A., Mendes, L., Cassiau, N., Kt´enas, D., Dryjanski, M., Pietrzyk, S., Eged, B., Vago, P. & Wiedmann, F. 2014. 5GNOW: non-orthogonal, asynchronous waveforms for future mobile applications. *IEEE Commun. Mag.*; 52:97–105.

Queseth, O., Schellmann, M., Schotten, H., Taoka, H., Tullberg, H., Uusitalo, M.A., Timus, B. & Fallgren, M. 2014. Scenarios for 5G mobile and wireless communications: the vision of the METIS project. *IEEE Commun. Mag.*; 52:26–35.

Jungnickel, V., Manolakis, K., Zirwas, W., Panzner, B., Braun, V., Lossow, M., Sternad, M. & Apelfr, R. & Svensson, T. 2014. The role of small cells coordinated multipoint, and massive MIMO in 5G. *IEEE Commun. Mag.* 52: 44–51.

Nam, W., Bai, D., Lee, J.& Kang, I. Advanced interference management for 5G cellular networks. 2014. *IEEE Commun. Mag.;* 52: 52–60.

Zhou, L., Rodrigues, J.J.P.C., Wang, H., Martini, M. & Leung, V.C.M. 2019.5G Multimedia Communications: Theory, Technology, and Application. *IEEE Multimedia*; 26(1): 8–9.

Liang, Y.N.J., Shi, X. & Ban, D. 2019. Research on Key Technology in 5G Mobile Communication Network. *International Conference on Intelligent Transportation, Big Data & Smart City (ICITBS), Changsha, China*; 199–201.

Gupta, A.& Jha, R.K. 2015. A survey of 5G network. Architecture and Emerging Technologies; 3:1206–1232.

Moysen, J.& Giupponi, L. 2018. From 4G to 5G-Self-organized network management meets machine learning. *Computer Communications*; 129: 248–268.

Shao-Yu, L., Chih-Cheng, T., Ingrid, M. & Leonardo, B. 2019. Recent Advances in 5G Technologies: *New Radio Access and Networking. Wireless Communications and Mobile Computing*; 1–2.

Lama, H., Ahmed, F., Abdelrahman, E. & Mahammad, S. 2020. 5G New Radio Prototype Implementation Based on SDR. *Communications and Network;* 12: 1–27.

Rabia, K., Pardeep, K., Jayakody, D.N.K.& Liyange, M. 2020. A Survey on Security and Privacy of 5G Technologies: Potential Solutions, Recent Advancements and Future Directions. *IEEE Communications Surveys & Tutorials*;22(1):196–248.

Recent Trends in Communication and Electronics – Sharma et al. (Eds)
© 2021 Taylor & Francis Group, LLC, ISBN 978-1-032-04572-6

Discovering inter-sentence relationship in medical text by using machine learning

Sarthak A. Kasturiwale
Department of Information Technology, PICT Pune, India

Sachin S. Pande
Department of Information Technology, Assistant Professor, PICT Pune, India

ABSTRACT: There has been significant growth in the health care sector in recent years. This has been instrumental in achieving a significant improvement in health care through the introduction of automated bots. Other text extraction approaches have been utilized for creating online chat bots and automated programs that interact with the patients and diagnose them. Numerous researches have been performed in this category and most of them have been found to have certain drawbacks or limitations such as low accuracy, improper detection etc. hence it is important to address them. Therefore, this paper describes an innovative approach towards determining inter sentence relationships in the medical text through the implementation of machine learning algorithms.

1 INTRODUCTION

The invention of the internet platform has also significantly contributed to the research space as researchers could effectively reduce the amount of time taken for exchanging information and collaborating on various new drugs and other preventive measures. This has been compounded by the fact that the internet has achieved an almost ubiquitous response from the citizens where most of the humans are now connected on this platform of a planet wide network. Therefore, various chat-bots and other online medical help portals were generated to help the general populace with their medical queries.

NLP or Natural Language Processing approach is a highly useful and one of the most popular approaches for extracting the semantics of a text. The NLP paradigm is aimed to understand the actual meaning behind the text rather than treating it as a collection of alphabets or characters. This is done through the implementation of term frequency and inverse document frequency along with noun identification and a bag of words implementation for a specific application on the biomedical text.

This NLP approach has been combined with a specialized bag of words that contains the biomedical text for effective and accurate identification. There is also the implementation of linear regression that effectively normalizes these results from the NLP module and represents effective regression between the attributes that are extracted from the biomedical text. These regression results are utilized further in and machine learning approach guided by an artificial neural network for the understanding of the biomedical context and the inter sentence relationship in the biomedical text corpus.

The utilization of random forest classification is an effective realization of the classification task for this biomedical text and inter sentence relationship extraction approach. The random forest approach is an extensive collection of decision trees that implement the if-then rules to

DOI: 10.1201/9781003193838-16

accurately classify the input data provided by the artificial neural network module. This classification is complete and results in the extraction of effective and highly precise relationship sets pertaining to the biomedical text given as input.

2 RELATED WORK

E. Atala et al. explains the paradigm of biological context and various biomedical texts that have events that have been defined through these approaches. This is a highly useful approach as biological event extraction can be utilized for the evaluation of various diseases and their progression in a large number of patients effectively. But due to the large size of the biomedical text corpus, it is highly difficult to manually guide and identify the biological context (Atala et al. 2019). Therefore the authors in this approach propose the utilization of various different classifiers through the use of natural language processing in extracted the biological context and cross-referencing it to the biochemical events for effective categorization of biomedical text.

P. Ren et al. discusses the utilization of various sentence decoration and text summarization approaches that have been implemented in a many fields (Ren et al. 2017). There has been a lot of improvement in these approaches but no as such improvement in the regression framework that can understand the inherent relationships which can be developed as features between the texts. Therefore the authors present a neural network that can effectively understand and extract the contextual features from the sentences.

T. Saha et al. elaborates on the paradigm of vector representation in sentences as it can provide various approaches that can be eventually utilized for ranking sentences, clustering, or classifying purposes (Saha et al. 2017). The authors have used bag of word approach for the implementation and a model has been retrofitted and regularized true context for effective sentence learning representation. The experimental results dictates that model can used for efficient ranking, clustering, and classification of the sentences.

D. Xue et al. introduces the paradigm of text classification and categorization that has been getting increasingly popular in various fields. There have been a lot of conventional approaches that have been utilized for this purpose to achieve text categorization effectively (Xue et al. 2015). But most of them have a lackluster performance for reduced efficiency. To improve this the researchers have proposed the implementation of a random forest algorithm for the purpose of text classification in a large corpus of text. The experimental results conclude the random forest has achieved significant improvements in text categorization.

M. Yousaf et al. introduces the concept of spatial information extraction through the use of existing natural language processing techniques to achieve an effective understanding of the text Corpus (Yousaf et al. 2017). There has been a large scale improvement in the spatial information extraction systems that have been utilized to extract place and descriptions. The researchers in this approach proposed the incorporation of contextual labels for spatial information extraction through the use of natural language place descriptions.

3 PROPOSED METHODOLOGY

The presented technique for the extraction inter sentence relationship in biomedical texts have been detailed in Figure 1 above.

3.1 *Step 1- dataset collection and preprocessing*

The biomedical text is contained in a Dataset that is obtained from this link https://ml4ai. github.io/BioContext/. The dataset acquired is in the document or .doc format. This doc file is provided to the system as an input for string generation. The string is transferred to the preprocessing module which is executed by using *Special Symbol Removal, Tokenization, Stop word Removal and Stemming.*

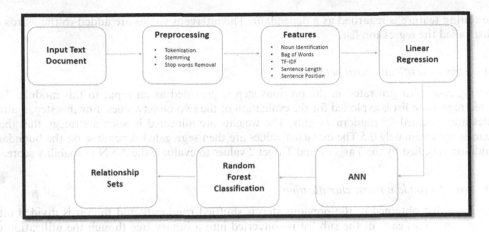

Figure 1. System overview diagram.

3.2 Step 2 - feature extraction –noun identification

The string pre-processed in the previous step is provided as an input in this module. This module uses the pre-processed string to segregate the individual words in the string. The segregated words that are extracted are then correlated using an Oxford Dictionary. If the word correlates to any of the words in the dictionary, it is classified as a noun which is provided with a score that is added along with the word at the end of a separate list.

Term Frequency and Inverse Document Frequency (TF-IDF) – In this step the identification of the important words is being performed through TF-IDF. A double dimension list is then created to store the lists of words along with their frequency in two columns. The logarithmic value of the documents is calculated by inversing the count obtained along with the ratio of the number of documents and the resultant value is stored as the Inverse document frequency (IDF). The extracted TF-IDF for a word is added to the previous list containing the other features that are extracted.

Bag of Words – The Bag of Words technique correlates all the words in the list with a medical dictionary. If the word occurs in the dictionary then a score of 1 is given. If the words that are correlated from the list are not detected in the dictionary, then a score of 0 is given to that specific word. The scores are added to the list of previous features next to that specific word on the list.

Sentence Length – The pre-processed string obtained in the previous step is utilized for the extraction of words. These individual words are the utilized to locate the original sentence. The length of the sentence is extracted and added to the respective word in the list. The information present in the length of the sentence is considerably more than that of the other sentences. Therefore, the length of a sentence is highly critical and must be regarded as an important feature.

Sentence Position – The position is capable of revealing the quality of the content that is comprised in the sentence. Therefore, for the measurement of the sentence position a score is given to the initial five sentences in the document. The scores provided as follows 1, 0.8, 0.6, 0.4, and 0.2 respectively. Any sentences after the fifth sentence are given a score of 0.

3.3 Step 3 - linear regression

After the completion of the training data, a test document is provided as an input to the system. This test document is utilized to perform all the steps in the feature extraction procedure. The extracted features from the testing data list and the training data list are correlated. For this purpose, the complementary trained feature is regarded as an independent value and the testing feature is regarded as a dependant. The intercept values are added to the words in a list called the regression list.

3.4 *Step 4 - artificial neural network*

The regression list generated in the previous step is provided as an input to this module. The sorted regression list is exploited for the evaluation of the two target values. For this step, 5 attributes are allocated 25 random weights. The weights are allocated in such a strategy that their mean is approximately 0.5.The extracted values are then segregated depending on the boundary condition specified by the Target 1 and Target 2 values to evaluate the ANN probability score.

3.5 *Step 5 - random forest classification*

The ANN list obtained in the previous step is shuffled randomly and then it is divided into N divisions. This each of the sub list is converted into a binary tree though the utilization of the probability factor. Then the words which conform to the particular levels in this tree are aggregated in the form of their respective clusters.

4 RESULTS & DISCUSSIONS

For the experimental evaluation, the proposed model uses a dataset from the URL: https://ml4ai.github.io/BioContext/. The dataset obtained from this URL is the corpus of biomedical papers from PubMed central. The proposed model uses some of these papers to train the model. On the other hand, some papers were used for testing purposes.

In the process of evaluation of entropy, each word that is used in the random forest is evaluated using the Shannon information gain. Each word 'W' is counted for its presence in all the sentences and it is counted as 'A'. 'C' indicates the total number of sentences and B gives the total number of sentences that don't contain the word, so B can be given as (C-A). Any gain value nearer to 1 represents the word is very important and if the values are nearer to 0 then it is not important. So this gain list is sorted in descending order and then its 80% of the data is considered as the model word bag, which is used for the evaluation of the sentence relationship using the precision and recall parameters.

$$IG = (-A/C)log(A/C) - (B/C)log(B/C) \qquad (1)$$

Where,
IG = Gain of the word

The precision is measured for the relative accuracy of the model and Recall can be denoted as the absolute accuracy of the model. If the obtained sentence relationships are present in the Information gain bag then it is considered as the accurately predicted or else it is considered as the inaccurate prediction.

On observing the Table 1 we come to know that average precision, recall and F-Score values are almost 64.28%. When we compare the obtained results with that of (Gu et al. 2017), which is mainly dealing with the extraction of chemical disease sentence relationships using the

Table 1. Precision, recall & F-Score readings.

Sr. No	Dataset File Name	X	Y	Z	Precision	Recall	F-Score
1	A Novel Neural	129	76	76	62.92	62.92	62.92
2	COTMAP3K8	152	86	86	63.86	63.86	63.86
3	Cells and the	68	35	35	66.01	66.01	66.01
4	Spatially restr	37	24	24	60.65	60.65	60.65
5	Structure of St	34	16	16	68.00	68.00	68.00

CNN. We found some facts in (Gu et al. 2017) like it yielded Average precision of 51.9%, Recall of 7.0% and F-Score of 11.7%. This is indicated that (Gu et al. 2017) performance is lower than that of our model for in-depth classification of the Inter the Inter- sentence relationship words. Whereas on the other hand, the authors of (Gu et al. 2017) mentioned in their research article that their model is needed to use rich neural networks. This eventually concludes that the proposed model works up to the mark in the first attempt of implementation.

5 CONCLUSIONS

The presented technique for the identification of inter sentence relationship in biomedical texts has been outlined in this research article. A dataset containing biomedical texts is furnished to the system for the purpose of extracting the NLP features. Initially the input dataset is preprocessed. After which the resultant string is provided as an input to the NLP module to extract the relevant features. The resultant data is then set as the training data and a test document is processed using the test data through linear regression. The ANN is fed by the regression list for effective neuron creation and hidden layer estimation. The probability list then classified using the Random Forest classification to extract the relevant inter sentence relationships. Extensive experimentation has been performed to achieve the performance metrics of the methodology. The results of the experimentation reveal excellent performance for a first time implementation of such a system for inter sentence relationship extraction from biomedical text corpora.

REFERENCES

Gu, J., Sun, F., Qian, L. & Zhoum, G. 2017 "Chemical-induced disease relation extraction via convolutional neural network", *Oxford University Press*, vol. 2017, April 2017.

Atala, E. N., Hein, P. D., Satish, S., Tumsi, S., Wong, Z., Wang X. & Sean, 2019. "Extracting Inter-sentence Relations for Associating Biological Context with Events in Biomedical Texts", *IEEE/ACM Transactions on Computational Biology and Bioinformatics*, March 2019.

Pennington, J., Socher, R. & Manning, C. D. 2014. "GloVe: Global Vectors for Word Representation.", *Proceedings of the 2014 Conference on Empirical Methods in Natural Language Processing, ACL*, pp. 1532–1543, October 2014.

Ren, P., Chen, Z., Ren, Z., Wei, F., Ma, J. & Rijke, M. 2017. "Leveraging Contextual Sentence Relations for Extractive Summarization Using a Neural Attention Model", *Proceedings of the 40th International ACM SIGIR Conference on Research and Development in Information Retrieval*, pp. 95–104, August 2017.

Kankaraj, M. & Kamath, S. 2014. "NLP based Intelligent News Search Engine using Information Extraction from e-Newspapers", *International Conference on Computational Intelligence and Computing Research, IEEE*, pp. 1–5, December 2014.

Saha, T., Joty, S., Hassan, N. & Hasan M. 2017. "Regularized and Retrofitted models for Learning Sentence Representation with Context", *Proceedings of the 2017 ACM Conference on Information and Knowledge Management, ACM*, pp. 547–556, November 2017.

Xue, D. & Li, F. 2015. "Research of Text Categorization Model based on Random Forests", *International Conference on Computational Intelligence and Communication Technology, IEEE*, pp. 173–176, February 2015.

Yousaf, M. & Wolter, D. 2017. "Spatial Information Extraction from Natural language Place Description for Incorporating Contextual Variables", *Proceedings of the 11th Workshop on Geographic Information Retrieval, ACM*, pp. 1–2, December 2017.

Selvi, T., Karthikeyan, P., Vincent, A., Abinaya, V., Neeraja, G. & Deepika, R. 2017. "Text Categorization using Rocchio Algorithm and Random Forest Algorithm", *Eighth International Conference on Advanced Computing (ICoAC), IEEE*, pp. 7–12, January 2017.

She, X. & Zhu, Y. 2018. "Text Classification Research Based on Improved SoftMax Regression Algorithm", *11th International Symposium on Computational Intelligence and Design (ISCID), IEEE*, pp. 273–276, December 2018.

Recent Trends in Communication and Electronics – Sharma et al. (Eds)
© 2021 Taylor & Francis Group, LLC, ISBN 978-1-032-04572-6

Heart detection and monitoring by IOT

Md. Shahbaz Alam
Assistant Professor, Department of Electronics and Communication Engineering, ABES Engineering College, Ghaziabad, India

Gaurav Dwivedi, Abhinav Gangwar, Sachin Rajput & Abdullah
B. Tech Student, ABES Engineering College, Ghaziabad, India

ABSTRACT: Below explained paper is related to heart pulse monitoring and heart attack detection system by Heart Beat Sensor and Arduino. Today we are having an increasing number of heart diseases which includes increasing in risk of heart attack. This device is implemented to an Arduino that helps measuring heart beat numbers, transmits it by a connection. According to the person he can fixed the standard value of measurement. By mapping further boundaries, this device examines the beat and when it goes more than standard value, it starts beeping the buzzer. All the reading of measurement will be displayed on the implemented led screen. With this advanced technology, nearby people can know about the danger and can help in securing the life of patients.

Keywords: IOT, heart beat sensing

1   INTRODUCTION

In this era of modernization, the advancement of electronic device, cellphones and laptops has become the crux in day-to-day livelihood. The future of today world is Internet, which established device, sensor, appliance, vehicle and many more. These devices may include the Radio Frequency Identification {RFID}, Phone, sensor, actuator and many more. By using Internet, we establish connection with everything, control from everywhere and anytime, effectively get many facilities and information about anything. Main purpose of Internet includes increasing the productivity of automatic controlling purpose, figure handling, permanent proper connectivity method. By use of Heartbeat sensors which is all time active and seeking data, all other systems would be connected local and globally (Ashrafuzzaman et al. 2013, Vaishnave, A. K., & Jenisha, S. T. 2013, Santos et al. 2020). The Internet system can give us a huge amount of data about mankind, things and there works. With joining of the present Internet technology, it gives us a huge amount of access and new services based on cheap products and wireless connectivity. Radio Frequencies Systems and Cloud computing increase compatibility with other network. It provides many methods for statistics collection, figure processing, system managements etc. Every system which try to connects to Internet needed a unique address or proper IPv6.Many people in this world are affected because of unproper access to medicals and health monitoring. With the help of newest technology, small wireless solutions which will connected to IoT could make it easy to keeping eye on patients remotely rather than visiting them or to the hospital. A system of different sensors which are connected to the patient can be used to collect the present health condition by easy and secure method, and the collected information will be improvised, analyzed {by using scientific algorithms} and sent to

DOI: 10.1201/9781003193838-17

the server using wireless transmission method. Any of the required professionals can get and view the data, can give suggestions accordingly to available services remotely with the Requirements of time and development of society; In the present time, people understand that healthy body and healthy lifestyle are much more important and helpful in promoting economic advancement, many believes that existing public health service and its present adoptability have been required to improve with respect to time. In present time every developed country is investing their huge amount on advancement of Internet of things, and many works include France Research System on IoT, Japan's a-Strategy planning, American advanced IoT research and India's Advanced health monitoring Project by Reliance Pvt Ltd.

2 OVERVIEW

The IoT applications in the field of health monitoring system will provide great chance to patients to use the better facilities available at less time, minimum cost and better service. Health monitoring is essential for our body and self to be monitored on every interval to check and maintain our body conditions and performance regularly (Sidheeque, et al. 2017, Ajitha, et al. 2017, Ashrafuzzaman, et al. 2013). Heart Rate of our body, Weight and Temperature measurements, Blood Pressure, ECG are the vital parameters observed for health checking. All the above-mentioned factors will give all the required information regarding our body health and performance, that is if someone having more than normal temperature having fever and different heart rates will indicate heart problems high temperatures indicates having heart problems. Remote patient monitoring is the best method for health Monitoring. This system works by collecting and resending the data to other station for displaying, interpreting and strafing for patient further use. Telemetry System can be used to capture a proper vital sign in such homemade system. With the help of such advancements visiting to patients and medical advisory can be efficiently used. Error in hospital monitoring system are one of the main cause of casualties. Systems should function better to create a better network with the various wireless monitoring systems. This is the connectivity of many physical devices to vehicles, buildings and other items in which different sensors are implemented which make it useful to collect and exchange data. It is defined as the infrastructure of information society by Global standards Initiative. By these methods objects can be controlled remotely from any places near the earth by wireless transmission. It creates huge opportunities in health and economic fields. It increases the accuracy and efficiency of present health monitoring system. These sensors and actuators help to improve the intelligence, better transport of information and better cyber physical systems connectivity which also encompasses technologies such as smart grids, smart cities etc.

3 WORKING DIAGRAM AND HARDWARE USED

3.1 *Working diagram*

The module is having mainly six embedded components: Electrocardiography,sensor, Heartbeat sensor, Infrared sensor, Global system for mobile communication (GSM) module, Arduino UNO and Liquid crystal Display (Patel N. et al., Patel, S. et al., Rani, S. U. et al.). Figure (1) is showing a working diagram.

3.2 *Hardware implementation*

To star the working of the system first we provide power supply to Arduino as it is the overall controlling unit. Figure (2) is presenting a hardware implementation of block diagram. Figure (3) shows the Arduino UNO board and Figure (4) presents an Arduino board. At the input end, there are various buttons like ECG and heart beat sensors, other manual buttons. The output is displayed on the Liquid crystal display at another end. However, the data is transmitted by the

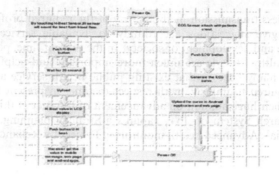

Figure 1. Working diagram.

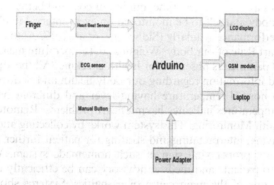

Figure 2. Hardware implementation of block diagram.

Figure 3. The arduino UNO board.

Figure 4. Arduino board.

GSM system into the cloud. The Push button is pressed to read the data after placed a finger on the Beat sensor. After this, the result is displayed on the LCD display. We can load the output data in the website/app and then text the message with the help of GSM by pressing another push button, Similar step is following with the ECG sensor by placing three electric pads. Due to too long characters of ECG result, the LCD display is not able to display it. For this the push button is pressed to send the data via GSM system which display the graph of ECG in the Website/app.

3.3 Pulse sensor unit

This unit is used to calculate beat (pulse) in real time and it can saved for later study. Figure (5) is showing a pulse sensor. It is a basic method to know the condition of our heart. This sensor can measure the heart beat by monitoring the blood circulation through the finger and display the results of the heartbeat after placing the finger. During working of this sensor, the beat LED glows/flashes with each heartbeat(pulse). The digital output is provided directly to measure the heartbeat in Beats per Minute (BPM). This working principle based on light modulation in each pulse.

3.4 ECG sensor

The Electrocardiography including disposable electrode connected with human chest to sense every heartbeat. Figure (6) is showing ECG Sensor. Electrode in this sensor change heart beats in the electricity. With light and small body which measures pulse accurately and represent the pulse rates. A chip of AD8232 is used to sense the electrical vibrations of the human heart (Yadav, Y. et al.). This electrical vibration can be plotted as an Electrocardiogram. Electrocardiography is useful to diagnose various heart conditions.

AD 8232 monitor include 9 connecting terminals in the integrated circuit. These are neutral, supply 3.3V, result, detect lead (LO -) and (LO+), (SDN) shut down, and RA,LA and RL are three inputs a,b,c respectively. It includes 3 cables.

4 SOFTWARE IMPLEMENTATION

In the implementation of software, the software part can be interface with the hardware part. In the hardware part there are many components as Arduino, Electrocardiography and Heartbeat Sensors, hardware of GSM and Liquid crystal display. We have to write program codes to implement the hardware part. In the program coding, it starts with the 7th pin of heart beat, heart beat button initiated with 6th pin, E C G pin is started at 2nd pin. Values of heartbeat with ECG are 0 at the starting. The SETUP method is used to set up the numbers of blocks in the LCD. button is used to take command. CALL BACK method is used when heartbeat does not interrupt.

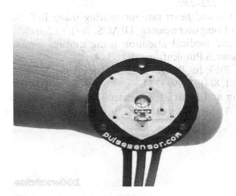

Figure 5.   Pulse sensor.

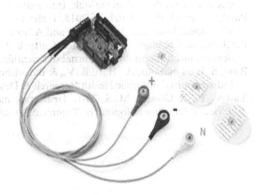

Figure 6.   ECG sensor.

```
setup Begin
  Input heart beat button, ECG button.
  Output lcd.print ("Heath Monitor")
  Delay for 4 seconds;
  callback used for not interrupt when heartbeat count for 20 seconds.
  callback method Begin
  n is the Timer counter.
  if (n>60) then
  n = 0;
  n++;
  end
```

Pulse count for 20 seconds, heart beat pin is high and display value. Heart beat value will be recorded for sixty seconds.

5 CONCLUSION

Sensors connected with controlling system read the patient, send it using Wi-Fi. These Systems set the heartbeat limits. Patient will get a alert notification whenever there heart beat will go above the limited heartbeat. This device also detects the lower heart beat. For checking these conditions, we are planning to propose the whole new paper.

REFERENCES

Sidheeque, A., Kumar, A., Balamurugan, R., Deepak, K. C., & Sathish, K. 2017. Heartbeat sensing and heart attack detection using Internet of Things: IoT. Int. J. Eng. Sci. Comput., 7(4), 6662–6666.

Ajitha, U., Aswathi, P. A., Sasidharan, A., Salman, V. A., Anand, V., & Arvind, A. 2017. Iot based heart attack detection and alert system. International Journal of Engineering and Management Research (IJEMR), 7(2), 285–288.

Ashrafuzzaman, M., Huq, M. M., Chakraborty, C., Khan, M. R. M., Tabassum, T., & Hasan, R. 2013. Heart attack detection using smart phone. International Journal of Technology Enhancements and Emerging Engineering Research, 1(3), 23–27.

Vaishnave, A. K., & Jenisha, S. T. 2019. IoT Based Heart Attack Detection, Heart Rate and Tempera-ture Monitor. International Research Journal of Multidisciplinary Technovation, 1(6), 61–70.

Mallick, B., & Patro, A. K. 2016. Heart rate monitoring system using finger tip through arduino and processing software. International Journal of Science, Engineering and Technology Research (IJSETR), 5(1), 84–89.

Santos, M. A., Munoz, R., Olivares, R., Rebouças Filho, P. P., Del Ser, J., & de Albuquerque, V. H. C. 2020. Online heart monitoring systems on the internet of health things environments: A survey, a reference model and an outlook. Information Fusion, 53, 222–239.

Patel, N., Patel, P., & Patel, N. 2018. Heart attack detection and heart rate monitoring using IoT. International Journal of Innovations and Advancements in Computer Science, IJIACS, 7(4), 612–615.

Patel, S., & Chauhan, Y. 2014. Heart attack detection and medical attention using motion sensing device-kinect. International Journal of Scientific and Research Publications, 4(1), 1–4.

Rani, S. U., Ignatious, A., Hari, B. V., & Balavishnu, V. J. 2017. Iot Patient Health Monitoring Sys-tem. Indian Journal of Public Health Research & Development, 8(4), 1329–1334.

Yadav, Y., & Gowda, M. S. 2016. Heart rate monitoring and heart attack detection using wearable device. International Journal of Technical Research and Applications, 4(3), 48–50.

Recent Trends in Communication and Electronics – Sharma et al. (Eds)
© 2021 Taylor & Francis Group, LLC, ISBN 978-1-032-04572-6

CT image denoising using bilateral and wavelet based thresholding

Rashmita Sehgal & Vandana Dixit Kaushik
Department of Computer Science, Harcourt Butler Technical University, Kanpur, India

ABSTRACT: The nature of Computed Tomography (CT) images assumes a significant job in clinical science. Thusly, this paper incorporates a denoising procedure where bilateral filter and wavelet parcel based thresholding are used. The results of the proposed calculation are analysed and furthermore tried with comparative existing ongoing calculations. The outcomes are assessed as far as visual investigation and furthermore utilizing execution measurements, for example, PSNR. From result examination, it was analysed that a large portion of the cases, the aftereffects of proposed plot is better in contrast with existing plans.

1 INTRODUCTION

In medicinal science, Computed Tomography (CT) is the most generally utilized innovation which includes checking the inner organs of the body utilizing X-Rays, (Buades et al.2005). It very well may be utilized to locate the bone and joint issues, for example, cracks and tumours. Computed tomography can without much of a stretch spot malignancy cells, coronary illness, and so on, (Massoumzadeh et al. 2010). In any case, the clatter of the CT image must be reduced with the goal that the edges are ensured. To secure the edges unmistakable edge preservation procedures are introduced by different researchers, (Tang et al. 2009) &(Diwakar et al. 2018). To handle the issue of boisterous edge, (Gu et al. 2005), proposed another estimation WNNM (Weighted Nuclear Norm Minimization) in which particular singular characteristics are delegated to different weight coefficients. Further, a denoising estimation subject to the discriminative WNNM (D-WNNM) methodology, which is specifically for the LDCT image, was in like manner introduced. In this, area entropy of the image is used to change the WNNM weight coefficient adaptively.

(Li et al. 2005) joined TV minimization using dictionary learning based requirements and it prompts the arrangement of the 3D dictionary learning technique. The significant point of this technique is to locate the dissipated portrayal of the info image as direct components called molecules and they form a dictionary. The 3-D dictionary is more effective than a 2-D dictionary as it could adequately catch spatial connections in each of the three measurements all the while. Wavelet thresholding is mostly used to filter the noise from the medical images such as surelet based thresholding, bayesthresholding and so on. In DWT based thresholding, DWT is utilized for the edge preservation while thresholding is for noise suppression. But this is not only sufficient. The accuracy on noise suppression over the DWT based thresholding must be increased. For that various other methods are also utilized. One of the popular concepts is mostly utilized for enhancing the existing algorithms that is method noise thresholding,(Xie et al. 2018)& (Arjovsky et al .2017).

With the inspiration of thresholding, a denoising plan is proposed so that edges ought to be increasingly saved. The remainder of paper is sorted out as: section 2 gives a short conversation on bilateral filter. Section 3 gives a conversation on proposed system. Section 4 gives result and conversation. At long last, section 5 gives the finishes of the paper.

DOI: 10.1201/9781003193838-18

A bilateral filter is a filter utilized for suppressing the commotion from the images by saving the edges and fringes of the images. It is a non-straight smoothing filter that trades every pixel of the image with a weighted normal of force esteems from the comparative close by pixel. The bilateral filter is defined as:

$$I^{filtered}(x) = \frac{1}{W_p}\sum_{x_i \in \Omega} I(x_i)v_r(||I(x_i) - I(x)||)b_s(||x_i - x||) \tag{1}$$

and normalization term, W_p, is designed as

$$W_p = \sum_{x_i \in \Omega} v_r(||I(x_i) - I(x)||)b_s(||x_i - x||) \tag{2}$$

2 PROPOSED WORK

In proposed methodology, the input noisy CT images are decomposed into wavelet transform to get approximation and detail parts. On approximation part, bilateral filter is performed to get sharp and smooth feature. In detail parts, Bayes thresholding is performed so that noise is suppressed from the high frequency coefficients. The images are further processed with method noise to recover the missing parts and reduce the noise. We perform wavelet bundle thresholding on the distinction images by utilizing round moving for better conservation of the edges and smothering the noise .Best threshold value can be obtained from the concept of circular shift. The important steps for proposed scheme are written below:

Step 1: Apply Discrete Wavelet Transform (DWT) over the noisy image(I) which divides the image into approximation part and detailed part.

Step 3: Apply bilateral filter on the approximation part.

Step 4: Apply soft thresholding on the detailed part using below equation:

$$\widehat{O'} = \begin{cases} sign(O)(|O| - \lambda), |O| > \lambda \\ 0, \ otherwise \end{cases}$$

Where, $\lambda = \frac{\sigma_\eta^2}{\sigma_y}$, $\sigma_\eta^2 = \left[\frac{median(|I(p,q)|)}{0.6745}\right]$, $\sigma_z^2 = max(\sigma_d^2 - \sigma_\eta^2, 0)$, $\sigma_d^2 = \frac{1}{n}\sum_{p=1}^m d_p^2$ and n = no of pixels in a selected block.

Step 5: Perform Inverse Discrete Wavelet Transform (IDWT) on results obtained from steps 3 and 4 and get denoised image R1.

Step 6: R2 = I - R1

Step 7: Apply Discrete Wavelet Transform (DWT) on R2 to get approximation part and a detailed part.

Step 8: On detailed part, perform soft thresholding.

Step 9: Perform Inverse DWT over step 5 and 6 to get denosied outcome as R3.

Step 10: Final denoised outcome can be obtained as: R = R3 + R1.

3 RESULTS AND DISCUSSIONS

The outcomes are assessed over the CT images as appeared in Figure 1, are gotten from http://www.via.cornell.edu/databases.

The images are affected with Gaussian White Noise and the proposed technique is applied on them for de-noising. The trial assessment is performed over the distinctive noise level (σ): 10, 15, 20, 25, 30. In Figure 2, the noise level of the CT images is 20.

To test the after-effects of the proposed works, the resultant images are contrasted and some already existing strategies. For looking at, strategies utilized are (Kang et al. 2013),

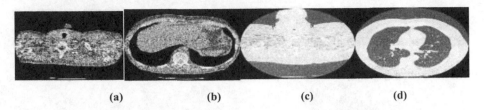

(a)   (b)   (c)   (d)

Figure 1.   Data set of original CT image.

(a)   (b)   (c)   (d)

Figure 2.   Noisy CT image data set.

(Chen et al. 2011), (Brenner et al. 2014), (Kumar et al. 2018), (Tang et al.2009) and (Kumar et al. 2019).

For relative outcome examination, visual outcomes are given as Figure 3: shows the consequence of, (Kang et al. 2013) Figure 4: shows the aftereffect of Phase-saving (Chen et al. 2011) Figure 5: shows the aftereffect of sparse portrayal (Brenner et al. 2014), Figure 6: shows the consequence of shrinkage (Kumar et al. 2018), Figure 7: shows the consequence of (Tang et al. 2009), Figure 8: shows the consequence of Hybrid denoising procedure (Kumar et al. 2019) and Figure 9: shows the consequence of Proposed strategy.

From Figures 3-9, it very well may be outwardly examined that in the majority of the cases the aftereffects of proposed strategy is better in contrast with existing strategies. Here, the visual outcomes are investigated as far as sharpness over the heterogeneous zone, perfection over the homogenous areas, differentiation and surface. The denoised images of the proposed strategy have greatest conservation of edges and noise concealment in contrast with existing techniques. In any case, just visual outcomes are not adequate to locate the best outcomes. In this manner, some significant presentation measurements are helpful to get the factual consequences of the denoising strategies. Here, PSNR is one of the well-known approaches to distinguish the best outcome in denoising strategies.

The previously existing techniques and the proposed structure are set next to each other with Peak Signal-to-Noise Ratio (PSNR). PSNR is a main consideration with the assistance of which we can perceive the image denoising quality. PSNR can be communicated as:

(a)   (b)   (c)   (d)

Figure 3.   Result of Shrink.

(a)   (b)   (c)   (d)

Figure 4.   Result of Phase-preserving.

Figure 5.   Result of sparse representation.

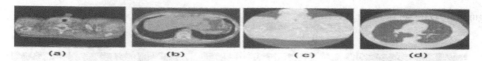

Figure 6.   Result of shrinkage.

Figure 7.   Result of total variation.

Figure 8.   Result of Hybrid denoising technique.

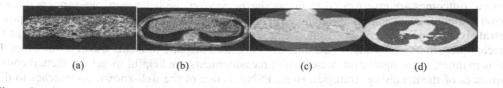

(a)   (b)   (c)   (d)

Figure 9.   Outcomes of Proposed Framework.

$$PSNR = 10 \times \log_{10}\left(\frac{255 \times 255}{MSE}\right)$$

Table 1 show to the arranged estimations of PSNR. Higher the estimation of PSNR results shows the best aftereffects of denoising strategy.

4   CONCLUSION

In this paper, proposed calculation is introduced which depends on thresholding and bilateral filter. We play out a half and half methodology in proposed calculation utilizing non-wavelet methodologies and wavelet strategies. The definitive results are broke down through PSNR and visual examination. Results procured are mind blowing the extent that edge protecting and image denoising. From result inspects, it is broke down that more often than not the aftereffects of proposed plot is better in contrast with existing plans. From basic investigation, power profile dispersion, it likewise shows that proposed strategy gives better results.

96

Table 1. PSNR of Noiseless CT images.

Σ	10	15	20	25	30	10	15	20	25	30
Input	**CT1**	**512×512**				**CT2**	**512×512**			
[23]	31.41	28.99	27.01	25.51	23.07	31.18	28.24	27.60	25.17	23.74
[20]	32.06	28.52	27.73	25.16	23.47	31.06	29.02	27.15	25.87	23.63
[24]	32.21	28.84	27.24	25.29	23.55	31.11	29.92	27.18	25.91	23.39
[21]	33.09	28.81	27.42	25.19	23.46	31.02	29.32	27.26	25.17	23.44
[22]	33.31	28.16	27.81	25.53	23.04	31.04	29.39	27.62	25.72	23.96
[19]	32.21	28.06	27.17	25.37	23.43	31.42	29.19	27.82	25.28	23.61
Proposed	**33.88**	**29.27**	**28.66**	**26.63**	**24.92**	**32.55**	**29.98**	**28.52**	**26.66**	**24.02**
Input	**CT3**	**512×512**				**CT4**	**512×512**			
[23]	31.12	28.64	26.12	25.62	23.27	31.55	29.08	27.97	25.27	23.74
[20]	31.77	28.45	26.99	25.33	23.06	31.11	29.57	27.73	25.82	23.07
[24]	31.93	28.53	26.33	25.43	23.03	31.17	29.82	27.03	25.42	23.74
[21]	31.72	28.93	26.43	25.05	23.79	31.32	29.06	27.43	25.94	23.94
[22]	31.64	28.27	26.90	25.42	23.96	31.42	29.33	27.52	25.62	23.66
[19]	31.61	28.17	26.20	25.23	23.63	31.21	29.35	27.23	25.22	23.16
Proposed	**32.33**	**29.65**	**27.67**	**26.02**	**24.46**	**32.47**	**29.99**	**28.28**	**26.34**	**24.78**

REFERENCES

Arjovsky, M., Chintala, S., & Bottou, L. 2017. "Wasserstein GAN." [Online]. AVALIABLE: https://arxiv.org/abs/1701.07875

Brenner, D .J., & Hall, E.J. 2014. "Computed tomography— An increasing source of radiation exposure," New England J. Med., vol. 357, no. 22, pp. 2277–2284.

Buades, A., Coll, B. & Morel Song, JM. 2005 .A review of image denoising algorithms, with a new one. SIAM J Multiscale Model Simul 4(2):490–530

Chen, Y. et al. 2011. "Improving low-dose abdominal CT images by weighted intensity averaging over large-scale neighbourhoods," Eur. J. Radiol., vol. 80, no. 2, pp. e42–e49.

Diwakar, M., & Kumar, M. 2018. CT image denoising using NLM and correlation-based wavelet packet thresholding. IET Image Processing, 12(5),pp.708–715.

Diwakar, M., Kumar, P., & Singh, A.K. 2020. CT image denoising using NLM and its method noise thresholding. Multimedia Tools and Applications, 79(21), pp.14449–14464.

Kang, D. et al. 2013. "Image denoising of low-radiation dose coronary CT angiography by an adaptive block-matching 3D algorithm," in Proc. SPIE Med. Imag., Lake Buena Vista, FL, USA, pp. 86692G1–86692G6.

Kumar, M. & Diwakar, M. 2019. "A new exponentially directional weighted function based CT image denoising using total variation" Journal of King Saud University-Computer and Information Sciences.

Kumar, M., & Diwakar, M. 2018. "CT image denoising using locally adaptive shrinkage rule in tetrolet domain." Journal of King Saud University-Computer and Information Sciences 30, no. 1: 41–50.

Massoumzadeh, P., Don, S., Hildebolt, K., Bae, K., & Whiting, B. 2010. "Validation of CT dose-reduction simulation," Med. Phys., vol. 36, no. 1, pp. 174–189.

Mishro, P. K., Agrawal, S., & Panda, R. 2019. Medical Image Denoising Using Spline Based Fuzzy Wavelet Shrink Technique. In International Conference on Computer Vision and Image Processing (pp. 185–194). Springer, Singapore.

Shreyamsha, Kumar BK. 2013. Image denoising based on non-local means filter and its method noise thresholding. Springer J Sig Image Video Process 7(6):1211–1227.

Tang, J., Nett, B.E., & Chen, 2009. "Performance comparison between total variation (TV)-based compressed sensing and statistical iterative reconstruction algorithms,". Phys. Med. Biol., vol. 54, no. 19, pp. 5781–5804.

Yu, G., Li, L., Gu, J., & Zhang, L. 2005. "Total variation based iterative image reconstruction," in Proc. 1st Int. Workshop Comput. Vis. Biomed. Image Appl., pp. 526–534.

Recent Trends in Communication and Electronics – Sharma et al. (Eds)
© 2021 Taylor & Francis Group, LLC, ISBN 978-1-032-04572-6

Assurance on data integrity in cloud data centre using PKI built RDIC method

V. Arulkumar
Asst.Prof, IT, Sri SivasubramaniyaNadar College of Engineering, Chennai

R. Lathamanju
Asso.Prof, ECE, SRM Institute of Science and Technology,Ramapuram

A. Sandanakaruppan
Asst.Prof, IT, Sri SivasubramaniyaNadar College of Engineering

ABSTRACT: Cloud computing has become a standard in many business applications. It is popular because cloud computing makes it possible to access large volume of data/ information ubiquitously. Due to attractive options, many organizations/associations prefer to store their data/information in cloud. However simultaneously a couple of ventures are not using cloud offices. The security dangers are the reasons why distributed computing isn't used by all. There are various kinds of security challenges in Cloud structure. Such risks are recognized as perils to secrecy, honesty, accessibility, responsibility and security. In this work, we center around the information/data trustworthiness as a significant quality of security. At the hour of putting away data to cloud, customer doesn't have even a thought where the data is put away. The security of the client's information should be ensured to preserve the integrity of the information.The proposed method based on PKI built RDIC method ensures the rightness of stored data in the cloud. It is confirmed with the assistance of outsider verifier.

Keywords: cloud computing, cloud security, data integrity

1 INTRODUCTION

Cloud computing, as a technology, is here to stay. With increased data size and requirement of such data to the user from anywhere at any time, it becomes easier and more pragmatic to keep the data in the cloud. In fact, many users are using the cloud without even being aware of it. It has also attracted wide interest from the research fraternity. Many organizations prefer cloud storage in place of IT infrastructure because of attractive economics. Consequently 96% of enterprises as of now utilize some type of the distributed storage. Many organizations influence almost five separate clouds and over 26% of ventures spend over $6 million every year on open cloud server farms.

1.1 *Cloud computing security*

Prior to cloud computing, users of cloud applications did not insist on the degree of security from service providers that they do today.Those saleable cloud applications were kept running inside the customer's edge behind the service providers firewall. Cloud security encompasses

DOI: 10.1201/9781003193838-19

Table 1. Types in physical security.

Personal security	Various information security worries to distinguishing nature and various specialists with cloud organization are typically managed
Privacy	Providers ensure that each and every essential datum is secured or encoded and that lone endorsed customers approach data totally.
PKI Management	Each try will have its own exceptional PKI the bosses' framework to control admittance to data and figuring assets.

a wide arrangement of approaches, applications, and the related frameworks of cloud are safe from external attack. It is a specialization of computer security, network system security, and more specifically data security. (Venkat Rao et al. 2013).

Virtual machines and code plans must be scripted so that no human intervention is required.

1.2 Cloud security controls

Framework for secure cloud environment must ensure that valid cloud security design is guarded and also the steps. A powerful security framework for cloud environment needs to ensure the issues that could escalate with countermeasures of the panel are kept in check. These approaches are to manage security breaches in cloud environment.

1.3 Cloud security and data security

There are numerous security issues in cloud as they facilitate resources management over web equally as follows. (Wang. et al 2009).

1.3.1 Physical security

Specialist cloud providers have to secure their environment against unapproved access, theft, and natural calamities. The providers must offer assurance to confirm that in any likelihood of interruption, the issues should be addressed immediately. The plan for this approach could be to keep their data centers safe. Some of the physical security management types are listed in Table 1.

1.3.2 Data security

Different security issues are connected with cloud information the executives with client's security threats, yet additionally with network sneaking around, interruption, and Denial answer in appropriated environment. There are extra distributed computing threats, for example, side channel attacks, virtualization weaknesses, and abuse of cloud organizations. Table 1 explains different issues with data security.

2 EXISTING AND PROPOSED METHOD

In existing framework, the data misfortune was obviously little in respect to the all-out data put away, yet any individual who execute a site a site can expeditiously perceive how terrifying a possibility any information adversity is. The current RDIC shows experience the evil impacts of the issue of an intricate organization, that is, they rely upon the public key framework (PKI), which may destroy the arrangement of RDIC practically speaking.

Proposed framework utilizes PKI-based RDIC protocol to diminish the system complexity in nature. This framework utilizes DriveHq as cloud specialist provider which is the finest for storage. The target objective is to check data integrity in cloud utilizing public auditing, PKI based remote data integrity trustworthiness checking ideas. By this, the proposal can guarantee the accuracy of data in the cloud condition.

3 LITERATURE REVIEW ON DATA INTEGRITY APPROACH

The literature study demonstrates that a wide range of methodologies for cloud data integrity checking utilizing auditing have been as of now actualized. Experts anticipated an issue just because that empowers data proprietors to check the integrity of remote data without unequivocal learning of the whole information. As of late, remote data integrity checking turns out to be increasingly noteworthy because of the advancement of distributed storage frameworks.

(Ateniese et al 2007) proposed technique verifiable information ownership which is uninhibitedly evident part grants not simply the proprietor yet to encounter the server everything considered CR Algorithm and it utilizes the holomorphicproperties. Various extemporization given reliant on this Appliance and help to creating various cloud environment. It give a correspondence cost of solicitation O(1). Considered dynamic PDP plot on the grounds that dependent on hash limits and symmetric key encryptions, which implies the information proprietor can capably invigorate their document after they store their data on the cloud worker. This arrangement is compelling yet has quite recently set number of requests and square addition can't explicitly be supported. This Scheme (Ayad et al 2012) is moreover do scrutiny information respectability in cloud. Implying Bi-linearity Property is to make a proof structure that is mixed close by a stamp that encourages encountering cloud server.

4 METHODS

The proposed method to ensure data integrity with zero knowledge of verifier using PKI built RDIC mechanism. In public cloud environment, users information been stored server which managed by cloud data center. Third party auditor has to monitor the data environment which owned by cloud provider. Key generation center authorized data owner based on login credential details. User can identified by PK which provided by key generation center. Data owner can ensure the data integrity through arising auditing request to TPA and auditing proof has to submit by service provider. This process get complete when auditing message received by providerand the same will shared to data owner in the form of report. In the intervening time if there is any intruders attack service provider environment that has to prevent by enquiring valid keys to ensure.

Table 2 contains the description of software framework that are required to setup the private cloud, deploy the web application over the cloud and how to use the client side application and ways to perform experimentation with user data.

DriveHQ (Drive Headquarters Inc.) is a cloud IT expert center generally in the undertaking administration market. DriveHQ's features join Cloud File Server, WebDAV Drive Mapping, Cloud Storage, Online Backup and FTP Server Hosting. DriveHQ was set up in 2003. As of Mar. 2016, DriveHQ has over 2.5 million enlisted customers. Among the greatest clients are Fortune 500 associations, for instance, Disney, Orange, Alstom, Cummins, etc. In 2012, DriveHQ pushed Cloud Recording and Surveillance organization Camera FTP.com.

The proposed method comprises the sections are user interface, Key generation, File uploadingProcess, Attacks generator, user challenges, proof generation and final check. The goal of

Table 2. Software description.

Programming languages	Scripts	Data base	Online Cloud Provider
Java, JSP	HTML, CSS, Glass fish 3.1.2.2	MYSQL	DriveHq

User interface (UI) configuration is to create a UI which makes it simple (clear as crystal), effective, and easy to use to work a machine in the manner in which creates the ideal outcome. The KGC (Key Generation Center) is a bit of a cryptosystem intended to diminish the danger trademark in exchanging keys. KGCs habitually work in structures inside which a couple of customers may have approval to use certain organizations at specific events and not at others. A common assignment with a KGC incorporates a sales from a customer to use some organization.

File uploading process includes DriveHq cloud server which we are utilizing to store the transferred information. Each client can make his/her very own record in DriveHq for their motivations. Client can transfer information from their nearby PC. Utilizing PKI component, client needs to encode the document utilizing private key and create hash code for the scrambled record. Subsequently, client needs to transfer the scrambled document to the cloud which serves security against the server. For guaranteeing information integrity, we have to show cloud attack to adjust the information put away in the cloud. For that reason, Phishing attack which is one of the cloud attacks is utilized. In user challenge, before the file to be transferred to the cloud, it must be scrambled utilizing client's private key and afterward hash value ought to be produced. Since the file is scrambled, the TPA ought to be ignorant of the document substance and hash code doesn't give any data about the record to TPA. In proof generation process involve with cloud server assumes a crucial job in proof generating process. After getting challenge from client, TPA will in general exchange the test solicitation to cloud for integrity checking purpose. TPA sends record id and file name to the cloud server. The cloud server finds the relating record utilizing filename. The TPA performs integrity checking process. For proof check process involves when the TPA gets cloud server computed hashcode, it does following methods.

Case 1: The TPA likewise keeps up the client created hashcode. It analyzes the hashcode from client and hashcode from cloud server .If the information isn't defiled by any external system, the returned value must be 1.That is, the information isn't ruined.

Case 2: Any external system, may ready to degenerate the client's data stored in cloud which is avail form of encoded format, there might be an opportunity to alter the encoded data. At the point when the information is modified, while TPA looks at the client's hashcode and cloud server's hashcode, the returned value must be 0.That is, the information is corrupted.

5 CONCLUSION

The cloud storage carries the helpful method to access files through various gadgets. In any case, one of the issues is security on the grounds that the file which is transferred could be stolen by unapproved people. Data protection from the outsider verifier is exceptionally fundamental since the cloud customers may store private or sensitive records to the cloud. Be that as it may, this issue has not been completely explored. The proposed PKI built RDIC system ensures correctness of data by giving assurance from the malignant worker and achieves zero information protection from a verifier. The clients encode the document before transferring and the decoding key is recollected by the client itself thus it gives the protection from cloud server and furthermore TPA doesn't have direct access to client's data.

REFERENCES

Ateniese R.C. 2007.Provable data possession at untrusted stores. ACM Conference on Computer and Communications Security: 598–609.
A. Juels 2007 Pors: proofs of retrievability for large files. ACM Conference on Computer and Communications Security: 584–597.

Ayad F. 2012. Integrity Verification of Multiple Data Copies over Untrusted Cloud Servers. Proceedings of the 12th IEEE/ACM ISCCG: 829–834.

C. Wang. 2009. Ensuring data storage security in cloud computing. In Proceedings International Workshop in Quality of Service: 1–9.

G. Ateniese 2008.Scalable and efficient provable data possession. In Secure Communication Fourth ICSPSN: 1–10.

Kan Yang. 2014. TSAS: Third-Party Storage Auditing Service in Security for Cloud Storage Systems, Springer Briefs in Computer Science: 7–37.

K. Venkat Rao. 2013. Data Integrity in Multi Cloud Storage. International Journal of Science Engineering and Advance Technology, Vol 1 issue 7.

Q. Wang. 2009. Enabling public verifiability and data dynamics for storage security in cloud computing. 14th European Symposium on Research in Computer Security: 355–370.

Y. Zhu. 2012. Efficient audit service outsourcing for data integrity in clouds Journal of system software: 1083–1095.

Recent Trends in Communication and Electronics – Sharma et al. (Eds)
© 2021 Taylor & Francis Group, LLC, ISBN 978-1-032-04572-6

A study on web mining classification and approaches

Santosh Kumar & Ravi Kumar
ABES Engineering College, Ghaziabad, India

ABSTRACT: Due to advancement in technology there is huge rise in structured as well as unstructured data which is growing at a very high pace. There is a great increase in users accessing web and all are sharing and posting data on web due to which the volume of data is increasing and making it complex to extract important information. In currant scenario there is certain need of extraction of potentially useful and valuable information from very large data available on web. This is done by the process of web mining. In this paper we have done the comparison and described the three categories of web mining will helpful in accessing the data from web in efficient way. Further, important parameters in web logs that are used in web usage mining have been described.

Keywords: Web Mining, Usage Mining, Content Mining, Structure Mining, Web Logs

1 INTRODUCTION

Due to tremendous growth in the area of computer technology, the cost of storage devices has been reduced. This has lead the business people to store their information in big databases. The web population is increasing in a gigantic manner. To fetch the required information from these large databases is a very crucial challenge. The web is distinguished as its used widely and have great potential to be applied to various business transactions (Barbara, P. & Yates, R. 2006). There is a certain need for the extraction of information effectively and systematically. We use search engines to find useful information but they are also overloaded with a very high pace of information that has been produced within friction of seconds. Web Mining is an approach through which we can extract the required information from hyperlinks between document usage of websites, usage logs, and web documents automatically by using various data mining techniques. By applying web mining techniques we can easily classify the web documents and identify the useful information contained on the web page (Ramakrishna et al. 2010) (Sharma et al. 2011). Thus it improves the power of the web search engine. Prediction of user behavior is also achieved moreover its useful in e-services and website page optimization. We can classify the techniques of web mining in three types web content mining (WCM), web usage mining(WSM) and web structure mining (WUM).

2 LITERATURE SURVEY

In (Alami et al. 2020) the author has talked about various research efforts and techniques used for key phrase extraction. Major approaches used AKPE like deep learning supervised learning; unsupervised learning other related techniques are discussed. Various problems and complexities faced by these approaches have been elaborated. At last, they have done comparative analysis on the performance of best performing techniques and analysis is done on why they are performing

DOI: 10.1201/9781003193838-20

well ten others. In (Berka, P. 2020) the author has stated about a cause-effect relationship and important relationship and the important correlation between learning time, learning perform-ance, and science inquiry competence exists. And analyzed exceptional shift contrasts in the learn-ing practices of learners with various learning performances and capabilities exist by using sequential lag and pattern extraction. In (Chen, C., & Wang, W. 2020), the author talked about an overview on the cutting edge in bunch recommender frameworks concerning different spaces. he has examined existing frameworks as for their accumulation and client inclination models. This association is extremely helpful to comprehend the complexities as for every domain. In (Dara, S., Chowdary, C.R. et al., 2020) the author has proposed a specific information mining process (to remove visit personal conduct standards) so as to uncover the densest periods, consequently extraction of recurrent sequential pattern is done. A period is viewed as dense in the event that it contains in any event one recurrentt sequential pattern for the group of users associated with the site in that period. In (Jabeen et al. 2020) the author has discussed about the basic components of association computation process prevalence of semantic measures. Evaluation methodologies in various fields that depends upon natural language semantics. Classification sources of background knowledge such as informal and formal knowledge sources and fundamental design models like network model, spatial and combinatorial which are applied in association computation process. Categorization of prevailing methods of association computation is done on the basis of applica-tion of sources of background knowledge into two main classes Knowledge Lean approaches and knowledge rich approaches. In (Massegliaet et al. 2008) the author presents a review of Web per-sonalization process can be considered as a use of information extraction needing support for all the stages of an average information extraction cycle. These stages incorporate pre-processing and information gathering, design disclosure and assessment, and finally applying the found informa-tion progressively to intervene between the user and the Web. In (Mobasher, B. 2007) the author has suggested an optimized framework that detects communities on the anticipated edges over network. Using probability of community affiliation and probability of unseen links simultan-eously to upgrade the quality of community detection. Community detection and link prediction are jointly strengthen to produce better results of both the task. In (Spiliopoulou, M. 1999) the author has discussed information extraction from web. Classification of relevant search is done as: Web text mining that focuses on information gaining by users, web usage mining that concen-trates on analyzing the search habits of users of web; user modeling that utilizes knowledge and data about users available over web. In (Zhang et al. 2020) the author investigated the diverse web mining methods and tools, utilized for extracting data from WWW. The various strategies with their advantages and downsides for WCM, WSM, WUM have been discussed.

3 CLASSIFICATION OF WEB MINING

Classification of different types of web mining is given in Table 1.

3.1 Web content mining

In web, content mining analysis of contents of web documents is done and useful information is extracted. Designing of web page is in such a manner that it may contain a group of facts called content data. Web content is composed of multimedia documents, images, audio, video, text, and other types of data.

3.1.1 Approaches to web content mining

There are two content mining approaches: First Agent-based Approach: In this, there is min-imum human involvement. It uses the automatic system and depends on automatic bots that analyze the web pages and find relevant websites and gather useful information. Three types of agents are there namely intelligent search agents, personalized web agents, knowledge filter-ing agents. Second, Data-based Approach: In this different techniques are applied to store the data on web in structured fashion. To analyze the data we use standard data mining tech-niques and database queries.

Table 1. Comparison of web mining categories.

| Parameters | Categories of Web Mining | | |
	Web Content Mining	Web Structure Mining	Web Usage Mining
Visualization of data	Unstructured docs and Structured Docs	Collection of interconnected web pages Hyperlink structure	Client/Server Interactions statistics
Source data	HyperText Documents Text Documents	Hyperlink structure	Logfile of server Log of Browser
Architecture	Relation between the contents of two different web pages	Way in which web pages are linked to each other	Way in which there is variation user behavior for websites listed.
Representation	Bag of words n-grams Term phrases Concept or ontology Relational	graph which is Edge label Relational	Graph having Relational Table
Working Model	Statistical Machine Learning	Page Rank algorithms	Association Rules Classification Clustering
Usability	web page relevance similarity in text paragraphs of segments	Personalization in Web Site Modeling of Management and adjustment	users behavior Mining Finding outliers Categorization
Association	Social based Filters or Collaborative	Reputation-based filters	Content-based filters

3.1.1.1 UNSTRUCTURED DATA MINING

The web contains large set of unstructured mainly text, by data. Knowledge discovery process we used to extract important information.

3.1.1.2 WEB DATA MINING LANGUAGE

Through WDML techniques to tranform unstructured data to structured data i.e data stored in table form and provide information to end users.

3.1.1.3 OBJECT EXCHANGE

Model Different techniques are applies on semi-structured data and important is extracted and saved in Object Exchange Model (OEM)

3.1.1.4 WEB CRAWLER

The hypertext web structure is traversed by software programs and required information . Web crawlers are utilized by search engines to obtain web page information.They are classified into two types internal and external web crawlers.

3.2 Structure mining

While surfing web pages user used to crawl through a sequence of inter-related web pages and mining significant information that distinguishes the page and utilize that information to give the page a rank build on given rules. Regardless, a set or foundation of website pages is given, an application known as a crawler will cross these pages and concentrate the required data from them. The data we are keen on is the hyperlinks contained inside the page. Mining this data should be possible utilizing regular Expressions. The major problem in applying these regular expressions is the assumption of code within web pages abides by various standards. After the extraction of hyperlinks contained in the web page, and the replication been removed the crawler used to crawl WebPages whose hyperlinks have been known since there is no need to crawl some page again.

3.2.1 *Page rank*
To find the importance of any web page the Page Rank is a significant factor as it determines the number of pages linked to it also known as the back links. Back links are formed from key page higher weightage assigned to link as compared to the links directed from less significant pages.

3.2.2 *Hypertext Induced Topic Search (HITS)*
There are two types of pages identified from a hyperlink structure the hubs containing a good source of content and high quality content web pages. Authorities and hubs are find using hits algorithm. The algorithm conduct the Sampling and iterative steps, sampling collection of co-related pages is performed i.e. for a given query a sub graph of G is recovered that holds acute influence pages. The algorithms starts with the root set B on the basis of query a set S is acquired that contains high relevant pages as well as the S is comparatively small and have good authorities

3.2.3 *Hubs*
It is a node having directed edges towards authorities a hub may have many edges directed to the authority.

3.2.4 *Authorities*
The web pages which are linked to hubs.

3.3 *Web usage Mining*
The prediction of users' interest can be determined by applying WUM as it uncovers the information hidden about user's interest while surfing on the internet. The main aim of UM is to model, store, and analyze behaviors, profiles of users' and patterns interacting with a website. It contains the important information like agent logs, referral logs, server logs, ratings of users the click streams, user profiles cookies transactions related to databases, site structure, and page content. Process of WSM is shown in Figure 1.

3.3.1 *Pre processing*
The main aim of applications of data mining is to create the dataset on which statistical and data mining approaches may be applied. In this combination of data collected from distinct sources is done and this data is converted to the format of input to data mining applications.

3.3.2 *Modeling of data*
The output of pre-processing are the page views on which these data mining tasks are applied. There is weights associated with the page views which shows the importance of the web page.

3.3.3 *Analysis and discovery*
The pre defined parameters are used like days, visitors, domains and we use statistical techniques on data to extract information about the user accessing the web pages.

3.3.4 *Navigational and sequential pattern analysis*
In this analysis of sequence of access of web pages is done to know the patterns of inter session. By gaining this knowledge future visit patterns can be determined.

3.3.5 *Web logs*
When user access any site logs are created which are text files which contains the information regarding the request to server. They are the real time files. These logs may provide information regarding transfer of files from any specific website. These logs are very useful to extend business as information provided by them helps us that in which zone we need to do better

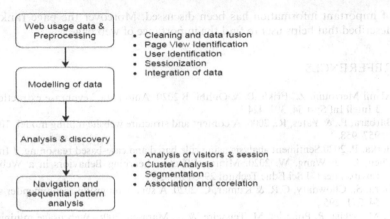

- Cleaning and data fusion
- Page View Identification
- User Identification
- Sessionization
- Integration of data

- Analysis of visitors & session
- Cluster Analysis
- Segmentation
- Association and corelation

Figure 1. Process of web usage mining.

Table 2. Parameter of web access log.

CLIENT IP	ACCESS DATE & TIME	METHOD	URL STEM	PROTOCOL	STATUS	BYTES	BROWSER
201.58.170.90	[18-AUG -2002:01:53:23-0500]	"GET	FRAMES. HTM	HTTP/1.1" 20011631	200	11631	MOZILLA/ 4.0
201.58.170.90	[18-AUG -2002:01:53:23-0500]	"GET	FRAMES. HTM/ CS.JSP	HTTP/1.1" 20011631	200	1153	MOZILLA/ 4.0
201.58.170.90	[18-AUG -2002:01:53:23-0500]	"GET	FRAMES JSP	HTTP/1.1" 20011631	200	782	MOZILLA/ 4.0
201.58.170.90	[18-AUG -2002:01:53:23-0500]	"GET	FRAMES/STYLE. CSS	HTTP/1.1" 20011631	200	1255	MOZILLA/ 4.0

and in which we are having success. It provides the summary of the access to our website. Parameters of access log are shown in Table 2. There are three types of log files agent Log, access log and refer logs.

These are composed user log i.e information of user like page referenced by him, ip address, and amount of time spent by the user on a particular page. Application server data: These are used to analyze different types of business events. Like visits to your website, most visited pages, keyword searches etc. These are logs in application servers like story server, web logic etc. Unique visitors: This information is identified by analyzing the ip address of the user. If the users have distinct ip address they are unique and if same ip access five contents then that is counted once.

4 CONCLUSION

Web mining deals with wide range of issues for extraction of information from large size database of web. This paper provides extensive literature review in the area of WM. Classification of the techniques of web mining is done which are quite useful for accessing important information in an efficient manner. A comparative summary of these approaches and different types of web content mining approaches has been elaborated. The whole process of extraction of important information has been discussed. Moreover the page rank algorithms has been described that helps user to find the importance of web page.

REFERENCES

Alami Merrouni, Z., Frikh, B. & Ouhbi, B.2020. Automatic keyphrase extraction: a survey and trends. J Intell Inf Syst 54, 391–424.

Barbara, P., & Yates, R., 2006. A content and structure website mining model. 10.1145/1135777.1135963, 957–958.

Berka, P. 2020.Sentiment analysis using rule-based and case-based reasoning. J Intell Inf Syst 55, 51–66.

Chen, C., & Wang, W. 2020. Mining Effective Learning Behaviors in a Web-Based Inquiry Science Environment. J Sci Educ Technol 29, 519–535.

Dara, S., Chowdary, C.R. & Kumar, C. 2020. A survey on group recommender systems. J Intell Inf Syst 54, 271–295

F. Masseglia, P. Poncelet, M. Teisseire, & A. Marascu, 2008. Web usage mining: extracting unexpected periods from web logs, Data Min Knowl Disc, 16, pp. 39–65.

Jabeen, S., Gao, X. & Andreae, P. 2020. Semantic association computation: a comprehensive survey. Artif Intell Rev 53, 3849–3899.

Mobasher, B. 2007. Data Mining for Web Personalization. In: Brusilovsky, P., Kobsa, A., Nejdl, W. (eds.) The Adaptive Web: Methods and Strategies of Web Personalization. LNCS, vol. 4321, pp. 90–135. Springer, Heidelberg.

Ramakrishna, M., T, Gowdar, L., K, Havanur, M., S, & Swamy, B., M, 2010.Web mining: Key accomplishments, applications and future directions, in Proc. Intl.Conf. on Data Storage and Data Engineering, pp. 187–191.

Sharma, K., Shrivastava, G., and Kumar, V. 2011. "Web mining: Today and tomorrow," in Proc. 3rd International Conference on Electronics Computer Technology, pp. 399–403.

Spiliopoulou, M. 1999. The laborious way from data mining to Web mining, Intl. J. of Computer Systems, Science, and Engineering, Special Issue on Semantics of the Web i4, 113–126.

Zhang, S., Li, C. & Lin, S. 2020. A joint optimization framework for better community detection based on link prediction in social networks. Knowl. Inf. Syst.

Recent Trends in Communication and Electronics – Sharma et al. (Eds)
© 2021 Taylor & Francis Group, LLC, ISBN 978-1-032-04572-6

Analyzing the impact of variation in hole block layer thickness on OLED performance

Shubham Negi
*Department of Electronics and Communication Engineering, Graphic Era Deemed to be University,
Dehradun and Department of Electronics and Communication Engineering, Tulas Institute, Dehradun*

Poornima Mittal
Department of Electronics and Communication Engineering, Delhi Technological University, New Delhi

Brijesh Kumar
*Department of Electronics and Communication Engineering, Madan Mohan Malviya University of
Technology*

ABSTRACT: The modern display market is dominated by the Organic Light Emitting Diodes (OLEDs). Therefore, impact of architectural and material changes on its performance is very important. The present article analyzes the Double Hole Block Layer (DHBL) architecture of OLED with respect to impact of thickness variation of the two hole block layers (HBLs) involved. Two analyses are performed. First is related to variation of thickness ratio of the two HBLs keeping their overall dimensions same. Second is related to increasing the thickness of these HBLs. The results of the analyses highlights that increasing the overall thickness of HBLs reduces the device performance. Further, changing their ratio improves the performance when ratio favours better performing HBL. However, still, changing the ratio of HBLs in device have poor performance as compared to original device. The reasons for these variations in performance are also discussed.

1 INTRODUCTION

Display is an important integral part of all the electronic display devices. It is the component which makes the first impression of any device and makes the device worth purchasing. At the same time, it is the component which is a source of power consumption in the device and one which gets easily damaged (Shin et al. 2011). Hence, it plays an important role in all the devices as it dictates its cost, efficiency, reliability and quality. Organic Light Emitting Diodes in the present time are being considered as the prime candidate for the display devices may it be portable like mobile phones or cumbersome as a 50" TV (Negi et al. 2020). The main reason behind this is its superior color quality (Negi et al. 2019). Apart from this, it exhibits a far reaching viewing angle with a slight blue shift (Fan et al. 2016) which human eyes cannot detect as compared to its predecessors. To add on, as it is composed of organic materials (Negi et al. 2019), therefore, it has the potential to compose a fully flexible (Negi et al. 2019; Kumar et al. 2014), robust (Negi et al. 2016; Mittal et al. 2015) and low power (Wee et al. 2012) handling devices. At the present time a lot of attention is being paid the world over for improving the performance of OLED. OLED consists of a layered architecture unlike conventional LED (Negi et al. 2018). Therefore, there are a number of ways that researchers propose to enhance it performance (Negi et al. 2019) and each researcher advocates their technique.

DOI: 10.1201/9781003193838-21

However, in the recent past the double hole block layer (DHBL) OLED architecture was proposed (Negi et al. 2018; Yang et al. 2006). This architecture shows a lot of potential to enhance the OLED performance. Therefore, in this article different analysis on DHBL OLED architecture has been performed. These analysis mainly focuses on the thickness of the two hole block layers (HBL) incorporated in the architecture. Some previous analyses performed by us on the HBL thickness are also included in this article. The article is divided in five sections. It includes the present introductory section followed by a brief of previous work presented by us in section two. In section three analyses with respect to the change in thickness of OLED are presented. Thereafter, in section four consists of results and discussion. It includes internal device analysis to support the results. Finally, in section five, concluding remarks of the paper are provided.

2 PREVIOUS WORK

The hole block layers (Hirata et al. 2011) is an excellent addition to the OLED architecture. These layers make a carrier confinement structure which helps in the following two purposes: First, these layers confine the holes movement further away from the emission layer (Its primary task). Thus, these help in enhancing the recombination rate according to the Langevin's theory (Yang et al. 2006). Secondly, if selected effectively, HBLs help in improving the electron injection (Secondary function) (Negi et al. 2018). For this latter function, HBL has to be selected such that they have the LUMO level matching the adjacent layers or the adjacent cathode work function. Therefore, these can be utilized as electron injection layer (EIL) as well. The previous study performed on the thickness of the double HBL architecture is related to the change in ratio of thickness for the two HBL (Negi et al. 2018). DHBL architecture is taken which consists of BAlq and BPhen as the two HBLs as shown in Figure 1. The architecture of the device consisted of ITO as the anode followed by m-MTDATA and NPB as hole injection and hole transport layers, respectively. Thereafter, QAD, which is sandwiched between the Alq_3 layers, is used as emission layer. It is followed by two HBL, BAlq and BPhen, and then the Alq_3 layer as electron injection layer. Finally, there is a bilayer cathode Al/LiF (Negi et al. 2018). All the dimensions are given in Table 1. Now, the analysis of single hole block layer architectures showed that BAlq is a better layer as compared to BPhen. Therefore, it was thought that varying the thickness of the two layers in DHBL architecture can reveal some useful results. Consequently, the thickness of the two HBLs is varied and the analysis is undertaken.

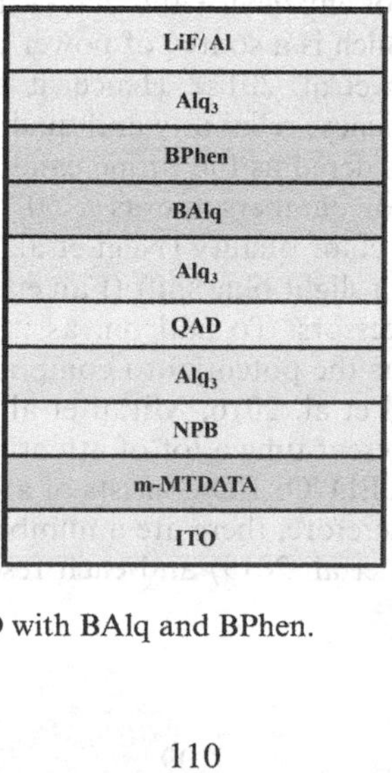

Figure 1.   Structure of DHBL OLED with BAlq and BPhen.

Table 1. Dimensions of the layers of DHBL OLED used for analysis.

Sr. No.	Name of the Layer	Dimension (in nm)
1	ITO (Anode)	50
2	m-MTDATA (HIL)	45
3	NPB (HTL)	10
4	Alq$_3$	5
5	QAD (EML)	0.1
6	Alq$_3$	10
7	BAlq (HBL)	8
8	BPhen	8
8	Alq3 (ETL)	44
9	LiF (Cathode)	1
10	Al (Cathode)	50

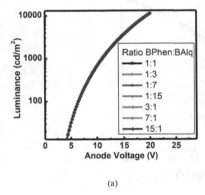

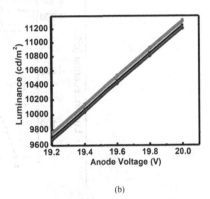

(a)   (b)

Figure 2.   (a) Plot for the luminance versus the anode voltage for the different ratios of double block layers, (b) The zoomed version of the plot in (a) to show the effect of ratio change.

The different ratios that were taken for the two HBLs is 1:1, 3:1, 7:1, 15: 1, 1:3, 1:7 and 1:15. The ratio is with respect to the BPhen: Balq layer (Negi et al. 2018). After the variation in thickness the device luminescence is analyzed and the results are given in Figure 2. The analysis results show that there is increase in the luminescence of the OLED whenever the ratio of BAlq layer is more than that of the BPhen layer. However, on increasing the thickness of BPhen layer, the luminescence decreases. Thus, it again proves that BAlq is a better hole blocking layer. But the role of BPhen layer cannot be ignored and the reason behind this is that the performance of DHBL OLED is always higher than OLED with single HBL, even though it is BAlq only OLED. Further, though there is a change in the performance of OLED with change in variation of DHBL thickness, but the change is not too significant.

3   EFFECT OF USING THICKER HOLE BLOCK LAYERS

The analysis work in this article is related to the thickness of the HBLs and their effect on the performance of OLED. Until now, it is noticed how the change in ratio of thickness of OLED affects its performance. Therefore, in this section the study is focused on the thickness of DHBLs. Before this, all the devices utilizing the HBLs had the same thickness of hole block layers i.e. single and double HBL had the same thickness. This means if the total thickness of

Table 2. Dimensions of the layers of OLED 2 used for analysis.

Sr. No.	Name of the Layer	Dimensions (in nm)
1	ITO (Anode)	50
2	m-MTDATA (HIL)	45
3	NPB (HTL)	10
4	Alq_3	5
5	QAD (EML)	0.1
6	Alq_3	10
7	BAlq (HBL)	16
8	BPhen	16
8	Alq_3 (ETL)	44
9	LiF (Cathode)	1
10	Al (Cathode)	50

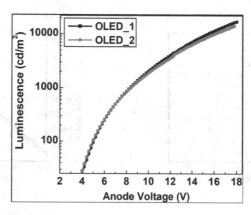

Figure 3. Luminescence comparison of OLED 1 and OLED 2.

HBL in single layer architecture is 16 nm then for the DHBL architecture the total thickness of DHBL is also 16 nm with individual HBL having only 8 nm thicknesses. The reason behind this is that the overall thickness of the OLED should not increase.

Now for the first time, study is focused on increasing the dimension of the OLED by utilizing the two HBL of same size as are used in single HBL OLED architecture. So analysis will be with respect to as to how the thickness of these layers affects the OLED device. So, two OLEDs are analyzed: OLED 1 with dimension same as given in Table 1. The other is OLED 2, in which the dimension of both the HBLs is doubled (16 nm) i.e. same as that in individual single layer architecture. Its dimensions are given in Table 2. The results of analysis are shown in Figure 3. The analysis results show that OLED 1 having the HBL of 8nm have a better luminescence as compared to OLED 2 with 16 nm HBL thickness. In the next section possible reasons for this have been discussed along with the internal device analysis.

4 RESULTS AND DISCUSSION

The analysis results in the previous section shows that the OLED 1 with HBL 8 nm was more effective as compared to OLED 2 with HBL of 16 nm thickness. The reason behind this could be possibly the defects introduced with an increase in thickness of HBLs or position of recombination zone. Different internal analysis has been performed on the device to validate this point with the help of Silvaco Atlas (Atlas manual 2014) device simulator. The first internal device analysis starts with hole and electron concentration in the different devices. As the results in Figure 4 show that there is a slight difference in electron and hole concentration, with OLED 1 consisting of slightly more charge carrier concentration as compared with OLED 2. It is followed by comparison with respect to Langevin's recombination rate as shown in Figure 5. It is clear from Figure 5, that recombination is higher in OLED 1 as compared to OLED 2. Thus is the main reason for improved luminescence of OLED 1. Reason for poor performance of OLED 2 is that the electron and hole concentration within the emission layer. Because of thicker HBLs the emission layer for OLED 2 is at137.1 nm as compared to 121.1 nm for OLED 1. As a result high electron and hole concentration are not within the emission layer rather in its nearby Alq$_3$ layer. Alq3 also shows luminescence properties but lower than QAD and therefore, Langevin's recombination observed in OLED 2 is lower. These analysis results, conclude that the HBL thickness do not help much in blocking the holes as both OLEDs have the same hole concentration.

Thereafter, electric field in the device is analyzed. The results for the electric field are as shown in Figure 6.The results for electric field show that the OLED 1 has a higher electric field as compared to OLED 2. Because of the higher electric field for the same applied voltage in OLED 1, it suggests that there are less number of traps in OLED 1 as compared to OLED 2.

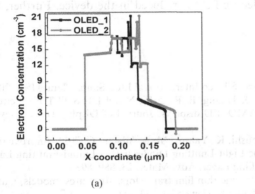 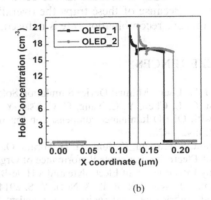

(a)   (b)

Figure 4.   Charge carrier concentration of OLED 1 and OLED 2: (a) Electron Concentration and (b) Hole Concentration.

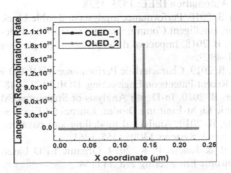

Figure 5.   Langevin's recombination rate for OLED 1 and OLED 2.

113

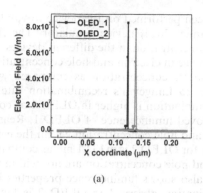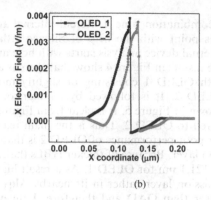

(a)   (b)

Figure 6. (a) Comparison of electric field in the in OLED 1 and OLED 2. (a) Comparison of electric field in the different layers of the two OLED.

5 CONCLUSION

The results in the article suggest that the increasing thickens of the HBL in the OLED does not help in improving the blocking of holes. On the contrary, as we increase the thickness of these layers, the performance of OLED decreases. The reason behind the decrease in the thickness of OLED is because as the HBL thickness increases there is increase in traps in the device. Because of these traps the overall electric field is reduced in the device. Further, the Langevin's recombination rate is reduced.

REFERENCES

ATLAS User's Manual Device Simulation Software. Silvaco International Ltd., Santa Clara, USA, 2014.

Fan, C. L. Chen, Y. C. Yang, C. C. Tsai, Y. K. & Huang, B. R. 2016. Novel LTPS-TFT pixel circuit with OLED luminance compensation for 3D AMOLED displays. Journal of Display Technology 12 (5): 425–428.

Hirata, S. Kubota, K. Jung, H. H. Hirata, O. Goushi, K. Yahiro, M. & Adachi, C. 2011. Improvement of Electroluminescence Performance of Organic Light-Emitting Diodes with a Liquid-Emitting Layer by Introduction of Electrolyte and a Hole-Blocking Layer. Adv. Mater. 23: 889–893.

Kumar, B. Kaushik, B. K. & Negi, Y. S. 2014. Organic thin film transistors: structures, models, materials, fabrication, and applications: a review. Polymer Reviews 54(1): 33–111.

Mittal, P. Negi, Y. S. & Singh, R. K. 2015. An analytical approach for parameter extraction in linear and saturation regions of top and bottom contact organic transistors. Journal of Computational Electronics 14(3): 828–843.

Negi, S. Baliga, A. K. Pandey, Y. Mittal, P. & Kumar, B. 2016. Performance analysis of dual gate organic thin film transistor through analytical modelling. 2016 International Conference on Computing, Communication and Automation IEEE: 1533–1538.

Negi, S. Mittal, P. & Kumar B. 2018. Performance analysis of double block layer OLED and variation in ratio of double block layer. Intelligent Communication, Control and Devices Springer: 123–128.

Negi, S. Mittal, P. & Kumar, B. 2018. Impact of different layers on performance of OLED. Microsystem Technologies 24(12): 4981–4989.

Negi, S. Mittal, P. & Kumar, B. 2019. Characteristic Performance of OLED Based on Hole Injection Transport and Blocking Layers. Recent Patents on Engineering. **DOI**: 10.2174/1872212113666190409151647: 13.

Negi, S. Mittal, P. & Kumar, B. 2020. In-Depth Analysis of Structures, Materials, Models, Parameters, and Applications of Organic Light-Emitting Diodes. Journal Of Electronic Materials 49: 4610–4636.

Negi, S. Mittal, P. & Kumar, B. 2019. Analytical modelling and parameters extraction of multilayered OLED. IET Circuits, Devices & Systems 13(8): 1255–1261.

Negi, S. Mittal, P. Kumar, B. & Juneja, P. K. 2019. Organic LED based light sensor for detection of ovarian cancer. Microelectronic Engineering 218: 111154.

Shin, D. Kim, Y. Chang, N. & Pedram, M. 2011. Dynamic voltage scaling of OLED displays. 48th Design Automation Conference, ACM: 53–58.

Wee, T. K. & Balan, R. K. 2012. Adaptive display power management for OLED displays. First ACM international workshop on Mobile gaming: 25–30.

Yang, H. Yi, Z. Jingying, H. & Shiyong, L. 2006. Organic light-emitting devices with double-block layer. Microelectronics Journal 37(11): 1271–1275.

Recent Trends in Communication and Electronics – Sharma et al. (Eds)
© 2021 Taylor & Francis Group, LLC, ISBN 978-1-032-04572-6

Simulative study of InGaAs and AlGaAs VCSEL in multimode fiber link

Sanket Malik
Department of Physics, GCW, Rohtak (India)

Abhimanyu Nain
Department of ECE, GJUS&T, Hisar (India)

ABSTRACT: The optical links have been most extensively used transmission medium for wired and wireless based communication systems because of its extremely high bandwidth and low loss features. To utilize the maximum out of available bandwidth along with minimum transmission losses, various types of Laser sources are proposed and employed by researchers and engineers. Vertical cavity surface emitting Lasers (VCSEL) are among the few promising light sources. In this paper, a simulative investigation of InGaAs and AlGaAs based VCSEL is discussed. It is reported that AlGaAs provides BER of 10^{-40} & Q factor of 16 which is significantly better compared to InGaAs.

Keywords: VCSEL, InGaAs, AlGaAs, multimode fiber

1 INTRODUCTION

The optical fiber communication has transformed the communication systems significantly by providing enormous bandwidth along with large distance transmission capabilities. High speed internet, multimedia services, various interactive services, Machine to machine communication, Internet of Things (IoT) etc are the examples of next generation wireless services which are a reality today due to the integration of optical fiber links and wireless RF transmissions. Different optical wireless technologies such as Radio over Fiber (RoF), Fiber to the Home (FTTH) etc are being deployed commercially to support the ultra fast data services (Nain 2018).

RoF is the most extensively utilized technology that caters to the different demands of billion of subscribers. It basically transports RF signal through optical fiber at the back end; remote antenna units are installed at the front end to provide mobile wireless service to the subscriber (Kumar 2017). To minimize the losses, highly coherent and monochromatic Lasers are used as light source. VCSEL is one of the probable options of Laser which can be used as light source. It has various advantages like easy manufacturing, smaller size, simple modulation, narrow divergence, high efficiency, excellent coupling to fiber, low operation power etc (Iga 2013). Different materials may be used to grow the VCSEL (Takahashi 1996, Michalzik 2013).

In this paper, InGaAs and AlGaAs based VCSELs are simulated in an optical link and their output performance are estimated on the grounds of BER & Q factor. The paper is organized in different sections which describe the simulation setup along with various simulation parameters and results are discussed in subsequent section. Finally a conclusion is drawn based on the obtained results.

DOI: 10.1201/9781003193838-22

2 SIMULATION SETUP

To compare the performance of InGaAs and AlGaAs based VCSEL, a simple link utilizing multimode fiber has been designed on Optsim simulation software. Figure 1 presents the designed link. The link comprises of a data generator and NRZ based modulator driver which converts logical data into corresponding electrical signals. The electrical signal thus obtained is fed into two parallel branches of AlGaAs and InGaAs based VCSELs which are operating on the wavelength of 1550 nm and linewidth of 10 MHz due to excellent dispersion tolerance property at this wavelength. The laser is followed by a linear multimode fiber whose transmission distance is varied from 0.5 to 1.5 km with attenuation of 0.2 dB/km. The dispersion of the fiber is set at 120 ps/nm/km. The avalanche photodiodes having quantum efficiency of 70% & avalanche gain of 25 dB are used to detect the transmitted optical signal. The electrical output is filtered using raised cosine 4th order electrical filter. The output is observed in terms of BER and Q factor and for this purpose BER estimator and Q meter are deployed in the link.

3 RESULTS & DISCUSSION

The system performance of the designed optical link is estimated on the grounds of BER and Q factor. The Figure 2 shows the behaviour of BER against fiber transmission distance for both the VCSELs. It can be seen that as the fiber transmission is varied from 0.5 to 1.5 km, the BER increases from 10-40 to 10-1 for AlGaAs and 10-18 to 10-1 for InGaAs respectively. It is also observed that AlGaAs provides quality transmission around in the range of 0.8 km to 1.2 km while InGaAs performs well around 0.6 km only.

The variation in Q factor with respect to transmission distance is presented in Figure 3. It reveals the same trend as shown in previous figure. The Q value decreases from 18.5 to 12 & 18.5 to 6 for InGaAs & AlGaAs respectively when distance is varied from 0.5 to 1.5 km. The AlGaAs offers higher Q value and performs better then InGaAs in the transmission range of 0.9 to 1.1 km and output quality deteriorates upon reaching the length of 1.5 km.

4 CONCLUSION

This paper presents a simulative investigation and comparison of InGaAs and AlGaAs based VCSELs by employing them in multimode fiber based optical link. The VCSELs are set at 1550 nm over the distance of 1.5 km. It is found that AlGaAs outperforms InGaAs VCSEL by providing better BER of 10-40 against 10-1 of InGaAs. Hence it is concluded that AlGaAs supports quality transmission in the range of around 1 km utilizing multimode optical fiber.

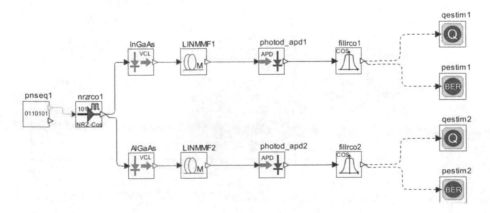

Figure 1. Simulative setup of the designed simulative optical link.

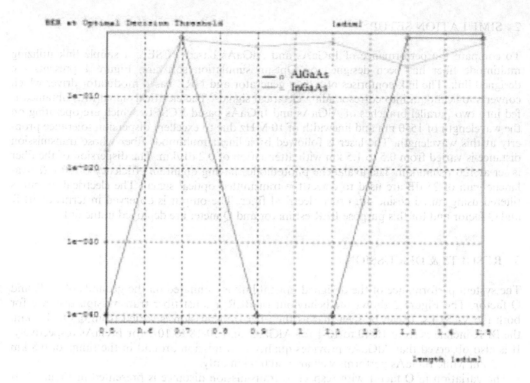

Figure 2. BER vs Distance for AlGaAs & InGaAs VCSELs.

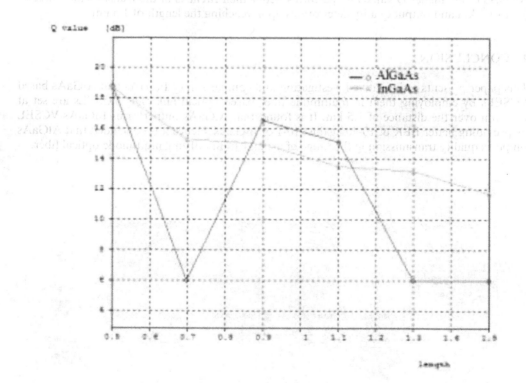

Figure 3. Q factor vs Distance for AlGaAs & InGaAs VCSELs.

REFERENCES

Iga, K. 2013. Vertical-Cavity Surface-Emitting Laser (VCSEL). Proceedings of IEEE 101(10), 2229–2233.

Kumar, S., Sharma, D., Nain, A. 2017. Evaluation of sub carrier multiplexing based RoF system against non-linear distortions using different modulation techniques. International Journal of Advance Research in Computer Science Software Engineering (IJARCSSE) 7 (6): 454–461.

Michalzik, R. 2013. VCSELs: A Research Review. Springer Series in Optical Sciences 166, 3–642.

Nain, A., Kumar, S. 2018. Performance Investigation of different Modulation Schemes in RoF Systems under the influence of Self Phase Modulation. Journal of Optical communication 39(3): 343–347.

Takahashi, M, Vaccaro, P., Fujita, K., Watanabe, T., Mukaihara, T., Koyama, F., and Iga, K., 1996. An InGaAs-GaAs vertical-cavity surface-emitting laser grown on GaAs (311) A substrate having low threshold and stable polarization. IEEE Photonics Technology Letters, 8 (6), 737–739.

Recent Trends in Communication and Electronics – Sharma et al. (Eds)
© 2021 Taylor & Francis Group, LLC, ISBN 978-1-032-04572-6

Implementation of sobel edge detection using MATLAB

N.K. Singh
KIET Group of Institutions, Ghaziabad, India

ABSTRACT: Digital image processing deals with digital images with the help of a digital computer. It is used in the medical field, remote sensing, machine vision, image sharpening and restoration, color processing, pattern recognition and microscopic imaging. It comes under the field of signals and systems but focuses partially on images. In image processing, edge detection is used to acquire data from the frame to extract its features. It divides an image into a matrix of pixel values which are compared with their respective neighbour values to examine a significant discontinuity. This paper proposes the algorithm for the implementation of sobel edge detection using Matlab thus reducing the complexity of edge detection concept.

1 INTRODUCTION

Image processing is a very vast area in which many things are still unexplored. It is used at many places like in medical, surveillance, robotics, automobiles, etc. Edge detection is one of the image processing (Maitra et al. 2018) technique. The abrupt changes of discontinuities in a picture are called edges. Generally, edges are of 3 types which are horizontal, vertical and diagonal edges. It is the first step to recover the information from the image. They mostly occur at the boundary of the image. The steps involved in edge detection include Filtering, Enhancement and Detection. Research on edge detection have run parallel and are intermittent from the last 40 years (Mittal et al. 2019). These approaches are divided into two general categories, such as using the first order and second order of the derivative (Taslimi et al. 2020). The problem with second-order derivative is that it has a lot of complicated equations so methods established on first-order derivatives are used more often. The classical Sobel operator is a commonly used gradient amplitude detection operator in edge detection (Wang et al. 2020).

2 SOBEL EDGE DETECTION

It is a very widely used algorithm along with canny and Prewitt for edge detection in input processing. It uses the second change of the luminance level to detect edges within an image (Jin et al. 2019). First, the image is handled in the horizontal and vertical directions, respectively. This will result in the outcome of an image which is the addition of horizontal and vertical edges. The approach for this processing is the calculation of the gradient of image intensity of every pixel in an image. The gradient vector (∇f) given in (Jin et al. 2019):

$$\nabla f = \begin{pmatrix} G_x \\ G_y \end{pmatrix} = \begin{pmatrix} df/dy \\ df/dy \end{pmatrix} \tag{1}$$

The magnitude and phase of ∇f can be calculated as given in (Cho et al. 2013):

DOI: 10.1201/9781003193838-23

$$|f| = \sqrt{G_x^2 + G_y^2} \tag{2}$$

$$R(f) = \tan^{-1}G_x/G_y \tag{3}$$

The operator first uses a 3×3 convolution template to perform weighted averaging or neighborhood averaging on the detected image (Wang et al. 2019), afterwards identifying the edge by first order differential calculation. Each point contains the horizontal as well as vertical derivative approximations, and respectively, the kernels as given in (Yusoff et al. 2018) are as follows:

Z_1	Z_2	Z_3		-1	0	1		-1	-2	-1
Z_4	Z_5	Z_6		-2	0	2		0	0	0
Z_7	Z_8	Z_9		-1	0	1		1	2	1

| (a) | (b) | (c) |

Figure 1.   Sobel Mask. (a) Pixel positions in matrix size of 3*3. (b) G_y. (c) G_x.

Sobel mask in horizontal direction (Gx) and Sobel mask in vertical direction (Gy) is calculated as by the mathmetical expression given in (Kuljic et al. 2008):

$$Gx = (Z_7 + 2 Z_8 + Z_9) - (Z_1 + 2 Z_2 + Z_3) \tag{4}$$

$$Gy = (Z_3 + 2 Z_6 + Z_9) - (Z_1 + 2 Z_4 + Z_7) \tag{5}$$

3   IMPLEMENTATION USING MATLAB

3.1   *Image captured by the camera and conversion into a grayscale image*

The image whose edge detection is to be taken out can be of any extension. It can be read using the inbuilt imread() function. The image is converted to the grayscale image so that less information needs to be provided for each pixel. A grayscale image M pixels tall and N pixel wide is represented as a matrix of double datatype of size M×N (Abdullah et al. 2016). The image can be converted to gray image simply by using inbuilt rgb2gray() function. Figure 2 shows the grayscale images.

3.2   *Creating a double-precision array of image*

The image is changed over into double to protect the exactness of the picture, rescaling the information if essential. It is done on the grounds that procedure on pictures are simpler when the images are represented in floating-point. It is entirely expected to change from integer number representation to floating-point representation which can be done by using inbuilt double() function.

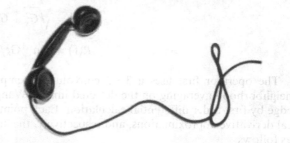

Figure 2. Grayscale image.

3.3 *Filtering*

As noise is observed in the image, a 2-d averaging digital filter is used to decrease noise in the image. After utilizing the filter, we can get rid of the noise, which can reduce the accuracy of the edges of the image and alter the result of our observation. It can be done using the following code

"out=filter2(fspecial('average',3),image_doubling)/225;"

image_ doubling contains the double-precision array of image created in previous step.

3.4 *Sobel mask*

Sobel operator is similar to other image processing operators such as Prewitt operator. It is utilized to recognize two sorts of edges in image mainly horizontal and vertical. This mask works precisely equivalent to the Prewitt operator having just a single distinction that it has "2" and "- 2" digits in the focal point of first and the third segment. When it is used on the picture the mask shall hone the vertical edges. Used Sobel mask G_y = [-1 0 1; -2 0 2; -1 0 1], G_x = [-1 -2 -1, 0 0 0, 1 2 1].

3.5 *Finding number of rows and columns*

The quantity of rows and columns of a picture can simply be found by using inbuilt size() function. The size is calculated so that its value can be used in loop depending upon which the pixels in the image will be processed.

3.6 *Applying sobel mask and result*

Sobel mask is implemented to complete image in both directions, that is vertical and horizontal. Two loops are used, one for operation in rows and the other for operation in columns. We will operate the loop up to $(r-3)^{th}$ row position and $(c-3)^{th}$ column position as the Sobel mask is a 3*3 matrix. The sobel mask is applied by multiplying its values with the matrix containing pixels. Finally, sum/combine all the horizontal and vertical edges. Figure 3 represents the final image which shows accurately the edges of the object in the image. It can be done using the code:

```
for q = 1:(r-3); for p = 1:(c-3)
gr_image = out(q:(q+2),p:(p+2));
res = sobel_maskx .* gr_image;
res2 = sobel_masky.*gr_image;
f(q,p) = sum(sum(res));
o(q,p) = sum(sum(res2));
```

Figure 3.   Sobel Edge Detection image.

end; end; output = f+o;

'r' represents rows, 'c' represents columns, 'out' contains filtered image, 'sobel_maskx' and 'sobel_masky' represents sobel mask in horizontal and vertical direction.

4   CONCLUSION

The project depends on Sobel edge detection. The project is coded in Matlab, which end up being a decent stage to inspect every single change in the image during the entire cycle. The algorithm can be used at various places like car plate number detection during surveillance, medical imaging, robot vision, etc. The main advantage of Sobel edge detection is its simplicity because of approximate gradient calculation whereas other edge detection techniques has greater computational complexity and time consumption.

REFERENCES

Abdullah, A., Palash, W., Rahman, A., Islam, K., & Alim, A. 2016. Digital image processing analysis using Matlab. *American Journal of Engineering Research (AJER)*, 5(12): 2320–0936.

Cho, J., Suh, J. W., Jeon, G., & Jeong, J. 2013. Surface modeling-based segmentalized motion estimation algorithm for video compression. *IEICE transactions on communications*, 96(4): 1081–1084.

Jin, S., Kim, W., & Jeong, J. 2008. Fine directional de-interlacing algorithm using modified Sobel operation. *IEEE Transactions on Consumer Electronics*, 54(2): 587–862.

Kuljic, B., Janos, S., & Tibor, S. 2008, September. Determining distance and shape of an object by 2D image edge detection and distance measuring sensor. In *2008 6th International Symposium on Intelligent Systems and Informatics*: 1–4. IEEE.

Maitra, I. K., & Bandhyopadhyaay, S. K. 2018. Adaptive Edge Detection Method towards Features Extraction from Diverse Medical Imaging Technologies. In *Computer Vision: Concepts, Methodologies, Tools, and Applications*: 1245–1278. IGI Global.

Mittal, M., Verma, A., Kaur, I., Kaur, B., Sharma, M., Goyal, L. M., ... & Kim, T. H. 2019. An efficient edge detection approach to provide better edge connectivity for image analysis. *IEEE Access*, 7: 33240–33255.

Taslimi, S., Faraji, R., Aghasi, A., & Naji, H. R. 2020. Adaptive Edge Detection Technique Implemented on FPGA. *Iranian Journal of Science and Technology, Transactions of Electrical Engineering*: 1–12.

Wang, X., Fang, Y., Li, C., Gong, S., Yu, L., & Fei, S. 2019. Static gesture segmentation technique based on improved Sobel operator. *The Journal of Engineering*, 2019(22): 8339–8342.

Yusoff, N. M., Halim, I. S. A., & Abdullah, N. E. 2018, August. Real-time Hevea Leaves Diseases Identification using Sobel Edge Algorithm on FPGA: A Preliminary Study. In *2018 9th IEEE Control and System Graduate Research Colloquium (ICSGRC)*. 168–171. IEEE.

Recent Trends in Communication and Electronics – Sharma et al. (Eds)
© 2021 Taylor & Francis Group, LLC, ISBN 978-1-032-04572-6

A comprehensive analysis of obstacles detection techniques using convolutional neural networks for visually impaired persons

Anamika Maurya & Prabhat Verma
HBTU Kanpur, India

ABSTRACT: The aim of this paper is to give a comprehensive survey in the field of object detection using deep learning techniques. Convolutional Neural Network(CNN) helps to classify more than one objects in the environment with great accuracy. The paper will highlight the features of CNN variations with time that help the researcher to incorporate the CNN in developing Electronic Travel Aid for the Visually Impaired-(VI) person. This study is an effort to report the research community about the capabilities of these neural network algorithms and emphasize the improvements over time. Recent advancement in CNN approach such as You Look Only Once(YOLO) and CSP-Net backbone could bring the ETA more compact, accurate and fast that would be easily affordable to VI.

1 INTRODUCTION

India is the country where the number of blind people is large as compare to other country. With passing year the population of India increases and reached to 130 crore. In which almost 9 crore people are visually impaired and approx. 8 crore people cannot travel independently(World Health Organization 2016). The inability of visualization imposes major challenges in performing daily life activities to the blind people. Without any assistance, it becomes a crucial task for them to go outside or to do any house chores. In recent years, many solutions have been proposed for navigational assistance. Sensor based approaches, Boosted classifier, Deformable Grid, Dictionary based approaches, Computer vision and Machine learnings are the most common approaches that have been used in object detection in ETAs but the issue of speed and accuracy is still a challenge for developers (Dakopoulos et al. 2010). In Deep Learning, Neural Network based obstacle detection approaches shown substantial improvement. The concept of CNN was described by LeCun in 1998 and the continuous research in CNN resulted its speed as being hailed as a milestone in object detection field. Researcher's rigorous effort in developing ETAs resulted many solutions in past but there is no device that is portable, low-cost, fast and accurate. The primary goal of this paper is to explain the work done in the field of CNN in such a handy way that could help the researcher to analyse and decide in which direction the scope of research is heading with the strengths and weaknesses of each in order to analyze the previous researches done in the area and what is the prospect for the further research in the area. The paper is organized as follows: Section II comprises the overview of CNN based approaches proposed by the researchers. Section III discuss the feasibility of CNN approaches for ETAs.

DOI: 10.1201/9781003193838-24

2 CNN AND ITS VARIATION

2.1 R-CNN

The research progress in the field of obstacle detection slow down in the period of 2010-2012 with minor variation in existing methods. Introducing the concept of region proposals with CNN named as R-CNN (Girshick et al. 2014). R-CNN is successful approach that consists of three modules. Module one gives region proposals that is independent of category and module two is large CNN to produce fix length feature vector and module three consist of multiple linear support vector machines that starts with a pre-trained ImageNet classification network and refine the Convolutional Network from end-to-end. R-CNN uses slow method i.e. selective search method and generate approx. 2000 region proposals for each image classification. For each region proposals, image classification is done through CNN. At the end regression method is used to refine each bounding box. Classifying 2000 region proposals for each image is a time taking process since it requires time to train the network and using fixed algorithm makes the R-CNN slow thus not feasible for real time applications.

2.2 SPPnet

The Spatial Pyramid Pooling SPPnet approach (He et al. 2014) lies between the hybrid models and R-CNN. The approach eliminates the requirement of fixed size input of image that sound impractical and trade off with object recognition accuracy. Irrespective of image size, SPPnet network structure generate fixed size of it. Unlike RCNN, SPPnet uses convolutional layers to extract all features of image. These convolutional features are not dependent on region proposals and shared by all regions. Like RCNN, SPPnet uses a region-wise multi layer perceptron(MLP) for image classification. In the SPP-based object detection method features are pooled on convolutional feature maps and convolutional feature is not computed again and again and supplied to the fully-connected layers for classification.

2.3 Fast R-CNN and Faster R-CNN

SPPnet is further enhanced in the Fast R-CNN and Faster RCNN, which perform better than the hybrid methods. In terms of speed, R-CNN is not as good as Fast R-CNN (Girshick et al. 2015), because in RCNN one has to feed region proposals to the convolutional neural network each and every time while Fast R-CNN omit this step. Like Fast R-CNN, Faster R-CNN take the image as an input and convolutional feature map is generated through convolutional network. Faster R-CNN Model has 2 modules: Region Proposal Network(RPN) and Fast R-CNN Module (Ren et al. 2015).

The RPN Module proposes regions and the second module used that regions since first module give the instruction to second module at which region to be work upon. Both modules will be trained independently and the target of author is to developing an algorithm that allow them to share convolutional layer. The author used four step alternating training. In Faster R-CNN, shared features are used in proposing regions. Despite the improvements, previously explained systems use MLP for region-wise classifiers. Sharing of convolutional features make the approach less costlier and detection rate of objects is 5-16 frame per second. i.e. faster than existing variation of RCNN.

2.4 Single Shot MultiBox Detector(SSD)

The proposed SSD object detector approach uses a single deep neural network (Liu et al. 2016). The single shot refer object localization and classification that were done in a single forward pass of the network whereas MultiBox technique used for bounding box regression. The network works as object detector that also classify the detected objects. The approach runs a convolutional network on input images only once and computes the feature map. They introduce SSD, a single-shot detector for different categories that is faster than YOLO (Redmon et al. 2016) and significantly more accurate than previous explained techniques that perform direct region proposals and pooling. High detection accuracy can be achieved by producing the predictions of different scales from feature maps and by separating the predictions on the basis of its aspect ratio.

2.5   RFCN

The proposed region-based Fully Convolutional Networks (RFCN) for efficient and accurate object detection (Dai et al. 2016). Unlike existing region-based detectors that apply a costly per-region sub network 100 of times, their region-based detector is fully convolutional with maximum computation shared on entire image. In R-CNN, Fast R-CNN, and Faster R-CNN (Ren et al. 2015) region proposals are generated by RPNs first. Then Region Of Interest(RoI) pooling is done and feed to fully connected layers for classification and bounding box regression. The fully connected layers are not shared among region of interest which increases time and makes RPN approaches slow compared to others. The fully connected layers increase the number of connections which also raise the complexity. R-FCN is even faster than Faster R-CNN.

2.6   FCN-NOC

In CNN, feature extractor and object classifier are two important components. SPPnet, Fast/ Faster R-CNN and many other systems focuses on feature extractor and significant research efforts have been done in the field of feature extractor but object classifier still does not get much attention of researchers and they just use simple multi layer perceptron for classification. FCN-NOC approach (Ren et al. 2016) focused on this component and designed deep network for classifiers. They focused on region-wise classifier architecture that use shared, region-independent convolutional features and called it: Networks on Convolutional feature maps(NOCs). Carefully designing region wise classifiers increases accuracy over multi layer perceptron. Using per region classifier with ReseNet and GoogLeNets resulted high detection accuracy.

2.7   Mask RCNN

Along with object detection, Mask RCNN generates mask for each instance (He et al. 2017). Adding pixel to pixel alignment between input and output in Mask RCNN is an extra feature added to Faster RCNN. A class label and a bounding box offset, these two outputs are generated by Faster RCNN, Mask RCNN added one more output of object mask with a very small overhead and running at 5fps.

2.8   YOLO and its version

The previous algorithms use regions to locate the object in the image. The complete image was not passed through the network but only parts of the image having high probabilities of carrying the object send to network. You Only Look Once(YOLO) is an object detection algorithm that is different from the above region based algorithms (Redmon et al. 2016). In this method, single convolutional network foresee the bounding boxes and the class probabilities for the predicted boxes. They take the image as input and split into SxS grid and consider m bounding boxes in each grid. For each bounding box, the network gives a class probability along with offset values. The bounding boxes with class probability above a certain point is choose to locate the object in the image. YOLO is faster approach as compare to other object detection algorithms but struggles with small objects in group and unusual aspect ratio object. The algorithm also suffer with localization error i.e. significantly higher than R-CNN variants.

To overcome the limitation of YOLO, second version of YOLO is introduced as YOLO9000.

YOLO9000 (Redmon et al. 2017) uses batch normalization, higher resolution classifier and anchor boxes that improves the mAP by 2% and fine-grained features and multi scale training help to detect the small objects of different configuration and dimension. YOLOv2 is further

Table 1. Comparison between CNN variants for object detection.

RCNN	Year	DATASET	Speed(s)	FPS	Layer	mAP%
Deep CNN	2013	Cross-Street			4	83.3
R-CNN	2014	VOC2007	47			63.0
SPP-Net	2014	VOC2007	4.3			66.0
Fast RCNN	2015	VOC2007	2.3	5-16		68.1
Faster RCNN	2015	VOC2007	0.2	7		73.2
YOLO	2016	VOC2007		45	26	63.4
SSD	2016	VOC2007		59		74.3
R-FCN	2016	VOC2007	0.17		101	83.6
CNN-NOC	2017	VOC2007				68.8
Mask R- CNN	2017	MS COCO		5		
YOLOv2(9000)	2017	MS COCO		40	32	78.6
YOLOv3	2018	MS COCO	0.005	30	106	57.9
YOLO LITE	2018	VOC2007		21		33.81
CSP-Net	2019	Image-Net				
YOLOv4	2020	MS COCO		65		43.5AP

improved and author named the new version 3 as incremental improvement[8] that claimed the improved accuracy and speed in real time object detection. Using logistic classifier, version3 gives the multi-label classification. To make accurate prediction YOLOv3(Redmon et al. 2018) uses featured pyramid networks. Deep learning approaches needed high computation power thus existing YOLO version requires GPU. A newer version YOLO-LITE (Huang et al. 2018) was introduced to run on a desktop or mobile phone that do not need GPU but the lightweight architecture compromised with accuracy. CSP-Net (Wang et al. 2019), a new backbone was designed to avoid the costly hardware dependency of CNN approaches and reduces the computation time as well as memory use.

Training and testing on conventional GPU with a large mini batch size is still a challenge for the CNN. Yolov4 combined the features such as Weighted-Residual-Connections (WRC), Cross mini-Batch Normalization (Cm), Cross-Stage-Partial-connections (CSP), Self-adversarial-training (SAT) and Mosaic data augmentation, Dish-activation, Drop Block regularization, Chou loss and CSPDarknet53 backbone as the feature-extractor model and became the fastest obstacle detection techniques with accurate prediction (Bochkovskiy et al. 2020).

The table 1 summarizes the performance of different approaches evolved till date. Since all the approaches are not using the same dataset, same architecture and do not run in the same environment, a true comparison can not be made. The researcher may compare the performance of these CNN variants using the speed, processing of frames using Frames per second-(FPS) and to measure the accuracy of these techniques, Mean average precision(mAP) is calculated that is also referred as average precision(AP).

3 ELECTRONIC TRAVEL AID WITH CNN

3.1 ETA features

The Electronic Travel Aids should follow the standard criteria such as the device should not occupy the important body parts of VI so that he/she would be able to interact with surrounding. Detection of objects up-to fixed distance so that he/she would be able to take decision accordingly and cost of the device should be moderate so that it could reach to maximum needy people and depending upon the severity(size, movement and depth) of object, feedback should be different or alarming. Based on users and developers experience, we found the challenges that should be considered while designing the ETAs are compactness, low in cost, range of device, able to customize, user friendly, multifunctional and most important features are accuracy and speed.

127

3.2 Integration of CNN variants with ETAs

Reviewing the neural network research work, we noticed that to classify and recognize object, the use of convolutional neural network is elite. But the dependency of deep learning on system configuration and GPU makes the research a bit costlier that turn ETAs with CNN costly.

R-CNNs are a two step family of technique that requires high computation power. Whereas YOLO and its versions are one step technique for object recognition with excellent speed and real-time use. The later version of YOLO can be trained on conventional GPU that makes easy to incorporate CNN for ETA with real time response of obstacle and much higher accuracy.

4 CONCLUSIONS

After reviewing the various obstacle detection approaches and the most trending convolutional neural network with their pros and cons, we concluded that CNN improves the accuracy day by day with enhancing concepts incorporating within it. The detection speed is also one of the major required features needed in ETA can be solved with CNN enabled ETA devices. CNN and and its variants outperforms as compare to their older version in respect of accuracy and speed. CNN advancement may remove the hurdle of developing the device that meet with user requirements. Depending upon users requirement, the developer may choose the CNN model for ETAs. If we need high detection rate and accuracy can be compromised up to certain extent the tiny version of YOLOv4 also serves the purpose but YOLOv4 is still unbeatable in terms of speed and accuracy in detection of obstacles.

REFERENCES

Bochkovskiy, A., Wang, C. & Liao, H. 2020. YOLOv4: Optimal Speed and Accuracy of Object Detection https://arxiv.org/abs/2004.10934

Dakopoulos, D. & Nikolaos, G. B. 2010. Wearable obstacle avoidance electronic travel aids for blind: a survey *IEEE Transaction On Systems, Man, And Cybernetics—Part C: Application And Reviews, Vol. 40.*

Dai J.& Li, Y., 2016. R-FCN: object detection via region-based fully convolutional networks, *30th Conference on Neural Information Processing Systems(NIPS).*

Girshick, R. & Donahue, D. 2014. Rich feature hierarchies for accurate object detection and semantic segmentation in *Proc. IEEE Conf. Comput. Vis. Pattern Recognit.,pp. 580–587.*

Girshick, R. 2015. Fast R-CNN in *Proc. Int. Conf. Comput. Vis., pp. 1440–1448.*

He, K. & Zhang, X. 2014. Spatial pyramid pooling in deep convolutional networks for visual recognition in *Proc. 13th Eur. Conf. Comput.Vis., pp. 346–361.*

He, K. & Gkioxari, G. 2017. Mask r-cnn in *IEEE International Conference on Computer Vision, pages 2980–2988*

Huang, R., Pedoeem, J. & Chen, C. 2018. YOLO-LITE: A Real-Time Object Detection Algorithm Optimized for Non-GPU Computers *IEEE International Conference on Big Data (Big Data), Seattle, WA, USA pp. 2503-2510.*

International Agency for Prevention of Blindness. 2009 [Online]. Available: http://www.iapb.org/

Liu, W. & Anguelov D. & Erhan D. 2016. SSD: single shot multiBox detector", *Eur. Conf. Comput.Vis.*

Redmon, J. & Divvala, S. & Girshick, R. & Farhadi, A. 2016. You only look once: unified,real-time object detection *Comput. Vis. Pattern Recognit.*

Redmon, J. & Farhadi, A. 2017. YOLO9000: better, faster, stronger In *Proceedings of the IEEE Conference on Computer Vision andd Pattern Recognition (CVPR), pages 7263– 7271.*

Redmon, J. & Farhadi, A. 2018. YOLOv3: An incremental improvement. *arXiv preprint arXiv:1804.02767, p.p. 2, 4, 7, 11*

Ren, K. H. & Girshick, R. & Sun, J. 2015. Faster R-CNN: towards real-time object detection with region proposal networks *Advances Neural Inf. Proc. Syst., pp. 91–99.*

Ren, S. & He, K. & Girshick R. 2016. Object Detection Networks on Convolutional Feature Maps *IEEE Transaction On Pattern Analysis And Machine Intelligence, VOL. 39, NO. 7.*

Wang, C. & Liao, H. M. & Yeh, I., Wu, Y., Chen, P., Hsieh, J. & Wang, C. 2019. CSPNet: A New Backbone that can Enhance Learning Capability of CNN *arXiv:1911.11929*

World Health Organization. 2016. Visual impairment and blindness [Online]. Available: http://www.who.int/mediacentre/factsheets/fs282/en/.

Recent Trends in Communication and Electronics – Sharma et al. (Eds)
© 2021 Taylor & Francis Group, LLC, ISBN 978-1-032-04572-6

Impact of wind energy based DG placement on congested electrical network under deregulated competitive power market

A. Agrawal & L. Srivastava
Electrical Engineering Department, Madhav Institute of Technology & Science, Gwalior, India

S.N. Pandey
Electrical Engineering Department, Dr. Bhim Rao Ambedkar Polytechnic, Gwalior, India

ABSTRACT: Deregulation has created the competitive environment in electrical power market. Day by day increasing demand and addition of new and renewable energy sources has made the power system operation more complex and insecure. In this competitive environment the profit maximization tendency of market participants has created overloading of transmission lines. This may lead to insecure and expensive operation of power system. For making power system operation smooth, secure and economic it is very necessary to make the transmission lines congestion free. Optimal placement of Wind Power Plant (WPP) as a Distributed Generation at the bottleneck location provides the required electric energy can effectively reduce the overloading of transmission corridors. In this paper Artificial Intelligence (AI) technique based Salp Swarm Algorithm (SSA) is proposed for getting optimal size and location of WPP based DG. The effectiveness of the proposed algorithm is compared with Genetic Algorithm (GA) at IEEE 30 bus system in MATLAB software environment.

Keywords: Salp swarm Algorithm, Wind Energy Source, Congestion Management

1 INTRODUCTION

Profit maximization environment in restructured power market may lead to bad effect not only on electricity prices but on the system security and reliability also. System security has threatened by new market operation methods (Mirjalili et al. 2017, Tiwari et al. 2019, Bigerna et al. 2015). Now a day's system security or overloading of the line has become a challengeable task for the system operator (Jain et al. 2013). Many researchers have proposed congestion management methods (Agrawal et al. 2020a) (Agrawal et al. 2020b). Strategic placement of small power generating units at or near customer loads provides incremental capacity to generate for highly reliable deregulated electricity markets (Gill et al. 2008, Agrawal et al. 2020c, Afkousi et al. 2010). Number of congestion management methods has been proposed (Kunz et al. 2015, Kunz et al. 2015) by the researchers in deregulated power market. Size and location of wind energy based DG play a key role to make the system operation smooth and economic (Ramandi et al. 2016, Agrawal et al. 2020d). Currently wind energy source based plants provide a large portion of electric energy which supports the development of country (Dhillon et al. 2014). Energy prices for the electric energy generated by the wind energy source based plants can also be at completive level and can reduce system operation cost. Wind energy source based plants does not always turns down electric energy price,

DOI: 10.1201/9781003193838-25

its perfect location and sufficient amount of energy generated are very crucial parameters for successful operation of wind energy source based plants (Ben-Moshe et al. 2015).

Most of the techniques in the literature were aimed to optimize either location or capacity and to estimate voltage improvement and loss reduction. This paper proposes an AI technique based optimal sizing and sitting of wind energy source based plants for congestion management and also reduces the energy prices.

2 POWER GENERATION FROM WIND ENERGY

Wind energy is under the category of renewable energy. The running cost of wind generator is not so much high as well as it eliminates the variations in energy price. Wind generators do not create any pollution for the environment. This attractive property of wind generator has made it more common and fastest growing energy source in worldwide. The electric energy generated by wind turbine can be shown by the following equation

$$P_{wind} = \frac{1}{2}\rho_{wind} A \gamma V_{wind}^3 \tag{1}$$

Where $'\rho_{wind}'$ is the air density factor, $'A'$ is the swept area of wind turbine, γ overall efficiency of wind power plant and V_{wind}^3 is the wind speed for the given height. Wind speed is not fixed and its electrical power output is variable all the times. The speed forecasted data can and cost of electric power generated by wind turbine 3.75 Rs/MW for 1 hour are taken from (Tiwari et al. 2019). Distributed Generator is generating real power only it also having generation limit.

$$0 \le P_{wind} \le P_{wind}^{Max} \tag{2}$$

3 SALP SWARM ALGORITHM

Salp swarm algorithm is a swarm based metaheuristic algorithm proposed by (Mirjalili et al. 2017). This algorithm work on swarm behavior of a population of salps. Like other swarm based other algorithm SSA population also having leader as a guide and other salps are the followers. The search space is multi dimensional. The position of all salps is stored in two-dimensional matrix and having a food source is the target of salps swarm. In each step only leader updates its position according to food position and having capacity of exploration and exploitation of solutions.

$$x_i^1 = \left\{ \begin{array}{l} f_i + c_1((ub_i - lb_i)c_2 + lb_i) \; c_3 \ge 0 \\ f_i - c_1((ub_i - lb_i)c_2 + lb_i) \; c_3 < 0 \end{array} \right\} \tag{3}$$

Where x_i^1 is the location of leader salp in ith dimension, f_I is the target food, ub_I and lb_I are the upper and lower bounds in i^{th} dimension C_1, c_2 and c_3 all are the random numbers. C_1 plays very important role of exploitation and exploration and can be expressed as follows

$$c_1 = 2e^{-\left(\frac{-4k}{K}\right)^2} \tag{4}$$

Where k and K are the current and total number of iterations, c_2 and c_3 are in such a manner these lies between [0,1]. Followers salp update their position with respect to leader salp only and cab expressed as

$$x_i^j = \frac{1}{2}\left(x_i^j + x_i^{j-1}\right) \tag{5}$$

Where x_i^j is location of jth follower in i^{th} dimension *and* $i \geq 2$.

4 PROBLEM FORMULATION

The object of the proposed work is to make the power system operation secure and smooth by mitigation congestion of transmission lines. Wind energy based energy source provides energy at the congested location and hence reduce the overloading of transmission lines, losses in the network system and system operation cost.

4.1 Objective function

The objective function having basic three components. Generation dispatch cost of thermal generators, congestion cost of transmission lines and cost of wind power generator.

$$\min\left(\sum_{i=1}^{N_G} C_i(P_{G_i}) + \sum_{k_i=1}^{T_L} TCC_{k_i} + C(P_{DG_i})\right) \tag{6}$$

4.2 Equality constraints

The system network having equality constraints in terms of active and reactive power available at all nodes. Sum of active and reactive power at each will be zero respectively. Power balance equations at each node:

$$P_i - P_{G_i} + P_{D_i} + P_{DG} = 0, \ i = 1, 2 \ldots . N_i \tag{7}$$

$$Q_i - Q_{G_i} + Q_{D_i} = 0, i = 1, 2 \ldots \ldots N_i \tag{8}$$

4.3 Inequality constraints

Generation capacity limits: Generating plants having its maximum and minimum active and reactive power limits due to economical and technical reasons.

Real and reactive power generation, bus voltage and transmission line capacity limits:

$$P_{G_i}^{Min} \leq P_{G_i} \leq P_{G_i}^{Max} \tag{9}$$

$$Q_{G_i}^{Min} \leq Q_{G_i} \leq QP_{G_i}^{Max} \tag{10}$$

$$V_i^{Min} \leq V_i \leq V_i^{Max}, i = 1, \ 2 \ldots . N_i \tag{11}$$

$$T_{k_i} \leq T_{k_i}^{Max}, k = 1, \ 2 \ldots . T_L \tag{12}$$

4.4 Cost function of thermal generators

$$C_i(P_{G_i}) = a_i(P_{G_i})^2 + b_i(P_{G_i}) + c_i \tag{13}$$

5 RESULT & ANALYSIS

In this section optimal placement of Salp swarm algorithm based wind energy source DG is discussed. IEEE 30 bus (Mirjalili et al. 2017) system having 41 lines, 24 loads buses and 6 generators. Transmission line rating is considered 32 MVA. Bus system is simulated in three cases. Generation dispatch, operation cost with congestion cost and system losses for all cases are shown in table no 1. MVA power flows on transmission system are presented in figure no 1.

a) Generation Dispatch without DG

From the figure no. 1 it is clear that MVA power flow on the line no 10 (between buses 6 and 8) is 32.4624 MVA and on line no. 29 (between buses 21and 22) is 32 MVA. Both lines are overloaded.

b) Generation Dispatch wind based DG with GA

The optimum location and size of GA based wind energy source DG is bus number 8 with 18.3529 MW capacities. Wind power generation cost is 61.8493 Rs. Now MVA power flow on line no. 10 and 29 are 32 MVA and 29.4207 MVA respectively. Optimal sizing and siting of DG with GA has made the power system operation economic but power flow on the line is still equal to its line rating.

Table 1. Generation dispatch, power loss and system operation cost.

Generation Dispatch/Losses/Cost	Base case without WPP	WPP with GA	WPP with SSA
PG1 (MW)	33.0359	43.052	40.3793
PG2 (MW)	45.4571	57.2004	54.0749
PG3 (MW)	25.5697	22.8844	21.8363
PG4 (MW)	40	30.6423	26.0389
PG5 (MW)	29.9909	16.3387	14.1232
PG6 (MW)	17.625	16.5596	114.2978
Total Generation Cost Rs	632.7243	574.547	569.434
Congestion Cost Rs	57.57	0.6147	00
Total Power Losses (MW)	2.4786	2.4772	2.1712

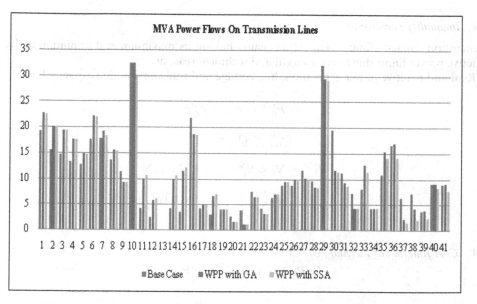

Figure 1. MVA Power flows on transmission lines.

c) Generation Dispatch wind based DG with SSA

Now simulation is carried out with SSA for finding out optimal size & location of DG. New location of DG is bus no 19 with capacity of 14.1359 MW. Generation cost for wind generator is 53.0096 Rs. MVA power flows on the line 10 and 29 are 31.2257 and 28.0422. It is clear from the results MVA power flows on lines are optimum with SSA based simulation as well as generation cost, congestion cost and system losses are less in this case.

6 CONCLUSION

In this paper optimal sizing and siting of WPP based DG is proposed to mitigate the overloading of transmission system. Optimal placement of DG has direct effect on reducing energy price and transmission congestion cost and its location and capacity must be optimized very carefully for secure and economic operation of power system. From the result it is clear that SSA provides perfect location and capacity for wind based DG placement with minimum generation and congestion cost. The proposed method can make the transmission system congestion free. Optimal placement of wind based DG avoids the load shedding minimizes rescheduling of generators and makes the system more reliable. Future scope includes implementation of more renewable energy sources with different scenarios.

REFERENCES

Afkousi Paqaleh, M., Fard, A. A. T., & Rashidinejad, M. 2010. Distributed generation placement for congestion management considering economic and financial issues. Electrical Engineering, 92(6), 193–201.

Agrawal A., Pandey S.N. and Srivastava L., 2020. A Review and Study on Market based Congestion Management Techniques in Restructured Power Market," International Conference on Advances in Systems, Control & Computing (AISCC-2020) Sponsored by Springer Under TEQIP-III MNIT Jaipur.

Agrawal A., Pandey S.N. and Srivastava L., 2020. Economic Benefits of implementing Demand Response in Deregulated Power Market. International Conference on Smart Communication & Imaging Systems (MEDCOM-2020) Sponsored by Springer GL Bajaj, Greater Noida Proc27-28 June., 2020

Agrawal A., Pandey S.N. and Srivastava L., 2020. Genetic Algorithm based Sizing & Siting of DG for Congestion Management in Competitive Power Market," International Conference on Advances in Systems, Control & Computing (AISCC-2020) Sponsored by Springer Under TEQIP-III MNIT Jaipur.

Agrawal A., Pandey S.N. and Srivastava L., 2020. Demand Response based Congestion Management Wind Energy Source in Competitive Power Market. International Conference on Smart Communication & Imaging Systems (MEDCOM-2020) Sponsored by Springer GL Bajaj, Greater Noida Proc27-28 June., 2020.

Ben-Moshe, O., & Rubin, O. D. 2015. Does wind energy mitigate market power in deregulated electricity markets. Energy, 85, 511–521.

Bigerna, S., Bollino, C. A., & Polinori, P. 2015. Marginal cost and congestion in the Italian electricity market: An indirect estimation approach. Energy Policy, 85, 445–454.

Dhillon, J., Kumar, A., & Singal, S. K. 2014. Optimization methods applied for Wind–PSP operation and scheduling under deregulated market: A review. Renewable and Sustainable Energy Reviews, 30, 682–700.

Gil, H. A., & Joos, G. 2008. Models for quantifying the economic benefits of distributed generation. IEEE Transactions on power systems, 23(2), 327–335.

Jain, A. K., Srivastava, S. C., Singh, S. N., & Srivastava, L. 2013. Bacteria foraging optimization based bidding strategy under transmission congestion. IEEE Systems Journal, 9(1), 141–151.

Kirthika, N., & Balamurugan, S. 2016. A new dynamic control strategy for power transmission congestion management using series compensation. International Journal of Electrical Power & Energy Systems, 77, 271–279.

Kunz, F., & Zerrahn, A. 2015. Benefits of coordinating congestion management in electricity transmission networks: Theory and application to Germany. Utilities Policy, 37, 34–45.

Mirjalili, S., Gandomi, A. H., Mirjalili, S. Z., Saremi, S., Faris, H., & Mirjalili, S. M. 2017. Salp Swarm Algorithm: A bio-inspired optimizer for engineering design problems. Advances in Engineering Software, 114, 163–191.

Ramandi, M. Y., Afshar, K., Gazafroudi, A. S., & Bigdeli, N. 2016. Reliability and economic evaluation of demand side management programming in wind integrated power systems. International Journal of Electrical Power & Energy Systems, 78, 258–268.

Singh, B., Mahanty, R., & Singh, S. P. 2015. Centralized and decentralized optimal decision support for congestion management. International Journal of Electrical Power & Energy Systems, 64, 250–259.

Tiwari, P. K., Mishra, M. K., & Dawn, S. 2019. A two step approach for improvement of economic profit and emission with congestion management in hybrid competitive power market. International Journal of Electrical Power & Energy Systems, 110, 548–564.

Tuan, L. A., Bhattacharya, K., & Daalder, J. 2004. A review on congestion management methods in deregulated electricity markets. power, 4, 6.

Recent Trends in Communication and Electronics – Sharma et al. (Eds)
© *2021 Taylor & Francis Group, LLC, ISBN 978-1-032-04572-6*

Forecasting industrial electric power consumption using regression based predictive model

R. Panchal
Research Scholar, Savitribai Phule Pune University, Maharashtra

B. Kumar
JSPM's Rajarshi Shahu College of Engineering, Pune, Maharashtra

ABSTRACT: Accurate electric power consumption forecasting plays key role in the decision-making, planning, executing and over all energy management. This paper presents a regression-based, model for accurate electric power consumption prediction by select industries in Ahmednagar city in Maharashtra state. The proposed model includes data cleaning, data smoothing and final data after preprocessing fed into regression-based model to predict industrial electric power consumption. The context features like calendar data, weather data, company data, and historical consumption of electricity are used as input for the regression-based model. In the conclusion, part accuracy of the proposed model is discussed as research.

Keywords: load forecasting, linear regression, energy management

1 INTRODUCTION

For the developing country, load forecasting is most concern task in electric energy generation, transmission and distribution. The primary purpose of load forecasting is decision on power market and overall energy management. The main challenge is to choose an effective & easy technique. Many factors have become dominant for energy management (Gupta & Pal 2017). Load forecasting classification shown in Table 1.

Above Table 1 elaborate the different types of load forecasting with their duration and purpose (Pandya & Parikh 2018).

2 REGRESSION BASED METHOD

It also referred as trend analysis or time-series analysis. Regression method based on the historical data to predict future as it is simplest method (Singh & Maini 2020).

This model can be employ to construct correlation between load (electric) and variables (external) such as;

Demand Data (Historical consumption of electricity in kWh.)

Calendar Data: (Month, Year, Season, etc.)

Weather Data: (Rainfall, Humidity, Temperature, etc.)

Industry Data: Industry Type, Types of electricity connection(HT/LT), Per unit electricity Price (Firsova 2018).

DOI: 10.1201/9781003193838-26

Table 1. Classification of different Load Forecasting.

Load Forecasting	Duration	Purpose
Very Short Term	Minute to an Hour	Used in energy management system
Short Term	Few Hour to Few week	Economic load dispatch for reliable study
Medium Term	Few week to Few months	To determine peak load
Long Term	1 Year to 20 year	Economic growth and planning of power system

With above data it will be simple to make association between input and output variable of model. Also to implement as well as handle easily (Singh & Maini 2020).

2.1 *Linear Regression (LR)*

The description of Linear Regression is to relate the independent and dependent variable. In other words, Linear Regression will calculate the Y value from X value.

Where, Y is Dependent variable and X is independent variable (Islam et al.2018).

The standard Linear regression equation (1) is as follows,

$$Y = a + bX \tag{1}$$

Where, a= intercept of line
      b= slope of line

This presents a simple straight line called regression line. This line reduces a squared deviation which was observed readings of Y (Kore et al. 2017).

2.2 *Performance measures*

For the performance measures, there are mainly two metrics are adopted for the validation: the mean absolute percentage of error called MAPE. The metric MAPE is a widely accepted measure for forecasting accuracy and the accuracy represents in percentage form (Zhang et al. 2018).

3 PROPOSED MODEL

Stage 1: Collection of Data

Stage 1 is the core of model that is "Data Collection" and for the current research proper data have been collected from MIDC Ahmednagar city, Maharashtra for the timeline of January 2011 to December 2018. Mainly two types of data collection as follows,

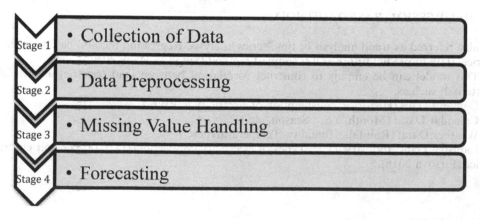

Figure 1. Industrial electric power consumption prediction model.

3.1 *Usage of electricity data*

The original document file consist of detail information like generation of power, consumption of power, loss of power, etc. for MIDC Ahmednagar city, Maharashtra. For the model these three parameters are transformed into three important variables,

✓ Total Power Generation,
✓ Total Power Consumption
✓ Total Power loss (Pandya & Parikh 2017).

3.2 *Weather data*

The secondly important Weather Data collected from online resources for January 2018 to December 2018. The weather related parameters are transformed into three important variables,

✓ Humidity in Fraction
✓ High Temperature in Celsius
✓ Low Temperature in Celsius

Stage 2: Preprocessing of Data
The pre-processing of actual data collected from stage 1 by converting into monthly average or in other words Mean value. And based on all above data processing stage CSV (Comma Separated Values) file was created for ultimate use.

Stage 3: Missing Value Handling
For the data mining data plays significant role. The data mining process improves data quality and better effectiveness. Due to few missing values, the real world data may be incomplete. The missing value analysis and Mean-Mode methods were applied (Voronkov 2014).

Stage 4: Prediction
The stage 4 is the key of whole research. After stage 2 pre-processing and stage 3 missing values handling, dataset is ready for prediction. Various data mining methods are used for prediction purpose, and the proposed model implement regression based prediction method for forecasting industrial electric power consumption (Goel 2016).

4 MODEL IMPLEMENTATION

4.1 *Multiple Linear Regression (MLR)*

Multiple Linear Regression is an expansion of regression method (simple linear regression). MLR can predict the value of one variable (target/dependent variable) derived from the value of two variables or more than two variables (input/independent variables).

The standard multiple regression equation (2) is as follows,

$$Y_i = \beta 0 + \beta 1 x 1_i + \beta 2 x 2 i + \beta 3 x 3_i + \ldots \ldots + \beta n x n_i + \varepsilon_i \qquad (2)$$

Where Yi=Dependent variable,
     xi =Independent variable,
      βi=Regression coefficient of xi
      εi =Random error.

The predicted response for the estimating model coefficients is as follows in equation (3).

$$\widehat{y}_i = b0 + b1 x 1_i + b2 x 2_i + b3 x 3_i + \ldots \ldots + b n x n_i \qquad (3)$$

Based on above equation, multiple regression model results are as follows,
Equation (4) shows that how to calculate the power consumption

Table 2. Average of monthly electricity data (demand data in KW) from January 2011 to December 2018.

Month	Jan	Feb	Mar	Apr	May	Jun	Jul	Aug	Sep	Oct	Nov	Dec
DemandData	1764	1484	1620	1945	2351	2443	2536	2418	2231	1909	1552	1634

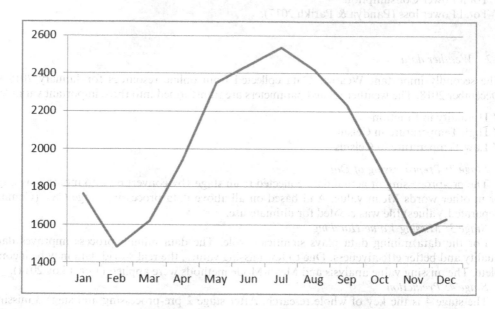

Figure 2. Plotting of average monthly electricity data (demand data in KW) from January 2011 to December 2018.

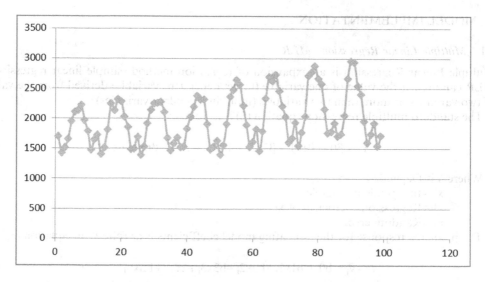

Figure 3. Electricity demand irregularity from January2011 to December 2018.

138

Table 3. Multiple Linear Regression Errors and Coefficient.

Sr. No.	Multiple Linear Regression MLR Errors	Error Values
1	Correlation Coefficient	0.9984
2	Mean Absolute Error (Average Absolute Error)	0.2621
3	Root Mean Square Error (RMSE) Standard Deviation of Prediction Errors.	0.7657
4	Relative Absolute Error	2.2025%
5	Root Relative Squared Error	5.3963%

$$Power\_Consumption = -1.6027 + 1.0186 * Power\_Generation$$
$$+ -1.0272 * Power\_Loss + 0.0579 \qquad (4)$$
$$*Lowest\ Temperature\ (Celcius)$$

5 RESULT AND CONCLUSION

It is evident that the proposed regression-based model would be accurate model to forecast industrial electric power consumption using Regression based predictive model for Ahmednagar city in Maharashtra state. There isn't any standard rule to prove the best model. However most suitable and appropriate the current proposed model has lower error. The Mean Absolute Error (Average Absolute Error) value is 0.2621. The RMSE Root Mean Square Error is 0.7657 also called standard deviation of prediction errors. Relative Absolute Error is 2.2025 % and Root Relative Squared Error is 5.3963 %. Research result proves that error rate has been reduced. More accuracy advanced deep learning algorithm would be implemented in this regards for the future perspective.

REFERENCES

Firsova, I. 2018. Energy Consumption Forecasting for Power Supply Companies. *International Journal of Energy Economics and Policy*, 9(1),1–6.

A. Goel, J. Gauta, and S. Kumar, "Real time sentiment analysis of tweets using naive bayes," in Proceedings of the 2016 2nd International Conference on Next Generation Computing Technologies (NGCT), IEEE, Dehradun, India, October 2016

Gupta, V., & Pal, S. 2017. An Overview of Different Types of Load Forecasting Methods and the Factors Affecting the Load Forecasting. *International Journal for Research in Applied* Science & Engineering Technology *(IJRASET)*, Volume 5 Issue IV, ISSN: 2321-9653.

Islam, M., & Moustafa, M.A. Regression-Based Predictive Models For Estimation Of Electricity Consumption. *International Conference on Sustainable Development (ICSD)*, (pp. 59–62).

Kore, S., & Khandekar, V. S. 2017. Residential Electricity Demand Forecasting using Data Mining. *International Journal of Engineering Trends and Technology (IJETT)*, Volume 49 Number 1.

Pandya, R., & Parikh, A. 2018. Data Mining for Agriculture Electricity Consumption Forecasting for Rural Area of Gujarat. *International journal of basic and applied research*, (P) 2278–0505.

Pandya, R., & Parikh, A. 2017.Classification of Data Mining Techniques for Electricity Load Forecasting. *International Journal on Future Revolution in Computer* Science & Communication Engineering, ISSN: 2454-4248, Volume: 3 Issue: 810–13

Singh, M., & Maini, R. 2020. Various Electricity Load Forecasting Techniques with Pros and Cons. *International Journal of Recent Technology and Engineering (IJRTE)*, ISSN: 2277-3878, Volume-8 Issue–6.

Voronkov, A. 2014. Keynote talk: EasyChair. In *Proceeedings of the 29th ACM/IEEE International Conference on Automated Software Engineering* (pp. 3–4). ACM. Retrieved from http://dl.acm.org/citation.cfm?id=2643085&dl=ACM&coll=DL

Zhang, X., & al, e. 2018. Forecasting Residential Energy Consumption Using Support Vector Regressions. *international conference on machine learning and application*.

Recent Trends in Communication and Electronics – Sharma et al. (Eds)
© 2021 Taylor & Francis Group, LLC, ISBN 978-1-032-04572-6

Retinal vessels extraction using maximum principle curvature

Preity & N. Jayanthi
Delhi Technological University, New Delhi, India

ABSTRACT: The extraction of blood vessels in retinal image is the fundamental step in the diagnosis of different ocular diseases like Diabetic Retinopathy, hypertensive retinopathy etc. Accurate analysis of vessel structure can make the diagnosis simpler for ophthalmologist. This paper proposed a novel framework to extract the retinal vessels accurately. The proposed method conducted in three stages (i) filtering of image using Gaussian filter & generation of mask using mathematical morphology,(ii) segmentation of vessels using Maximum Principle curvature followed by cleaning operation and (iii) comparison of segmented image with the ground truth image and evaluation of the performance parameters. MATLAB 2015a is used for implementation purpose and the work is tested with DRIVE database which is publicly available online. The proposed method efficiently extracts the retinal vessels with the accuracy of 96.7%.

1 INTRODUCTION

Vision impairment has been increased drastically these days and it will be a kind of disability in humans and the reason behind this is not only the ocular diseases but also due to some cardiovascular and pulmonary diseases. Eye is that part of human body which helps in visualization of everything and vision loss will bring various types of other abnormalities that affects people physically and mentally too. Now a days both the diseases are quite common in human beings which becomes the cause of Diabetic Retinopathy and Hypertensive Retinopathy and these ocular abnormalities leads to partial and full vision loss in human being which is one of the important matter of concern. Early detection of these kind of eye disease can prevent one from losing his vision also reduce the chances of getting the condition worsen. So for the screening of these eye disease, retinal features extraction plays a vital role. By segmenting the vessels it can be used at initial stage of diagnosis. If the retinal vessels have leakage or if there is clotting inside the vessels then those vessels are in the category of abnormal vessels and these leakage will cause severe damage in the retina if left untreated. So in short we can say that damaged retinal vessels are the main reason behind different abnormalities. Basically Exudates, hemorrhage and micro-aneurysms are the different abnormalities in retina. Hence retinal vessels extraction is a compulsory step that should be taken in diagnosis of various ocular disease like DR and MD. This novel proposed method is based on principle curvature technique. There are many different research works already done in this area using different image processing methods also some of authors presented their work using principle curvature also but still the results they obtained were not up to the mark. We have organized this paper in 5 sections, section 2 is about the related work section 3 describes the proposed methodology, section 4 &5 contains the results and conclusion respectively.

DOI: 10.1201/9781003193838-27

2 RELATED WORK

We have seen that the diagnosis of several eye diseases are based on analysis of anatomy of retina in which the retinal blood vessels are the prime feature that plays most important role in early detection. Numbers of researchers published their work in this area some of image processing based methods are mentioned here. Fuzzy c mean clustering used for segmentation of retinal vessels and this method was proposed by (Kande et al. 2010). The method was tested on STARE and DRIVE database used for testing the method and upto 93.85% accuracy is achieved. ROC curves for this were also determined. Matched filter was used for contrast enhancement purpose. Another method of auto segmentation of retinal vessel is proposed by (Bantan et al. 2016) and it was tested on HRF dataset and calculated 3 performance parameters accuracy, specificity and sensitivity. The author calculated error between the resulting image and ground truth image so the method was 94% accurate. (Dash et al. 2017) proposed an extraction technique for retinal vessels and it was based on mean c thresholding. STARE and DRIVE database were used for testing the method and obtained accuracy of 95%. Another methodology proposed by work (Bandara et al. 2017) based on image enhancement technique. An improved version of Tyler Coye algorithm and SUACE algorithm is used to reconstruct the retinal blood vessels hough line transform is also used. DRIVE database is used for performance testing for this methodology, accuracy of 94% was achieved by the author. There are some challenges confronting the existing methods like misdetection of blood vessels and not detecting the thinner vessels, complex methodology and requirement of large storage devices but in our work these problems have been solved and accuracy is increased.

3 PROPOSED METHOD

The proposed method is tested on DRIVE database which is publically available online. In this method we have used Maximum Principle curvature technique to extract the retinal vessels efficiently. Basically the proposed method is divided into three sections one by one explained below.

3.1 *Gaussian filtering and generation of mask*

Generation of mask of the input image we have taken is done by using morphological operation. Conversion of image into binary image is done firstly and creates diamond structured element and the size of that element is taken 20. A $\ominus$ B represents erosion of image A to the structuring element B which has origin at (x, y). And the new pixel value obtained by the equation given below.

$$g(x,y) = \begin{cases} x, & B\ hits\ A \\ y, & else \end{cases} \qquad (1)$$

In this proposed method we have used Gaussian filter. As de-noising is one of the important step which is to be done and here we have used Gaussian filter as it is more efficient in comparison to others. The value of standard deviation is responsible for smoothening of image. In this work the value we have chosen of standard deviation is 1.45.

3.2 *Principle curvature and finding Lambda function*

Hessian matrix is the fundamental of principle curvature as for the evaluation the principle curvatures and direction Hessian matrix is needed. Mathematically it is represented as

$$H = \begin{bmatrix} Ixx & Ixy \\ Iyx & Iyy \end{bmatrix} \qquad (2)$$

Hessian matrix is a symmetric matrix also a square matrix therefore I x y = I y x. The Eigen values are then calculate on this Hessian matrix and these Eigen values are represented by Lambda function [λ]. The maximum principle curvature is decided by the maximum value of lambda. The Eigen values gives the principle orientation. From this hessian matrix two Eigen values are calculated and these are λ_1 and λ_2. Where $\lambda_1 \leq \lambda_2$, convexity is measured by λ_1 and concavity is measured by λ_2. Basically in any surface principle curvature and principle direction is obtained by the Eigen values and Eigen vector of Hessian matrix. After the calculation of Eigen values and Eigen vectors at each pixels is completed then weights are assigned to the pixels of the image as Blood Vessels also the condition of $|\lambda 2.| > 1$ is to be satisfied. In the direction of Eigen vector the region is growing and this weights assigned pixels then it states that pixel belongs to the Vessel region if it satisfy the condition mentioned above otherwise the pixel is considered to be non-vessel part of the image. Now multiplication of a rescaled mask with the rescaled maximum principal curvatures of the Hessian id done and hence the region of growing of blood vessels is obtained. The mask takes two parameters namely (1) the binary form of input image and (2) diamond structuring element that takes into consideration only those pixels which will reach to the radius given as parameter and proceeds accordingly by convoluting to the maximum principal curvatures of the Hessian. By evaluating maximum principal curvatures function leads to effective extraction of vessels vasculature from the given input image.

3.3 Contrast enhancement and ISODATA

After finding the lambda function that is principle curvature of the image the image is gone through the contrast enhancement and thresholding process. For contrast enhancement adaptive histogram equalization is used. On different sections of image, individual histograms are computed in AHE so it is slightly different from histogram equalization. For better version of output contrast enhancement is a crucial step. After contrast enhancement ISODATA thresholding technique is employed. It is an iterative method use to find threshold by separating the background and foreground of an image. After this contrast enhancement and thresholding step final filtering of image and processed output data is obtained. The final output data is obtained after final filtering and applying some morphological operations. Here in this proposed method morphological opening is employed and then final segmented image is obtained. After segmentation we compare the image with ground truth image to and evaluated the performance parameters.

4 RESULTS AND DISCUSSION

The proposed method implemented using DRIVE database which consist of 40 images. The proposed method is an efficient method to extract the retinal vessel using principle curvature. The performance parameter calculated in this method are Accuracy, Sensitivity and Specificity. An excellent result is obtained by this method. An average accuracy of 96.6%, 75.43 % of Sensitivity and 98.9 % of specificity is achieved by the proposed method. Table 1 depicts the comparison of our method with the existing methods in terms of performance parameters. The implementation results of the proposed method is shown in Figure 1.

Table 1. Comparison of the proposed method with the existing methods.

Authors	Accuracy	Sensitivity	Specificity
Dash [2017]	95%	70.4%	97.1%
Bantan [2016]	94.7%	69%	97%
Roychaudhary [2014]	95.2%	75.2%	98.3%
Proposed method	96.7%	75.4%	98.9%

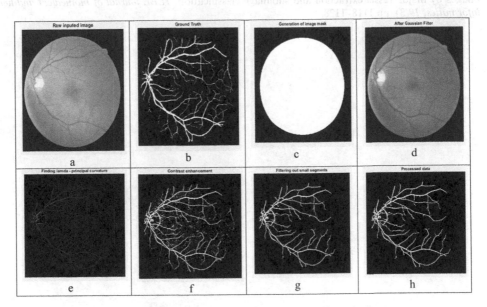

Figure 1. a. depicts input image b. is the ground truth image c. is the image mask d. is the image after Gaussian filtering e. depicts the principle curvature f. is the image after contrast enhancement g. filtered image h. is the final processed data.

5 CONCLUSION

If we see the current statistics approximately 80% of the people having any issue of diabetes and hypertension for past 15 years are more prone to ocular diseases specially Diabetic retinopathy and the detection of these retinal diseases are truly based on vessel extraction. Vessel extraction is the basic step for detection of any kind of eye disease. The advantage with this method is that the accuracy and specificity obtained by this method is 96.7% and 98.8% respectively which is excellent if we compare it with existing methods also it is quite easy to implement and segment fine vessels accurately. If we see the future aspects of this topic we can conclude that apart from image processing techniques, machine learning based approach may be a good option.

REFERENCES

Bandara, A.M.R.R. and Giragama, P.W.G.R.M.P.B., 2017, December. A retinal image enhancement technique for blood vessel segmentation algorithm. In *2017 IEEE international conference on industrial and information systems (ICIIS)* (pp. 1–5). IEEE.

Bantan, M.T., 2016, November. Auto-segmentation of retinal blood vessels using image processing. In *2016 4th Saudi International Conference on Information Technology (Big Data Analysis)(KACSTIT)* (pp. 1–6). IEEE.

Dash, J. and Bhoi, N., 2017. A thresholding based technique to extract retinal blood vessels from fundus images. *Future Computing and Informatics Journal*, *2*(2), pp.103–109.

Kande, G.B., Subbaiah, P.V. and Savithri, T.S., 2010. Unsupervised fuzzy based vessel segmentation in pathological digital fundus images. *Journal of medical systems*, *34*(5), pp.849–858.

Mapayi, T., Viriri, S. and Tapamo, J.R., 2015. Comparative study of retinal vessel segmentation based on global thresholding techniques. *Computational and mathematical methods in medicine*, *2015*.

Roychowdhury, S., Koozekanani, D.D. and Parhi, K.K., 2014. Blood vessel segmentation of fundus images by major vessel extraction and subimage classification. *IEEE journal of biomedical and health informatics*, *19*(3), pp.1118–1128.

Recent Trends in Communication and Electronics – Sharma et al. (Eds)
© 2021 Taylor & Francis Group, LLC, ISBN 978-1-032-04572-6

Classification of product review polarity using LSTM

Chiranjeevi Pandi & Ch. Vijay Kumar
Faculty of CSE, ACE Engineering College, Hyderabad, India

K Adi Narayana Reddy
Faculty of CSE, BVRIT College of Engineering for Women Hyderabad, India

Ramesh Alladi
Faculty of CSE, ACE Engineering College, Hyderabad, India

P. Sumithabhashini
SAMSKRUTI College of Engineering and Technology

ABSTRACT: In this rapid increasing technological world, the data can be taken from various social networking platforms such as Facebook, Twitter, LinkedIn etc. The data is mainly in the form of text. The Customer reviews play a very important role, while buying the products . In various machine learning techniques were used to extract the polarity from the reviews but those machine learning techniques failed to understand the semantics and context of the review. The sequential model capture the semantics of the product reviews. In this paper we propose LSTM (Long Short Term Memory) technique, which classifies the polarity of the product reviews automatically. The results indicate that the LSTM Technique outperform over the baseline Machine Learning Techniques.

Keywords: Polarity, Machine learning, Deep learning, LSTM, Neural networks, Opinion mining

1 INTRODUCTION

The day to day of the business strategies has improved a lot when compared to traditional business environment. The two main tasks should have been doing and maintained by business holders is that producing products/services with new innovation when it compared to the previous inventions, and the second task is to monitor their customer's needs and views of purchasing a product or acquiring a service, which changes unexpectedly and rapidly. The customer's review on a product appears to be in the text format of typed as Positive, Negative and Neutral. These reviews play major role while buying products/services. These have a capacity of changing mindset of people. The customer's attitude and viewpoint of buying products which are in the form of positive, negative and neutral comments are very important to the businesses to improve their features in products by which they can attract their customers. In this regard, this work is used to more precise method which is an improved one when compared to the other methods used in previously developed works. This paper deals with the polarity of customer's opinions by using deep learning techniques to fulfill the above-mentioned points. In the second part of the terms used in this opinion mining is defined and explained the

DOI: 10.1201/9781003193838-28

basic concepts. The third part discuss about the literature survey. The fourth part describes about the proposed method used in this work. The fifth part analyze the proposed model with the dataset of mobile phones and laptop. The last parts deals with conclusion and references of this work.

2 LITERATURE SURVEY

The studies on published work of this subject on finding the polarity of customer's reviews on goods and services they purchase, some of the articles used deep learning methods and some do not use deep learning methods. Below is the explanation some of them. Research(Hu, M. et al. 2004) used Natural Language Processing(NLP) techniques and supervised data mining methods. To take out the customer's opinion on product's review(Pandi, C., et al. 2018)and find polarity of opinions, after that the opinions are summarized. Research(Chiranjeevi, P. et al 2019) explained different existing opinion mining techniques and supervised data mining methods. To take out the customer's opinion on product's review (Balazs, et al. 2016) and find polarity of opinions, after that the opinions are summarized.

Research (Liu, B et al. 2010)This research used opinion dataset which contains total details of five products. They are, mobile phone, MP3 player, two digital cameras, DVD player. This work has divided opinion mining into two different tasks, finding product's features, finding opinions belongs to product's features(Crossley, et al. 2013), calculating the polarity of the opinions and placing them in order according to their strength. This work has used OPINE as unsupervised data extracting method which find opinions(Hochreiter, et al. 1997) to build a model using important features, qualities of products and comments on products. The method used in this work has given 22% greater efficiency in feature extraction when compared to other works. Research(Deng, L et al 2014) This research has used amazon e-commerce dataset. This work developed algorithm which creates a search engine to extract comments by using PoS-tagging (Popescu, et al. 2007) and data preprocessing techniques. This work not only creates search engine but also extract useful information from product's characteristics. The proposed work contain the following steps. They are (Liu, et al 2012), data pre-processing, opinion mining engine, opinion ranking algorithm, numbering the taken features, search engine developed and interacting(Eirinaki, et al. 2012) user interface. The proposed work not only categorize opinions as positive, negative and neutral but also take out features of opinion and also rank each opinion.

Research(Riaz, et al. 2019)used different approach compared to the ranking method of comments in social media. It analyzes the sentimental words (Mars, et al. 2017)] in comments and place them into a cluster of same emotions. Research (M. Marelli et al 2014) used many techniques such as text mining tools, Big Data, Machine learning (Day, et al. 2017) and also providing labels to the comments to easy identification of polarities. It used the dataset of electronic products such as Nokia 6610 mobile phone, Nikon camera J3, Nikon 4300. This research (Gupta, et al. 2015) has proposed a technique which is combination of both big data (Pandi, C. et al. 2018) and text mining. This proposed model contains these following steps. In first step it prepares a tree with positive and negative emotions which is called as lexical ontology. The second step contains collecting and storing data written by customers on social media sites. The third step consists of using techniques like Natural Language processing, PoS-tagging, information mining, text summarization, thematic models, clustering, classification Map Reduce technique is also used to find polarity of comments present in bis data collected by above steps.

3 DATASET ANALYSIS AND PROCEDURE FOR POLARITY EXTRACTION

The solution consists of finding the polarity of customer's reviews on mobile phones and laptop datasets by using deep learning technique.

3.1 Pre-processing of the data

Pre-processing of data is done after the tokenization is done on the data which removes all the unnecessary (Pang, et al. 2008) data form the dataset such as stop words, punctuation, pictures, spelling mistakes, links, (such as https, www). For example, if it wants the data should contains the sentences of length 200 then, we can use this padding technique(Hochreiter, et al. 1997), and this is used when we enter the data into neural network for training. Positive class is attributed as 1, Negative class attributed as -1 and Neutral class attributed as 0.

LSTM: After representing each word with its corresponding vector trained by the Word2Vec model, the sequence of words are input to the LSTM one by one in sequence. The work flow of LSTM is presented in the Figure 1 and the corresponding mathematical equations are

$$\text{input gate}: \mathrm{i}^{\mathrm{t}} = \sigma\left(W^i * h_{t-1} + I^i * x_t\right)$$

$$\text{forget gate}: \mathrm{f}^{\mathrm{t}} = \sigma\left(\mathrm{W}^{\mathrm{f}} * \mathrm{h}_{t-1} + \mathrm{I}^{\mathrm{f}} * x_t\right)$$

$$\text{output gate}: \mathrm{o}^{\mathrm{t}} = \sigma(\mathrm{W}^{\mathrm{o}} * \mathrm{h}_{t-1} + \mathrm{I}^{\mathrm{o}} * x_t)$$

$$\text{New memory cell}: c^t = \tanh\left(W^c * h_{t-1} + I^c * x_t\right)$$

$$\text{Final memory cell}: \mathrm{c}^{\mathrm{t}} = \mathrm{f}^{\mathrm{t}} * \mathrm{c}^{t-1} + \mathrm{i}^{\mathrm{t}} * c^t$$

$$\text{Final hidden state}: \mathrm{h}^{\mathrm{t}} = \mathrm{o}^{\mathrm{t}} * \tanh\left(\mathrm{c}^{\mathrm{t}}\right)$$

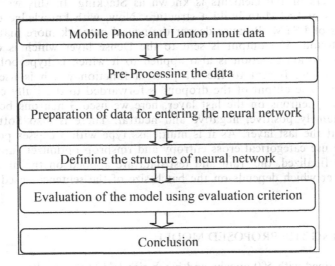

Figure 1. Procedural steps.

Table 1. Size(sentences) of the dataset used.

Domain	Training data	Testing data	overall
Mobile Phone	5732	360	6092
Laptop	5804	360	6164

147

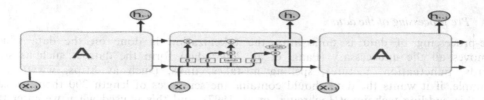

Figure 2. Work flow of LSTM from Colah's Blog.

Once the feeding of input to LSTM model is complete, the output of LSTM hidden layer is flattened and it is concatenated with the Gene and Variation features. The output layer classifies into nine classes.

4 IMPLEMENTATION

In this section the implementation of the research is explained. The main concepts used here to find the polarity of customer's reviews are Recurrent Neural Network (RNN) and Long Short-term Memory (LSTM). The input of this work is customer review on products/services, which is pre-processed and converted into numeric vectors to enter into the Neural Network. So, first the train data is pre-processed then classifier is constructed and tested. The parameters such as accuracy, recall, precision, F1 score are noted down to evaluate the results of different classifications and get the final results. An LSTM of 100 elements is taken to construct a network. Binding all such LSTM layers of 100 elements is known as Stacking. In this work we are using such 100 Blocks of LSTM to build a structure. Now, with knowledge of mathematics, we use 3 layers of LSTM. Because deeper the neural network more mathematical combinations are found. The output is sent to the Dense layer which is connected layer and here an activation function is also applied to it which is hyperbolic tangent. The output obtained here is sent to the Dropout, a function which is used to avoid the over fitting in it. The output of the dropout is forwarded to the Fully connected layers which contains 3 neurons on the last layer, here we used 3 neurons because there are three classes namely positive, negative and neutral. There is also Softmax activation function here at the last layer. As it is multiclass type with 3 classes positive, negative and neutral we use categorical cross entropy and rmsprop optimizer function. After the parameters are finalized, the train and test data are applied on the model in the form of epoch number which depends on the batch size of the sentences used in training the model.

5 EVALUATING THE PROPOSED MODEL

The model is trained with 500 apochs and batch size as 128 the results are presented in the Table 2. The model also includes rmsprop as optimizer function. The show that the proposed LSTM Technique is outperformed over the baseline Machine Learning based algorithms. The model is built on both laptop and phone datasets. The accuracy of the proposed model for laptop dataset is 96.72 and mobile dataset is 97.87, for both the dataset is the accuracy is more than the existing model.

Table 2.   The result analysis

Research/Model	Domain	Accuracy
[12]/SVM	Both	76.46
[12]/NB	Both	75.12
[12]/LSTM	Both	94.00
[13]/PSO	Mobile phone	78.48
	Laptop	71.25
[14]/Baseline	Mobile phone	64.30
	Laptop	51.10
[14]/All Positive	Mobile phone	64.20
	Laptop	52.10
[14]/Rule-Based	Mobile phone	77.80
	Laptop	67.70
[14]/SVM	Mobile phone	81.00
	Laptop	70.50
[15]/JU-CSE	Mobile phone	65.54
	Laptop	53.21
[16]/Sentitu	Mobile phone	78.70
	Laptop	79.30
Proposed Work	Mobile phone	97.87
	Laptop	96.72

Graphical Comparison:

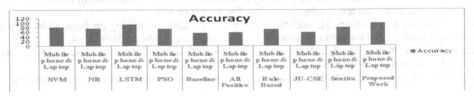

Figure 3.   Algorithms comparison.

6   CONCLUSION

In this work, the polarity of the product reviews are extracted using deep learning technique. LSTM is a sequential model, which captured the semantics of the product review. The proposed LSTM based technique is outperformed over the Machine Learning based algorithms. The proposed model is evaluated using accuracy. The accuracy of the model is far better than any other traditional machine learning algorithms.

REFERENCES

Balazs, J.A. and Velásquez, J.D., 2016. Opinion mining and information fusion: a survey. Information Fusion, 27, *pp.95-110*.

Chiranjeevi, P., Santosh, D.T. and Vishnuvardhan, B., 2019. Survey on sentiment analysis methods for reputation evaluation. In Cognitive Informatics and Soft Computing (pp. 53–66). Springer, Singapore.

Chung, J., Gulcehre, C., Cho, K. and Bengio, Y., 2014. Empirical evaluation of gated recurrent neural networks on sequence modeling. arXiv preprint arXiv:1412.3555.

Crossley, S.A., 2013. Advancing research in second language writing through computational tools and machine learning techniques: A research agenda. Language Teaching, 46(2), pp.256–271.

Day, M.Y. and Lin, Y.D., 2017, August. Deep learning for sentiment analysis on google play consumer review. In 2017 IEEE international conference on information reuse and integration (IRI) (pp. 382–388). IEEE.

Deng, L. and Yu, D., 2014. Deep learning: methods and applications. Foundations and trends in signal processing, 7(3–4), pp.197–387.

Elman, J.L., 1990. Finding structure in time. Cognitive science, 14(2), pp.179–211.

Eirinaki, M., Pisal, S. and Singh, J., 2012. Feature-based opinion mining and ranking. Journal of Computer and System Sciences, 78(4), pp.1175–1184.

Gupta, D.K., Reddy, K.S. and Ekbal, A., 2015, June. Pso-asent: Feature selection using particle swarm optimization for aspect based sentiment analysis. In International conference on applications of natural language to information systems (pp. 220–233). Springer, Cham.

Hochreiter, S. and Schmidhuber, J., 1997. Long short-term memory. Neural computation, 9(8), pp.1735–1780.

Hu, M. and Liu, B., 2004, August. Mining and summarizing customer reviews. In Proceedings of the tenth ACM SIGKDD international conference on Knowledge discovery and data mining (pp. 168–177).

Liu, B., 2010. Sentiment analysis and subjectivity. Handbook of natural language processing, 2(2010), pp.627–666.

Liu, B., 2012. Sentiment analysis and opinion mining. Synthesis lectures on human language technologies, 5(1), pp.1–167.

Mars, A. and Gouider, M.S., 2017. Big data analysis to Features Opinions Extraction of customer. Procedia computer science, 112, pp.906–916.

Pandi, C., Dandibhotla, T.S. and Bulusu, V.V., 2018. Corpus Linguistic Rules Based Review Sentence Selection for Opinion Targets Extraction and Opinion Orientation: A Distant Supervision Approach. International Journal of Intelligent Engineering and Systems, 11(5), pp.114–124.

Pandi, C., Dandibhotla, T.S. and Vardhan, V., 2018. Reputation based online product recommendations. ISI, 23(5).

Pang, B. and Lee, L., 2008. Foundations and Trends® in Information Retrieval. Foundations and Trends® in Information Retrieval, 2(1-2), pp.1–135.

Popescu, A.M. and Etzioni, O., 2007. Extracting product features and opinions from reviews. In Natural language processing and text mining (pp. 9–28). Springer, London.

Riaz, S., Fatima, M., Kamran, M. and Nisar, M.W., 2019. Opinion mining on large scale data using sentiment analysis and k-means clustering. Cluster Computing, 22(3), pp.7149–7164.

Shyam Chandra Prasad G, Adi Narayana Reddy K, "Sentiment Analysis Using Multi-Channel CNN-LSTM model", JARDCS, vol. 11, p. 489–494, 2019

Zhao, J., Liu, K. and Xu, L., 2016. Sentiment analysis: mining opinions, sentiments, and emotions.

Recent Trends in Communication and Electronics – Sharma et al. (Eds)
© 2021 Taylor & Francis Group, LLC, ISBN 978-1-032-04572-6

Driver's drowsiness identification using eye aspect ratio with adaptive thresholding

Swati Srivastava

Department of Computer Engineering and Applications, GLA University, Mathura, India

ABSTRACT: Driver's sleepiness is one of the most widespread cause for the road collisions. The prevailing problem across the world these days is the flourishing number of road mishap. Driver drowsiness monitoring is one of the extensive interest for many researchers. Collisions can be avoided by providing a warning to the driver after drowsiness level detection. Preliminary signs of exhaustion can be detected before a critical situation arises. This paper proposed a system that can precisely detect drowsy driving in real time and generate alarms accordingly. This work aims to prevent the drivers from drowsy driving and generate a safer driving environment.

Keywords: Drowsiness, eye aspect ratio, eye blinking, safety, eyes detection

1   INTRODUCTION

Transports are an essential part of human life nowadays. As we know, for any type of automobile, the safety should be on the highest priority. Protection of the traveler should be on the top priority while travelling. Road collisions are caused significantly due to driver's drowsiness and tiredness. It increases gradually years by years, the number of deaths and injuries globally. Therefore, researchers focused on refining the vehicle safety using computer vision incorporated with monitoring of eye blinks of driver. It is therefore imperative to incorporate latest technologies to devise and fabricate systems that are proficient to keep an eye on driver and to quantify the level of awareness during driving. The proposed system aims routine exposure of driver's drowsiness based on eyes detection and adaptive threshold. Proposed system detects face and eyes, estimate the eye state, measure the duration of eye closure and accordingly generate alarm.

There are assorted aspects which contribute to road collisions. Some of them are lost vehicle control, over speed, alcoholic condition of driver, driver drowsiness, weather environment, and vehicle conditions. Ordinarily, prior to driving a vehicle, a person do not concerned about the self physical and mental state. Consequently, it instigate road collisions, deaths, injuries and smash up to belongings. Thus, there is a need to propose a system which correctly identify the sleepiness of the driver which could help out the driver immensely by generating alarm. The proposed approach mainly focuses on the identification of driver drowsiness by integrating facial and eyes landmarks with the objective to provide essential alerts when the warning is required to driver. This helps the driver to make a decision whether to persist driving or not. Accordingly, he can catch a proactive approach to road collisions rather than a reactive approach.

Rest of the paper is organized as: Section 2 depicts the related work. Section 3 portrays the proposed work. Findings are discussed in Section 4. Finally, Section 5 presents the concluding remarks.

DOI: 10.1201/9781003193838-29

As evidenced from literature, wide-ranging sum of work has been done on drowsiness detection. Due to space limitation, only a few imperative and appropriate related literature are specified here. Drowsiness detection can be primarily classified into three facets: measures based on vehicle, physiological measures and behavioral measures. In vehicle based measures, sensors are placed on different vehicle constituents such as steering wheel and the accelerator. To measure the movement of steering wheel, a steering angle sensor is placed on steering. When the driver is drowsy, the figure of micro-corrections on the steering wheel trims down as compared to normal driving (Kang, 2013). To distinguish whether the driver is sleepy or not, small SWMs are used and provide an alarm accordingly (Feng et.al., 2009). SDLP (Sahyadehas et.al., 2012) is another vehicle based gauge to detect the sleepiness of driver. To track the position of lane, an exterior camera has been mounted. Enslavement on external factors such as street markings, illumination and climate surroundings is the major constraint of this method. Consequently, these driving performance assesses are not precised to the driver drowsiness (Sahyadehas et.al., 2012).

Physiological measures are defined as physical changes due to tiredness that arise in our body. (Chellappa et al. 2016) uses physiological as well as bodily indications for drowsiness recognition. The values from both the causes were given as input to the system. An amalgamation of these is used as a parameter. Physiological participation embraces body heat and pulse rate whereas physical contribution involves blinking and yawning. The physiological changes can be simply measured by ECG, EOG, EEG and EMG. ECG signal provides the critical parameters related to heart such as respiration rate or inhalation occurrence. Each of these are related to sleepiness (Ingre et al. 2006) (Awais, et al. 2017).

Yawning, eye closure and eye blinking are a number of the behavioral changes that occur during drowsing. The yawning rate in usual circumstances is less than that of in fatigue situation. Thus, it can be identified whether the driver is in exhausted state or not (Sun et al. 2011)) (Khunpisuth et al. 2016) (Dong et.al., 2010) by observing the yawning rate. Eye Aspect Ratio (EAR) and Template Matching approaches are the main techniques used for eye blink detection. Recently, researchers gave a survey on behavioral methods as they are invariant to light conditions and vehicle modifications (Ngxande et al. 2017). The drowsiness can also be detected by using model compression of deep neural network for embedded system (Reddy et al. 2017). Researchers also developed an android based application using deep neural network for drowsiness detection which is more convenient for drivers to use (Jabbar et al. 2018).

3 PROPOSED SYSTEM

To scrutinize the driver's condition such as drowsiness and yawning, this paper presents a real time system that alert the driver to avoid collision if driver closes eye or yawn for more than 4 sec. The proposed work is accomplished by a webcam that is mounted in front of the driver. It continually takes images of driver and detect face and eyes. Then with the help of EAR, the system is able to detect driver's drowsiness. The system computes EAR to detect driver's drowsiness based on adaptive thresholding (Gupta et al. 2018). The proposed driver fatigue detection system consists of a webcam which imprisons the pictures of a driver. The captured images are used for face and eyes detection. The eye area is detected using EAR which in each image will be calculated as:

$$EAR = \frac{\|p2 - p6\| + \|p3 - p5\|}{2\|p1 - p4\|} \tag{1}$$

where p1, p2, p3, p4, p5 and p6 are eye landmarks. The average EAR is measured as follows:

$$EAR_{avg} = \{L(EAR) + R(EAR)\}/2 \qquad (2)$$

where L(EAR) and R(EAR) is the EAR for the left and right eye respectively. The area of the eye is inversely proportional to EAR. The increased area of the driver's eye will decrease the level of drowsiness and vice-versa. The working of proposed system is given in Figure 1. The initial phase of the proposed system for driver protection and drowsiness identification is image acquisition.

3.1 Image acquisition

The camera is mounted in front of the driver. It is used to imprison the pictures of the driver. The captured pictures are then send to the face-eye detection stage which is the next phase of the proposed system. Face-eye detection phase detects the face and eyes of the driver.

3.2 Face-eye detection stage

This stage of driver wellbeing and drowsiness exposure system uses Haar Cascade Classifier to identify the precised locations of the face and eyes. This stage employs Haar cascade face detection and Haar cascade eye detection. These aforementioned processes are used to identify face and eyes respectively. The output of this stage is send to the next stage of the driver's drowsiness detection arrangement for further processing towards the objective of the proposed work.

3.3 Eye area estimation stage

The area of the eyes will be determined at this stage of the proposed arrangement. EAR is the measure that keeps track of eye lids and identifies whether the eyes are in open or closed state.

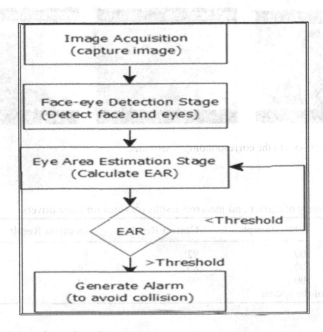

Figure 1. Proposed drowsiness detection system.

3.4 *Generate alarm*

Small value of EAR indicates that eyes are open. Large values of EAR indicates the eye closure which results in invisibility of the iris. The system generates alert if the eye closure percentage of the driver is greater than the specified threshold.

4 RESULTS

The proposed system has been tested on different real time images. The system accurately detected the face and eyes. Further, the eye closure has been successfully estimated using EAR. The average accuracy of the proposed system is 89.33%. Figure 2 shows the results obtained for the real time images. The results includes eye detection and the corresponding EAR value. If the EAR value is less than the defined threshold, the system generates an alarm to avoid road collision.

Experiments have been performed on three drivers. As shown in Table 1, for each driver, the process of drowsiness detection has been repeated 100 times. It is found that for driver 1, 92 times the correct results are obtained whereas 8 times the system wrongly interpreted. For second driver, 86 times correct results are obtained and 14 times the proposed system gave incorrect results. For the third driver, the system gave 90 correct results and 10 wrong outputs. Thus, accuracy is calculated by dividing the total number of correct results with total number of attempts. The average accuracy of the proposed system is 89.33% which evidences the efficiency of the proposed drowsiness identification system.

5 CONCLUSION

In this work, a real time system that monitors and detects the loss of attention of drivers of vehicles is proposed. The face of the driver has been detected by capturing facial landmarks

Figure 2.   Eye detection and the corresponding EAR value.

Table 1.   The summary of correct and incorrect results obtained for three drivers.

Input Images	Total Attempts	Correct Result	Incorrect Result	Accuracy(%)
Driver1	100	92	8	92
Driver2	100	86	14	86
Driver3	100	90	10	90
Average Accuracy of the system				89.33%

and warning is given to the driver to avoid real time crashes. The proposed approach uses Eye Aspect Ratio with adaptive thresholding to distinguish driver's drowsiness in real-time. This is useful in situations when the drivers are used to strenuous workload and drive continuously for long distances. The future work can include integration of the proposed system with globally used applications like Uber and Ola. The system, if integrated, can reduce the number of casualties and injuries that happen regularly due to these drowsy states of the drivers. The proposed approach also gives the comparable accuracy for the people wearing spectacles. Accuracy of the proposed system improves with the increase in brightness of the surrounding environment.

REFERENCES

Awais, M., Badruddin, N. and Drieberg, M., 2017. A hybrid approach to detect driver drowsiness utilizing physiological signals to improve system performance and wearability. *Sensors, 17*(9), p.1991.

Chellappa, Y., Joshi, N.N. and Bharadwaj, V., 2016, August. Driver fatigue detection system. In *2016 IEEE International Conference on Signal and Image Processing (ICSIP)* (pp. 655–660). IEEE.

Dong, H.Z. and Xie, M., 2010, December. Real-time driver fatigue detection based on simplified landmarks of AAM. In *The 2010 International Conference on Apperceiving Computing and Intelligence Analysis Proceeding* (pp. 363–366). IEEE.

Gupta, I., Garg, N., Aggarwal, A., Nepalia, N. and Verma, B., 2018, August. Real-time driver's drowsiness monitoring based on dynamically varying threshold. In *2018 Eleventh International Conference on Contemporary Computing (IC3)* (pp. 1–6). IEEE.

Ingre, M., Åkerstedt, T., Peters, B., Anund, A. and Kecklund, G., 2006. Subjective sleepiness, simulated driving performance and blink duration: examining individual differences. *Journal of sleep research, 15*(1), pp.47–53.

Jabbar, R., Al-Khalifa, K., Kharbeche, M., Alhajyaseen, W., Jafari, M. and Jiang, S., 2018. Real-time driver drowsiness detection for android application using deep neural networks techniques. *Procedia computer science, 130*, pp.400–407.

Kang, H.B., 2013. Various approaches for driver and driving behavior monitoring: A review. In *Proceedings of the IEEE International Conference on Computer Vision Workshops* (pp. 616–623).

Khunpisuth, O., Chotchinasri, T., Koschakosai, V. and Hnoohom, N., 2016, November. Driver drowsiness detection using eye-closeness detection. In *2016 12th International Conference on Signal-Image Technology & Internet-Based Systems (SITIS)* (pp. 661–668). IEEE.

Ngxande, M., Tapamo, J.R. and Burke, M., 2017, November. Driver drowsiness detection using behavioral measures and machine learning techniques: A review of state-of-art techniques. In *2017 Pattern Recognition Association of South Africa and Robotics and Mechatronics (PRASA-RobMech)* (pp. 156–161). IEEE.

Reddy, B., Kim, Y.H., Yun, S., Seo, C. and Jang, J., 2017. Real-time driver drowsiness detection for embedded system using model compression of deep neural networks. In *Proceedings of the IEEE Conference on Computer Vision and Pattern Recognition Workshops* (pp. 121–128).

Ruijia, F., Guangyuan, Z. and Bo, C., 2009, March. An on-Board System for Detecting Driver Drowsiness Based on Multi-Sensor Data Fusion Using Dempster-Shafer Theory. In *Proceedings of the International Conference on Networking, Sensing and Control* (pp. 897–902).

Sahayadhas, A., Sundaraj, K. and Murugappan, M., 2012. Detecting driver drowsiness based on sensors: a review. *Sensors, 12*(12), pp.16937–16953.

Sun, Y., Yu, X., Berilla, J., Liu, Z. and Wu, G., 2011. An in-vehicle physiological signal monitoring system for driver fatigue detection. In *3rd International Conference on Road Safety and Simulation Purdue University Transportation Research Board.*

Recent Trends in Communication and Electronics – Sharma et al. (Eds)
© *2021 Taylor & Francis Group, LLC, ISBN 978-1-032-04572-6*

Question-answer pairs generation from text segment

Amit Kumar Agarwal, Akhilesh Kumar Srivastava & Shubham Singh
ABES Engineering College, Ghaziabad, UP, India

ABSTRACT: The need of questions and answers is prompted for various purposes, e.g. self-study, academic assessment, and coursework. However, the conventional way to create question-answer pairs has been both tedious and time-consuming. In the present study the authors propose an automatic question generation for sentences from text passages in reading comprehension. Authors introduce a rule-based automatic question generation for the task, as well as implement statistical sentence selection and various configurations of named entity recognition. Three types of WH-questions ("What", "Who", and "Where") can be produced by this system. The system performs well on generating questions from simple sentences, but weakens on more complex sentences due to incomplete transformation rules.

1 INTRODUCTION

The translation of machine-readable, non-linguistic information into human language representation is the task of natural language generation (NLG), which is one of substudy of natural language processing (NLP). Question-answer generation (GAQ) from text is classified as NLG task focused on generating question-answer pairs from unstructured text. Basically, there are several NLG tasks related to question-answer generation, such as content determination (to decide which information should be involved) and linguistic realization (to apply grammar rules to produce valid sentences) (Reiter &Dale 2000). However, in order to be able to generate questions and answers from text, the system has to understand what it processes first. Then, natural language understanding (NLU) also should be included in question-answer generation to convert human language into representations that the machine understands (D. L. Lindberg,(Lindberg 2013). It has been suggested that question-answer generation has numerous possible, useful applications. QAG can be used to suggest frequently asked questions (FAQ) related to pre-defined documents or other media, to generate questions needed to assess deeper learning (self-learning help), and to provide question-answer pairs for tutoring purposes. There are also several applications of QAG that lie outside education field, such as to give suggestions about questions that might be asked in legal or security contexts (Lindberg 2013). In the present study, we aim to focus on generating question-answer pairs for academic purposes.

2 RELATED WORK

There are several approaches to accomplish question-answer generation. Du et al. introduced the usage of end-to-end fashion and deep sequence-to-sequence learning to generate questions for reading comprehension (Du et al. 2017). They implemented an attention-based sequence learning with two encodings, namely sentence encoding and

DOI: 10.1201/9781003193838-30

paragraph-level information. Their system, Neural Question Generation (NQG), is able to generate "what", "when", "where", "who", "why", and "how" questions (without answers). There is also an ongoing recent work about question generation by Sarvaiya, A., whose framework we adopt and develop into our system (Sarvaiya 2018). The strategy used in this study is selecting important sentences, finding candidate gaps in sentences, then forming the interrogative sentences.

3 QUESTION-ANSWER GENERATION

Formally, given a passage P, question-answer generation (QAG) system retrieves the most important sentence S from P . Then, QAG system produces a set of question-answer pairs {(Q, A)}, where each generated j, $_jA_j$ can be found in S, and its pair Q is the interrogative j version of S or a clause C from a set of clauses in k {C}k S, without A in it. As shown in Figure 1, here are four main j modules in our QAG system.(Sarvaiya 2018)

a. Preprocessing, which cleans the input passage P from unnecessary characters and shapes it into the desirable form (list of sentences).
b. Sentence Selection, which picks top-N most important sentences {S, ..., S} given . The 1 N P text summarization method used can be chosen between TextRank, multi-word phrase extraction (MWPE), and latent semantic analysis (LSA). The chosen method ranks the sentences in P and selects the top-N highest ranked sentences as the output.
c. Gap Selection, which selects phrases in S that can be used as answers {A} based on constituent tree j from syntactic parser and named entity recognition (NER).
d. Question Formation, which creates the interrogative version of S or C {C} in S k ∈ k without A to make a question for each answer j Qj in {A}. The final output of this module is j question-answer pairs {(Q, A} related to . jj P

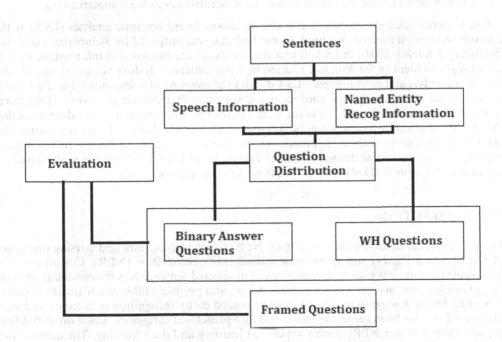

Figure 1.   System architecture in overview.

4 SENTENCE SELECTION

Sentence selection module was implemented using text summarization methods. One of the methods is TextRank (Mihalcea et al. 4), which is a graph-based ranking model for graphs extracted from natural language texts. TextRank makes an assumption that the most important sentences are the ones that are the most similar to every other sentence. The similarity can be computed using cosine distance or Jaccard distance. It utilizes PageRank algorithm by Google Search (Page & Brin 1998) to rank the importance of each sentence in the passage. The outline of TextRank is as follows.

– Each sentence in passage P is added as vertex in the graph
– Calculate similarity between every pair of sentences and use it as edge in the graph
– Normalize the similarity values
– Run PageRank algorithm until convergence
– Rank the sentences based on their scores
– Pick top-N sentences as the summarization

In the second method, multi-word phrase extraction (MWPE), we used TextRank to extract key phrases instead of the highest ranked sentences. Then, the occurrences of words that formed the key phrases were counted in each sentence of the passage. The most important sentences were the ones that had the highest number of occurrences. The following is the outline of multi-word phrase extraction.

– Each tokenized word is annotated with part-of-speech (POS) tags
– Filter the words (only leave those that are nouns or adjectives)
– Add the words to the graphs as vertices
– Add an edge between words that co-occur within window size W words of each other
– Run PageRank algorithm until convergence
– Rank the words based on their scores
– Pick top-K words as keywords
– Pair the keywords together as key phrase if they are adjacent in the graph
– Calculate the occurrence of words that form the key phrases in each sentence
– Pick top-N sentences with the highest number of occurrences as the summarization.

Our sentence selection module is also able to choose latent semantic analysis (LSA) as the text summarization method. We use LSA method that was proposed by Steinberger and Ježek (Steinberger &Ježek 2004). In LSA, it uses the context of the passage and information such as term usages to identify the semantic relation between sentences. It does not use syntactic relation, word order, and morphologies. LSA decides the meaning of a sentence using the word it contains, and the meaning of a word using the sentences that contain the word. Then, inter-relations between words and sentences were discovered with singular value decomposition (SVD). The outline of latent semantic analysis is as follows. Extract terms from sentences →Build input matrix with an approach on the terms, e.g. tf-idf or binary representation →Compute singular value decomposition (SVD) on input matrix →Compute ranks based on sigma and V T from SVD →Pick top-N sentences as the summarization

5 GAP SELECTION

The gap selection module utilizes two main NLP components: constituent parsing (using the PCFG Stanford Parser) and an in-house named entity recognition (NER). Constituent tree from syntactic parser is used to determine parts of selected sentence S as the candidate answers {a}. j Gap selection accepts a sentence (as a string), and produces fill-in-the-blanks Q-A pairs.

Named Entity Recognition: The purpose of named entity recognition is to locate and classify named entities from the unstructured text into predefined categories. There are two different approaches in our NER, namely supervised learning and deep learning. The dataset used to train the NER only consists of three useful tags, namely person, location, and organization.

To enable time tagging, we integrate another system (Agarwal et al.2018) with our NER, which uses a rule-based date time finder to extract date time entity from the text. Supervised learning approach in NER is done using several features and Conditional Random Field (CRF) classifier. CRF classifier is used because it outperforms other sequence modeling models like Hidden Markov Model (HMM) based on the previous work (Agarwal et al.2018).

Gap Selection: Utilizing the constituent tree and BIO encoding labels from NER, phrases from the sentence are selected as candidate answers. Candidate answers are noun phrases that act as a subject, an object, or a complement in the prepositional phrase of a declarative clause. Furthermore, there exists a possibility that a candidate noun phrase only consists of a part of a named entity. This is because each word is, by itself, a valid noun phrase. More specifically, if a candidate answer a is a part of named j entity and a does not include all corresponding words j needed to form a proper named entity, aj is eliminated from the set of candidate answers {a}. In the end of the process, j the candidate answers are passed to the next module as the answers {A} of. j S

Formation of Question: As with gap selection, question formation is mostly rule-based. Question formation uses the named entities (and their tags) and the constituent tree produced by gap selection, as well as a NLG module to convert fill-in-the-blanks question-answer pairs to proper, interrogative-form question-answer pairs. A detailed explanation of each step taken is follows.

Conversion to Interrogative Form: Depending on the role of the answer in its clause, there are three cases for converting a declarative sentence to its interrogative form.

- Answer as subject converting a statement to a question when the answer is its subject is simple: replace the subject answer with the question word, add a question mark to the end.
- Answer as object converting a statement to a question when the answer is its object is considerably more difficult. The constituent tree is traversed top-down with recursive descent and subtrees are used as input for the NLG module. The answer object is elided from being inputted to the NLG; instead, the NLG generates an interrogative sentence.
- Answer as prepositional phrase similarly, converting a statement to a question when the answer is its object is difficult. The constituent tree is traversed top-down with recursive descent and subtrees are used as input for the NLG module, depending on its constituent tag.

Determining Question Words: The question words for an interrogative sentence is determined by the NE tag of its answer. For answer as subject or answer as object, the only possible question words are "What" and "Who", as both roles only allow nouns. For answer as prepositional phrase, possible question words are "Who", "Where", "When", and "What", depending on if the answer is, a person, a location, a time preposition, or anything else.

6 EXPERIMENT

6.1 *Dataset*

We used two datasets, SQuAD2.0 for the question-answer generation system in general, and CoNLL2003 was specifically used for named entity recognition (NER) (Sang et al. 2003). Stanford Question Answering Dataset (SQuAD) is a dataset that includes questions generated by crowdworkers and the corresponding passages (Sang et al. 2003).

6.2 *Evaluation on sentence selection*

Summaries generated by system using different methods in sentence selection were evaluated by cosine similarity. The evaluation was done with 10 passages from SQuAD dataset, each of them was at least 10 sentences long, with three-sentence summaries. Based on averaged cosine similarity, it can be seen that almost all of the methods performed better when they considered the presence of stop words. Stop words are used to eliminate insignificant words from the processed sentences.

6.3 *Evaluation on named entity recognition*

The best model in supervised learning is achieved using tuned c1 and c2 hyper parameters with all the features explained in Section 5. During the feature importance experiment, it can be seen that all features give some contributions to the model, with the most impactful feature is bag-of-words feature. This is because the feature gives context to the model when it is classifying a word. The least impactful feature is position feature, because there are many words in one sentence and there is no pattern for each sentence in the dataset. Word2Vec CBOW, Word2Vec SG, and Fast Text are trained using train dataset of CoNLL2003.

7 RESULT

Figure 2 Given below shows sample question-answer pairs generated by our proposed system, as well as its context sentence. The first five question-answer pairs show successful question generation; the final pair, however, presents a failed question generation. Questions generated using our system successfully converted simple declarative statements into questions, but for more complex sentences, our devised rules show difficulties in converting more complex statements to questions. This is due to an implementation error—the rules to convert a statement to a question are made in-house, instead of implementing an already established rule-based question-generation system.

8 CONCLUSION AND FUTURE WORK

Authors use multiple methods to select noteworthy sentences in a paragraph, then use named entity recognition and constituent parsing to generate possible question-answer pairs. The sentence is then transformed to an interrogative form based on a set of rules and possible answers. The system performs well on simple sentences, but falters on more complex sentences. Our question-answer generation system is able to produce three types of WH-questions, e.g. "What", "Where", and "Who". Future improvements to the system include implementing a more robust, already established rule-based question-generation system.

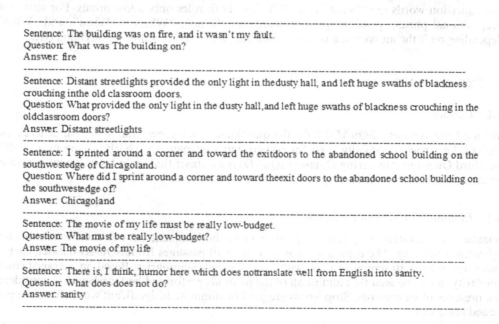

Figure 2.   Sentence-question-answer.

160

REFERENCES

Du, X. et al. 2017. Learning to Ask: Neural Question Generation for Reading Comprehension. Proceedings of the 55th Annual Meeting of the Association for Computational Linguistics (Volume 1: Long Papers).

LindbergD. L., 2013. Automatic Question Generation from Text for Self-directed Learning. DOI: 10.13140/RG.2.2.33776.92162

M. Agarwal et al. 2018. Comparative Analysis of the Performance of CRF, HMM and MaxEnt for Part-of-Speech Tagging, Chunking and Named Entity Recognition for a Morphologically rich language. Nov 2018.

Mihalcea, R. &Tarau, P. 2004. Textrank: Bringing order into text, Proceedings of the 2004 conference on empirical methods in natural language processing, 2004.

Page, L. &Brin, S. 1998. The PageRank Citation Ranking: Bringing Order to the Web.

Reiter, E. & Dale, R. 2000. Building Natural Language Generation Systems, Cambridge University Press.

Sarvaiya, A. 2018. Using Natural Language Processing for Smart Question Generation. Intel AI Academy.

Steinberger, J. & Ježek, K. 2004. Using Latent Semantic Analysis in Text Summarization and Summary Evaluation. Proceedings of the 7th International Conference ISIM, Jan. 2004.

Tjong Kim SangE. F. et al. 2003 E. Tjong Kim, and F. De Meulder, Introduction to the CoNLL-2003 shared task in Proceedings of the seventh conference on Natural language learning at HLT-NAACL 2003

Recent Trends in Communication and Electronics – Sharma et al. (Eds)
© 2021 Taylor & Francis Group, LLC, ISBN 978-1-032-04572-6

Low-energy consumption using Bayesian-Hidden Markov model in wireless sensor networks

Gauri Kalnoor

PhD Research Scholar, Department of Computer Science & Engineering, BMS College of Engineering, Bangalore, India

S GowriShankar

Professor, Department of Computer Science and Engineering, BMS College of Engineering, Bangalore, India

ABSTRACT: The network becomes vulnerable to various attacks because of its wireless nature. The natural environment with an inherent transmission is unreliable due to its characteristics. In this article, we attempted to design an intrusion detection mechanism using the novel model based on machine learning algorithms known as Bayesian Network Model and Hidden Markov Model. We evaluate the aspect of detecting abnormal/malicious nodes in a Wireless Sensor Network (WSN). This proposed model has enabled a hierarchical WSN that generates the trained trusted values among different sensor nodes. The performance of the mechanism is evaluated and the impact is analyzed based on threshold value (fixed and a dynamic) to identify the malicious nodes. Further evaluation of a sensor network is conducted. The results obtained by experiments shows that the proposed model encourages in detecting sensor nodes that are malicious. The Hidden Markov Model (HMM) is imposed with the Bayesian Model, to achieve the goal of minimum energy consumption. The network performance is evaluated and results are compared, based on the experiments conducted. In the recommended framework, the sensor nodes evaluate the detection rate and the obtained results while simulating are tabulated. Finally, an intrusion detection experiment is illustrated for various types of attacks. Based on the comparison made and the mathematical analysis shows that our proposed system in a distributed WSN makes an effective use of the communicating sources. Finally, we attain high accuracy in the detection of malicious nodes with minimum energy consumption.

Keywords: Bayesian network, WSN, IDS, Bayesian-Hidden Markov model, training set, clustering

1 INTRODUCTION

In Wireless Sensor Networks (WSN) is most favorable and motivating areas of technology since the past few years. Such wireless network comprises of numerous tiny, low cost, low-powered sensors and are very large collecting information regarding the environment's physical nature. The fault-tolerant, self-organized, flexible, highly sensed fidelity, rapid deployment and low-cost features of the WSN have formed many novel areas of applications that are sensing remotely. The proposed algorithm aims to determine the malicious or abnormal nodes within the range of sensing. The dynamic characteristic of the sensor nodes is the additional prior information that is required.

DOI: 10.1201/9781003193838-31

The mechanism of intrusion detection is developed by using the model known as Bayesian-Hidden Markov such that the trust values are evaluated for every node that are distributed hierarchically for a WSN. The proposed technique is able to detect the malicious nodes while choosing an appropriate threshold of trust value. The hierarchical structure of WSN is used to minimize the traffic of the network that is caused by communications between neighboring nodes.

The contributions of the proposed work are summarized below:

a. A trust-based mechanism of intrusion detection is developed based on the Bayesian network model to detect malicious nodes and computes trust values in WSN. These rely on a hierarchical structure which is scalable also included the cluster heads (CHs) and the sensor nodes (SNs). The SNs records the information of trust values initially during the communication of peer nodes. A CH gathers the trust values and its reports obtained from each and every SNs. Next the comprehensive trust value is calculated within the sensing range of all the nodes in the sensor network (for example, the clusters). The network's malicious or abnormal nodes arc then detected by choosing a suitable trust threshold.

b. Further simulation is conducted in WSN to detect and identify the malicious nodes by finding an appropriate trust threshold. Computation of trust values is carried out and then analyzed for 10 clusters. The trust threshold values are pointed out within a WSN which is highly dynamic in nature as compared to wired communication network.

c. The experiments are conducted and the performance of WSN where a Bayesian-Hidden Markov Model is implemented and computed using the static threshold value along with a dynamic threshold value. Thus, the results gained illustrates our proposed design that encourages the detection of abnormal activity of the nodes considered as malicious. This is done by choosing a trust value that is appropriate threshold value, along with an rate of false positive and false negative. Also, the current findings are compared and analyzed with the previous obtained results using different methodologies [1], and the possible overhead is presented in our developed detection mechanism.

The proposed work is discussed in detail in the following sections. The related work carried out is discussed in Section 2. The proposed Model known as Bayesian-Hidden Markov Model is explained in section 3 and a mathematical model is framed and discussed, followed by Results and Analysis in the 4rth Section. The inference of the work is defined in section 5.

2 RELATED WORK

The authors in [2] has explained and proposed a new scheme known as "frivolous Group-based Trust Management Scheme (GTMS)" which is designed based on the clustering in WSN. The GTMS is a scheme which evaluates the values of trust within a group of SNs and has been considered based on two types of topologies: intergroup topology where the approach of centralized trust administration was used and intragroup topology where the distributed approach of trust management was adopted. Then, the author in [3] proposed the establishment of a active trust and framework management for a hierarchical WSNs. The developed framework considers the direct as well as indirect (group) trust in evaluation. The energy was also evaluated by the authors that are related with the sensor nodes in the selection of services. The method carried out by the authors has also considered the dynamic nature of trust values by implementing the varying function of trust and is used for greater weight as compared with the recently obtained values of trust during the process of calculation. Additionally, the approach applied by the authors having the capability for considering mobile nodes f computation of rom one cluster of the network to the other cluster.

The other work by different authors explains about the protocols of trust management which is referred to [4], [5], [6] and [7]. As compared to the previous work, in the referred articles, our work mainly attempts to compute the sensor node's trust values by using Bayesian model and additional knowledge-based intrusion detection technique is used for a hierarchical WSN.

3 BAYESIAN-HIDDEN MARKOV MODEL

Intensions of the Proposed model: The designed system utilizes novel algorithm that includes both Bayesian model and Hidden Markov Model. The Detection is through with Hidden Markov Models (HMM) having maximum likelihood to estimate the algorithm with real time viterbi optimization. The model based on Hidden Markov is used for detection of misused patterns which improves the accuracy by applying the Entropy method.

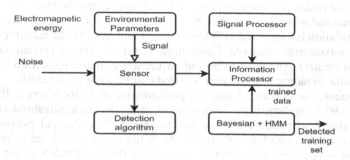

Figure 1. System architecture.

The procedure is described in the steps as mentioned below:
The procedure is clarified in the subsequent steps:
Step 1: Calculation of the sample set SS_t at constant breaks
Assuming sample set be: SS_t
$$k = \{\theta_j, S_i, j = 1, 2, 3, \ldots, i = 1, 2, 3 \ldots u\}$$
at t
Step 2: If the interval of time to obtain the abnormal or malicious nodes is not within the range that is not at constant intervals than need to go to step 4 for finding prediction.
Step 3: else if the interval of time to obtain the abnormal or malicious nodes is within the range that is at constant intervals then fix the sample set $\hat{S}\hat{S}_t$ by obtaining the signal strength of body in motion and its mentioned nodes via readers i.e.: $\hat{S}\hat{S}_t = \{S_i, \theta_j\}$
Go to step 5
Step 4: The calculated history has the trace of sample sets of last 10 breaks. Let the final set be:

$$\hat{S}\hat{S}_t = \{S_i, \theta_j\}$$

Step 5: obtain the Euclidean distance from signal asset of following object and reference nodes neared to the tracking object:
$E_j = \sqrt{(\sum n_i = 1(\theta_i - S_i)2)}$ where j = 1, 2 …..m.
Step 6: next stage the Euclidean distance is calculated, in reference to this distance obtained the set of k recommendation tags neared to the target.
Let $E_q = \{E_1, E_2, \ldots \ldots E_k\} where\ q = 1, \ldots .k$
Step 7: The weighting factors are predicted by next formula:

$$W_j = \frac{\frac{1}{E_i^2}}{\sum_{i-1}^{k} \frac{1}{E_i^2}}$$

Step 8: The co-ordinates of the tacking object is given
by: $(x, y) = \sum_{i-1}^{k} W_1(x_1, y_1)$

In statistical analysis, Bayesian Model (also known as *Bayesian inference*) is the methodology of implication in which the Bayes' rule is applied such that the probability estimation is updated for a normal theory as supplementary indication. The main aim of applying the *Bayesian Model* in our novel work is for calculating the trust values of sensor nodes (and also cluster heads) in a clustered WSN. This model proposed is grounded on a statement mentioned below:

Statement: All the packets are directed after a node which are independent from other nodes. Thus, if any packet is detected to be an abnormal/malicious packet, the probability of the next immediate packet being a suspicious malicious packet will remain ½.

This statement of probability theory shows that any kind of attacks may appear in different forms, either in a single packet or in large number of packets thus the trust values can be calculated and derived. Also, an assumption is made for N packets that are directed from a node among which k packets are proved to be the *normal packets*.

following, some terms are provided which are explained in the previous work [7].

$P(in : normal) = p$ (*means the probability of the ith packet is normal.*)

Vi (*means that the ith packet is normal.*)

$n(N)$ (*means the number of normal packets.*)

Based on analysis and the above assumption made, we also assume the "distribution of observing $n(N) = k$ is governed by a Binomial distribution2", as described below.

$$P(n(N) = k|p) = \binom{N}{k} p^k (1-p)^{N-k}$$

4 RESULTS AND DISCUSSIONS

The performing of our proposed model is assessed based on the experiments that are conducted and this experimental scenario was simulated by using network simulator- NS2.

Thus, it is observed that the proposed method when compared with the mainstream methods, has various advantages, mainly with respect to QoS parameters like detection

Table 1. Simulation scenarios for different attacks.

Experiment	Attack type
Experiment 1	Black hole
Experiment 2	Flooding
Experiment 3	Rushing
Experiment 4	Multiple-attacks
Experiment 5	Black hole
Experiment 6	Flooding
Experiment 7	Rushing
Experiment 8	Multiple-attacks

165

rate and FAR. However, the designed framework and methodology has achieved a stable performance as observed in every experimental scenarios.

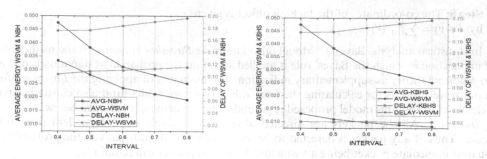

Figure 2.   Average energy consumption and delay.

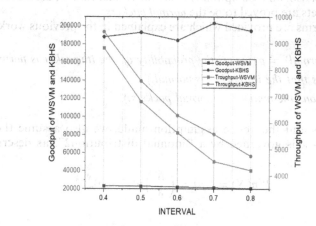

Figure 3.   Goodput and Throughput comparison. Note: True positive is represented as (_P) and true negative is represented as(_N) for all algorithms.

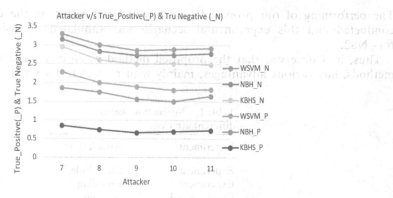

Figure 4.   Attacker v/s true positive and true negative.

Note: True positive is represented as (_P) and true negative is represented as(_N) for all algorithms

5   CONCLUSION

A novel approach is framed and designed to obtain better performance as compared with the mainstream algorithms. A naïve model called as Bayesian-Hidden Markov Model is proposed such that high detection rate of malicious packets is obtained with minimum energy consumption of sensor nodes. Hidden Markov Model (HMM) is the model used for calculating the probability and detecting the malicious nodes during the active states of sensor nodes. An HMM applicator is used during the training and testing stage of the data packets. The model is designed with Bayesian Model where a Baye's rule is applied to calculate the trust threshold values and then calculate the energy consumption of the sensor node. The IDS is created to find whether the traffic of the network is normal or intrusions have been detected. The model is run at the base station to detect the attack. Thus, high throughput and minimum delay is obtained with the proposed model and also detection rate, accuracy and other parameters are maximized when compared with the mainstream IDS techniques deployed in WSN.

REFERENCES

[1] D.He, et.al. 2015 "Robust anonymous authentication protocol for health-care applications using wireless medical sensor networks," Multimedia Systems, vol. 21, pp. 49–60.

[2] M.Li & H.J.Lin (2015), "Design and Implementation of Smart Home Control Systems Based on Wireless Sensor Networks and Power Line Communications," IEEE Transactions on Industrial Electronics, vol. 62, pp. 4430–4442.

[3] M.G.Ball, et.al. 2015 "A Review of the Use of Computational Intelligence in the Design of Military Surveillance Networks,"Recent Advances in Computational Intelligence in Defense and Security, vol.621, pp. 663–693.

[4] H.C. Qu, et. al. 2015 "Hybrid Computational Intelligent Methods Incorporating Into Network Intrusion Detection," Journal of Computational and Theoretical Nanoscience, vol.12, pp. 5492–5496.

[5] Huang J.Y., et. al. 2011, "Shielding wireless sensor network using Markovian intrusion detection system with attack pattern mining," Information Sciences, vol.231, pp.32–44.

[6] Patil S. & Chaudhari S., (2016) "DoS Attack Prevention Technique in Wireless Sensor Networks," Procedia Computer Science, vol.78, pp. 715–721.

[7] Su M.N. & Cho T.H., (2014) "A Security Method for Multiple Attacks in Sensor Networks: Against False-Report Injection, False-Vote Injection and Wormhole Attacks," Open Transactions on Wireless Sensor Network, vol.2, pp.13–29.

Recent Trends in Communication and Electronics – Sharma et al. (Eds)
© 2021 Taylor & Francis Group, LLC, ISBN 978-1-032-04572-6

COVID-19 detection using machine learning and deep learning techniques: A review

T. Aishwarya, V. Ravikumar & V. Megha
Vidyavardhaka College of Engineering, Mysuru, India

ABSTRACT: Frequently alluded to as Coronavirus disease, Covid-19 is a transmittable infection brought about by a newfound Coronavirus. There are neither definite treatments nor appropriate medicines for COVID-19 in the present scenario. Nonetheless, there are a few clinical investigations that are as of now testing potential treatments and immunizations. To detect Covid-19, Artificial Intelligence models can be adapted and leveraged with preliminary clinical understanding to address COVID-19 and the novel challenges associated with it. AI centered tools can be built and established for achieving the same. The availability of limited set of datasets and accessibility to those datasets is the biggest challenge for researchers. There is likewise less number of specialists for marking the information explicit to this new strain of infection. In this paper, we look into a few techniques of Machine Learning and Deep Learning that have been employed to analyze Corona Virus Data.

1 INTRODUCTION

The globally spreading infection of coronavirus has become an undermining threat to general well-being worldwide. This virus mimics behaviour like that of viral pneumonia and, if not immediately treated, often leads to an increasing death rate. However, since the signs do not occur until the early stages, it is difficult to detect and rapidly isolate individuals that have acquired the disease. This respiratory disease has made way to a pandemic globally, which has confirmed global cases in more than 72 countries as provided by the WHO data.

Machine Learning and Deep Learning techniques are playing a vital role in this task of Covid-19 detection. The Chest CT is an analytical and diagnostic tool used for the detection of pneumonia. It is found to be exceptionally helpful in distinguishing ordinary COVID-19 radiographic highlights, particularly with dainty slices. Some conditions such as swelling of lesion signs, bronchiectasis depicting the irregular widening of the bronchi and blurriness in CT images are demonstrated by CT images to easily examine COVID-19 (Gozes 2020). Ongoing studies show that RT-PCR has only 30 to 60 percent of moderately low rate of discovery and repeated tests are frequently required. With a moderately high critical proliferation number (R0) of 2.2, it can be transmitted from individual to individual and so far has no proficient medicines and control methodologies (Chan 2020).

As demonstrated by the occurrence and explosion of cases of corona disease, patients may be divided into different stages of disease, such as mild, moderate, severe and critically ill. If there is some change in patient symptoms, the progressive resolution of consolidation in computed tomography images may be observed. Nevertheless, despite the advances, detection of COVID-19 in CT images from common cases of pneumonia remains a pronounced obstacle in screening and identification. Incorrect negative results can be developed especially during

DOI: 10.1201/9781003193838-32

the mild stage of the disease and this process is often time consuming. Therefore, it is very desirable to use computerised investigation and analysis methods to extricate Commonly Acquired Pneumonia (CAP) from Covid-19.

2 MACHINE AND DEEP LEARNING TECHNIQUES AGAINST COVID-19

We discuss the different techniques of machine learning and deep learning to detect Covid-19 that have been developed and used by various researchers.

2.1 *Deep CNN for detection of covid-19 cases from chest x-ray*

The deep learning technique proposed by Wang & Wong (2000) combines a machine-human design approach utilizing a network called the COVID-Net. Network architecture is created which is customized and tailor-made in discovering COVID-19 cases from images of CXR, Chest XRay. A publicly accessible benchmarked dataset was created and it is referred to as "COVIDx". Five different data repositories were combined and modified to create this dataset. COVIDx, which was employed to train and assess the suggested COVID-Net, encompassed approximately 13,870 patient cases with a total of 13,975 images of CXR. In this methodology, a network design prototype has been constructed, preliminarily to mark one among the three subsequent predictions: i) normal, which is no infection, ii) infections that are not COVID-19 and iii) positive COVID-19 communicable infection. The logic and motivation for opting these probable predictions is because it can assist the doctors and clinicians to enhance the decision as to who should be ranked for Polymerase Chain Reaction but also in addition to what kind of treatment should be utilized depending upon the reason for infection. This is because both non-COVID-19 diseases and COVID-19 disease necessitate varied plans in treatment.

The next stage of the machine-human collective plan for design which is engaged in developing the suggested COVID-Net is the design of the Machine-driven exploration stage. Precisely, in this particular stage, the primary prototype of the network design, data, behaves as a monitor to a design exploration strategy besides the human specific design requirements. This is supportive in identify and learning the ideal micro-architecture and macro-architecture designs. This is in turn used to create the ultimate Deep Neural Network architecture that is custom made. It is also possible to have a considerably superior flexibility that is probable through the manual human-driven design of the architecture. This can be still done while confirming that the domain-specific operational requirements are fulfilled by the Deep Neural Network architecture. Sensitivity to cases of COVID-19 is substantial in bounding the missed cases of COVID-19 to a great extent. It is of utmost importance for the design of COVID-Net. They combine the machine-driven design exploration strategy as generative synthesis. This is centred on a sophisticated interaction amongst a pair of inquisitor-generator which works in line to gather intuitions and understandings. By achieving 93.3% testing accuracy, COVID-Net put out a good and decent accuracy. This accentuates the potential of leveraging a collaborative human-machine design strategy to produce highly customized architectures of deep neural network.

2.2 *Ensemble and classical machine learning algorithms*

The detection of this disease with the aid of different AI methods may be one of the remedies for managing the current havoc. In this technique proposed by Khanday (2020), various conventional and ensemble based ML algorithms were utilized. Techniques such as Bag of Words (BOW), Report length and Term Frequency/Inverse Document Frequency (TF/IDF) were used for Feature extraction. Collecting data, data refinement, pre-processing, feature extraction and classification were employed sequentially in this work. The dataset used for this process encompassed a total of 24 attributes. Attributes like patient_id, age, sex, offset, went_icu, modality, temperature, needed_supplemental_O2, extubated, neutrophil count, lymphocyte count, date, location were used. There were approximately 212 reports that were used and their length was estimated.

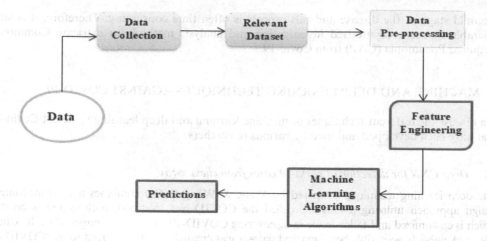

Figure 1. Methodology proposed (Khanday 2020).

Because the study was based on text mining, relevant clinical notes and findings were extracted. The text was present in clinical notes while the attribute finding contained the labels of the corresponding text. The methodology used is depicted in Figure 1.

With unigrams and bigrams also being extracted, a simplified representation called the Bag of Words was also taken into account. In order to classify the text into four distinct virus groups, classification was carried out. These distinct groups of virus chosen to be classified were SARS, COVID, ARDS and both COVID-ARDS. This task of classifying text was performed using algorithms such as SVM-Support Vector Machine, MNB-Multinomial Naive Bayes, Logistic Regression, Decision Tree, Random Forest and Adaboost. The class of numerical variable is predicted by Logistic Regression centered on its association with the label. SVM proceeds with 'n' number of features for the specific text with the specified label (Zhang 2012). Multinomial Naïve Bayes (Xu 2017) uses the Bayes rule to calculate class probabilities of a given document. The Decision Tree divides the space recurrently, conferring to the inputs and classification occurs at the nethermost of the tree. The text is classified by the leaf nodes into four classes. Splitting criterion is a critical feature that must be considered when constructing a decision tree. This function determines exactly in what way the data ought to be divided to maximize and optimize output. Random Forest (Katuwal & Suganthan 2018) is an Ensemble machine learning algorithm that is employed for the purpose of classification and this algorithm operates similar to a decision tree. To train a random forest algorithm, a bootstrap aggregation technique is deployed. The overall predictions can be done by averaging the predictions of all distinct regression trees. Adaboost is an ensemble based learning algorithm that works with weighted dataset instances.

The results of this study are shown in Table 1. It can be observed that Multinomial Naive Bayes Algorithm and Logistic Regression show superior results when compared to other

Table 1. Comparative results of the classical and ensemble based algorithms.

Algorithm	Precision	F1 Score	Recall	Accuracy(%)
Multinomial Naïve Bayes	0.94	0.95	0.96	96.2
Support vector machine	0.82	0.86	0.91	90.6
Logistic Regression	0.94	0.95	0.96	96.2
Decision Tree	0.92	0.92	0.92	92.5
Adaboost	0.85	0.88	0.91	90.6
Random forest	0.93	0.93	0.94	94.0

algorithms by having an accuracy of 96.2%, recall of 96%, F1 score of 95% and precision of 94%. Good results were also shown by other algorithms such as random forest and decision trees having an accuracy of 94% and 92.5% respectively.

2.3 *DeTrac deep convolutional neural network*

A significant system that can offer a favorable solution by the transmission of information from general object recognition task to domain definitive task is the Transfer Learning. A deep CNN referred to as DeTrac is employed in detecting Covid-19 (Abbas, Abdelsamea & Gaber 2020). DeTrac is an acronym for Decompose, Transfer and Compose is very beneficial in classification of chest X-ray images. Examining its class boundaries with the mechanism called class decomposition; DeTraC can cope with any anomalies in the CXR dataset. When Transfer Learning is considered, it can be executed by three important settings (Li 2014). They are Shallow tuning, Deep tuning and Fine tuning. Only the final classification layer is adapted to the fresh task in Shallow tuning. Without training, it then freezes the remaining layer parameters. Deep tuning aims at retraining every pre-trained network parameters throughout. By tuning the learning parameters, Fine-tuning intends to progressively train extra layers before a substantial performance improvement is achieved. In the course of the adapting and training a pre-trained CNN model called ImageNet, shallow tuning was employed using the CXR image dataset collected. The DeTrac model used in this technique is shown in Figure 2.

The deep CNN architecture that is constructed on the feature of class decomposition, referred to as DeTraC - Decompose, Transfer and Compose model is adapted to adequately enhance the performance capability of pre-trained models. To aid this process, a class decomposition layer is supplemented to these models. This decomposition layer of the class attempts to partition individual class into several sub-classes within the image dataset. It then assigns new labels to the new set, where each sub-set is treated as independent classes. Finally, these sub-sets are convened together to generate the ultimate predictions. The DeTraC model typically involves of three stages. The pre-trained model of DeTraC is trained in the first stage to mine deep local features from every image considered. Applying class-decomposition layer of DeTraC follows next. This is done to facilitate simplification of the local structure of the distributed data. During the subsequent stage, training is carried out by means of a sophisticated method of gradient descent optimization. A layer of class composition is introduced in the third stage to perfect the last classification of CXR images. In classifying CXRs images from regular to critical respiratory syndrome cases, this model achieved a good accuracy of 93.1% working with a sensitivity of 100%.

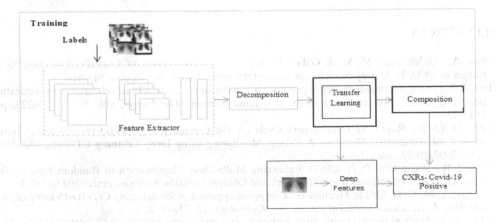

Figure 2. Depicting the DeTrac- Decompose, Transfer and Compose model employed (Abbas et al. 2020).

Table 2. Outline of the techniques reviewed.

Title	Techniques	Data Type	Accuracy
1. Deep CNN for Covid-19 Detection	COVID-Net	COVIDx Chest XRays	93.3%
2. Ensemble and classical ML Algorithms	- Multinomial Naïve Bayes		96.2%
	- Support vector Machine	Textual Clinical	90.6%
	- Logistic Regression	Reports	96.2%
	- Decision Tree		92.5%
	- Adaboost		90.6%
	- Random forest		94%
3. DeTrac CNN	Decompose, Transfer and Compose Model	Chest XRay Images	93.1%

3 DISCUSSION

The key aim of this particular work was to expound prime methods of deep learning and machine learning to detect Covid-19. The summary of the works considered in this paper is presented in Table 2.

4 CONCLUSION AND FUTURE WORK

Machine learning and Deep learning techniques are playing a significant role in battling the COVID-19 disease. Researchers are employing these techniques to identify people, to test potential treatments and to examine the impact of this disease on general public. We presented a survey of techniques in ML and DL for detecting Covid-19. Classical and ensemble based ML algorithms were used to classify clinical reports with good accuracy. Deep learning techniques using COVID-Net and DeTrac models were also expounded. It was observed that Deep Convolutional Neural Network based models were quiet efficient for the task of Covid-19 detection. AI strategies regularly require a lot of information for computational models. With the availability of data, this detection task can be improvised on by using Internet of Things applications powered by AI and numerous deep learning networks like Capsule Network. Nevertheless, we envision the quantity of COVID-19 based AI studies to grow significantly in the coming time as more COVID-19 subtleties like radiographs and CT become more accessible.

REFERENCES

Abbas, A., Abdelsamea, M. M. & Gaber, M. M. (2020). Classification of COVID-19 in chest X-ray images using DeTraC deep convolutional neural network. *Applied Intelligence*. Springer.
Chan, J. F. (2020). A familial cluster of pneumonia associated with the 2019 novel coronavirus indicating person-to-person transmission: a study of a family cluster. *The Lancet*, vol. 395, no. 10223, pp. 514–523.
Gozes, O. (2020). Rapid AI Development Cycle for the Coronavirus (COVID-19) Pandemic: Initial Results for Automated Detection & Patient Monitoring using Deep Learning CT Image Analysis. arXiv:2003.05037.
Katuwal, R., Suganthan, P. N. (2018). Enhancing Multi-Class Classification of Random Forest using Random Vector Functional Neural Network and Oblique Decision Surfaces. arxiv:1802.01240v1.
Khanday, A. M. U. D. (2020). Machine learning based approaches for detecting COVID-19 using clinical text data. *International journal of information Technology*, 12, 731–739.
Li, Q. (2014). Medical image classification with convolutional neural network. 13th international conference on control automation robotics & vision (ICARCV), *IEEE*, pp. 844–848.

Narinder, S. (2020). Covid-19 Epidemic Analysis Using Machine Learning and Deep Learning Algorithms. medRxiv:2020.04.08.20057679.

Wang, D., et al. (2020). Clinical characteristics of 138 hospitalized patients with 2019 novel coronavirus–infected pneumonia in Wuhan, China. *Jama*, 323(11):1061–1069.

Wang, L. & Wong, A. (2020). Covid-Net: A Tailored Deep convolutional Neural Network Design for detection of covid-19 cases from chest x-ray. arXiv:2003.09871v4 [eess.IV].

Xu, S., et al. (2017). Bayesian Multinomial Naïve Bayes Classifier to Text Classification. In J. Park, S. C. Chen, K. K Raymond Choo (eds), *Advanced Multimedia and Ubiquitous Engineering*. Singapore: Springer.

Zhang, Y. (2012). Support Vector Machine Classification Algorithm and Its Application. In C. Liu, L. Wang, A. Yang (eds), *Information Computing and Applications*. Berlin: Springer.

Comprehensive review of publications related to Covid-19 in science and engineering journals

Anil Kumar Dubey, Mala Saraswat & Santosh Kumar
ABES Engineering College Ghaziabad, India

ABSTRACT: The current pandemic linked to novel coronavirus: Covid-19 offered a new era of research for prediction and diverse challenges to control. The presently, every one heard and discuss for corona virus (covid-19) that's plays an important role to changing the daily activities of not only individuals life, but also society as well as country routine. United State of America and India are major country impacted with Covid-19.Several laboratory already confirmed that continues increment of coronavirus cases alarming to peoples. Numerous misinformation, false reports and unwanted doubts in this regards have been regularly spread since the occurrence of COVID-19. To response such theatres, we Illustrate a detailed review for all major aspects of Covid-19 from various reliable sources.

1 INSTRUCTIONS

On 11[th]March 2020, World Health Organization (WHO) confirmed that the quick transmission of the Covid-19 (novel Coronavirus) is pandemic situation for world [1].Since then the investigators as well as associated industry, working on its prediction, diagnosis and possible treatment. As of 22th Aug 2020, there have been up to 22,813,065Confirmed cases in 216 Countries, areas or territories with cases (2,975,701India) and 7,95,134Confirmed deaths (55,794India)[1]. The Coronaviruses (CoV) is the subfamily member of Corona virinae, within family of Coronaviridae in order to Nidovirales. This types of virus seems like crown as projections on surface, so it's named "Corona" means crown. [2]. Covid-19 are sphere-shaped positive sense RNA viruses with diameter from 600Å to 1400Å [3], having spikes proteins protruding from surface. The significant epidemic infectious disease arises the most health emergency, researcher use several techniques to examine and forecast disease trends [5].

In this pandemic, the classes, training, workshops and other meeting organized through online mode, due to this mobile become very important. Governments plan to prevent the virus spread through confine the peoples stay to their home [6]. The significant support to fighting this pandemic situation is to discontinuity the infection chains of Covid-19. Not only in bounded region of a country, but also whole world's feel the financial belongings of this pandemic force [8, 9]. V Chamola et al., [4] elaborate the review of contribution towards Covid-19 with several perspective with role of current technology. As per WHO records, fever, fatigue, dry cough, and sputum production are most common symptoms, while myalgias, rhinorrhea, chest pain, and gastrointestinal symptoms are less common symptoms [7, 10, 11]. In December 2019, a novel Coronavirus disease befell an eruption in Wuhan, China. This eruption originated from human seafood market at Wuhan. In this patients have unknown pneumonia disease and had the travel history to seafood market. Day to day, the number infection cases began to rise with diverse characteristics as not travel history of seafood market, specify the possibility of virus transmission from human-to-human. [2, 12]. G Pascarella et al., [13] present a comprehensive review on diagnosis and management of

DOI: 10.1201/9781003193838-33

Covid-19. V M Mahalmani et al., [12] illustrate a review to highlights the diverse characteristics of Covid-19 like etio-pathogenesis, epidemiology, diagnosis, & prevention to be adopted to fight such pandemic situation. Author's focus on challenges and ethics preparation for developing country (like India) during such eruption, also emphases on several approved approaches to develop effective therapeutic policies included drugs, convalescent plasma therapy, and vaccine therapy. Tanu Singhal [3] presented are view of Covid-19, and discuss its origin, spreads, about epidemiology and pathogenesis, medical topographies, differential diagnosis methods and treatments. Syed A Hassan et al.,[14] present a review towards medical topographies, diagnosis, and treatments of on Covid-19. Authors describe the general outline, medical topographies, assessment, &several adopted treatments for Covid-19 patients. S Chauhan [15] illustrates the comprehensive review of Covid-19. Author delivers an update of infection rapidly evolving worldwide.

Need of Contribution: To our knowledge there are no such review articles currently available that summarized all the associated contribution towards COVID-19 for publication. Here we present the Covid-19related contribution by investigator in different scenario of individuals, organizations, country and citation of publication.

Research Question: The primary goal of this review is to find the solution of the following research questions (RQ). These are given as

RQ1 Which publisher published highest publication of Covid-19 during this pandemic?

RQ2 Which subject mostly concerned by top most Science & Engineering publisher in publication for Covid 19 articles?

Resources used: Due to latest impact on society, we concern to write the article, it is suggested that to review these diverse research databases, that can extract for existing work, and figure out the contribution toward Covid-19. For this study global databases are searched from search engine to find the existing publication contributed by publisher in this domain. Major contribution selected with standard of science and engineering publisher from search databases as: Google Search Engine, IEEE, Elsevier, Sage, Springer, Tailor and Francis, MDPI, ACM, Hindawi, Wiley etc.

Contribution: The goal of this study is to outline the publisher concern towards publication for Covid-19 articles among top scientific and engineering publisher. We also highlight the publisher concern to publication for different tactics of Covid-19 as prediction, diagnosis, treatments, and Impacts.

2 RELEVANT STASTICS

All text, On 22nd August 2020, the WHO provides the statistics of Covid-19 to worldwide with infection and death statistics. As medical disease the small percent of infectious peoples shows the major confrontation as its serious problem. Therefore each and every person infection and death we consider from all population worldwide. As of statistics there have been 22,813,065 confirmed cases detected in whole words, and 15,142,021 are recovered, while 7,95,134 confirmed deaths. From currently active infected Covid-19 patients, up to 99% in mild condition, and only 1% are critical. From closed cases, about to 95% patient recovered from Covid-19 and only 5% patient deaths. As per statistics of WHO for Covid-19 and map them with total population for infections for top five populated country worldwide, we find as, China is most populated country in the world having up-to 1.4 billion of population, in which only 90.1 thousand peoples are infected with 0.06% and .003% deaths. While united states of America have the highest infection population up to 18.33% and 0.56% deaths. [1]

As per record 795134 patients were dead worldwide due to Covid-19, in which 46.57% of cumulative death occurred in top five listed county. From all cumulative death, United State of America having the highest percent of death as 21.76% death, 14.12% in Brazil, and 7% in India. On 22th Aug 2020, death newly reported in last 24 hours shows that 60.06% of death covered by top five country, in which Brazil newly reported death for day as 20.31% is the highest death percent from whole world. In USA the newly reported death is 17.96%, 15.94% death in India. The results demonstrate that newly reported death is more than double and highest increment as 8.93% in India from cumulative death percentage. Similarly Brazil has up to 6.19% increment in newly reported death from cumulative death percentage. [1]

3 PUBLICATION STATISTICS

This section of paper describes the contribution of researcher and publisher to published several publication associated to Covid-19. Contributions focuses on several attributes of Covid-19 includes understanding concepts, study, approaches used to identify the disease, effects of disease to human health, treatments, and medicine stages of testing, etc. As not only medical organization working on Covid-19, but also other organization contributed the works using their technologies for prediction, diagnosis, analysis, and treatment approaches. Author studies several available resources for publication and manually compute the publication contributed by publisher for publishing the Covid-19 associated articles.

The above figure illustrates that the top most engineering & science publisher publishing the contribution of Covid-19 work. Among these Elsevier have the top most publication as published near about to five thousand publications related to Covid-19.Covid-19 is the disease, therefore in these above mentioned journals limited work is published, while in medical science associated journals, total published articles on Covid-19 are 106529 and 81 patents as per Dimension data [16]. As per statistics up to 82081 articles in medical associated journal are published for Covid-19 disease, diagnosis, treatment and analysis. In addition to these articles, 21985 articles preprinted, 1216 book chapters, 450 edited books, 424 proceedings, and 373

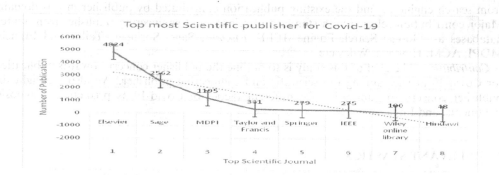

Figure 1.   Top most known Science & Engineering publisher of Covid-19 Work.

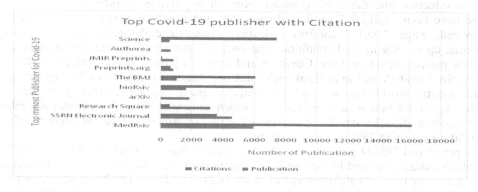

Figure 2.   Covid-19 top publisher with citation [Data source- Dimension [16]).

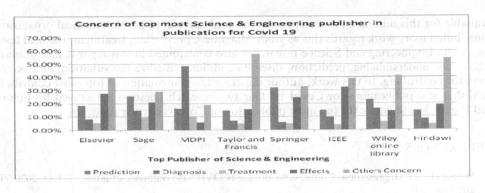

Figure 3. Top most Science & Engineering publisher of Covid-19 concern in published articles.

monograms are published for Covid-19. Most of the researcher published the articles as it's up to 80 percent of total publication.AS more than one lacs paper is published for Covid-19 from all over world, where top five country has contributed up to 28 percent of publication (in which 41.23% by United States of America, 21.2% by China, 17.07% by United Kingdom, 13.47% by Italy, and 7% by India).

As in the worlds several organizations contributed towards Covid-19 publication, human desire to know the top contributed organization worldwide during this panic situation. As per statistics of publication, Harvard University, United States of America is the top most contributors for publication in Covid-19 as six hundred sixteen among all institution of worlds. Huazhong University of Science and Technology, China contributed five hundred ten publications towards Covid-19 and reached to second position from whole world with relevant contributing publication. University of Oxford, UK, Johns Hopkins University, USA, University of Toronto, Canada, are respectively published 499, 480, and 477 papers towards covid-19 and situated at third, fourth and fifth positive.

Several journals are publishing the Covid-19 related work for readers support. In this figure, author illustrates the top publisher of Covid-19 work among all existing publisher with their citation. Figure shows that MedRxiv is the top most publisher of Covid-19 contribution and published 5950 publication with 16137 citations. The Science publisher focus on citation as well as top ten publishers, having up to 12% of Citation, as highest Citation among the top ten publishers of Covid-19.As per Dimension data [16], Viroj Wiwanit kit from Dr. D.Y. Patil Vidyapeeth, Pune, India, is the top most contributed author for Covid-19 work as 109 publications. Kwok-Yung Yuen from University of Hong Kong, China, have the highest citation as 128.01 from top ten contributed author of Covid-19. Abi Rimmer is the top ten contributed publishing author having lowest percentage of citation 1.12% among all top ten authors. Here fifty percent of author having less than 10% of citations, and only one author have more than 100% citation, among top ten authors. Elisabeth Mahase is published ninety five Covid-19 papers with approximately sixty percent of citation.

The above figure illustrates that the top most engineering & science publisher publishing the Springer concern about to thirty two percent of publication for corona viruses is the highest among all top most publisher. MDPI mainly focus on publication of diagnosis of Covid-19 disease and published up to forty eight percent as highest among top publisher for diagnosis concern. Thirty two percent as highest publication for Covid-19 effects by IEEE.

4 CONCLUSIONS

The novel conoravirus 2019 create the pandemic situation in the environments to everyone personally as well as globally. Due to this panic situation researchers of diverse era focusing to possible approaches for contribution towards the Covid-19. Therefore, in a very short period of time, more than one lacks publication are published and most of them are freely available for this duration. As its types of virus and harmful to human, medical organization contributed more work regards this as study, diagnosis, prevention, treatment, medical testing stages etc. Engineering and Science researcher also contributed in this regards and published articles for understanding, prediction, disease condition, visualize the situation, and possible ways for controlling. In this work, author illustrate the contribution of top most contributor, publisher and publications for Covid-19. Our two research question is resolved by Figure 1 and Figure 3 as illustrate the top most publication of Covid-19 articles concern.

REFERENCES

1. World Health Organization, Coronavirus disease (COVID-19) outbreak situation. WHO; 2020 [cited 2020 Apr 24].
2. Chen, Y., Liu, Q. and Guo, D., 2020. Emerging coronaviruses: genome structure, replication, and pathogenesis. *Journal of medical virology*, *92*(4), pp.418–423.
3. Singhal, T., 2020. A review of coronavirus disease-2019 (COVID-19). *The Indian Journal of Pediatrics*, pp.1–6.
4. Chamola, V., Hassija, V., Gupta, V. and Guizani, M., 2020. A Comprehensive Review of the COVID-19 Pandemic and the Role of IoT, Drones, AI, Blockchain, and 5G in Managing its Impact. *IEEE Access*, *8*, pp.90225–90265.
5. Zheng, N., Du, S., Wang, J., Zhang, H., Cui, W., Kang, Z., Yang, T., Lou, B., Chi, Y., Long, H. and Ma, M., 2020. Predicting covid-19 in china using hybrid AI model. *IEEE Transactions on Cybernetics*.
6. Romero-Rodríguez, J.M., Aznar-Díaz, I., Hinojo-Lucena, F.J. and Gómez-García, G., 2020. Mobile Learning in Higher Education: Structural Equation Model for Good Teaching Practices. *IEEE Access*, *8*, pp.91761–91769.
7. Zhou, P., Yang, X.L., Wang, X.G., Hu, B., Zhang, L., Zhang, W., Si, H.R., Zhu, Y., Li, B., Huang, C.L. and Chen, H.D., 2020. A pneumonia outbreak associated with a new coronavirus of probable bat origin. *nature*, *579*(7798), pp.270–273.
8. Yee, J., Unger, L., Zadravecz, F., Cariello, P., Seibert, A., Johnson, M.A. and Fuller, M.J., 2020. Novel coronavirus 2019 (COVID-19): Emergence and implications for emergency care. *Journal of the American College of Emergency Physicians Open*, *1*(2), pp.63–69.
9. Bekeschus, S., Kramer, A., Suffredini, E., Von Woedtke, T. and Colombo, V., 2020. Gas Plasma Technology—An Asset to Healthcare During Viral Pandemics Such as the COVID-19 Crisis?. *IEEE Transactions on Radiation and Plasma Medical Sciences*, *4*(4), pp.391–399.
10. Quayson, M., Bai, C. and Sarkis, J., 2020. Technology for social good foundations: A perspective from the smallholder farmer in sustainable supply chains. *IEEE Transactions on Engineering Management*.
11. Shikuku, K.M., 2019. Information exchange links, knowledge exposure, and adoption of agricultural technologies in northern Uganda. *World Development*, *115*, pp.94–106.
12. Mahalmani, V.M., Mahendru, D., Semwal, A., Kaur, S., Kaur, H., Sarma, P., Prakash, A. and Medhi, B., 2020. COVID-19 pandemic: A review based on current evidence. *Indian Journal of Pharmacology*, *52*(2), p.117.
13. Pascarella, G., Strumia, A., Piliego, C., Bruno, F., Del Buono, R., Costa, F., Scarlata, S. and Agrò, F.E., 2020. COVID-19 diagnosis and management: a comprehensive review. *Journal of Internal Medicine*.
14. Hassan, S.A., Sheikh, F.N., Jamal, S., Ezeh, J.K. and Akhtar, A., 2020. Coronavirus (COVID-19): a review of clinical features, diagnosis, and treatment. *Cureus*, *12*(3).
15. Chauhan, S., 2020. Comprehensive review of coronavirus disease 2019 (COVID-19). *Biomedical Journal*.
16. https://app.dimensions.ai/discover/publication?search_mode=content&search_text=Covid-19&search_type=kws&search_field=full_search

Recent Trends in Communication and Electronics – Sharma et al. (Eds)
© 2021 Taylor & Francis Group, LLC, ISBN 978-1-032-04572-6

Study on enhancing the accuracy of fingerprint recognition system

Abhishek & Ajai kumar Gautam
Delhi Technological University, Delhi, India

ABSTRACT: With the rapid advancements of the technology everyone is socially connected with each other by means of internet. This will also give rise to the problem of theft of each other identities whether there is a swipe card transaction, border crossing, time attendance, driving license, personal computers, personnel digital assistants to access the confidential files. So unique impression of each individual which can't be duplicated is thus required. Therefore in order to avoid such kind of theft, bio-metrics plays a dominant role, which include face recognition, iris recognition and fingerprint recognition among all of these fingerprint provides high level of abstraction to the user security. In this paper we have done comparison of various image enhancements methods and discussed classification of fingerprint recognition with some machine learning techniques like Random Forest, SVM, Naive Bayes, Radial Basis Function network algorithms along with some deep learning algorithms like Convolutional neural network.

Index Terms—Fingerprint Recognition, Random Forest, SVM, Naive bayes, Radial Basis Function (RBF), Convolutional Neural Network(CNN)

1 INTRODUCTION

In the era of growing information and communication technologies, security became the serious concern for all of us that's why demand of security system has also increased. Fingerprint recognition systems act as a bridge to this demand since every individual has their own unique fingerprint, which reduces the chances of unauthorized access. Along with that it offers the least cost, fast and reliable way of providing security. Fingerprint recognition systems has wide variety of applications including law enforcement, access control, IT systems security, border management system airport [1]. Fingerprint consists of ridges and valley and every individual has unique pattern of ridges. Generally pattern of ridges can be classified into 3 types i.e. arch, loop and whorls [2]. The major steps involved in the fingerprint recognition system are acquiring the data it can be either online or offline, followed by pre-processing, feature extraction, and classification and then pattern identification, and matching. The acquisition or enrollment can be done via optical finger reader or through ink impression on the paper the later method is known as offline method and former called online method. The major challenging part of this process comes when the fingerprint image is noisy or corrupted due to poor skin condition or dirt accumulated on the finger. So the fingerprint image is subjected to pre-processing and enhancement step which is then followed by binarization and thinning which would be necessary to clearly identify the ridge structure. Generally in most of the researches done earlier Gabor filter is widely used in the enhancement process whether there it is Texture Segmentation [3], Face recognition [4], Fingerprint enhancement [5], although Fast Fourier Transform (FFT) were also used So the fingerprint image is subjected to pre-processing and enhancement step which is then followed by binarization and thinning which would be necessary to clearly identify the ridge structure. earlier The popularity of Gabor enhancement is due to its characteristics such as the frequency and inclination representation

DOI: 10.1201/9781003193838-34

which is similar with the visual system of humans. The function for Gabor filter in complex form can be described as follows:

$$G(A, B; \mu, A, \Phi, \Gamma) = EXP(-(A^2 + \Gamma^2 B^2))EXP(I(2\Pi A/\mu + \Phi)) \quad (1)$$

Where, $A = A \cos \alpha + B \sin \alpha, B = A \cos \alpha - B \sin \alpha$

The filter has the following characteristics:

- Wavelength (μ): it represents the number of cycles per pixel.
- Inclination (α): Angle of the normal to the parallel strips of Gabor function
- Phase angle (φ): it is the offset of the sinusoid.
- Spatial aspect Ratio (γ): it represents the ellipticity which is generally less than 1.

This paper is described in different sections. Section 2 represents the Literature review and comparison of various enhancements methods and discuss the accuracy of different machine learning and deep learning techniques applied. Finally in section 3 we have made the observations and conclusions.

2 LITERATURE REVIEW

Fingerprint recognition process has gone through series of steps fingerprint enrollment, preprocessing, feature extraction, then fed to the classifier. Preprocessing is required to deal with noisy image so to avoid any error in the results and it can be further divided into five types normalization, enhancement, binarization, filtering and thinning [6].

2.1 Normalization

The main purpose of normalization is to compress the pixel values of the fingerprint image so that it comes under the desired range which makes the computations efficient Normalization function can be written as

$$N(x, y) = M_0 + \sqrt{(V0) * (\ln(x, y) - M)}^{2/\sqrt{V}} \quad (2)$$

If ln(x, y)>M

$$N(x, y) = M_0 - \sqrt{(V0) * (\ln(x, y) - M)}^{2/\sqrt{V}} \quad (3)$$

If ln(x, y)<M
N(x, y): Normalized image function
M_0: Desired mean
M: Mean of the image pixels
V_0: Desired variance
V: Variance of input image pixels
In(x, y): Intensity function of the pixels at a Particular point x, y.

2.2 Enhancement

In order to extract the feature from the image there could be the problem of scars, blurs and incipient ridges so to minimize such undesirable effects and to ex-tract the certain quantities of good features it can be ridge bifurcation, ridge ending or can be singularities likes cores and delta image enhancement methods are required. There have been many enhancement methods discussed. Each has their own merits and demeritsso based upon the comparison suitable technique can be chosen properly.

2.3 Binarization

In this process a grey level image is transformed into black and white image i.e during the process of binarization if the pixel value of fingerprint image is more than threshold it is treated as 1(white) otherwise 0(black) so after this images has having the ridges and valleys. Ridges corresponds to black and valley as white.

2.4 Thinning

It is a morphological operation and its primary objective is to reduce the pixels so that it becomes only one pixel by eliminating the redundant pixels present in the image.

Table 1. Comparison of enhancement methods.

Method	Demerits	Merits
Histogram equalization	This method sometimes enhances the background noise in the image	This method straight works on the finger print image pixels
Bandpass filtering	This technique fails if the image has huge noise	To some extent noise removal to retain the true structure and ridges of fingerprint
Gabor filtering	This method will not produce good results if the image contains noises	This method produce good results when the parameters of anisotropic and low pass filter are combined
Binarization and thinning	This method sometimes results in deviated lines for empty medial lines of fingerprint	fingerprint parameters and ridges connections is maintained in this method
2D fourier Transform	This method drops the accuracy as it is countinuously accepting the frequency	This is fast and categorizes the location to 16 directions
Wavelet based Transformation	Since its wavelets are divisible therefore not suited for diagonal parameters	Fast and efficient in denoising fingerprint images
Wave atom Transform		This method hasn't any disadvantages until now as with the other methods

Figure 1. Fingerprint.

Figure 2. Enhanced Fingerprint.

Figure 3.   Binarized Fingerprint.

Figure 4.   Thinned Fingerprint.

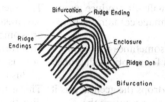

Figure 5.   Fingerprint minutiae.

2.5 *Feature Extraction*

Feature extraction in the fingerprint recognition system can be done via two methods pattern based approach and minutiae based approach in pattern based [7]. Generally in pattern based methods fingerprint pattern (arches, loops and whorls) are used to classify the candidate fingerprint and stored template where as in minutiae based approach minutiae(fine details) are used for the identification purpose. Although there are 150 various different local characteristics which can be singular points (core and delta), ridges, valleys etc. but most prominently ridge endings and ridge bifurcations are used because of their stability and robustness. So in order to identify these features cross number(CN) method is used in which 3*3 mask is moved on the pixels of the image and if the central pixel in the mask is 1 and if their at least 3 neighborhood pixels are exactly 1 then it will be considered as ridge bifurcation if the central pixel in the mask is 1 and if their is only one neighborhood pixels is exactly 1 then it will be considered as ridge termination.

2.6 *Classification/identification*

The main aim of this step is to classify the fingerprints of the different owners and training the algorithm so that when learning or training is completed the algorithm will enable identification. In the paper discussed [9] by A. Ali they evaluated the accuracy of fingerprint database fvc2002 using different machine learning algorithms such as multilayer perceptron, Random forest, Naive Bayesian and RBF network and found the maximum accuracy of 58.75 percent is achieved in case of random forest In the approach discussed [6] where ridge bifurcation is used as feature extraction for the Artificial neural network classifier using 3 different activation functions hyperbolic, elliot and sigmoid and it is found that the maximum accuracy of 81.25 percent is achieved using sigmoid activation function. Deep learning classifier CNN is used in the text described [10] by B. Pandya preprocessing method such as Gabor enhancement and fingerprint thinning are also used and achieved the accuracy of 98.21%. An effective preprocessing helps in reducing the training time without compromising the classification accuracy [11] by using this approach J.M Shrein achieved. 95.9 percent accuracy on the NIST-DB4 database. In the recent times finger vein recognition also becoming popular in the approach discussed [12] by W. Liu uses CNN with 5 convolutional layers and 2 fully connected layers and achieves the maximum accuracy of 99.53% with minimum equal error rate (ERR) of 0.8%.

3 OBSERVATIONS AND CONCLUSIONS

Observations are made after going through the papers that fingerprint recognition using deep learning algorithm provide much better classification then the machine learning algo- rithms with effective preprocessing technique combined with deep learning can reduce the training time without hampering the accuracy. Although fingerprint recognition accuracy can be affected due to several factors such as environmental conditions, age, illness, personal cause etc. as discussed [13]. The finger vein recognition on the contrary does not encounter any such problem and could achieve a very high accuracy of more than 99% [12] Since fingerprint using deep learning can enhance the ac- curacy but has some anomalies as discussed so finger vein technology on the other hand prevent such limitations and also enhance the accuracy to a much great extent.

REFERENCES

1 Babatunde, I. G., Kayode, A. B., Charles, A. O., Olatubosun, O. (2012). Fingerprint image enhancement: Segmentation to thinning.

2 Borra, S. R., Reddy, G. J., Reddy, E. S. (2016, March). A broad survey on fingerprint recognition systems. In 2016 International Conference on Wireless Communications, Signal Processing and Networking (WiSP- NET) (pp. 1428–1434). IEEE.

3 Weldon, T. P., Higgins, W. E., Dunn, D. F. (1996). Efficient Gabor filter design for texture segmentation. Pattern recognition, 29(12), 2005–2016.

4 Lampinen, J., Oja, E. (1995). Distortion tolerant pattern recognition based on self-organizing feature extraction. IEEE Transactions on Neural Networks, 6(3), 539–547.

5 Hong, L., Wan, Y., Jain, A. (1998). Fingerprint image enhancement: algorithm and performance evaluation. IEEE transactions on pattern analysis and machine intelligence, 20(8), 777–789.

6 Oulhiq, R., Ibntahir, S., Sebgui, M., Guennoun, Z. (2015, October). A fingerprint recognition framework using artificial neural network. In 2015 10th international conference on intelligent systems: theories and applications (SITA) (pp. 1–6). IEEE.

7 Ali, M. M., Mahale, V. H., Yannawar, P., Gaikwad, A. T. (2016, Febru- ary). Fingerprint recognition for person identification and verification based on minutiae matching. In 2016 IEEE 6th International Conference on Advanced Computing (IACC) (pp. 332–339). IEEE.

8 Zhou, J., Chen, F., Gu, J. (2008). A novel algorithm for detecting singular points from fingerprint images. IEEE transactions on pattern analysis and machine intelligence, 31(7), 1239–1250.

9 Ali, A., Khan, R., Ullah, I., Khan, A. D., Munir, A. (2015, October). Minutiae based automatic fingerprint recognition: Machine learning approaches. In 2015 IEEE International Conference on Computer and Information Technology; Ubiquitous Computing and Communications; Dependable, Autonomic and Secure Computing; Pervasive Intelligence and Computing (pp. 1148–1153). IEEE.

10 Pandya, B., Cosma, G., Alani, A. A., Taherkhani, A., Bharadi, V., McGinnity, T. M. (2018, May). Fingerprint classification using a deep convolutional neural network. In 2018 4th International Conference on Information Management (ICIM) (pp. 86–91). IEEE.

11 Shrein, J. M. (2017). Fingerprint classification using convolutional neural networks and ridge orientation images. In 2017 IEEE Symposium Series on Computational Intelligence (SSCI) (pp. 1–8). IEEE.

12 Liu, W., Li, W., Sun, L., Zhang, L., Chen, P. (2017, June). Finger vein recognition based on deep learning. In 2017 12th IEEE Conference on Industrial Electronics and Applications (ICIEA) (pp. 205–210). IEEE.

13 Rathod, V. J., Iyer, N. C.,Meena, S. M. (2015, October). A survey on fingerprint biometric recognition system. In 2015 International Con- ference on Green Computing and Internet of Things (ICG- CIoT) (pp. 323–326). IEEE.

Recent Trends in Communication and Electronics – Sharma et al. (Eds)
© 2021 Taylor & Francis Group, LLC, ISBN 978-1-032-04572-6

Image fusion based on multiscale transform

Shubham Shweta
Delhi Technological University, New Delhi, India

ABSTRACT: Image fusion is a method to improve the quality of two or more than two fused images. There are so many techniques used to perform this method such as DWT, Multiscale transform. In this paper we use an image fusion method in which a fusion process is based on MST using DWT and compare its result with simple using DWT and Lifting Wavelet Transform. In this process we first decompose each input source image into its sub-bands using the MST -DWT method which further divide each subband into low-pass and for high pass band we use maximum absolute (max-abs) rule. In this method we calculate the inverse MST at last to get the fused images back. There are so many methods which we use in MST LWT lifting wavelet transform, MST -DWT, DTCWT dual-tree complex wavelet transform and NSCT non-sub-sampled contourlet are used. We also perform this technique with different method like simply discrete wavelet transform and calculate difference for real-life images. By using the proposed method, we observe that it improves the quality of image like the contrast, clarity and visual information of the fused results as well as the energy of the images.

1 INTRODUCTION

Generally, in the MST based fusion techniques, firstly we decompose the given source images into their sub-bands respectively. After that the fusion method we use from the above mentioned fusion rules we merge these sub bands. At last we apply the inverse MST to reconstruct the fused image on the sub bands which is merged earlier. As we mentioned that each sub-band has some type of salient features of the images. Therefore, we divide the image into the sub bands having the low pass and high-pass bands which are used as the approximation and details of the given source image. In this process generally, we fuse the low-pass bands with the help of averaging rule although the high-pass band is fused by using max abs or maximum absolute values. Thus, on the basis of the max-abs rule, we calculate the activity level by using the absolute value. While with high-absolute values, we measure the sharp edges and fine details coefficients. As we know that from the low pass theory the Low-pass band consists of significant image energy. Therefore, we could not apply the averaging rule as by doing this we can lose some significant energy from the given source images. The problem arises in multi-modal fusion where the source images are taken using different camera. Here the sensor captures the different features from the same imagery. Therefore, the source image may contain the different levels of aspects such as brightness due to this color of the image may differ. For example, like in multimodal images in which if we get the high energy differences then we select the max-$L2$-norm rule while their energy difference is not high then max-$L1$-norm is chosen. Usually we apply the max-abs rule on the high pass bands of sub bands. As we know that there are few fields in which we can apply this MST based fusion for real-life image are as follows such as medical image like CT and MRI of images, multimodal of IR and visual image for military purpose. Here, we divide the activity levels into three process as absolute value, $L1$-norm or $L2$-norm depend on its specific criteria. We described our process of fusion using MST transform in general by following diagram.

DOI: 10.1201/9781003193838-35

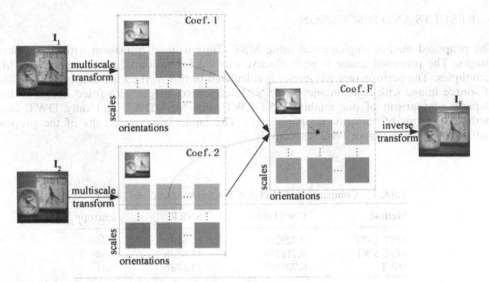

Figure 1.   MST for image 1 and 2 into fused image.

2   PROPOSED METHOD

This method is divided into four steps which we described below; for better understanding of the process we consider only two images. The four steps are mentioned as follows:

1) In the first step we decompose the both source image I1 and I2 independently with the help of MS method in which we divide each image into sub bands of low-pass L1 and L2 respectively and the high-pass into H1 and H2 .
2) As we divide each low-pass band of L1and L2 into image patches of size $n \times n$ with the help of the running of the sliding window method. Thus, we obtain the image patches having image size of M × N.
3) Since the high-pass bands of the source image consists of information of image details like edges etc. As we know that the coefficients with more information always have higher absolute value. Therefore we apply the max-abs or maximum absolute value for the fusion rule for high pass bands.
4) Finally, we apply the inverse MST on the low-pass and high-pass bands which are calculated from the earlier process of step 2 and 3 for final fused image.

2.1   Advantages of proposed method over individual MST

As far we discussed earlier, we understand the MST based fusion technique has many drawbacks like the averaging of low pass band due to which it lost maximum amount of the image energy. So, this problem get solved by using MST with our proposed method. Further for its clarity we have taken an example of multi-modal images which include the visible and MMW images pair are represented. In this process the source images represented by the multi-modal fusion are captured from different sensors in which different aspects of physical features of the source image is represented from the same scene. Since the visible image will represent edges and textures of the source image although the minute detail of the metallic object can be also find out in the smallest type of millimeter image as a brighter part. From this we conclude that the larger part of grey-levels represents the higher energy level which also represent the brighter areas. As we know that, the averaging rule will not be suitable for high differences values or levels.

185

3 RESULTS AND DISCUSSION

The proposed method implemented using MST lifting wavelet transform with two different images. The proposed image is very effective method for fusion of two images using MST techniques. The performance parameter is calculated in this method are Entropy, correlation of source image with fused images and SNRR of source images with fused image. Table 1 Depicts comparison of our method MST-LWT with MST-DWT and only DWT fusion method in terms of performance parameters. The implementation results of the proposed method is shown in Figure 2 & 3 respectively.

Table 1. Comparison of MST-LWT with DWT and MST-SWT.

Method	Correlation	SNRR	Entropy
MST-LWT	0.729219	17.68db	7.36
MST-SWT	0.719219	17.68db	7.34
DWT	0.709495	17.58db	7.31

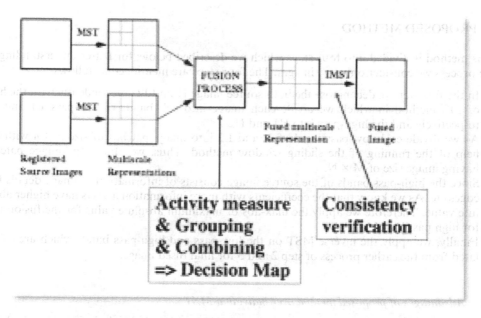

Figure 2. Block diagram of MST transform method.

a.   b.   c

Figure 3. a. Image 1 before fusion b. Image 2 before fusion c. Fused image of image 1 & 2.

4 CONCLUSION

In this process for calculating the activity level, we divide it into two steps firstly we calculated max-$L1$- norm and max-$L2$-norm. Further for the measurement of the desirable fusion rule, then we took into account that the source images energy. In this calculation if we conclude the large energy gap as we represented in the case of multi-modal images captured from multiple sensors, if there was not so large energy gap then we took 'max-$L2$-norm' into consideration vice versa. In this process we concluded that this rule arises a large amount of energy transfer due to improve the contrast for further clarification of the fused image ultimately. We also compare our work with simple discrete wavelet transform or DWT and lifting wavelet transform LWT and we also compare our proposed method with MST-DWT image fusion method. Thus we concluded that the method we used in this Image fusion enhanced the information details in the context of the fused image as well as the image energy.

REFERENCES

Lee, H., Yeom, S., Guschin, V.P., et al., 2009. Image fusion of visual and millimeter wave images for concealed object detection. Proc. IEEE Int. Conf. on Infrared and Millimeter Wave, Busan, September 2009, pp. 1–2.

Li, X., He, M., Roux, M., 2010. Multifocus image fusion based on redundant wavelet transform. IET Image Process., 2010, 4, (4), pp. 283–293.

Jiang, Y., Wang, M., 2014. Image fusion using multiscale edge-preserving decomposition based on weighted least squares filter. IET Image Process., 2014, (3), pp. 183–190.

Sahu, D. K., Parsai, M. P., 2012. Different image fusion techniques. Int. J. Mod. Eng. Res., 2012, 2, (5), pp. 4298–4301.

Hossny, M., Nahavandi, S., Creighton, D., et al., 2010. Towards autonomous image fusion. 11th IEEE Int. Conf. on Control Automation Robotics and Vision (ICARCV), 2010, pp. 1748–1754

Wald, L., 1999. Some terms of reference in data fusion. IEEE Trans. Geosci. Remote Sens., 1999, 37, (3), pp. 1190–1193.

Liu, Y., Wang, Z., 2014. Simultaneous image fusion and denoising with adaptive sparse representation. IET Image Process., 2014, 9, (5), pp. 347–357.

Li, H., Manjunath, B., Mitra, S. 1995. Multi-sensor image fusion using the wavelet transform. Graph. Models Image Process., 1995, 57, (3), pp. 235–245.

Modified DiEfficientNet for road extraction from aerial imagery

P. Das & S. Chand
Jawaharlal Nehru University, New Delhi, India

ABSTRACT: Extracting road networks from high-resolution aerial imagery significantly impacts many socio-economical applications such as urban planning, updating navigation system, rapid disaster response, etc. Researchers are still struggling to find an end to end method to resolve issues like occlusion, class imbalance, and high inter-class similarity. In this study, we present an end-to-end encoder-decoder architecture, modified DiEfficientNet. This model combines sophisticated spatial features along with semantic features from all the levels. We combine layers of pre-trained EfficientNet with dilated convolutions for extracting fine-grained multiscale features. EfficientNet scales up depth, width, and resolution of an input image simultaneously with less number of parameters and minimum computational cost. We introduce a customized loss function, an integrated loss that handles class imbalance, and boundary loss problem. All the experiments are conducted on publicly available Massachusetts road data, and the results with 0.89 mIoU demonstrate the efficiency of the proposed model.

1 INTRODUCTION

For the past few years, remote sensing images are being widely used by commercial organizations for a large variety of applications like urban planning & management, automated driving, updating navigation systems, emergency rescue operations in a disaster like emergencies, etc. Due to rapid growth in the GIS industry, a substantial amount of remote sensing imagery is available nowadays. Yet few challenges like different weather conditions, terrain features, and high inter-class similarity still exist. Traditional machine learning methods face difficulty in handling these critical issues. Also, there are some issues in datasets, e.g., imbalanced class distribution, occlusion, shadows, and wrong ground truth annotations. Dealing with all the issues with machine learning methods is labour-intensive as well as error-prone. Therefore researchers were willing to explore efficient ways of automatic road extraction methods. In recent times, most computer vision tasks like classification, detection, localization, and segmentation are majorly solved by convolutional neural networks (CNNs) – most actively used architectures of deep learning method (K. He et al., 2016), (Krizhevsky et al., 2012), (Sermanet et al., 2014) (Simonyan & Zisserman, 2015), (Zeiler & Fergus, 2014). Road map detection and extraction from remote sensing is similar to a semantic segmentation task where each pixel is labeled with its respective class enclosing the object or region. For the segmentation task, CNN is considered one of the most suitable choices among the computer vision community in recent years. Some CNN-based segmentation models are FCN (Long et al., 2015), U-Net (Ronneberger et al., 2015), DeepLab v3 (Chen et al., 2017), Segnet (Badrinarayanan et al., 2017), LinkNet (Chaurasia & Culurciello, 2017). Satellite imagery often carries too much information in each image; it becomes computationally expensive to process them due to the presence of rich and sophisticated features. However, aerial imagery is readily available due to the growing demand in the market for real-time applications. Many GIS-related organizations prefer aerial imagery, which is time and labour efficient and requires minimum hardware support. A CNN based model,

DOI: 10.1201/9781003193838-36

modified dilated EfficientNet, is proposed, which is also being actively used for extracting the road maps from aerial imagery and evaluate its performance using the Massachusetts road dataset that consists of diversified aerial imagery.

2 RELATED WORK

A wide range of diverse road extraction approaches has been discussed in recent years; however, most of them fail to address the main challenges entirely because they used traditional machine learning methods. Here, we review some important road extraction methods.

In (Wei et al., 2017), a road structure refined CNN model along with fusion layers is discussed that considers both spatial correlations as well as geometric information of road structure. In (Zhang et al., 2018), a deep residual U-Net model having residual units and skip connections is discussed that eased the model training and enhanced information propagation respectively at the same time. In (Gao et al., 2018), a multiple feature pyramid network model with a weighted balance loss function is discussed. Two new modules were added, feature pyramid and tailored pyramid pooling module, to utilize the multi-level semantic features. In (Eerapu et al., 2019), a dense refinement residual network (DRR Net) along with multiple DRR modules is discussed. A hybrid loss function is introduced, integrating two very effective loss functions: BCE (binary cross-entropy) and Lovasz. A U-Net-based model, along with ASPP (atrous spatial pyramid pooling), is discussed in (Wulamu et al., 2019) that utilizes multiscale features. A hybrid loss function integrating dice coefficient, BCE, and Lovasz hinge loss is used. In (Yuan et al., 2019), a U-net like model named WRAU-Net (wide-range attention unit) is discussed that employs an attention approach channel-wise by including a partially dense connection branch. In (H. He et al., 2019), ASPP (atrous spatial pyramid pooling) combined with the U-Net model having customized loss aggregating both structural similarity (SSIM) and BCE is discussed. In (Xin et al., 2019), a DenseUnet model having dense units in and U-Net features in the basic building blocks with lesser parameters is discussed. In (Ding & Bruzzone, 2020), a direction-aware residual network (DiResNet) with two new sub-networks added, segmentation network DiResSeg, aims at structural completeness a refinement network DiResRef that optimizes the road extraction results is discussed.

3 PROPOSED WORK

Extracting the features of only one object, e.g., roads, is far simpler than extracting multiple objects' features. So in this paper, we adopt a lightweight architecture, modified dilated EfficientNet (modified DiEfficientNet) with less number of parameters and computational complexity. Training a model from scratch may take significant time and effort. Therefore we use layers of the EfficientNet (Tan & Le, 2019) model as our encoder for faster convergence of the model. This model outperforms standard CNN baseline models such as ResNet (K. He et al., 2016) and MobileNet (Sandler et al., 2018). It suits perfectly for high-resolution aerial imagery as it jointly scales up the width, depth, and resolution of an image. The key components of EfficientNet are MBConv (Mobile inverted Bottleneck convolution) (Tan & Le, 2019), which includes depth-wise convolutions (DW Conv) and SE-blocks (Squeeze and Excitation block). The DW convolution optimizes the proposed model by reducing the computations significantly in which each channel is given attention separately. The SE-blocks extract only influential features and discard nonessential features. These blocks reduce the computational burden of the model without affecting the performance. However, the deeper networks are also more challenging to train due to the vanishing gradient problem. On the other hand, wider networks help capture fine-grained features for which they can easily be trained. Thus, EfficientNet simultaneously scales up all three dimensions: depth, width, and resolution of the input image by producing better results faster. Then we modify the proposed network by adding dilation between each encoder-decoder layer connecting them. The dilation rates used are 1, 2, 4, and 8 that captures the multiscale features from different layers. We follow the decoder structure of the LinkNet model (Chaurasia & Culurciello, 2017) with transposed convolutions as the main operation for upsampling features. The intuition behind using a deeper convolution network is that they capture more complex and sophisticated features. The proposed model maintains a perfect balance between the prediction of road maps and inference speed by successfully combining spatial information of local and global features.

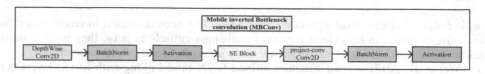

Figure 1. Components of MBConv blocks.

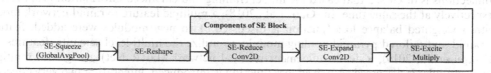

Figure 2. Components of squeeze and excitation block.

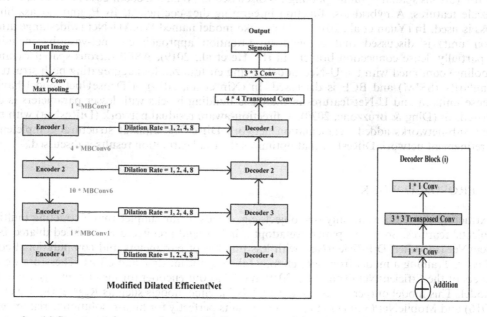

Figure 3. (a) Structure of proposed model, (b) Components of decoder.

4 EXPERIMENTAL ANALYSIS

4.1 *Dataset*

To evaluate all the experiments conducted in this paper, we use the publicly available Massachusetts road dataset (Mnih, 2013) covering nearly 2600km^2. Two classes are present: road and non-road. The dataset consists of 1108, 14 and 49 imagery for training, validating and testing respectively. The 3-band RGB imagery have size 1500 x 1500 pixels with a ground resolution of 1.2m^2.

4.2 Evaluation metric

The datasets having class imbalance problems may need a better loss function than cross-entropy as it gives equal weightage to classes having both less and large samples. This reduces the model's generalization capability, which often shows biasness towards class, having large samples. The Massachusetts road dataset is highly imbalanced, with nearly 96%, 4% of non-road class and road class, respectively. So the model shows biasness towards the background class during training. So to deal with this challenge, we present an integrated loss function to evaluate all the experiments. The Dice coefficient (DC) loss is being integrated with Binary cross-entropy (BCE) loss. The dice loss handles the class imbalance problem well because it gives equal importance to the features of the classes with minimum spatial representation. It usually penalizes the predictions with low confidence with a normalization effect. The dice loss always considers the context of loss both locally as well as globally.

$$\text{Dice coefficient} = \frac{2 \times |A.B|}{|A| + |B|} \quad or \quad \frac{2 \times TP}{(TP + FP) + (TP + FN)} \quad (1)$$

Where $|A.B|$ denotes the common area between ground truth mask and predicted map, and $|A| + |B|$ represents the total area in both ground truth mask and predicted map separately. TP (true positives), FP (false positives), FN (false negatives).

$$L_{integrated} = \alpha \times (L_{BCE}) + (1 - \alpha) \times (L_{DC}) \quad (2)$$

Here α is a weighted parameter that supervises the values of loss functions. We also evaluate precision and recall metrics for a better understating of the results.

4.3 Implementation details

Tensorflow framework with Adam optimization algorithm is used to implement our proposed method. Default values of exponential decay rates are used that is set ($\beta1 = 0.9$ and $\beta2 = 0.999$).

We conduct all the experiments on NVIDIA GeForce 940MX GPU, having on-board memory of 4 GB. We use the batch of size 4 and get the optimal results at around 50 epochs. The value of learning rate for initial layers is fixed at 0.000. For making model training easier, we resize the image size to 1024 x 1024 pixels. We rotate the images (90,180, 270 degrees) and flip them both horizontally and vertically. Data augmentation enhances the diversity of training data and regulates the overfitting problem of our model.

4.4 Quantitative and visual comparison of results

Here, we analyze and study the outcomes of the modified DiEfficientNet model with other recent methods. The models like DRR Net, WRAU-Net, and DiResNet fail to properly extract narrow roads partially occluded by the shadows of buildings or trees, as shown in the 2nd row of Figure 4. The proposed model successfully extracts the roads. The WRAU-Net model doesn't perform well in the dense urban scenario as they have to process a lot of features accumulated via the concatenation of feature maps. Other methods have ignored the ambiguous boundary lines in multi-lane road maps, which is very crucial information for complex urban roads. The newly introduced integrated loss function alleviates the inaccurate predictions and mostly retains the connectivity of road links. It can be seen in the 1st row of Figure 4 that the proposed model precisely separates the boundary lines between multi-lane roads. The model performs significantly and produces very better results by extracting different types of roads both in urban and rural areas while preserving the road connectivity. So, our model has resolved the issue of narrowness, diversity, and sparsity faced by most of the road extraction methods.

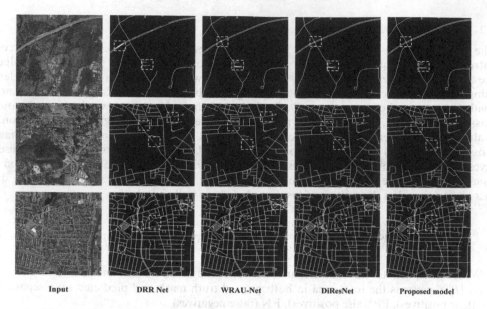

| Input | DRR Net | WRAU-Net | DiResNet | Proposed model |

Figure 4. Visual results of DRR Net, WRAU-Net, DiResNet, and proposed model Modified DiEfficientNet. Zoom in for better visualization.

Table 1. Performance comparison of different models using OA (overall accuracy), Precision, Recall, and mIoU evaluation metrics.

Model	OA	Precision	Recall	mIoU
DRR Net	0.7794	0.9805	0.9792	0.8330
WRAU-Net	0.7563	0.7450	0.8290	0.8050
DiResNet	0.9813	0.8038	0.7941	0.8582
Proposed Model	0.8765	0.9368	0.9455	0.8945

However, it fails to retain the width size of the roads. From Table 1, we observe the performance of the states of the art methods as well as our proposed model. Our model correctly differentiates between roads and roads like other classes because the values of recall and precision are quite reasonable compared to other methods. The mIoU and overall accuracy values of the modified DiEfficientNet demonstrate how significantly it surpasses other recent methods in terms of completeness and coherence.

5 CONCLUSION

In this study, we have discussed an improved encoder-decoder segmentation network, modified dilated EfficientNet, having less number of parameters, and reduced inference speed. A newSly introduced loss function, integrated loss, handle the imbalanced class distribution problem, and produces excellent results compared to other methods. The proposed model integrates the localization information from initial layers with semantic features from intermediate layers to address the challenges faced by the road extraction task.

REFERENCES

Badrinarayanan, V., Kendall, A., & Cipolla, R. (2017). Segnet: A deep convolutional encoder-decoder architecture for image segmentation. *IEEE Transactions on Pattern Analysis and Machine Intelligence*, *39*(12), 2481–2495.

Chaurasia, A., & Culurciello, E. (2017). Linknet: Exploiting encoder representations for efficient semantic segmentation. *2017 IEEE Visual Communications and Image Processing (VCIP)*, 1–4.

Chen, L.-C., Papandreou, G., Kokkinos, I., Murphy, K., & Yuille, A. L. (2017). Deeplab: Semantic image segmentation with deep convolutional nets, atrous convolution, and fully connected crfs. *IEEE Transactions on Pattern Analysis and Machine Intelligence*, *40*(4), 834–848.

Ding, L., & Bruzzone, L. (2020). Direction-aware Residual Network for Road Extraction in VHR Remote Sensing Images. *ArXiv Preprint ArXiv:2005.07232.*

Eerapu, K. K., Ashwath, B., Lal, S., Dell'Acqua, F., & Dhan, A. N. (2019). Dense refinement residual network for road extraction from aerial imagery data. *IEEE Access*, *7*, 151764–151782.

Gao, X., Sun, X., Zhang, Y., Yan, M., Xu, G., Sun, H., Jiao, J., & Fu, K. (2018). An end-to-end neural network for road extraction from remote sensing imagery by multiple feature pyramid network. *IEEE Access*, *6*, 39401–39414.

He, H., Yang, D., Wang, S., Wang, S., & Li, Y. (2019). Road extraction by using atrous spatial pyramid pooling integrated encoder-decoder network and structural similarity loss. *Remote Sensing*, *11*(9), 1015.

He, K., Zhang, X., Ren, S., & Sun, J. (2016). Deep residual learning for image recognition. *Proceedings of the IEEE Conference on Computer Vision and Pattern Recognition*, 770–778.

Krizhevsky, A., Sutskever, I., & Hinton, G. E. (2012). Imagenet classification with deep convolutional neural networks. *Advances in Neural Information Processing Systems*, 1097–1105.

Long, J., Shelhamer, E., & Darrell, T. (2015). Fully convolutional networks for semantic segmentation. *Proceedings of the IEEE Conference on Computer Vision and Pattern Recognition*, 3431–3440.

Mnih, V. (2013). *Machine learning for aerial image labeling.* Citeseer.

Ronneberger, O., Fischer, P., & Brox, T. (2015). U-net: Convolutional networks for biomedical image segmentation. *International Conference on Medical Image Computing and Computer-Assisted Intervention*, 234–241.

Sandler, M., Howard, A., Zhu, M., Zhmoginov, A., & Chen, L.-C. (2018). Mobilenetv2: Inverted residuals and linear bottlenecks. *Proceedings of the IEEE Conference on Computer Vision and Pattern Recognition*, 4510–4520.

Sermanet, P., Eigen, D., Zhang, X., Mathieu, M., Fergus, R., & LeCun, Y. (2014). OverFeat: Integrated Recognition, Localization and Detection using Convolutional Networks. *ArXiv:1312.6229 [Cs].* http://arxiv.org/abs/1312.6229

Simonyan, K., & Zisserman, A. (2015). Very Deep Convolutional Networks for Large-Scale Image Recognition. *ArXiv:1409.1556 [Cs].* http://arxiv.org/abs/1409.1556

Tan, M., & Le, Q. V. (2019). Efficientnet: Rethinking model scaling for convolutional neural networks. *ArXiv Preprint ArXiv:1905.11946.*

Wei, Y., Wang, Z., & Xu, M. (2017). Road structure refined CNN for road extraction in aerial image. *IEEE Geoscience and Remote Sensing Letters*, *14*(5), 709–713.

Wulamu, A., Shi, Z., Zhang, D., & He, Z. (2019). Multiscale road extraction in remote sensing images. *Computational Intelligence and Neuroscience*, 2019.

Xin, J., Zhang, X., Zhang, Z., & Fang, W. (2019). Road Extraction of High-Resolution Remote Sensing Images Derived from DenseUNet. *Remote Sensing*, *11*(21), 2499.

Yuan, M., Liu, Z., & Wang, F. (2019). Using the wide-range attention U-Net for road segmentation. *Remote Sensing Letters*, *10*(5), 506–515.

Zeiler, M. D., & Fergus, R. (2014). Visualizing and understanding convolutional networks. *European Conference on Computer Vision*, 818–833.

Zhang, Z., Liu, Q., & Wang, Y. (2018). Road extraction by deep residual u-net. *IEEE Geoscience and Remote Sensing Letters*, *15*(5), 749–753.

Recent Trends in Communication and Electronics – Sharma et al. (Eds)
© 2021 Taylor & Francis Group, LLC, ISBN 978-1-032-04572-6

BASIC: Blockchain application and smart intelligent contract

Kanika Gupta
ABES Engineering College, Ghaziabad, India

Rahul Johari
SWINGER (Security, Wireless IoT Network Group of Engineering and Research) lab, USICT, GGSIP University, Sector-16C, Dwarka, Delhi, India

ABSTRACT: As well known, a blockchain is said to be a database of distributed ledger or records in which all transactions or series of events that have been committed are distributed among several entities. The series of transactions in the public ledger is checked by consensus of a several participants in the whole system. The events can never be tampered once it is entered. The blockchain technology creates a system by establishing a consensus in the online world of digital era. In the current research work, emphasis has been laid on the concept of Smart Contract in BlockChain Technology, for which a BASIC: BlockChain Application and Smart Intelligent Contract based Algorithm has been proposed.

Keywords: Blockchain, Bitcoin, Smart Contract

1 INTRODUCTION

Around 2008, some unknown group(s)/person(s) were there behind the evolution of Bitcoin, enumerated how the technology of Blockchain, a linked peer-to-peer distributed technology could be utilized to eliminate the problem of establishing transaction orders and to prevent the problem of double spending.

Developers and researchers exploit the capabilities of the BlockChain technology and explore various applications across a vast array of areas. Based on the target audience, three generations of blockchain have now surfaced: Blockchain v1.0 that includes digital cryptocurrency transactions applications, Blockchain v2.0 that includes the combination of applications beyond transactions of cryptocurrency; and Blockchain v3.0 that includes the applications which are beyond the scope of above two mentioned categories such as health, IoT, Government and Science. It is a set of interconnected nodes or blocks which provide some implementation-specific features to the infrastructure. At the bottom level of the infrastructure, various transactions which are signed exists between different peers. These types of transactions indicate the contract agreement between two participants, who are involved the completion of task, transfer of digital or physical assets etc. Atleast one participant can sign the valid series of events and it can be distributed to its neighbors. Specifically, any party which is associated with blockchain is known as node. Therefore, all the points that checks various rules and specifications of blockchain are known as full nodes. Various types of nodes packs various events into several blocks and these are held responsible for determining whether the series of events are committed, and to be put in that particular set of blockchain or not.

A series of events is said to be valid, for example: Darth accepted one Bitcoin from Bob. However, Bob tried to put together the same currency, as this is known as asset which is digi-

DOI: 10.1201/9781003193838-37

tal in nature, to Carol. Furthermore, all the blocks of the particular blockchain must confirm to the state that which events are to be kept in the blockchain to assure that there will be no divergences and corrupted layers. Different consensus mechanisms exist depending on the type of blockchain. The most popular mechanism of consensus is the Proof-of-work (PoW). It needs the formulation of complex process of computation, like different pattern hashes to ensure verifiability and authentication. Proof-of-stake (PoS) protocols divide the blocks which are proportionally divided. In this manner, the choice is transparent and it avoids any of the party from suppressing the particular network. Various Blockchains like Ethereum, have shifted to the framework of PoS due to reduction in consumption of power and its scalability is improved.

2 RELATED WORK

To give a scientific, open and useful literature review of several applications based blockchain, the systematic process has been adopted. It has been seen through some study that approximately 260 papers have been published between April 2014 and April 2018. The aim of the research papers analysis is manifold like:- (1) It provides us with the very interesting insights of research trends in the blockchain technology (2) It helps to visualize various multidisciplinary research approaches adopted by blockchain technology (3) It helps to identify the correct classification structure of blockchain technology. There are two key criteria for descriptive analysis: (a) distribution of publications over theme area and time (b) distribution on the basis of publication over the period of time.

As well known, the blockchain was introduced through Bitcoin [1] as its core technology, and it took various years by the researchers to be fully known about the block chain's potential and to take advantage of other features of blockchain as well.

There are several domains which have been identified various applications which are blockchain based from the analysis. Large portion is of business-oriented applications of all the available applications followed by Health, IoT, Governance and data management. In the last couple of years, health-care oriented applications received much attraction from all communities of world. Blockchain applications can be classified into non-financial and financial ones since various cryptocurrencies represent a portion of the blockchain which is existing in enabled networks. In other way round, we can also classify according to the different versions of blockchain. A detailed categorization of the various available applications which are blockchain-enabled based on the survey of various research papers is presented as follows: -

i) Financial Applications

Presently, emerging technology of blockchain is enabled to wide array of sectors in fields of finance which includes financial assets settlement, business services, prediction markets and various transactions of finance. It is known that it will play an important part in the global economy growth and giving corner-edge advantages to the customers, to the existing system of banking and to entire society, in common. The economy is discovering the ways in which applications, which are blockchain enabled can be used such as flat money, securities and derivative contracts [2]. Blockchain enabled applications presents a great change to markets of capital for performing various transactions like digital payments, securities and derivatives transaction, financial auditing, loan management schemes and general banking services. Blockchain based implementations of this system can be circulated through Viacoin, a source cryptocurrency which performs faster transactions than Bitcoin.

ii) Verification of Integrity

One of the most recent area of blockchain field is verification of integrity [3]. These blockchain enabled applications preserves transactions and information related to the generation and lifecycle of services and products. The several areas can be: (i) Insurance (ii) Intellectual property (IP) management (iii) Provenance and counterfeit. The work presented describes the summary to preserve and manipulate the information in an automatic manner in the context of integrity of information and provenance.

iii) Governance

All the Governments throughout year are dealing with holding and managing service records of both enterprises and citizens. Applications which are blockchain enabled changed the way the government at state level/local level operates by disintermediating record keeping and transactions. The safety, automation and accountability that the blockchain proposes for record handling eventually prevents the corruption and the various services of government more efficient. This technology could serve as a platform which is secure in nature for communication and integration of social, physical and various business infrastructures in context of smart city [4].

iv) Management of healthcare

This could play an essential part in the management of industry of healthcare with numerous applications in several areas like health automation claims, patient access online, sharing patient's medical data, healthcare management in public [5], healthcare records, clinical trials, drug counterfeiting and precision medicine. Management of patients Electronic Healthcare Records (EHR's) is most common area with the highest development.

v) Internet of Things

Enormous amount of data, approximately 90% globally has been produced in past three years. The resultant of this is due to: (a) The advent of IoT (b) Population growth. While the possibilities of their expansion, Blockchain and IoT technologies [6], are already a greater area to talk upon and their symbiotic relationships. For example, the distributed wireless sensor networks, inspite of its disadvantages, are one of the strengths of human and technological evolution, demonstrates that technology of blockchain and its architecture may improve IoT by minimizing its disadvantages and maximizing the strength.

In [7] author(s) explored the potential and contribution of Blockchain technology in smart cities and provide a novel solution for easy and automated consumer utility payments based on smart contracts. In [8] author(s) presents an insight into concept of blockchain, highlights the initiatives undertaken by Government of India and a presented a real time simulation showcasing MOLE: Multiparty Open Ledger Experiment, based on BlockChain Technology using IIT-MHRD'S Virtual lab. In [9] author(s) discuss various applications such as: finance, medical sciences, academics, Intellectual Property Rights of Blockchain Technology and describe how they are impacting the lives of common people in day- to-day-life.

3 SMART CONTRACTS IN BLOCKCHAIN

Smart Contract [9] is the set of terms and conditions signed and secured using Message Digest 5 (MD5) Algorithm or Secure Hash Algorithm (SHA). It helps to detect and prevent multiple type of attacks especially the men in middle attack. It can allow anyone in a blockchain network to trade currency, property, tangible or intangible assets or any valuable thing in a simple, easy to use and hassle-free manner. The user can compare it with a concept of C-Language that is: "if-else statement". If one condition is fulfilled, then a particular condition will take place and if a particular condition is not met then, some other task/condition would take place. The smart contracts are more like a set of Committed Codes [10]. Smart Contracts aims to establish a mutual agreement between the parties involved in a commercial deal so that smooth, trusted, secure and harmonious relationship between the Party A and Party B.

4 ALGORITHM FORMULATION

The BASIC Algorithm have been designed to handle the concept of design and deployment of Smart Contract between Company and Client.

> BASIC Algorithm: BlockChain based Smart and Intelligent Contract Agreement between the Company and Client
> NOTATION:

Smart Contract/Agreement	: S_c
Transaction	: T_i **(value ranging from T_i for i=1 to n)**
Company	: C_o
Company Address	: CO_{ADD}
Company GST Number	: CO_{GST}
Company Email ID	: CO_{EM}
Company Phone number	: CO_{PNO}
Client Name	: C_{LN}
Contract Name	: C_N
Contract Purpose	: C_P
Terms of Contract	: C_{TEM}
Contract Duration	: C_D
Client Age	: C_A
Client Address	: C_{ADD}
Client AADHAR Number	: C_{AN}
Client Email ID	: C_{EM}
Client Phone number	: C_{PNO}
Client Personal Details	: C_{PERS}
Client Professional Details	: C_{PROF}
Cryptographic Hash Algorithm	: CH_A
[MD 5 Algorithm ‖ SHA Algorithm]	

> **Trigger: Client Approaches Company for signing Smart Contract**

1. Input the Client Details and create a single composite record C_{PERS}

$$C_{PERS} \leftarrow \{C_{LN} + C_A + C_{ADD} + C_{AN} + C_{EM} + C_{PNO}\}$$

2. Input the Company Details and create a single composite record C_{PROF}

$$C_O \leftarrow \{CO_{ADD} + CO_{GST} + CO_{EM} + CO_{PNO}\}$$

3. Validate: If Client's Record/data is Non-Numeric String Record and it's length is 96 (16x6) bit long && Company Data is 64 bits(16x4)bit long.

$$X \leftarrow \{Valid(C_{PERS})\}$$
$$Y \leftarrow \{Valid(C_O)\}$$

If (Valid() ← Fails ‖ *SizeOf* (C_{PERS}) < 96 ‖ *SizeOf* (C_O) < 64}}
{
throw runtime exception;
}*else*
{
goto step 4;
}

4. if (X ‖ Y is NULL)
{

197

```
print "Error in Received Client Details";
print "Try Again";
break;
}
else
{
goto step 5;
}
```

5. Initiate the steps towards creation and signing off the Agreement

```
if (Ti between Client and Company are Valid)
{
Initiate Signing of Sc of input CN
If (CP is served)
{
Secure Contract using CHA
}

}
```

5 CONCLUSION

The Primary objective in the current research work was to design, develop and deploy the Smart Contract Algorithm in BlockChain Technology which was successfully met.

REFERENCES

M. Alharby and A. van Moorsel, "Blockchain-based smart contracts: A systematic mapping study," arXiv preprint arXiv:1710.06372, 2017.

V. Buterin, "A next-generation smart contract and decentralized application platform.," Available online at: https://github.com/ethereum/wiki/wiki/White-Paper/

K. Petersen, R. Feldt, S. Mujtaba, and M. Mattsson, "Systematic mapping studies in software engineering.," in EASE, vol. 8, pp. 68–77, 2008.

S. Nakamoto, "Bitcoin: A peer-to-peer electronic cash system," 2008. A. Lewis, "A gentle introduction to smart contracts.," Available online at:https://bitsonblocks.net/2016/02/01/a-gentle-Introduction.to-smart-contracts/ [Accessed19/07/2018].

G. Wood, "Ethereum: A secure decentralized generalized transaction ledger," Ethereum Project Yellow Paper, vol. 151, 2014.

Johari, Rahul, Kanika Gupta, A. S. Parihar "Smart Contracts in Smart Cities: Application of Block-Chain Technology." In 2nd International Conference on Innovations in Information and Communication Technologies (IICT-2020), Springer, October 2020 (In-Print).

Gupta, Kanika, Rahul Johari, Suyash Jai "MOLE: Multiparty Open Ledger Experiment, Concept and Simulation using BlockChain Technology". In ICAAAIML International Conference on Advances and Applications of Artificial Intelligence and Machine Learning, Springer, October 2020 (In Print).

L. Luu, D.-H. Chu, H. Olickel, P. Saxena, and A. Hobor, "Making smart contracts smarter," in Proceedings of the 2016 ACM SIGSAC Conference on Computer and Communications Security, pp. 254–269, ACM, 2016.

I. Nikolic, A. Kolluri, I. Sergey, P. Saxena, and A. Hobor, "Finding the greedy, prodigal, and suicidal contracts at scale," arXiv preprint arXiv:1802.06038, 2018.

Johari R., Gupta K., Jha S.K., Kumar V. (2020) CBCT: CryptoCurrency Based Blockchain Technology. In: Batra U., Roy N., Panda B. (eds) Data Science and Analytics. REDSET 2019. Communications in Computer and Information Science, vol 1230. Springer, Singapore.

Block chain: IoT security, privacy and resource challenges

Prabira Kumar Sethy
Department of Electronics, Sambalpur University, Burla, India

Chanki Pandey & Mohammad Rafique Khan
Department of Electronics and Telecommunication Engineering, GEC Jagdalpur, India

Santi Kumari Behera
Department of Computer Science and Engineering, VSSUT Burla, India

Sharad Chandra Rajpoot
Department of Electrical Engineering, GEC Jagdalpur, India

ABSTRACT: The purpose of the paper is to provide a vague overview of the latest block-chain use cases in the information technology industry, in particular, the Internet of Things (IoT). Respective use cases have been discussed in current research articles, Master Theses, business white papers, and industry expert blogs. The paper also provides an overview of the technical aspects of the blockchain and IoT. This paper offers a comprehensive overview of the history and current situation of IoT security, privacy, and resource challenges centred on blockchain technology. It helps understand the need for a comprehensive blockchain feasibility model to be created.

Keywords: Blockchain, Security, Privacy, Internet-of-Things, Resource, Challenges

1 INTRODUCTION

The technology of information and communications is rising rapidly. Recent advancements in electronic components and communications technologies enable interaction of devices using internet as medium. These technologies allow the computer and the machines to interact with human beings. "One such trend can be measured in several words, including IoE, IoT, Internet-of-Vehicles (IoV), IoMT and IoBT (Banerjee *et al.* 2018)". Usually, modern devices includes sensor for sensing and interpreting different data from the surroundings. The observed information is further processed for review and retrieval by different applications in centralized cloud storage. Data are susceptible to multiple threats in the unified cloud. "Blockchain is a shared network where both parties hold a record of each transaction (Nakamoto 2008)". Transactions are transparent and it is quick to spot any adjustments. For instance, an intelligent society with real-time displays of parking spaces to users. The centralized database is updated when the sensors detects a free parking space. The server administrator who manages this database will add a car park to himself without displaying a slot for others. In this situation, the quality of the sensor data is compromised.

The goal is to prevent use by third parties by the built-in system blockchain network and to ensure that the sensor's real-time data can be accessed without alteration by any network node. Furthermore, blockchain allows IoT networks to instantly connect and determine.

DOI: 10.1201/9781003193838-38

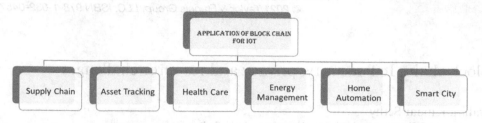

Figure 1. Application of Block chain for IoT.

Decentralization of the IoT network has a variety of advantages, including reducing the expense of operating the unified IoT data base and increasing security and privacy, which eliminates expectations of third parties. However, it is uncertain how IoT will incorporate these functions. This is primarily attributed to the constraints of processing speed, control, and energy of IoT computers. This is why the cryptocurrency blockchain specification is not ideal for IoT implementations. Related IoT implementations are illustrated which can gain from blockchain in Figure 1. This covers supply chain administration, health services, business security, and automation of home facilities, electricity storage, and asset monitoring.

There is little traceability and transparency of conventional supply chain management. The price of products may be produced artificially. "Without an intermediary, Blockchain will allow the supply chain industry to manage false ledgers and monitor products. (Abeyratne and Monfared 2016, Borah *et al.* 2020)". That also enables better accountability and eliminates supply chain abuse. In healthcare, IoT and blockchain combinations quickly capture, track and safely store patient data in real-time (Simic *et al.* 2017). Home equipment and IoT can be automated in intelligent cities by blockchain, which enables devices to communicate between equipment. Technology is going to blockchain because it is capable of growing prices and negative impacts on the environment (Andoni *et al.* 2019). Blockchain can assist in asset monitoring by the collection of open, encrypted, and accountable information from asset-associated IoT devices. Energy conservation is also one of the key problems to be solved as by adopting the combination of blockchain and IoT.

Current approaches for IoT implementations are strongly clustered, raising multiple security problems, such as one-faith, anonymity, and point of failure. It also restricts their optimization and hence suggests the need for a decentralised IoT trust scheme. The Blockchain provides a cryptographic methodology of trust without a central authority. Several blockchain-based IoT technologies have recently gained attention thanks to their potential to improve security and privacy. A recent Juniper analysis study (Ethereum 2020) forecasts that a mix of IoT and food business blockchain will save billions of dollars by that retailer's prices, promoting regulatory enforcement and combating fraud. "Giants from the food business such as Carrefour, Nestle, and Cermaq have already begun utilizing hyper ledger Fabric, an IBM-developed blockchain technology (Carrefour 2020, Cermaq 2020, and Nestlé Global 2020)."

2 SECURITY AND PRIVACY ASPECTS FOR IOT THROUGH BLOCKCHAIN

Cloud infrastructure IoT devices are linked through a cloud server. The data submitted and retrieved by the devices is processed and stored. Cloud-connected computers are therefore susceptible to multiple threats. The IoT block is a limitation or a possible failure point. (Swan 2015). The cloud paradigm may be exploited. In Flint, Michigan, for example, intelligent water metres was adopted to in order to monitor water quality. The authorities insisted on safe drinking of water in the region, while CNN Article claimed that officials had to adjust sample data to minimize the lead level of water. It reported that two of the obtained samples were rejected by the officers. (Research 2020).These malpractice forms may be prevented by introducing IoT blockchain. The explanation is that the data produced by the sensors cannot be updated.

Devices are based on a smart contract to exchange blockchain messages. A digital signature of a letter with the private key of the author guarantees the reply was given by the owner himself. The risk of centre, recycling or other types of attacks is therefore minimized. This reduces the risk of people in the center, replay, and other forms of attacks (Swan 2015).

A few of the strengths of IoT's blockchain are:

- Lowered costs 8.4 billion IoT products were reported by Gartner (Gartner 2017) in 2017, which was 31 percent higher relative to 2016. The disc and network space for these devices has been dramatically improved. Blockchain allows computers to connect and to conduct acts automatically. Therefore, cloud management and administrative personnel would not be expected to manage cloud storage (Sun *et al.* 2016).
- Single failure point Each IoT architecture entity's roles are different. Therefore a malfunction of any system will produce a single failure point. In a blockchain, every node is linked and all transactions are copied to another network node; thus a failure of one device does not affect any device activity.
- IoT systems immune to malicious attacks are susceptible to several forms of attacks because of their unified design. Examples of threats involve a remote server interruption, annoyance assault, and data stealing. This can be stopped with the IoT blockchain design though blockchain is susceptible to certain other kinds of assaults.
- Safety and secrecy Due to IoT's centralized infrastructure, information is likely to be managed when computers are connected and hashed in blockchain. Therefore, data processing on a specific computer cannot be extended to other blockchain systems.

While several attacks are reported for blockchain, majority of them are not significant in practice (Buccafurri *et al.* 2017). In the literature, some of the assaults contain;

- Malware: "The hierarchical existence of blockchain design contributes to malware dissemination. With the advent of modern protocols and the ability to store and quantify data, the blockchain can preserve malicious data (Cermeño 2016)". Malware impact on the blockchain devices can contribute to its dissemination to other blockchain nodes. This will trigger the nodes to crash.
- DDOS: The (Vasek *et al.* 2014) reported that 7.4% of bitcoin-related providers had DDOS experience. There are more chances of assault in eWallets, financial services, and mining pools. Just as in a standard wallet, the bitcoin wallet must also be secured. Two-factor authentication is suggested to secure the bitcoin wallet. For an extra authentication measure, the wallet and backup can be secured.
- Phishing attacks: "In 2018, on Bitcoin wallets, numerous phishing attacks were released on Bitcoin wallets and blockchain.info (Information. n.d.)". Hackers developed and attempted to steal the wallet details on a website close to blockchain.info. In another scenario, hackers have described legal receivers and convinced customers to submit bitcoin. If the bitcoin has been submitted, it cannot be retrieved.
- Majority Attacks: This form of attack is also known as a 51 percent invasion. The group of miners will decide which transactions should or should not be tolerated if many of the mining resources of the network can be handled. This will encourage them either to deny their purchases or to double them. That could be accomplished with the growth of mining pools if the blockchain network is open and available. The assault does not, however, provide complete ownership over the Bitcoin network. In the same manner, proof of work would be applied in a private or registered blockchain under the supervision of the Regulator, thus the Regulator will have the right to regulate the network (Cermeño 2016).
- Sybil Attack: Sybil attack (Douceur 2002) is utilizing several identities to exploit a peer-to-peer network. A single agency produces several bogus network control personalities. If an intruder can monitor most mining nodes in the blockchain, a fraudulent transaction can be generated and inserted.
- Eclipse attack (Heilman *et al.* 2015): This was a coordinated assault on the distributed infrastructure where a malicious intruder segregated a certain node and removed both

inbound and outbound links with his peers. Attackers are attempting to achieve 51% of mining capacity by isolating some of the mining nodes.

3 RESOURCES ASPECTS FOR IOT THROUGH BLOCKCHAIN

IoT networks have minimal device, connectivity, and storage capacity whereas Blockchain systems need enormous capital. Includes under 10 kB of data memory and less than 100 kB of programme memory for Low Power IoT platforms (Bormann *et al.* 2014), and GBs memory includes the Blockchain Node (BitcoinCore 2020). Moreover IoT device capacities with limited low power tools are far beyond the computational requirements of consent algorithms, including proof of work. Contemporary Blockchain technologies are thus unadapt since their resources are needed for such low-power IoT applications. Figure 1 demonstrates that the Blockchain processes cannot be managed by network nodes and client computers and that the cloud layer is ideal for modern Blockchain technologies. A central IoT solution could bind to a decentralised network of the blockchain, which is meant to act as a point of entry to a blockchain network.

4 CONCLUSION

Blockchain technology has now rendered a huge influence on the use of crypto currencies. For the IoT and supply chain management applications, the core building blocks of blockchain technologies – the digital ledger, consensus processes, and public-key cryptography. This paper gives a quick summary of blockchain technologies to illustrate the technological problems in utilizing blockchain for IoT.

REFERENCES

Abeyratne, S.A. and Monfared, R., 2016. Blockchain ready manufacturing supply chain using distributed ledger. *International Journal of Research in Engineering and Technology*, 05 (09), 1–10.

Andoni, M., Robu, V., Flynn, D., Abram, S., Geach, D., Jenkins, D., McCallum, P., and Peacock, A., 2019. Blockchain technology in the energy sector: A systematic review of challenges and opportunities. *Renewable and Sustainable Energy Reviews*.

Banerjee, M., Lee, J., and Choo, K.K.R., 2018. A blockchain future for internet of things security: a position paper. *Digital Communications and Networks*, 4 (3), 149–160.

BitcoinCore, 2020. Running A Full Node - Bitcoin [online]. *2020*. Available from: https://bitcoin.org/en/full-node# minimum-requirements [Accessed 18 Sep 2020].

Borah, M.D., Naik, V.B., Patgiri, R., Bhargav, A., Phukan, B., and Basani, S.G.M., 2020. Supply Chain Management in Agriculture Using Blockchain and IoT. *In*: *Kim* S., *Deka* G. *(eds) Advanced Applications of Blockchain Technology. Studies in Big Data*. Springer, Singapore, 227–242.

Bormann, C., Ersue, M., and Keranen, A., 2014. *Terminology for constrained node networks*. Internet Requests for Comments, RFC Editor, RFC 7228,.

Buccafurri, F., Lax, G., Nicolazzo, S., and Nocera, A., 2017. Overcoming limits of blockchain for IoT applications. *In*: *ACM International Conference Proceeding Series*. New York, NY, USA: Association for Computing Machinery, 1–6.

Carrefour Group. Carrefour launches Europe's first food blockchain. [online], 2020. Available from: https://www.carrefour.com/current-news/carrefour-launches-europes-first-food-blockchain [Accessed 18 Sep 2020].

Cermaq | Cermaq contributes to traceability with blockchain [online], 2020. Available from: https://www.cermaq.com/wps/wcm/connect/cermaq/news/mynewsdesk-press-release-2945012/[Accessed 18 Sep 2020].

Cermeño, J.S., 2016. Blockchain in financial services: Regulatory landscape and future challenges for its commercial application. *BBVA Working Papers*, (December), 33.

Douceur, J.R., 2002. The sybil attack. *In*: *Lecture Notes in Computer Science (including subseries Lecture Notes in Artificial Intelligence and Lecture Notes in Bioinformatics)*. Springer Verlag, 251–260.

Ethereum, 2020. Home | ethereum [online]. Available from: https://ethereum.org/en/[Accessed 18 Sep 2020].

Gartner, I., 2017. Gartner Says 8.4 Billion Connected 'Things' Will Be in Use in 2017, Up 31 Percent From 2016 [online]. Available from: https://www.gartner.com/en/newsroom/press-releases/2017-02-07-gartner-says-8-billion-connected-things-will-be-in-use-in-2017-up-31-percent-from-2016 [Accessed 18 Sep 2020].

Heilman, E., Kendler, A., Zohar, A., and Goldberg, S., 2015. *Eclipse Attacks on Bitcoin's Peer-to-Peer Network.*

Information., C.N. and I.S., n.d. Bitcoin Phishing Attack Hacking Methods Used for Cryptowallets. [online].

Nakamoto, S., 2008. Bitcoin: A Peer-to-Peer Electronic Cash System [online]. Available from: https://bitcoin.org/en/bitcoin-paper [Accessed 18 Sep 2020].

Nestlé breaks new ground with open blockchain pilot Nestlé Global [online], 2020. Available from: https://www.nestle.com/media/pressreleases/allpressreleases/nestle-open-blockchain-pilot [Accessed 18 Sep 2020].

Research, C.E., 2020. Flint Water Crisis Fast Facts - CNN [online]. Available from: https://edition.cnn.com/2016/03/04/us/flint-water-crisis-fast-facts/index.html [Accessed 18 Sep 2020].

Simic, M., Sladic, G., and Milosavljević, B., 2017. A Case Study IoT and Blockchain powered A Case Study IoT and Blockchain powered Healthcare. *In: The 8th PSU-UNS International Conference on Engineering and Technology (ICET-2017).*

Sun, J., Yan, J., and Zhang, K.Z.K., 2016. Blockchain-based sharing services: What blockchain technology can contribute to smart cities. *Financial Innovation*, 2 (1), 1–9.

Swan, M., 2015. *Blockchain: Blueprint for a new economy (NOTE! Poor quality).* O'Reilly Media, Inc.

Vasek, M., Thornton, M., and Moore, T., 2014. Empirical analysis of denial-of-service attacks in the bitcoin ecosystem. *In: Lecture Notes in Computer Science (including subseries Lecture Notes in Artificial Intelligence and Lecture Notes in Bioinformatics).* Springer Verlag, 57–71.

Recent Trends in Communication and Electronics – Sharma et al. (Eds)
© 2021 Taylor & Francis Group, LLC, ISBN 978-1-032-04572-6

Analysis of varied architectural configuration for 7T SRAM bit cell

B. Rawat & P. Mittal
Delhi Technological University, Delhi, India

ABSTRACT: In this paper three different bitline configuration of 7T SRAM bit cell – 7TDE, 7TSE and 7THE - are designed for 32 nm technology node and the simulation results are analyzed. The static noise margin obtained for 7TDE, 7TSE and 7THE for hold operation are - 75, 75 and 87 mV respectively, while for the read operation are 30, 75 and 87 mV respectively. The bit cell is a part of a larger circuit and embedded circuits are subjected to temperature variation during its course of operation. So the bit cell are analyzed for temperature variation from 25 °C to 110 °C. This analysis highlights that the 7THE bit cell has higher temperature tolerance for read and hold operation whereas the 7TSE has a better write operation temperature tolerance. While 7TDE shows inferior performance for all operation modes.

1 INTRODUCTION

The increasing popularity of portable devices with longer battery life, has pushed designers to continuously scale the feature size of the varied circuit elements. Reducing the feature size for the circuit also reduces the area occupied by the circuit. It also increases the density for the integrated circuit, thereby making it more economical. But like all good things, scaling has its own trade-offs. As the feature size for a device approaches the tens of nanometer vicinity, tunneling effect, process variation and short channel effects start to take their toll on the performance of the device characteristics (Karmakar, 2018). A digital memory block that has found its way into almost every processor is the static random access memory (SRAM) bit cell. Its performance and compatibility has made it one of the most researcher memory element. But the steep trends observed in technology scaling and voltage reduction have made it quite challenging for SRAM designers to keep up the performance of the bit cell.

The conventional 6T SRAM bit cell has been the basic building block for cache memory development in various System on Chip (SoC). But this cell is losing its charm with the progressive scaling in the technology node and reduction in the supply voltage available to the bit cell (Mittal et al. 2016). As a result the performance of the bit cell has degraded in terms of read stability, write ability and delay (Paul et al. 2007, Lin et al. 2010 and Patel et al. 2018). Also with the downscaling of the technology node process variations such as random dopant fluctuations and line edge roughness have become major concern in the design of robust SRAMs (Mishra et al. 2019). To curb these limitation of 6T SRAM bit cell various but cell have been reported. 7T SRAM bit cell is also a popular choice against 6T. There are various configuration for a 7T SRAM bit cell. The additional transistor can be used to break the mutual feedback or even to insert an additional port for the bit cell. In this work three different 7T SRAM schematics are analyzed for their performance. The rest of the paper is structured as follows – Section II details the varying architectural configurations for 7T bit cell. In section III the result for static noise analysis, write margin and Temperature variation analysis for the bit cell are analyzed and section IV concludes the paper.

DOI: 10.1201/9781003193838-39

2 ARCHITECTURAL VARIETY FOR 7T SRAM BIT CELL

In this work a comparative analysis of the different 7T SRAM bit cells is performed. This is done to understand the implication of varying port configuration on the performance of a SRAM bit cell composed of seven transistors only. The 7T SRAM bit cells used for the analysis is given in Figure 1. The first 7T SRAM bit cell represented in Figure 1 (a) was reported by Aly et al in 2008. This 7T bit cell uses the same set of bit line for the read and write operation and has a differential read operation so it is referred for as 7TDE. The second bit cell is a 7T SRAM bit cell with a single bit line configuration (7TSE). This bit cell relies on a single bit fine for the read as well as the write operation for the bit cell. Another bit cell the combines the differential and single ended approach to access the 7T SRAM bit cell is given in Figure 1 (c). This bit cell is therefore referred to as a hybrid 7T SRAM (7THE) bit cell. The details for the working and operation for the bit cell is given in Figure 1 are given in the subsequent sub section.

2.1 Differential 7T SRAM bit cell (7TDE)

The 7TDE SRAM bit cell comprises of two inverter pairs formed by PUL-PDL and PUR-PDR. An additional NMOS transistor (DL) is used to establish a connection and disconnection within the feedback connection of the two inverters. The DL transistor is controlled by the signal W. When, W = 1, the DL transistor is in ON state and the mutual feedback connection is maintained. This configuration of the bit cell is used to perform the read operation for the bit cell. While for W = 0, the DL transistor is in OFF state. In this scenario the positive feedback between the memory core is disconnected. This cascaded inverter pair configuration is used to perform the write operation for the 7TDE SRAM cell.

2.2 Single port 7T SRAM bit cell (7TSE)

A single ended 7T (7TSE) bit cell configuration was reported by Yang et al. 2016. The schematic for this cell is given in Figure 1 (b). This cell has a single ended single port architecture. Transmission gates; TN and TP, are used for disconnection of the feedback between the node Q and the input node for the right-side inverter. The cell is accessed through transistor; ACL. Four control signals are used to steer the bit cell into different modes of operations. These control signals are, WL, WWLA, WWLB and BL.

2.3 Hybrid port 7T SRAM bit cell (7THE)

A configuration for SRAM bit cell that entails an isolated read port by inserting an additional transistor into the bit cell to perform the read operation was reported by researchers Liu et al. 2017. The additional read port is connected at the QB node of the memory core of the bit cell. This additional NMOS transistor is referred to as R2 in the schematic shown in Figure 2 (c). The gate of R2 is controlled by the QB node of the memory core, the source is connected to read bit line, RBL and drain to read word line, RWL. Prior to the read operation RBL is precharged and RWL is turned low. The RBL is discharged through R2 to RWL when the QB node stores 1.

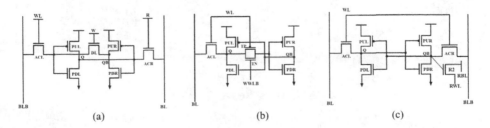

Figure 1.   Schematic for (a) 7TDE, (b) 7TSE and (c) 7THE SRAM bit cells.

3 SIMULATION RESULTS

The simulation in the paper are performed using predictive technology models for 32 nm technology node. The environment temperature for the simulation in maintained at 27 °C for noise margin and write margin analysis.

3.1 *Static noise margin*

The first parameter for evaluation for an SRAM bit cell is its ability to withstand noise. The measurement for the same is done in terms of the static noise margin (SNM). The conventional method to measure SNM, which is defined as the maximum dc noise voltage required to flip the bit cell data (Kang, 2015). It is obtained by graphically superimposing the voltage transfer characteristics of the two inverters resulting in two-lobes curve (Kumar et al. 2014), called the butterfly curve. The SNM is measured as the maximum possible square embedded into the lobes of the butterfly curve. The values of SNM obtained for the hold operation for 7TDE, 7TSE and 7THE bit cells 75, 75 and 87 mV respectively. For the ease of comparison the data for SNM is presented graphically in Figure 2. It can be observed from Figure 2 that the hold SNM values for the three bit cell are similar but the highest value is obtained for 7THE. While during the read operation the SNM values for the 7TDE, 7TSE and 7THE bit cell is 30, 75 and 87 mV respectively. The read SNM for 7TDE cell is considerably less in comparison to 7TSE and 7THE. This degradation in the SNM for a differential cell is a result of vulnerability experienced by the bit cell when it is accessed from both ends. Consequently increasing the probability for a destructive read operation in a differential cell, while for a single ended SRAM bit cell, the possibility of a destructive read operation is less (Ahmad et al. 2017).

3.2 *Write margin*

The write ability of an SRAM cell is a measure of its ability to flip the data stored in its storage nodes. This is measured in terms of write margin (WM). By definition, WM is estimated as the voltage difference between VDD and WL when the storage nodes Q and QB flip (Ahmad et al. 2017). A balanced WM is essential for a effective write operation for an SRAM bit cell. A very high value for WM makes it easier to write into the cell but increases the noise susceptibility of the cell. While a low write margin makes it difficult to write into the bit cell (Metrelliyoz et al. 2008) The WM obtained for 7TDE, 7TSE and 7THE are 225, 187 and 185 mV respectively. The graphical comparison of WM for the discussed bit cell is presented in

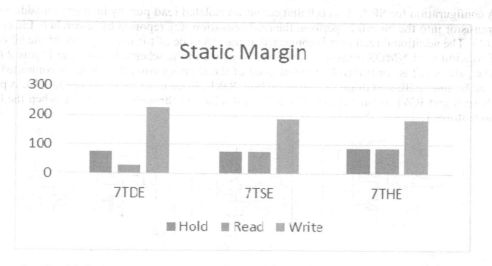

Figure 2. Graphical representation for the SNM and WM for different 7T SRAM bit cells.

Figure 2. It is observed that the WM for 7TDE is quite high, while for 7TSE and 7THE is similar. This implies that the write ability for single ended and hybrid ended is balanced in comparison to differential cell.

3.3 *Temperature variation analysis*

SRAM circuits form a major part of SoC. All SoCs are designed to function under a wide range of temperature. Due to continuous operation or uncontrolled environmental conditions, the device temperature may surge. Therefore, to account for the temperature variation encountered by an SRAM bit cell, a temperature variation analysis is performed for SRAM bit cells discussed before. The analysis is done for the temperature range from 25 °C to 110 °C.

3.3.1 *Static noise margin*

In the previous subsection, the SNM for the hold and read operation were analyzed. The SNM results for the varying 7T SRAM bit cells were analyzed and a few conclusions were drawn. But the differentiator lies in the temperature tolerance of each circuit. The temperature variation analysis results for SNM for the bit cell are given in Figure 3. It can be observed from the curves given in Figure 3 that maximum variation in performance with respect to temperature variation is registered for 7TDE, while the minimum is for 7THE. While the 7TSE bit cell showcased an intermediary temperature tolerance. Thus implying that hybrid ended cell respond better to temperature variation that the circuit might encounter.

3.3.2 *Write margin*

In the previous section it was established that the 7TSE bit cell has better performance than 7TDE and 7THE in terms of SNM for temperature variation. The other mandatory operation

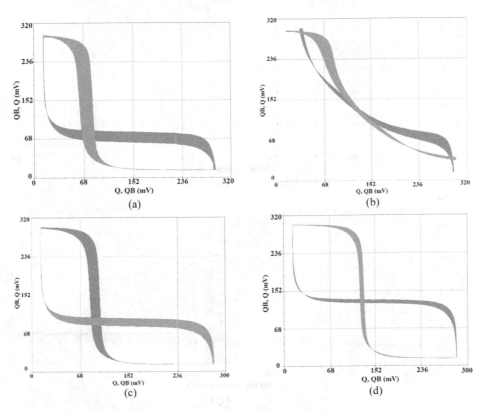

Figure 3. Temperature variation analysis for (a) 7TDE hold operation, (b) 7TDE read operation, (c) 7TSE hold and read operation and (d) 7THE hold and read operation.

than an SRAM bit cell preforms is write operation. The graphs obtained for WM for the three SRAM bit cells in consideration are given in Figure 4. As can be observed from Figure 4 the least variation in the performance is by 7TDE while the most variation is observed for 7THE. Here as well the performance of the 7TSE bit cell is intermediary.

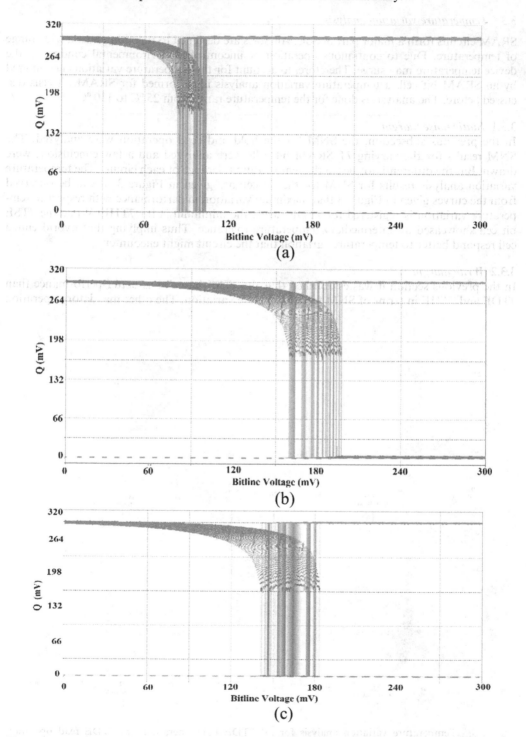

Figure 4. Temperature variation analysis for write margin for (a) 7TDE, (b) 7TSE and (c) 7THE SRAM bit cell.

208

4 CONCLUSION

SRAM bit cells are a significant part of a processor designing. With the decrease in size and increase in the popularity of portable devices, memory designers are experimenting with different configuration to create and SRAM bit cell that will cater to the needs of the semiconductor industry. In this paper three different bit line configuration for 7T SRAM bit cells were designed for 32 nm technology node. The bit cells are simulated for 300 mV supply voltage and the results so obtained are analyzed. The bit cells are analyzed for their static noise margin for read and hold operation. While for the write operation the write margin is analyzed. This static analysis highlights that the 7TDE bit cell is a bit inferior to 7TSE and 7THE bit cell. To identify the superior bit cell amongst 7TSE and 7THE the static parameters are analyzed for temperature varying from 25 ºC to 110 ºC. This analysis highlights that 7THE has a better performance for read and hold SNM under temperature variation. While the write margin is better for 7TSE.

REFERENCES

Ahmad, S. Gupta, M.K. Alam, N. Hasan, M. 2017. Low Leakage Single Bitline 9 T (SB9T) Static Random Access Memory. *Microelectronics Journal* 62: 1–11.

Aly, R.E. & Bayoumi, M.A. 2007. Low-Power Cache Design Using 7T SRAM Cell, *IEEE Transactions on Circuits and Systems II: Express Briefs* 54(4): 318–322.

Kang, S. Leblebici, Y. & Kim, C. (3rd) 2015. *CMOS digital integrated circuits*. New York; McGraw-Hill.

Karmakar, S. 2018. Ultra-Fast SRAM Using Spatial Wave-Function Switched FET (SWSFET). *International Journal of Electronics Letters* 7(1): 40–49.

Kumar, B. Kaushik, B.K. & Negi, Y.S. 2014. Design and analysis of noise margin, write ability and read stability of organic and hybrid 6-T SRAM cell. *Microelectronics Relability* 54(12): 2801–2812.

Kumar, B. Raj, K. & Mittal, P. 2009. FPGA implementation and mask level CMOS layout design of redundant binary signed digit comparator. *International Journal of Computer Science and Network Security* 9(9): 107–115.

Lin, S. Kim, Y.B. & Lombardi, F. 2008. A highly-stable nanometer memory for low power design. *Proc. IEEE International Workshop Design Test of Nano Devices, Circuits Systems*: 17–20.

Lin, S. Kim, Y.B. & Lombardi, F. 2010. Design of a CNTFET-based SRAM cell by dual- chirality selection, *IEEE Trans. Nanotechnology* 9(1): 30–37.

Meterelliyoz, Mesut, 2008. Thermal Analysis of 8-T SRAM for Nano-Scaled Technologies. *Proceeding of the Thirteenth International Symposium on Low Power Electronics and Design - ISLPED 08*.

Mishra, N. Mittal, P & Kumar, B. 2019. Analytical Modeling for Static and Dynamic Response of Organic Pseudo All-p Inverter Circuits. *Journal of Computational Electronics* 18(4): 1490–1500.

Mittal, P. Negi, Y.S. & Singh, R.K. 2016. A Depth Analysis for Different Structures of Organic Thin Film Transistors: Modeling of Performance Limiting Issues, *Microelectronics Engineering* 150: 7–18.

Patel, P.K. Malik, M.M. & Gupta, T.K. 2018. Reliable high-yield CNTFET-based 9T SRAM operating near threshold voltage region. *J Comput Electron* 17: 774–783.

Paul, B. C. Fujita, S. Okajima, M. Lee, T. H. Wong, P.H.S. & Nishi, Y. 2007. Impact of a process variation on nanowire and nanotube device performance, *IEEE Trans. Electron Devices* 54(9): 2369–2376.

Rathod, A.P.S. Lakhera, P. Baliga, A.K. Mittal, P. & Kumar, B. 2015, Performance Comparison of Pass Transistor and CMOS Logic Configuration based De-Multiplexers, *IEEE International Conference on Computing Communication and Automation (ICCCA- 2015)*: 1433–1437.

Yang, Y. Jeong, H. Song, H. Wang, S.C. Yeap, J. Jung G. & S.-O. 2016. Single Bit-Line 7T SRAM Cell for Near-Threshold Voltage Operation With Enhanced Performance and Energy in 14 nm FinFET Technology, *IEEE Transactions on Circuits and Systems I: Regular Papers*, 63(7): 1023–1032.

Recent Trends in Communication and Electronics – Sharma et al. (Eds)
© 2021 Taylor & Francis Group, LLC, ISBN 978-1-032-04572-6

Improved performance of on chip communication fabric using different arbitration techniques

A. Musala Venkateswara Rao, B. Kaja Hemanth & C. Shaik Razia
A. Assistant Professor, Department of ECE, B.U.G Student & C. Associate Professor, Department of CSE
A.C. Koneru Lakshmaiah Education Foundation, Vaddeswaram, India, & B.GMR institute of Technology, Rajam, India

ABSTRACT: The Network on Chip (NoC) is an effective practice for the interfacing of Application on Chip (AoC)s in development of embedded architectures. The AoC uses various processors, embedded devices, and soft cores. Over the years, the demand for expertise in the implementation of efficient arbiters has increased due to the emerging of on-chip interconnections. Which requires low latency at the time of message transmission. The aim of the communication channel is to provide the better performance for the present scenario like increase in volume of processing elements. arbitrators are recommended for a fair outcome in NoC with minimal power, area, time, and router control paths consist of buffers, arbiters, and the allocators at every resource point. In this approach we proposed an arbiter with less power, area, and arbitration delay.

Keywords: Network on Chip (NoC), Application on Chip (AoC), arbiter, power, area, delay, Musala Venkateswara Rao

1 INTRODUCTION

In most instances, arbitration is required since multiple sources ask for access to a common resource. This is the normal case for most interconnected architectures. Arbiter will assist in settling certain demands for better communication performance. This is commonly used in bus-based systems of several masters and slaves. When the resource is shared through a buffer, channel or input/output ports, arbiters plays a major role. Which makes it easier to find which information is most important to grant and provide access to the desired request. Crossbar switches are used by arbiter's dependent on switching in scheduling algorithms McKeown (1999), Gupta and McKeown (1999), Li et al. (2001).

Arbiters are mainly used in complex computer systems like multiprocessors and in system on chip embedded systems. When the quantity of centers utilized is little, a brought together exchanging fabric is used Yalala et al. (2006). Arbiters are mostly used in routing logic, virtual-channel allocation, and crossbar scheduler Nawathe et al. (2008). One of the fundamental components of multi-core System-on-chips (MCSoCs) are memory cores, interconnections, and interfaces, these are found in memory modules with multiple ports Adiletta et al. (2002), regular transport-based interconnections Reed and Manjikian (2004), Zitouni and Tourki (2008). and Virtual-Channel (VC) as shown in the Figure 1 and switch allocators of present day on chip switches.

DOI: 10.1201/9781003193838-40

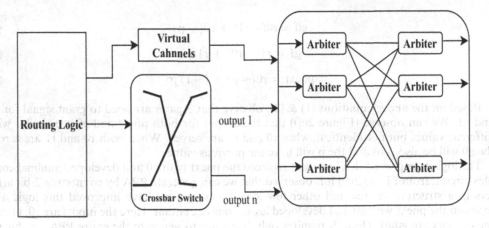

Figure 1.   Abstract architecture of a NoC.

2   DESIGN AND ITS FUNCTIONALITY

Observations about arbiters in related works proved that arbiters are used to decrease the arbitration delay Dimitrakopoulos et al. (2008). Main aim is to reduce the complexity of arbiter and make a secured network on chip. The proposed arbiter is based on the round robin scheduling algorithm, the recommended logic will be able to assess and information will be transferred among the different layers of Network on chip Passas et al. (2014). To permit a reasonable designation of assets and to accomplish high execution switch activity Das et al. (2004) Chao et al. (1999).

The general PPA will consists of mainly two parts. One is to programmable arbitration logic that is used to decide for which input need to be given a highest priority and for accessing the data and other one is programmable logic which helps to promote the input. The programmable arbitration logic which helps to scan the inputs in a cyclic manner is shown in Figure 2(a) which consumes more area compared to proposed circuit. In the Figure 2(a) r0, r1, and p are the inputs, go,g1, and pnext are outputs as described below.

Ping Pong Arbiter

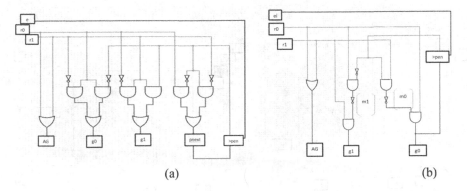

(a)   (b)

Figure 2.   Logic diagram for existing (Figure 2. (a)) and Proposed (Figure 2. (b)) Ping Pong Arbiter (PPA).

211

$$g0 = r0(\sim r1) + r0(\sim p) \tag{1}$$

$$g1 = r1(\sim r0) + r1(p) \tag{2}$$

$$pnext = r0(\sim p) + (\sim r1)p \tag{3}$$

Based on the above equations (1) & (2), observe that 3 gates are need to grant signal for g0 and g1. We can observe (Figure 2(a)) that the output for both pnext and g0 are similar with different values but not identical, when r0 and r1 are zeroes. When both r0 and r1 are zeroes the g0 will be also zero and the p will have the previous value.

The logic is improved this logic and replaced the pnext with g0 and developed minimal complex circuit, from Figure 2(b) it is observed that we can generate PPA by connecting 2-bit arbiters in a structure of tree and other few logic gates. So, we have improved this logic and replaced the pnext with g0 and developed lesser complex circuit. Here the inputs are r0, r1 and the outputs are g0,g1. Here, it requires only five gates to generate the entire PPA. In this we introduce a signal called m (m1 and m0) which is used to mask the signals of arbiters and to give grant. The outputs for the Figure 2(b) is as follows:

$$pnext = g0 \tag{4}$$

$$m0 = \sim (r1.p) \tag{5}$$

$$m1 = \sim (r0.p) \tag{6}$$

$$g0 = r0.m0 \tag{7}$$

$$g1 = r1.m1 \tag{8}$$

The AG signal refers to the signal of 2-bit arbiter enable signal.

A. Ping lock arbiter

B. In ping lock arbiter a new updating policy is used by following round robin architecture. So that this will provide a fair arbitration. By servicing each and every input before the higher priority input is updated, fair arbitration can be achieved. In PLA 2-bit arbiters relate to few other logic gates which is similar to PPA. 1-bit output signal is generated from the improved 2-bit arbiter, which is passed through AG signal.

The Figure 3(a) shows that that l0 is generated when both requests are active. The inputs and outputs of the of the Figure 3(a) are r0, r1 and g1, l0, g0.

The equation of the output l0 is as shown as

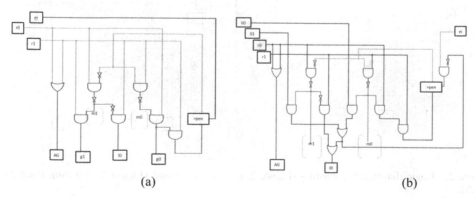

| (a) | (b) |

Figure 3. Logic diagram for existing (Figure 3.(a)) and Proposed (Figure 3.(b)) Ping Lock Arbiter (PLA).

$$l0 = r0.r1.(\sim p) \qquad (9)$$

The priority vector should not be updated till the highest input request is granted. The request will be granted by the root arbiter Ugurdag and Baskirt (2012). If the input AG signal is granted its respective l0 signal will be asserted. It is observed (Figure 3(b)) that it is a modified arbiter of 2-bit from the previous ping lock arbiter. In these two lock signals will be added which are named as li0 and li1. In this for every request of the input, the output logic. When any request is granted by the arbiter and its respective li signal will be active, the priority value will not change. The signal lp is used to verify all the above conditions and generates the output by using following equations.

$$lp = li0.g0 + li1.g1 \qquad (10)$$

$$lp = li0.r0.m0 + li1.r1.m1 \qquad (11)$$

$$lp = li0.m0 + li1.m1 \qquad (12)$$

The signal m is used (Figure 3(b)) to mask the arbiters signal and required to grant the request. When there are two requests, the highest priority will be only for the first request. Hence, the equation will be as follows:

$$l0 = lp + r0.m1 \qquad (13)$$

C. Proposed-X arbiter

In this proposed arbiter it eliminates multiple no. of requirements of FPAs and new circuit is designed which handles a cyclic nature of priority input signals. The Figure 4 shows the proposed-X arbiters with the inputs GP1, GP2, GR1, GR2 and the outputs X, Y. In this the GP defines group priority and GR defines group propagate. From the Figure 4 it is observed that this defines a different group of propagate and priority Dimitrakopoulos et al. (2012). This arbiter is generated by using only 3 gates which has less complexity and has area efficiency. The output equations of the above Figure 4 follows as below.

$$X = GP1 + (GR1.GP2) \qquad (14)$$

$$Y = GR1.GR2 \qquad (15)$$

D. Matrix arbiter

When compared with other arbiters the matrix arbiter gives you a fair arbitration. The matrix arbiter is most efficient arbiter since it requires less inputs and is area efficient. It provides a fast and fair arbitration. It is less cost and easy to implement. Figure 5 represents a matrix arbiter with less inputs. The inputs are denoted as gi, gj and the outputs are wij, w

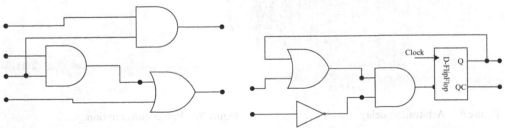

Figure 4. The proposed-X arbiter.

Figure 5. Logic diagram for the matrix arbiter.

213

(ij+1). In this a **D** flip-flop is used. The flip-flop helps to reset the values. In every stage of wij a D flip-flop is set synchronously. The output of the and gate is given as the input of the D flip-flop.

3 EXPERIMENT RESULTS AND ANALYSIS

It represents results which include the delay of arbitration. We observed the fairness of arbitration between different types of arbiters. Different RTL codes are developed for the PPA, PLA, proposed-X arbiter, and matrix arbiter in Vivado. The analysis of the results represents the latency, area efficient, power consumption, arbitration delay, cost, and the fairness of the arbitration. We have compared arbitration delays of all the arbiters and the arbitration policy of almost all the arbiters have the same, but the fairness of arbitration is more in matrix arbiter since it has less inputs and easy to implement. In this we have developed 2 bit arbiters of PPA and PLA. The proposed-X arbiter follows the groups of priority and propagates. The matrix arbiter follows the triangular approach. It consists of a flip-flop to set the grant of rows and columns in synchronously. Existing [16] refers Monemi et al. (2017).

The Figure 6 represents the arbitration execution delay among the existing arbiter, proposed-X arbiter, and proposed matrix arbiter. When compared there is a considerable improvement in the reduction of arbitration execution delay. In this 2-bit,4-bit,6-bit,8-bit have been taken and observed the delays of all the arbiters and proved that the proposed arbiters have a fair arbitration. The Figure 7 represents the utilization of the power among the existing arbiters, proposed –X arbiter and proposed matrix arbiter. It has been noticed that the consumption of the power is less in the proposed arbiters when compared to the existing arbiters.

Figure 8 Signifies about the consumption of the area among the arbiters Khan and Ansari (2012). It is also noticed that the area efficiency is high in the proposed arbiters. The work focuses on the minimization of the arbitration execution delay, power consumption and to improve the area efficiency. So, we have taken 2-bit,4-bit,6-bit and 8-bit respectively.

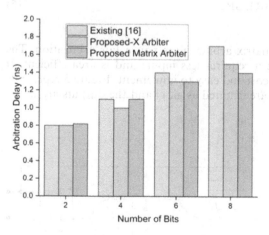

Figure 6.   Arbitration delay.

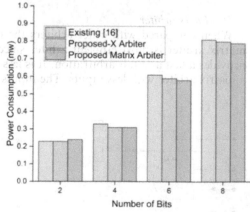

Figure 7.   Power consumption.

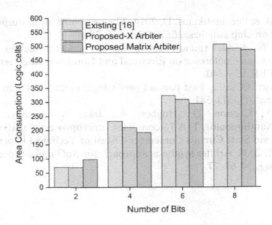

Figure 8. Area consumption.

4 CONCLUSION

The main goal of this work is to reduce the arbitration delay and area efficiency. In this paper we compared between the existing and proposed arbiter's arbitration delay, power consumption and the area efficiency. To reduce the complexity of the existing arbiters we have developed the modified arbiters. We have proposed a modified version of the arbiters to decrease the arbitration execution delay. When compared to the other arbiters PPA has more arbitration policies.

REFERENCES

Adiletta, M. J., Wheeler, W., Redfield, J., Cutter, D. & Wolrich, G. 2002. SRAM controller for parallel processor architecture including address and command queue and arbiter. Google Patents.

Chao, H. J., Lam, C. H. & Guo, X. A fast arbitration scheme for terabit packet switches. Seamless Interconnection for Universal Services. Global Telecommunications Conference. GLOBECOM'99.(Cat. No. 99CH37042), 1999. IEEE, 1236–1243.

Das, S., Fan, A., Chen, K.-N., Tan, C. S., Checka, N. & Reif, R. Technology, performance, and computer-aided design of three-dimensional integrated circuits. Proceedings of the 2004 international symposium on Physical design, 2004. 108–115.

Dimitrakopoulos, G., Chrysos, N. & Galanopoulos, K. Fast arbiters for on-chip network switches. 2008 IEEE International Conference on Computer Design, 2008. IEEE, 664–670.

Dimitrakopoulos, G., Kalligeros, E. & Galanopoulos, K. 2012. Merged switch allocation and traversal in network-on-chip switches. *IEEE Transactions on Computers*, 62, 2001–2012.

Gupta, P. & Mckeown, N. 1999. Designing and implementing a fast crossbar scheduler. *IEEE micro*, 19, 20–28.

Khan, M. A. & Ansari, A. Q. 2012. Area-efficient programmable arbiter for inter-layer communications in 3-D network-on-chip. *Central European Journal of Computer Science*, 2, 76–85.

Li, Y., Panwar, S. & Chao, H. J. On the performance of a dual round-robin switch. Proceedings IEEE INFOCOM 2001. Conference on Computer Communications. Twentieth Annual Joint Conference of the IEEE Computer and Communications Society (Cat. No. 01CH37213), 2001. IEEE, 1688–1697.

Mckeown, N. 1999. The iSLIP scheduling algorithm for input-queued switches. *IEEE/ACM transactions on networking*, 7, 188–201.

Monemi, A., Ooi, C. Y., Palesi, M. & Marsono, M. N. 2017. Ping-lock round robin arbiter. *Microelectronics Journal*, 63, 81–93.

Nawathe, U. G., Hassan, M., Yen, K. C., Kumar, A., Ramachandran, A. & Greenhill, D. 2008. Implementation of an 8-core, 64-thread, power-efficient sparc server on a chip. *IEEE Journal of Solid-State Circuits*, 43, 6–20.

Passas, G., Katevenis, M. & Pnevmatikatos, D. 2014. The combined input-output queued crossbar architecture for high-radix on-chip switches. *IEEE Micro*, 35, 38–47.

Reed, J. & Manjikian, N. A dual round-robin arbiter for split-transaction buses in system-on-chip implementations. Canadian Conference on Electrical and Computer Engineering 2004 (IEEE Cat. No. 04CH37513), 2004. IEEE, 835–840.

Ugurdag, H. F. & Baskirt, O. 2012. Fast parallel prefix logic circuits for n2n round-robin arbitration. *Microelectronics Journal*, 43, 573–581.

Yalala, V., Brasili, D., Carlson, D., Hughes, A., Jain, A., Kiszely, T., Kodandapani, K., Varadharajan, A. & Xanthopoulos, T. A 16-core RISC microprocessor with network extensions. 2006 IEEE International Solid State Circuits Conference-Digest of Technical Papers, 2006. IEEE, 305–314.

Zitouni, A. & Tourki, R. 2008. Arbiter synthesis approach for SoC multi-processor systems. *Computers & Electrical Engineering*, 34, 63–77.

216

Recent Trends in Communication and Electronics – Sharma et al. (Eds)
© *2021 Taylor & Francis Group, LLC, ISBN 978-1-032-04572-6*

Computational analysis of Diabetic Retinopathy detection

Deepti Seth, K.P Mishra & Abhishek Gupta
KIET Group of Institutions, Ghaziabad

ABSTRACT: Diabetic Retinopathy (DR) is a typical retinal ocular infection related with diabetes. It is a significant reason for visual impairment in center just as more established age gatherings. Subsequently early identification through ordinary screening and ideal mediation will be exceptionally helpful in successfully controlling the advancement of the infection. Since the proportion of individuals compared with the number of eye expert which diagnosis demand greater, hence the demand of robotized indicative framework corresponding to diseases corresponding to diabetics in the eye with the goal which lone unhealthy people has to be alluded to the master for additional intercession and treatment. Different angles and phases of retinopathy are dissected by inspecting the shaded retinal pictures. Image processing process can be utilized for robotized identification of these different highlights and phases of Diabetes Retinopathy and can be alluded to the pro likewise for intercession, consequently making it an extremely successful instrument for compelling screening of Diabetic Retinopathy patients.

1 INTRODUCTION

Diabetic retinopathy is a visual ailment that can cause vision incident and visual debilitation in people who have diabetes. This is when high glucose levels cause mischief to veins in the retina. These veins can develop and spill. Or then again they can close, keeping blood from experiencing. Sometimes strange new veins create on the retina. These movements can take your vision. Diabetic retinopathy has two principle stages: nonproliferative and proliferative. "Proliferative" implies whether there is neovascularization (sporadic vein advancement) in the retina. The abnormality in the retina is said to be(NPDR). As the disease rapidly grow this abnormality may be named as PDR.

Hyperglycemia achieves mischief to retinal vessels. This cripple the dividers of slim and results in small thin thread type abnormalities named as microaneurysms. Microaneurysms is situated inside the retina near the location of internal membrane, Their physical appearance is like dot named as hemorrhages. The weakened vessels also become broken, causing fluid to soak the retina. Fluid declaration under the macula, or macular edema, interferes with the macula's run of the mill limit and is a commonplace explanation behind vision hardship in those with DR. Neovascularisation (strange vein development) in diabetic retinopathy happens in light of retinal ischaemia (absence of blood stream to the retina). New vessels may develop on the optic circle (where the optic nerves enter the eye) or somewhere else in the eye. Since the veins are strange, they may seep into the retina or glassy liquid, causing spots of blood in the eye that square vision. Proliferative retinopathy is a sight-compromising condition.

2 METHODOLOGY

2.1 *Segmentation of blood vessels*

There are four stages of DR. Primary it is the place little regions of inflatable like growing show up in there retina's minuscule veins named as microaneurysms. At the point

DOI: 10.1201/9781003193838-41

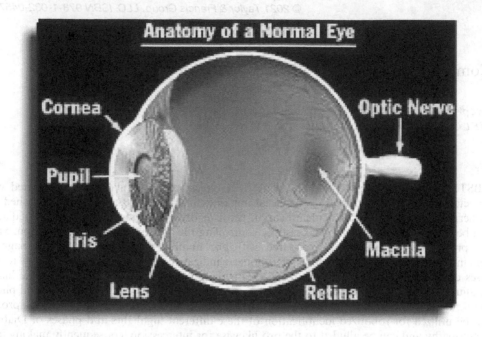

Figure 1. Anatomical structure.

where the situation become active large ammount of veins will be hindered, confining the progression of blood to a few territories of retina. These denied regions impart signs to the body to develop fresh blood vessels for sustenance. In this situation the new blood vessels grows. These vessels will be develop on the inner region of the retina and boundary of the vitreous retina.

2.2 Segmentation of optical disc

The vein is situation in the region of optic plate. For a robotized framework, limitation as well as division should be significant Undertaking. In this way the center is to naturally identify the area of optical disc.

2.3 Segmentation of red lesions

For microaneurysms and hemorrhages set up, we consider only lesions. By identify these any one can easily recognize the disease like diabetics. A programmed as well as basics recognition of red injuries will help the patient not losing the visual perception. These will be named as the basic stages for identification of disease, the basic procedure is concealing rectification of veins must be taken out .In this paper we have utilized picture procurement and picture handling method to discover different highlights from the picture of an eye. The optical arrangement of fundas gives an upstanding, in this paper, the proposed strategy has 4 phases, In first we have identified and removed the highlights of diabetic retinopathy. We had a picture of eye fundus which is a RGB. Shading design as information. From this picture significant highlights, for example, clustered vessels, blood vessels, optic disc, hard exudates and microaneurysms are removed. These highlights can be utilized to group, whenever input picture has diabetic retinopathy or not. Positively whenever input picture doesn't hard exudates, grouped vessels and microaneurysms or a portion of these that of picture can be considered to have diabetic retinopathy. For such reason we have utilized an integral asset MATLAB 2020a. Image processing tool compartment is required for such reason. This toolboxes worked in MATLAB 2020a.

3 RESULT

A fast and simple computational technique for the automatic localisation of optic disc in the fund as retinal pictures we have take some typical pictures and some strange pictures as contribution in the wake of preparing and investigating we have did measurable examination by plotting histogram subsequently by ascertaining and breaking down mean middle standard deviation and so on of the two picture.

According to information determined it has been discovered that for typical pictures the estimation of the estimation of mean and middle are roughly comparative and be around of (0.2152 - 2 0.2165) for mean (8 - 15) for middle. A similar pattern has been seen in the standard deviation for ordinary pictures the qualities are practically comparative with the range 0.12 - 0.13 for X estimations of plot and 30.58-30.48241 Y estimations of plot.

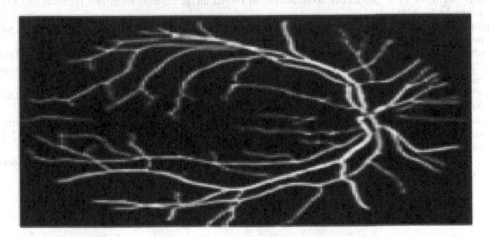

Figure 2.   Vessels detection for finding Diabetic Retinopathy.

Figure 3.   Optic disc detection for finding Diabetic Retinopathy.

Figure 4.   Micron Detection for finding the Diabetic Retinopathy.

REFERENCES

Amrutkar, Nikhil. et.al April 2013. Retinal blood vessels segmentation algorithm for diabetic retinopathy and abnormal detection using image substruction. International journal of Advance research in Electrical Electronics and Instrumentation Engineering 2 (4), 2320–3765.

Gargeya, R. et al Elsevier. 2017. Automated identification of diabetic retinopathy using deep learning. 124 (7):962–969

Gupta, Ankita. et al 2018. Diabetic Retinopathy Present and Past. Procedia Computer Science 132: 1432–1440.

K, Manju. et al July 2015. A Survey on Segmentation Techniques for Detection of Diabetic Retinopathy International Journal of Engineering Research & Technology 5(1): 2278–0181.

Li Huiqi et.al 2001. Automatic location of optic disc in retinal images. In Proceedings of the International Conference on Image Processing 837–840.

Mai S, Mabrouk et al. September 2006. Survey of retinal image Segmentation and registration" GVIP Journal 6(2): 1–11.

Philip, S. et.al 2007. The efficacy of automated disease/no disease grading in a systematic screening program. In British journal of Ophthalmology. 2: 1512–1517.

Pradhan, Sandeep. et al 2008. An Integrated Approach using Automatic seed generation and Hybrid Classification for the detection of red lesions in Digital Fundus Images. IEEE International Conference on Computer and Information Technology Workshops, CIT 2932–2935.

Singh, Neera. et al October 2010. Automated Early Detection of Diabetic Retinopathy Using Image Analysis Technique. International Journal of Computer Applications 8(2): 0975–8887.

Walter, T. et.al October 2002. A contribution of Image Processing do the diagnosis of Diabetic retinopathy detection of Exudates in Color Fundus Images of the Retina. IEEE Trans. Med. Imag, 21(7): 1010–1019.

Liu, Z, et al 1997. Automatic Image Analysis of fundus photograph proceedings of the International Conference of IEEE Engineering in Medicine and Biology Society 2(1): 524–525.

Recent Trends in Communication and Electronics – Sharma et al. (Eds)

Survey on road traffic-flow prediction: Recent trends in India

Kranti Kumar & Bharti
School of Liberal Studies, Dr. B. R. Ambedkar University Delhi, Delhi, India

ABSTRACT: Efficient and fast transport has become the backbone of modern economy. Transport has a significant role not only in the development of economy but also in the day to day life of a common person. Not only developing but also developed countries are facing the problems caused by transport related activities like congestion, accidents, air and noise pollution. Intelligent transportation systems (ITS) are seen as a crucial component in traffic management. Extensive research has been carried out by various researchers in area of traffic flow. This paper reviews various short term as well as long term methods for road traffic prediction in India. Traffic approaches were subdivided in two categories: Mathematical/Statistical and artificial intelligence based approach. It was found that neural network based models are more popular for traffic flow prediction.

Keywords: traffic flow, model, artificial neural network, intelligent transportation system

1 INTRODUCTION

Nowadays, road traffic is a major concern for everyone. Everyone needs to do some planning before commuting in order to avoid wastage of time in traffic jams. According to WHO (2018) report, a road user dies in every 20 seconds in the world due to road accidents. In India 151.4K deaths by road accidents were reported by Ministry of Road Transport and Highways (MoRTH) in 2018. Figure 1 shows the number of deaths due to road accidents in India during 2014-2018.

TomTom NV, a Dutch multinational company engaged in developing and creation of location technology prepared traffic congestion index of 416 cities covering 57 countries of the world (TomTom NV, 2020). According to this index 4 out of 10 severely congested cities in the world are in India which includes Bengaluru, Delhi, Mumbai and Pune. Bengaluru was found most severely congested city in the world. Study shows that average Bengaluru driver waste 243 hours in traffic jams every year. Main reason is continuously increasing population of India over the years shown in Figure 2, since larger population leads to slower mobility as well as more congestion. As per MoRTH Annual report 2019-20, there are 253.3 million motor vehicles registered in India which has quadrupled since 2001 shown in *Figure 3*. On the other hand increment in the number of buses which form a significant part of public transport is extremely low as compared to two wheelers and other vehicles. Out of all registered vehicles there are 102.6 lakh in Delhi followed by Bengaluru (68.33 lakh), Chennai (52.99 lakh) and Mumbai (30.53 lakh). Lack of adequate infrastructure, increasing population and number of vehicles are making traffic congestion problem of Indian cities worse day by day. Hence, there is a strong need for implementation of new efficient traffic management and control systems. One of the prime necessity of these systems is to predict the traffic flow precisely well in advance so that the congestion can be reduced. These systems are based upon traffic flow models.

In order to design efficient traffic flow models it is necessary to study various characteristics of traffic flow. Traffic flow models are generally classified into three categories namely

DOI: 10.1201/9781003193838-42

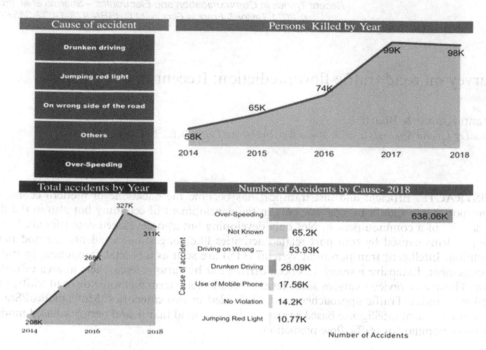

Figure 1. Death by road accidents due to various factors in India.
Source: morth.nic.in

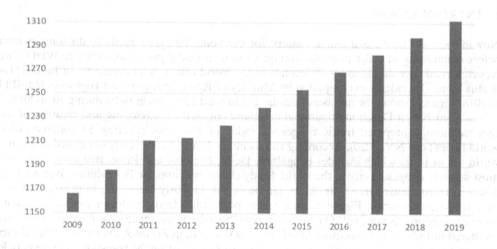

Figure 2. Population of India (in million).
Source: Ministry of Statistics and Program Implementation (MOPSI)

microscopic, macroscopic and mesoscopic models. Microscopic model describe vehicles as individual particles forming traffic flow which widely includes car-following models and cellular automata. These models also describe the behavior/reaction of every driver depending on the surrounding traffic. Macroscopic models describes traffic flow similar to liquid or gases in motion and are often called "hydrodynamic models".

Fundamental macroscopic traffic variables are density(ρ), speed(v) and flow(q). Significant time has been spent to understand these characteristics and interrelation between them.

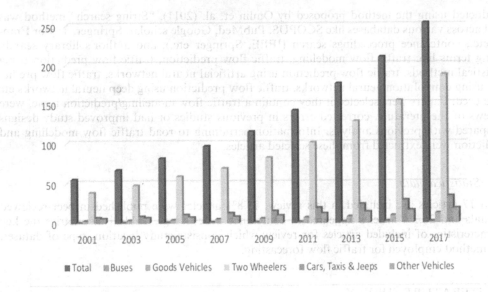

Figure 3.   Registered motor vehicles in India (in million).
Source: morth.nic.in

Predicting these variables can help in comprehending the changes that might occur in road-traffic networks and eventually help in reducing congestion, accidents and increasing traffic flow. Mesoscopic models combine both microscopic and macroscopic approaches which results in a hybrid model. Traffic flow modelling started with Greenshields studies in 1935. Lighthill, Whitham (1955) and Richards (1956) proposed a continuum model for vehicular flow called LWR model which was based on fluid flow and set the foundation for macroscopic traffic flow modeling. The LWR model is formulated by the conservation equation

$$\frac{\partial \rho}{\partial t} + \frac{\partial q(\rho)}{\partial x} = 0$$

where $q = \rho.v$ is flow rate, ρ is density, and v is speed of traffic and all are functions of space(x) and time(t). Car-following models were developed in 1960s, with many experiments conducted by different researchers. But each model has its shortcomings. Several attempts have been made to model the traffic flow of homogeneous as well as heterogeneous nature. In developing countries like India mixed traffic conditions are observed i.e. vehicles do not always abide by the rules and disruptive lane changing can be observed. Hence, models developed under homogeneous and lane disciplined conditions are not suitable for heterogeneous traffic and cannot be applied directly. At present complex and more precise models for traffic flow prediction are developed and employed. This study aims to do a systematic review of traffic flow models. This review will be focused on following broad questions:

– Which techniques/models are used for traffic flow prediction?
– To what extent work has been done?
– Which models are suitable for different traffic flow conditions in India?

2   MATERIALS AND METHODS

To have an idea about the past and current road traffic flow modelling and prediction studies con- ducted by various researchers, literature survey from the last 20 years (2000 to 2020) was conducted using the method proposed by Omlin et. al. (2011). "String search" method was used across various databases like SCOPUS, PubMed, Google scholar, Springer, Taylor Francis etc., conference proceedings search (IEEE, Springer etc.), and author's library search. String terms like traffic flow modelling, traffic flow prediction, traffic flow prediction using statistical methods, traffic flow prediction using artificial neural networks, traffic flow prediction using convolution neural networks, traffic flow prediction using deep neural networks etc were used. Papers were selected if they contain a traffic flow modelling/prediction topic, were reviews of the literature, corrected errors in previous studies or had improved study designs compared with previous analysis. Information pertaining to road traffic flow modelling and prediction were extracted from these selected articles.

2.1 Statistical data

Total 17 articles are included in this review. 73.8% articles were published in peer-reviewed journals and 26.2% articles appeared in conference proceedings. Table 1 summaries the key characteristics of included articles for review which consists study location, type of dataset, and method employed for traffic flow forecasting.

3 LITERATURE SURVEY

Abundant work in the area of road traffic flow prediction has been done in India by different researchers. Various approaches have been used by them to predict the traffic flow. For review purpose these approaches have been divided in two parts: Statistical/Mathematical based and artificial intelligent approaches.

Table 1. Summary of literature.

No.	Study	Location	Dataset	Method
1	Anveshrithaa & Lavanya (2020)	California	Performance Measurement System	LSTM
2	Tsuboi & Yoshikawa (2020)	Ahmedabad	CCTV	Data Envelopment Analysis (DEA)
3	Poonia & Jain (2020)	Jaipur		LSTM
4	Rajendran & Ayyasamy (2020)	Bangalore	Toll data	Local weighted learning
5	Bhatia et al. (2019)	Ahmedabad	Simulated traffic	SDN enabled VANET
6	Chavhan & Venkataram (2020)	Metropolitan area	Simulated	Agent technology and emergent intelligence
7	Miglani & Kumar (2019)	Punjab	Video	Deep learning (review)
8	Giraka & Selvaraj (2019)	Vellore	Video	SARIMA
9	Sharma et al. (2018)	Roorkee	Video survey	BP ANN
10	Kumar (2017)	Vellore	Video survey	Kalman filtering technique
11	Omkar & Kumar (2017)	Vellore	Video survey	Multiplicative decomposition
12	Mohan & Ramadurai (2017)	Madras	Video	Extended AR model
13	Kumar & Vanajakshi (2015)	Chennai	Collect-R camera	SARIMA
14	Kumar et al. (2013)	Muzaffarnagar	Video survey	ANN
15	Theja & Vanajakshi (2010)	Madras	Video	Support vector machines
16	Arasan & Koshy (2005)	Chennai	Video survey	Simulation in C++
17	Khare & Author (2001)	Delhi	Video survey	Gaussian plume model

3.1 Statistical/Mathematical approach

Khare & Sharma (2001) developed a model to predict empirically modified traffic for Delhi. They derived the model from gaussian plume model to account for local conditions. Traffic amount calculated hourly can be utilized to limit the entry of vehicles to a specific amount, for example a short-term control measure. This model was used to create hourly predictions of traffic volume for a major traffic intersection. Arasan & Koshy (2005) proposed a model for highly heterogeneous traffic flow. Simulations were done in such way that there was absence of lane discipline in heterogeneous traffic flow. Model was implemented and validated using C++. It is applicable for any level of traffic mix, including the intense automobiles traffic streams condition. It is impractical to use ARIMA in cases wherever adequate data is unavailable. Data-driven approach using Box-Jenkins ARIMA model demands a solid database for model construction. Therefore, the availability of data creates an issue which was solved by a prediction scheme with Seasonal ARIMA (SARIMA) model by Kumar & Vanajakshi (2015), Kumar (2017). They also proposed and evaluated a method supported by Kalman filtering technique (KFT), which does not require large data set. Data was collected for two days and then converted into passenger car units (PCUs). It was finally aggregated into the specified amount of time interval. KFT was utilized for predicting the following day traffic flow together with the specified accuracy, and mean absolute percentage error (MAPE). Mohan & Ramadurai (2017) developed a model for heterogeneous traffic flow which is very common in India. ARIMA requires sufficient data for prediction purpose. To overcome this, Omkar & Kumar (2017) used multiplicative decomposition. This method segregates the elements of time series into trend cycle and seasonal factors. By further calculating moving average (MA) and centered moving average (CMA) randomness was removed from the data. Results show that model performs well with lesser data sets. Chavhan & Venkataram (2020) proposed a traffic management system that makes use of static and mobile agents. They use agents and emergent intelligence (EI) techniques because it is adapted to the changing behavior in distributed environments. A region/particular area can be predicted by this model with low, moderate, and high traffic densities. In another study, Omkar, & Selvaraj (2019) developed a model using R software to forecast the direction wise turning volumes based on SARIMA at an unsignalized intersection. Tsuboy & Yoshikawa (2020) analyzed the traffic flow quantitively using Data Envelopment Analysis (DEA) for the very first time in Ahmedabad city. DEA is a fluid dynamics based concept. Authors have explained how this concept is used in traffic flow using viscous and turbulent flow. Traffic-flow equations based on Greenshields model were derived for each road by assigning unique traffic parameters. Speed ratio was used to understand and compare the congestion in a selected area. Rajendran & Ayyasamy (2020) used locally weighted learning (LWL) along with structure pattern and regression for short term traffic flow forcast. Developed model was used to predict both upstream and downstream traffic. Proposed method was found more accurate and efficient than normal LWL model.

3.2 Artificial intelligence based approach

Short term prediction of traffic flow has attained a considerable amount of attention in India with the recent interest in applications of Intelligent Transportation Systems (ITS) like advanced traveler information systems (ATIS) and advanced traffic management systems (ATMS). Theja & Vanajakshi (2010) proposed a method for the prediction of short-term traffic flow variables in heterogeneous and less lane disciplined situations. Machine learning technique known as support vector machines (SVM) was used which is blend of pattern classification and regression technique. Sensitivity analysis was done to search the best parameters of support vector regression (SVR) concerning running time and accuracy of this model. Results of developed model were compared with a multi-layer feed forward NN with backpropagation. It was found that optimized results can be obtained by the blend of various optimal algorithms, therefore varied strategies are useful for traffic flow forecasting. Nagare & Bhatia (2012) presented a review of three methods for traffic flow forecasting, which were based on Back-propagation NN (BPNN), Simulated Annealing Genetic BPNN and Radial Basis Function (RBF) NN Optimized by Particle Swarm Optimization (PSO). All these methods were compared on various performance parameters. Kumar, Parida & Katiyar (2013) used ANN for short term traffic flow prediction. Input variables were taken distinctively for every class of vehicles. It was shown that even with the increasing forecasting time interval, the model performed consistently well. Model was found suitable for mixed traffic conditions such as non-urban Indian highways also. Multi-lane undivided highways with mixed traffic conditions constitute a considerable portion of roads in India. Sharma, Kumar, Tiwari, Yadav & Nezhurina (2018) proposed a short term traffic flow prediction model using backpropagation ANN for heterogeneous traffic on two-lane undivided highway. Sensitivity analysis was performed and results were compared with SVM, k-nearest neighbor classifier, random forest, multiple regression models, and regression tree. It was shown that proposed model performs better than other approaches used in the study. It has been observed that investigation on autonomous vehicles has changed from the underlying statistical models to adjustable machine learning methods. Miglani & Kumar (2019) presented a detailed review of traffic flow forecasting in autonomous vehicles using deep-learning techniques including parametric and non-parametric techniques. An artificial intelligence scheme inspired by data-driven approach for traffic behavior prediction was proposed by Bhatia, Dave & Bhayani (2019). Flexibility, measurability, and adaptableness influenced by the SDVN (software-defined vehicular networks) architecture was combined together with the machine learning algorithms. An insightful approach to seek out congestion sensitive areas within the VANET (Vehicular Ad-Hoc Network) employing clustering algorithm was introduced. It predicted the upcoming density of traffic for every spot by recurrent neural network (RNN). A long short term memory (LSTM) NN that overcomes the problem of back-propagated error decay through memory blocks for spatiotemporal traffic prediction was employed. It was shown that the model forecasts real-time density of traffic with 97% accuracy. Anveshrithaa & Lavanya (2020) integrated big data frameworks like Apache, Spark, and Kafka into a deep-learning method LSTM, in order to process large amount of data for traffic flow forecasting. Real-time predictive analysis of the traffic data was carried out using Apache Kafka and Spark streaming. It was found that model performed well in real time data analyis and vehicular traffic flow prediction. Poonia & Jain (2020) used LSTM for momentary traffic stream forecast. Model was tested using collected data and it performed well.

4 DISCUSSION AND POLICIES

There are various models outlined with different techniques for the road traffic flow prediction in the literature. Most of the methods are statistical/mathematical and artificial intelligence based. In this paper, a range of techniques used by different researchers were reviewed. It was interesting to note that how these methods are improving over time. It has been noticed that there is huge inclination towards AI based techniques in the recent years. ANN is a part of artificial intelligence which is simplifies model of biological neural network. It simulates characteristics of the human brain like adaptability, learning ability, fault tolerance, flexibility, and many more with the assistance of different algorithms such as backpropagation, perceptron learning algorithm, forest algorithm, etc. Due to these abilities ANN is used in many fields including traffic flow modeling. Hence, there is a rigorous shift in traffic flow modelling through various mathematical or statistical models to different AI and hybrid models that includes deep learning.

Good traffic flow models are needed to sort the traffic problems like traffic jams and accidents with the use of recent technology based upon these real world models. In fact there is a large gap between the building and testing of these models with real traffic data. Various factors are responsible for poor traffic conditions in Indian cities. These include not only scientific/technical factors but also administrative, management, infrastructure, policy, planning, enforcement and research and development related issues which need to be addressed in a coordinated and effective way. Following points describe some important factors which are responsible for better traffic flow movement and other associated traffic related problems.

I. **Population:** Undoubtedly human population is an important factor in management of transport related operations. As the population of India is increasing with a high rate and in turn the numbers of vehicles are also increasing (MOSPI, 2018 and MoRTH, 2018). Most of the Indian cities are bearing the pressure of handling large amount of population and in turn increased traffic flow often exceeding their carrying capacity.

II. **Funding:** Funding is most important aspect for improvement in Indian transport sector. Although, in past two decades large amount and funds have been allocated in this sector by central and state Governments and in turn roads and highway network along with metro rails have expanded tremendously (MoRTH, 2018). To accommodate the needs of increasing population, large amount of funds are required not only form Government but also from private sector in the form of investments in transportation sector.

III. **Infrastructure and Public Transport Facilities:** Robust infrastructure is primary requirement of all the sectors including transportation one. When adequate public transport facilities in proportion to the population/commuters increase are not available, it results in increase of private vehicles plying on roads. When number of vehicles exceeds the capacity of existing roads then it appears in form traffic jams, increased travel time, accidents or deteriorated traffic flow conditions (Pucher, Korattyswaropam, Mittal & Ittyerah (2005)). We need to focus more on public transport to improve the traffic flow conditions and reduce private vehicle use. India has witnessed great expansion in metro rail and highway infrastructure, yet a lot still needed to be done.

IV. **Planning and Implementation:** Unplanned urbanization growth is a big problem with which not only big but also small cities are suffering in India. If a new model/project is developed and implemented after a lot of efforts, it's not necessary that the model will work successfully. One such example is the failure of BRT (Bus Rapid Transit) system in most of the Indian cities. Though it was partially successful in Ahmedabad, India, but a huge failure in the capital Delhi. Reports by the International Association of Public Transport (2020) suggests that this type of arrangement requires dedicated lanes that carries only busses, swift movement of buses with disciplined frequency, a matching height of the station, and a lot more. BRT failure shows that better planning, implementation and coordination among different government agencies and departments is essential for success of any important project.

V. **Traffic Management and Rules Enforcement:** Traffic rules enforcement is a crucial factor for better management of traffic. In the absence of proper traffic rules followed by travellers no modern traffic management system or technology can solve the traffic flow related problems.

VI. **Research and Development:** Proper research is needed to address the transport related problems in India. Another issue comes with the implementation of research outcomes which needs coordination between researchers and field implementing agencies. Inadequate research funding, old data collection equipment's/methods and coordination among different sectors are the primary problems associated with transportation research. We can observe that in India, traffic data collection itself is a daunting task, most of the time it is being done by using video cameras and followed by manual traffic parameters (such as traffic volume, speed) extraction.

VII. **Use of Technology** Use of modern technological systems like advance traffic management system (ATMS), AI/deep learning based softwares, mobile apps, high tech systems of smart cameras, traffic counters/detectors which are beneficial to manage huge traffic flow, route diversion and to avoid unnecessary traffic jams.

5 CONCLUSION

This paper reviews traffic flow prediction studies conducted in India during past two decades. Research studies dealing with short as well as long term traffic flow prediction found in literature were included for review purpose. Different approaches were applied by various researchers for their studies. These approaches were broadly categorized in two parts namely Statistical/Mathematical and ANN based. Depending upon traffic flow characteristics, data availability and local conditions different traffic flow prediction models were designed/developed by researchers. Review analysis suggests that in recent time artificial intelligence based models like ANN, deep learning and hybrid models are more popular and are extensively used.

ACKNOWLEDGEMENT

Financial support to carry out this study from University Grants Commission (UGC) through the start-up grant research project "Modelling and simulation of vehicular traffic flow problems" through the grant No. F.30-403/2017(BSR) is thankfully acknowledged.

REFERENCES

Anveshrithaa, S. & Lavanya, K. 2020. Real-Time Vehicle Traffic Analysis using Long Short Term Memory Networks in Apache Spark. *International Conference on Emerging Trends in Information Technology and Engineering, ic-ETITE 2020*: 1–5.

Arasan, V. T. & Koshy, R. Z. 2005. Methodology for modeling highly heterogeneous traffic flow. *Journal of Transportation Engineering*, 131(7): 544–551.

Bhatia, J., Dave, R., Bhayani, H., Tanwar, S. & Nayyar, A. 2020. Sdn-based real-time urban traffic analysis in vanet environment. *Computer Communications*, 149: 162–175.

Chavhan, S. & Venkataram, P. 2020. Prediction based traffic management in a metropolitan area. *Journal of Traffic and Transportation Engineering (English Edition)*. Elsevier Ltd.: 447–466.

Giraka, O. & Selvaraj, V. K. 2019. Short-term prediction of intersection turning volume using seasonal ARIMA model. *Transportation Letters. Taylor & Francis*: 1–8.

Greenshields, B.D., Channing, W. & Miller, H. 1935. A study of traffic capacity. *In Highway research board proceedings. National Research Council (USA), Highway Research Board.*

Institute for Transportation and Development Policy (ITDP). 2020. Challenges of a Bus-Rapid-Transit System in Indian Cities: The Rainbow case study. <https://www.itdp.in/challenges-of-a-bus-rapid-transit-system-in-indian-cities-the-rainbow-case-study/>.

Khare, M. & Sharma, P. 2001. An Empirically Modified Traffic Forecasting Model for Delhi. *The Transport Asia Project-On Land Use, Transportation and Environment, Organized by Harvard University at Pune, India*: 3–4.

Kumar, K., Parida, M. & Katiyar, V. K. 2013. Short Term Traffic Flow Prediction for a Non Urban Highway Using Artificial Neural Network. *Procedia - Social and Behavioral Sciences. Elsevier B.V.*, 104: 755–764.

Kumar, S. V. 2017. Traffic Flow Prediction using Kalman Filtering Technique. *Procedia Engineering*: 582–587.

Kumar, S. V. & Vanajakshi, L. 2015. Short-term traffic flow prediction using seasonal ARIMA model with limited input data. *European Transport Research Review*, 7(3): 1–9.

Lighthill, M. J. & Whitham, G. B. 1955. On kinematic waves II. A theory of traffic flow on long crowded roads. *Proceedings of the Royal Society of London. Series A. Mathematical and Physical Sciences*, 229(1178): 317–345.

Miglani, A. & Kumar, N. 2019. Deep learning models for traffic flow prediction in autonomous vehicles: A review, solutions, and challenges. *Vehicular Communications. Elsevier Inc.*, 20: 100184.

Ministry of Road Transport and Highways (Morth). 2020. Annual Reports. <https://morth.nic.in/annual-report>.

Ministry of Statistics and Programme Implementation. 2018. *Projected Population*. Statistical Yearbook of India. <http://mospi.nic.in/statistical-year-book-india/2018/171>.

Mohan, R. & Ramadurai, G. 2017. Heterogeneous traffic flow modelling using second-order macroscopic continuum model. *Physics Letters, Section A: General, Atomic and Solid State Physics. Elsevier B.V.*, 381(3): 115–123.

Nagare, A. & Bhatia, S. 2012. Traffic Flow Control using Neural Network. *International Journal of Applied Information Systems*, 1(2): 50–52.

Omkar, G. & Vasantha Kumar, S. 2017. Time series decomposition model for traffic flow forecasting in urban midblock sections. *International Conference On Smart Technology for Smart Nation, SmartTech-Con*: 720–723.

Poonia, P. & Jain, V. K. 2020. Short-Term Traffic Flow Prediction: Using LSTM. *International Conference on Emerging Trends in Communication, Control and Computing (ICONC3), IEEE*: 1–4.

Pucher, J., Korattyswaropam, N., Mittal, N. & Ittyerah, N. 2005. Urban transport crisis in India, *Transport Policy*, 12(3): 185–198.

Rajendran, S. & Ayyasamy, B. 2020. Short-term traffic prediction model for urban transportation using structure pattern and regression: an Indian context. *SN Applied Sciences. Springer International Publishing*, 2(7): 1–11.

Richards, P. I. 1956. Shock Waves on the Highway. *Operations Research*, 4(1): 42–51.

Sharma, B., Kumar, S., Tiwari, P., Yadav, P. & Nezhurina M.I. 2018. ANN based short-term traffic flow forecasting in undivided two lane highway. *Journal of Big Data. Springer International Publishing*, 5(1): 48

Tang, J., Gao, F., Liu, F., & Chen, X. 2020. A Denoising Scheme-Based Traffic Flow Prediction Model: Combination of Ensemble Empirical Mode Decomposition and Fuzzy C-Means Neural Network. *IEEE*: 11546–11559.

Theja, P. V. V. K. & Vanajakshi, L. 2010. Short term prediction of traffic parameters using support vector machines technique. *Proceedings - 3rd International Conference on Emerging Trends in Engineering and Technology, ICETET 2010*: 70–75.

TomTom International BV. Traffic Index Ranking, 2019. <https://www.tomtom.com/en_gb/traffic-index/ranking/>.

Tsuboi, T. & Yoshikawa, N. 2020. Traffic flow analysis in Ahmedabad (India). *Case Studies on Transport Policy. Elsevier*, 8(1): 215–228.

WHO Global Status report on Road Safety 2018. Accessed September 10, 2020. <https://extranet.who.int/roadsafety/death-on-the-roads/>.

Recent Trends in Communication and Electronics – Sharma et al. (Eds)
© 2021 Taylor & Francis Group, LLC, ISBN 978-1-032-04572-6

Study on road traffic congestion: A review

Manoj Kumar & Kranti Kumar
School of Liberal Studies, Dr. B. R. Ambedkar University Delhi, India

Pritikana Das
Department of Civil Engineering, Maulana Azad National Institute of Technology Bhopal, India

ABSTRACT: Traffic congestion is one of the most visible, pervasive, and immediate transport problems plaguing not only India's but also most of the cities of the world on a daily basis. It affects all modes of transportation especially roads and all socioeconomic groups. Rapid population growth, increasing urbanization, inadequate/unplanned transport infrastructure, poor public transport systems and the rising number of personnel vehicles are some of the primary causes of congestion. This article reviews the findings of studies based on road traffic congestion. Various traffic congestion measurement metrics have been discussed. These metrices categorized into three parts (1) Travel time based, (2) speed based and (3) level of service-based. Also, congestion data collection techniques employed in different studies have been discussed. Findings of the study indicate that improved traffic management and control, better public transport services, increase in the funding for transport infrastructure, use of modern technology and overall coordination of transport and land-use policies are important parameters to reduce the congestion.

Keywords: Traffic congestion, Traffic flow, Congestion measures, Data collection techniques

1 INTRODUCTION

Traffic congestion has become a major issue with the metropolis facing the most. Due to urbanization, population increase, and the high use of personnel vehicles; the problem of traffic congestion is increasing day by day. Inadequate infrastructure, unplanned urban growth, and poor traffic management have contributed to a great extent to raise it further. Traffic congestion in urban or nonurban areas increases travel time and energy consumption, increase pollution and stress, reduces production, and increase transportation costs also. The problem of traffic congestion is an open challenge for metropolitan cities as well as medium and small cities. According to TomTom (2020), India is the most affected country by traffic congestion. 4 of the 10 most crowded cities in the world are in India, namely Bengaluru (71%), Mumbai (65%), Pune (59%), and Delhi (56%). According to the Boston Consulting Group report-2018, daily commuters during peak traffic hours on an average spend 1.5 hours or more in Delhi, Mumbai, Bengaluru, and Kolkata. Traffic congestion measured (Stipancic et al., 2017) by using the congestion index (CI) during peak hours shows that congestion is directly correlated with accident frequency. World Road Statistics (2018) released a report on road accidents and deaths due to road accidents comprising of 199 counties. In this report, India got 1st rank in the number of road accident deaths. It is believed that identifying the characteristics of the congested portion of the road is the initial step for such efforts as it is the necessary guidance for the selection of suitable measures. Various researchers and research organizations have provided several types

DOI: 10.1201/9781003193838-43

of definitions of traffic congestion but they are all based on some traffic parameters such as volume and capacity (or density), travel time (or delay), and speed. However, the uses of these definitions are dependent on collected data, and literature consists of several types of data collection techniques. It is found that there does not exist a unique definition of congestion in the literature (Anthony, 2004). Aftabuzzaman (2007) critically analysed various traffic congestion measures and defined criteria for a congestion measure. It was suggested that a congestion measure should have clarity, simplicity, continuity, and comparability.

The objective of this paper is to discuss different kinds of congestion measures considering quantitative and qualitative indicators. A systematic review has been done on various congestion measurement methodologies used at national and international levels and it also illustrates different types of data collection techniques. It will help in selecting an appropriate congestion measure as well as applying suitable data collection methodology for practitioners, planners, and policy decision-makers.

2 MEASURES OF TRAFFIC CONGESTION

To tackle the problem of traffic congestion, various measures have been developed for the identification and quantification of traffic congestion by various researchers. These measures can be helpful for finding the degree of traffic congestion and the performance of the roadway. We have categorized these measures into three parts: Travel time based, speed-based, and level of service based. It is shown in Table 1.

Table 1. Traffic congestion measures.

Travel time based: Travel time is the time taken to cross a section of a road by a vehicle. This time has been used as a parameter in traffic congestion studies. Urban Link's performance evaluation was conducted based on travel time. Some measures related to travel time are listed below.	
Delay	Delay is used to quantify traffic congestion. Delay is defined as extra travel time taken during a journey against the expectations. Lomax et al. (1997) represented delay as the difference between free-flow travel time (FFTT) and average travel time (ATT) i.e. delay can be calculated by using Equation (1). $$\text{Delay} = \text{ATT} - \text{FFTT} \qquad (1)$$
Planning time index (PTI)	PTI is the ratio between free-flow travel time and 95th percentile travel time (TT) (Karuppanagounder and Muneera, 2017). As travel time increases, PTI also increases. Thus, PTI should have the minimum value for better traffic operation, and it can be calculated by using the Equation (2). $$\text{PTI} = \frac{95\%\text{TT}}{\text{FFTT}} \qquad (2)$$
Congestion index (CI)	CI measure is the ratio between delay and FFTT. Here delay is the difference between ATT and FFTT. For better traffic operation, ATT for commuters should be minimum (Karuppanagounder and Muneera, 2017), and index can be calculated by using the Equation (3). $$\text{CI} = \frac{\text{Delay}}{\text{FFTT}} \qquad (3)$$
Travel time index (TTI)	TTI index is the ratio between peak period travel time (PPTT) and FFTT (Lomax and Schrank, 2005). This measure has been used in both recurring and incident traffic congestion conditions (Rao and Rao, 2012), and it can be calculated by using the Equation (4).

(Continued)

231

Table 1. (*Continued*)

$$TTI = \frac{PPTT}{FFTT} \qquad (4)$$

Travel time index (TTI)

TTI is the ratio between ATT and FFTT (Karuppanagounder and Muneera, 2017). For better traffic performance, TTI should be minimum. Index value can be calculated by using the Equation (5).

$$TTI = \frac{ATT}{FFTT} \qquad (5)$$

Buffer time index (BTT)

BTI is extra "buffer" time required to be on time 95th percent of the time. It represents extra time generally travellers add to ATT while planning their journey. Buffer time index (Nakat et al., 2014) is defined as the Equation (6).

$$BTI = \frac{95\%TT - ATT}{ATT} \times 100\% \qquad (6)$$

Speed based: Speed is the most commonly used measure of performance for roadway and traffic congestion. Speed can be calculated in several ways, first, based on average travel time combined with the length of the corridor under study. Average travel speed can be calculated by the Equation (7).

$$Average\,travel\,speed = \frac{Length\,of\,the\,corridor}{Average\,travel\,time} \qquad (7)$$

Second, the average speed of the stream is the weighted average of recorded spot speeds of vehicles using radar guns over the count period. On that spot, it is calculated using the following Equation (8) (Jain and Jain, 2017),

$$V_s = \frac{\sum_{i=1}^{n} u_i \times q_i}{\sum_{i=1}^{n} q_i} \qquad (8)$$

Where, V_s = average speed of the stream, u_i = average spot speed of ith category of vehicle, q_i = flow of the ith vehicle type of total n types.

Congestion Index (CI)

Speed based congestion index has been used (Stipancic et al., 2016) to estimate road congestion from GPS data. It provides information from both in view microscopic or macroscopic of network performance. Dias, et al. (2009) calculated traffic congestion by the Equation (9).

$$CI = \begin{cases} \frac{V_F - V_A}{V_F}, & if\ CI < 0 \\ 0, & if\ CI\,0 \end{cases} \qquad (9)$$

Where, V_F = free-flow speed, and V_A = actual speed.

Speed Reduction Index (SRI)

Speed reduction index is the rate of vehicle speed reduction due to congestion. This rate allows comparing the degree of traffic congestion for different types of transport services by using a constant as a scale to distinguish between different congestion classes at different levels (Kukadapwar and Parbat, 2015). It can be calculated by using the Equation (10).

$$SRI = \frac{Non - peak\,flow\,speed - Peak\,flow\,speed}{Non - peak\,flow\,speed} \qquad (10)$$

(*Continued*)

232

Table 1.   (*Continued*)

Very-low-speed Index (VLSI)	Very low-speed index is the ratio between the time-traveling at a much slower speed and the total travel time. It has been used (Kukadapwar and Parbat, 2015) in traffic congestion evaluation on the urban roadway. Thus, it can be calculated by using the following Equation (11).

$$VLSI = \frac{\text{Time spent in delay}}{\text{Total travel time}} \qquad (11)$$

Corridor Mobility Index (CMI)	CMI is used to calculate the capacity of corridors. Par values for freeway and arterial operation are combined with the speed person volume to generate a corridor mobility index (Lomax, 1990). Equations (12) and (13) were used to calculated CMI values for freeway and arterial roads respectively. For freeway high-occupancy-vehicle lanes

$$CMIF = \frac{\text{Travel speed(mph)} \times \text{Peak} - \text{Hour person volume per lane}}{100000} \qquad (12)$$

For arterial high-occupancy-vehicle lanes

$$CMIA = \frac{\text{Travel speed(mph)} \times \text{Peak} - \text{Hour person volume per lane}}{20000} \qquad (13)$$

Travel speed rate	TSR is the ratio between the reduction in speed during congestion and the speed in free-flow conditions (Hamad and Kikuchi, 2002). The travel speed rate can be calculated by Equation (14).

$$TSR = \frac{V_F - V_{Av}}{V_F} \qquad (14)$$

Where V_F = free-flow speed, and V_{Av}= average speed.

Level of Service (LOS) based: LOS has been used as a qualitative measurement that explains the operational status of traffic stream and its perception by drivers and commuters. LOS is used for the identification of traffic performance as athreshold. Generally, LOS is categorized into six levels from A to F (Manual, 1985). In which the free-flow speed is represented by LOS A and congestion by LOS F (i.e. stop and go flow).

Roadway Congestion Index (RCI)	RCI combines as a ratio of daily vehicle-mile travel (DVMT) per lane-mile for freeways and principal arterial street systems (PASS) that compares the existing DVMT with the determined DVMT values in congested conditions. It is calculated by using the Equation (15).

$$RCI = \frac{\left(\text{freeway}\frac{DVMT}{\text{lane}} - \text{mile}\right) \times \text{freewayDVMT} + \left(\text{arterial}\frac{DVMT}{\text{lane}} - \text{mile}\right) \times \text{arterialDVMT}}{13000 \times \text{freewayDVMT} + 5000 \times \text{arterialDVMT}}$$

$$(15)$$

where, 13,000 vehicles per lane per day for freeways and 5,000 vehicles per day for principal arterial roads were used, again on an area basis for congestion limits.

Congestion Severity Index (CSI)	CSI is an index of freeway delay in urban areas, which calculates delays travel per million vehicle kilometers (km). CSI uses 1985 HCM calculation and local freeway traffic volume distribution to estimate both recurring and non-recurring delays of each hour of a - typical day. An analysis has been done by Lindley (1987) using this index. Threshold for congestion (v/c ratio of 0.77 or greater) on a section of a freeway was found.
Lane Mile Duration	LMDI was created as a measure of recurring freeway congestion in urban networks (Cottrell, 1991). This index was used to calculate an AADT/C based on the Highway

(*Continued*)

Table 1. *(Continued)*

Index (LMDI)	Performance Monitoring System (HPMS) data to the freeway of the urban section. LMDI can be calculated by using the Equation (16).

$$\text{LMDI} = \sum_{j=1}^{N} \left[\text{congestedlanemines}_j \times \text{congestionduration}_j\right] \quad (16)$$

where, j = Individual segment of N total number of segments in an urban area.

3 LITERATURE SURVEY ON VARIOUS APPROACHES FOR CONGESTION STUDY

3.1 *International status*

Levinson and Lomax (1996) defined congestion and described various methods for congestion measurement. A congestion index was developed by the concept of the position of the vehicle to describes the time and speed, they found a tool of analytical calculation for judging issues and policies and is simply accessible to the non-technical public. For determining the level of congestion on arterial roadways, a fuzzy inference-based measure was proposed by Hamad and Kikuchi (2002). Dias et al. (2009) investigated the occurrence of traffic accidents by using the Bayesian Belief theory on traffic information such as traffic density, volume, and congestion index. They claimed when density and CI increase then the corresponding road section is getting congested, chances for an accident also increase. Lee and Hong (2014) proposed the Traffic Congestion Score (TCS) based concept on the spatiotemporal aggregation method. TCS measures the capacity of the existing road which found a range from 0 to 100 percent by using an approximation ratio of the speed limit. They investigated the traffic congestion of Busan in South Korea during February 2014. Ye et al. (2013) defined three new indicators of congestion to estimate urban road congestion based on travelers' feelings such as travel time satisfaction (TTS), transportation environment satisfaction (TES), and traffic congestion frequency and feeling (TCFF).

A comparative study by using several popular networks such as SVM, PART, J48, ANN, and KNN was done by using the past data for traffic speed, volume, and occupancy to predict whether the road section is congested or non-congested for a future time (Morris et al., 2016). They determined that the J48 network algorithm out of these networks had the best performance and also able to predict the congestion up to 6 minutes and 10 minutes of time with the highest and good performance respectively. Stipancic et al. (2016) proposed measures to represent the level of congestion for spatiotemporal in the Quebec City of Canada using the GPS data based on smartphones of drivers. Congestion measured using the CI during PM-peak periods was found to be positively correlated with an accident frequency (Stipancic et al., 2017).

A resilience-oriented approach for quantitatively assessing recurrent spatiotemporal congestion on urban roads was proposed by Tang and Heinimann (2018). Signs of the congestion index not only describe the intensity of the overall congestion but also indicate the discharge process after its formation. It was found to be applicable in both arterial and freeway cases. In signal-controlled traffic, the metric performs well enough. A congestion measure was proposed by Stipancic et al. (2019) to estimate and visualize traffic congestion in levels using a congestion index (CI). CI was divided into three classes; high, moderate, and low. Using the extreme learning machine (ELM) theory, the symmetric ELM cluster technique was proposed (Yuan and Yang, 2019) as a new fast learning methodology to predict large scale traffic congestion. Ranjan et al. (2020) constructed a hybrid network by adding the convolutional neural network (CNN), long short-term memory (LSTM), and transpose CNN to predict the city-wide traffic congestion using the TOPIS map data in South Korea. With the city scale congestion analysis, Gong et al. (2020) discovered congestion key points in traffic jams using the Baidu map data in the cities of China.

3.2 *Indian status*

Joseph and Nagakumar (2014) investigated the capacity and LOS of roads in Banglore city. Required data like traffic volume, speed, and road geometry were collected through field studies. It was found that mid blocks were congested during peak hours. Level of service F was found for the entire stretch. A quantification of congestion on the urban mid-block sections for Ahembdabad City by including both operational and volume characteristics of traffic movement was done by Patel and Gundaliya (2016). It was shown through this study that defining LOS quantitatively could be a better option. Effect of the width of a carriageway on the level of congestion was also determined. It was concluded that by increasing one lane or widening the carriageway, congestion can be reduced up to 30–50%. Fuzzy inference approach was proposed for measuring the degree of congestion with greater accuracy and low error margin on the major arterial road networks (Kukadapwar and Parbat, 2015). Developed model was demonstrated by using the real-time traffic data of road networks in Nagpur city. Patel and Mukherjee (2015) investigated the assessment of congestion in road networks rather than link flow congestion at Ranchi city, of Jharkhand State, India with the help of Traffic Congestability Value. It was shown that if TCV value was lower, then congestion was higher in the spatial zone and vice versa. In the study of Inner Ring Road, Delhi, Rao, and Rao (2016) established the thresholds of congestion on arterial roads to identify the traffic congestion in terms of speed using floating car data. 57 km/h was used as a free speed measured on these roads. Stream speed observed under normal conditions was between 27 km/h to 33 km/h and the speed of the congested stream was between 18 km/h to 22 km/h. From this study, it was observed that in congested conditions 50 percentile of the vehicles were running below the speed of 19 km/h. Congestion speed was found to be 19 km/h for these roads. Kamble and Kounte (2020) identified road congestion by predicting the traffic speed using the Gaussian process in machine learning based on the GSP vehicle trajectory data.

Jain and Jain (2017) proposed a congestion index for estimating congestion on Delhi roads using categorized vehicular traffic data. Collected Data involved three traffic parameters such as traffic volume, spot speed, and travel time. and These parameters were extracted by manual counting, radar gun, and moving car method respectively in the intervals s of 15 minutes. Using different methodologies on Travel Time Index (TTI), Taxes Transportation Institute (Systematics, 2005) determined the factors affecting congestion. Henry and Koshy (2016) used the travel time index to estimate the congestion at a road segment in Kottayam, Kerala. By analyzing the hourly variance of congestion, it has been observed that vehicles take 81% to 133% more time to cross the segment.

A comparative study of the performance evaluation on the urban link under heterogeneous traffic conditions using the travel time-based indices such as delay, travel time index, congestion index, and planning time index was reported (Karuppanagounder and Muneera, 2017). It was estimated that the performance indicators have higher values when the value of travel time was high on that link. Chikaraishi et al. (2020) investigated the applicability of several types of machine learning models for the prediction of non-recurrent traffic congestion, in which deep neural network models got the highest prediction accuracy.

4   CONGESTION DATA COLLECTION TECHNIQUES

Data collection methods have been classified into two classes, quantitative and qualitative data. For traffic congestion studies reported in literature one or the other of these techniques were used by researchers. Various quantitative data collection techniques have been discussed in Table 2.

Table 2. Quantitative data collection techniques.

Trajectory data	Cross-sectional data	Floating car data
Data based on trajectory provide space-time profiles of all traffic vehicular in a selected location of road segment (Treiber et al., 2013) In this method, the selected segment of traffic can be viewed directly above a tall building by cameras or an airplane. Extracting the data over time from video footage using various software to track the position of each vehicle. Traffic density, lane changes can be measured using this data collection method. A novel traffic congestion estimation approach on roadways in the open-world scene observed from TV cameras placed on poles or buildings was proposed by Li et al. (2008). Palubinkas et al. (2008) presented an approach to estimate congestion on temporal data collected by optical digital camera images. This approach is used to detect the vehicles on a segment of the road by changing detection between two images with a short time interval. Hinz et al. (2007) gave a theoretical overview for upcoming dual-channel radar satellite missions to the surveillance of road networks from space and illustrated the potentials and limitations of real data.	Single-vehicle data or microscopic data: Microscopic data is collected on the specific road section, data can be captured from radar or light barriers by using pneumatic tubes to lying across on the road. Commonly, induction loops are deployed beneath the road surface for microscopic data collection. A single-loop detector may be used to automatically measure at a fixed point on a road section, passing the time of vehicle, traffic volume, and speed when vehicle length is known. The double-loop detector is made up of two (or more) induction loops separated by a fixed segment of road, for example, 1m. Thus, this technique can provide traffic volume, speed, etc. In this technique, we need two inductive loop detectors mounted at many places of the road that is used to classify the vehicle category depending on axle spacing. (Coifman and Cassidy, 2002). This method of data collection is expansive and provides travel time of vehicles between detector stations. Wireless magnetic sensor networks for traffic measurements on freeways and intersections provide an attractive, low-cost alternative to inductive loops	Floating car data (FCD) provide information about a single-vehicle only. These vehicles are specially equipped. In contrast to trajectory data, floating data is captured inside a vehicle. In this method, we can collect georeference coordinates by GPS-receivers that are then "map-matched" to a road on the traffic map. It can determine the travel time and speed from two GPS points in road space. (Taylor, 1992). Other types of sensors (such as radar) are deployed on the probe vehicle to record the travel distance and speed of the leading vehicle (Stipancic et al., 2017; Stipancic et al., 2019). In this way, augmented floating-car data is referred to as extended floating-car data. With the low speed of many vehicles like taxis, van, bus, and trucks of commercial transport agencies or companies present on roads, we are not able to describe the entire traffic by using FCD. Fortunately in free-flow speed and congested conditions where difference does not matter then FCD information becomes relevant: and then this bias vanishes.

Camera-based techniques consist of complex and error-prone methodologies, these methodologies demand automated and robust algorithms to track the vehicles, and thus these techniques are costly for data collection. Also, the camera-based methods give information only at a few hundred meters away from a road segment.

(Cheung et al., 2005). It is used to obtain the number of vehicles, speed, and occupancy. These sensors are used to classify the category of vehicles based on the non-axle spacing of vehicles that cannot be derived from standard loop data.

Aggregated data or macroscopic data: Macroscopic data can be obtained by averaging the aggregate microscopic single-vehicle data over fixed time intervals. Time interval varies between 20s to 5min generally, but most common interval being 60s. Ranjan et al. (2020) proposed another way of collecting macroscopic data. Seoul Transportation Operation and Information Service (TOPIS) is an online open-source web service in South Korea. It is going to use an efficient and inexpensive data acquisition method across the city by taking a snapshot of traffic congestion maps. Similarly, several types of traffic maps in different countries like Google map, Bing map, and Baidu map, are used to determine city-wide congestion levels in real-time.

5 DISCUSSION

Congestion occurs due to various reasons for instance environment, mechanical, human, and infrastructure. Quantitative and qualitative both types of congestion indices have been discussed, suggested by the afore-mentioned researchers. Different congestion data collection methods have been explained which will be useful in the selection of a suitable technique for Indian conditions. Congestion measurement criteria can be adopted based on available budget by adopting speed based or LOS based metrices.

Ridesharing emerges as an important tool with high efficiency with the use of existing assets such as personal vehicles. Recently the role of ridesharing has increased in most of the cities. Ridesharing is a new point-to-point transport model, and has seven features: (1) dynamic pricing, (2) dynamic routing, (3) smart dispatching, (4) customer network effect, (5) demand pooling, (6) feedback collection, and management system and (7) flexible supply base. From ridesharing, we can find four advantages that can automatically reduce traffic congestion by 17%-31% and optimize infrastructure investment by (1) Accelerating public transportation, (2) Providing alternatives to car ownership, (3) Supplementing incomes, (4) Optimising infrastructure timing and location (Chin et al., 2018).

A mobile application based on a new high-tech system linked to smart cameras will be launched in Delhi (Aijaz, 2020). It will be deployed at traffic signals to help ease the chaotic roads. Signals will change to red or green in real-time based on traffic volume and not at fixed intervals.

It will enhance traffic management capabilities and efficiency as well as manpower reduction since there will be no requirement to manage the traffic physically by traffic cops. This system is being adapted based on the studies of best international practices in the cities such as Sydney, Singapore, and Amsterdam. A similar system based on real-time traffic is functional at Closure Home, Coimbatore, Tamil Nadu, India.

6 CONCLUSION

This paper reviews the traffic congestion-based studies along with congestion measures and congestion data collection techniques. Different kinds of congestion measures used by researchers were discussed. Broadly congestion measures were categorized into three parts based on travel time, speed, and level of service. Quantitative traffic congestion data collection techniques reported in literature namely trajectory, cross-sectional, and floating car was also described. It has been observed that the selection of a congestion measure depends on various factors like data type, study/measurement location, traffic characteristics, and availability of equipment for data collection. Study results suggest that artificial intelligence-based traffic congestion prediction models are extensively applied in recent times. It can be concluded that traffic congestion prediction models and techniques based on them can play an important role in mitigating congestion.

7 FUNDING

Financial support to the first author (Manoj Kumar) by the University Grants Commission (UGC), New Delhi, India as a Junior Research Fellowship (JRF) is thankfully acknowledged.

REFERENCES

Aftabuzzaman, M. 2007. Measuring traffic congestion-a critical review. *30th Australasian Transport Research Forum.*
Aijaz, R. 2020. The Smart Cities Mission in Delhi, 2015-2019: An Evaluation. Observer Research.

Anthony, D. 2004. *Still Stuck in Traffic: Coping with Peak-Hour Traffic Congestion.* Brookings Institution Press Washington, DC.

Cheung, S. Y., Coleri, S., Dundar, B., Ganesh, S., Tan, C.-W. & Varaiya, P. 2005. Traffic measurement and vehicle classification with single magnetic sensor. *Transportation Research Record*, 1917(1), pp 173–181.

Chikaraishi, M., Garg, P., Varghese, V., Yoshizoe, K., Urata, J., Shiomi, Y. & Watanabe, R. 2020. On the possibility of short-term traffic prediction during disaster with machine learning approaches: an exploratory analysis. *Transport Policy.*

Chin, V., Jaafar, M., Subudhi, S., Shelomentsev, N., Do, D. & Prawiradinata, I. 2018. Unlocking Cities: The impact of ridesharing across India. Boston Consulting Group. Available at: http://image-src.bcg. com/Images/BCG

Coifman, B. & Cassidy, M. 2002. Vehicle reidentification and travel time measurement on congested freeways. *Transportation Research Part A: Policy and Practice*, 36(10), pp 899–917.

Cottrell, W. D. 1991. Measurement of the extent and duration of freeway congestion in urbanized areas. *Institute of Transportation Engineers Meeting.*

Dias, C., Miska, M., Kuwahara, M. & Warita, H. 2009. Relationship between congestion and traffic accidents on expressways: an investigation with Bayesian belief networks. *Proceedings of 40th annual meeting of infrastructure planning (JSCE).*

Gong, K., Zhang, L., Ni, D., Li, H., Xu, M., Wang, Y. & Dong, Y. 2020. An Expert System to Discover Key Congestion Points for Urban Traffic. *Expert Systems with Applications.* 113544.

Hamad, K. & Kikuchi, S. 2002. Developing a measure of traffic congestion: Fuzzy inference approach. *Transportation research record*, 1802(1), pp 77–85.

Henry, S. & Koshy, B. 2016. Congestion Modelling for Heterogeneous Traffic. *International Journal of Engineering Research & Technology (IJERT)*, 5(02), pp 114–119.

Hinz, S., Meyer, F., Eineder, M., Bamler, R. 2007. Traffic monitoring with spaceborne SAR—Theory, simulations, and experiments. *Computer Vision and Image Understanding*, 106(2-3), pp 231–244.

Jain, S. & Jain, S. S. 2017. A methodology for modelling urban traffic congestion based on ITS. *2017 2nd IEEE International Conference on Intelligent Transportation Engineering (ICITE).* IEEE, 295–299.

Joseph, E. N., Nagakumar, M. S. 2014. Evaluation of capacity and level of service of urban roads. *International Journal of Emerging Technologies and Engineering*, 2(85–91.

Kamble, S. J. & Kounte, M. R. 2020. Machine Learning Approach on Traffic Congestion Monitoring System in Internet of Vehicles. *Procedia Computer Science.* 171(2235–2241.

Karuppanagounder, K. & Muneera, C. 2017. Performance Evaluation of Urban Links under Heterogeneous Traffic Condition. *European Transport*, (65), pp 1–10.

Kukadapwar, S. R. & Parbat, D. 2015. Modeling of traffic congestion on urban road network using fuzzy inference system. *American Journal of Engineering Research*, 4(12), pp 143–148.

Lee, J. & Hong, B. 2014. Congestion score computation of big traffic data. *2014 IEEE Fourth International Conference on Big Data and Cloud Computing*, IEEE, 189–196.

Levinson, H. S. & Lomax, T. J. 1996. Developing a travel time congestion index. *Transportation Research Record*, 1564(1), pp 1–10.

Li, L., Chen, L., Huang, X. & Huang, J. 2008. A traffic congestion estimation approach from video using time-spatial imagery. *2008 First International Conference on Intelligent Networks and Intelligent Systems*, IEEE, 465–469.

Lindley, J. A. 1987. Urban freeway congestion: quantification of the problem and effectiveness of potential solutions. *ITE journal*, 57(1), pp 27–32.

Lomax, T., Turner, S., Shunk, G., Levinson, H., Pratt, R., Bay, P. & Douglas, G. J. T. 1997. NCHRP report 398: quantifying congestion. *National Research Council, Washington, DC.*

Lomax, T. J. 1990. Estimating transportation corridor mobility. *Transportation Research Record*, 1280.

Lomax, T. J. & Schrank, D. L. 2005. The 2005 urban mobility report.

Morris, B., Paz-Cruz, A., Mirakhorli, A. & de la Fuente-Mella, H. 2016. Traffic congestion classification using data mining techniques. *Economics, Social Sciences.*

Nakat, Z., Herrera, S. & Cherkaoui, Y. 2014. the Government of Egypt, P. 2014. Cairo traffic congestion study. *Report prepared by ECORYS Nederland BV and SETS Lebanon for the World Bank and the Government of Egypt, Phase.*

Palubinskas, G., Kurz, F. & Reinartz, P. 2008. Detection of traffic congestion in optical remote sensing imagery. *IGARSS 2008-2008 IEEE International Geoscience and Remote Sensing Symposium.* IEEE, II-426-II-429.

Patel, J. & Gundaliya, P. J. 2016. Estimation of Level of Service through Congestion–A Case Study of Ahmedabad City. *International Research Journal of Engineering and Technology.*

239

Patel, N. & Mukherjee, A. B. 2015. Assessment of network traffic congestion through Traffic Congestability Value (TCV): a new index. Bulletin of Geography. *Socio-economic Series*, 30(30), pp 123–134.

Ranjan, N., Bhandari, S., Zhao, H. P., Kim, H. & Khan, P. J. I. A. 2020. City-Wide Traffic Congestion Prediction Based on CNN, LSTM, and Transpose CNN. *IEEE Access*. IEEE. 8(81606–81620).

Rao, A. M. & Rao, K. R. 2012. Measuring urban traffic congestion-a review. *International Journal for Traffic & Transport Engineering*, 2(4), pp.

Rao, M. & Rao, K. R. 2016. Identification of traffic congestion on urban arterials for heterogeneous traffic. *Transport Problems*, 11

Shrank, D., Turner, S. & Lomax, T. 1993. Estimates of Urban Roadway Congestion. *College Station, TX: Texas Transportation Institute (Texas A&M University)*.

Stipancic, J., Miranda-Moreno, L. & Labbe, A. 2016. Measuring Congestion Using Large-Scale Smartphone-Collected GPS Data in an Urban Road Network. *Proceedings of Transportation Association of Canada Annual Conference, Toronto, Ontario, Canada*.

Stipancic, J., Miranda-Moreno, L., Labbe, A. & Saunier, N. 2019. Measuring and visualizing space-time congestion patterns in an urban road network using large-scale smartphone-collected GPS data. *Transportation Letters*, 11(7), pp 391–401.

Stipancic, J., Miranda-Moreno, L. & Saunier, N. 2017. Impact of congestion and traffic flow on the crash frequency and severity: application of smartphone-collected GPS travel data. *Transportation Research Record*, 2659(1), pp 43–54.

Systematics, C. 2005. Traffic congestion and reliability: Trends and advanced strategies for congestion mitigation (No. FHWA-HOP-05-064). United States. Federal Highway Administration.

Tang, J. & Heinemann, H. R. 2018. A resilience-oriented approach for quantitatively assessing recurrent spatial-temporal congestion on urban roads. *PLoS One*. 13(1), pp e0190616.

Taylor, M. 1992. Exploring the nature of urban traffic congestion: concepts, parameters, theories, and models. *Proceedings 16th ARRB conference, 9-13 Nov 1992, Perth, Western Australia*. Vol 16, Part 5.

TomTom. 2020. TomTom Traffic Congestion Index Report-2020.

Treiber, M., Kesting, A. 2013. *Traffic flow dynamics*. Traffic Flow Dynamics: Data, Models and Simulation, Springer-Verlag Berlin Heidelberg.

Ye, L., Hui, Y. & Yang, D. 2013. Road traffic congestion measurement considering impacts on travelers. *Journal of Modern Transportation*. 21(1), pp 28–39.

Yuan, C. & Yang, H. 2019. Research on K-value selection method of K-means clustering algorithm. *Multidisciplinary Scientific Journal*. 2(2), pp 226–235.

Recent Trends in Communication and Electronics – Sharma et al. (Eds)
© 2021 Taylor & Francis Group, LLC, ISBN 978-1-032-04572-6

Literature review: Sleep stage classification based on EEG signals using artificial intelligence technique

Satyam Mishra & Rajesh Birok
Delhi Technological University, New Delhi, India

ABSTRACT: Psychiatry and neurology are one of the most important diagnostic approaches. Sleep specialists have taken up time consuming and challenging activities. However, pre signal processing has always been helpful in terms of accuracy. Sleep analysis is centered on a comprehensive EEG study and can only be read and interpreted by a specialist in this area. Computational intelligence approaches have shown a positive outcome in different sleep stage classifications. Sleep classification plays a valuable role as it can help identify different sleep associated disorders such as anxiety, restless leg syndrome (RLS), parasomnia, and several more. While reviewing this sleep classification concept based on EEG signals, I have categories methods in supervised learning, unsupervised learning and deep learning. Specific algorithms for all stages of sleep function differently. As every procedure has its own different classifiers. Having said that SVM, k-means and CNN are the algorithms often used and deliver decent results if not better.

1 INTRODUCTION

Sleep Human sleep has been classified on the basis of rapid eye movement. Non-rapid eye moment (NREM) and rapid eye moment (REM). Further NREM is classified into 4 stages, namely, W(wakefulness), N1, N2, and N3. Adult human being has 60-70% sleep in only N3 and REM stage. There are different characteristics of all sleep stages and generates different sleep waves. For example, beta waves are generated by the human brain in the wakefulness(W) stage, low amplitude mixed frequency (LAMF) (Mishra et al 2020) in sleep stage N2. Sleep classification is important as suppose if a person misses REM sleep or N3 and REM sleep stage together, sleep deprivation can occur and it will be severe for that person's health. Sleep deprivation for a long time can increase a toxic protein called beta-amyloid which shuts the memory door to create any new memory and, in some cases, it could cause dementia later. While recording the EEG signal human brain which is active also gives the reading of muscle movements, eye moments and other body organ actively giving signal which the brain processes accordingly, which makes sleep classification a more tedious task to tackle. Hence the signals should be pre-processed and this review has considered only that papers where signal acquisition is been done properly. In any technique it is very important to get the required attributes to classify signals on the basis of mathematical grounds (Nicola.M 2019).

Generally, sleep groups are based on EEG waveforms. But it is a more common technique that does not offer a smooth score. The waveform and frequency can be seen, but it is not worthy only to get frequency for many sleep diseases. A variety of papers in the field of sleep classification are published and the positive and weak aspect of each technique are considered in this review paper. Papers which are published are majorly classified based on the classification methodology adopted by the researchers namely supervised learning, unsupervised learning, and finally reinforcement learning.

DOI: 10.1201/9781003193838-44

2 IMPLEMENTED METHODS

In artificial intelligence, classifying a problem that is fundamentally repetitive and complex becomes a little simpler if not better. It is difficult to mathematically deal with non-stationary signals such as PSG signals. In order to deal with such complex issues Artificial Intelligence requires an imperative structure, an architecture. In general, the training and test module is common. Most of the problem about classification is iterative in nature. We will consider all three machine learning techniques used in sleep classification based on EEG signals to carry out a thorough analysis.

2.1 Supervised learning

In simple terms, by knowing the training set and predicting the result or outcome is called supervised learning. In sleep classification, support vector machine (SVM) is often used. There are various version of SVM such linear regression, Multi-class SVM and Dendrogram SVM. (Huang et al. 2013) and (Sharma et al. 2018) uses multi-class SVM and shown good promise. Random forest is also a supervised learning technique that has been used (Fraiwan et al.2012) There accuracy has been summarized in different stages are given below:

Table 1. Classification accuracy of various sleep stages reported by different researchers employing supervised learning techniques.

Authors	Classifier	W	N1	N2	N3
Huang [2013]	Multi-SVM	76.85%	34.24%	72.39%	88.67%
Sharma [2018]	Multi-SVM	95.41%	17.39%	76.38%	57.11%
Fraiwan [2012]	Random Forest	93.33%	43.22%	84.76%	68.37%

2.2 Unsupervised learning

Unsupervised learning requires learning from results, but without the aim of estimation. This is because perhaps the dataset is not supplied with the desired response variable (label) or one prefers not to show the answer. It can also be used as a pre-processing stage for supervised learning. In supervised algorithms linear classification are easy to implement. However, when the given boundary is non-linear in nature, classification becomes difficult. So, unsupervised learning is suited best. K- means clustering, Gaussian mixture model (GMM), agglomerative clustering, neural networks are some common example which are used in sleep classification. Some uses K-means clustering which shows exceptional performance in class N3. (Sors, et al 2018) uses bragging agglomerative clustering. However, using neural network shows great results in REM class and N3 class, which requires separate discussion. Accuracy of discussed methods are summarized in the table given below:

Table 2. Classification accuracy of various sleep stages reported by different researchers employing unsupervised learning techniques.

Authors	Classifiers	W	N1	N2	N3	REM
Shuyuan [2015]	K-mean	76.14%	11.76%	69.94%	97.12%	94.44%
Sors [2018]	Bragging	96.6%	27.48%	82.93%	76.92%	69.57%
Rodriguez [2014]	J-mean	84%	15%	91%	59%	38%

2.3 Neural network

The neural network is a collection of algorithms that are designed to recognize the links underlying the data through a process that mimics the functioning of the body nervous system. Throughout this way, neural networks are applied to biological or artificial neuron structures. Neural networks should be tailored to evolving inputs such that the network should achieve the best outcome possible without redefining the performance criterion. Neural networks, embedded in artificial intelligence, are rapidly becoming common in trading systems growth. Several neural network combinations are used to give the N3 and REM sleep class a strong classification. Feedforward network (CNN) is used for the recurrent neural network (RNN). The feedforward network is back-propagated neural networks. The best classification is accomplished by using CNN and various RNN variants, including vanilla RNN, long-term memory (LSTM) and graded recurring unit (GRU).

Table 3. Classification accuracy of various sleep stages reported by different researchers employing neural network.

Authors	Classifier	W	N1	N2	N3	REM
Hsu [2013]	ElmanRNN	70.8%	36.70%	97.30%	89.70%	89.50%
Sors [2018]	CNN	91.4%	34.92%	89.24%	85.08%	85.82%
Wei [2017]	CNN	92.7%	26.66%	87.4%	87.05%	82.74%
Michielli [2019]	LSTM RNN	95.29%	61.09%	89.48%	91.66%	83.76%
Mishra [2020]	CNN RNN	90.74%	46.02%	89.4%	87.49%	86.92%

3 DISCUSSION

We understand that in-depth learning allows the new aspect to the classification of sleep. Furthermore, the iterative character of different hidden layers indicates that the fine line between sleep stages will definitely deplete by more refined datasets and the repetition of various hidden layers. The findings revealed that the method could be used on different EEG channels (F4-EOG (left), Fpz-Cz, and Pz- Oz) without modifying the software design and the training algorithms. In comparison with state-of- the-art hand-engineered procedures for both the Sleep-EDF and MASS datasets, the total exactness and macro F1 score have been similar, with different features, such as sampling rates, entropy and non-linear attributes (AASM) and R&K standards (shuyuam et al 2015). The results show that the spatial data obtained from the extraction of features leads to better performance. RNN is classified as awake to the

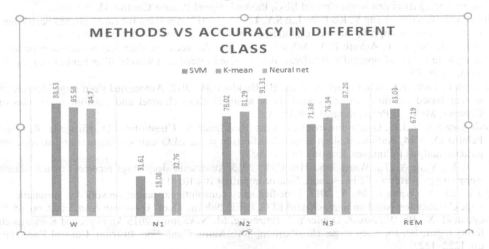

Figure 1. Comparative analysis of different method with its accuracy.

NREM N1 deep sleep since this transition from alertness to sleep N1 phase is so seamless that differentiation requires a backward radiating neural network.

Now, we need to consider which procedures are to be used, since different sleep stage classifications require different methods. Comparing the accuracy of various grades of different approach.

As we can see different sleep stage has different accuracy with various method. After this thorough study we can for class,W, N2, SVM and its variation works better, while REM vs N3 will be classified well in deep learning method, whereas for class N2, N3 we can use k-mean and its variation.

4 CONCLUSION

Pre-processed EEG signals make classification a bit less tedious work. After reading different methods for classification on a problem which is different for all other humans, it is safe to say that no single method will be able to classify all the sleep stages, even after so many iterations it not evident that the algorithm can be helpful in classifying all stages. One must understand that supervised learning, unsupervised learning, and deep learning will work efficiently on some stages, certainly on not all stages. Its combination can be lethal while dealing with such a mathematically complex and exhausting process. To classify in stage W, N1, N2, we may say that supervised learning or unsupervised learning would be appropriate, as classification is not that tiring and ease of implementation is not a crucial factor. SVM, K-mean and its variation algorithms, which do not have REM or N3 stages, should be implemented.

The limitation of sleep stage classification is with increasing accuracy, it will be compensated by cost. Sleep is an indispensable part of human life. Sleep analyzes enable us identify the most "profound" pacing of our sleep and quality sleep at the same time. Sleep problems can develop after not having a good quality of sleep for a period of time. Different sleep cycles are created by different sleep disorders, so understanding the sleep classification will help identify the condition.

REFERENCES

Mishra S, Birok R, July 2 nd -4 th, 2020. "Sleep Classification Using Cnn And Rnn On Raw EEG Single- Channel" in IEEE Industry Applications society(USA) & IEEE Kolkata section 2020 international conference on Computational performance Evaluation organized by department of biomedical engineering, North-Eastern Hill University (NEHU) Shillong, Meghalaya.

Nicola M, Rajendra U, Acharya, Molinari. F 2019. "Cascaded LSTM recurrent neural network for automated sleep stage classification using single-channel EEG signals" computers in biology and medicine 106.

Sors A, Bonnet. S,Mirek .S, Vercueil.L.J.-, Payen, F. 2018 A convolutional neural network for sleep stage scoring from raw single-channel EEG, Biomed. Signal Process Control 42, 107–108.

Huang C.S, Lin C.L, Yang V, Ko L.W, Liu S.Y, Lin C.T. 2013, Applying the fuzzy cmeans based dimension reduction to improve the sleep classification system, IEEE Int. Conf. Fuzzy Syst, pp. 1–5.

Sharma .M, Goyal D., Achuth P. V. Acharya U.R, 2018 An accurate sleep stages classification system using a new class of optimally time-frequency localized three-band wavelet filter bank, Comput. Biol. Med. 98 58–75.

Fraiwan.L, Lweesy.K, Khasawneh.N, Wenz H., Dickhaus.H, 2012, Automated sleep stage identification system based on time–frequency analysis of a single EEG channel and random forest classifier, Comput. Methods Progr. Biomed. 108,10–19.

Rodríguez-Sotelo J.L., Osorio-Forero A., Jiménez-Rodríguez A., Cuesta-Frau D., Cirugeda- Roldán E., Peluffo D. 2014, Automatic sleep stages classification using EEG entropy features and unsupervised pattern analysis techniques, Entropy 16,6573–6589.

Hsu Y.-L., Yang Y.-T.,. Wang J.-S,. Hsu C.-Y, 2013 Automatic sleep stage recurrent neural classifier using energy features of EEG signals, Neurocomputing 104,105–114.

Wei L., Lin Y., Wang J., Ma Y. 2017, Time-frequency convolutional neural network for automatic sleep stage classification based on single-channel EEG, IEEE 29th Int. Conf. Tools with Artif. Intell, pp. 88–95.

Shuyuan H. X., Bei W., Jian Z. Qunfeng, Z. Junzhong, M. Nakamura, 2015 An improved K-means clustering algorithm for sleep stages classification, 54th Annu. Conf. Soc. Instrum. Control Eng. Japan, pp. 1222–1227.

Recent Trends in Communication and Electronics – Sharma et al. (Eds)
© 2021 Taylor & Francis Group, LLC, ISBN 978-1-032-04572-6

Medical face mask detection using modified MobileNetV2

Vikrant Choudhary & Krishna Kant Singh
Department of ECE, KIET Group of Institutions Delhi-NCR, Ghaziabad, India

Akansha Singh
Department of CSE, ASET, Amity University, Noida, Uttar Pradesh, India

ABSTRACT: In this paper, a simple and computationally efficient approach as per the complexity has been presented for Face mask detection using a Deep Learning architecture called MobileNetV2 including some additional specifications for the improvisation of the results. The secondary objective is to keep the pre/post-processing of the images minimal. The presented model is trained on images from the RMFD dataset which includes masked and non-masked people images. The aim is to achieve an accuracy above 95% with a minimum number of parameters after evaluation.

Keywords: computer vision, artificial intelligence, coronavirus, convolution neural network

1 INTRODUCTION

In December 2019, some extreme instance of pneumonia came into account in Wuhan, China and within two months they spread drastically throughout the globe. The World Health Organization further declared this outbreak as pandemic [1]. As of now no specific medicine or vaccination is available to prevent or cure COVID-19. Also, Fever, dry cough and exhaustion are the most common symptoms of COVID-19 [2]. Government is continuously insisting on using medical masks to slow down COVID-19 transmission, the most widely recognized method of transmission of SARS-CoV-2 is likely through respiratory droplets of size 5 µm to 10 µm ejected in the surroundings when speaking, sniffing or coughing [3].

In biometric knowledge, Automatic human face detection is one of the most appealing fields which is widely accepted by the technology community. The objective of this technique is to detect faces in a digital image or videotape in presence of varieties, for example, present, light, impediment and expression [4]. The main advantage of this face recognition in this COVID period Is that it does not require any kind of physical contact from an individual. Face detection has improved a lot in terms of accuracy and with the introduction of the Viola-Jones object detection framework (Haar cascade classifier), the computation time is also reduced significantly [5]. the algorithm is treated on a lot of true images (images of faces) as well as False images (images without faces) to train the classifier.

In this research paper, a deep learning model for detection of Face mask is proposed whose architecture is inspired by MobileNetV2 [6]. the networks take Images as raw input from our RMFD (Real-World Masked Face Dataset) Figure 1. [7] and maps it to the desired output, the network uses depth separable CNN at the core [8]. We have compared the MobileNetv2 performance with a Resnet 50 [9], Resnet101 and MobileNet1 [10] to see how it outscored all other models.

DOI: 10.1201/9781003193838-45

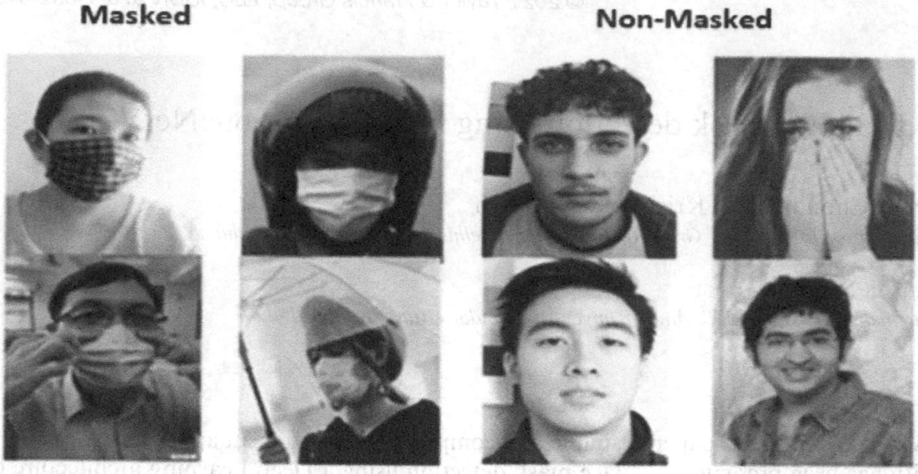

Masked   **Non-Masked**

Figure 1.   Sample images from the dataset.

2   PROPOSED METHOD

In the proposed method, we have trained MobilenetV2 model. Images from the dataset are loaded along with the labels, image Normalization is performed in which converts the images into an array and dividing them by 255 is done. It helps in describing the range of the image on a scale of 0.0 -1.0. after this we split the data into 80:20 ratio, 80% for the training and rest 20 for validation. The augmented training set images are fed into the convolutional neural network with all parameters set, Softmax optimizer is applied which outputs the prediction of classes in terms of probabilities. In the last step, performance evaluation is done. The process flow is in Figure 2.

2.1   Model

MobileNetV1[10] is a deep neural network specially designed to overcome the issues of high computational power and large memory storage without affecting original accuracy. Mobile-NetV2 is an upgradation in accuracy as well as compactness, built at Google Inc. [5]. Inverted Residuals MobileNetV2 follows inverted approach compress->enlarge->compress. Initially, a 1x1 convolution enlarges the neural network because the count of parameters had been already compressed by a 3x3 depth wise convolution Figure 3. Then after one more 1x1 convolution is applied so that no of initial and final channels will be equal. ReLU6 is the non-linearity used because unlike ReLU, It performs very well when used with low-accuracy computation. [7]

2.2   The architecture

The proposed model initially contains 32 filters fully connected convolution layer, further followed by 19 residual bottlenecks shown in Table 1. Kernel of size 3 × 3 along with batch normalization and dropout is used at every single stage during training.

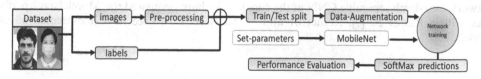

Figure 2.   The proposed method with all steps.

246

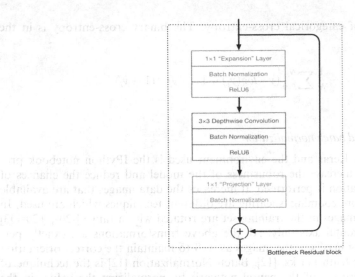

Figure 3. Bottleneck Residuals.

Table 1. Detailed network architecture.

Input	Operator	t	c	n	s
$224^2 \times 3$	conv2d	-	32	1	2
$112^2 \times 32$	bottleneck	1	16	1	1
$112^2 \times 16$	bottleneck	6	24	2	2
$56^2 \times 24$	bottleneck	6	32	3	2
$28^2 \times 32$	bottleneck	6	64	4	2
$14^2 \times 64$	bottleneck	6	96	3	1
$14^2 \times 96$	bottleneck	6	160	3	2
$7^2 \times 160$	bottleneck	6	320	1	1
$7^2 \times 320$	conv2d 1×1	-	1280	1	1
$7^2 \times 1280$	avgpool 7×7	-	-	1	-
$1 \times 1 \times 1280$	conv2d 1×1	-	k	-	

they used a factor of 6. t denotes channels expansion rate. c denotes the total input channels and n represents how frequently the block is used again. s shows first repetition of a block with stride of 2 for the down sampling process.

3 TRAINING METHOD

Training procedure has been carried out on 80% of the images available out of the 3843 images in the dataset. The actual number of images will be more because of the new transformed images that will be added after the image augmentation process. As per the architecture of the network the total trainable parameters 164,226 out of 2,422,210 the non-trainable parameters are 2,257,984.

3.1 Loss function

In the model loss function is binary cross-entropy [11]. Here, only two possible outcomes (Masked face, non-masked (0 or 1)) are there. So, here binary cross-entropy is used as

247

the loss function instead of categorical cross-entropy. The binary cross-entropy is in the below form:

$$L(y, \hat{y}) = -\frac{1}{N} \sum_{i=0}^{N} (y * log(\hat{y}_i) + (1 - y) * (1 - \hat{y}_i)$$

3.2 *Image augmentation and batch normalization*

The implementations are in Keras and the environment used is the IPython notebook provided by Google Colab. to increase the robustness of the model and reduce the chances of overfitting, Image Augmentation is performed. It increases the data images that are available in the dataset. Image rotation, zooming, shearing are different techniques which are used. In the image rotation, all the images in the training set are rotated with a range [-20°, +20°] [3] and flipped along the horizontal axis only. All the above transformations are exactly performed on the corresponding masks of the images as well to maintain the correct orientation of feature images with their truth masks. [12]. Batch Normalization [12] is the technique of fastening up the learning process of the neural network by normalizing the values in the hidden layers similar to the principle behind the normalization of the features in the data or activation values. In proposed the batch normalization layer is present after every convolution layer.

4 EXPERIMENTAL DESIGN AND RESULTS

4.1 *Dataset*

The RMFD (Real-World Masked Face Dataset) contains 3843 images belonging further to two classes either face with a mask or without a mask [7]. The non-mask section contains 1918 normal faces with various illumination, pose, occlusion etc. For the training dimension of each image is fixed to 224 x 224 before feeding it into the network. batch size of 32. The dataset has been provided and sponsored by the National Engineering Research Center for Multimedia Software (NERCMS), School of Computer Science, Wuhan University.

4.2 *Performance evaluation*

We are using precision and recall as evaluation metrics, defined as follows.

$$precision = \frac{True\ Positives}{True\ Positives + False\ Positives}$$

$$recall = \frac{True\ Positives}{True\ Positives + False\ Negatives}$$

4.3 *Result and analysis*

The performance of MobileNetV2 is compared to various other models including Mobile-NetV1, ResNet50[9] and ResNet101, the accuracies for both masked and non- masked faces along with total no of parameters is given in Table 2.

The curves include the loss curve and the accuracy curve with respect to the epochs along the horizontal axis. Initially, for the MV2 test set the loss is above 0.121 which gradually declined and reached 0.021 while the accuracy began 0.96 and ended up at 0.99.

248

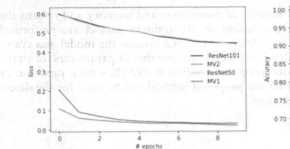
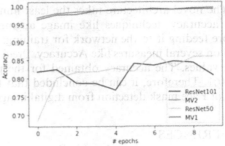

Figure 4. The testing curves (a) losses (b) accuracy.

Table 2. Performance statistics after testing for 10 epochs.

| Model Name | With Mask | | Without Mask | | No of Parameters |
	Precision	Recall	Precision	Recall	
Mobile V2	.99	.99	.99	.99	2.4 M
Mobile V1	.99	.99	.99	.99	3.3 M
ResNet50	.90	.81	.82	.91	23.8 M
ResNet101	.92	.68	.74	.94	42.9 M

Figure 5. Live camera results.

4.4 Application based results

We have implemented the model on OpenCV using haarcascade frontal classifier [5]. The following are the results Figure 5(a). Detecting multiple masked faces with different distances and in 5(b) Mask predictions in a helmet with covered with front glass. An accurate prediction in these unusual circumstances shows how well the model is performing. In 5(c) face covered with hand shows no mask and these are some of the actual circumstances which the model will definitely encounter in the real world.

5 CONCLUSION AND DISCUSSION

Most of the bio-metrics require physical contact with the individual for any kind of verification or detection but during this covid crisis its highly unsafe, to overcome this challenge our proposed method uses a vision-based technique which does not require any kind of physical touch. which can be further integrated with surveillance cameras or hardware like Raspberry pi. The novel architecture takes the least number of parameters and memory and returns the best accuracy. techniques like image augmentation on the training dataset are performed before feeding it to the network for training. After the training process the model was evaluated on several measures like Accuracy, Precision, Recall and number of parameters for statistical values. The accuracy obtained for the proposed method is 99.61% with a precision of 99.00%. Therefore, it can be concluded that the proposed method can be used for the detection of Face mask detection from digital images.

REFERENCES

1. Dataset: https://github.com/X-zhangyang/Real-World-Masked-Face-Dataset by National Engineering Research Center for Multimedia Software (NERCMS), School of Computer Science, Wuhan University.
2. Gondauri D, Mikautadze E, Batiashvili M. Research on COVID-19 Virus Spreading Statistics based on the Examples of the Cases from Different Countries. Electron J Gen Med. 2020;17(4):em209. https://doi.org/10.29333/ejgm/7869
3. He K, Zhang X, Ren S, Sun J arXiv:1512.03385v1 [cs.CV] 10 Dec 2015 *"Deep Residual Learning for Image Recognition"* Microsoft Research
4. Howard A, Zhu M, Chen B, Kalenichenko D, Wang W, Weyand T, Andreetto M, Hartwig A; arXiv:1704.04861v1 [cs.CV] 17 Apr 2017 *"MobileNets: Efficient Convolutional Neural Networks for Mobile Vision Applications"* Howard A, Zhu M, Chen B, Kalenichenko D, Wang W, Weyand T, Andreetto M, Hartwig A.
5. Howard, J.; Huang, A.; Li, Z.; Tufekci, Z.; Zdimal, V.; van der Westhuizen, H.; von Delft, A.; Price, A.; Fridman, L.; Tang, L.; Tang, V.; Watson, G.L.; Bax, C.E.; Shaikh, R.; Questier, F.; Hernandez, D.; Chu, L.F.; Ramirez, C.M.; Rimoin, A.W. Face Masks Against COVID-19: An Evidence Review. Preprints 2020, 2020040203 (doi: 10.20944/preprints202004.0203.v1).Navabifar, Emadi, Yusof, Khalid *"A short review paper on Face detection using Machine learning"*.
6. Howard, J.; Huang, A.; Li, Z.; Tufekci, Z.; Zdimal, V.; van der Westhuizen, H.; von Delft, A.; Price, A.; Fridman, L.; Tang, L.; Tang, V.; Watson, G.L.; Bax, C.E.; Shaikh, R.; Questier, F.; Hernandez, D.; Chu, L.F.; Ramirez, C.M.; Rimoin, A.W. Face Masks Against COVID-19: An Evidence Review. Preprints 2020, 2020040203 (doi: 10.20944/preprints202004.0203.v1).Navabifar, Emadi, Yusof, Khalid *"A short review paper on Face detection using Machine learning"*.
7. Ho Y, Wooke S; 27 December 2019 **DOI**: 10.1109/ACCESS.2019.2962617 *"The Real-World-Weight Cross-Entropy Loss Function: Modeling the Costs of Mislabeling"*.
8. Ioffe, Szegedy, 11 February 2015. "*Batch Normalization: Accelerating Deep Network Training by Reducing Internal Covariate Shift*," [Online]. Available: https://arxiv.org/abs/1502.03167.
9. Padilla R, Filho C.F.F and Costa M.G.F; International Science Index, Computer and Information Engineering Vol:6, No:4, 2012 waset.org/Publication/2910; *"Evaluation of Haar Cascade Classifiers Designed for Face Detection"*.
10. Sandler M, H Andrew, Zhu M, Zhmoginov A, Chieh L Google Inc; arXiv:1801.04381v4 [cs.CV] 21 Mar 2019 *"MobileNetV2: Inverted Residuals and Linear Bottlenecks"*.
11. Shorten, C., Khoshgoftaar, T.M. A survey on Image Data Augmentation for Deep Learning. *J Big Data* **6**, 60 (2019). https://doi.org/10.1186/s40537-019-0197-0
12. WHO, "*WHO |Coronavirus*," World Health Organization, 2020[Online] available at: https://www.who.int/health-topics/coronavirus#tab=tab_2
13. Zhang R, Zhu F, Liu J and Liu G; August 2019 IEEE Transactions on Information Forensics and Security PP(99):1 DOI: 10.1109/TIFS.2019.2936913 *"Depth-wise separable convolutions and multi-level pooling for an efficient spatial CNN-based steganalysis"*.

Recent Trends in Communication and Electronics – Sharma et al. (Eds)
© 2021 Taylor & Francis Group, LLC, ISBN 978-1-032-04572-6

Monitoring system for health

Shahbaz Alam
Department of Electronics and Communication Engineering, ABES Engineering College, Ghaziabad, UP, India

Chhavi Puri, Priyanshu Tyagi, Sia Saini & Yash Kumar
ABES Engineering College, Ghaziabad, UP, India

ABSTRACT: In this paper, we are proposing an advancement of the already designed health monitoring system by using a heart beat rate, respiratory rate and ECG monitor system. It offers the advantage of portability. The paper focuses on: how microcontroller board is used to analyze different data/input from patients in real-time, how to use data from different sensors, heart beat rate, respiratory rate and ECG and fire an alarm in case of any emergency or abnormality faced by the patient. This is very useful for future analysis and review of a patient's health condition. This system can also be useful in controlling and monitoring a patient's health and/or athletic people's health for a long time period. The program reads, stores and analyses the rate of heart beat signals, respiratory rate and ECG signals continuously. Hardware and software architecture is tailored to a single-chip microcontroller-based device, thus minimizing scale. Also all the process parameters within an interval selectable by the user are recorded online. The first tests were encouraging.

Keywords: Health, Heart Beat, Respiratory system

1 INTRODUCTION

In the field of health surveillance systems, the most crucial and critical people are aged 40 years and above. The people aged above 40 years tend to show more health issues than users below 40 [1-6]. But continuous monitoring of critical parameters, commonly, heart rate, respiratory rate and ECG from a remote location is very difficult. In a hospital, the health care persons and professionals such as nurses and/or doctors have to check and monitor the health of each and every patient physically, due to which continuous monitoring is not possible. Thus, any critical situation is very difficult to be found unless the health care persons and professionals check the patient's health continuously. This may be an issue for the doctor, who has to take care and provide health services to many in the medical facility. To keep alert, connected and to track critical and emergency health conditions, a health monitoring system of heart rate, respiratory rate and ECG is studied and drawn up in this paper. In the next category, the proposed system is explained with the help of block diagram. In the next to next category the hardware is presented for the above-mentioned set-up. The implementation is mentioned in section IV. Then results are mentioned in section V. Conclusions and justification is drawn at the end.

DOI: 10.1201/9781003193838-46

2 PROPOSAL

The proposal presents a methodology of health monitoring which gives people continuous service. The health surveillance system of heart rate, respiratory rate and ECG consists of three sections, one for the patient section, one for the operating system section, and the third for the communication unit. Heart rate, Respiratory rate and ECG signals are calculated with the help of the heart rate sensor, respiratory sensor and ECG sensors respectively, and afterwards processed by a microcontroller, AT89S52 [7-9]. The information is used by the microcontroller to alert the hospital person by firing an alarm in case of crisis or abnormal situation. It can help the doctors in diagnosis of serious health issues and can improve the efficiency, efficacy and quality of health management. When the measured heart rate, respiratory rate or ECG exceeds the normal range or if the pulse calculated is unusual or irregular, it activates an alarm.

3 HARDWARE

This paper proposes a model of Patient Health Monitoring System, with different components like fall detection and sleep pattern analysis. The sensors utilized as a part of this project are Accelerometer and Gyroscope (MPU6050), Heart beat sensor, Body temperature sensor, and blood oxygen level (MAX30100), and Proximity sensor (KY032). These sensors work autonomously of each other. The measured reading from the sensor is broke down for the patient and is made accessible to the specialist or to any concerned individual in the type of the web or smart phones. This web interface and additionally versatile application serves as the user interface for this model. The other element added to this application is examination of the information in past to caution visualizing the latest and the current reading of the exposure of the patients monitored, along with the display of graph. Another element added to this application is investigation of the information in past to caution the specialist and patient about huge changes event, or make an alarm to specialist or any concerned individual related with the patient when it sees any probability of therapeutic crisis. The interfacing between the equipment and the product part is done on the stage of AWS IoT. The readings are sent to AWS IoT through. The Modular diagram of the patient monitoring system is described in Figure 1.

3.1 *Respiratory rate sensor*

The breathing rate sensor contains the novel piezoelectric technology of PMD while the lobe contains the electronics and the rechargeable batteries. When applied to the skin of the patient, the single use Sensor translates and outputs the breathing deflections into the Lobe as

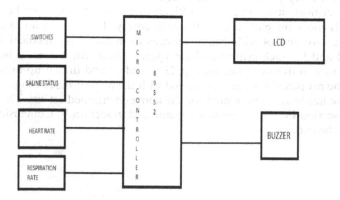

Figure 1. Flow chart.

Figure 2.   Respiratory rate Sensor.

a varying low voltage signal [10]. The groundbreaking Sensor Engineering of PMD Solution contributes to a range of clinical applications and is a cure.

3.2   *Heart beat sensor*

The Pulse Sensor provides an easy means of studying heart rate. This sensor tracks blood flow through a clip, which can also be used between a thumb and index finger on a fingertip or on the skin [11]. Heart rate varies from person to person. For an adult person at rest 72 pulse per minute is accepted. Athletes who are active have usually lower heart rate compared to less active citizens.

Kids tend to have faster heart beat rates (about 90 per minute), but there are major variations too. In totality, the heartbeat sensor can be used just like any other sensor connected to a device.

3.3   *Microcontroller (AT89S52)*

The microcontroller in use is AT89S52. The Microcontroller reads different sensors, here, heart beat sensor, respiratory sensor and ECG signals. The processed output is then used for analysis, if any sensor reading is not in the normal range then an alarm triggers ON. The same data is also sent to the LCD screen. The programming of the microcontroller is performed using Embedded C, a language of middle level controller modules.

High-efficiency, Low-power Microchip, 8-bit Automatic Voltage Regulator (AVR) reduced instruction set computer (RISC) microcontroller blends 256 KB of ISP flash memory, 8 KB of SRAM, 4 KB of EEPROM. It has 86 Input/output pins for general purpose, 32 registers for general purpose, real-time clock, 6 versatile mode timer and counters, PWM, 4 USARTs, 2-wire byte-oriented serial interface, 16-channel 10-bit Analog/Digital converter, and on-chip debugging

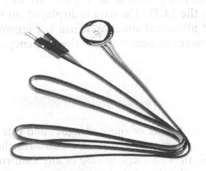

Figure 3.   Heart Beat Sensor.

Figure 4. Mictrocontroller-AT89S52.

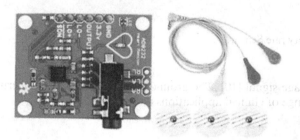

Figure 5. ECG Sensor.

JTAG port. At 16 MHz the system reaches a 16 MIPS throughput and operates between 4.5-5.5 volts.

3.4 *ECG sensor*

The module AD8232 cuts out 9 IC connections to which pins, wires, or other connectors can be soldered. The necessary pins to connect with a microcontroller or other development board are SDN, LO+, LO-, OUTPUT, 3.3V, GND to run this display. Custom sensors can be mounted and used with the help of RA (Right Arm), LA (Left Arm), and RL (Right Leg) pins. A Driven indicator light is also present which pulsates to a heartbeat rhythm.

4 IMPLEMENTATION

This project presents a system that provides a continuous service for people to monitor their health. Heart rate, respiratory rate and ECG is measured with the help of the heart rate sensor, respiratory rate sensor and ECG sensor respectively and are then processed by a microcontroller, AT89S52. The processed data is then studied by a microcontroller. Finally the analysed data is sent to the LCD. The data is displayed on the LCD continuously and in case of any emergency and abnormal situation alarm is triggered. It can help the doctors in diagnosis of serious health issues and can improve the efficiency, efficacy and quality of health management.

5 CONCLUSION

In this paper, through the hard work of our colleagues and our respectable mentor, we have shown a health monitoring system, which not only measures the different parameters of the body but also displays it and fires an alarm in case of any abnormal situation. We have implemented the hardware and have analyzed the output.

REFERENCES

Aboobacker, Arith, Balamurugan, Deepak, Sathish "Heartbeat Sensing and Heart Attack Detection using Internet of Things: IoT" International Journal of Engineering Science and Computing April 2017.

Ajitha, U., et al. "IOT Based Heart Attack Detection and Alert System." International Journal of Engineering and Management Research (IJEMR) 7.2 (2017): 285–288.

Ashrafuzzaman, Md, et al. "Heart attack detection using smart phone." International Journal Of Technology

Gowrishankar, S., M. Y. Prachita, and Arvind Prakash. "IoT based Heart Attack Detection, Heart Rate and Temperature Monitor."

Mallick, Bandana, and Ajit Kumar Patro. "Heart rate monitoring system using finger tip through arduino and processing software." International Journal of Science, Engineering and Technology Research (IJSETR) 5.1 (2016): 84–89

Manisha, Mamidi, et al. "Iot on heart attack detection and heart rate monitoring." International Journal of Innovation in Engineering and Technology (IJIET).

Mayur, Suraj, Shubham, Nikhil "International Journal For Engineering Applications And Technology".

Patel, Shivam, and Yogesh Chauhan. "Heart attack detection and medical attention using motion sensing devicekinect." International Journal of Scientific and Research Publications 4.1 (2014).

Riazul Islam S. "The Internet of Things for Health Care: A Comprehensive Survey", Date of publication June 1, 2015, DOI 10.1109/ACCESS.2015.2437951.

Rani S. U. "IOT patient health monitoring system" Indian journal of public health research and development 2017.

Yadav, Yashasavi, and Manasa Gowda. "Heart Rate Monitoring and Heart Attack Detection using Wearable Device." International Journal for technical research and Application (2016).

Recent Trends in Communication and Electronics – Sharma et al. (Eds)
© 2021 Taylor & Francis Group, LLC, ISBN 978-1-032-04572-6

A comparative study and analysis for image fusion techniques

Vineeta Singh & Vandana Dixit Kaushik
HBTU Kanpur, India

ABSTRACT: With the evolution of new image processing techniques day by day, several image fusion techniques have been proposed. In this research article we have studied several image fusion techniques evolved specially in multi-focus domain and analyzed accordingly. Comparatively image fusion schemes have been studied and elucidated here in the research paper.

Keywords: image fusion, multi-focus image fusion, comparative study, fusion techniques

1 INTRODUCTION AND BACKGROUND OF THE WORK

As we have seen many research papers published during these times in the field of image fusion technique. Image fusion is accompanied in different ways: multi-sensor image fusion, multi-spectral image fusion, multi focus image fusion and many more fusion techniques. All these techniques have further many algorithms developed by the researchers during recent times to fulfil the purpose of image quality enhancement. See Figure 1. Broader categorization for image fusion techniques is in two domains namely: transform domain as well as spatial domain. A series of images are generated for a single scene in image processing system. But for the human perception and machine interpretation these series of images don't fulfil the purpose. To accompany that we have concept of image fusion, where two or more than two images are fused to form a combined image for the analysis purpose as proposed by Parekh et al. (2014).

Transform Domain Methods undergo a three-step procedure:

1. Transformation of input images into transform domain is accompanied and respective coefficients for transform domain are generated.
2. Coefficients for transform domain are combined together as per fusion scheme.
3. Perform inverse transform on combined coefficients to get the fused image.

Spatial Domain Methods undergo the fusion process directly in spatial domain and further it is categorised in mainly two categories Pixel Level Methodology as well as Feature Level Methodology.

In static scenes it is a little bit easier to capture the images but in dynamic scene there are issues in capturing scenes due to motion related things like dynamism in scenes movement of the camera positions. Ellmauthaler et al. (2013) proposed multi-scale image fusion with the help of UDWT. In this research paper spectral factor as well as non-orthogonal factor was used with the UDWT. Zhu et al. (2013) proposed sparse based image fusion model. Shen et al. (2012) proposed in their research a volumetric medical image fusion. Do et al. (2005) discussed Contourlet transform in their research paper, contourlet transform leveraged with efficient as well as directional and multi-resolution image representation effectively. Cunha et al. (2006) discussed, Nonsubsampled Contourlet Transform with its' design characteristics, qualities and applications as well. Krishnamoorthy et al. (2010) & Kaur et al. (2016), devised

DOI: 10.1201/9781003193838-47

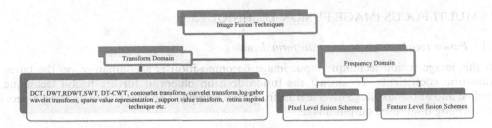

Figure 1. Image fusion techniques.

a research paper on comparative study of different image fusion methods and different fusion methods were developed, as a part of the research work. Manu et al. (2012) proposed a novel statistical fusion rule and it was compared with wavelet method as well as with non-subsampled contourlet transform. Biswas et al. (2015) proposed a scheme for remote sensing image fusion and it was accompanied via PCNN model. Wang et al. (2014) has devised a research paper for fusing multi-spectral images with panchromatic images with the help of sparse representation as well as local autoregressive model. Cheng et al. (2015) devised a framework to fuse remote sensing images via wavelet transform as well as sparse representation. Li et al. (2019) devised Infra-red and visible image fusion. Jiang et al. (2014) proposed image fusion with morphological component analysis. Manviya et al. (2020) & Vijana et al. (2020) devised in their research work with detailed discussion on several image fusion models.

Refer Table 1 to see comparative study of different image fusion techniques.

Table 1. Comparative study of image fusion methods.

S.No.	Fusion Techniques	Pros	Cons
1	DWT Method	Better PSNR and good quality fused output image	Shift variance and additive noise drawback
2	DCT Method	Efficient output	Blurring problem
3	RDWT Method	Shift variance problem of DWT resolved	Some additive noise issue
4	SWT Method	Yields good results at level 2 decomposition	Time consuming method.
5	Pyramid Based Methods	Good visual quality	False information and false edge issue
6	Averaging Based Method	Simple method.	Presence of noise degrades quality of fused image.
7	PCA	High spatial quality of fused output images	Spectral degradation is issue
8	HIS	Good results for remote sensing.	Noise and colour distortion issue
9	Brovey Method	Simple and fast method.	Spectral distortion issue
10	Sparse Based Fusion	Spectral model errors were reduced	Less sensitive in case of panchromatic image
11	Novel Cross Scale Fusion	Versatile and effective fusion rule	Implementation is complex
12	UDWT	Robustness for fusion	Not exclusive for multi-sensor images
13	Unsupervised Change Detection	For detection of changes, method is better than previous methods	Additive noise is drawback
14	Kalman Fusion	Better results for multi-sensor satellite data	Not very exciting results but satisfactory up to some extent
15	Wavelet Based Fusion	At decomposition levels, detection is done efficiently	More computational time required, colour distortion issue

2 MULTI-FOCUS IMAGE FUSION TECHNIQUES

2.1 Fusion techniques based on transform domain

In this image fusion technique input image decomposition is accompanied on the basis of transform coefficients. To extract the fusion decision blueprint further fusion technique is applied and at the last stage inverse transformation is applied on that fusion decision blueprint for forming final fused output image.

2.1.1 Discrete Wavelet Transformation (DWT)
DWT is used to break an image in high level and low-level frequency sub-bands for further image processing and finally inverse DWT is performed to get back the enhanced result.

2.1.2 Discrete Cosine Transformation (DCT)
In DCT image decomposition takes place into non-imbricating N*N sized blocks where for every single block coefficient of DCT are calculated and further fusion schemes are developed for obtaining final fused DCT coefficients.

2.1.3 Redundant DWT
Shift variance shortcoming of DWT is overcome by Redundant DWT. Three levels of decomposition takes place with the help of daubechies filters in redundant DWT in input images for generation of an approximate wavelet band.

2.1.4 SWT
SWT follows the concept of translation invariant and its working principle is similar to DWT.

2.1.5 Pyramid based method
Pyramid based methods are made up of band pass or low pass instance of an image. Every image instance denotes a varied scale pattern details. In pyramid method every single level is a factor of two shorter as the predecessor as well as higher level focuses over smaller partial frequencies. At this place pyramid does not include all information respective to re-construct the source image.

2.2 Fusion techniques based on spatial domain

Spatial Domain fusion techniques are straight away applied to the source images.

2.2.1 Averaging method
Every corresponding pixel of input image is undergoes for averaging to generate fused output image.

2.2.2 Principal Component Analysis (PCA)
PCA is a kind of statistical method.

2.2.3 Intensity Hue Saturation (IHS Method)
IHS model overcomes the limitations of RGB model. Here intensity refers to the whole portion of the light that falls over the eye. Here hue means wavelength in the visible-light spectrum where the energy output out of the source is maximum; saturation may be explained as pureness of whole portion of white light of a colour.

2.2.4 Brovey method
Brovey method exhibits a colour transform method. It surpassed the drawbacks in multiplicative method. Brovey method is the simplest method for integrating the various sensor data. Arithmetic operation is applied in this method. Spectral bands are used in this method, but before using spectral bands, normalization of spectral bands take place, then further multiplication takes place with panchromatic image. Here respective spectral feature of every pixel is

preserved and further transformation of whole information of luminance is done to a high-resolution panchromatic image.

3 MODERN FUSION TECHNIQUES

3.1 *Sparse based fusion technique*

Compressive sensing theory-based pan-sharpening method. Pan sharpening method includes two kinds of images one is low resolution multi-spectral input picture. While another one is panchromatic image of high resolution. The main aim is to increase input image spatial resolution while producing multispectral image.

3.2 *Novel cross scale fusion*

Method used for decomposition based on multi-scale. Intra and inter scale consistencies are considered and used for volumetric medical images. To preserve the source image details revealing the artefact fusion model passes information at every decomposition level. At starting step, information flows from lower level to higher level and every fused coefficient calculation is fulfilled via flowing information there. The value of membership is used to guide the details of coefficient selection.

3.3 *Un-Decimated Wavelet Transform (UDWT)*

UDWT fusion scheme divides the image decomposing procedure in two filtering processes via spectral factorization of analysis filters. Use of filter bank is done in UWT. Decomposition of one dimensional signal is accompanied by filter bank to the coefficients of high pass. Here this technique is better than conventional fusion model and overcomes the issues where fused image was supposed to reduce the undesired coefficient values spread throughout imbrications image singularities which ultimately results into complex process for feature selection which leads to the 'reconstruction errors' in the fused final image. Such procedure reduces the ringing artefact like side effects while fusing the final image rebuilding process.

3.4 *Unsupervised change detection*

Change detection algorithm has an important part to play for versatile applications use for example motion detection, medical diagnosis and surveillance etc. Unsupervised change detection methodology is suggested on the behalf of k-means clustering image fusion kernel for getting better result of radar images with synthetic aperture. Here for fusion purpose, mean ratio as well as log ratio methodologies, DWT are used. False alarm rate is minimized via non-linear clustering thus accuracy of clustering process increases.

3.5 *Kalman fusion*

For interpreting an image information content in an intended way, the image quality enhancement process is accompanied. Image fusion is one of the techniques to fulfil its objective. Let us talk about remote sensing domain, where image acquisition can be accompanied either from various sources or from original source but in several acquiring contexts, proper data fusion is essential for getting nicer and consistent explanation of the image to extract the whole perception from a phenomenon. To enhance all such shortcomings kalman filter works for image fusion at pixel-level. The formation of final fused image is accompanied via fusing optimal estimate coefficients of pixel values to get the nice image quality for information interpretation.

259

3.6 *Wavelet based fusion*

In medical science tumour spotting with real size and position is yet a demanding task. For fulfilling this purpose, a wavelet-based algorithm was proposed in International Journal of Computer Applications (0975-8887) Volume 109 – No. 6, January 2015. Here this algorithm provided the redundant as well as complementary information out of image data. This algorithm works effectively to consume the captured information out of source image to produce combined fused output image which ease out the tumour detection problem up to some extent. In this procedure, fusion is accompanied with the consideration of highest coefficient values corresponding to changes of sharper brightness, yields in the pertinent features presented. In multi-scale based transformation or in the case of multi resolution analysis, fragmentation of images are accompanied into different frequency sub-bands such as low-high (LH), low-low (LL), high-low (HL), high-high (HH). This way picks out the suitable fragmentation levels by fusing the main characteristics of input pictures to conquer the demerits of the adjacent characteristics imbrications of the minimum band signals.

4 COMPARATIVE STUDY OF MULTI-FOCUS IMAGE FUSION TECHNIQUES

5 CONCLUSION

Here in this research paper, several techniques based on multi-focus image fusion have been discussed, out of which some traditional fusion schemes have been discussed and some advanced fusion schemes have been discussed and concise description with their merits and drawbacks is listed in Table 1.

REFERENCES

Biswas, Biswajit & Sen, Biplab Kanti et al. 2015. Remote Sensing Image Fusion using PCNN Model Parameter Estimation by Gamma Distribution in Shearlet Domain", Procedia Computer Science, vol. 70, pp. 304.

Cheng, Jian & Liu, Haijun & Liu, Ting et al. 2015. Remote sensing image fusion via wavelet transform and sparse representation. ISPRS Journal of Photogrammetry and Remote Sensing. 104. 10.1016/j.isprsjprs.2015.02.015.

Cunha, A. L., Zhou, J., & Do, M. N. 2006. The nonsubsampled contourlet transform: theory, design, and applications. IEEE Transactions on Image Processing, 15(10),3089–3101.

Do, Minh et al. 2006. The Contourlet Transform: An Efficient Directional Multiresolution Image Representation. IEEE transactions on image processing: a publication of the IEEE Signal Processing Society. 14. 2091–106.

Ellmauthaler, Andreas et al. 2013. Multiscale Image Fusion Using the Undecimated Wavelet Transform With Spectral Factorization & Nonorthogonal Filter Banks. Image Process., IEEE Transactions on. 22. 1005–1017. 10.

Jiang, Yong & Wang, Minghui. 2014. Image fusion with morphological component analysis. Inf. Fusion 18 (July, 2014), 107–118. DOI: https://doi.org/10.1016/j.inffus.2013.06.001.

Kaur, Gurpreet & Kaur, Prabhpreet. 2016. Survey on multi-focus image fusion techniques. Electrical Electronics and Optimization Techniques (ICEEOT) International Conference on, pp. 1420–1424.

Krishnamoorthy, S., & Soman, K. P. 2010. Implementation and comparative study of image fusion. International Journal of Computer Applications, 9(2),25–35.

Li, H. & Wu, X. 2019. DenseFuse: A Fusion Approach to Infrared and Visible Images. IEEE Transactions on Image Processing, vol. 28, no. 5, pp. 2614–2623. Doi: 10.1109/TIP.2018.2887342.

Manu, V. T. et al. 2012. A novel statistical fusion rule for image fusion & its comparison in non-subsampled contourlet transform domain & wavelet domain. The Inter. Journal of Multimedia & Its Applications, 4(2),69–87.

Manviya, Monica & Bharti, Jyoti. 2020. Image Fusion Survey: A Comprehensive and Detailed Analysis of Image Fusion Techniques. 10.1007/978-981-15-2071-6_53.

Parekh, Pramit & Patel, Nehal & Macwan, Robinson et al. 2014. Comparative Study and Analysis of Medical Image Fusion Techniques. International Journal of Computer Applications. 90. 12–16. 10.5120/15827-4496.

Shen, Rui & Cheng, Irene & Basu, Anup. 2012. Cross-Scale Coefficient Selection for Volumetric Medical Image Fusion. IEEE transactions on bio-medical engineering. 60. 10.1109/TBME.2012.2211017.

Vijana, Anish & Dubey, Parth & Jain, Shruti. 2020. Comparative Analysis of Various Image Fusion Techniques for Brain Magnetic Resonance Images. Procedia Computer Science. 167. 413–422. 10.1016/j.procs.2020.03.250.

Wang, Wenqing & Jiao, Licheng et al. 2014. Fusion of multispectral and panchromatic images via sparse representation and local autoregressive model. Information Fusion. 20. 73–87. 10.1016/j.inffus.2013.11.004.

Zhu, Xiao & Bamler, Richard. 2013. A Sparse Image Fusion Algorithm With Application to Pan-Sharpening. IEEE Transactions on Geoscience and Remote Sensing. 51. 2827–2836. 10.1109/TGRS.2012.2213604.

261

Recent Trends in Communication and Electronics – Sharma et al. (Eds)
© 2021 Taylor & Francis Group, LLC, ISBN 978-1-032-04572-6

Performance evaluation of two-dimensional cyclic shift coding technique for Optical Code Division Multiple Access System

Navya Siringi & Gurjit Kaur
Delhi Technological University, New Delhi, INDIA

ABSTRACT: This paper proposed design and implementation of wavelength and phase Optical Code Division Multiple Access System (2D-OCDMA) where Cyclic Shift (CS) coding technique is used. Previously this cyclic shift encoding was done in spectral domain but we have proposed the encoding technique in the spectral as well as phase domain. The proposed 2-D Cyclic shift code with spectral as well as phase domain increases the number of users by splitting the optical source to two orthogonal phase signals without the requirement to increase the code length as well as number of LEDs. Thus, the wavelength required for our proposed encoding structure is also reduced. In this paper, this encoding technique has been tested for four users. The performance is analyzed with increase in number of users. The proposed 2D is compared with the previously implemented 1D CS SAC-OCDMA code. Performances of the system designed show better result than the previously designed SAC-OCDMA Cyclic Shift code. The results of the proposed system in terms of parameters like bit error rate is within the range of 10^{-9}.

1 INTRODUCTION

OCDMA is a spread spectrum multiple access technique in which we use orthogonal codes for multiple users to access the spectral band without any interference. Codes with low cross-correlation and high autocorrelation are required for viable communication. Several codes are being developed for high capacity in terms of data, users, and range of the network. Spectral amplitude coding (SAC), which is one of the techniques used in optical code division multiple access (OCDMA), each user is assigned a unique codeword in the spectral domain. There are several SAC-OCDMA codes like Cyclic Shift, Hadamard Coding, Optical Orthogonal Code (OOC), Modified Frequency Hopping (MFH), Random Diagonal (RD), Dynamic Cyclic Shift (DSC) etc. These codes are characterized by code length; number of users supported and code weight. As the users increase the code length gets to increase making the system more complex. Spreading in one dimensional (1D) spectral-domain optical codes requires ultra-short optical pulses for coding; this makes such systems practically difficult especially for systems with long code lengths. 2D codes can be formed using combinations of wavelength, time, space, and polarization. For 2D OCDMA, some are based on algorithm while some use different combinations of 1-D sequences to construct new 2-D codes.

This paper proposes a 2D OCMA. The proposed 2D coding is done in the spectral and phase domain. The 2D coding technique increases the number of assigned users without an increase in code length. The cyclic Shift coding technique is used for encoding in the spectral domain. One dimensional coding set is used with a polarized optical source, which is by splitting the optical broadband source into two orthogonal phases, this reduces the wavelength requirement. Cyclic Shift coding technique is characterized by zero cross-correlation, reduced

DOI: 10.1201/9781003193838-48

number of filters required, this reduces the components required in the system designed to reduce the cost of the system designed. Cyclic Shift code for SAC-OCDMA is developed for two users.

For the proposed system, we construct the cyclic shift code for two users we form a matrix as shown below:

$$
\begin{array}{ccccc}
\text{User 1} = & 1 & 1 & 0 & 0 \\
\text{User 2} = & 0 & 0 & 1 & 1 \\
& \downarrow & \downarrow & \downarrow & \downarrow \\
& 1550 & 1550.8 & 1551.6 & 1552.4
\end{array}
$$

Each '1' is assigned a wavelength as shown in the above matrix.

Let $C_K(j)$ is the jth item of the Kth Cyclic Shift code sequence. For two sequence the code is presented as

$$
\sum_{j=1}^{L} C_K(j) C_N(j) = \begin{cases} W, & \text{for } K = N, \\ 0, & else \end{cases}
\tag{1}
$$

The optical source is not polarized and with flat spectrum over the range of $\left[f_0 - \frac{\Delta V}{2}, f_0 + \frac{\Delta V}{2} \right]$ where Δv is the bandwidth f source and f_0 is central frequency. After the encoding of optical signals, the binary data signal transformed to electrical RZ signal which is then modulated. At the receiver section, the Power Spectral Density (PSD) is given with the expression 2.

$$
r(v) = \frac{P_{sr}}{\Delta v} \sum_{k=1}^{K} d_k \sum_{j=1}^{L} C_K(j) \left\{ u \left[v - f_0 - \frac{\Delta v}{2L}(-L + 2j - 2) \right] - u \left[v - f_0 - \frac{\Delta v}{2L}(-L + 2j) \right] \right\}
\tag{2}
$$

Photocurrent at the receiver for the polarized states, is given in equation 3.

$$
I = \frac{ne}{hv} \int_0^{-\infty} G(v) dv = \frac{ne}{hv} \cdot \frac{P_{sr}}{L} \quad \text{if } \theta = 90^\circ \text{ or } \theta = 0^\circ
\tag{3}
$$

The Signal to Noise Ratio (SNR) is given in equation 4.

$$
SNR = \frac{I^2}{\sigma^2} = \frac{\left(\frac{RP_{sr}W}{L} \right)^2}{eBR \frac{P_{sr}W}{L} + \frac{4KT_n B}{R_L}}
\tag{4}
$$

and Bit Error Rate (BER) is given in equation 5.

$$
BER = .5 \, erfc \left(\sqrt{SNR/8} \right)
\tag{5}
$$

Where K is the Boltzmann's constant, e is the electron charge, T_n is the temperature in kelvin, B denotes the electrical bandwidth, R_L is the load resistance and P_{ST} is the power received.

2 DESIGN OF TWO-DIMENSIONAL OCDMA SYSTEM

The system is designed for data rate of 622Mbps for a distance of 20Km in OptiSystem version 17.

In the transmitter section, Light Emitting Diode (LED) used as a broadband source is split into two orthogonally polarized signals. Each polarized signal is encoded with previously proposed 1D cyclic shift code for 2 users. WDM mux is used as encoder. The same set of coding is used for orthogonally polarized signals. The optically modulated signals of the users are combined through power combiner and transmitted using optical fiber. At the receiver section the received signal is split using power splitter. For detection of received signal, direct detection technique is used. At the receiver the signals are first decoded then they are phase shifted using polarizer. The schematic of the system is shown in Figure 1.

The polarized state of optical source signal as given by the polarization meter is presented in Table 1.

The parameters used in designing the system are formulated in Table 2.

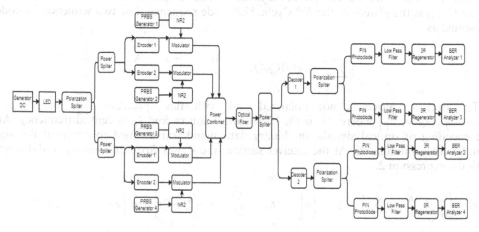

Figure 1.   Proposed system design.

Table 1.   Parameters of optical source polarization.

Polarization Meter	For user 1 and 2	For user 3 and 4
Degree of Polarization	100%	100%
Signal Power	11.9 dBm	12.12 dBm
Signal Phase	1	-1

Table 2.   Parameters used in system design.

	Parameters	Values
1	Data rate	622Mbps
2	No of users	4
3	Wavelengths used	1550-1552.4
4	Wavelength spacing	0.8nm
5	LED source wavelength	1551nm
6	Fiber optic length	20km
7	Attenuation	25dB/Km
8	Photo detector gain	3
9	Uniform FBG bandwidth	0.3nm
10	Polarization device angle	0°
11	Photo detector gain	3
12	Responsivity	1A/w
13	Low pass filter cutoff	$7.5e^{+009}$

3 RESULTS AND DISCUSSIONS

The results of the proposed 2D OCDMA are compared with one dimentional SAC-OCDMA. The electrical signal transmitted and received for first user as observed in the oscilloscpe visualizer is shown in below Figure 2.

The oscilloscope visualizer shows the received signal is similar to the transmitted signal. The performance prameters of four users for the proposed 2-D Cyclic Shift SAC-OCDMA with polarization.

As the number of users are increased, the Quality factor and and Bit Error Rate (BER) starts degrading. When a single user is transmitted in the system the received signal has high Q-Factor and BER. As the number of users are enlarged the performance deteriorates. However the minimum values required for the system is maintained. The performance for received signal for 3rd user with the increasing users is shown in Figure 3.

The Eye diagram of the received signal for third user is shown in Figure 4.

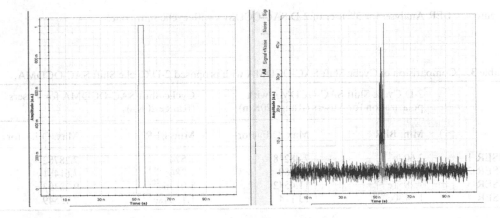

Figure 2.   Transmitted and received signal for 3rd user.

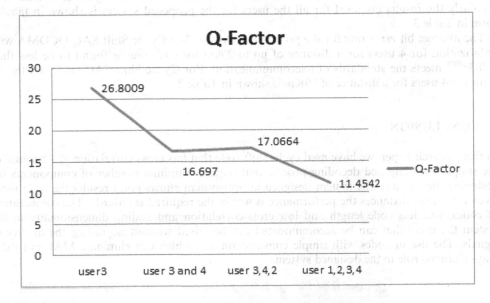

Figure 3.   Performance with the increase in number of users.

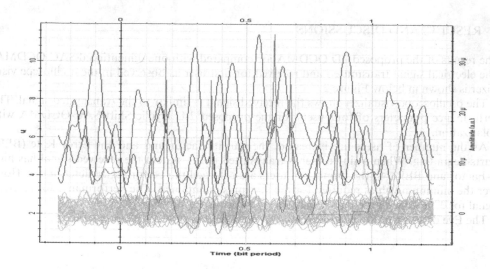

Figure 4.   BER Analyser for 3rd user of 2-D SAC-OCDMA with cyclic shift code.

Table 3.   Comparision of Cyclic Shift SAC-OCDMA with proposed 2-D Cyclic Shift SAC-OCDMA.

	2-D Cyclic Shift SAC-OCDMA with polarization for 4 users (Range 20Km)		Cyclic Shift SAC-OCDMA for 4 users (Range 20Km)	
	Min. BER	Max. Q Factor	Min. BER	Max. Q Factor
USER 1	$3.607e^{-014}$	7.48218	$1.57479e^{-013}$	7.28782
USER 2	$97.15582e^{-054}$	15.480	$1.286e^{-014}$	7.61481
USER 3	$7.2851e^{-031}$	11.4542	$8.0196e^{-021}$	9.28393
USER 4	$1.3430e^{-039}$	13.114	$1.13548e^{-015}$	7.92549

Performance of the system is studied in terms of parameters like bit error rate and quality factor along with the system designed for four users in one dimentional Cyclic Shift OCDMA. Similarly the results obtained for all the users for the proposed system is shown in tabular form in Table 3.

The average bit error rate for the proposed system of 2-D Cyclic Shift SAC-OCDMA with polarization for 4 users for a distance of up to 20km for each user is found to be less than $2.783e^{-008}$ meets the standards of telecommunication. For Cyclic Shift SAC-OCDMA the bit error for 4 users for a distance of 10Km is shown in Table 2.

4   CONCLUSION

In this research paper, we have used cyclic shift code that has cross-correlation ≤ 1 because of the simple encoding and decoding process that requires minimal number of components for designing the system. The system designed in opti-system shows good results the code used. Even for longer distances the performance is within the required standards. The combination of codes with less code length and low cross-correlation and adding dimensionality to the system the users that can be accommodated can be raised without degrading the quality of signals. The use of codes with simple construction and which can eliminate MAI and PIIN plays a crucial role in the designed system.

ACKNOWLEDGEMENT

The authors would like to thank the "Research Project grant to the faculty of DTU" under Delhi Technological University for its financial support of this work under file No. DTU/Council/BOM-AC/Notification/31/2018/5738.

REFERENCES

Mostafa, S., Mohamed, A. E. N. A., El-Samie, F. E. A., & Rashed, A. N. Z. (2019). Cyclic Shift Code for SAC-OCDMA Using Fiber Bragg-Grating. *arXiv preprint arXiv:1904.00373*.

Salehi, J. A., & Brackett, C. A. (1989). Code division multiple-access techniques in optical fiber networks. II. Systems performance analysis. IEEE Transactions on communications, 37(8), 834–842.

Z. Wei, "Unipolar Codes With Ideal In-Phase Cross-Correlation for Spectral Amplitude-Coding Optical CDMA Systems," IEEE Transactions on communications, vol. 50, no. August, pp. 1209–1212, 2002.

Smith, E. D., Blaikie, R. J., & Taylor, D. P. (1998). Performance enhancement of spectral-amplitude-coding optical CDMA using pulse-position modulation. *IEEE Transactions on Communications*, *46*(9), 1176–1185.

Fadhil, H. A., Aljunid, S. A., & Ahmad, R. B. (2009). Performance of random diagonal code for OCDMA systems using new spectral direct detection technique. *Optical Fiber Technology*, *15*(3), 283–289.

Kakaee, M. H., Seyedzadeh, S., Fadhil, H. A., Anas, S. B. A., & Mokhtar, M. (2014). Development of multi-service (MS) for SAC-OCDMA systems. *Optics & Laser Technology*, *60*, 49–55.

Anas, S. A., Abdullah, M. K., Mokhtar, M., Aljunid, S. A., & Walker, S. D. (2009). Optical domain service differentiation using spectral-amplitude-coding. *Optical Fiber Technology*, *15*(1), 26–32.

Koonen, T. (2006). Fiber to the home/fiber to the premises: what, where, and when?. *Proceedings of the IEEE*, *94*(5), 911–934.

Fathallah, H. (2006). Optical CDMA communications and the use of OFCs. *Optical Fiber Components: Design and Applications*, 201–43.

Garadi, A., Bouazza, B. S., Bouarfa, A., & Meddah, K. (2018). Enhanced performances of SAC-OCDMA system by using polarization encoding. *Journal of Optical Communications*, *1* (ahead-of-print).

Recent Trends in Communication and Electronics – Sharma et al. (Eds)
© 2021 Taylor & Francis Group, LLC, ISBN 978-1-032-04572-6

Variations in the human consciousness - An explanation on the basis of model theory (RKT) and its applications

K. Reji Kumar

Department of Mathematics, N. S. S. College, Cherthala, India

ABSTRACT: Consciousness is the end result of all information processing activities in our mind. Human beings show significant differences due to the variation in the processing of information. Conscious actions of human being are controlled by mental processes and hence it is subjected to variations. Consciousness can be explained in terms of the models processed in the mind. In this paper, we study the reasons of personal variations in performing different actions on the basis of the models processed in the mind.

1 INTRODUCTION

In the review article of Perlovsky (Perlovsky 2006), the possibility of a scientific study of human consciousness is discussed in detail. It also gives us a picture of the research in this area up to 2006. A scientific study of consciousness demands a collaborative effort of researchers from various disciplines such as philosophy, mathematics, physical sciences, social sciences, biological science, etc. All fundamental concepts must be scientifically defined. Explaining consciousness is one of the unsolved problems of the modern world. Every scientific study uses mathematical modeling methods to give accuracy, objectivity, and reliability to give the concepts explained. A scientific study of consciousness also demands the use of mathematical modeling at all levels.

We proceed to discuss some prominent theories proposed by researchers using the methods of mathematical modeling. The *Global Workspace Theory* (GWT) proposed by Baars (Baars 1997, Baars & Franklin 2009) argues that consciousness is computational. It tries to explain the phenomenon based on an interaction between bottom-up and top-down intentional modulation mechanisms. Another important theory is the *representation theory,* which is mainly of two types. The *first order* and *higher order*. According to this theory, our consciousness is directly associated with mental representations. It also argues that consciousness has no connection to the physical states. The first order representation theory, which is advocated by Dretske (Dretske 1995), Tye (Tye 2000), Mehta & Mashour (Mehta & Mashour 2013) is on the assumption that the perceptual representations formed in the sensory regions cause the conscious mental state. The thoughts of Locke and Kant have motivated the higher-order representation theory.

Tononi's *Integrated Information Theory* (IIT) (Balduzzi & Tononi 2008, Edelman & Tononi 2000, Oizumi, Albantakis & Tononi 2014) measures the level of consciousness of a system using some mathematical methods. *Quantum theories of consciousness* attempt to explain consciousness using the methods of quantum mechanics. The four-dimensional approach to consciousness is an attempt to explain conscious experiences based on space-time intervals (Sieb 2016). Koehler's mathematical approach is also an important step forward in this area (Koehler 2011).

DOI: 10.1201/9781003193838-49

A theory of consciousness based on models was proposed by Reji Kumar in the year 2008 (Kumar 2009). The basic assumption of the theory is that the fundamental units of consciousness are models. Models are representations of reality. All models are not of the same nature and type. In the context of the study, the models are classified into three broad sets. It is based on the complexity associated with it. The very fundamental or basic models are named α - models. More complex (middle-level) models are named β – models and the models with a higher level of complexity are called the γ – models. Models as fundamental units of consciousness give a mathematical framework to the mathematical modeling of consciousness. A set of axioms of consciousness were suggested (model axioms) to give the theory a firm foundation (Kumar 2010b, Kumar 2011).

A comparison is a mental activity that generates different levels of consciousness. Complex models are formed using simple models and used for representing reality (Kumar 2010a). The problem of combinatorial complexity was explained using the model axioms and models as the fundamental units (Kumar 2015a). How the comparison of models leads to variations of consciousness in two different individuals is based on this study. Modeling theory (Reji Kumar's Theory or RKT) can be used to study various forms of knowledge and further explain the difference between the various knowledge forms such as Mathematics, Science, etc (Kumar 2009).

Subjectivity is a character of consciousness, which is closely related to our experiences. Experience also contains objective content. Subjectivity is feeling or experiencing a common thing differently by different persons. For example, the experience of the night is different for different people but the object of experience is the same for all. So the night is an objective content even though all experience it very subjectively. Thus objective content and subjective consciousness cannot be treated separately. They are the two sides of a coin. The subjectivity of consciousness is mathematically explained in (Kumar 2011, Kumar 2016). The scientific explanation of a phenomenon is a kind of representation, that can be called a model. Different consciousness can represent a particular phenomenon using different models. Consciousness can be defined as the totality of all representations in the mind of a person. Furthermore, a phenomenon can have different representations on different occasions in consciousness.

All living organism do activities, which is either conscious or unconscious. We can plan, experience, and control our conscious activities, while all our unconscious activities are out of our control. In this study, our focus is only on conscious activities and we attempt to explain its organization in the mind using the method of models and operations associated with it.

The following definitions are relevant to our discussion, which are adopted from the theory of models (RKT) (Kumar 2015b). We denote the collection of all models that constitute the consciousness of an individual x by Cx, A model in the consciousness id represented by the small letter m. The collection of all models, which generates a model m is denoted by m_c. If $m_1, m_2, \ldots, m_n$ are the models that make a model m, then we write $[m_1, m_2, \ldots, m_n]$. The sub-model and super-model relations are defined as follows. The model m_i is a sub-model of the model m, if it is extended to make a new model m, and the model m contains the whole m_i without any change. It is denoted by $m_i \leq m$. At the same time, The model m is a super-model of m_i, which is denoted by $m \geq m_i$. We say that the model m is a generalization of m_1 and m_2 if $m \leq m_1$ and $m \leq m_2$. The null models and the universal model are defined to perform special functions. Null models $\emptyset$ can represent any reality and the Universal model U containing all models in a consciousness. Usually, a collection of models is denoted by capital letters.

In the language of model theory, a simple collection of models is also a model. But we can define different types of relations among the members of a collection of models and the collection of models along with the relations can become a new model. Moreover, relations of various types can be defined simultaneously among the models in a collection. Thus the field of model theory is very vast and demands intense research at a deeper level.

2 CONSCIOUSNESS AND HUMAN ACTIONS

A mind can process almost infinitely many models. A developing mind contains γ - models that are very complex. Comparison is the fundamental activity of the mind based on which it derives conclusions and decisions. Conclusion and decisions are also models in the language of model theory. The procedure of comparison of a model with another model is explained in (Kumar 2015a) . When the number of sub-models associated with a model is very large in number the comparison procedure faces the problem of combinatorial complexity. The problem of combinatorial complexity can be explained as follows. Consider two models $M_1 = [m_1, m_2, \ldots, m_p]$ and $M_2 = [n_1, n_2, \ldots, n_q]$. By comparison of the models M_1 and M_2 we mean each sub-model of M_1 is compared to each sub-model of M_2. More clearly, m_1, is compared to $n_1, n_2, \ldots, n_q$, m_2 is compared to $n_1, n_2, \ldots, n_q$ and so on. As the numbers p and q becomes large the procedure becomes extremely complex and impossible to complete. This is the combinatorial complexity of comparison. But human consciousness easily overcomes this complexity. This is by avoiding many steps of comparison by making the assumption that the sub-models compared are the same or similar. Moreover, the comparison procedure may stop when sufficiently level of disagreements between sub-models. The number of comparisons made by two different consciousness varies significantly that can cause a difference in the level of consciousness.

Now we apply the above-discussed procedure of comparison to the performance of conscious actions of a human being. We make the assumption that our conscious actions are controlled by the two relevant areas in our brain. One is for planning the activities and the other is for recording the completed actions. These areas can store and process enumerable models and the models are subjected to comparison in a continuous fashion. Let the collection of sub-models of the planning area be represented by M_P and the collections of sub-models belonging to the area of completed actions be represented by M_A. Continuously comparison of the models and sub-models are carried out and the results of the comparisons are used for improving the performance of the actions.

A possible method of comparison can be explained using two example models. Consider $M_P = [m_1, m_2, \ldots, m_r]$ and $M_A = [n_1, n_2, \ldots, n_r]$. Since the comparison is between the collections of models representing the planning and performance of the same action, we also assume that the number of sub-models in the collections is the same. The evaluation process can be explained by defining a function $f : M_p {}^{\circ} M_A \rightarrow [0, 1]$ such that $f(m_i, n_i) = 1$ if the models m_i and n_i perfectly agree and $f(m_i, n_i) = 0$, if the models completely differ. Other levels of similarities and agreements between the models are represented by a suitable real number between 0 and 1. The chance of committing errors in the comparison also cannot be avoided.

3 A CASE STUDY OF VARIATION OF ACTIONS

To measure the variation of comparison of consciousness we introduce a value of comparison $V = \sum f(m_i, n_i)$. This value is named the *value of the agreement*. If the value of the comparison is[i] very high the agreement between the models is very high. On the other hand, if the value is very low then the agreement is very leek and the models will be modified accordingly. This study is independent of the effect of any external factors or physical factors and that influence the activities of a person. We focus only on the variations due to the variations in consciousness in this study. We use the data in Table 1 and Table 2 to illustrate the situation to explain the comparison. Corrective measures are taken by the consciousness to improve the performance of the action depending on the values obtained, by comparison. In this example, we include only five sub-models for simplicity.

We can compute the value of the agreement for the first comparison equal to 4.3and that for the second comparison equal to 0.7. So the first evaluation indicates very high similarity between the models compared to that of the second comparison. Considering the low value of

Table 1. Data showing high level of similarity.

	n_1	n_2	n_3	n_4	n_5
m_1	0.8	0	0	0	0
m_2	0	0.7	0	0	0
m_3	0	0	0.8	0	0
m_4	0	0	0	1	0
m_5	0	0	0	0	1

Table 2. Data showing a very low level of similarity.

	n_1	n_2	n_3	n_4	n_5
m_1	0.1	0	0	0	0
m_2	0	0.25	0	0	0
m_3	0	0	0	0	0
m_4	0	0	0	0.2	0
m_5	0	0	0	0	0.15

comparison, if the conscious make the necessary changes in the models, then the performance of the consciousness can be improved.

The conscious can continue the steps of comparison and modification in the models up to a level decided it. The degree and the extent to which the procedure continues is highly important and that itself decided the variations in the level of consciousness. The idea presented in this paper can be applied in the study of artificial intelligence and robotics. It can guide us towards the artificial creation of conscious activities and explain effectively the variations shown in the performance of similar activities by different individuals.

4 CONCLUSION

In this paper, we briefly describe a mathematical model to explain the performance of conscious activities of human beings using the ideas of the Model Theory of consciousness. The highly subjective nature of consciousness has kept scientists away from proposing a scientific theory of consciousness for centuries. But now the situation has started changing. We are in a position to define and explain the unanswered questioned and unexplained phenomena in a very scientific and reliable manner. Mathematics can play a great role in these changes. The study presented here can shed light on the areas of research in consciousness.

REFERENCES

Baars, B.J. 1997. In the theatre of consciousness: Global workspace theory, A rigorous scientific theory of consciousness, Journal of Consciousness Studies.

Baars, B.J. & Franklin S. 2009. Consciousness is computational: The LIDA model of global workspace theory. International Journal of Machine Consciousness, 1(1), 23–32. https://doi.org/10.1142/S1793843009000050.

Balduzzi D. & Tononi, G. 2008. Integrated information in discrete dynamical systems: Motivation and theoretical framework. PLoS Computational Biology, 4(6).

Dretske F. 1995. Naturalizing the Mind, Cambridge, MA: Bradford/MIT Press.

Edelman, G. & Tononi, G. 2000. A Universe of Consciousness. New York: Basic Books.

Koehler, G. 2011. Q-Consciousness: Where is the Flow? Nonlinear Dynamics, Psychology, and Life Sciences, 15(3), 335–357.

Mehta, N. & Mashour, G.A. 2013. General and specific consciousness: A first order representationalist approach. Frontiers in Psychology. https://doi.org/10.3389/fpsyg.2013.00407.

Oizumi, M., Albantakis, L. & Tononi, G. 2014. From the phenomenology to the mechanisms of consciousness: Integrated Information Theory 3.0. PLoS Computational Biology, 10 (5). https://doi.org/10.1371/journal.pcbi.1003588 [doi]

Perlovsky, L.I. 2006. Towards physics of the mind, concepts, emotions, consciousness and symbols, Phys. Life Rev. 5, 23–55.

Reji Kumar, K. 2009. Modeling of consciousness: Mathematics, Science and other forms of knowledge, Manuscript.

Reji Kumar, K. 2010a. Mathematical modeling of consciousness, Project report submitted to DRDO, India.

Reji Kumar, K. 2010b. Modeling of consciousness - Classification of models, Advanced Studies in Biology, 3, 141–146.

Reji Kumar, K. 2011. Modeling of consciousness - A model based approach, Far East Journal of Applied Mathematics, 1, 1–14.

Reji Kumar, K. 2015a. How does consciousness overcome combinatorial complexity?, Advances in Intelligent systems and Computing Springer India, 325, 167–172.

Reji Kumar, K. 2015b. Mathematical Foundation of Information Processing: A Consciousness Based Study, I J C T A (International Science Press), 8 (5), 1989–1995.

Reji Kumar, K. 2016. Mathematical Modeling of Consciousness: Subjectivity of Mind, Proceedings of the International Conference on Circuit, Power and Computing Technologies [ICCPCT].

Sieb, R.A. 2016. Human Conscious Experience is Four-Dimensional and has a Neural Correlate Modeled by Einsteins Special Theory of Relativity. NeuroQuantology, 14, 630–644. https://doi.org/10.14704/nq.2016.14.3.983.

Tye M. 2000. Consciousness, Color, and Content. Cambridge: MIT Press.

Recent Trends in Communication and Electronics – Sharma et al. (Eds)
© 2021 Taylor & Francis Group, LLC, ISBN 978-1-032-04572-6

The IoT based digital postal services

Kuldeep Singh
Kabadi Techno Pvt. Ltd., Agra, India

Pushpa Singh
CSE, Delhi Technical Campus, Gr. Noida, India

Gayatri Arora, Anuj Kumar & Bikash Dutta
IEC College of Engineering & Technology, Gr. Noida, India

ABSTRACT: A well-structured postal service, in addition to promoting national cohesion, provides the essential armature for the proliferation of industry and economy. Postal operators should offer new products and services that reflect the evolving insistence to bind the nation together in a new world where people are perceptibly communicating digitally. In this regard, the present paper aims at the postal sector development in the condition of the digital economy by devising an IoT based digital post machine (Digi Post). The proposed prototype of Digital Post machine will use the ESP32, an open source electronics platform to direct the working of the hardware. The ultrasonic sensors, one of the hardware component, measure the volume of the digital post machine, simultaneously the machine's metadata is transmitted over the Internet, and the collector is notified at a certain threshold. It recuperates the postal sector by reducing the last mile distance covered by the user.

Keywords: postal service, iot, sensors, blockchain, digi post

1 INTRODUCTION

The upsurge in the use of Internet Services and digital transformation has indicated an extensive repercussion on the mandate for postal services. It allows people to send anything to any part of the world smartly and intelligently. The postal services are still in existence and ensuring the right to communicate through exchanging messages and goods, and presently require rapid and significant changes. The emerging technology such as IoT, Machine Learning, Cloud computing and Blockchain are essentially changing the world of communications, business, and commerce. The postal sector is no exception in this regard. In the conditions of the digital economy, postal services need to revolutionize their role to accommodate the digital age. Postal operators should offer new products and services that imitate the growing mandate to bind worldwide, where people are gradually collaborating digitally (Otsetova 2019). The postal operators must understand their competitive advantages and alter the key aspects of their business in a new direction with the use of IoT and other emerging technology. The basic postal facilities such as parcel distribution, money transfer, insurance, and pension schemes offered by Indian post offices. The IoT performs sensing, actuating, data gathering, storing, and processing by connecting physical or virtual devices to the Internet (Gunde et al. 2020). IoT offers the best-connected network application (Singh & Agrawal 2018). Nowadays, with the business embracing digital technology at high speed rate, customers are gradually expect-

DOI: 10.1201/9781003193838-50

ing to interact directly with the postal operators through digital channels. The postal operators significantly extended the kind of their activities, which currently include financial, insurance, e-services and other benefits besides the more traditional postal facilities (RARC Report 2016). Very few texts have been reported to enhance the services of traditional postal services. In this regard, the present paper aims at the postal sector development in the condition of the digital economy by devising an IoT based digital post machine.

In this paper, we discuss IoT based digital post machine for automating postal services. The proposed system is based on the IoT with ultrasonic sensors to measure the volume of posts in containers, simultaneously transmits the machine's information over the Internet, and notifies the collector at a certain threshold. The rest of the paper is structured into the subsequent sections. Section 2 denoted the IoT based proposed digital post machine. Section 3 discusses the result and discussion. Finally, concluding remarks are given in section 4.

2   PROPOSED IOT BASED DIGITAL POST MACHINE

We have offered to revolutionize the traditional postal services through an intelligent and digitalized solution that is an IoT based Digital Postal Machine. The Postal machine is a fully functional and automated model. The proposed prototype is built using ESP 32 other interfacing components. The ESP 32 board is prepared with sets of digital and analog input/output pins that may be interfaced with a Laser Sensor (TOF10120), Stepper Motor (NEMA 17), Motor Driver (DRV8834), a weight sensor (HX711), and an android mobile. The specific module used here is ESP32-WROOM-32E with the embedded chip ESP32-D0WD-V3.the administrator monitors the entire system. Arduino IDE & Android studio software are required. The workflow of the Postal Machine is based on the following steps, which are represented in Figure 1.

1) The customer or user is an entity who wants to send their post, courier, and parcel. The customer is required to register himself with either the web portal or the application.
2) The user can locate the nearby Digi Post machines using the 'TRACK' option in the software system, which uses Google API.
3) The customer then navigates to the machine and fills in the essential credentials regarding the courier. Eventually, a QR code is generated, encrypting the above information and permits the parcel over the weighing machine.
4) The Digi Post accordingly displays the cost and provides a QR code for payment. Then the machine generates a receipt with a barcode that he/she is instructed to attach on his courier or parcel item.
5) After the payment is successful the inlet gate opens, the user drops the packet into the machine which is further delivered by the concerned authorities

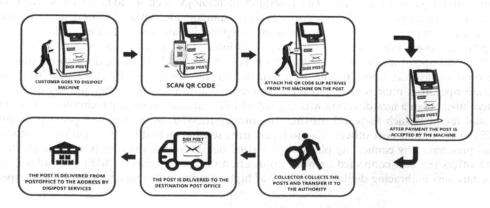

Figure 1.   Work flow of digi post.

6) The collector (postman) will collect and deposit all the parcels to the warehouse (Post office), which is further delivered to the destination.

2.1 *Algorithm for the customer*

The user installs the Digi Post software on his device and gets him registered with his credentials. The person intending to send the post enters the senders and receivers' address along with the essential specifications of the post. If the customer wishes to know its approximate price beforehand, he can do it using the price calculator feature provided in the software. In order to post the shipment, he proceeds by confirms the details entered by him. Eventually, a QR code is generated, which will be used during further processing. The machine authenticates the user as a customer using the QR code. If the authentication fails, an error message is displayed over the screen. The Digi Post machine instructs the customer to place his shipping commodity over the machine. The display screen shows the specification of the shipment comprising of its weight and the total incurred cost, and an option to proceed for payment. The user proceeding with the payment process gets a QR Code displayed over the screen to pay the specified cost. On successful payment, a confirmation message is displayed over the screen and user received receipt. The overall working is shown in algorithm 1.

2.2 *Algorithm for the collector*

The collector installs the Digi Post software on his device and is registered as an authorized collector after the administrator verifies his credentials. He is responsible for periodically evacuating the Digi Post machine and surrendering the post at the hub from where they will be dispatched. The collector is required to vacate the machines in an order which is based on a predefined criterion. The criterion is implemented through an algorithm that is devised to help him reach the machine which needs to be evacuated at the earliest. The collector approaches the Digi Post and scans the collector specific QR code from the inbuilt mobile device installed at Digi post. The overall working is shown in algorithm 2. Both the algorithm 1 and 2 is represented in Table 1.

Table 1. Algorithm 1 and Algorithm 2.

Algorithm 1: Working of Digi Post for customer	Algorithm 2: Working of Digi Post for Collector
Step 1. If customer authenticated using QR Code	Step 1. If collector authenticated
Step 2. Digi Post displays, "Place the Shipment over the weighing machine."	Step 2. Collector receives one-time-password on his registered mobile number
Step 3. Digi Post displays the Shipment details and the proceed option.	Step 3. The one-time password is entered by him on the Digi Post screen
Step 4. If the customer chooses the proceed option, the payment QR Code is displayed over the screen.	Step 4. If validation is successful the outlet gate 2 opens
Step 5. If the payment is successful, a confirmation message is displayed over the screen.	Else step 1 is recapitulated.
Else- the session is expired, and the machine resets itself	Step 5. The collector collects the post and scans the QR code on each shipment from his mobile device.
Step 6. After successful payment, receipt generated.	Step 6. If the collector press the close button on-screen the gate 2 closes.
Step 8. The user presses the Finish Button to acknowledge	Else the Gate closes after 10 minutes automatic- ally.
Step 9. The shipment is disposed into the machine	Step 8. The Digi Post level is updated in the database
Step 10. end	Step 9. end

2.3 Maximum priority algorithm for the collector

The collector is accountable for evacuating the Digi Post Machines intermittently based on pre-defined criteria. The machine which has reached or is about to surpass the threshold (maximum allowed capacity) possesses the highest priority. To optimize the entire process, the collector uses the maximum priority algorithm to help the collector locate Digi Post in non – decreasing order of priority. The maximum priority algorithm is a variant of the Dijkstra Algorithm (Amaliah et al. 2016). The maximum priority algorithm explains the single-source shortest-paths problem on a weighted, directed graph (G) with set of vertexes (V) and edges (E). The weights of all the edges are non-negative. In this paper, we assume that w(x,y)≥0 for each edge (x,y) belongs E. The maximum priority algorithm maintains a set S of vertices whose final maximum-ratio from the source (s) has already been determined. The algorithm repeatedly selects the vertex x belongs V-S with maximum ratio adds x to S and relaxes all edges leaving x. In the following implementation, we use a max-priority queue (Q) of vertices, keyed by their ratio values. The ratio is computed using the given formula (1). Where capacity is the capacity of node and distance is distance between two nodes.

$$ratio = \frac{(capacity + distance)}{capacity}$$

2.4 Block diagram of DIGI POST

In this proposed model, a 24x7 monitoring system is designed to dispatch the shipment smartly. The customer's device initiates the entire work process of the Digi post. ESP-32 is an open-source electronics platform, a system on chip (SoC) series with integrated Wi-Fi and Bluetooth used to interact with hardware (sensors, microcontrollers, etc.) and software (Arduino IDE) (Aghenta et al. 2019). Here ESP32-WROOM-32E is deployed which integrates ESP32-D0WD-V3, with higher stability and safety performance. ESP32-WROOM-32E is a dominant, generic Wi-Fi, Bluetooth, BLE MCU component that targets an extensive range of applications, ranging from low-power sensor networks to the most challenging tasks. It is the prime mover of all the components. The proposed system intakes user's information then updates the database. Simultaneously, the Bluetooth directs the ESP32 to initiate the working of the hardware. If the container of the Digi Post machine is filled with 80%, then the laser sensor captures this level and sends an alert message to the mobile application of admin or concerned authority. Collector will collect the post or courier according to the time duration of the collection process like 6, 8, 12 hours. Eventually, the Post or courier is collected by the collector immediately and deposit to the warehouse for the next process of delivery that shipment. The primary circuit diagram of the proposed model is represented in Figure 2.

3 RESULT & DISCUSSION

The proposed framework relies on the accuracy of distance measures by ultrasonic sensors as threshold distance is taken 80% of the overall length of the Digi Post container. If the system exceeds this limit, then the system sends an alert message to the admin and also updates the database. Admin monitors the machine location and nearest collector to evacuate the Digi Post container. The database server also maintains the customer registration record with demographic and account details to generate the QR code. When the customer disposes of his courier along with the QR code, the collector disposes it to the warehouse. The entire process is done under complete supervision, which ensures assured delivery and increased user reliability. The digital postal machine offers customers seamless access to more than one utility within the same online transaction. The Digi Post offers a more flexible and advanced system as compared to the conventional means and also saves time, efficiency and customer satisfaction.

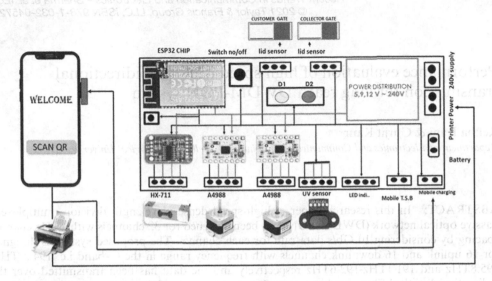

Figure 2.   Digi post basic circuit diagram.

This work could be extended with Blockchain in future. The technology of Blockchain will permits authentication without any third-party. Across every individual node (that are participants), the ledger is disseminated in the Blockchain. Thus, it is distributed in nature. Blockchain protects the data, and transactions processed are recorded chronologically (Singh & Singh 2020).

4   CONCLUSION

The Digi Post aims at the digitalization of smart cities. IoT defines a methodology where items in the physical world, and sensor associated with these items, are connected to the Internet via wireless and wired connections. A smart IoT-based solution is devised comprising user-friendly hardware that collects shipments from the customer without paperwork and helps the collector dispatch it to the destination as soon as possible using the maximum priority algorithm. The Printed circuit board uses ESP-32 as one of its components and helps achieve the mentioned objective effortlessly as compared to other IC's available in the market. This work could be extended with the help of Blockchain in future.

REFERENCES

[1]   Aghenta, L. O., & Iqbal, M. T. 2019. Low-cost, open source IoT-based SCADA system design using thinger. IO and ESP32 thing, *Electronics* 8 (8),822.
[2]   Amaliah, B., Fatichah, C., & Riptianingdyah, O. 2016. Finding the Shortest Paths among Cities in Java Island using Node Combination Based on Dijkstra Algorithm. *International Journal on Smart Sensing & Intelligent Systems*: 9(4).
[3]   Gunde, S., Chikaraddi, A. K., & Baligar, V. P. 2017. IoT based flow control system using Raspberry PI, In *2017 International Conference on Energy, Communication, Data Analytics and Soft Computing (ICECDS)*: 1386–1390. IEEE.
[4]   Otsetova, A. 2020. Postal Sector Development Perspective and Challenges.
[5]   RARC Report. (2016), Riding the Waves of Postal Digital Innovation, available online https://www.uspsoig.gov/sites/default/files/document-library-files/2016/RARC-WP-16-014.pdf
[6]   Singh, P., & Agrawal, R. 2018. A customer centric best connected channel model for heterogeneous and IoT networks, *Journal of Organizational and End User Computing (JOEUC)*, 30(4): 32–50.
[7]   Singh, P. & Singh, N. 2020. Blockchain with IoT & AI: A Review of Agriculture and Healthcare, *International Journal of Applied Evolutionary Computation (IJAEC)* 11(4):13–27.

Performance evaluation of high speed 320Gb/s bidirectional transmission for long reach DWDM-PON system

Rekha Rani & Gurjit Kaur

Department of Electronics and Communication Engineering, Delhi Technological University, Delhi

ABSTRACT: In this research paper, a high-speed dense wavelength division multiplexed-passive optical network (DWDM-PON) has been designed for 32 channels with 1THz channel spacing by considering 10 Gb/s data rate for each channel. The proposed system is designed for 16 uplink and 16 downlink channels with frequency range in the C–band i.e. 194.3 THz-195.8THz and 191.1THz-192.6THz respectively and the data has been transmitted over the bidirectional optical fiber simultaneously. The performance of the proposed system has been analyzed in terms of maximum quality factor (Q-factor), minimum bit error rate (BER), maximum optical to signal noise ratio (OSNR) by varying the optical fiber length for maximum transmission reach. We have successfully transmitted 320Gb/s data over bidirectional optical fiber length up to 140 km for DWDM-PON network.

1 INTRODUCTION

Recent communication network uses fiber optics in system to transmit data over the network by using light signals. It is a promising technology which facilitates us by providing optical channels with more number of users and huge data capacity keeping high speed and enormous reliability (Iwatsuki, 2010). In DWDM-PON FFTH system for 32 channels with a data rate of 10Gbps has been designed which can carry data up to 129km by (Ali et al., 2019). The researchers designed16 channels 50 GHz spaced DWDM-PONs system by considering Kerr nonlinearities (Vilcane et al., 2020). In (Kachhatiya and Prince, 2016), the authors have demonstrated a WDM-DWDM PON network capable of transmitting 10Gb/s data 2.5Gb/s for 16 downstream and 32 upstream channels by using power splitters. The comparison of 8 channels WDM PON with TDM PON with data rate 10Gbps has been presented in (Usman, 2018). In (Zhang et al., 2014), a WDM-PON scheme has been described with 40Gb/s data for each downstream channel and 10Gb/s for upstream channel with EDFA and self wavelength controlled tunable filter. The researchers revealed that a WDM PON exhibits better performance than a TDM PON. The experimental setup has been designed in (Artiglia et al., 2016)for the 40Gb/s WDM-PON system by providing a very high power budget and 110 km reach. A PON network has discussed for 10Gb/s data transmission rate over a distance of 130km of optical fiber using DFB laser and coherent receiver (Corsini et al., 2015). WDM-PON system based on Fabry-Perot laser diode with AWG has been presented for 10Gb/s downstream and 2.5Gb/s for upstream channels (El-nahal et al., 2015). DWDM-PON has been analyzed with different modulation schemes, four wave mixing (FWM) and hybrid optical amplifier (Apoorva et al., 2020; Sharda et al., 2008)(Kaur et al., 2016).

It is critical to increasing a further number of channels with a maximal reach of the DWDM-PON system in a cost-effective manner. By increasing the number of channels to some extent over a single fiber or increase the data rate per channel, intersymbol interference between the signals takes place. Due to this, the quality of the signal is degraded and the corresponding bit error rate is increased. So, there is a need to design a system in such a way

DOI: 10.1201/9781003193838-51

that quality factor should be equal to and greater than 7 to tolerate a maximum BER of recommended by ITU-T. To meet these requirements, we have designed the DWDM-PON system with 320Gb/s downlink-uplink data rate over a single bidirectional optical fiber length up to 140 km effectively.

The paper is organized as follows. Section 2 explains the architecture of DWDM-PON with simulated parameters. Results are discussed in section 3 and conclusion is reported in section 4.

2 DESIGN OF DWDM-PON SYSTEM

DWDM-PON architecture for 16 downlink and 16 uplink channels with total data rate 320 Gb/s is shown in Figure 1. In this system, we have used 16 downlink signal frequencies vary from 191.1THz-192.6THz THz that are uniformly spaced 0.1THz are generated by continuous wave (CW) laser sources with 0dBm input power.

Each pseudorandom non-return to zero (NRZ) bit sequence of length 1024 bits having data rate10Gb/s is modulated by each frequency of CW laser source with the help of Mach-Zehnder modulator (MZM) of extinction ratio 20dB. Multiplexer combines these signals and transmit over the bidirectional optical fiber for transmission. Multiplexed DWDM data is shown in Figure 2. Erbium-doped fiber amplifier (EDFA) is used to

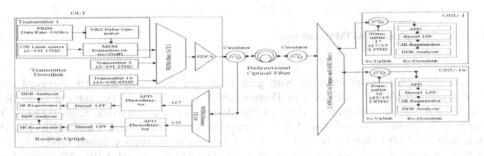

Figure 1. Proposed architecture of DWDM-PON.

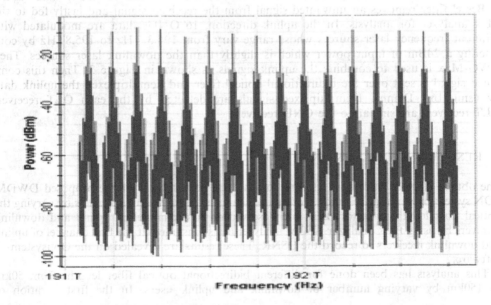

Figure 2. Multiplexed DWDM downlink signal.

279

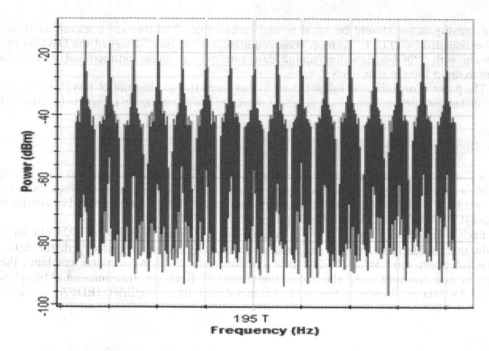

Figure 3. Multiplexed DWDM uplink signal.

increase the signal power strength with gain 10dB and noise Figure 4dB. Arrayed wave-guide gratings (AWGS) acts as 1:16 Demux which is installed at the remote node (RN). It is routed the de-multiplexed data to the desired optical network units (ONU). The reference frequency is used for AWG Demux is 191.1 THz with bandwidth 100GHz. Each port of AWG Demux is connected to a particular frequency tuned ONU. A single ONU is equipped with an avalanche photodiode (APD), Bessel LPF, 3R regenerator, and BER analyzer. APD is used to provide high gain and large SNR for long-distance communication whose responsivity and dark current are set to 1A/W and 10nA. A Bessel filter removes an unwanted signal from the received signal and lastly fed to the BER analyzer for analysis. In the uplink direction, 16 ONUs data are modulated with different frequency laser sources whose range vary from 194.3 THz to 195.8THz by con-sidering a 2dBm of input power which is slightly than the downlink laser sources. Then AWG-Mux is used to combine the uplink signals as shown in Figure 3. Then this com-bined signal is sent over the bidirectional optical fiber and demultiplexed the uplink data by using 16:1 Demux. Demultiplexed signals are detected by the each OLT receiver. OLT receivers are similar to the ONU receivers.

3 RESULT AND DISCUSSIONS

The obtained results have been used to compare the performance of the proposed DWDM PON system by varying the number of uplink channels and downlink channels by varying the optical fiber length. Q-factor, BER are observed on first channel of uplink and downlink receivers by using BER analyzer. WDM analyzer is also installed at the first channel of uplink and downlink receivers to record the OSNR. These results are revealed on the Optisystem-17 software.

This analysis has been done for different bidirectional optical fiber lengths from 50km to 150km by varying number of downlink and uplink users. In the first iteration of

50 km, we have received a high Q factor and maximum OSNR with minimum BER. After that, we observe gradual degradation of signal takes place with addition in optical fiber length. The same pattern is also observed for other curves. We have received a signal with high Q-factor of 12.0906 and 12.8410 for 2 downlink and uplink users and 11.5 and 10.3788 for 16 downlink and uplink users as shown in Figure 4 and Figure 5. In last iterations, Q-factor decreases from 6.9380 to 6.6031 and 8.44 to 7.5037 at distance 150km for downlink and uplink users respectively. Figure 6 shows that increases in BER with increase in number of users. BER increases from $4.60*10^{-34}$ to $2.69*10^{-31}$ at distance 50 km and $1.88*10^{-12}$ to $9.87*10^{-11}$ at distance 150km with increasing the number of downlink users from 2 to 16 respectively. Similarly, BER also increases for uplink channels as optical fiber length varies from 50 km to 150 km. We have recorded value of BER from $3.6*10^{-38}$ to $1.05*10^{-25}$ at 50 km and $1.08*10^{-16}$ to $2.73*10^{-14}$ at 150 km by varying the number of uplink users from 2 to 16 which is shown in Figure 7. As number of users increases other parameter i.e OSNR also reduces. By varying downlink users from 2 to 16, OSNR decreases from 52.0644 to 42.9101 and 51.5001 to 41.7000 for distance 50 km

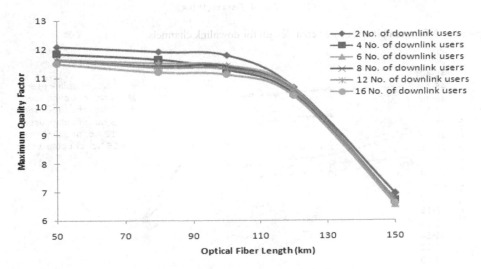

Figure 4.   Maximum quality factor versus length for downlink channels.

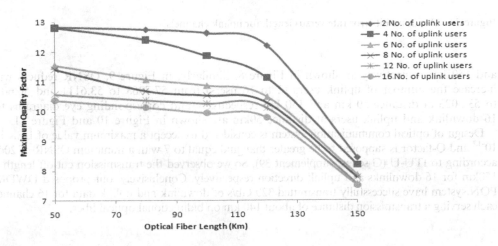

Figure 5.   Maximum quality factor versus length for uplink channels.

281

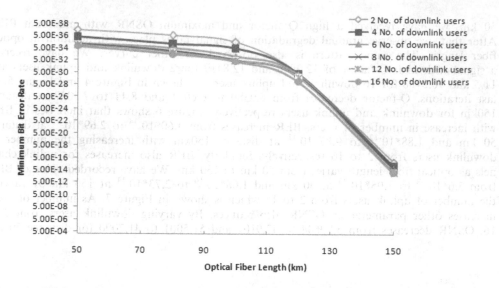

Figure 6. Minimum bit error rate versus length for downlink channels.

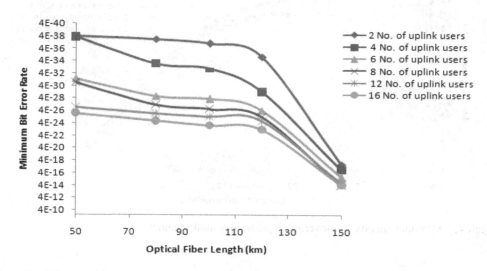

Figure 7. Minimum bit error rate versus length for uplink channels.

and 150 km respectively as shown in Figure 8. Similarly, in Figure 9, OSNR reduces with increase the number of uplink users (2 to 16 users) from 57.7968 to 53.6111 and 56.6010 to 53.2023 at distance 50 km and 150 km respectively and corresponding eye diagrams for 16 downlink and uplink users at distance 50km are shown in Figure 10 and Figure 11.

Design of optical communication system is considered to accept a maximum value of BER i.e. 10^{-12} and Q-factor is supposed to be greater than and equal to 7 with a minimum OSNR of 20dB according to ITU-U (G-series, supplement 39), So we observed the transmission cut-off length at 150km for 16 downlinks and uplink direction respectively. Conclusively, our proposed DWDM-PON system have successfully transmitted 320 Gb/s of downlink and uplink data for 16 channels each serving a transmission distance of about 140 km on bidirectional optical fiber.

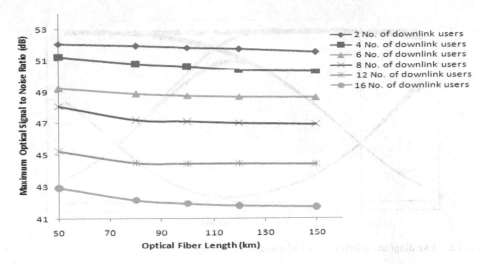

Figure 8.　Maximum optical signal to noise ratio (dB) versus length(km) for downlink channels.

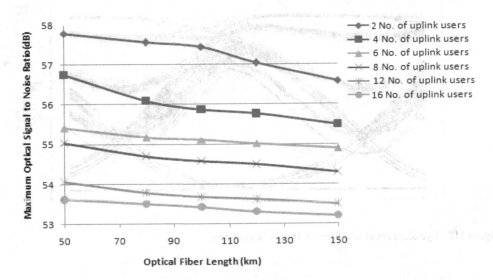

Figure 9.　Maximum optical signal to noise ratio (dB) versus length(km) for uplink channels.

4　CONCLUSION

We have proposed DWDM-PON system with 2, 4,8,16, 24, and 32 users at 10 GB/s data rate of each channel for a transmission distance from 50 km to 150km in uplink and downlink direction simultaneously. It can be accomplished that quality factor and optical signal to noise ratio decreases while bit error rate increases by increasing the optical fiber length as well as number of users. Further to increase the performance of this system, different modulation schemes and dispersion compensation techniques will be used.

ACKNOWLEDGEMENT

The authors would like to thank the "Research Project grant to the faculty of DTU" under Delhi Technological University for its financial support of this work under file number DTU/

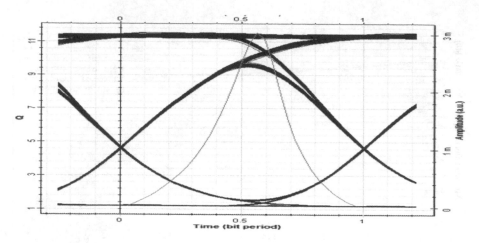

Figure 10.   Eye diagram at first users of 16 downlink channels.

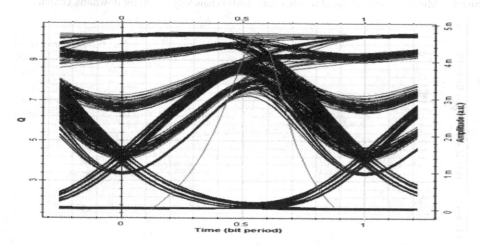

Figure 11.   Eye diagram at first users of 16 uplink channels.

Council/BOMAC/Notification/31/2018/5738. The authors also would like to thank the anonymous reviewers for the feedback.

REFERENCES

Ali, M.H., Almufti, A.M., Abu-Alsaad, H.A., 2019. Simulative analyzing of covering suburban areas with 32 × 10 Gbps DWDM-PON FTTH using different dispersion and power. *J. Commun.* 14: 381–389. https://doi.org/10.12720/jcm.14.5.381-389

Apoorva, Priyanka, Wadhwa, H., Kaur, G., 2020. Performance analysis of hybrid optical amplifiers for 32 channel WDM system at 10 gbps bit rate for wan applications. *J. Opt. Commun.* 41: 23–29. https://doi.org/10.1515/joc-2017-0139

Artiglia, M., Bottoni, F., Corsini, R., Presi, M., Rannello, M., Valvo, M., Ciaramella, E., 2016. 4×10 Gb/s coherent WDM-PON system over 110 km single mode fibre and with 55 dB ODN power budget. 2016 Opt. Fiber Commun. Conf. Exhib. OFC 2016 1: 3–6. https://doi.org/10.1364/ofc.2016.w2a.66

Corsini, R., Presi, M., Artiglia, M., Ciaramella, E., 2015. 10-Gb/s Long-Reach PON System With Low-Complexity Dispersion-Managed Coherent Receiver. *IEEE Photonics J.* 7. https://doi.org/10.1109/JPHOT.2015.2486678

El-nahal, F.I., Alhalabi, M., Husein, A.H.M., 2015. Wavelength Division Multiplexing Passive Optical Network (WDM - PON) technologies for future access networks 2: 34–40.

Iwatsuki, K., 2010. Application and technical issues of WDM-PON. *Broadband Access Commun. Technol. IV*: 7620, 76200C. https://doi.org/10.1117/12.846598

Kachhatiya, V., Prince, S., 2016. Wavelength division multiplexing-dense wavelength division multiplexed passive optical network (WDM-DWDM-PON) for long reach terrain connectivity. *Int. Conf. Commun. Signal Process. ICCSP 2016*: 104–108. https://doi.org/10.1109/ICCSP.2016.7754518

Kaur, G., Yadav, R., Zaman, M., 2016. Design of Dpsk Modulator and Direct Detection Receiver for Dwdm Based Optical Communication System. *ICTACT J. Commun. Technol.* 7: 1321–1325. https://doi.org/10.21917/ijct.2016.0194

Khaksar, S., 2015. A survey on challenges in SOA Maintenance. *Comput. Eng. IT* 21: 1–6.

Sharda, A.K., Kaur, G., Gupta, N., 2008. Performance comparison of 2.5GBPS FWM based DWDM system for enhanced number of users. Proc. *CAOL 2008 4th Int. Conf. Adv. Optoelectron. Lasers*: 137–139. https://doi.org/10.1109/CAOL.2008.4671912

Usman, M., 2018. Performance comparison of a WDM PON with TDM PON At 10 GBPS Performance Comparison of a WDM PON with TDM PON At 10 GBPS.

Vilcane, K., Matsenko, S., Parfjonovs, M., Murnieks, R., Aleksejeva, M., Spolitis, S., 2020. Implementation of Multi-Wavelength Source for DWDM-PON Fiber Optical Transmission Systems. *Latv. J. Phys. Tech. Sci.* 57: 24–33. https://doi.org/10.2478/lpts-2020-0019

Zhang, Z., Sun, Y., Chen, X., Wang, L., Zhang, M., 2014. 40-Gb/s Downstream and 10-Gb/s Upstream Long- reach WDM-PON Employing Remotely Pumped EDFA and Self Wavelength Managed Tunable Transmitter: 280–283.

Recent Trends in Communication and Electronics – Sharma et al. (Eds)
© 2021 Taylor & Francis Group, LLC, ISBN 978-1-032-04572-6

Analysis of noise reduction techniques for electrooculography signals for automatic sleep stage scoring application

Ravi Raja
Research Scholar, ECE Department, SRMIST, Ramapuram, Chennai, Tamil Nadu, India

Phani Kumar Polasi
Professor & Head, ECE Department, SRMIST, Ramapuram, Chennai, Tamil Nadu, India

ABSTRACT: Sleep stage scoring is an important task for analyzing the sleep pattern and for timely diagnosis of sleep disorders and sleep related studies. Polysomnography (PSG) is re-quired for conventional sleep-related studies, which uses different biological signals like Electrooculography (EOG), Electromyography (EMG), Electroencephalography (EEG), Electronystagmography (ENG), Electrocardiography (ECG). Acquisition and analysis of all these signals requires expertise. Sleep stage scoring can also be accomplished by a single channel EOG. In this connection, noise reduction of EOG signal gained importance, and acquiring these signals doesn't require corresponding field expert and sophisticated lab setup. In this paper, different suitable wavelet noise reduction techniques are used for reducing the noise in EOG signal. The EOG signal dataset is taken from publicly available ISRUC-Sleep database. Performance of each technique is analyzed and presented in terms of SNR and MSE.

1 INTRODUCTION

According to AASM-American Academy of Sleep Medicine (Iber 2007) the sleep is an essential component for human survival, but work stress and busy life is causing sleep deprivation and causes insomnia, but many people doesn't know that they are depressed or affected by insomnia (Nutt 2008). The symptoms of disturbed night time sleep in people with depression/ work stress have caught many researcher's eye in both clinical and epidemiological studies. Even though the sleep related studies started in early 1968 (Rechtschaffen 1968), it gained importance in 2007 because of sleep staging rules changed by AASM (Iber 2007). In 1968, Rechtschaffen A, Kales A, proposed and gave standard guidelines for classification of sleep into 6 stages [wake-W, Sleep Stage 1-S1, Sleep Stage 2-S2, Sleep Stage 3-S3, Sleep Stage 4-S4, Sleep Stage-REM (Rapid Eye Movement). In 2007, AASM modified these 6 sleep stages into 5 sleep stages [Wake-W, Sleep Stage 1-N1, Sleep Stage 2-N2, Sleep Stage 3-N3 and Sleep Stage REM-R]. The Sleep Stage S3 and S4 are combined and formed Sleep Stage-N3. The rules for scoring these stages are given through an AASM manual (Iber 2007) and every year they are updating and releasing the manual according to new researches (Silva 2014).

The sleep stage scoring requires PSG signals which contains different biological signals like EOG, EMG, EEG, ENG, ECG. Acquisition and analysis of all these signals require expertise in those corresponding fields, require a huge lab setup and also time-consuming process. M. M. Rahman et. al., 2018 (Rahman 2018) used single channel EOG for classification of sleep stages and proved scoring accuracy of 92.6%. EOG signal acquisition is flexible when com-pared to other biological signals. EOG signals gained importance in the application area of sleep scoring, with careful examination of single channel EOG, sleep scoring can be done.

DOI: 10.1201/9781003193838-52

But while acquiring EOG data from the test subject, noise might get induced (Bulling 2007). Even though these signals are recorded during sleep, subject's movements and external factors may induce noise into the signal. There are several noise removal techniques available (Dasgupta 2017) for biological signals but signal peaks play important role in sleep stage scoring (Crofft 2005). These methods smoothen the signals and hence there is a possibility of losing sleep stage estimates. So, care must be taken while choosing a noise reduction procedure.

The (Rajesh 2015) Stationary Wavelet Transform (SWT) is useful for noise reduction purpose, but all these biological signals are non-stationary in nature and also thresholding procedures used in these references remove the required features for sleep scoring. The idea of implementing Discrete Wavelet Transforms (DWT) for the reduction of noise in EOG signals with biorthogonal (Naga 2012) Daubechies and Haar wavelet basis functions produced suitable results for different applications (Mala 2016). The analysis shows good noise elimination at higher thresholds, but these higher thresholds are not suitable for sleep scoring, since they elim-inate EOG signal peak values. In this paper, analysis made on Haar, Symlets, Fejer-Korovkin, and Biorthogonal wavelets for noise reduction in EOG signal for sleep stage scoring application. MSE and SNR parameters are used for performance calculation.

For the analysis of sleep stage scoring the dataset is taken from ISRUC-SLEEP Dataset (Khalighi 2016). This dataset is made publicly available for researchers and the data is recorded ac-cording the new rules formed by AASM (Iber 2007). These datasets are recorded from hu-man adult subjects, including healthy, and subjects suffering from different sleep disorders. The recordings will be in the format of PSG, These PSGs will contain most of the biological signals as indicated earlier and all these datasets associated with each subject (Khalighi 2013), are visually scored and analyzed by experts and results are made available for validation purposes. For this study healthy subject's EOG dataset is taken into consideration.

2 METHODOLOGY

As EOG is a nonstationary signal, Discreet Wavelet Transform (DWT) is one of the best tools available to analyze the signal (Yansong 2011). Wavelet refers to a small wave, Haar, Symlets, Daubechies etc., Some of the wavelet families are useful for analysis of a signal (Mali 2011). Wavelet transform can have both time and frequency resolution using scaling and dilation operations. Therefore, DWT is used to decompose the EOG signal. Decomposed signal consists of various frequency components. So that noise reduction method can be applied at any frequency. Finally, reconstruction of noise free signal can be obtained by using inverse DWT. The above said process is depicted in Figure 1.

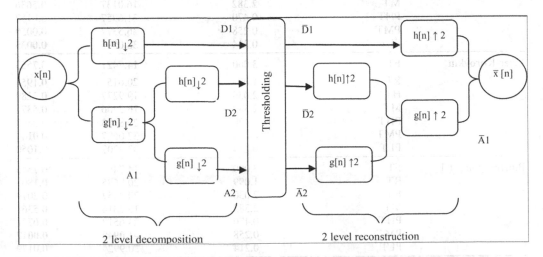

Figure 1. Two level discrete wavelet decomposition and reconstruction model.

In Figure 1, x[n] represents the EOG input signal, h[n], g[n] represents high pass and low pass filter responses respectively, and x̄[n] is estimate of input signal received at the output, in 2 level decomposition structure ↓2 represents down sampling by a factor of 2 and in 2 level reconstruction structure ↑2 represents up sampling by a factor of 2. In this fashion each low pass and high pass filter creates coefficients (A1, D1, A2, D2) that represent original signal information in compact form. Ā1, D̄1, Ā2 and D̄2 represents coefficients generated after thresholding. According to Yansong W 2011, performance of the wavelet decomposition depends on the similarity of input signal and wavelet shape. So, the Haar, Symlets, Biorthogonal, Fejer-Korovkin basis functions are similar to EOG signal, and these are used for decomposition of signal into 4 levels, and soft thresholding procedure used, which includes Fixed form threshold (FT), Rigorous sure threshold (RT), Heuristic sure threshold (HT), Minmax threshold (MT), Penalize high threshold (PHT) Penalize medium threshold (PMT), Penalize low threshold (PLT). For each output SNR and MSE are calculated and results are presented in Table 1.

The procedure followed for this study involves four steps.

i. DWT is applied on EOG signal with a specified wavelet as basis function and decomposition done to 4 levels.
ii. Soft thresholding is applied for each decomposed level.
iii. Reconstruction of decomposed signal x[n] to obtain denoised EOG signal x̄[n].
iv. Calculation of SNR, MSE

Table 1. Comparison of SNR and MSE values obtained from various wavelet decompositions.

Wavelet	Thresholding Method	Threshold Value	SNR in dB	MSE
Haar	FT	3.000	14.7469	0.7545
	RT	1.099	20.9685	0.1801
	HT	2.030	20.3457	0.2079
	MT	2.382	16.2314	0.5361
	PHT	0.470	24.6517	0.0771
	PMT	0.258	29.6556	0.0244
	PLT	0.214	30.9924	0.0179
Symlets	FT	3.000	14.5086	0.7971
	RT	1.099	21.8618	0.1466
	HT	2.030	20.9299	0.1817
	MT	2.382	16.0137	0.5636
	PHT	0.470	31.6957	0.0152
	PMT	0.258	36.5377	0.0050
	PLT	0.214	**38.0595**	**0.0035**
Fejer-Korovkin	FT	3.000	14.7421	0.7553
	RT	1.099	20.615	0.1954
	HT	2.030	19.9277	0.2289
	MT	2.382	16.2206	0.5374
	PHT	0.470	26.8283	0.0467
	PMT	0.258	32.0697	0.0140
	PLT	0.214	33.3207	0.1050
Biortho-gonal 1.1	FT	3.000	14.7469	0.7545
	RT	1.099	20.9685	0.1801
	HT	2.030	20.3457	0.2079
	MT	2.382	16.2314	0.5361
	PHT	0.470	24.6517	0.0771
	PMT	0.258	**41.0949**	**0.0017**
	PLT	0.214	30.9924	0.0179

2.1 Estimation of SNR and MSE

The SNR and MSE values are estimated between the reference EOG signal and denoised EOG signal and formulae is given in equations (1) and (2) respectively.

$$SNR = 10 \log_{10} \frac{\sum_{i=1}^{N} x(i)^2}{\sum_{i=1}^{N} (x(i) - \bar{x}(i))^2} \qquad (1)$$

$$MSE = \frac{1}{N} \sum_{i=1}^{N} (x(i) - \bar{x}(i))^2 \qquad (2)$$

where x(i) is the input EOG signal (Figure 1), x̄ (i) is the denoised estimate of the input EOG signal (Figure 2), i represents instance, N is the EOG signal's length.

3 RESULTS AND DISCUSSION

In this paper, for analysis the dataset is taken from publicly available ISRUC-Sleep Data (Kha-lighi 2016) and it consists each subject's data in the form of epochs. Each epoch consists of 30 seconds data sampled at 200 samples/sec and each epoch consists of 6000 samples. A total of 960 epochs is stored for each subject for 8 hours of sleep. For the analysis purpose 10 seconds data is considered consisting of 2000 samples as shown in Figure 1.

The EOG signal is decomposed into 4 levels and undergone through 7 different soft thresholding methods as specified in methodology. The results are tabulated in Table 1. It shows the SNR and MSE values with corresponding thresholding method and its value. Since the main requirement here is the transition states of EOG signal must not be filtered out, for the application of sleep stage scoring. Hence, care is taken while selecting thresholds to preserve the transition states so that each stage can be identified in future part of the work. Even though for some wavelets the data is filtered out as shown in Figure 3. The best SNR and MSE values obtained through each applied wavelet procedure are tabulated in Table 2.

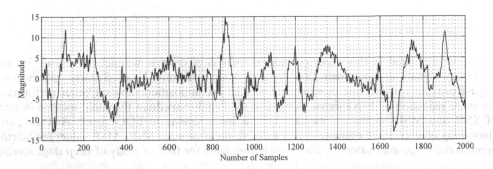

Figure 2. Electrooculography Signal consisting data of 10 seconds.

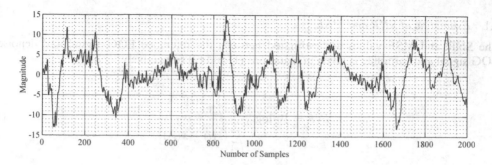

Figure 3. Denoised Electrooculography Signal using Biorthogonal 1.1 wavelet with Penalize-Medium Threshold.

Table 2. Best SNR value and corresponding MSE from each wavelet decomposition.

Wavelet Method	Signal to Noise Ratio (SNR) in dB	Mean Square Error (MSE)
Haar	30.9924	0.0179
Symlets	38.0595	0.0035
Fejer-Korovkin	33.3207	0.1050
Biorthogonal (1.1)	41.0949	0.0017

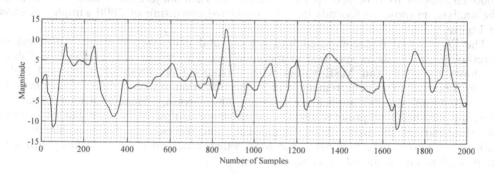

Figure 4. Denoised Electrooculography Signal using Haar wavelet with Fixed Form Threshold.

4 CONCLUSIONS

The work presented in this paper provides usefulness of discrete wavelet transforms for non-stationary signals in noise reduction process. After analyzing Haar, Symlets, biorthogonal, Fejer-Korovkin wavelets the biorthogonal wavelet decomposition shows a better performance for noise reduction of EOG signal along with the Penalize medium threshold method in terms of SNR and MSE because of wavelet shape is more similar to EOG signal. And also, over smoothing of sleep stage estimate is also reduced, the computed SNR, MSE values for biorthogonal wavelet decomposition are extremely encouraging for further study of sleep stage scoring.

REFERENCES

Bulling, A. 2007. Automatic artefact compensation in EOG signals. *2nd European Conference on Smart Sensing and Context*: 12–13.
Croft R J, 2005. EOG correction: A comparison of four methods. *Psychophysiology* 42(1): 16–24.

Dasgupta A. 2017. A two-stage framework for denoising electrooculography signals. *Biomedical* Signal *Processing and Control* 31: 231–237.

Iber C. 2007. The AASM manual for the scoring of sleep and associated events: rules, terminology and technical specifications. *American Academy of Sleep Medicine*: 1–86.

Khalighi. 2013. Automatic sleep staging: a computer assisted approach for optimal combination of features and polysomnographic channels. *Expert Systems with Application* 40: 7046–7059.

Khalighi. 2016. ISRUC-Sleep: A comprehensive public dataset for sleep researchers. *Computer methods and programs in biomedicine* 124: 180–192.

Mala S. 2016. Electrooculography de-noising: Wavelet based approach to reduce noise. *International Journal of Advanced Engineering Technology*: 482–487.

Mali, R. D. 2011 Removal of 50Hz PLI using discrete wavelet transform for quality diagnosis of biomedical *ECG* signal. *International Journal of Computer Applications*: 1–6.

Naga R. 2012. Denoising EOG signal using stationary wavelet transform. *Measurement Science Review* 12(2): 46–51.

Nutt, 2008. Sleep disorders as core symptoms of depression, *Dialogues in clinical neuroscience* 10: 329–36.

Rahman M. M. 2018 Sleep stage classification using single-channel EOG. *Computers in Biology and Medicine* 102: 211–220.

Rajesh N. 2015. A segmental approach with SWT technique for Denoising the EOG signal. *Modelling and* Simulation *in Engineering*.

Rechtschaffen A. 1968. A manual of standardized terminology, techniques and scoring system of sleep stages in human subjects. *Los Angeles: Brain Information Service/Brain Research Institute*.

Silva Da. 2014. An EOG-Based Automatic Sleep Scoring System and Its Related Application in Sleep Environmental Control. *Physiological Computing Systems: First International Conference*: 71–88.

Yansong W. 2011. Discrete Wavelet Transfom for Nonstationary Signal Processing. *Discrete Wavelet Transforms Theory and Applications IntechOpen*.

Recent Trends in Communication and Electronics – Sharma et al. (Eds)
© 2021 Taylor & Francis Group, LLC, ISBN 978-1-032-04572-6

Real time estimation of calories from Indian food picture using image processing techniques

U. Anitha, R. Narmadha & G.D. Anbarasi Jebaselvi
Associate Professor, Sathyabama Institute of Science and Technology, Chennai, India

S. Lavanya Kumar & P. Raja Shekhar Reddy
Student, Sathyabama Institute of Science and Technology, Chennai, India

ABSTRACT: Due to the mass transformation in people'sfood habits and standards of living, obesity rates are increasing at an alarming speed, and this will ultimately reflecting people's health. People need to avoid the fast foods and control their daily calorie intake by eating healthier foods, which is the most important thing to be followed. Though food packaging comes with nutrition and calorie labels, it's still not very convincing the people to refer those in an effective way. The static food picture/photo usually contains various kinds of ingredients used in the dishes. To recognize the food images with various dishes, it needs to detect every single ingredient in an image. In recent years, the accuracy of object detection has been improved radically by means of using neural network. In this paper, a neural network-based object detection method is proposed to identify the various dishes from a food-image and label its ingredients. In addition, this food detector has been applied to food calorie estimation for each ingredient in a food picture/photo of multiple dishes. The dataset has food picture/photo of many dishes and its ingredients with food calorie of each ingredients. In this research work, counting of food calories has been done by combining the quantities of each ingredient present in that dish.

Keywords: Food picture/photo, GLCM algorithm, object detection, neural network, calories estimation, image processing, mobile application

1 INTRODUCTION

Now days, increase in the rate of obesity cases influence people's health. To avoid the risk of being obesity and its ill-effects, people need to know their daily food calorie intake. The main aim of food items recognition is to provide a nutritional information of the food to be consumed is through mobile phone applications. In general, dietetic information are look up in official database/catalogue of top-rated hotels or food processing industries and it is time consuming long process. A mobile based food calorie estimator will become a good alternate for the conventional method. In 2009, Wu, Wen & Jie Yang have started the work using count-based matching (CBM) method, it shows the low accuracy in identifying food items individually. A compute descriptive statistical analysis has been tried by McAllister, P., et.al. 2017. It also has a drawback that there is a major difference between sum of calorie estimate of each item and its overall estimate. Local Variation Segmentation (LVS), Multi-Kernel SVM and Deep Convolutional Neural Networks (DCNNs) methods also applied on the food samples. However, Ege, Takumi, et.al. 2019 and Aslan, et.al. 2018, have concluded that the calorie estimation is a difficult task associated with the diet management system. In 2018, Minija, et.al.,

DOI: 10.1201/9781003193838-53

developed a method called Weakly-supervised Segmentation Based Calorie Estimation which requires 3D food volume data to develop the system and is very expensive and time consuming.

Therefore, the crises of food identification by determining what type of food present in the input photo, and that food is the focus of the image. Liu, Chang, et.al. 2016 and Sudo, Kyoko, et.al. 2014, were assuming the background is plain, like a plain surface. Kawano, et.al. 2013, Yunus, Raza, et al. 2018 and Ciocca, Gianluigi, et.al. 2015, evident that with such a controlled problem domain, food recognition is a very tough. Object recognition has been more effective, for the images having car, face but food is very difficult and vague. The definition of any food has nothing to do with shape, need its ingredients and method of preparation. Ingredients and method of preparation are evident themselves in its features like shape, colour, and texture. However, each food has inflexible shape and structure, sometimes the cheese is hidden inside the bun, etc. Fuchs, Klaus, et al. and Knez, Simon & Luka Sajn. 2020, proved the real time food calorie estimate using mobile phone via headset is in demand and it need high accuracy. It also faces a drawback that the algorithm needs to be updated for all local languages. In this paper, all problems were addressed by trying to recognize the ingredients that constitute a food item. Humans express food items and their variation in terms of the ingredients they contain and the way the ingredients are arranged, so it makes sensitive that may be able to collect more information by extricating features at an intermediary ingredient level. After categorizing the ingredients in the image, extract features that describe the relative's quantities of all ingredient.

2 PROPOSED METHODOLOGY

Initially 2D input image which was taken from the database is fed into pre-processing unit as shown in Figure 1. It converts the input image in to a gray scale image. The resize along with filtering of the noise in preprocessed image is executed using discrete wavelet transform (DWT), then forwarded to next step. Here the image is enhanced by the gray level co-occurrence matrix (GLCM) features followed by GLCM algorithm in feature extraction procedure. The texture features calculated by GLCM are energy, entropy, contrast, correlation coefficient and homogeneity. The extracted features for the food identification are stored in the database extended to the list of ingredients and its calorie chart. Next to the feature extraction, classification of new feature extracted image by using Neural Networks has been

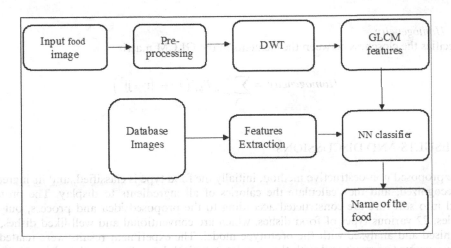

Figure 1. Proposed Methodology.

executed by comparing it with the image of the database. The name of the food and calorie value is shown as a result in the GUI environment using MatLab software.

2.1 *Texture features*

Image texture parameters are used to determine the perceived texture of object in an image. The parameters such as energy, entropy, contrast, correlation coefficient and homogeneity are used for this application as shown in equation (1) to (5).

2.1.1 *Energy*

It is a measure the smoothness of the image and It is also suitable to detect the disorder in texture image.

$$Energy = \sum_{i,j=0}^{N-1} (P_{ij})^2 \tag{1}$$

2.1.2 *Entropy*

Entropy gives a measure of density of the image. Complex textures tend to have higher entropy.

$$Entropy = -\sum_{i=1} \sum_{j=1} P(i,j) log(P(i,j)) \tag{2}$$

2.1.3 *Contrast*

It is to measure shadow depth on texture in a gray scale image.

$$Contrast = \sum_{i,j=0}^{N-1} P_{ij}(i-j)^2 \tag{3}$$

2.1.4 *Correlation coefficient*

It measures the mutual probability occurrence of the specified pixel pair.

$$Correlation = \sum_{i,j=0}^{N-1} P_{ij}(i-\mu)(j-\mu)/\sigma^2)) \tag{4}$$

2.1.5 *Homogeneity*

It specifies the closeness between the elements in the GLCM matrix.

$$Homogeneity = \sum_{i,j=0}^{N-1} P_{ij}/\left(1 + [i-j]^2\right) \tag{5}$$

3 RESULTS AND DISCUSSIONS

In the proposed non-destructive method, initially the food type is classified, and its ingredients are recognized, and then calculate the calories of all ingredients to display. The prototype model into software are constructed according to the proposed idea and process, out of 72 samples, 22 various types of food dishes, which are conventional and well-liked dishes, were recognised and analysed with the prototype model. The experiment results were related with the destructive test result and with the average statistical data of destructive assessment available in the open websites. The results show the proposed method possess more accurate calorie values than the average statistical data of destructive assessment appearing on the common websites. This proves that the food samples for this work and destructive test at the lab both are similar, but the destructive statistical evaluation results in the common website are not exactly matches with the test data samples. This says that the proposed software successfully calculates food calories in each food. The performance of the software is evaluated by comparing obtained calories with the calories set up by the traditional destructive method showing error as 3.57% for the given dataset.

Figure 2 have shown total 22 typical types of food dishes, which are known as the highly popular dishes in India, were tested with the proposed software model of the food to measure the calorie in it using image process technique. In Figure 3, the finding of the calorie of the food the first step is to browse the image sample from the database by using the food ingredients detection system.

In Figure 4 is an image pre-processing output for further processing. In Figure 5, feature parameters are extracted which is used for image classification. Neural network is trained using those values and the new test image parameter values are compared with the existing database to find the food type. Figure 6 represents completion of Neural Network training phase.

In Figure 7, it will display the name of the food by using the data base. In food type categorization, which is the first procedure of the proposed method, is completely considered as a most important factor for the further procedures so that the results must be extremely consistent. Even though the results are not absolutely 100%, they are relatively close to 100% in all food types. However, few samples were missed unfortunately for the classification was

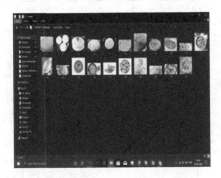

Figure 2.   Image samples.

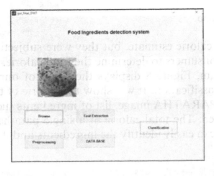

Figure 3.   Browsing the Image in GUI.

295

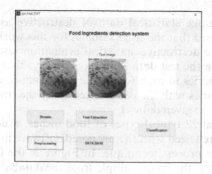

Figure 4.   Pre-processing of the food.

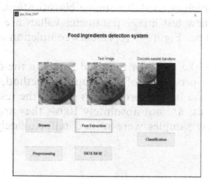

Figure 5.   Feature Extraction.

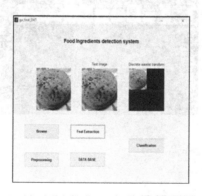

Figure 6.   Classification of Food.

likely to produce errors in calorie estimate, but they were subjectively tolerable in practice as given in guidelines to the consumers to determine the food calorie.

By using testing algorithm, Figure 8 displays the result of our project. After all this pre-processing, simply apply classification. It will show the calorie of the food and its ingredients. For the given single input PARATHA image, list of ingredients such as maida, oil, salt, water, and its calories are tabulated. The total calorie of a single paratha is 1074 which is displayed here. By this method, user can easily identify the ingredients and its calorie value.

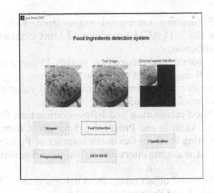

Figure 7.   Name of an India Food (Paratha).

Ingredients	Calories
Maida	190
Refined Oil	884
Salt	0
Water	0
Total	1074

Figure 8.   Total Calorie of the food displayed in the MATLAB window.

4   CONCLUSION

The Gray-Level Co-Occurrence Matrix (GLCM) based features to characterize food ingredient provides better performance than other methods which loses lots of essential information about food. The proposed work helps for the easy identification and useful for medical diet. This method gives a sizable upgrade in categorization accuracy over standard methods, but it takes approximately 1 or 2 minutes to compute the result. If that time delay is tolerable, then this model would be easy to incorporate into a mobile phone application. In Future, this system may be implemented by using hardware model for nutrition and calorie estimation along with mass calculation. The MATLAB software and hardware are interfacing with the help of controller for measuring the mass using a high megapixel camera and a precision sensor to obtain results for liquid food such as milk, sauce, tea, juices, etc.

REFERENCES

Wu, Wen & Jie Yang. 2009. Fast food recognition from videos of eating for calorie estimation. IEEE International Conference on Multimedia and Expo.
McAllister, P., et.al. 2017. Automated adjustment of crowdsourced calorie estimations for accurate food image logging. 2017 IEEE International Conference on Bioinformatics and Biomedicine (BIBM).
Ege, Takumi, et.al. 2019. Image-Based Estimation of Real Food Size for Accurate Food Calorie Estimation. 2019 IEEE Conference on Multimedia Information Processing and Retrieval (MIPR).
Aslan, et.al. 2018. Semantic food segmentation for automatic dietary monitoring. IEEE 8th International Conference on Consumer Electronics-Berlin (ICCE-Berlin).
Minija, et.al. 2018. Image processing based Classification and Segmentation using LVS based Multi-Kernel SVM. 2018 International Conference on Smart Systems and Inventive Technology (ICSSIT).
Liu, Chang, et.al. 2016. Deep food: Deep learning-based food image recognition for computer-aided dietary assessment. International Conference on Smart Homes and Health Telematics. Springer.

Sudo, Kyoko, et.al. 2014. Estimating nutritional value from food images based on semantic segmentation. Proceedings of the 2014 ACM International Joint Conference on Pervasive and Ubiquitous Computing: Adjunct Publication.

Kawano, et.al. 2013. Real-time mobile food recognition system. Proceedings of the IEEE Conference on Computer Vision and Pattern Recognition Workshops.

Yunus, Raza, et al. 2018. A framework to estimate the nutritional value of food in real time using deep learning techniques. IEEE Access 7: 2643–2652.

Ciocca, Gianluigi, et.al. 2015. Food recognition and leftover estimation for daily diet monitoring. International Conference on Image Analysis and Processing. Springer, Cham.

Fuchs, Klaus, et al. 2020. Supporting food choices in the Internet of People: Automatic detection of diet-related activities and display of real-time interventions via mixed reality headsets. Future Generation Computer Systems.

Knez, Simon & Luka Sajn. 2020. Food object recognition using a mobile device: Evaluation of currently implemented systems. Trends in Food Science & Technology.

Recent Trends in Communication and Electronics – Sharma et al. (Eds)
© 2021 Taylor & Francis Group, LLC, ISBN 978-1-032-04572-6

Study of aging effect on solar panel performance using LabVIEW

Salim
KIET Group of Institutions, Delhi-NCR, Ghaziabad, India

J. Ohri
Electrical Engineering, National Institute of Technology, Kurukshetra, India

ABSTRACT: In today's solar energy situation, the general motto is "set it and forget it". This argument derives from thorough analysis and reliability studies into photovoltaic panels. This ensures there is no need for repairs and daily monitoring. Yet several things can go wrong to cause the real output to deviate from the planned output. If defects and unanticipated problems of deterioration go undetected, they would result in reduced production of energy. The degradation of solar modules was also evident with the rise in series resistance of the module and consequent degradation of the output electrical power of the modules. This research paper explores the precedent of ageing among modules currently in use, with the objective of evaluating the durability of photovoltaic modules. This paper discusses comparative product life analysis of PV modules. Analysis of all solar PV modules is accomplished with LabVIEW software.

Keywords: solar photo voltaic (pv), current and power-voltage characteristics, LabVIEW

1 INTRODUCTION

Increasing worries about air pollution and global warming, sustainable green energy sources are expected to play a significant or important part in the global strength of the future. If we have to get the 24x7 energy motive for every person in the world, the medium-term solution seems to be solar power. Geographically, India has perfect solar energy characteristics. This will cut energy demand by more than 50%.Owing to the spontaneous nature of their occurrence, the climatic conditions are uncertain. These complexities contribute to either over-or underestimation of the energy output of PV modules. Modeling PV modules is one of the core components responsible for the proper operation of PV systems. One diode model studied for simulation and modeling of PV cell [1]. Modeling is a means to consider the current, voltage, and power relationships of PV modules. The calculation of the models however is influenced by multiple intrinsic and extrinsic variables, which essentially impact the action of current and voltage. The R_p- model studied for solar PV [2]. Perfect modeling is therefore important for estimating the efficiency of PV modules under various environmental conditions. Real time monitoring studied and [3][4]. In addition, the author also found that the model estimate could be further improved either by adding two parallel diodes with an individual fixed saturation current or by considering the diode quality factor as a variable parameter instead of a fixed value such as 1 or 2.The method of extraction of solar PV parameters tested [5][6]. Advanced MPPT technology observed [7]. One diode and two diode variants studied [8]. The method of modeling and evaluating solar PV carried out [9]. Pure silicon recovered from old PV modules [10].

DOI: 10.1201/9781003193838-54

The effect of MPPT controller based on incremental conductance algorithm on Shell SM50H PV modulehas been studied [12]. Real time monitoring using data dash board in LabVIEW has been studied [13]. Different LabVIEW models studied [14]. With the aid of LabVIEW, comparison of solar PV module at different life span has been shown in this paper. Solar panel input and output parameters are compared and analyzed at different time period. Mathematical modeling of solar panel is studied and developed for the simulation of the solar PV module characteristics.

2 MATHEMATIC MODEL OF SOLAR PANEL

One diode model created in MULTISIM with series resistance is shown in Figure 1. The power given by the PV module depends on its electrical characteristics (current and voltage) and its external atmospheric conditions.

PV cell current output:

$$I = N_p \times \left[I_{photo} + I_0 - I_0 \times \exp\left(\frac{V \times N_p + N_s \times I \times R_s}{n \times V_t \times N_p \times N_s} \right) \right] - I_{sh} \qquad (1)$$

3 PARAMETRIC DATA OF 100W SOLAR PANEL

Figure 2 shows a LabVIEW VI model of the solar PV panel. Using LabVIEW simulation tool characteristics of solar panel are studied. Input and output parametric data of solar panel at different time period are shown in Table 1.

4 SIMULATION AND RESULTS

4.1 *Outcomes of solar rradiance change*

Solar panel photocurrent depends on irradiance (G) and surrounding temperature (T) as shown in Equation 1.I_{photo} is directly proportional to G. Solar panel power versus voltage characteristics at different isolation level (1000 W/m^2, 800W/m^2)and (600W/m^{2}and400W/m^2)are delineated in Figure 3, Figure 4 respectively. During this simulation surrounding temperature (T) is kept constant at $25^\circ C$.

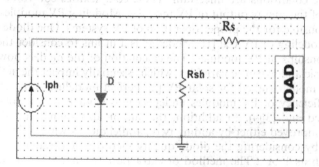

Figure 1.   Single diode identical illustrative for photovoltaic cell.

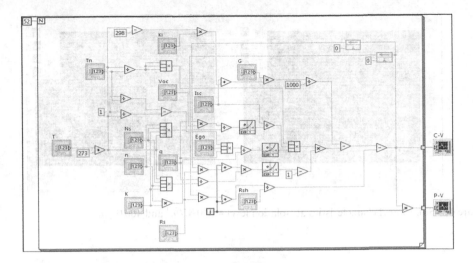

Figure 2.   Developed LabVIEW model of the PV module.

Table 1.   Comparison of input and output parameters of PV module at different time period.

S. No	Parameters	Symbol	New Panel	After 1 Decade	After 2 Decade	After 25 Year
1	Open Circuit Voltage (V)	V_{oc}	45	43	37	32
2	Short circuit current (A)	I_{sc}	2.22	1.82A	1.5	1.1
3	Band Gap energy(eV)	E_{g0}	1.12	1.12	1.12	1.12
4	Boltzmann's constant (J/ K)	k	1.38×10^{-23}	1.38×10^{-23}	1.38×10^{-23}	1.38×10^{-23}
5	Charge of electron (C)	q	1.602×10^{-19}	1.60×10^{-19}	1.60×10^{-19}	1.60×10^{-19}
6	Reference Temperature (K)	T_r	298	298	298	298
7	Ideal Factor	n	1.96	1.82	1.53	1.40
8	No. of cell in series	N_s	54	54	54	54
9	No. of cell in Parallel	N_p	1	1	1	1
10	Temperature Coefficient	K_i	0.0032	0.0032	0.0032	0.0032
11	Series Resistance (Ω)	R_s	0.00272	0.00472	0.0988	0.12
12	Shunt Resistance (Ω)	R_{sh}	3050	2550	1350	1050
13	Rated Power (W)	P_m	100	90.3	55.5	35.2

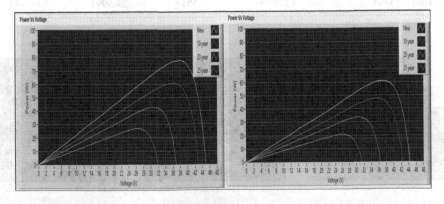

Figure 3.   Solar Panel characteristics at isolation level = $1000 W/m^2$ and $800 W/m^2$.

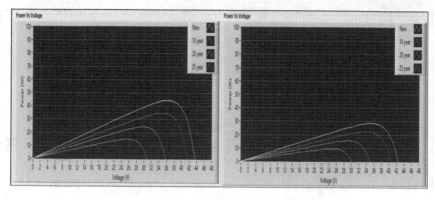

Figure 4. Solar Panel characteristics at isolation level = $600\,W/m^2$ and $400\,W/m^2$.

Table 2 contains comparison of solar panel power at different illumination level at different time span.

4.2 *Effect of varying cell temperature*

In mathematical modeling of solar panel temperature plays an important role. Solar panel current depends on temperature. So, it is important to study solar PV characteristics with varying temperature. Solar PV power and voltage characteristics at different temperature ($5\,°C, 15°C,\ 25°C,\ 35°C\ and\ 45°C$) are shown in Figure 5 and Figure 6 respectively. During this simulation irradiance level is kept constant at $1000w/m^2$.

Table 2. Comparison of maximum Power at different irradiance and time span at constant temperature = $25°C$.

Irradiance (W/m^2)	1000	800	600	400
New Cell	77.4371	60.9169	44.6003	28.709
After 10 year	70.3421	55.2258	40.5162	26.079
After 20 year	57.7969	44.8508	32.9102	21.1686
After 25 year	42.8403	33.5817	24.5991	15.744

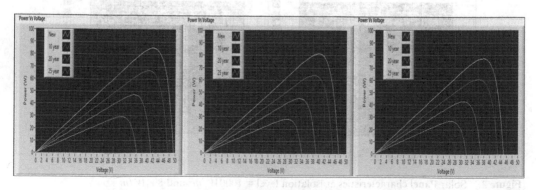

Figure 5. Solar Panel characteristics at temperature level = $5°C$, $15°C$ and $25°C$.

302

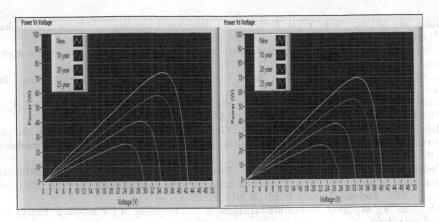

Figure 6. Solar Panel characteristics at temperature level = $35°C$ and $45 °C$.

Table 3. Comparison of Power of PV module at different temperature and different time span.

Temperature ($°C$)	5	15	25	35	45
New Cell	85.2744	81.4533	77.4371	73.3696	69.2291
After 10 year	77.4523	73.9563	70.3421	66.5546	62.6054
After 20 year	63.5639	60.7684	57.7969	54.7877	51.6743
After 25 year	47.2058	45.1023	42.8403	40.4276	37.8743
After 30 year	29.3853	28.0597	26.5891	25.027	23.3594

Under standard test conditions characteristics of solar panel at different temperature level is illustrated in above figures. Around 65 percent of power output of solar panel decreased in 25 years as illustrated in Table 3.

5 CONCLUSION

By using appropriate MPPT controller and proper maintenance life of a solar panel and its output efficacy can be improved. MPPT controller can produce maximum output from the panel at any instant of time. With the population growth, need of electricity rises. To fulfill these needs advance renewable energy sources are required. Solar energy is heart of today energy demand. So solar panel health monitoring is required. This paper presents a comparative simulation study of 100W solar panel. From simulation results it is concluded that after installation of solar panel, there is a 29% decrease in voltage, 50% decrease in current and 65% decrease in output power of solar panel. As year goes there is a 9.16 percent decrease in output of solar panel first decade and 17.83 percent decrease in second decade. By using proper recycling process of solar panel and appropriate MPPT controller problem of clean energy has been solved.

REFERENCES

[1] Chatterjee, A., Keyhani, A., & Kapoor, D., 2011. Identification of photovoltaic source models. IEEE Trans. Energy Convers. 26(3): 883–889.
[2] Yetayew, T. T. & Jyothsna,T. R., 2013. Improved single-diode modeling approach for photovoltaic modules using data sheet. 2013 Annu. IEEE India Conf. INDICON 2013.
[3] Preethi, G., Lavanya, D., Sreesureya, V., & Boopathimanikandan, S., 2019. Real time monitoring and controlling of solar panel using labview. Int. J. Sci. Technol. Res., 8(8): 1747–1752.

[4] Aung, W., M., M., Win, Y., & Zaw, N., W., 2018. Implementation of Solar Photovoltaic Data Monitoring System. Int. J. Sci. Eng. Technol. Res., 7(8): 2278–7798.

[5] Rahman, S., A., Varma, R., K., & Vanderheide, T., 2014. Generalised model of a photovoltaic panel. IET Renew. Power Gener., 8(3): 217–229.

[6] Salim, Ohri, J., & Naveen, 2013. Speed Control of DC Motor using Fuzzy Logic based on LabVIEW. International Journal of Scientific and Research Publications. 3(6):1–5.

[7] Salim & Ohri, J., 2019. Controlling of Solar Powered S.E DC Motor using IMC Controller. WSEAS Transactions on Power Systems. 14: 65–69.

[8] Hejri, M., Mokhtari, H., Azizian, M., R., Ghandhari, M., & Soder, L. 2014. On the parameter extraction of a five-parameter double-diode model of photovoltaic cells and modules. IEEE J. Photovoltaics. 4(3): 915–923.

[9] Yadav, Y., Roshan, R., Umashankar, S., Vijayakumar, D., & Kothari, D., P., 2013. Real time simulation of solar photovoltaic module using labview data acquisition card. 2013 Int. Conf. Energy Effic. Technol. Sustain. ICEETS 2013. 512–523.

[10] Fthenakis, V., M., 2000. End-of-life management and recycling of PV modules. Energy Policy, 28 (14): 1051–1058.

[11] Caracas, J., V., M., Farias, G., D., C., Teixeira, L., F., M., & Ribeiro, L., A., D., S., 2014. Implementation of a high-efficiency, high-lifetime, and low-cost converter for an autonomous photovoltaic water pumping system. IEEE Trans. Ind. Appl. 50(1): 631–641.

[12] Bai, J., Liu, S., Hao, Y., Zhang, Z., Jiang, M., & Zhang, Y., 2014. Development of a new compound method to extract the five parameters of PV modules. Energy Convers. Manag. 79: 294–303.

[13] Salim S., & Ohri, J., 2020. Monitoring and Analysis of Solar PV based on GUI. 2020 First IEEE International Conference on Measurement, Instrumentation, Control and Automation (ICMICA), Kurukshetra, India. 1–4, doi: 10.1109/ICMICA48462.2020.9242687.

[14] Salim & Ohri, J., 2020. Fuzzy based PID controller for speed control of D.C. motor using LabVIEW. WSEAS Transactions on Systems and Control. 10: 154–159.

[15] Kolhe, M., Joshi, J., C., & Kothari, D., P., 2004. Performance analysis of a directly coupled photovoltaic water-pumping system. IEEE Trans. Energy Convers. 19(3): 613–618.

Combination of load balancing and colouration mechanism for inverse of broadcasting in WSN's for getting higher performance

K.M. Veeresh
Computer Science & Engineering, JSSATE, Noida, India

B. Ramesh Naik
Computer Science & Engineering, GITAM Deemed University, Bangalore, India

ABSTRACT: In this paper we present a routing scheme, which provides enhancement to greedy geographical routing known as adaptive load balancing, along with additional colouration feature, RTS and CTS are used. Main goal is to showcase our scheme as an energy-efficient which achieves remarkable performance. Here we present adaptive load balancing scheme, a novel scheme for sensor networks based on geographic routing, channel contention, and load balancing. Relay selection mechanism avoids nodes that are heavily loaded, and is able to successfully handle higher traffic loads than comparable solutions. We also present colouration, an additional feature of load balancing scheme that makes it possible to solve the problem of dead ends, this feature together solves the problem of routing around dead ends. More importantly, unlike other solutions currently available in the literature, load balancing with colouration is very efficient. Broad set of simulations claim that this scheme outperforms similar solutions.

1 INTRODUCTION

Improvements in embedded microprocessors and wireless technology made new production of sensor networks suitable for commercial applications. The very typical sensing parameters of sensor node are temperature, humidity, pressure, and acceleration, light, radioactive. Size and cost constraints on sensor node results in corresponding constraints on resources such as memory computational speed and communication bandwidth. A Wireless Sensor Network (WSN) fit in Wireless Personal Area Network (WPAN) among different wireless networks. The technology which is employed here is low power and low data rate technology. Sensors sense the physical parameter of environment and they generate the signals and for this we have some special devices known as sensor nodes. One important characteristic of WSN is, it is mainly application oriented. Sensor networks are collective systems, so there is a chances of occurring dead end nodes since nodes are collaborative [1]. Paper deals with how to deal with load balancing in WSN and these dead ends. Important requirement of these networks is power must be utilized in most efficient way so maximization of network lifetime is most important objectives of network design.

Load balancing in wireless sensor network is very important criteria where by achieving load balancing energy utilization of network can be minimized. Load balancing in this project is achieved by calculating the (QPI) queue priority index and by calculating GPI (geographic priority index) parameters for relay characterization. The dead ends which are occurring in a network can be efficiently avoided and packets are routed around connectivity hole by colouration feature. Sensing of data and data gathering are

DOI: 10.1201/9781003193838-55

the key ingredients of WSN's. The sensor nodes perform collection of data, and the data is transmitted to a sink via multihop wireless routes [1]. Many schemes, fail to address routing around connectivity holes and efficient route selection. Existing solutions and algorithm minimizes the emerge of dead ends, but could not remove them, like localization error results in packet loss during packet transmission, flooding results in a more power consumption of sensor nodes and Unable to forward the packets or receive. Therefore packets will not reach destination.

In this paper, we propose an approach to the problem of load balancing and routing around connectivity holes that is suitable for any connected topology. The combination of these, results in an integrated solution for convergecasting in WSNs. In this adaptive load balancing mechanism we are going to achieve load balancing among the nodes by choosing the node which is having lower loads on it as relay for packet transmission from sender to sink. Along this mechanism we are also using one feature called colouration, which avoids dead end creation and also makes routing of packets around connectivity holes. This is achieved by giving colours to nodes to differentiate between normal nodes and node which is a connectivity hole.

2 RELATED WORK AND SURVEY ON GEOGRAPHIC ROUTING

The position information of each node is often available at the node in geographic routing algorithms [2]. In geographic forwarding scheme the node which is having packet to transmit to sink will forward a packet to the next node which is in direction of sink and provide best advancement towards the sink With respect to first and simplest formulation, geographic routing algorithm will make use of position information [1]. It will not maintain routing information. So that geographic routing becomes attractive for WSN's. GFG Routing [3] in WSN, reach the destination in every step by choosing neighbor which is nearer to the sink as a next hop. One of the disadvantages of greedy forwarding is that it fails in reaching a node that is closer its neighbors. In many geographic routing, planarization and face routing [2] are used which is not required in proposed algorithms. GeRaF [5] is one of the most popular algorithms for convergecasting in cross layer integration manner [7]. In this nodes will alternatively awake/asleep according to schedule with fixed duty cycle. Integrated routing and IRIS [6] is a non-geographic based on hop count routing. Many of the existing algorithms will fail to address some of the important issues of geographic routing in wireless sensor networks like routing around dead ends, efficient relay selection and resilience to localization error. Our proposed algorithms will give efficient performance in terms of these disadvantages of existing algorithms.

3 EXISTING SYSTEM

With respect to first and simplest formulation, geographic routing algorithm will make use of position information for making packet forwarding decision [1]. GFG Routing [3] in WSN, reach the destination in every step by choosing neighbor which is nearer to the sink as a next hop. In many geographic routing, planarization and face routing [2] are used which is not required in proposed algorithms. GeRaF [5] is one of the most popular algorithms for convergcasting in cross layer integration manner [7]. Integrated routing and Interest Dessimination system (IRIS) [6] is a non-geographic based on hop count routing. Many of the existing algorithms will fail to address important issues with respect to geographic routing in wireless sensor networks like routing around dead ends, efficient relay selection and resilience to localization error. Our proposed algorithms will give efficient performance in terms of these disadvantages of existing algorithms.

4 PROPOSED SYSTEM

A. Adaptive Load Balancing Geographic Routing

The algorithm we proposed is a solution for convergcasting in WSN. Each and every relay selection will be done by considering two important parameters; they are Queue priority index (QPI) and Geographic priority index (GPI). QPI is calculated as follows:

$$\text{Equation1} : QPI = Q + NB/M \tag{1}$$

Where, Q = packets in queue NB = packets requested to be transmitted in burst M = average packets the relay is capable to transmit without an error.

The nodes having less QPI are selected as relay, so which will in turn minimizes the latency by maintaining load between forwarders. Depending up on the location of each node and the geographic location of the sink, each node will calculate the value of the GPI. Figure 1 represents the example of how the values of QPI and GPI are calculated, this is done as follows. The node which is having packet to transmit to the sink is denoted by sender S, the cross marks and small circles (white) will represent asleep and awake neighbor nodes of sender S respectively. Awake nodes are those which are available for S to receive its RTS signal. The light gray color area will represent advancing area for that nodes. In the Figure 1 we have considered NB is 2 packets. In awake nodes, A has no packets in its queue, but with worst advancement (M=1), Therefore its QPI is 2. M=4 for both the nodes C and B, C has Q=2, its QPI is 1, the B with Q=5 have QPI=2. The sender will consider first the node which is having QPI value less. If the QPI values of two nodes are find the same then they will consider GPI value that is nodes with less GPI value to broadcast the RTS signal.

B. Coloring Feature

For more detailed explanation we will consider a Figure 2 in which the node x is a node having packet to transmit and the transference region of x is divided as F and Fc which composed of all the neighbors of x which will offer positive advancement and negative advancement towards the sink. When x has packet to transmit it will choose relay in the region either in F or in Fc depending on its colour Ck, chosen from colours set C0, C1......... Ck coloured nodes where K > 0, can can send packets to nodes with colour Ck or Ck+1. Ck coloured nodes where K >0, search for Ck-1 and Ck. In conclusion the C0 coloured nodes will search for C0 colors only to establish relay. This rainbow colouring is shown in Figure 3. Iinitially all the nodes will be in the C0 color and routing happens according to adaptive load balancing algorithm in the absence of dead ends so no color will change in the nodes. In the case dead end if it's not possible to select relay in Nhsk attempts, they conclude that they may actually be dead end and increase their color to C1. Then C1 node will forward packet through c0 or c1 nodes only throughout process.

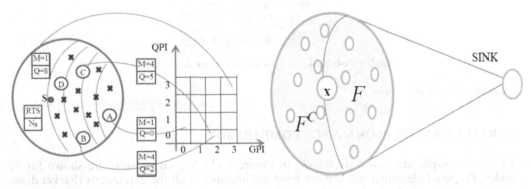

Figure 1.   Calculating QPI and GPI values.   Figure 2.   The F and F^C region.

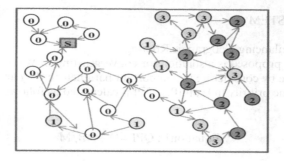

Figure 3.   Rainbow coloring.

5   DESIGN OF SYSTEM

Figure 4 Will represent the overall design of the adaptive load balancing scheme and colouration scheme.

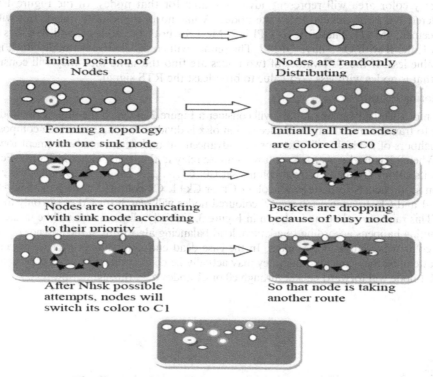

Initial position of
Nodes

Nodes are randomly
Distributing

Forming a topology
with one sink node

Initially all the nodes
are colored as C0

Nodes are communicating
with sink node according
to their priority

Packets are dropping
because of busy node

After Nhsk possible
attempts, nodes will
switch its color to C1

So that node is taking
another route

Finally coloring will happen with whole topology

Figure 4.   Design of a system.

6   RESULT AND PERFORMANCE COMPARISION

The result graphs are shown as follows in Figure 5 (A) & 5 (B), results are shown for 40 nodes. Proposed algorithm will achieve good performance in all the parameters (Packet delivery, energy consumption) compared to existing algorithms.

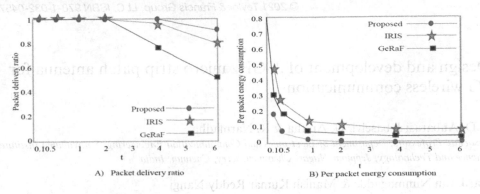

Figure 5 (A) & 5(B).   Result graphs.

7   CONCLUSION

In this paper, we have proposed mechanism; adaptive load balancing based geographic routing in combination with colouration feature, together it will provide a very efficient converg-casting in WSNs in terms of complementing load at the nodes and forwarding a packet around dead end. Colouration feature will make sure that the packet reached the destination. Our results will show that adaptive load balancing geographic routing with colouration feature is an energy efficient and achieves good performance in terms of packet delivery ratio, end-to-end latency, throughput and energy consumption. Performance evaluation of proposed algorithm is done with existing algorithm like GeRaF and IRIS, which concludes that proposed algorithm will outperform existing in terms of packet delivery ratio, energy consumption.

REFERENCES

[1] Akkaya K, and Younis M, "A survey on routing protocols for wireless sensor networks and adhoc networks 3,3 (may 2005), page no. 325–349.
[2] "On the effect of localization error in geographic face routing in sensor network" by K. Seada, A. Helmy and R. Govindam, proc. IEEE/ACM Third Int/symp .Information processing in sensor networks (IPSN 04), pp 71–81, apr 2004.
[3] I Stojmenovic, "Position based routing in adhoc networks ", IEEE comm, magazine, vol. 40, no.7, pp. 128–134 july 2002.
[4] "Survey on efficient routing methods via geographic routing with connectivity holes in WSN's", IJCST, volume 2 Issue 5, sept-oct 2014, page no, 75–79.
[5] "GeRaf: A new contention based MAC protocol for geographic forwarding in adhoc and sensor networks", by Michele Zorzi, August 29, 2003.
[6] Hands on IRIS: Lesson learned from implementing a cross layer protocol stack for WSN's, by Alessandro Camillo and Chiara Petrioli, Globecom 2012- Adhoc and sensor networking symposium, page no. 157–163.
[7] "Efficient beaconless geographic cross layer routing protocol: CoopGeofor optimal relay selection in wireless Adhoc networks "by Srimathy,IJECE, ISSN 0974-2166 volume 5, number 3(2012) pp. 257–266.

Recent Trends in Communication and Electronics – Sharma et al. (Eds)
© 2021 Taylor & Francis Group, LLC, ISBN 978-1-032-04572-6

Design and development of 33GHz micro strip patch antenna for 5G wireless communication

G.D. Anbarasi Jebaselvi, U. Anitha & R. Narmadha
Associate Professors, Department of Electronics and Communication Engineering, Sathyabama Institute of Science and Technology, Jeppiaar Nagar, Chemmencherry, Chennai, India

Harikiran Nimmagadda & Manish Kumar Reddy Nangi
UG Students, Department of Electronics and Communication Engineering, Sathyabama Institute of Science and Technology, Jeppiaar Nagar, Chemmencherry, Chennai, India

ABSTRACT: Since wireless/mobile communication systems have been urbanized rapidly in recent years, re-configurable antennas with scalable quantities are necessary and have to be developed and their wave propagation should be improved. Being a front component of communication systems, an antenna should possess a few major qualities like wide band, good radiation pattern and a quick switchable ability. In addition, well designed antenna relaxes much complexity and improves the performance of the receiver. The specific configuration, its dimension depends on the application and the operating frequency. In this research paper, a micro strip antenna which can be extended for 5G wireless communications with an operating frequency of 33GHz has been fabricated and tested by relevant software. The fundamental parameters like return loss, directivity and Voltage Standing Wave Ratio (VSWR) of the proposed antenna are computed and optimized values are obtained using High Frequency Structural Simulator (HFSS). 3D polar plot gain and 3D polar plot directivity at the operating frequency of 33GHz have been obtained and it is observed from the simulated results that the proposed antenna has a return loss of -16.7dB, gain as 1.54dB and VSWR as 1.3 at 33GHz which has a minimum reflection coefficient.

1 INTRODUCTION

1.1 *Review on reconfigurable antennas*

Re configurable antennas find their suitability and applications in wireless communication and radar systems where object detection, secure communication, multi-frequency communication are involved and the vehicle speed tests. Rectangular or half-circular shaped slots are preferably used as patches in antennas so that it can be reconfigured properly. As the size of the micro strip patch antenna becomes much small, the frequency increases and hence micro strip patch antennas found very much useful in ultra-high frequency signals. These antennas are widely used in mobile communication devices and are becoming more profitable due to their low cost and adaptable designs. In addition, micro strip antennas render many attractive characteristics like most economical, light weight, miniature size, and conformability to fit in planar and non-planar surfaces, stiff and trouble-free installation.

1.2 *Applications and access technologies used*

Slotted micro strip patch antenna is competent enough to sense frequencies lower than microwave frequency. This antenna has a operating frequency of 33GHZ and this is a wide band

DOI: 10.1201/9781003193838-56

range of frequency and this range is suitable for WIMAX, telematic, WLAN and wireless communication etc.,

This proposed antenna with an operating frequency of 33GHz shows a less return loss and very good radiation pattern. The radiation pattern has circularly polarized waves with omni directional radiation in wider azimuth planes. In this the dielectric substrate taken is fr4epoxy with a thickness of 3mm. This antenna is light in weight and low cost and has a low profile.

2   COMPUTATIONS AND DESIGN DETAILS

2.1   *Formula used*

Wireless communication systems have been enforced towards the fifth generation, 5G (Chen & Zhao, 2014) due to the demands of compact, high-speed, and large bandwidth systems. The proposed antenna has the patch of an E-shape. A substrate Fr4epoxy which has a dielectric constant of ε_r=4.4 is used, the operating resonant frequency being (fr =33GHz) and the thickness of the substrate is (h=3mm). The dimension of the substrate is being 21.7mm × 13.2 mm × 3mm and the patch dimensions are 21.7mm × 13.2mm and works at the frequency range between 25-35 GHz, this is the frequency range of wide band, at this band, the future 5G communication falls. The dimensions of the patch are calculated by using the formula written below in equations viz. (1) to (7) which is taken from the literature survey related to rectangular patch antenna (Goyal & Sharma, 2015 and 2016).

$$w = \frac{c}{2f_r}\sqrt{\frac{2}{\varepsilon_r+1}} \tag{1}$$

$$L_{eff} = \frac{c}{2f_r\sqrt{\varepsilon_{eff}}} \tag{2}$$

$$\varepsilon_{eff} = \frac{\varepsilon_r+1}{2} \tag{3}$$

The effective patch length is,

$$\frac{\Delta L}{h} = 0.412\frac{(\varepsilon_{eff}+0.3)}{(\varepsilon_{eff}-0.258)}\frac{\left(\frac{w}{h}+0.264\right)}{\left(\frac{w}{h}+0.8\right)} \tag{4}$$

$$L = L_{eff} - 2\Delta L \tag{5}$$

$$L_g = 6h + L \tag{6}$$

$$W_g = 6h + L \tag{7}$$

W_g - width of the ground plane
ε_{eff} - All the notations used here namely
w- width, L- length, Leff - effective length-effective dielectric constant
L_g - change in effective patch length h - substrate thickness
ΔL- length of the ground plane;c and fr are the velocity of the light and desired frequency respectively.

Three dimensional views of the proposed patch antenna and micro strip feeding antenna in the form of co-axial probe shown in red colour using HFSS is depicted below in Figure 1.

| (a) | (b) |

Figure 1a and 1b. View of proposed patch antenna and feeding microstrip antenna in 3D using HFSS.

2.2 *Simulation*

The dimensions of both patch and the substrate are same and the design of the proposed antenna is made by high frequency structure simulator (HFSS) and simulated. HFSS uses finite element method and solves for parameterization of any device with electromagnetic structures,user friendly and enables the design and optimization of devices to be operated over wide range of frequencies.

There are 6 main steps to create and solve a proper ANSOFT HFSS simulation. They are:

1. Creation of an optimized suitable design/geometry
2. Allocation of flexible boundaries
3. Consign excitation
4. Proper set-up
5. Solution to the problem
6. Post process of results

From this patch, the radiation pattern obtained is in the form of circularly polarized waves and the strength of the radiation can be seen in wider azimuth planes as shown in Figure 2.

It is to use large radiation box to obtain error free results in the first run during simulation. In general, there exist four types of feeding used in antenna namely micro strip line, coaxial probe, aperture coupled and proximity coupled (Hang Wong et al, 2012). From this proposed design, inside the radiation box, the waves are circularly polarized which are in omni direction (360 degrees). In this proposed antenna the patch and the base is made up of copper, air box is in vacuum and the substrate is made up of fr4epoxy which has the properties like high loss, low gain, cheap and easy availability. There is one more material called RT DOURIOD 6002 which has low loss, high gain, high cost and the availability of the material is very difficult and hence not preferred (Haraz, et al 2015). A new technique, tapered H-shaped ground, is proposed to reduce the thickness of a magneto-electric (ME) dipole antenna. By employing this technique, the height of the ME dipole antenna is reduced from 0.25 to 0.11 $\lambda 0$ (where $\lambda 0$ is the wavelength of 5.5 GHz) and this new ructure can be easily realized by multi-layer PCB technology (Hau Wah Lai & Hang Wong (2015).

Figure 2. Circular polarized waves in the radiation box.

3 RESULTS AND DISCUSSIONS

3.1 *Return loss*

The S11 parameter is called as return loss and used to determine the power of the signal returned/reflected by a transmission line or an optical fiber with discontinuity in its signal transmission. This discontinuity is because of the mismatch brought in by either a terminating load or by an active device which is inserted in a line and the value is found to be -18.88dB. This value is obtained when the frequency is of 33GHz and getting another value of -13.678dB at a frequency of 29 GHz(m2, measurement2) which is deliberately shown in Figure 3. The consideration of the frequency is 33 GHz since the return loss is less at -10dB which is depicted as m1 in the graph (Zoom in and can be seen) and is taken as the best base value for the return loss.

3.2 *VSWR*

The voltage standing wave ratio (VSWR) is also known as reflection coefficient, describes the power reflected from the antenna. This ratio should be small so that the antenna can be better matched to the corresponding transmission line and hence more power is delivered to the antenna. The tolerable level of VSWR for the wireless applications must be less than 2.5dB. In this proposed antenna the value is 1.257dB for the operating frequency 33 GHz(m1) and 1.522dB at the frequency 29 GHz(m2). It is safe to consider the smallest value for getting more power, so the desired value of 1.257dB at a frequency of 33 GHz is chosen. Figure 4. shows the plot for VSWR.

3.3 *Gain and directivity*

The directivity of an antenna is defined as the ratio of maximum power density to its average value over a sphere as observed in the far field of an antenna. The gain of the transmitting antenna tells how well the antenna converts the input power into effective radio waves in a specified direction. In the receiving antenna, the gain of the antenna describes how well it

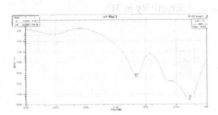

Figure 3. Plot for return loss (s11).

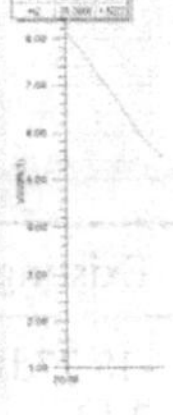

S.No	Types of parameters	Simulated values of the Proposed Antenna	Existing values

Figure 4. Plot for VSWR.

converts radio waves into electric power. When there is no direction specified, the peak value denotes the gain. In this specific antenna, a gain value of 1.5246 dB and the directivity value of 3.7779 dB are obtained. Figure 5(a) and Figure 5(b) show the plot of 3D view for gain and directivity respectively.

3.4 *Radiation pattern*

The major types of radiation patterns available in antennas are: (i) Omni-directional pattern (also called non-directional pattern) which has a doughnut shape (ii) Pencil beam (iii) Fan beam. The beam which has a sharp pencil shaped directional pattern is called pencil beam pattern. The beam which has fan shaped pattern is called fan beam pattern. The beam which is non-uniform and without having a proper pattern is known as doughnut shaped beam pattern. Here, in this proposed antenna the radiation pattern has doughnut shaped beam pattern. The omni directional radiation pattern of the proposed antenna in 2D is shown below in Figure 6.

The gain depends on the type of radiation pattern and if the pattern changes then the gain also changes as both are inter-related with each other (Sunakshi Puri et al 2014). These are very similar to the Omni directional dipole, however, the gain is increased which in turn, reduces the beam width. When installed on a tower, high gain omni antennas typically to provide coverage outward. Table 1 shows the simulated values of the major parameters of the proposed micro-strip antenna.

(a)   **(b)**

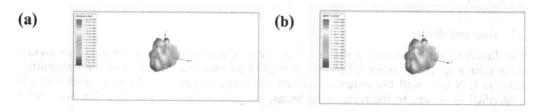

Figure 5a and 5b. Plot for gain and directivity in 3D.

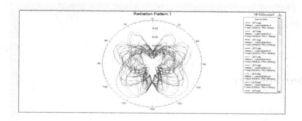

Figure 6. Plot for Radiation Pattern of the proposed antenna in 2D view.

Table 1. Optimized parameter values of proposed antenna.

S.No	Types of parameters	Simulated values of the Proposed Antenna	Existing values
1.	S-11 parameter	-18.88dB	-18.27dB
2.	VSWR	1.25	2.13
3.	Gain	1.5246 dB	4.4887 dB
4.	Directivity	3.7779 dB	7.388 dB

314

4 CONCLUSION

Here by, it is concluded from the simulated results obtained for the design and development of micro strip antenna apt for 5G applications has the resonating frequency of 33GHz. At this frequency this antenna can be used for communication purpose and other Wi-Fi, Wi-max, telematic and WLAN applications. At this frequency it shows the good return loss of value -18.88dB and it also has a good VSWR value 1.25 dBwhich is near to 1 at the same frequency. It has a strong radiation pattern with a good gain and directivity values 1.5246 dB and 3.7779 dB respectively. This antenna has a low profile because of its dimensions 21.7mm x 13.2mm, for this reason it is easily integrated in devices and recommended for 5G applications.

ACKNOWLEDGMENT

The authors are indebted to **Dr.M.Sugadev**, Assistant Professor, Department of ECE, Sathyabama Institute of Science and Technology, helped in getting the results in simulation using HFSS software and **Ms.L.Magthelin Therase**, Assistant Professor, Department of ECE, Sathyabama Institute of Science and Technology, helped in assessing the optimized parameters of the proposed antenna during fabrication.

REFERENCES

Chen.S & Zhao.J. 2014. The requirements, challenges, and technologies for 5G of terrestrial mobile telecommunication, in IEEE Communications Magazine, vol. 52, no.5, pp. 36–43.

Goyal R.K. & Sharma K.K. 2015. Multi-Ring Shaped Microstrip Patch Antenna at 60 GHz for Point-tp-Point Communication, IJSER, Volume 6, Issue 12, ISSN No.-2229-5518.

Goyal R.K. & Sharma K.K. 2016. T-Slotted Microstrip patch antenna for 5G Wi-Fi network. IEEE Explorer, pp 2684–2687,ISBN no.978-1-50902029-4.

Hang Wong et al. 2012. Small antennas in Wireless Communications, Proceedings of the IEEE Journal, vol. 100, no. 7, pp. 2109–2121.

Haraz, O.M et al. 2015. Single-band PIFA MIMO antenna system design for future 5G wireless communication application in Wireless and Mobile Computing, Networking and Communications (WiMob), 2015 IEEE 11th International Conference on, vol., no., pp.608–612.

Hau Wah Lai & Hang Wong. 2015. Substrate Integrated Magneto- Electric Dipole Antenna for 5G Wi-Fi, in Antennas and Propagation, IEEE Transactions on, vol.63, no.2, pp.870–874.

Ka Ming Mak et al. 2014. Circularly Polarized Patch Antenna for Future 5G Mobile Phones, in Access,. IEEE, vol.2, no., pp.1521–1529,2014.

Rappaport, T.S et al. 2013. Millimeter Wave Mobile Communications for 5G Cellular: It Will Work in Access, IEEE, vol.1, no., pp.335–349.

Sunakshi Puri et al. 2014. A Review of Antennas for Wireless Communication Devices. International Journal of Electronics & Electrical Engineering, vol 2, no. 3, pp. 199–201.

Goyal R.K. & Sharma K.K. 2015.Slotted Microstrip Patch Antenna at 60 GHz for point to point Communication. IEEE Explorer, pp: 371–373, ISBN: 978-1-5090-0051-7.

Recent Trends in Communication and Electronics – Sharma et al. (Eds)
© 2021 Taylor & Francis Group, LLC, ISBN 978-1-032-04572-6

Throughput and delay comparison of 4G & 5G mobile communications

Himanshu Sharma, Umang Agarwal, Vikash Singh, Vattsal Singhal & Aditya Pandey
ECE Department, KIET Group of Institutions, Delhi-NCR, Ghaziabad, U.P., India

ABSTRACT: The 5th generation (5G) cellular network has already been deployed in developed countries like China, South Korea & the United States of America (USA) till 2020. But in developing countries like India, it will take another next five years for complete 5G network deployment. The advantages of 5th generation (5G) New Radio cellular system is larger bandwidth, better signal strength, better network connectivity, very high throughput (Mbps) and very low latency (µs) resulting in higher data download speeds. In this paper, a comparative analysis of throughput & latency of 4G & 5G cellular network systems is performed. From the simulation results, it is proved that a 5G networks consisting of 100 cellular nodes has a throughput of 2.43 Mbps per user as compared to 4G networks which has only throughput of 1.62 Mbps per user. Furthermore, the latency is also reduced from 0.35s in 4G to 0.022s in 5G cellular networks.

Keywords: 4G, 5G, wireless mobile communication, delay, throughput

1 INTRODUCTION

The 5th generation cellular system will have hundred times more devices, virtually zero latency, faster response time, very high bandwidth, ubiquitous connectivity and throughput speed up to 10 Gbps (Xiang & Bian 2018). Ideally, the 5G mobile networks will approach the Shannon limit on data communication rates (Bai & Heath 2015). The Figure 1 shows the basic block diagram of 5 G mobile cellular network. It consists of all Internet protocol (IP) networks so that all types of Radio Access Networks (RAN) such as 2G, 3G, 4G and Wi-Fi can be directly connected to the Nano core of the 5G cellular network (Sharma & Kumar 2016).

2 SIMULATION PARAMETERS

Simulation parameters	4G parameters	5G parameters
Channel Noise	AWGN	AWGN
Grid Length	1 km	1 km
Outdoor Scenario	Urban_Macro	Urban_Macro
No. of mobile nodes	20 -140	20 - 140
Transport Protocol	UDP	UDP
Channel bandwidth	20 MHz	100 MHz
Modulation Technique	QAM -64 bit	QAM-256 bit
Simulation Time	1 Hour	1 Hour

DOI: 10.1201/9781003193838-57

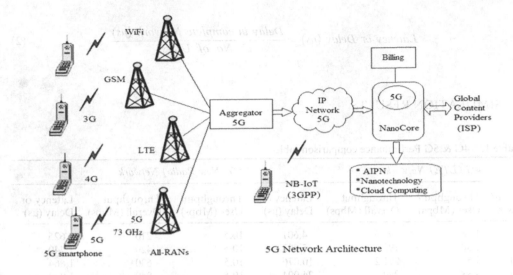

Figure 1. Block diagram of 5 mobile cellular network.

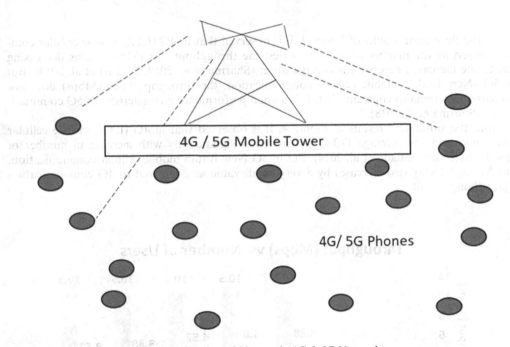

Figure 2. Scenario for video downloading of 20 users in 4G & 5G Networks.

3 PERFORMANCE PARAMETERS

The main performance parameters of mobile cellular communication are its Throughput/user (Mbps) (Lee & Suh 2020) and delay or latency (μs) (Bag et al. 2019).

$$Throughput\ (Mbps)/User = \frac{Complete\ Network\ Throughput\ (Mbps)}{No.\ of\ Users} \qquad (1)$$

317

$$Latency \ or \ Delay \ (\mu s) = \frac{Delay \ in \ complete \ Network \ (\mu s)}{No. \ of \ Users} \quad (2)$$

4 SIMULATION RESULTS

Table 1.   4G & 5G Performance comparison table.

	4G (LTE) Network			5G (New Radio) Network		
No. of Users	Throughput/ User (Mbps)	Throughput Overall (Mbps)	Latency or Delay (μs)	Throughput/ User (Mbps)	Throughput Overall (Mbps)	Latency or Delay (μs)
20	4.88	97.6	4,601	10.5	210	1,625
40	4.80	192	5,893	10.5	420	1,740
60	4.52	271.2	10,076	10.5	630	1,864
80	3.88	230	24,094	10.5	840	1,960
100	3.62	362	35,322	10.5	1,050	2,225

From the simulation results of Figure 3, it is observed that in 4G (LTE) mobile cellular communication as the number of users increases, the throughput/user (Mbps) starts decreasing due to the increased network load on the tower (Sharma et al. 2019, Sharma et al. 2018). But in 5G (New Radio) mobile cellular communication, the throughput/user (Mbps) does not decreases and remains constant. So this is a good performance characteristic of 5G communication (Sharma et al. 2018).

From the simulation results of Figure 4, it is observed that in 4G (LTE) mobile cellular communication the Average Delay (μs) increases significantly with increase in number of mobile users (Buenestado et al. 2014). But in 5G New Radio mobile cellular communication, the Average Delay (μs) increases by a very small value as compared to 4G communication (Ijaz, Zhang, 2016).

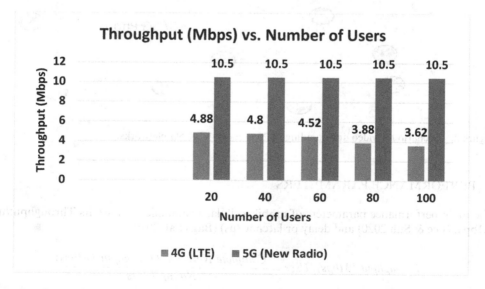

Figure 3.   Comparison of throughput (Mbps) versus Number of users in 4G & 5G.

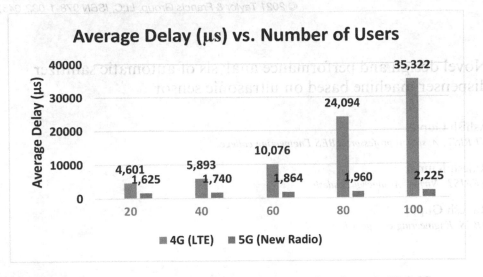

Figure 4.   Comparison of latency or Average delay (µs) versus Number of users in 4G & 5G.

5   CONCLUSION

In this paper, the comparison of 4G & 5G wireless networks performance is performed using NETSIM simulation. Out results show that the 5G (New Radio) mobile cellular communication has more throughput (Mbps) and less delay as compared to 4G (LTE) communication.

REFERENCES

Xiang C., Bian J., 2018. A Survey of 5G Channel Measurements and Models, *IEEE Communications Surveys & Tutorials*, volume 20, Issue 4.

Bai T., Heath R.W., 2015. "Coverage and Rate Analysis for Millimetre-Wave Cellular Networks", *IEEE Transactions on Wireless Communications*, Volume: 14, Issue: 2.

Sharma T., Kumar R., 2016. "Analogous study of 4G and 5G", *3rd International Conference on Computing for Sustainable Global Development* (INDIACom).

Lee W., Suh E., 2020. "Comparative Analysis of 5G Mobile Communication Network Architectures", *Journal of Applied Sciences*, MDPI, .volume 10, no. 7.

Bag, Garg T., Shaik S., 2019, "Multi-Numerology Based Resource Allocation for Reducing Average Scheduling Latencies for 5G NR Wireless Networks", *European Conference on Networks and Communications (EuCNC)*, Valencia, Spain.

Sharma H., Haque A., Jaffery Z. A., 2019. "Maximization of Wireless Sensor Networks Lifetime using Solar Energy Harvesting for Smart Agriculture Monitoring", *Adhoc Networks Journal, Elsevier*, Vol. 94, Netherlands, Europe.

Sharma H., Haque A., Jaffery Z. A., 2018. "Solar energy harvesting wireless sensor network nodes: A survey", Journal of Renewable and Sustainable Energy, *American Institute of Physics (AIP), USA, vol.10, no.2*, pp.1–33.

Sharma H., Haque A., Jaffery Z. A. 2018, "Modelling and Optimization of a Solar Energy Harvesting System for Wireless Sensor Network Nodes", *Journal of Sensor and Actuator Networks, MDPI, USA, vol. 7, no. 3*, pp.1–19.

Buenestado V., Ruiz-Avilés J. M., 2014. "Analysis of Throughput Performance Statistics for Benchmarking LTE Networks", IEEE Communications Letters, Volume: 18, Issue: 9.

Ijaz A., Zhang L., 2016. "Enabling Massive IoT in 5G and Beyond Systems: PHY Radio Frame Design Considerations", IEEE Access, Volume 4.

Recent Trends in Communication and Electronics – Sharma et al. (Eds)
© 2021 Taylor & Francis Group, LLC, ISBN 978-1-032-04572-6

Novel design and performance analysis of automatic sanitizer dispenser machine based on ultrasonic sensor

Ashish Gupta
NERIST, Assistant professor, ABES Engineering college

Rajesh Kumar
NERIST, Nirjuli, Arunachal Pradesh

Rakesh Gupta
ABES, Engineering college, Ghaziabad

ABSTRACT: Viruses such as COVID-19 are transferrable through touch and contact. There are health guidelines to clean or sanitize hands regularly to reduce the risk of infection. Sanitizer dispensing using bottle and its storage would require manual intervention which may cause spread of pathogens. In this paper we propose a novel design of touch-less sanitizer machine to reduce the risk due to contact. we have analyzed its performance by installing in various location. The system can sense the proximity with the help of ultrasonic sensor and sends signal to micro-controller. The controller processes the sensor data & actuates the pump and solenoid valve. The sanitizer liquid dispenses through mist nozzle. Mist nozzle helps to optimize the use of dispensing liquid and save wastage of sanitizer liquid.

Keywords: automatic, ultrasonic proximity sensor, sanitizer machine, pump, mist nozzle

1 INTRODUCTION

Hygiene is an important aspect to remain healthy as cleanliness helps to keep away viruses and bacteria which can cause disease in human body. There are various aspects of hygiene like clean environment, clean body, clean clothes and healthy food. A clean hand is one of them. Hands generally are touched at various surfaces like door handles and can be exposed to direct contamination. Cleaning hands at regular interval is recommended by various health organizations including IMA, government of India and WHO. Due to COVID-19 pandemic spread hand cleaning is nowadays regarded as one of the most important components of hygiene which can help in infection control activities [1–3]. Scientific research has shown the evidence which supports the observation that Hand hygiene when properly implemented, alone can reduce significantly the risk of infection cross-transmission in healthcare amenities (HCFs) [Boyce JM, Pittet et al.].

There are also sufficient researches that suggests hand sanitization lessens the spread of pathogens which are associated with healthcare and related occurrence of HCAI (healthcare associated infections) significantly [6]. Now-a-days, sanitizers based on alcohol are increasingly used instead of soap and water for sanitization of hands as hygiene practice in healthcare industry as it does not require to rinse water and make separate arrangement for washing hands and installing basin. Poor or inadequate hand washing is known to be challenging in hospital settings as there are various points where hand touch becomes necessary, and is

DOI: 10.1201/9781003193838-58

a major cause of infections contracted while patients are admitted to a hospital. It is known by some researches that there is need to place hand washing stations and hand sanitizer dispensers throughout medical facilities including in examination rooms, hallways, lobbies, and even patient rooms. However, those systems are purely mechanical and are incompetent of providing touchless and automated means of establishing accountability of good hygienic practices [7].

The Severe Acute Respiratory Syndrome Coronavirus 2 (SARS-CoV-2) is the virus which was initially reported in Wuhan, China on December 31, 2019, and later was declared as a pandemic by the World Health Organization on March 11, 2020. The need of touch-less automatic dispenser is identified after observing that it is the point of contact for contamination. In this paper we present a novel design of automatic hand sanitizer dispenser. The circuit includes a ultrasonic sensor SC-04. The sensor senses the proximity of hands under the machine. The machine is designed for wall mount at a height of 4ft such that anyone can reach to get sanitizer dispense. The sensor send signal to the microcontroller and the controller takes decision to actuate the pump an valve simultaneously to dispense the liquid sanitizer through a mist nozzle. This dispenser can be wall mounted at entry gates of society, schools, colleges, factories, and offices.

2 PROPOSED SYSTEM

We have designed a sanitizer dispensing machine in a plastic cabinet as shown in Figure 1. Proposed system comprises of proximity sensor based on ultrasonic principle. The sensor used in the system is SR04 to sense the hands are under the machine or not. The cabinet design was originally fabricated for water RO system and has been modified for the purpose of sanitizer dispensing action. The sanitizer storage section is on the front side upper region. Filters have been removed and the water dispensing tap has also been removed. Mist nozzle has been added at the bottom side of the cabinet. The pump is used to suck the sanitizer and pump it with a pressure to the nozzle. The solenoid valve has also been used to control the opening of nozzle and to facilitate to control the dispensing of liquid sanitizer. Pipes and attachments helped to make it easy to fabricate.

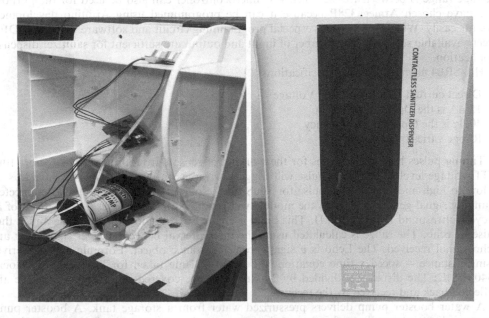

Figure 1. Original image of the proposed design in plastic cabinet a) back side b) front side.

Table 1. Comparison with existing systems.

Sl. no.	System existing/proposed	Features	Remarks
1	Existing system [8]	Sanitizer dispensers with compliance verification	Not for hand sanitization. It is an automated washing system with compliance verification, user carries or wears an RFID tag (or other automatically detectable identification to monitoring whether a consumable item used in the sanitizer dispenser.
2	Existing system [9]	Automated dispenser for disinfectant with proximity sensor	The system was not designed for alcohol based sanitizer dispensing. It is Automated dispenser for disinfectant with proximity sensor. Dispensers for soap, It was reported for soap based dispenser. Electrical control is used for the dispensing mechanism
3	Existing system	IR based sanitizer machine	IR sensor does not properly function in external environment where temperature is normally high. It may give false actuation also.
4	Existing system	Foot based mechanical system	Need to press by foot and exert force, inconvenient for children and not adjustable as height is generally fixed

2.1 Hardware and Software

Atmega-328 microcontroller IC is an AVR family microcontroller with 28 pin DIP. The microcontroller has 6 analog input pins for interfacing with analog signals and 13 digital input/output pins for digital input and output device. The AVR architecture helps to reduce power consumption up to picowatts. The controller is 8- bit RISC(reduced Instruction set computer) microcontroller with 1024 bytes of EEPROM and 2KB SRAM (static Random-access Memory) and 32 general purpose working registers for arithmetic and logic operations. Total inline serial flash memory ISP is 32KB. The watchdog timer with internal oscillator is programmable and helps to avoid processor hang conditions due to overload and also have five selectable power saving modes which can be configured by software. The device operating voltage range is between 1.8-5.5 volts. PIC microcontroller can also be used for the purpose. We have chosen Atmega328P because it can be programmed using Arduino development boards easily. Whereas PIC needs especial programming circuit and software. Arduino IDE is freely available software. The number of input and output are sufficient for sanitizer dispenser application.

HC-SR04 ultrasonic sensor specifications

- Direct current 5V is Operating Voltage
- 15mA is the Operating Current
- 40Hz is the Operating Frequency
- Range varies from 2cm to 4m

Timing pulses have three pulses for the working of sensor. It requires supply a short 10uS TTL voltage level pulse. These pulse will trigger the input. In programming we have to send a low to high and high to low pulse for 10uS on the pin at which the trig pin is connected. Ranging signal gets started.Then the module will generate an output pulse in a pattern of an 8-cycle ultrasound burst at 40 kH. This burst is created by the ultrasonic module and it then raise its echo. The range is calculated using the time interval between trigger signal sent and echo signal received. The Echo is a signal from a distance object. Formula can be derived using distance = speed x time equation. the range = Time taken for level High * velocity (340M/S)/2; The distance is divided by 2 as the distance covered is transmit to receive and reflected back and received by the receiver.

A water booster pump delivers pressurized water from a storage tank. A booster pump increases low water pressure for the nozzle spray and also increases the flow. It provides the extra enhancement needed to fetch your water pressure to the desired level. Most water booster pump comprise the similar core components in most of the manufactured units, as follows:

- Motor
- Impellers
- Inlet and outlet
- Pressure or flow sensing device

Solenoid Valve only operates when it is supplied with DC Voltage. This is a 2 way Solenoid Valve. So, it has 2 connections, one of them is liquid inlet and other is outlet. It accepts DC voltage of 24V. It can also works on 12V DC. When there is input voltage on the Solenoid terminals, it operates, and opens up, allowing the liquid to flow from inlet to outlet. The Solenoid has push fit connectors on each side. So, this connector will allow to directly push the pipe or nozzle into the solenoid without worrying of leaking. This also helps to make system modular.

2.2 Specifications

Operates on voltage ranging from 12V to 24V
    Current Requirement: 400 mA
    Connections type: Push Fit
    Nozzle Diameter: 6.3 mm
    It is a Normally Closed type of valve
Solenoid valve controls the flow of liquid from inlet to the outlet. It has electromechanical components to achieve the desired function of opening or closing an orifice in a valve body. This may which either allows or prevents flow through the valve. The solenoid coil when energized in a normally open valve position, the plunger covers off the orifice, which in turn thwarts flow. 12 volt solenoid valve is having 15 watt coil draws a current of 1.25 amps. When it is connected to a battery, it will have a substantial power trough and will need icing up accordingly. Amps = watts/12 volts.
    Amps: current consumption in the coil
    Watts: power consumption of coil
All mist type of systems are comprised of a nozzles placed in a line like a series. When they are attached to high-pressure pumps, water is pushed through nozzles. The high pressure supports the formation of tiny dorplets which evaporate into mist when they come in contact of the outdoor air. Mist nozzle material is generally made up of Brass, Stainless Steel, Aluminium & much more. It can withstand the pressure up to 50 Bar with a spray angle of 45-120 degree. The connection Type: 1/4 BSPP and the size is 0.3 inch.

2.3 Circuit description

The PCB is designed in proteus and the circuit is simulated in the same software. The PCB is printed using CNC machine. It is single sided PCB with thorugh hole type components and soldered on other side. The schematic is shown in Figure 2.

3 WORKING OF PROPOSED SYSTEM

System block diagram is shown in Figure 3. In the figure we can see placement of sensor to detect the proximity of hands when placed under the machine. It works on ultrasonic waves reflection principle. Module HC - SR04 Ultrasonic based sensor provides 2cm - 400cm range of non-contact measurement function. It generally have an accuracy of 3mm. The modules includes ultrasonic emitter, detector and control circuit.
    Test distance = (high level time×velocity of sound (340M/S)/2.

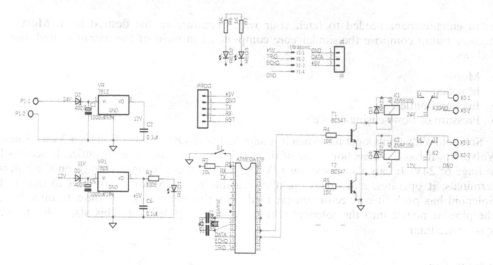

Figure 2. Schematic of the sanitizer dispensing machine circuit.

Figure 3. Schematic of the sanitizer dispensing machine circuit.

As the controller receive High signal from the sensor module it triggers the pump to pull water from storage area and send to the nozzle in mist form. The program runs the pump for 3 seconds. It has been seen during testing 3 seconds are sufficient to sanitize the hands with mist spray. Even we can change the time as per user need through program.

4 RESULT AND CONCLUSION

A automatic sanitizer dispensing machine designed and developed. The machine is wall mount at entrance gates of society, schools, colleges or any commercial building. It can spray 40 times with 100 ml liquid and is effective in optimize use of liquid sanitizer. The machine is tested for 24hour operation for more than a week and is working fine. It helped to reduce the contact for getting sanitizer and also reduce man power employed to spray sanitizer with a spray bottle.

The power consumption is very low. For each spray the maximum current consumption is 2 ampere at 24 V. It consumes 48W if run continuously for 1 hour. The estimated cost of the system is about 3000/INR with a 10 litre capacity storage cabinet. Major cost is incurred for plastic cabinet, pump and power adaptor. The systems are installed at various location such as gate of society and offices on wall at a height of 4 ft. The system worked fine. At the time of rain due to humidity the circuit trigger by itself so it was programmed again with slight calibration of sensor. The device is used by more than 100 people daily and successfully worked at the location. It was advised that the system should not be placed under direct sunlight and also do not expose to rain directly.

ACKNOWLEDGMENT

The Author would like to thank Mr. Anoop pandey for helping and providing moral support. He have extended support while installing pump and making mechanical attachments. Testing was done and the sanitizer dispenser machine worked successfully.

REFERENCES

[1] Guide to implementation of the WHO multimodal hand hygiene improvement strategy. [accessed on August 24, 2010]. Available from: http://www.who.int/patientsafety/en/

[2] WHO Guidelines on Hand Hygiene in Health Care. First Global Patient Safety Challenge. Clean Care is Safer Care. [accessed on August 24, 2010]. Available from: http://www.who.int/patientsafety/en/

[3] Boyce JM, Pittet D. Guideline for Hand Hygiene in Health-Care Settings. Recommendations of the Healthcare Infection Control Practices Advisory Committee and the HICPAC/SHEA/APIC/IDSA Hand Hygiene Task Force. Morb Mortal Wkly Rep. 2002; 51:1–44. [PubMed] [Google Scholar].

[4] Daniels IR, Rees BI. Handwashing: simple, but effective. Ann R Coll Surg Engl. 1999;81:117–8 [PMC free article] [PubMed] [Google Scholar].

[5] Sickbert-Bennett EE, DiBiase LM, Willis TM, Wolak ES, Weber DJ, Rutala WA. Reduction of Healthcare-Associated Infections by Exceeding High Compliance with Hand Hygiene Practices. Emerging Infect. Dis. 2016 Sep; 22(9):1628–30. [PMC free article] [PubMed].

[6] T Prodanovich, SJ Heim Sanitizer dispensers with compliance verification, US Patent 8,085,155, 2011.

[7] TR Winings, R Samson, Automated dispenser for disinfectant with proximity sensor, US Patent 5,695,091, 1997.

[8] AJ Johnson, TA Robertson, TA Casper, Automated wetted or dry sheet product dispensers, US Patent App. 16, 2020 - Google Patents.

Recent Trends in Communication and Electronics – Sharma et al. (Eds)
© 2021 Taylor & Francis Group, LLC, ISBN 978-1-032-04572-6

Automated vehicle detection and classification methods

Shreya Kaul, Gaurav Joshi & Akansha Singh
Department of CSE, ASET, Amity University, Noida, Uttar Pradesh, India

ABSTRACT: Recognition, Localization and Classification of objects, especially vehicles, have been an area of immense research in Computer Vision (CV). Automated Vehicle Detection and Classification on real world traffic and over various terrains has found its use in autonomous cars, intelligent parking systems, automated petrol systems, intelligent vehicle tracking systems, security systems and many more CV based applications. There has been a considerable evolution in the methods adopted before to detect and classify vehicles in real world scenario to those being employed today. Vehicle Detection and Classification (VDC), today, involves prediction of the coordinate-based location of particular category of vehicles in a given input image by means of bounding boxes using Deep Learning and Neural Networks (NNs). Feature calculation carried out by Deep learning-based vehicle detectors are more in use due to its better performance, ease and higher accuracy compared to the earlier employed machine learning techniques involving SVMs and Regression to identify patterns in vehicle frames. Background Detection methods and use of Gaussian Mixture Models for detection and classification are among the oldest attempts made in this problem. This paper reviews some of the crucial methods, by highlighting the different works done so far by researchers across the globe, to process real-time traffic images and videos to eliminate the need for manual surveillance and thus gradually introducing the idea of smart traffic management tools. This paper further points to the history of CV in VDC problem along with the future possibilities.

1 INTRODUCTION

Computer Vision is the sphere of Artificial Intelligence that works to incorporate machines with the sensitivity and intricacies of human vision. To process visual data like images, videos etc and distinguish objects in visual data similar to humans. It equals the study and analysis of visual data attempting to recreate human vision in machines (Shanahan & Dai, 2020). Majority of bits flying around the World Wide Web is visual data. The availability of visual information has exploded to a huge degree in the last couple of years. This is largely due to the development and deployment of sensors all across the globe. Today, from our laptops, cameras to mobile phones, all have more than one visual data sensors. Experiments conducted by M. Cho (Cho et al, 2015) used binary classification method to detect and classify vehicles mounting Scala sensor. The sensor, placed at the front of the vehicle, was capable to scan the data twice-once in upper and then in the lower direction-all in one single period (Cho et al, 2015).

A 2011 study conducted by CISCO estimated that by 2015 roughly 674 days of video data will flow in every second over the internet (https://tinyurl.com/y4ap3pv6, 2011). Approximately, 80% of all traffic on the internet would be video. Another statistic examined that every second of clock time, five hours of visual data is uploaded. It is difficult to analyze

DOI: 10.1201/9781003193838-59

visual data using algorithms and thus leading to it being called the dark matter of internet. Developing algorithms to catalogue, analyze, annotate, manage and utilize visual data is crucial. Algorithms restructure the complexity of human vision to dive in and automatically understand the content of visual data, eliminating the manual study of enormous degree of visual data.

Major problems in CV include image classification, object detection, image segmentation, image captioning. Due to the rapid advancement in hardware and software, automated surveillance systems are being deployed to detect vehicles, track their movements and even classify these (Tian et al, 2017). These require high end video cameras and techniques to monitor and manage traffic movement, replacing the laborious manual traffic monitoring using CCTV. The VDC problem solved under the field of CV along with Deep learning forms a critical part of the intelligent traffic surveillance systems (Tian et al, 2017). It also finds its application in intelligent parking space detection, autonomous vehicles, vehicle counting systems, etc. In this study, we have presented a review of the existing works in VDC problem and discussed the evolution of methods developed for vehicle detection and classification consuming various computer vision techniques.

2 VEHICLE DETECTION AND CLASSIFICATION METHODS

Automatic vehicle detection and classification software comes under the Computer Vision problem of Object Detection and Classification. Computer vision decodes features and patterns present in the visual data. Objects or vehicles in our case differ in their physical attributes like body type, axle number, colour, number of wheels and category of vehicle and so on. There are numerous challenges such as variation in pose, difference in resolution, occlusion, etc., to 24/7 video monitoring of real-time traffic as discussed by B. Tian in their study (Tian et al, 2017). In object detection, the various smaller sub-images within given image are focused. Sub-images comprise the required object (vehicle) enclosed within bounding boxes. This determines the coordinates i.e., location of the object within the image. Object detection consists of object localization whereas object classification determines and places these localized objects into categories or classes. In short, we identify real world entities like vehicles, trees, person etc, within real-time data stream input to the machine. Figure 1 illustrates a general framework to solve VDC problem using existing CV techniques. Earlier ways of object detection in CV consisted of coding each line by the developers and involved tons of manual effort with just restricted output performance level. Automation was very less with a wide range of errors.

Background detection or foreground detection is the Image Processing (IP) technique used with the blob detector to identify vehicle (object) and eventually classify the detected object results from background subtraction methods using a classification algorithm (Benezeth et al, 2010). Background subtraction has been used with MOG (Mixture of Gaussians) models to locate pixels of not only the foreground objects but also their shadows (Zivkovic, 2004). This model has been further employed in videos to detect moving people (Arinaldi & Fanany, 2017). Yet another study uses the concept of capturing spatio temporal frames combined with object oriented fuzzy concept based classification. This study not only helped in dealing with one of the most common problems encountered in several vehicle classification and detection systems, that is, occlusion by using spatial time frames. Also, using object oriented classification techniques the complexity was greatly reduced as compared to the prior works (Sharma et al, 2019).

Simpler way to recognize and classify vehicles using machine learning techniques, which generated more accurate results and better performance, was introduced. The statistical tools like logistic regression, Support Vector Machines (SVM), linear regression, decision trees were used to recognize patterns without manually coding each rule into the software application. Machine learning is also combined with low level features such as SIFT or Scale-Invariant Feature Transform (Lowe, 2004), SURF or Speeded Up Robust Feature, etc.

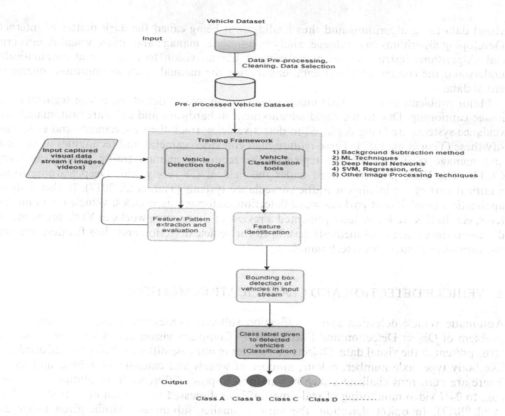

Figure 1. General flow of solving VDC problem.

In comparison to machine learning, training computer to analyze visual data using deep learning techniques to perform vehicle detection and classification is easier, effective and faster (Soin & Chahande, 2018). Deep neural networks have general purpose function which can be used to solve problems that are represented by examples. If we feed the deep neural network with labeled example images of certain types of objects, it searches for common patterns and features in the example images corresponding to those given labels to formulate a mathematical equation or algorithm that indentifies and classifies different objects in the future test images. Deep Reinforcement Learning gives a whole new angle to this problem.

3 COMPARATIVE STUDY

"Chen et al (2017) aimed at using CNN for developing a framework robust enough for vehicle type classification solely based on captured rear-side images. This paper solely focuses on the rear–view images. The three basic structures of the CNN are convolution layer, pooling layer which specifically made use of the adopted maximum pooling and the fully connected layer. Softmax classifier was used to generate the output using probabilistic models. ReLU was used as the activation function for enhancing the expressive ability of the network."

"Filho et al (2009) proposed a less expensive infrared-based system for classification in urban setting. A template dataset was also created for matching with the real time captured images. Dynamic Time Warping or DTW method is utilized for making the comparisons. A down looking pulsed infrared beam through LED's was radiated which bounces back from the vehicle and is sensed by a photodiode which produces an electronic signal. The system developed here resonated with an emitter and a receiver mechanism. The electronic signal is then passed to a band pass filter where an amplitude is generated which is used by the envelop detector to detect the vehicle profile, dynamic time warping (DTW) is put to use for classifying the detected vehicles which matches with the existing template database."

"Peng et al (2014) presented a model for classification of vehicle frames, constructed using binary coding. This paper focuses on analysing the object in motion from the principal (key) frames. The model deploys spatial pyramid matching with binary coding in which there is no analysing of the relative specifications of the vehicle, for instance breadth, edge, stature, length, and so on. Utilizing a hash function dense gradient-based feature was programmed. The classification task was then executed by employing the SVM classifier with spatial pyramid kernel. No pre or post pre-processing was required."

"Jo et al (2018) proposed a classification system using transfer learning and convolutional neural network (CNN) which was trained on a huge dataset. The vehicle was detected from the roadway using Haar-like features secondly; the transfer learning-based vehicle classification was implemented by using GoogLeNet which proved to be beneficial for classifying the vehicle models. The several methods used in this paper are detecting the vehicle using the Haar-like features."

"Pena-Gonzalez & Nuno-Maganda (2014) presented their study which used a CV based framework for recognizing, tracking and categorizing vehicles in motion. A HD-RGB camera comprised the data acquisition system and selected classification and clustering algorithms were used to process information. The study employs an HD-RGB camera, signifying a traditional vision based approach to detect the vehicle, the input frames were then passed for grayscale image conversion, followed by the region of interest selection, temporal difference, though in this paper pixel to pixel difference was not calculated, furthermore, the ROI was divided into n square segments, merging into a temporal object (MO), which forms a part of clustering process, followed by yet another clustering process to form a refined object RO, all overlapping RO were merged together to form the moving objects MO. Go pro hero-3 camera was employed to capture the vehicle images."

"Kaewkamnerd et al (2010) deployed a sensor board having two AMR sensors, having a microprocessor, accountable for the arrival and departure of the vehicles, their identification, speed, distance (length of vehicle) assessment and signal transmission. The extracted features used in the classification are relative length of vehicle, length of signal, number of Hill-pattern peaks and differential energy parameters. These extracted features are then used to build the classification hierarchal tree. They used four vehicle classes (motorcycle, car, van and pickup) and classification was carried out by building up a classification hierarchal tree using low computational extracted feature combination."

A review of the above mentioned works is summarized, along with their respective advantages and challenges, in Table 1 as depicted below.

4 CONCLUSION AND FUTURE SCOPE

The various Computer Vision techniques involving deep learning to solve VDC problem is already being integrated in the form of applications or software modules and even being sold as products to private and government organizations in major developed countries. The solution to VDC problem has a huge prospect in autonomous driving applications facilitating it to make sense of their surrounding traffic. The software collects video input from different angles and distances via high end cameras and detects obstacles and vehicles around it to ensure a safe drive. Figure 2 contains a bar graph that depicts the average number of works in some vehicle detection and classification techniques between the years 2010 to 2020.

Deep Neural Nets, brought into picture in the years 2013 and 2014, generative models were introduced. Moreover, the advent DRL is ensuring machines learn the same way as humans by enabling the agent to take decisions with consideration to its environment. Equipping all these improvements into Deep learning based vehicle detectors and special eye on challenges

Table 1.   Review of existing works.

S.No.	Title	Authors	Approach	Advantages	Challenges
1.	Vehicle Type Classification based on Convolutional Neural Network (Chen et al, 2017)	Yanjun Chen, Wenxing Zhu*, Donghui Yao, Lidong Zhang	Convolutional Neural network applicability was put to test against the various transformations in the image such as rotation, scaling and translation.	The CNN which takes the raw input directly was found to be much more computationally feasible as compared to the various traditional methods employing techniques such as HOG and SIFT. The CNN was found to be robust enough to the various transformations. The proposed method was able to outperform existing traditional pattern recognition methods.	Offers only limited application as it was applicable on only three types of vehicles classes, reason being the unavailability of data sources.
2.	Infrared-based system for vehicle classification (Filho et al, 2009)	Antonio Carlos Buriti da Costa Filho, Joao Pereira de Brito Filho, Renato Evangelista de Araujo, Clayton Augusto Benevides	Exploring the usage of infrared light emitted from LED's and Dynamic Time Warping for efficient comparison of vehicle from existing template database.	Infrared-based system was low cost.	The various problems faced were contributed by factors such as the distance between the vertical equipment and the incoming vehicles, different heights of vehicle parts having different reflectance.
3.	Binary Coding-Based Vehicle Image Classification (Peng et al, 2014)	Peng Yishu, Yan Yunhui, Zhu Wenjie, Zhao Jiuliang	Analysing image near the key frames captured from the video.	Was able to give high efficiency and proved to be low on storage reason being the fact that it used only 32 bits for feature description.	Similar-sized vehicles were sometimes not classified correctly; this was observed in the case of a van and a sedan.
4.	Transfer Learning-based Vehicle Classification (Jo et al, 2018)	So Yeon Jo, Namhyun Ahn, Yunsoo Lee, Suk-Ju Kang	Using GoogleNet as a deep neural network pre-trained on ILSVRC-2012.	Better accuracy was obtained when using transfer learning instead of the conventional methods; it was found to have an	a)Limited scale dataset b) The range of layers was not fixed which affected the performance of the method badly.

(Continued)

Table 1. (*Continued*)

S.No.	Title	Authors	Approach	Advantages	Challenges
				accuracy of 0.983.	
5.	Computer vision based real-time vehicle tracking and classification system (Pena-Gonzalez & Nuno-Maganda, 2014)	Raul Humberto Pena-Gonzalez, Marco Aurelio Nuno-Maganda	The study relies entirely on the vision based system. The data acquisition system consisted of a HD-RGB camera.	Go pro hero-3 camera provides tall HD video streams and sharper images, high accuracy.	Not robust enough in different weather and illumination conditions.
6.	Vehicle Classification Based on Magnetic Sensor Signal (Kaewkamnerd et al, 2010)	Saowaluck Kaewkamnerd, Jatuporn Chinrungrueng, Ronachai Pongthornseri and Songphon Dumnin	AMR magnetic sensors were used to extract features of low complexity.	The overall accuracy was found to be 81.67% which was higher than that obtained in the previous study (77%).	The accuracy of classification faced a major setback whenever the speed was not estimated accurately, which was observed especially in the case of high speed vehicles.

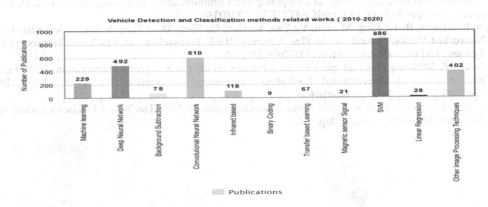

Figure 2. Bar graph on VDC methods.

like occlusion, vehicle overlapping, illumination, weather conditions while devising the solution is a crucial future task.

REFERENCES

Arinaldi, A. & Fanany, M.I. 2017. Cheating video description based on sequences of gestures, in: 5th International Conference on Information and Communication Technology (ICoIC7), pp. 1–6. doi: 10.1109/icoict.2017.8074679. [Google Scholar]

Benezeth, Y.; Jodoin, P. -M.; Emile, B.; Laurent, H. & Rosenberger, C. 2010. Comparative study of background subtraction algorithms. Journal of Electronic Imaging 19(3), 033003 (1 July 2010). https://doi.org/10.1117/1.3456695. [Google Scholar]

Chen, Y.; Zhu, W.; Yao, D. & Zhang, L. 2017. Vehicle type classification based on convolutional neural network. 2017 Chinese Automation Congress (CAC). doi: 10.1109/cac.2017.8243078.

Cho, M.; Choi, B.; An, J. & Kim, E. 2015. Vehicle detection and classification in the Scala sensor by using binary classification. 2015 15th International Conference on Control, Automation and Systems (ICCAS). doi: 10.1109/iccas.2015.7364700.

Filho, A. C. B. da C.; Filho, J. P. de B.; de Araujo, R. E. & Benevides, C. A. 2009. Infrared-based system for vehicle classification. 2009 SBMO/IEEE MTT-S International Microwave and Optoelectronics Conference (IMOC). doi: 10.1109/imoc.2009.5427528.

Jo, S. Y.; Ahn, N.; Lee, Y. & Kang, S.-J. 2018. Transfer Learning-based Vehicle Classification. 2018 International SoC Design Conference (ISOCC). doi:10.1109/isocc.2018.8649802.

Kaewkamnerd, S.; Chinrungrueng, J.; Pongthornseri, R. & Dumnin, S. 2010. Vehicle classification based on magnetic sensor signal. The 2010 IEEE International Conference on Information and Automation. doi:10.1109/icinfa.2010.5512140.

Lowe, D. G. 2004. Distinctive Image Features from Scale-Invariant Keypoints. International Journal of Computer Vision 60: 91–110. https://doi.org/10.1023/B:VISI.0000029664.99615.94. [Google Scholar]

Pena-Gonzalez, R. H. & Nuno-Maganda, M. A. 2014. Computer vision based real-time vehicle tracking and classification system. 2014 IEEE 57th International Midwest Symposium on Circuits and Systems (MWSCAS). doi:10.1109/mwscas.2014.6908506.

Peng, Y.; Yan, Y.; Zhu, W. & Zhao, J. 2014. Binary coding-based vehicle image classification. 2014 12th International Conference on Signal Processing (ICSP). doi:10.1109/icosp.2014.7015138

Shanahan, J. G. & Dai, L. 2020. Introduction to Computer Vision and Real Time Deep Learning-based Object Detection. KDD '20: Proceedings of the 26th ACM SIGKDD International Conference on Knowledge Discovery & Data Mining: 3523–3524. https://doi.org/10.1145/3394486.3406713. [Google Scholar]

Sharma, P.; Singh, A.; Raheja, S. & Singh, K. K. 2019. Automatic vehicle detection using spatial time frame and object based classification. Journal of Intelligent & Fuzzy Systems, 1–11. doi:10.3233/jifs-190593.

Soin, A. & Chahande, M. 2018. Moving vehicle detection using deep neural network. 2017 International Conference on Emerging Trends in Computing and Communication Technologies (ICETCCT), Dehradun, 2017; pp. 1–5. doi:10.1109/ICETCCT.2017.8280336.

Tian, B.; Morris, B. T.; Tang, M.; Liu, Y.; Yao, Y.; Gou, C.; Shen, D. & Tang, S. 2017. Hierarchical and Networked Vehicle Surveillance in ITS: A Survey. IEEE Transactions on Intelligent Transportation Systems, 18(1): 25–48. doi:10.1109/TITS.2016.2552778.

Zivkovic, Z. 2004. Improved adaptive gaussian mixture model for background subtraction, in: Proceedings of the 17th International Conference on Pattern Recognition; pp. 28–31. doi:10.1109/icpr.2004.1333992. [Google Scholar]

2011, June 01. Global Internet Traffic Projected to Quadruple by 2015. The Network: Cisco's Technology News Site. https://tinyurl.com/y4ap3pv6.

Recent Trends in Communication and Electronics – Sharma et al. (Eds)
© 2021 Taylor & Francis Group, LLC, ISBN 978-1-032-04572-6

Precoding techniques to minimize PAPR for OFDM system

Vanita B. Kaba & Rajendra R. Patil
Department of ECE, Godutai Engineering College for Women Kalaburagi, India

ABSTRACT: Orthogonal frequency division multiplexing (OFDM) is a multicarrier modulation scheme which utilizes the spectrum effectively, simple design complexity and less exposure to echoes. As OFDM-system provides these benefits, it is adopted in 4th generation wireless scheme because of these major advantages. The major requirement for OFDM system is to lower the peak to average power ratio (PAPR) to meet the needs of 5G network. Different types of modulation have been recommended for OFDM in order to satisfy the above requirement such as precoding, subband filtering and pulse shaping. In this paper, uniform amplitude with polyphase sequences known as Zadoff Chu (ZC) sequences is used for MIMO-OFDM system which low of out of band radiation. ZC sequences bears admirable correlation features and are used to precode the transmitting data to minimize the peak to average power ratio (PAPR) of MIMO-OFDM system. Simulation represents the ZCT precoding technique reduces the ratio of PAPR significantly.

1 INTRODUCTION

The intense demands and employment of wireless multimedia applications has precede for the progress of diverse multicarrier modulation techniques to fulfill the high bit rate requirement. A 4G system offers a widespread and secured communication for multimedia applications [2]. The 4th generation wireless scheme uses Orthogonal Frequency Division Multiplexing (OFDM) and Multiple Input Multiple Output (MIMO) [3]. OFDM is a multiple carrier modulation scheme which segments the available spectrum into multiple carriers. In OFDM, channels are spaced much closer together to improve the spectral efficiency, and all carriers are orthogonal to each other to avoid intervention between the nearly spaced carriers. These overlapped signals lead to more PAPR, which drives the power amplifier from linear region to nonlinear region [1].

The design of 5G networks requires huge connectivity over high throughput and increased spectral proficiency. The conventional OFDM does not satisfy the need of 5G because of its PAPR, which is high, synchronization between the users is not as typical of 5G networks [2]. To deal with this issue in 5G networks, distinct modulation techniques have been recommended, like as pulse shaping, precoding and subband filtering to lower the out of band exposure of OFDM signals. Filtered OFDM is based on subband filtering that minimizes the coincide between carriers with the well-designed filter, the leakage over the stopband can be significantly suppressed. However, there is a bit error rate (BER) performance degradation in this method. Pulse shaped OFDM have also been designed for 5G networks, which is also observed as a subcarrier based filtering, can successfully lower the PAPR.

Distinct PAPR minimization techniques have been studied with tradeoff between PAPR minimizing efficiency and loss in bit rate, deterioration of BER achievement, increase in signal power and the increase in computational difficulties. The precoding is a simple linear technique that does not require any additional information for implementation. In precoding tech-

DOI: 10.1201/9781003193838-60

niques, the data are coded or transformed before transmission to bring, which saves time and ensure high speed transmission. The precoding and inverse precoding cancel the effect of each other and provides a mechanism to reduce PAPR by reshaping the data. In this work Zadoff chu transformation precoding technique is proposed for out of band radiation reduction in OFDM system and the results are compared with WHT precoded OFDM system [5].

The precoding is done before IFFT on the transmitter side and inverse pre-coding is done before FFT on the receiver side. The precoding and inverse precoding cancel the effect of each other and provides a mechanism to reduce PAPR by reshaping the data. The precoding techniques like Walsh Hadamard Transform briefly discussed in the section 2 of the manuscript. The technique to reduce the PAPR is being discussed in Section 3 Zadoff Chu transform and Section 4 represents simulation results, Section 5 concludes.

2 PROPOSED SCHEME

2.1 *Zadoff Chu sequence*

The Zadoff Chu sequences are complex valued mathematical sequence, which has uniform magnitude and polyphase. Prime length sequences possess a zero autocorrelations it means that the autocorrelation is not equal to zero at zero lag toward the cyclic shifted version [8]. ZC sequence is by

$$Z(I) = \begin{cases} e^{-i\frac{2\pi}{L} * ql\frac{(l+2k)}{2}} \text{ when } L \text{ is even} \\ e^{-i\frac{2\pi}{L} * ql\frac{(l+1+2k)}{2}} \text{ when } L \text{ is odd} \end{cases} \tag{1}$$

Where L =Sequence Length q =Sequence root index which comparably prime with respect to L k =any integer $l = 0, 1 \ldots L - 1$ $i = \sqrt{-1}$ .We can use a modified expression for ZC series which is authentic for odd length sequence and as well as for even length Zadoff Chu sequences.

The auto correlation properties of Zadoff Chu sequence are useful to lower the out of band radiation of the MIMO OFDM system. The ZC sequences allow to generate sequences (both orthogonal and non-orthogonal), which greatly increase the spreading sequences compared to traditional orthogonal code based spreading [7].

2.2 *ZCT precoding based OFDM system*

The proposed block diagram attic representation of the proposed precoded OFDM system shown in Figure 1, the transmitter modulated data streams with QAM modulator followed by precoding matrix.

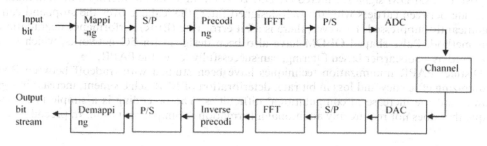

Figure 1.   Block diagram of precoded OFDM.

The QAM mapped input $X = [X_0, X_1, \ldots X_{L-1}]^T$ is given to the precoder before the of IFFT which transforms the input symbol to precoded symbols of length N that can be represented as $Y = PX = [Y_0, Y_1, \ldots Y_{L-1}]^T$ where P is precoding matrix

$$P = \begin{bmatrix} p_{0,0} & p_{0,1} & \cdots & & p_{0,L-1} \\ p_{1,0} & p_{1,1} & \cdots & & p_{1,L-1} \\ p_{N-1,0} & p_{L-1,1} & \cdots & p_{L-1,L-1} \end{bmatrix} \tag{2}$$

The precoding transfers the input symbols into complex space and reduces Peak to average power ratio of transmitting symbol compare to the conventional OFDM system. At the reception precoder matrix is employed for the demodulation.

2.3 Walsh hardmard transform

This transform is a non-sinusoidal and linear transform [5]. WHT does not increase design complication of the system. Compared to the conventional OFDM system, the proposed scheme is capable to reduce the peak power significantly. The auto correlation of the input is lowered using WHT to solve the occurrence of high PAPR and this technique does not require the any supplementary information need to be transmitted to the receiver [6]. The core of Walsh Hardmard transform can be written as $H_1 = [1]$.

$$H_2 = \frac{1}{2}\begin{bmatrix} H_1 & H_1 \\ H_1 & -H_1 \end{bmatrix} = \frac{1}{2}\begin{bmatrix} 1 & 1 \\ 1 & -1 \end{bmatrix} \tag{3}$$

$$H_{2N} = \frac{1}{2N}\begin{bmatrix} H_N & H_N \\ H_N & H_N^{-1} \end{bmatrix} \tag{4}$$

Where H_N^{-1} is the binary complement of H_N. The correlation between the Walsh sequence and its shifted version is maximized when there is zero lag when used as synchronization.. The correlation is zero, if there exists any lag in between these sequences.

3 ZCT PRECODED MIMO OFDM

The primary concern of the multiple user communication system is the multiple reception and multiple transmission with the usage of Mr and Mt antennas respectively. The orthogonal frequency division based MIMO operates with the unique property of transmitting sequences of data blocks in association with multiple antennas which are correlated to corresponding multiple collateral symbols and performs the operation of block modulation by using the operator IFFT to construct respective OFDM block for transmission. The block diagram of proposed

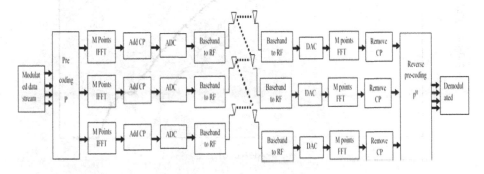

Figure 2.   Block diagram of the ZCT precoded MIMO OFDM transreceiver.

335

OFDM based MIMO transreceiver is as shown in Figure 2, the transmitter modulated data streams with QAM modulator followed by precoding code book which is reshaped to form a precoding matrix. The combination of antenna is done by generating the various code generated from code books by employing non-orthogonal codes.

4 SIMULATION RESULTS

4.1 *ZCT precoded OFDM system*

The PAPR performance of the Zadoff Chu Transform precoded OFDM system, its performance has been compared with the performance of Walsh Hadamard codes. To show the effect of these techniques, an OFDM system has been considered with M-QAM modulation. Figure 3 shows the ratio of peak to average power of conventional OFDM system, Walsh precoded OFDM and ZCT precoded OFDM system with M=16 modulation. Results show that ZCT based PAPR reduction technique perform better as compared to Walsh technique.

4.2 *Zadoff Chu transformation MIMO OFDM precoding*

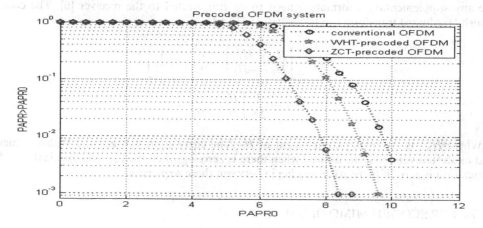

Figure 3. Precoded OFDM System.

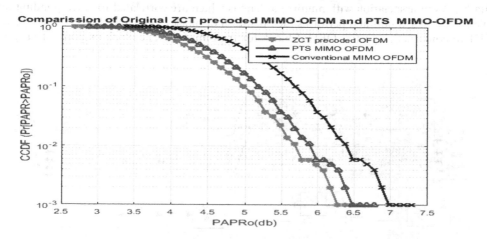

Figure 4. ZCT precoded MIMO OFDM and PTS MIMO OFDM.

Figure 4 shows the ratio of peak to average power of conventional MIMO OFDM system, PTS MIMO OFDM and ZCT precoded MIMO OFDM system. Results show that ZCT based PAPR reduction technique perform better as compared to PTS technique.

5 CONCLUSION

Here we have proposed non-orthogonal spreading sequence known as Zadoff Chu series, which are classes of multiphase sequences. Precoding matrix is form by rearranging the ZC sequence. To lower the out of band radiation Zadoff Chu transformation is performed. Conventional PAPR reduction techniques are less efficient and in most of these techniques a tradeoff between bandwidth and PAPR reduction is required. Coding techniques can be explored for PAPR reduction but these requires computational complexity. So, in this paper, a precoding technique has been used for PAPR reduction of an MIMO-OFDM system. From simulation results, shows that peak to average power ratio of Conventional MIMO-OFDM is more compare to ZCT pre-coded MIMO-OFDM system.

REFERENCES

[1] Lili Wei, Rose Qingyang Hu, Yi Qian, and Geng Wu, 2014. Key Elements to Enable Millimeter Wave Communications for 5G Wireless Systems. *IEEE Wireless Communications*.
[2] Tsung- WeiWu, Char-Dirchung, 2014. Spectrally Precoded DFT-Based OFDM and OFDMA With Oversampling. *IEEE Transactions on Vehicular Technology*, Volume: 63, Issue 6.
[3] Sen Hung Wang,Chih-Peng Li 2015. A Novel Low-Complexity Precoded OFDM System with Reduced PAPR. *IEEE transactions on signal processing*, volume 63.
[4] Dumitrel Loghi; Shaofeng Cai 2019. The Disruptions of 5G on Data-Driven Technologies and Applications. *IEEE Transactions on Knowledge and Data Engineering* Volume: 32, Issue: 6.
[5] LOU Xiaocui 2009. Reducing PAPR with Novel Precoding Method in OFDM System. *IEEE Second International Symposium on Information Science and Engineering, Shanghai*, China.
[6] Shankar T, Nithya Ramgopal 2017. Hadamard based SLM using genetic algorithm fo PAPR reduction in OFDM systems. *Innovations in Power and Advanced Computing Technologies (i PACT)*, Vellore, India.
[7] Miin-Jong Hao and Chiu-Hsiung Lai 2008.Pulse Shaping Based PAPR Reduction for OFDM Signals with Minimum Error Probability. *International Symposium on Intelligent Signal Processing and Communication Systems* Swissôtel Concorde Le, Bangkok, Thailand.
[8] Imran Baig and Varun Jeoti 2008, A New ZCT Precoded OFDM System with Pulse Shaping: PAPR Analysis. *IEEE Asia Pacific Conference on Circuits and Systems*, Kuala Lumpur, Malaysia.
[9] Jian Dang, Minghao Guo, Zaichen Zhang 2017.Imperfect Reconstructed Filter Bank Multiple Access System Using Wide-Banded Subbands. *IEEE 85th Vehicular Technology Conference (VTC Spring)*, Sydney, Australia.
[10] Zhuyan Zhao1, Deshan Miao1 2016, Uplink Contention Based Transmission with Non-Orthogonal Spreading. *IEEE*
[11] Imran Baig and Varun 2010. A New ZCT Precoded OFDM System with Improved PAPR, *Asia Pacific Conference on Circuits and System*.
[12] D. C. Chu 1972. Polyphase codes with good periodic correlation properties. *IEEE Trans. on Inf. Theory*, vol. IT-18, pp. 531–532,1972.
[13] Srdjan Budisin 2010, Decimation Generator of Zadoff Chu Sequences, *Sequences and their applications SETA*.
[14] Imran Baig and Varun, 2010. PAPR Reduction in OFDM Systems: Zadoff Chu Matrix Transform Based Pre/Post-Coding Techniques.
[15] Mohammad M. Mansour 2009, Optimized Architecture for Computing Zadoff Chu Sequences with Application to LTE, Department of Electrical and Computer Engineering. American University of Beirut Beirut, Lebanon.

Recent Trends in Communication and Electronics – Sharma et al. (Eds)
© 2021 Taylor & Francis Group, LLC, ISBN 978-1-032-04572-6

Is India ready for COVID-19: A survey during lockdown period

Arun Kumar Tripathi, Niranjan Darshan & Priyanka Patel
KIET Group of Institutions, Delhi-NCR, Ghaziabad

ABSTRACT: In December 2019, Wuhan City of China starts suffering from an unforeseen virus, later this virus was named as Coronavirus Disease 2019 (COVID-19). It is a highly transmittable and contamination disease, which affects the human respiratory system and causes for human death. To stop man-to-man transmission in the third stage of corona, most of the countries implemented Lockdown. India is the second most populated country in the world and to stop third stage of coronavirus transmission, it has applied lockdown in the early stage. The lockdown breakups the transmission chain of COVID-19 in India. Lockdown suffers from some serious issues such as migration of people from workplace to hometown, decrease in food supply process, social distancing etc. Although government of India arranged food parcels distributions to needy people during the lockdown to prevent migration and announced several relief packages such as "Aatmanirbhar Bharat scheme" to the economy recover from the impact of COVID-19. This article provides an ephemeral overview of the COVID-19 epidemic, symptoms, and routes of transmission of infection. Furthermore, the paper reviews systematic investigation of lockdown and impact of unlock in India during COVID-19 on basis of number of COVID-19 tests, infected persons, number of COVID-19 hospitals, availability of beds in these hospitals, availability of ventilators with respect to patients.

1 INTRODUCTION

Now a days the entire world is suffering from a severe respiratory disease named as COVID-19, subsequently, named Severe Acute Respiratory Syndrome Corona Virus-2 (SARS-CoV-2) [1]. Earlier, in November-2002, first time the SARS virus-1 was found in China. In January-2020, the China officially announced about the COVID-19 and source of spreading as the seafood market of Wuhan city. Initially, it was expected that, the persons who have consumed the infected animals as a food or visited the seafood market are suffering from COVID-19. Furthermore, during the investigation, it is found that it is a transmissible disease and COVID-19 symptoms were also found in the patients, who have never visited the seafood market. The main symptoms [2] of COVID-19 are cough, sneezing and breathing problem etc. As the number of COVID-19 patients were increasing exponentially, the China declared lockdown in the Wuhan city. Later, the corona virus was spread out in Italy, Spain and United State of America in large scale and forced the government for lockdown. At present India is ranked as second most populated country in the world and does not have medical facilities as of China, Italy or America. In India, first COVID-19 patient was reported on 30-January-2020. As the number of patients were increasing daily basis, the government of India has taken the epidemic seriously and announced the complete lockdown-1.0 on 24-March-2020. Subsequent lockdown is announced in month of April and May in India. This is should be an extremely patient factor while designing the epidemic response suited to the country. In gen-

DOI: 10.1201/9781003193838-61

eral, COVID-19 affects the respiratory system and may have one or more symptoms like fever, cough, body ache, headache, shortness of breath etc. To overcome from COVID-19 requires an incubation period six-to-seven days. But an ideal incubation period is referred as fourteen days. Most of the severe case of COVID-19 may causes pneumonia, Severe Acute Respiratory Syndrome (SARS), kidney disorder, or even may cause death.

2 METHOD

The paper analyzes the effect of COVID-19 [3], during the lockdown in the India on day to day basis. In this paper, we have conducted a retrospective observational study on the data from first week of March'2020 to first week of June-2020. For analysis, the data is fetched from the official website of government of India, online newspapers, statista website etc. We analyzed total active cases, confirmed cases, recovered cases deaths etc. from the available data on daily basis. India is one of the most populated country and active cases may depends on population. The population in India is also not distributed normally. Therefore, number of active cases, recovered cases vary from one geographical location to another. Moreover, to know how much India is ready to fight with COVID-19, we have surveyed quantitively for availability of resources in India

3 RESULTS AND DISCUSSION OF COVID-19 IN THE INDIA

Italy, America, and Brazil were most affected counties after china in March-2020. To control the transmission rate of coronavirus in India, government of India imposed lockdown. Along with this, government also suggested guidelines for lockdown to reduce the transmission rate of coronavirus. In this section, we have analyzed the transmission speed of coronavirus during the specified period of lockdown in India. Along with this, various medical facilities available to deal with coronavirus in the states of country.

3.1 *Analysis of total cases in India per day*

During the specified interval, although the third stage of virus transmission i.e. community transmission in the India has not started, still there is exponential growth in the number of COVID-19 patients [3]. As depicted in graphical representation, there is measurable increase in active cases of corona virus as unlock phase started from first week of June'2020. Figure 1 depicts the growth of active cases in India per day basis.

As the COVID-19 patients [3] are increasing daily basis. Figure 2 shows the cumulative growth of coronavirus patient in India. During the period 10-March-2020 to 31-March-2020

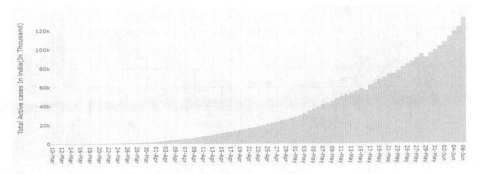

Figure 1.   Day-wise representation of COVID-19 active cases.

339

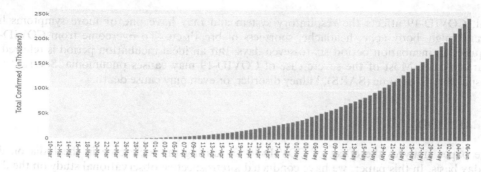

Figure 2.   Cumulative confirmed cases of COVID-19.

only 0.11 people were infected 1 lakh people. While it is increased by 1.54 during the period 01-April-2020 to 22-April-2020.

3.2 *State-wise analysis of COVID-19*

In this section, we have analyzed the state-wise confirmed, active, recovered and death cases [4], [5] in India. The number of COVID-19 patient increases gradually as the number of COVID-19 are increased. In India, Maharashtra, Tamil Nadu, and Delhi were most affected from coronavirus during the specified time slot taken into consideration. The main cause of it was migration from workplace and not follows the social distancing. Figure 3 shows state-wise analysis of COVID-19 confirmed, active, recovered and death cases in India.

3.3 *Reason for spreading of COVID-19 in India*

During the survey, there are four main factors migrants [6], less awareness about COVID-19, tourists and habitants are considered as basic cause for spreading of COVID-19 in India. After the lockdown, a huge migration of workers from one state to another is responsible for spreading of COVID-19. Figure 4 shows that migrant is one of basic cause of spreading of COVID-19 in India.

3.4 *State-wise resource analysis for treatment COVID-19 patients*

In this section we will discuss, the state-wise resource [7]. [8], [9], [10], [11] availability in India. We have considered the physical resources such as number of hospitals Intensive Care Unit (ICU) and availability of ventilators state-wise in India.

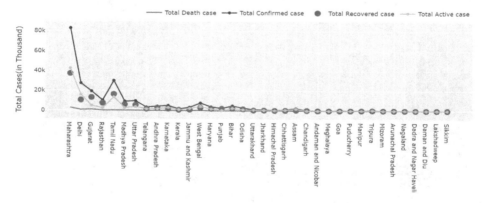

Figure 3.   State-wise analysis of COVID-19.

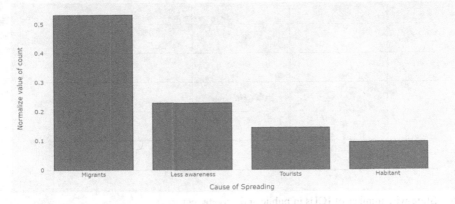

Figure 4.　Reason for spreading of COVID-19.

3.4.1 *State-wise availability of hospitals*
As the number of COVID-19 patients are increasing exponentially. During the lockdown, as per Government order, for treatment of COVID-19 the patient must be admit in the hospital. We have analyzed the number of available hospitals state-wise. As per government record, Uttar Pradesh has largest count in hospitals. Figure 5 shows state-wise number of hospitals in India.

3.4.2 *State-wise availability of ICU in hospitals*
An old age or a critical patient, who is suffering from COVID-19 may need to admit in ICU of hospital. We have graphically represented the availability of ICUs in public and private hospitals in Figure 6. As per analysis, 0.0075 ICU beds are available on as per people in India. In a survey, during the specified period, approximately, 27317 patients i.e. 15.34% need to admit in ICUs with oxygen support.

3.4.3 *State-wise availability of ventilators in hospitals*
If a patient is suffering from severe acute respiratory syndrome or kidney disorder due to COVID-19 may need ventilator support. In a survey, during the specified period, at least 7423 patients i.e. 4.16% required ventilator support across the India for the active COVID-19 patients. Figure 7 shows the state-wise availability of ventilators in India to deal with COVID-19 patients.

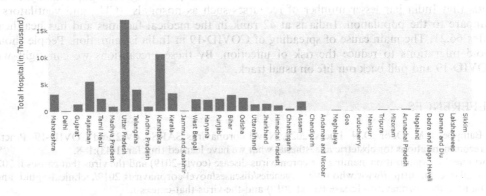

Figure 5.　State-wise availability of hospitals in India.

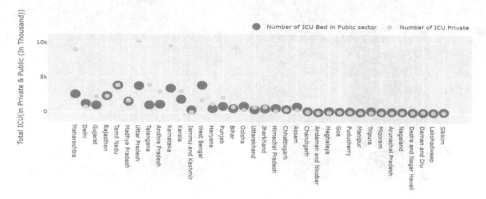

Figure 6.   State-wise number of ICUs in public and private sector.

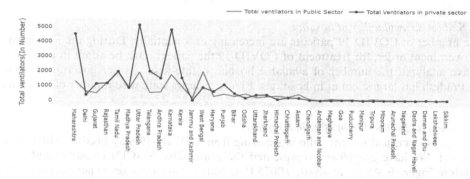

Figure 7.   State-wise availability of ventilators in public and private sector.

4   CONCLUSION

India is one of most populated country the world and huge chance of community transmission in the densely populated regions. Hence, people should follow the guidelines suggested by government of India such as wearing of face masks, frequently sanitizing of hands and surroundings maintaining the social-distancing and avoid social gathering. The paper analyzes and finds that India has lesser number of recourses such as hospitals, ICUs, and ventilators in compare to the population. India is at 42 rank in the medical facilities and has healthcare index 66.21. The main cause of spreading of COVID-19 in India is migration. People should avoid migrations to reduce the risk of infection. By these precautions we can fight with COVID-19 and pull back our life on usual track.

REFERENCES

S. Bampoe, P.M. Odor, D.N. Lucas, 2020, Novel coronavirus SARS-CoV-2 and COVID-19. Practice recommendations for obstetric anaesthesia: what we have learned thus far. 1(43):1–8.
world health organization. naming the coronavirus disease (covid-2019) and the virus that causes it 2020. available at: https://www.who.int/emergencies/diseases/novel-coronavirus-2019/technical- guid ance/ naming-the-coronavirus-disease-(covid-2019)-and-the-virus-that-causes-it.
Ministry of health and family Welfare, Government of India. https://www.mohfw.gov.in/
https://www.statista.com/statistics/1104054/india-coronavirus-covid-19-daily-confirmed-recovered-death-cases/
https://www.statista.com/statistics/1103458/india-novel-coronavirus-covid-19-cases-by-state/

Influx of migrant workers, https://www.hindustantimes.com/india-news/bihar-witnesses-influx-of-migrant-labourers-over-2-000-test-positive-top-10-covid-19-updates-from-bihar/story-9plij4xi2djrvkiytlnuyp.html

Chakraborty PS Shamika Ravi, and Sikim, 2020, COVID-19 | Is India's health infrastructure equipped to handle an epidemic? https://www.brookings.edu/blog/upfront/2020/03/24/is-indias-health-infrastructure-equipped-to-handle-an-epidemic/

COVID-19 in India: potential impact of the lockdown and other longer-term policies. in: center for disease dynamics, economics & policy. https://cddep.org/publications/covid-19-indiapotential-impact-of-the-lockdown-and-other-longer-term-policies/

https://health.economictimes.indiatimes.com/news/industry/out-of-indias-total-covid-19-cases-only-4-16-pc-of-patients-required-ventilator-support-govt-official/76615799

https://www.hindustantimes.com/india-news/only-4-16-of-covid-19-patients-in-india-require-ventilator-support-official/story-isrjjrco3mgdk8ihtd4w3h.html

Recent Trends in Communication and Electronics – Sharma et al. (Eds)
© 2021 Taylor & Francis Group, LLC, ISBN 978-1-032-04572-6

Efficient design of perovskite photovoltaic cell using $CH_3NH_3PbI_3$ as perovskite and PCBM as ETM

P. Ojha
Research Scholar, School of Renewable Energy and Efficiency, National Institute of Technology, Kurukshetra

A. Dahiya
Department of Electrical Engineering, National Institute of Technology, Kurukshetra

ABSTRACT: In recent years because of high percentage of power conversion efficiency, the Perovskite Solar have become more popular. It has also left behind the organic solar cells and dye-sensitized photovoltaic cells and in terms of performance. In terms of structure a perovskite solar cells have mainly five layers consisting of an anode, a coating of transparent conducting oxide, a metal-based cathode layer, an electron transport material layer, an absorber layer, and finally a hole transport material layer. This paper has proposed a structure of perovskite photovoltaic cell with CuI as a HTM, $CH_3NH_3PbI_3$ as an absorber layer and phenyl C61 butyric acid methyl ester as an ETM. For mathematical of the designed model of the cell SCAPS-1D tool is used. The PCE of the optimized structure is found to be above 26% with O.C. voltage of 1.14 V, Current density of 25.72 mA/cm² and the fill Factor of 86.38%.

Keywords: perovskite, SCAPS, fill factor, photovoltaic

1   INTRODUCTION

An article was published in Springer Nature about a perovskite photovoltaic cell having an efficiency of more than 10 percent. From that day the perovskite solar cells have gained attention among the scientists and manufacturers as their performance is high, cost of fabrication is low and they have a great potential for commercialization [1]. In case of perovskite solar cell, there is a sharp rise in the efficiency from 3% - 23% in a very short span of time i.e. within 10 years. Such a fast increment has not been achieved in any other photovoltaic cell technology as on date which is very astonishing [2–4]. This success is achieved because of several semiconducting properties of perovskite materials which are very favorable. These properties include high tolerance to defects, long minority carrier lifetime, tunable bandgap, high absorption coefficient and long diffusion length [5]. Due to the presence of these materials on earth in abundance attracts low-cost manufacturing due to ease in fabrication processes. The general formula of a Perovskite material is **ABX₃** which is similar to calcium titanate (CaTiO3) in terms of structure. Here, **A** is an inorganic or organic monovalent cation e.g. CH_3NH_{3+}, **B** is a divalent small metallic cation, e.g. Pb_{2+}, X is a monovalent halide anion e.g. I, Br, Cl. In a crystalline structure, **A** ion is neighboured by eight 3-D structure of a corner which shares octahedral BX6 units [6]. $CH_3NH_3PbX_3$ is the most popular perovskite absorber material whose optical bandgap ranges from 1.5 to 2.3 eV [7]. For further enhancement in performance of a photovoltaic cell a study is carried out with tin doped methylammonium lead chloride perovskite cell experimentally [8]. The lead based perovskite photovoltaic cells are toxic and a lot of research is going to find its alternatives to widespread the

DOI: 10.1201/9781003193838-62

commercialization of this technology [5]. Tin has got the very good efficiency among all the materials tested in B-site in composition of Perovskite cell [9,10]. As proposed by Shockley-Queisser limit, tin has achieved a bandgap of 1.3 eV [11]. The combination of tin base perovskite solar cell with suitable ETM and HTM provides a gateway to fabricate highly efficient solar cells. In this work $CH_3NH_3PbI_{3-x}Cl_X$ is used as perovskite absorber layer, PCBM as ETL layer and CuI as HTL. The performance of the proposed device is studied using SCAPS simulator.

2 DEVICE SIMULATION AND MATHEMATICAL MODELLING

The simulation of the proposed design is done with the help of SCAPS simulator. This simulation software solves the Poisson's and carrier continuity equation. [12,13]

$$D_n \frac{\partial^2 n(x)}{\partial x^2} + \mu \, E_n(x) \frac{\partial n(x)}{\partial x} + G_n(x) - R_n(x) = 0 \qquad (1)$$

$$D_p \frac{\partial^2 p(x)}{\partial x^2} - \mu \, E_p(x) \frac{\partial p(x)}{\partial x} + G_p(x) - R_p(x) = 0 \qquad (2)$$

Where, n(x) shows the concentration of electrons and p(x) shows the concentration of holes. Here $G_p(x)$ and $G_n(x)$ represents the photo generation rates and $R_p(x)$ and $R_n(x)$ the recombination of holes and electrons respectively. [15] Figure 1 represents the structure of the proposed device.

Table 1 represents the simulation parameters of different layers used in the designed solar cell structure. The proposed cell structure consists of transparent conductive oxide as front

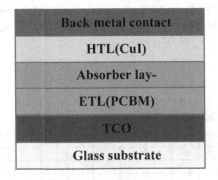

Figure 1. Basic structure of a perovskite photovoltaic cell.

Parameters	TCO	PCBM	Defect layer 1	$CH_3NH_3PbI_3$	Defect layer 2	CuI
Thickness (nm)	500	20-60	10	400	10	400
Band gap (eV)	3.5	2	1.3	1.55	1.55	3.1
Affinity of Electron (eV)	4	3.9	4.2	3.9	3.9	2.1
Relative permittivity	9	4	10	6.5	6.5	6.5
Effective density of conduction band, (cm^{-3})	2.2×10^{18}	1×10^{21}	1×10^{18}	2.2×10^{18}	2.2×10^{18}	2.2×10^{18}
Effective density of valance band, (cm^{-3})	1.8×10^{18}	2×10^{20}	1×10^{19}	1.8×10^{18}	1.8×10^{18}	1.8×10^{18}
Electron/hole mobility, μ_n/μ_p $(cm^2V^{-1}s^{-1})$	20/10	0.01/0.01	2/2	2/2	2/2	44/44
Donor/Acceptor Concentration (cm^{-3})	2×10^{19}	1×10^{21}	1×10^{14}	1×10^{13}	1×10^{13}	3×10^{18}
Defect density, N_t (cm^{-3})	10^{15}	10^{17}	10^{21}	2.5×10^{13}	1×10^{10}	1×10^{15}

contact superseded by PCBM as ETL, interface layer 1, absorber $CH_3NH_3PbI_{3-x}Cl_X$, interface layer 2, CuI as HTL and Ag as a back contact. The parameters of the various layers are represented in Table 1.

3 RESULTS AND DISCUSSION

3.1 *Current – voltage curve*

The thickness of absorber layer is taken as 400 nm with the defect density of 2.5×10^{13}. The doping level of ETM and HTM are taken as 3×10^{18} and 1×10^{21} respectively. The current density-voltage(J-V) curve is represented in Figure 2. The open circuit (O.C) voltage is 1.14, current density(J) is 25.72 mA/cm^2, fill factor is 86.38 % and percentage power conversion efficiency(PCE) is 26.21.

3.2 *Influence of width of the absorber layer on current and voltage*

The thickness of the absorber layer is an important parameter in determining the efficiency of a perovskite solar cell. [18] The absorber layer is set at exquisite thickness for the absorption of maximum number of electrons to generate more electron-hole pairs. The width of the absorber layer is varied from 100 nm - 600 nm. When the thickness of the absorber layer is increased, a sufficient number of electron-hole pairs are produced due to longer illumination wavelength. When the width of the absorber layer is reduced, the fill factor and efficiency are reduced as the distance between the depletion layer and back contact has reduced and due to this the electrons participating in generation process will recombine [19,20]. Therefore, the optimized thickness of the absorber layer is 350nm.The PCE at this thickness is 26.1 %, open circuit voltage is 1.14 V, current density is 25.61 mA/cm2, and FF of 85.1%.

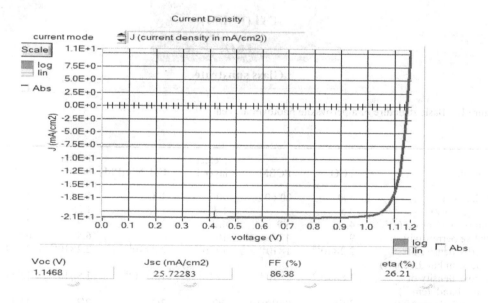

Figure 2.   Result of the simulated model in SCAPS simulator.

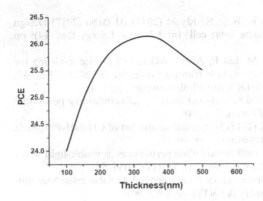

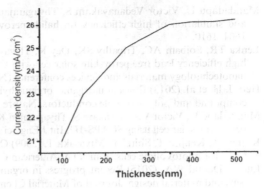

Figure 3a. PCE of the proposed cell versus thickness of layers.

Figure 3b. Current density of the proposed cell versus thickness of layers.

4 CONCLUSION

The performance of the proposed cell having architecture TCO/PCBM/CH3NH3PbI3/CuI is studied using SCAPS simulator. The effect of thickness of absorber layer and the influence of defect densities are also investigated for the performance of the designed cell. The effect of doping concentration of hole transport layer and electron transport layer are also analysed. It was also observed that absorber with low defect densities are with moderate thickness shows better performance. After the analysis of all the parameters and keeping the thickness and defect density of absorber layer 400nm and 2.5×10^{13} cm^{-3} respectively, the PCE of the optimized structure is found to be above 26% with open circuit voltage of 1.14 V, Current density of 25.72 mA/cm^2 and the fill Factor of 86.38%. Further by adding the interface defects suitably, the efficiency of the proposed model can be enhanced.

REFERENCES

Snaith, H. J. Perovskites: (2013) The emergence of a new era for low-cost, high- efficiency solar cells. J. Phys. Chem. Lett. 4, pp 3623–3630.

Miyasaka T, Kojima A, Teshima K, Shirai Y: (2009) Organometal halide perovskite as visible-light sensitizer for photovoltaic cells. J Am Chem Soc 131: pp 6050–6051.

Ganose AM, Savory CN, Scanlon DO (2017) Beyond methylammonium lead iodide: prospects for the emergent field of ns2 containing solar absorbers. Chem Commun 53: pp 20–24.

Saliba M, Matsui T, Seo JY, Domanski K, Correa-Baena JP, Nazeeruddin MK, Zakeeruddin SM, Tress W, Abate A, Hagfeldt A, Gra¨tzel M (2016) Cesium-containing triple cation perovskite solar cells: improved stability, reproducibility and high efficiency. Energy Environ Sci 9: pp 1989–1994.

Shi Z, Jayatissa AH (2018) Perovskites-based solar cells: a review of recent progress, materials and processing methods. Materials 11: pp 729–734.

Wang, Z. L. & Kang, Z. C. (1998). Functional and Smart Materials. New York: Plenum Press.

Jeon NJ, Noh JH, Yang WS, Kim YC, Ryu S, Seo J, Seok SI (2015) Compositional engineering of perovskite materials for high-performance solar cells. Nature. pp 476–480.

Sarkar P, Srivastava A, Tripathy SK, Baishnab KL, Lenka TR, Menon PS, Lin F, Aberle AG (2020) Impact of Sn doping on methylammonium lead chloride perovskite: an experimental study. J Appl Phys 127:125110 pp 1–11.

Konstantakou M, Stergiopoulos T (2017) A critical review on tin halide perovskite solar cells. J Mater Chem A 5: 11518. pp 55–65.

Anwar F, Mahbub R, Satter SS, Ullah SM (2017) Effect of different HTM layers and electrical parameters on ZnO nanorodbased lead-free perovskite solar cell for high-efficiency performance. Int J Photoenergy, Vol. 2017, Article ID 9846310, pp 1–10.

Shookley W, Queisser IIJ (1961) Detailed balance limit of efficiency of p–n junction solar cells. J Appl Phys 32: pp 510–516.

Mandadapu U, Victor Vedanayakam S, Thyagarajan K, Raja Reddy M (2017) BJ Babu (2017) Design and simulation of high efficiency tin halide perovskite solar cell. Int J Renew Energy Res 7(4): pp 1604–1610.

Lenka TR, Soibam AC, Tripathy SK, Dey K, Thway M, Lin F, Aberle AG (2019) Device modeling for high efficiency lead free perovskite solar cell with Cu2O as hole transport material. In: 2019 IEEE 14th nanotechnology materials and devices conference (NMDC), Stockholm, Sweden, pp 1–4.

Heo, J. H. et al. (2013) Efficient inorganic–organic hybrid heterojunction solar cells containing perovskite compound and polymeric hole conductors. Nature Photon. 7, pp 487–492.

Mandadapu U, Victor Vedanayakam S, Thyagarajan K (2017) Numerical simulation of $CH_3NH_3PbI_3$-xClx perovskite solar cell using SCAPS-1D. Int J Eng Sci Invention 2: pp 40–45.

Kojima A, Kenjiro T, Shirai Y, Miyasaka T. (2009) Organometal halide perovskites as visible-light sensitizers for photovoltaic cells. Journal of American Chemical Society.; 131(17): 6050–51.

Luo S, Daoud WA. (2015) Recent progress in organic-inorganic halide perovskite solar cells: Mechanisms and material design. Journal of Material Chemistry A.; 3(17): pp 8992–9010.

Tessema G. (2012) Charge transport across bulk-heterojunction organic thin film. Applied Phy-sics. A, Materials Science and Processing.; 106(1): 53–7.

Wayesh Q, Yesmin AJ, Gloria MD, Tashfiq M, Hossain MI, Islam SN. (2015) Optical analysis in $CH_3NH_3PbI_3$ and $CH_3NH_3PbI_2Cl$ based thin-film perovskite solar cell. American Journal of Energy Research.; 3(2): pp 19–24.

Yin WJ, Shi TT, Yan YT. (2014) Unusual defect physics in $CH_3NH_3PBI_3$ perovskite solar cell absorber. Applied Physics Letters A.; 104 (6): 063903, pp 1–4.

Design of AGC circuits for oscillator using current conveyor based translinear loops

Shahbaz Alam

Department of Electronics and Communication Engineering, ABES Engineering College, Ghaziabad, UP, India

ABSTRACT: With the increasing demand of portable and battery driven devices a low voltage operating device has become necessary and for this current mode techniques are ideally suited. This thesis is focused on current conveyor based translinear (TL) loops in which a Second generation current conveyor (CCII) based on CMOS technology is studied. Using the same a new circuit of an integrator is implemented using log-domain principle. Further, three circuits for automatic gain control are proposed and implemented. All the proposed circuits are verified for the functionality using PSPICE and 0.18μm TSMC CMOS technology parameters.

1 INTRODUCTION

Current mode circuits are gaining acceptance in variety of applications such as data acquisition and processing, testing, filter design and image sensory processing to name a few [8]. This is due to the many advantages offered by current mode circuits such as high speed, large bandwidth and ability to work under lower power supply voltage [6]. Another important consideration is that while voltage mode circuits require the use of complex techniques to lower power dissipation, current mode circuits can do the same in standard VLSI CMOS technology [2]. To attaining maximum dynamic range in the traditional circuit implementation of op-amp, MOS-C, transconductance-C and switched capacitors, there was a restriction on the bases of the supply voltage.

2 OVERVIEW

However, most of the AGC circuits have been developed in voltage-mode configuration [1,5]. In this thesis the work is presented on the current mode configuration to implement the AGC circuits. The proposed AGC circuit which includes an integrator circuit has been conceived from the building blocks of multiplier and rectifier with low-pass filter circuits which had been proposed earlier in the research work available in literature [4]. All circuits can be realized using Current Conveyors. Consequently, the circuit is very suitable to assemble with the various portable equipment like hearing instrument and communication devices. The simulation results are done by PSPICE. In this I present the background of Translinear (TL) circuits, the concept of Translinearity and Translinearity principle is studied. It is further extended to the study of log-domain principle and √-domain principle. The concept of all three generations of current conveyors are also studied namely CCI, CCII and CCIII. Using these CCs the three types of Automatic gain Controllers is proposed and implanted.

DOI: 10.1201/9781003193838-63

3 TRANSLINEARITY

Basically Translinearity could be considered as Transconductance which is linear with respect to current or voltage. So, what can be the devices can have transconductance linear with respect to current or voltage? Think of it as a synthesis procedure, so, if one has to call for a device what should be its properties and characteristics. Its relation between current and voltage could be given as equation 1, 2 & 3:

$$I_0 = f(V_i) \tag{1}$$

$$\text{Transconductance, } g_m = \partial I_o / \partial V_i \tag{2}$$

Where I_0 is output current and V_i is input voltage. If g_m is linear with respect to current, then

$$g_m = \partial I_o / \partial V_i = kI_0 \tag{3}$$

The principle applies where a number of junctions (V_{be}) are connected in loop in forward bias condition. Depending upon the direction of current the transistor loop may in clockwise or in anticlockwise. The transistor could be NPN or PNP, but the number of clockwise junctions should be equal to the number of anticlockwise junctions. To satisfy this condition the numbers of V_{be} are even (equation 4).

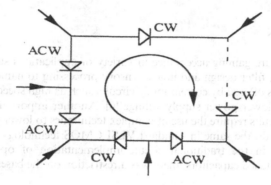

Figure 1.   Loop with j npn Vbe and k pnp Vbe junction in each direction.

Consider a loop in which there are:
J= number of junctions of NPN V_{be} in each direction
K= number of junctions of PNP V_{be} in each direction

$$\Sigma V_{bej} + \Sigma V_{ebk} = \Sigma V_{bej} + \Sigma V_{ebk} \tag{4}$$

Icj and Ick represents the collector current associated with each Vbe junction in the loop. The currents Isn and Isp represents the reverse saturation current in npn and pnp transistors respectively. These currents can be represented in terms of there current densities:

$$(Isn = JsnAjandIsp = JspAp) \tag{5}$$

3.1   Dynamic translinear principle

The basis of the dynamic translinear circuits or log domain circuits, are the capacitor currents as shown in Figure 2 which can be derived by time derivative of equation given below[3],

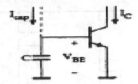

Figure 2.   BJT Translinear circuit.

Where all the symbols have usual meaning.

$$Ic = Ise^{V_{BE}/V_{Th}}$$ (6)

$$I_{Cap} = CV_{th}(\dot{I_c}/I_C)$$ (7)

Where C is the capacitance value and dot shows the time derivative.

The equation represents a non linear relation between I_c and I_{cap}. On multiplying both sides of equation by I_c a direct relation can be achieved. The dynamic translinear principle thus states *"the time derivative of a current* is *equivalent to a product of currents"*. This product can be implemented using translinear principle. As there is a key role or translinear principle or its preferably known as "Dynamic translinear principle" [7].

4   THE BASICS OF CURRENT CONVEYOR

The general equation of a typical current conveyor is given the following matrix equations.

$$\begin{bmatrix} I_Y \\ V_X \\ I_Z \end{bmatrix} = \begin{bmatrix} 0 & A & 0 \\ B & R_X & 0 \\ 0 & C & 0 \end{bmatrix} \begin{bmatrix} V_Y \\ I_X \\ V_Z \end{bmatrix}$$

The block diagram of a typical current conveyor is shown below having two input terminal and one output terminal.

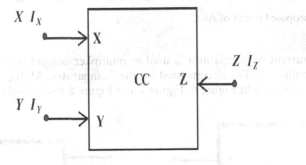

Figure 3.   Typical symbol of CC.

The CMOS implementation of Second generation current conveyor is given below.

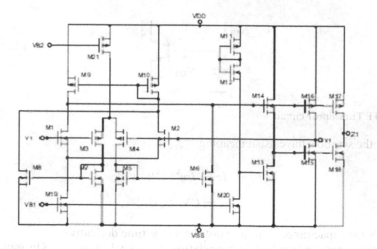

Figure 4.   Translinear second generation current conveyor (CCII).

5   AUTOMATIC GAIN CONTROL (AGC)

The block diagram of a proposed Automatic Gain Control (AGC) is given, in which there are mainly three types of circuits- Multiplier, precision rectifier and Comparator.

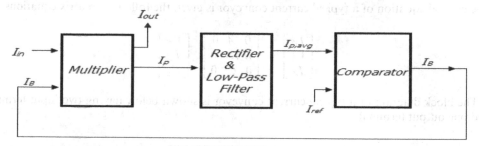

Figure 5.   The proposed circuit of AGC.

The output current of comparator is used as multiplier control current to control the gain of the current multiplier. The circuits used for the Comparator, Multiplier and Half wave precision rectifier are shown in Figure 6, Figure 7 and Figure 8 respectively.

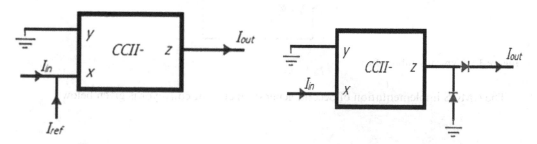

Figure 6.   Comparator.

Figure 7.   The half wave precision rectifier.

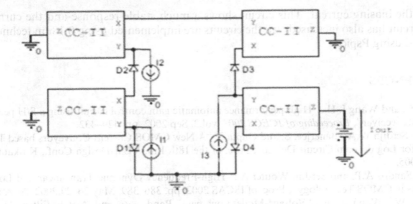

Figure 8.   The multiplier.

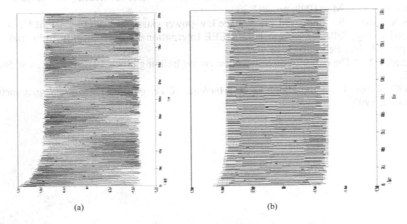

(a)   (b)

Figure 9.   The output current of AGC when input current is (a) increased (b) decreased.

The output response of this automatic gain control circuit is as shown in the Figure 9. The circuits are set as to give an output response fixed at 1uAmp. The multiplier circuit consists of the only forward biased diode; hence the circuits only entertains the positive half of the circuit. To avoid that an additional D.C. biasing is required in the circuit such that on application of input the diodes are all forward biased, which is removed from the circuit after multiplier action is done. But still it can be seen from the wave-form that this circuit is able to modulate only the positive half of the waveform, which turns out to be a drawback for the circuit. This drawback is removed in the next pro-posed circuit.

6   CONCLUSION

In this paper a new AGC is introduced which is implemented using current conveyor as a translinear element. This integrator exploits the log domain principle even working with the MOS transistors in strong inversion region. Three new AGC circuits has been proposed. The first one is a single multiplier circuit with a comparator output as its biasing current. This cir-cuit is able to control the gain of the +ve envelop of the input waveform only. The second AGC is a double multiplier circuit with comparator output as there biasing current. This cir-cuit overcome the drawbacks of the first circuit. The third circuit is an enhanced version of the second with a double multiplier and an integrator after the comparator whose output is used as the biasing current. This circuit shows a much stable response and the current range of the circuit has also increased. All the circuits are implemented using 0.18um technology and simulated using Pspice.

REFERENCES

Chow H.C. and Wang I. H., "High performance automatic gain control circuit using a S/H peak detector for ASK receiver", *Proceeding of ICECS 2002*, vol.2, Sep 2002, pp. 429–432.

Dutta, D, Serdijn W.A., Banerjee, S. and Gupta S. "A New CMOS Current Conveyors based Translinear Loop for Log domain Circuit Design": Proc. of the 18th Intl. VLSI Design Conf., Kolkata. pp. 850–853, 2005.

Haddad, Sandro A.P. and Serdin, Wouter A.: "High-Frequency Dynamic Translinear and Log-Domain Circuits in CMOS Technology", Proc. of ISCAS 2002, pp. 386–389, May 26–29,2002, Arizona.

Horng J. W., Wang Z. R. "Voltage-Mode Low pass, Band pass and Notch Filters Using Three Plus-Type CCIIs", Circuits and Systems, Volume 2, pp. 34–37, January 2011.

Martinez J. S. and Salcedo J. S. "A CMOS automatic gain control for hearing/aid devices", *Proceeding of ISCAS'98*, vol. 1, May 1998, pp. 297–300.

Moustakas K., Siskos S. "Improved low-voltage low-power class AB CMOS Current Conveyors based on Flipped voltage follower", Cape Town, IEEE International Conference on Industrial Technology (ICIT), pp. 961–965, February 2013.

Smith K., Sedra A., "The current-conveyor-a new circuit building block," IEEE Proc., vol. 56, pp. 1368–69, 1968.

Toumazou C., Lidgey F. J., and Haigh D. G. *"Analogue IC design: the current-mode approach"*, Peregrinus, London, 1990.

Recent Trends in Communication and Electronics – Sharma et al. (Eds)
© 2021 Taylor & Francis Group, LLC, ISBN 978-1-032-04572-6

Impact of technology on human behavior during COVID-19

Arun Kumar Tripathi & Shweta Singh
KIET Group of Institutions, Ghaziabad, India

Nupur Pandey
Invertis University, Bareilly, India

ABSTRACT: Coronavirus Disease (COVID-19), a pandemic, has enforced a global emergency worldwide. More than 200 countries are affected by this pandemic. Different strategies (like, quarantine, isolation, work from home approach, positive leadership, etc.) are practiced globally to recover from pandemic. COVID-19 has endangered the economic health in various counties. The crisis in economic health, globally, has initiated a behavioral and attitudinal change in the human being. Due to longer lockdown duration, the negative impact on human behavior is dominating than the positive one and as a result, people feels to be irritated, frustrated and addictions of unnecessary competition, at times. Nowadays, technological development and availability of gadgets have reformed human lives and behavior. The article is based on a survey on a list of technological parameters which is responsible for bringing behavioral changes in any individual during pandemic. The perspectives include social, cultural, environmental factors, etc. A detailed questionnaire was made and distributed to people of different age groups, their responses are then analyzed. A detailed survey will allow researchers and people to better understand both negative and positive effects of different perspectives, and how these factors can initiate a behavioral change and extreme psychological burdens to an individual.

1 INTRODUCTION TO COVID-19

Past year, i.e. December 2019, an unknown disease was recognized in extremely populated Wuhan City, Hubei province of China. It is observed that most of the patients were suffering from severe clinical symptoms like fever, bilateral lung in-filtrates, dyspnea, and dry cough on imaging. During this critical observation of patients, it is found that major cause of it is coronavirus. During the initial investigation of China, it is linked to the seafood wholesale market of Wuhan. This market is famous for trading numbers of different live animal species that includes bats, fish, poultry, snakes, and marmots, mainly. Soon, this pneumonia was stated as Severe Acute Respiratory Syndrome Coronavirus 2 (SARS-CoV-2). Further, it was entitled as COVID-19 (Li-sheng et al. 2020) through World Health Organization (WHO). Initially, most of the infected patients of COVID-19 are found with mild symptoms like fever with sore throat and dry cough.

Furthermore, majority of the infected patients were treated and their infection level was resolved, but few generated fatal complications including severe pneumonia with following symptoms life septic shock, organ failure, Acute Respiratory Distress Syndrome (ARDS), and pulmonary edema. These fatal complications were seen in people who are older in age and required intensive care support and who faced health issues earlier including cerebrovascular, digestive, cardiovascular, endocrine, and respiratory disease.

DOI: 10.1201/9781003193838-64

The COVID-19 has been declared that it transmits from one human to another when comes in contact with any infected human. For its treatment, it is critically recommended to keep in separate incubation sections and to continue their treatment, while staying in quarantine. Diagnosis of COVID-19 can be done by taking respiratory tract specimen for SARS-CoV-2 RNA. For instance, if SARS-CoV-2 tests are reported positive, patient is kept in isolation in anticipation of two consecutive SARS-CoV-2 tests are reported negative.

2 A WIDE CONSEQUENCES OF COVID-19 IN REFERENCE TO LOCKDOWN

On concerning the great threat for countries with susceptible health systems, the WHO affirmed this Chinese outbreak of COVID-19 as pandemic, on January 2020. Including emergency committees of different countries and WHO, it has been stated that further propagation of COVID-19 need to be stopped by following a set of procedure such as early detection, prompt treatment, isolation of infected person with from others. This outbreak has brought an economic breakdown in the whole world. Since, to stop its propagation and recurrence, several isolation and quarantine processes are encouraged to manage the immediate effects of COVID-19.

Different strategies, such as lockdown, being in quarantine, isolation, different mobile applications for medical related help, practicing social and personal distance-related activities, etc. are practiced around the world. Technology (Divya et al. 2018) during lockdown, in some way or the other, encourages both positive and negative impacts on the majority. The most faced challenges in lockdown through technology as aspects of behavioral change can be, such as, Threat perception, Social context, Science communication, Individual and collective interest, Leadership, Stress and copying. The incidences of the perspectives discussed above are directly dependent on an individual's mind-set. An individual's mind-set is made through a sequence of stages from early childhood to the older age.

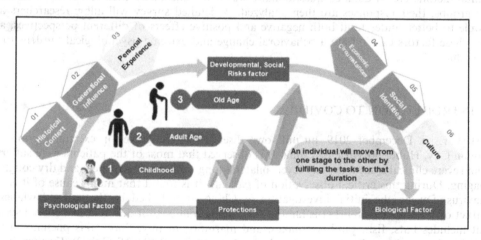

Figure 1. The Human Behavioral Developmental Model.

3 RESEARCH METHODOLOGY

Digital information is given relevance for several aspects, such as for news update, COVID-19 update, medical diagnosis and treatment at small level, etc. These aspects are promoting imprints on human behavior.

The present area of study relies on the following specific objectives:

a. Generating the factors that trigger behavioral change in an individual during pandemic.
b. Furthermore, considering those factors for data analysis and to depict the utmost significant factors out of them, and justify their inter-connectivity build healthy mind-sets.

3.1 Sample size and target

The investigation is performed on suspects in North-East Delhi region. The target sample size at initial was set to 430 out of which approximately 398 respondents validated with appropriate responses.

3.1.1 Pilot testing

Pilot testing (Etchegaray et el. 2011) is utilized to attain the lucidity in questionnaire and the reliability of variables. For this, several pre testing methods need to be performed. These pre-testing surveys are performed so that questions that are hard to understand or confusing can be easily rectified, and questionnaire is to be non-confusing.

3.1.2 Sampling procedures

Research relies on a survey technique to gather information via questionnaires distributed to various suspects in North-East Delhi region. These suspects were mixed who were ranging between 12 years and above of age. An appropriate sampling technique was utilized while investigation.

3.1.3 Response rate

A count of 440 questionnaires was circulated out of which 396 satisfactory responses are received. Though a satisfactory response rate of 90% is attained, rest mentioned issues such as lack of interest, busy schedules, or other personal reasons.

3.1.3.1 INSTRUMENT AND MEASURES

While designing the questionnaire, total count of 20 factor dimensions was taken at initial. The overall process has resulted in mainly 100 items in questionnaire consisting of both generalized and proposed scenario related questions.

3.1.3.2 DATA ANALYSIS

Exploratory Factor Analysis (EFA) (Yong et al. 2013) was performed to cut-off the count of proposed items to some level. It is a two-way approach, mainly.

3.2 Reliability and validity test

This section concentrates more on analyzing the reliability and validity tests on the factors considered while making questionnaire. Afterwards, Cronbach's Alpha is measured for each of them, so that the factors with highest loading factor are discovered.

3.2.1 Reliability

An IBM SPSS Statistics software (version 20.0) (George et al. 2011) is used to perform reliability test on 100 items in questionnaire. Table 1 depicts the alpha coefficients calculated for each factor involved. To improve the scales, those adapted from previous studies, Cronbach's alpha coefficient (Tavakol et al. 2011) and EFA were applied.

As recommended by testing research theory (Dziuban et al. 1974), a cut-off level is to be fixed at 0.7. Thus eliminating the factors, those are unsatisfactory in level of reliability. Optimized item-to-total correlation (Dziuban et al. 1974) is facilitated that helps to choose which item is to be removed.

3.3 Exploratory Factor Analysis (EFA)

EFA is used basically to cut-off the count of items in questionnaire, and also to evaluate validity for the construct. The very famous, Kaiser-Meyer-Olkin (KMO) and Bartlett's Test

357

(Galgalikar 1994) are opted by most researchers to validate the robustness for each factor analysis and sampling adequacy (Nunnally et al. 1994) procedures. Table 2 depicts KMO calculate of sample adequacy as (0.855) which is approximately 1. Furthermore, while applying Bartlett's Test of Sphericity, a considerable value (p=0.000) is received i.e. approximately 0.05 (such that p-value <0.5).

Table 1.   Reliability test for factors in relevance to impact of technology on Human Behavior during COVID-19.

S. No.	Factors Considered	Cronbach' Alpha	Calculated Alpha coefficients for factors namely 2, 4, 10, 11, 12, 13, 14, 15, 16, 18, 19 and 20 has met the considerable range of reliability that lies in between 0.701 and 0.872. Though, factors like 1, 3, 5, 6, 7, 8, 9 and 17 didn't met the minimum reliability with loads as 0.685, 0.690, 0.678, 0.590, 0.673, 0.634, 0.599 and 0.696. Thus, these indicate inadequate level of reliability in relation to role of technology in behavioral change of a human being in a pandemic. To achieve constancy, unsatisfactory factors were removed, and was restricted to further load the factors with minimum of 0.7.
1	Threat	0.685	
2	Emotion and risk perception	0.722	
3	Prejudice and dicrimination	0.690	
4	Disaster and panic	0.737	
5	Trust and compliance	0.678	
6	Identity and leadership	0.590	
7	Ingroup elevation	0.673	
8	Zero-sum thinking	0.634	
9	Moral decision making	0.599	
10	Cooperation	0.814	
11	Conspiracy theories	0.812	
12	Fake news	0.857	
13	Persuasion	0.701	
14	Social norms	0.769	
15	Social inequality	0.820	
16	Culture	0.752	
17	Political polarization	0.696	
18	Social isolation and connection	0.801	
19	Intimate relationships	0.795	
20	Healthy mind-sets	0.872	

Table 2.   KMO and Bartlett's Test.

Test		Adequacy
Kaiser-Meyer- Olkin Measure of Sampling Adequacy		0.855
Bartlett's Test of Sphericity	Chi- Square	10752.627
	Significant Value	0.000

4 SURVEY FINDINGS

It can be concluded from Table 1 & 2 that we are now able to find the optimized count of total factors (12 factors) after applying reliability tests on set of factors. It is found that factors named Healthy Mind-sets, Fake News and Social Inequality has received greatest loading factor.

Figure 2 and 3, concludes the summary report as per the questionnaire distributed to various respondents. It can be stated clearly from the Figure 3 that respondent has validated the proposed scenario.

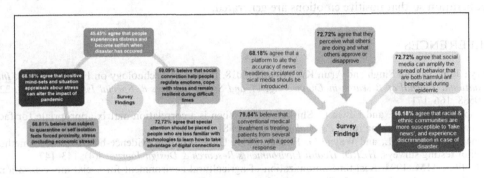

Figure 2.   Survey Summary.

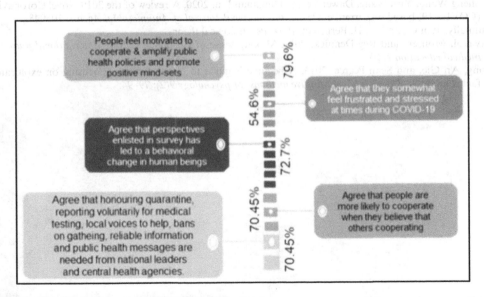

Figure 3.   Survey Summary to validate the conclusion.

5 CONCLUSION AND FUTURE SCOPE

In the paper, a survey is performed on various factors that contribute in finding the impact (both positive and negative aspects) of technology on human beings, i.e. how technology is acting as a mediator for behavioral change in human beings. Furthermore, a primary questionnaire was made, to which Cronbach' Alpha factor is applied to justify utmost significant factors from proposed list of factors set. From the overall study, it can be stated that people, nowadays, relies more find themselves isolated and feels stressed during pandemic, which triggers a behavioral on technological or digital aspects. A strong sense of trust, respect, and shared identity is required so that people are not threatened in pandemic, so that strategies are routed effectively. The purpose behind this survey is to provide basis for governmental and non-governmental authorities that promote a sense of shared identity and motivate people for cooperation so that positive emotions are generated.

REFERENCES

Divya Jain, Shweta Singh and Arun Kr. Tripathi. 2018. Impact of Technology on Human Beings. *International Journal of Research in Engineering*, IT *and Social* Sciences *8, Special Issue.* ISSN No. 2250-0588: 166–171

Dziuban, Charles D., and Edwin C. Shirkey. 1974. When is a correlation matrix appropriate for factor analysis? Some decision rules. *Psychological bulletin* 81(6): 358

Etchegaray, Jason M., and Wayne G. Fischer. 2011. Understanding evidence-based research methods: Pilot testing surveys. *HERD: Health Environments Research & Design Journal* 4(4): 143–147.

Galgalikar MM. 1994. Real-time automization of agricultural environment for social modernization of indian agricultural system. *In Computer and Automation Engineering (ICCAE)*, The 2nd International Conference 1: 286–288.

George, Darren, and Paul Mallery. 2011. IBM SPSS statistics 23 step by step: A simple guide and reference. *Routledge.*

Li-sheng Wang, Yiru Wang, Dawei Ye and Qingquan Liu. 2020. A review of the 2019 Novel Coronavirus (COVID-19) based on current evidence. *International Journal of Antimicrobial Agents*: 105948.

Nunnally, Jum C., and Ira H. Bernstein. 1994. Psychological theory.

Tavakol, Mohsen, and Reg Dennick. 2011. Making sense of Cronbach's alpha. *International journal of medical education* 2: 53.

Yong, An Gie, and Sean Pearce. 2013. A beginner's guide to factor analysis: Focusing on exploratory factor analysis. *Tutorials in quantitative methods for psychology* 9(2): 79–94.

Implementation of energy efficient multi-hop protocol employing wireless sensor networks for precision agriculture

Manish Zadoo
Department of Electronics & Communication Engineering, ABES Engineering College, Ghaziabad, India

Amit Choudhary
Department of Electronics & Communication Engineering, Jamia Millia Islamia, New Delhi, India

Meenakshi Sharma
Department of Electronics & Communication Engineering, IPEC, Ghaziabad, India

ABSTRACT: In Precision agriculture, a Wireless Sensor Network (WSN) is utilized for collection of data related to atmospheric conditions. A wireless sensor network consists of Sensor nodes installed at multiple points in a greenhouse for monitoring soil properties like moisture, pesticide levels, air temperature and humidity level etc. Sensor nodes transmit the measured parametric digital information to a sink node. Sink further transmits the sensed data to a Decision Support System (DSS). On the basis of a crop growth model, the DSS precisely control irrigation, fertilization and air conditioning systems for controlling climate temperature and humidity levels of green house. In this manner, proper inputs are applied to crops leading to better crop health and yield. Reliable transmission data is a crucial design objective for WSN in precision agriculture as the presence of foliage in transmission channel causes significant attenuation, of the radiated waves. Additionally it may cause scattering and diffraction of signals as well. A dynamic data routing protocol selects the optimum data paths for node data transmission in a WSN. This paper presents a novel energy efficient multi-hop data routing protocol for WSN precision agriculture applications. The critical parameters for the selection of intermediate forwarder nodes are residual energy of the node, distance of main node from sink and average distance of the node from other sensor nodes. Forwarder collects data packets of source nodes and retransmits them to sink. This aids in reducing the packet loss rates of distant nodes. The proposed protocol utilizes a cost effective methodology employing a Multi Attribute Decision making algorithm that makes a parametric comparison of network nodes and selects the forwarder node as an optimized node. The suggested protocol renders superior performance in terms of, stability, throughput and network lifetime as compared to the existing protocols.

Keywords: Network lifetime, Precision Agriculture, Wireless Sensor Network, Routing protocol

1 INTRODUCTION

Wireless sensor networks have proven themselves very helpful for precision agriculture applications (Kone et al. 2015, Imam et al. 2015). Multiple sensor nodes, capable of sensing various soil and air characteristics are grounded to farmland or green house. These node sense farmland condition parameters like moisture, pesticide levels, air temperature and

DOI: 10.1201/9781003193838-65

humidity level etc. and send data to base station. Base station computer after necessary processing, upload this data to internet cloud using IOT techniques (Garcia et al. 2009, Beckwith et al. 2004). Farmer can access this data regarding soil and atmospheric conditions on his smart phone using internet connection from anywhere in the world. For this Taylor made android application can be designed (Imam et al. 2017). Based on this real time information availability, farmers become aware regarding soil and atmospheric conditions and can provide required inputs to their crop like proper irrigation, and fertilization to their crops (Figure 1).

Nodes in a wireless sensor network have tiny battery which gets consumed in a short duration of time due to data communication and nonstop measurement of parameters at high sampling rates (Akyildiz et al. 2002).Reliable transmission data is a crucial design objective for WSN in precision agriculture as the presence of foliage in transmission channel causes significant attenuation, scattering, diffraction, and absorption of the radiated waves (Garcia et al. 2004).High energy efficiency along with high throughput and low latency are significant design goals for wireless sensor networks for precision agriculture application.

A routing protocol continuously searches shortest, energy efficient data paths from source to sink node. In case of multi-hop routing protocol, a source node transmits its data packet to a nearby relay (forwarder) node. Relay node retransmit the data packet to sink node (Pantazis et al. 2013). This aids in decreasing packet loss rates of the nodes along the periphery of the network.

A wide range of research works done in the field of WSN application based precision agriculture is available in existing literature. Below is a concise review of some of them. (Beckwith R. 2004) proposed the design of wireless sensor network for large scale deployment in agriculture. Similarly (Garcia et al. 2009) studied the benefits obtained from recent developments in the field of different sensor and wireless communication technologies like ZigBee based WSN, passive, semi-passive & active RFID in the segment of Agriculture and Food Industry. A popularly known clustering routing protocol, as proposed by (Heinzelman et al. 2008) is LEACH. LEACH elects next cluster forwarders for every transmission cycle. SEP (Stable Election Protocol) by (Smaragdakis et al. 2004) is an extension of LEACH for heterogeneous networks.

The current work presents the design of a new energy efficient WSN multi-hop routing protocol for green house farming applications. In the proposed work, selection of Intermediate forwarder nodes is done on the basis of node residual energy and distance of reference node from sink and distance of the node from other sensor nodes. The proposed protocol employs a cost effective Multiple Attribute Decision Making algorithm which makes a parametric comparison of network nodes based on their performance and selects the node consuming minimum energy as a forward transmission node (Smaragdakis et al. 2004). It is observed after the analysis that the overall Protocol performance (Lifetime of Network, Network-stability and Network-throughput) in comparison with the current protocols is drastically improved.

Section-2 of the paper emphasizes on the designed WSN system model while section-3 elaborates the proposed protocol. Simulation results have been represented in section 4. Section 5 gives the concluding remarks besides the future scope of work.

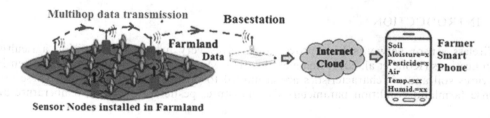

Figure 1.   Use of WSNs in precision agriculture.

2 SYSTEM MODEL

WSN system used in the present work contains eight sensor nodes that have been placed at different indoor locations of a 20 m X 20 m green house. The green house has been partitioned into eight grids implanted with the tomato crop. Each grid is provided with a senor.

2.1 *WSN architecture*

Sensors 1& 8 communicate data directly to Sink. Remaining nodes report data to sink via forwarder node. (Figure 2).

2.2 *Energy model*

First order radio model as given in (Heinzelman et al. 2008) model node data transmission energy losses. Equation-1 models node transmission energy for communication of data word of L-bits length to a node at a distance-X away. Node energy consumed in receiving same data word is given by Equation-2

$$E_{Tx}(,D) = E_{TX-elect} \times L + E_{Amp} \times L \times X^2 \tag{1}$$

$$E_{RX}(W) = E_{RX-elect} \times L \tag{2}$$

$E_{Tx\text{-}elect}$ /$E_{Rx\text{-}elect}$ represents the energy consumed in transmitting each bit of information for driving the transmitter or the receiver circuit and E_{Amp} is the energy consumed by the power amplifier of transmitter circuit.

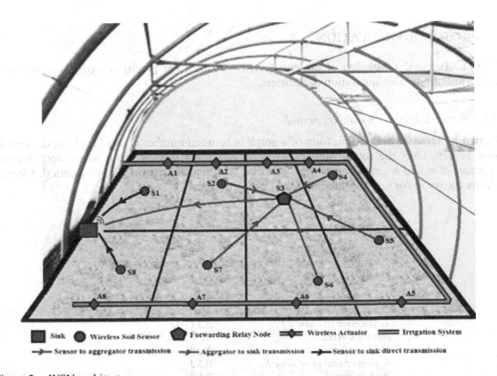

Figure 2.   WSN architecture.

3 PROPOSED PROTOCOL

Proposed protocol works in following phases.

3.1 *Forwarder selection phase*

Forwarder selection is sink assisted and is carried out in following phases.

Sink applies a cost function based MADM method for forwarder node selection as follow. This method ranks each candidate node on basis of their current energy level $E_R(i)$; $i \ldots$ (1, 2...N,), their average distance from other sensor nodes $D_{avg}(i)$; $i \ldots$ (1, 2...M), and their distance from sink node $D_{toSink}(i)$; $i \ldots$ (1, 2...M) for their appropriateness for forwarder job. In this case the Sinknode assignsthe calculatedcost function value for every sensor node using equation (3). Here M is the number of sensor nodes.

$$CF(i) = \frac{E_{Res}(i)}{D_{avg}(i) + D_{toSink}(i)} ; i \in (1,2..M) \tag{3}$$

The candidate sensor node with highest cost function is selected as the forwarder head.

3.2 *Transmission and sensing of data*

In this procedure, sensor nodes observe their assigned soil or air characteristics. During this period node transceivers are off. Starting from node-S2, sensor nodes throw their sensed data words to forwarder on one by one basis. When a node is communicating, other nodes remain in sleeping condition with their transceiver off. Then Forwarder transmits its packet to sink. Then sensor-S1transmits its data to sink.Then sensor-S8 transmits its data to sink.

4 PROTOCOL SIMULATION

MATLAB based simulation of presented WSN protocol for precision agriculture is carried out. Table 1 contains simulation parameters.

4.1 *Network lifetime & stability period*

Network lifetime is obtained in form of a graph in between number of dead nodes and transmission round. Thus the round number when all nodes get died is taken as network lifetime. Number of round when very first node gets died is termed as network stability period. Figure-3 shows the result for network lifetime & stability period.

Table 1. Simulation parameters.

Simulation parameter	Value
$E_{TX-elect}$, $E_{RX-elect}$ & E_{Amp}	15.7, 35.9 & 1.97 nJ/bit
Data packet size (W)	4000
Transceiver frequency f	2.5 GHz
Network Area	20 x 20 m^2
Nodes utilized	Nodes=8,Sink=1
Reference energy of node E_o	0.5 J
Peak energy E_{Thr}	0.1 J

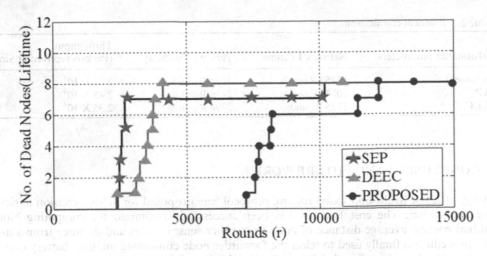

Figure 3.   Representation of network lifetime&network stability period.

4.2   *Network throughput*

Network throughput is the number of successfully transmitted information packets to sink. Figure-4 shows the result for network throughput.

The suggested protocol shows better performance compared to the current protocol in terms of network lifetime, stability of network and throughput. Table 2 compares the proposed protocol with existing SAP and DEEC protocol.

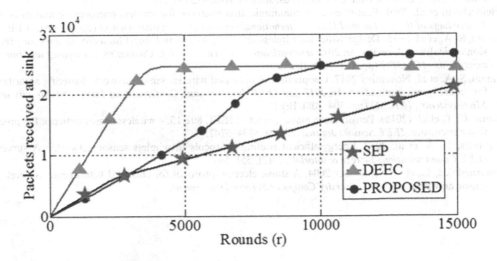

Figure 4.   Network throughput.

Table 2. Protocol comparison.

Performance parameters	Network Lifetime	Period of Stability	Throughput (Packets received at Sink)
(Proposed)	12450 Rounds	7564 Rounds	2.75×10^4
SAP	10345 Rounds	2456 Rounds	2.45×10^4
DEEC	5125 Rounds	2296 Rounds	2.55×10^4

5 CONCLUSION AND FUTURE WORK

Energy efficient Multi-hop based routing protocol was proposed for WSN precision agriculture application. The cost function has been successfully optimized for computing Node residual energy, average distance of node from other sensor nodes and distance from sink . This procedure is finally used to select the forwarder node consuming minimal battery energy thereby extending the network lifetime. It is established that the suggested protocol displays a superior performance in terms of network lifetime, stability and throughput as compared to current protocol. In near future, the suggested protocol can be designed with higher number of sensor nodes in a real time environment.

REFERENCES

Akyildiz, I. F. et al. 2002. Wireless sensor networks: a survey. *Computer networks*, *38*(4), 393–422.

Beckwith, R. et al. 2004. Unwired wine: Sensor networks in vineyards. In *SENSORS, 2004 IEEE* (pp. 561–564). IEEE.

Garcia, L. et al. 2009. A review of wireless sensor technologies and applications in agriculture and food industry: state of the art and current trends. *sensors*, *9*(6), 4728–4750.

Heinzelman et al. 2000. Energy-efficient communication protocol for wireless microsensor networks. In *Proceedings of the 33rd annual Hawaii international conference on system sciences* (pp. 10–pp). IEEE.

Imam, S. A. et al. 2015. Design issues for wireless sensor networks and smart humidity sensors for precision agriculture: A review. In *2015 International Conference on Soft Computing Techniques and Implementations (ICSCTI)* (pp. 181–187). IEEE.

Imam, S. A. et al. November 2017. Cooperative effort based wireless sensor network clustering algorithm for smart home application. In *2017 2nd IEEE International Conference on Integrated Circuits and Microsystems (ICICM)* (pp. 304–308). IEEE.

Kone, C. T. et al. (2015). Performance management of IEEE 802.15. 4 wireless sensor network for precision agriculture. *IEEE Sensors Journal*, *15*(10), 5734–5747.

Pantazis, N. A. et al. 2012. Energy-efficient routing protocols in wireless sensor networks: A survey. *IEEE Communications surveys & tutorials*, *15*(2), 551–591.

Smaragdakis, G. et al. September 2004. A stable election protocol for clustered heterogeneous wireless sensor networks. *Boston University Computer Science Department*.

Artificial intelligence to predict consumer behaviour: A literature survey

Gautam Srivastava & Narendra Singh
GL Bajaj Institute of Management and Research, Greater Noida

ABSTRACT: Artificial intelligence application has been extending in marketing. Marketers use AI to predict the buying behaviour of consumers. This study reviews the research paper of consumer behaviour and artificial neural network and explores the opportunities to develop a framework of consumer behaviour using artificial neural network. Framework based on an artificial neural network can interpret consumer behaviour more precisely. Forecasting the future action of the consumers is not a facile task. Consumer behaviour changes continuously. It also explores the possibility of developing a machine learning model through AI to monitor and forecast the consumption pattern of the consumer.

Keywords: artificial intelligence, consumer behavior, machine learning, predictor

1 INTRODUCTION

Today the world is moving fast towards automation. The automation system works on artificial intelligence. AI system predicts the future by analyzing the past data. Every sector like manufacturing, banking, real state, service sector etc. are using AI. Its scope is not limited to automated robotics system. Its applications are widely accepted in marketing also. AI can also be useful to predict the buying trends of the consumers (Nadimpalli 2017). We can develop the artificial intelligence system to analyze the consumers. AI itself will detect the future response of the consumers toward products and services (Davenport, et al. 2020). So, there is a need to develop a model based on artificial intelligence to predict the buying behavior of the consumers. The proposed model can be supervised or unsupervised learning model. This review paper focused on the scope of the paper-based on AI to develop the classification model. An intensive literature review has been to determine the research had been done related to this area and identified that rare any research conducted to develop the machine learning model to predict the future purchasing behaviours of the consumers based of artificial intelligence.

2 LITERATURE REVIEW

In digitalized world consumer express their need and wants on a social networking platform like Twitter, Facebook, blogs, videos etc. Analyzing the buying behavior of the consumer is an extremely complex task (Kietzmann et al. 2018). It is also very important to identify factors that affect the buying behaviour of consumer (Yadav et al., 2018; Singh et al., 2018). The recent era is called the era of artificial intelligence.

DOI: 10.1201/9781003193838-66

Artificial intelligence has been bringing drastic changes in marketing. It became easier for the marketer to analyse the buying behavior of the consumers. Artificial intelligence is a better tool to do predictive analysis in comparison to other traditional tools (Abiodun, et al., 2017). Effective market segmentation is requisite for marketing strategy. Artificial intelligence works on artificial neurons. The way of doing shopping is changing continuously. Businesses organization should have the past data of the consumers to develop the machine learning model. These models help to classify consumers (Badea, 2014). Due to globalization, growing information and communication technology, the complex data of consumer buying habits need more complex problem-solving techniques. Artificial intelligence used to estimate, classify, data conceptualization, solving problem-related to the buying behaviour of consumers (Staub, et al. 2015). Machine Learning techniques are used to predict valuable customer in order to retain them (Singh & Agrawal, 2018).

In marketing artificial intelligence can be used to predict the buying behaviour of the consumers. Artificial intelligence has an artificial neuron on which it works. Neurons decide based on past trends of the consumers. Banking sectors using an automation system to segment the eligibility of the consumers for plastic money. Similarly, big retail outlet also uses this system to give the credit point to their consumers and try to retain them (Dumitriu and Popescu 2019). In the present scenario, consumers spent on an average of 5 to 6 hrs daily on social networking sites or search engine which generates billion of data daily. These data can be used to develop the machine learning model which segments the consumers through an automationsystem.

Authors	Findings
Martínez-López & Casillas, 2009	Machine learning model help to develop the marketing intelligence system which support marketing decision.
Furaji, Łatuszyńska, Wawrzyniak, &Wąsikowska, 2013	This study determines the effectiveness of advertising on male and female consumers and keep the database of the consumers which help the companies to retain them by launching separate promotional strategy for different clusters.
Khade, 2016	Implementation of big data driven technology play a significant role to monitor the buying behaviour of theconsumers.
Singh & Agrawal, 2019	Implement KNN for customer retention.
Srivastava, 2019	Consumer attitude, consumer perception and consumer satisfaction are the three parameters to classify the behaviour of theconsumers.
Singh et al., 2020a	Customer Segmentation by using various machine learning techniques such as decision tree, multi-layer perceptron (MLP) etc.
Ameen, Ameen, &Hosany, 2020	Virtual reality, artificial intelligence and wearable technology has proven successful for effective marketing strategy.
Blasco-Arcas, Reyes, Raya, &Kastanakis, 2020	Differentiated between implicit and explicit data to understand the consumer behaviour.
Kachamas, Akkaradamrongrat, Sinthupinyo, & Achara, 2019	Artificial intelligence model based on naïve Bayes give the best result to predict the behaviour of the consumers.
Verma &Bandi, 2019	Artificial intelligence is being used to solve the complex recruitment and retention process of HR.
Pillai &Sivathanu, 2020	AI has positive impact of talent acquisition. HR is adopting machine learning system to hunt the talented employees.
Gubbi, Buyya, Marusi, &Palaniswami, 2013	Cloud based model increase the availability of internet of things (IOT) in remote areas.

3 CONCEPTUAL MODEL

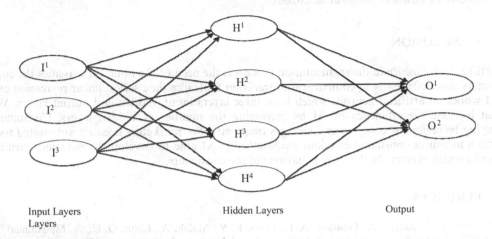

Figure 1. Artificial neural network.

The architect of an artificial neural network is based on three layers: input layers, hidden layers, and output layers. The input layers in the first layer in ANN and consist of input variables. The second layer is the hidden layer and it processes the data. The hidden layer generates artificial neurons by an activation function. An activation function is a mathematical function that determines the output of the artificial neural network. Output layers give the result of an artificial neural network. This ANN model could be used for measuring and analysing the buying behaviour pattern of consumers. Consumer's behaviour is dynamic, and it changes frequently, so an advanced statistical model is needed to predict the buying behaviour of the consumers more accurately.

4 DISCUSSION

The taste and preferences of consumers change continuously. To measure the need and expectations of the consumers, marketers continuously do the market survey. Market segmentation is the result of the study. Segmentation is very much needed for the classification of consumers based on similar characteristics. No business enterprises can fulfill the demands of consumers by producing unique products. Consumers are different on the parameters of economic, social, culture and psychological behaviour. Companies need to produce different product length for a different segment of consumers. Segmentation is an ongoing process. Marketers required some automated system to do the segmentation continuously. AI may be one of the solutions. Research is needed to develop the framework for the classification of consumers by using artificial intelligence.

5 FINDINGS

This study reviews the research paper of reputed journal to explore the research that has been done in analyzing consumer behaviour using artificial neural network. Determining the buying behaviour pattern of consumers is one of the most complex tasks. This research tries to find out the possibilities of using an artificial neural network to develop a framework for consumer behaviour. An artificial neural network is an advance deep learning tool having a wide application. It can be used in the marketing area to measure and develop an appropriate framework for buying behaviour of the consumers. Further, this work can be extended with distributed artificial intelligence in the prediction of consumer behaviour in electronic commerce or business (Singh et al. 2020b).

6 CONCLUSION

Artificial intelligence in the segmentation process is the need of the hour. AI classifies the consumers more efficiently in comparison to tradition tools like time series, linear regression etc. AI works on artificial neurons which have three layers input, hidden and output layers. We can increase the efficiency of AI by increasing the number of hidden layers. Consumers buying behaviour is complex and changes frequently. So, marketers need an automated tool which measures consumerbehaviour automatically. AI able to develop the machine learning model which predict the dynamic behaviour of the consumers.

REFERENCES

Abiodun1, O. I., Jantan, A., Omolara, A. E., Dada, K. V., Malah, A., Linus, O. U., … Mahammad, a. (2017). Comprehensive Review of Artificial Neural Network Applications to Pattern Recognition. *IEEE Acess*, 2–9.

Ameen, N., Ameen, N., & Hosany, a. S. (2020). Consumer interaction with cutting-edge technologies . *Computers in Human Behavior*.

Badea, L. M. (2014). Predicting Consumer Behavior with Artificial Neural Networks .*Procedia Economics and Finance*, 238–246.

Blasco-Arcas, L., Reyes, A., Raya, M. L., & Kastanakis, a. M. (2020). Leveraging User Behavior and Data Science Technologies for Management .*Journal of Business Research*.

Davenport, T., Guha, A., Grewal, D., & Bressgott, a. T. (2020). How artificial intelligence will change the future of marketing. *Journal of the Academy of Marketing Science*, 24–42.

Dumitriu, D., & Popescu, a. M.-M. (2019). Artificial Intelligence Solutions for Digital Marketing. *Procedia Manufacturing*, 630–636.

Furaji, F., Łatuszyńska, M., Wawrzyniak, A., & Wąsikowska, a. B. (2013). Study on the influence of advertising attractiveness on the purchase decisions of women and men. *Journal of International Studies*, 20–32.

Gubbi, J., Buyya, R., Marusi, S., & Palaniswami, a. M. (2013). Internet of Things (IoT): A vision, architectural elements, and future directions. *Future Generation Computer Systems*.

Kachamas, P., Akkaradamrongrat, S., Sinthupinyo, S., & Achara, a. (2019). Application of Artificial Intelligent in the Prediction of Consumer Behavior from Facebook Posts Analysis. *International Journal of Machine Learning and Computing*, 91–97.

Khade, A. A. (2016). Performing Customer Behavior Analysis using Big Data Analytics. *Procedia Computer Science*, 986–992.

Kietzmann, J., Paschen, J., & Treen, a. E. (2018). Artifcial Intelligence in Advertising: How Marketers Can Leverage Artifcial Intelligence Along the Consumer Journey. *Journal of Advertising Research*, 263–267.

Martínez-López, F. J., & Casillas, a. J. (2009). Induction approach based on Genetic Fuzzy Systems. *Industrial Marketing Management*.

Nadimpalli, M. (2017). Artificial Intelligence – Consumers and Industry Impact. *International Journal of Economics & Management Sciences*, 6(4).

Pillai, R., & Sivathanu, a. B. (2020). Adoption of artificial intelligence (AI) for talent acquisition in IT/ITeSorganizations .*Benchmarking: An International Journal*.

Singh, P., & Agrawal, R. (2018). Prospects of Open Source Software for Maximizing the User Expectations in Heterogeneous Network. International Journal of Open Source Software and Processes (IJOSSP), 9(3), 1–14.

Singh, N., Gupta, M., & Dash, S. K. (2018). A study on impact of key factors affecting buying behaviour of residential apartments: a case study of Noida and Greater Noida. International Journal of Indian Culture and Business Management, 17(4), 403–416.

Singh, P., & Agrawal, V. (2019).A Collaborative Model for Customer Retention on User Service Experience.In Advances in Computer Communication and Computational Sciences (pp. 55–64). Springer, Singapore.

Singh, N., Singh, P., Singh K.K., & Singh, A., (2020a). Machine Learning based Classification and Segmentation Techniques for CRM: A Customer Analytics, *Int. J. of Business Forecasting and Marketing Intelligence*, 6(2), pp. 99–117, DOI: 10.1504/IJBFMI.2020.10031824.

Singh, P., Singh, R., Singh, N., Singh, MK. (2020b) Distributed Artificial Intelligence:The future of AI, Distributed Artificial Intelligence: A Modern Approach, CRC Press. DOI: 10.1201/9781003038467-16

Srivastava, G. (2019). Buying behaviour of consumers towards Green Products. RESEARCH REVIEW International Journal of Multidisciplinary, 4(1), 477–481.

Staub, S., Karaman, E., Kaya, S., KarapÕnar, H., & Güven, a. E. (2015). Artificial Neural Network and Agility. *Procedia Social and Behavioural Sciences*, 1477–1485.

Verma, R., & Bandi, a. S. (2019). Artificial Intelligence & Human Resource Management in Indian It Sector. *Digital Strategies for Organizational Success*, (pp. 962–967).

Yadav, N. S., Gupta, M., & Singh, P. (2018). Factors affecting Buying Behavior & CRM in Real Estate Sector: A Literature Survey. Asian Journal of Research in Business Economics and Management, 8 (6), 32–39.

Recent Trends in Communication and Electronics – Sharma et al. (Eds)
© 2021 Taylor & Francis Group, LLC, ISBN 978-1-032-04572-6

IoT based smart car parking system for smart cities

Himanshu Sharma, Shruti Talyan, Shambhavi kaushik & Kartikeya dwivedi
ECE Department, KIET Group of Institutions, Delhi-NCR, Ghaziabad, U.P, India

ABSTRACT: Today in this modern world, most of the urban cities are facing problems like traffic jam, limited car parking space, poor road safety conditions, costly transportation of goods from one city to other etc. In this paper, we have addressed the problem of smart car parking system using Internet of Things (IoT) for smart cities. Here, the wireless IoT sensor nodes provides remote access to the car driver to keep a track of the availability of the parking area. In this paper, the driver's efforts & time both are reduced by providing an instant notification message about the free parking slots area using RFID technology. The developed smart car parking is simple, low cost, and efficient a system which can be used in Municipal Corporation authorized car parking stations at busy places such as Shopping Malls, Bus Stand Car parking, Metro Train car parking, airports etc.

Keywords: smart car parking, arduino, IoT, RFID

1   INTRODUCTION

Now a days, all new automobile vehicles are equipped with Internet of Things (IoT) sensor nodes for smart car parking, route navigation, accident detection on road, vehicle theft safety & connectivity with owner using smart mobile phone at all the time (Awaisi & Abbas 2019). An embedded system connected to the internet is called Internet of Thing (IoT) device, for example, a computer, laptop, smart phone or camera etc (Sharif & Guo 2019). The IoT has enabled the smart car parking stations in busy places such as Shopping Malls, Bus Stand Car parking, Metro Train car parking, Airports etc (Sharma et al. 2019). The IoT consists of wireless sensor networks (WSN) nodes for sensing, measurement & control applications (Sharma et al. 2018). Now a day, the energy harvesting technique is also used in IoT to prolong the network lifetime of sensing devices (Sharma et al. 2018). The major car parking problems are comfortable car parking, loss of time & fuel for searching parking place (Kannan & Mar, 2020).

In this paper, we have designed a smart car parking system as shown in Figure 1. It consists of four parking slots (Slot1-Slot4), an Arduino microcontroller, a router, IoT server, and smart mobile phone (Denis et al. 2020). As shown in the flow chart, when a driver arrives at the parking gate a parking request is made from outside itself (Shabasy & Abdellatif 2020). If an empty slot is available for car parking, then the driver is allowed to take the car inside the parking stations.

2   COMPONENTS & WORKING ALGORITHM

The various components of a smart car parking system are shown in Figure 1 as follows:

- Parking Sensors:(Passive Infrared (PIR) Sensors
- Processing Unit: Arduino ATmega 128 Microcontroller

DOI: 10.1201/9781003193838-67

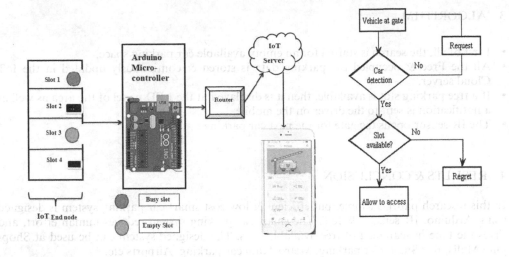

(a) Architecture of Smart Car Parking   (b) Flow Chart

Figure 1.   IoT based smart car parking system.

- Wi-Fi Module: ESP 8266
- Android Mobile Application

In Figure 1(a) the available empty car parking slots are shown by green color & busy car parking slots are shown by red color. The remaining slots are reserved. All car parking slots are continuously being monitored by the IoT servers (Mahindra & Sonoli 2017). The Arduino microcontroller uses various types of sensors like, Camera, Infrared (IR) etc . to sense the nearby objects. The Arduino microcontroller sends the data to the router. From the router, the data is sent to the IoT server & then finally to the smart phone of the user (Praveen & Harini 2019). As shown in Figure 1(b) flowchart, when the vehicle arrives at the gate, then the driver checks his smartphone for any empty slot. If the slot is available, then he is allowed to enter in the parking area (Meenaloshini et al. 2019). When user finds a free available parking, then the IR sensors and Servo motor are used to find the distance between other cars (Saarika et al. 2017) as shown in Figure 2. In Figure 2(A) The Arduino connections with the servo motor & IR sensor are shown. In Figure 2(b) The operation of IR sensor is explained. Here as the object comes closer to the IR sensor mounted on the backside of car, the reflected waves trigger the Buzzer and an alert sound is produced. This alerts the car driver while care parking. Some possible mounting positions of IR sensor are highlighted in Figure 2(c).

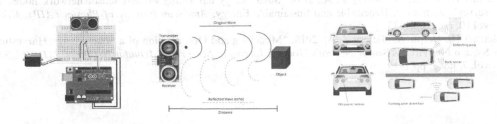

(a) Arduino & IR Sensor   (b) IR sensor operation   (c) Car parking using sensors

Figure 2.   Car parking using IR sensor & Servo motor.

3 ALGORITHM

- First of all, the search is started for an empty available car parking space.
- All the Free & occupied car parking data is stored & continuously updated in the IoT Cloud server.
- If a free parking slot is available, then it is displayed at the LED panel of the area as well as a notification is sent to the driver on the mobile.
- The IR sensor opens the gate for the next car parking.

4 RESULTS & CONCLUSION

In this research paper a simple, but efficient & low cost smart car parking system is designed using Arduino, IR sensors & IoT. The smart car parking system reduces human effort, and saves the time in searching of free available slots. The designed system can be used at Shopping Malls, Bus Stand Car parking, Metro Train car parking, Airports etc.

REFERENCES

Awaisi K. S., Abbas A. 2019. "Towards a Fog Enabled Efficient Car Parking Architecture", *IEEE Access Journal*, Volume 7, no. 159100.

Denis A., Tiwari A., Jirge V., 2020. "Smart Parking System using IoT Technology", *IEEE International Conference on Emerging Trends in Information Technology and Engineering (IC-ETITE)*, IEEE, USA.

Kannan M., Mary L. W., 2020. "Towards Smart City through Virtualized and Computerized Car Parking System using Arduino in the Internet of Things" *IEEE International Conference on Computer Science, Engineering and Applications (ICCSEA)*, IEEE, USA.

Mahendra B. M., Sonoli S. 2017. "IoT based sensor enabled smart car parking for advanced driver assistance system". *2nd IEEE International Conference on Recent Trends in Electronics, Information & Communication Technology (RTEICT)*.

Meenaloshini M., Ilakkiya J., Sharmila P., Sheffi J.C., Nithyasri S., 2019. "Smart Car Parking System in Smart Cities using IR", *IEEE 3rd International Conference on Computing and Communications Technologies (ICCCT)*.

Praveen M., Harini V. 2019. "NB-IOT based smart car parking system", IEEE International Conference on Smart Structures and Systems (ICSSS), 2019.

Saarika P., Sandhya K., Sudha T. 2017. "Smart transportation system using IoT", International Conference on Smart Technologies for Smart Nation (SmartTechCon).

Shabasy N. E., Abdellatif M., 2020. "IoT for Smart Parking", IEEE International Conference on Advances in the Emerging Computing Technologies (AECT), IEEE, USA.

Sharif A., Guo J. 2019. "Compact Base Station Antenna Based on Image Theory for UWB/5G RTLS Embraced Smart Parking of Driverless Cars", *IEEE Access Journal*, Volume 7, no. 180898.

Sharma H., Haque A., Jaffery Z. A. 2019. "Maximization of Wireless Sensor Networks Lifetime using Solar Energy Harvesting for Smart Agriculture Monitoring", *Adhoc Networks Journal, Elsevier*, Vol. 94, Netherlands, Europe.

Sharma H., Haque A., Jaffery Z. A., 2018 "Solar energy harvesting wireless sensor network nodes: A survey", Journal of Renewable and Sustainable Energy, *American Institute of Physics (AIP), USA*, vol.10, no.2, pp.1–33.

Sharma H., Haque A., Jaffery Z. A., 2018. "Modelling and Optimization of a Solar Energy Harvesting System for Wireless Sensor Network Nodes", *Journal of Sensor and Actuator Networks, MDPI, USA*, vol. 7, no. 3, pp.1–19

Recent Trends in Communication and Electronics – Sharma et al. (Eds)
© 2021 Taylor & Francis Group, LLC, ISBN 978-1-032-04572-6

An enhanced ENCIPHER to encrypt large text & image using basic arithmetic and logic operation with substitution-transposition

D. Prasad, S. Kumar & V. Bharti
School of Computing (CSE), DIT University, Dehradun, India

ABSTRACT: In this digital era of every possible computation, information sharing, classic to modern device inter-connectivity, of threats and attacks, of cybercrimes and ethical fraud, there is hardly any area where security is of no concern. The information is power and it has to be controlled while flow through a cascade of networks. This tall reliance of both organizations and people on the data makes it very fundamental. This, in turn, has driven to a increased mindset for secure information including ensuring the genuineness of information and messages, and to secure frameworks from network-based assaults. Thus cryptography has developed itself to give its best effort to eliminate all the vulnerable attacks as part of network security. In this work we are proposing an enhanced cipher that uses the basic capability of its previous version but now with more challenges including large text character base and image. This proposed method makes cryptanalysis indeed more troublesome using "Random Number as key" and slightly different concept to arrange of encryption rounds and keys for ultimate scrambling the plain text. This time Unicode character set is used which eliminates the limitation of encryption to only a small size data. Also makes algorithm suitable for image encryption as image is just an array of bytes. It also allows key exchange between receiver and sender automatic by adding key with the message itself at random locations (like in the middle of two halves decided again on the basis of random number).

Keywords: text encryption, cipher, encryption-decryption, information security, substitution-transposition, hybrid encryption, logical NOT operation

1 INTRODUCTION

The Organizations have experienced two major changes for the requirement of information security. Before the use of data processing tools, the security of the valuable information of any organization was provided primarily by physical and administrative means. For example in order to store the sensitive documents, the strong cabinets with a combination lock were used. The example for latter is the use of personal screening procedure in the hiring process.

The use of automated methods in order to protect the sensitive information on the computer has become evident in the digital era, especially in the case of time-sharing system. The need is more important for the systems accessible over a telephone network, or the internet. So the collection of methods and tools designed for the protection of the data, sensitive information and to prevent hacking is known as computer security.

Information of the distributed system and the network and communication methods for carrying the data between users has a major role in the security. Data should be protected during the transmission by using the network security measures. Since all Business, government and academic corporations are integrated with tools for data processing and have inter-

DOI: 10.1201/9781003193838-68

connected group of such small networks which we refer as internet, so the term internet security can be used in place of network security.

There are no clear boundaries between these two shapes of security. For example, the most
common type of attack i.e. the computer virus may be introduced in to a system physically
through optical disk or it may also arrive over an internet. In both the cases, when the computer
system is infected with a virus, detection and recovery is necessary through the internet computer security tools.

2 SHORTCOMINGS OF EXISTING ALGORITHM

- The existing algorithm uses a random number just to make a decision to select order of
  substitution and transposition and not for key. This way, it is easy to break as key generation part is very simple.
- The algorithm use auto key generation and putting key in message itself although in
  encrypted form makes decryption easy but this makes algorithm less secure since place of
  key and of random number notation is fixed at either end.
- The algorithm can only encrypt and decrypt small range of ASCII characters that makes it
  not suitable for image and binary files encryption.
- The algorithm has too many decision steps and entire message treated as a whole that
  result in great latency in large file and image encryption.

3 OUR CONTRIBUTION

In this enhanced algorithm proposed here the key generation part is focused very much. Previous algorithm uses random number not for its proper utilization. Randomness always gives
security. Here in this, random number serves as a key. Other keys further are generated from
this random number only. Encryption-Decryption are secure if key transfer between both the
parties i.e. sender & receiver is done in a highly secure manner. The beauty of the previous
algorithm to lighten the burden of sharing the key is retained here also but this time with
more enhanced and secure feature. The key is added in message itself but at random place. No
sign to indicate what was the key where it could be is missing now. Also the key is also
encrypted along with the message. To improve the performance, message is divided into
halves of different sizes determined with a sense of randomness and then a confusing order of
substitution and transposition (rail-fence) is applied with each. This time domain of substitution character set has been improved to support a Unicode character which was ASCII character set previously. This significant improvement makes this algorithm suitable for image
encryption and binary file encryptions. For each half a new key is generated and main key i.e.
random number is put between the two halves. As how halves are created is difficult to guess
hence main key can't be guessed. This part is the significant improvement in previous algorithm. Finally, when each half is encrypted as per the algorithm, logic NOT operation is
applied to complement each character of each half to get final cipher text character. To facilitate decryption the notation for half creation is attached at the end of cipher text followed by
length of main key. Both the information will be in encrypted form and know to two parties
only and will be secured just we do for pin number of ATM cards. Now, the final message can
be flowed through the network.

To the other side of coin, decryption, will take place after a hint on number to create the
halves. Since length of key only is known at the end of cipher text, key recovery is the first activity. Once keys and parts are separated, the reverse of encryption is applied starting with logic
NOT operation.

4 ENCRYPTION ALGORITHM

Step 1: Random(R) generation.
Step 2: Generate another Random number R1 to guess how two halves created.
Step 3: Create two halves
Step 4: Count Length of message
Step 5: if length is even, go to **Step** 6. **Else** go to **Step** 10.
Step 6: This is for first half which we call part1
Step 7: calculate Key (K1)=R % Length(part1)
Step 8: Do Substitution by formula C= (P+K1) mod 65535; where C,P & K1 are Cipher, Plain and Key text respectively.
Step 9: Do Rail Fence transposition on **Step** 8 result and do same for other half as mentioned in Setp 10 to **Step** 12.
Step 10: Apply rail-fence transposition to the 2nd half which we call part3
Step 11: calculate Key (K2)=R % Length(part3)
Step 12: Do Substitution by formula C= (P+K3) mod 65535; where C,P & K3 are Cipher, Plain and Key text respectively.. Do same for part1 as mentioned in step 6 to Step 9.
Step 13: Create one part to store R which we call part2
Step 14: Apply logical NOT with each individual parts ie. part1, part2 & part3.
Step 15: merge three parts in the order –part1, part2 & part3.
Step 16: Append notation for creating halves & length of main key to cipher text.
Step 17: Get final cipher text after applying Logical NOT on merged result.
STOP

5 ENCRYPTION AND DECRYPTION RESULT

Example:

   Let's our plaintext message "A T T A C K @ C y b e r ?". The Encryption and Decryption results obtained from Algorithm are as follows-

- Main key i.e. Random Number is generated. Assuming this main key be 62290. Another random number to guess the partition method is generated. Let's this indicate that partition will be N/3, 2N/3 where N is length of message.
- Till this, step, we have part1 length=ceiling(13/3)=4; part1={A,T,T,A}. part2 length=-number of digits (R) =number of digits(62290)=5; part2={6,2,2,9,0}. Part3 length=ceiling(2*13/3)=9; part3={C,K,@,C,y,b,e,r,?}.
- Count the length of message which is 13, odd, hence, 2nd part will be taken first and apply rail-fence then substitution. Then after, for the 1st part we apply substitution then rail-fence.
- After applying rail-fence on part3 it will become; part3={C,@,y,e,?,K,C,b,r}.

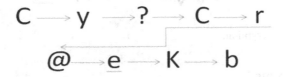

Figure 1.   Apply transposition as rail-fence.

- key3=Length (part3)=62290%9=1; Using ASCII(1), we do substitution to get jumbled part3 by C=(p+key3)%65535.
- After round 1 of encryption with part3, and applying we get –

ㅂㅊ口ㅏㅌㅍㅏ리ㅎ

- Similarly, after round 1 of encryption with part1, we get-

ㅎ래ㅌ래ㅎ

- Part2 will become; □ ◆ ? □

- The final cipher text after merging all three parts, adding key length and notation for partition method.

- Cipher Text= ㅎ래ㅌ래ㅎ □ ◆ ? □ ㅂㅊ口ㅏㅌㅍㅏ리ㅎ

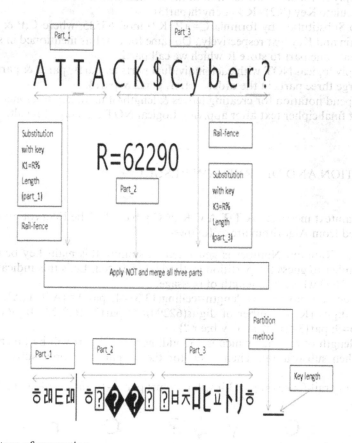

Figure 2.   One stage of encryption.

6   RESULT ANALYSIS

Following are two tables displaying characteristics of different algorithms and results obtained through a program implemented in C# and .NET environment. The hardware configuration utilized is mentioned as below-

Table 1. Different characteristics/parameters of few algorithms including proposed algorithm.

Algorithm	Designers	Key-Size	Block-Size	Algorithm Structure	Rounds
Blow fish	Bruce Schneier	32-448 Bits	64 Bits	Feistal network	16
Two fish	Bruce Schneier	128,192 or 256 Bits	128 Bits	Feistal network	16
RSA	Rivest, Shamir & Adleman	1024-4096 Bits	Any Byte length	——	1
Modified ENCIPHER	Devendra et. Al.	Variable	Variable	Hybrid	1

Table 2. Estimated processing time.

Size of I/P file (KB)	Processing Time RSA (ms)	Processing Time Blow Fish (ms)	Processing Time Two fish (ms)	Processing Time Proposed Algo (ms)
1	7.4578	7.122	6.5963	5.2031
5	9.1236	6.3216	6.3125	6.3269
10	9.8496	6.8956	7.1256	6.8969
15	10.1250	8.2369	7.8269	8.1063
20	11.1000	9.1256	8.1256	9.6236

- Intel Core i5 8the Gen @1.60 GHz
- 8 GB RAM

We call the overall processing time as throughput including time to encrypt & decrypt.

Observations reveal that proposed algorithm gives almost similar processing time as of Blowfish. Also in the case of proposed algorithm high throughput resulted is in high speed but with less time used.

7 ADVANTAGES OF ALGORITM

- Less time complexity as message is divided into two halves.
- Easy to understand and implementable
- Uses classic and basic arithmetic and logic operations
- Efficient & secure key generation technique utilizing random numbers.
- Unicode characters are used. This makes it suitable for image encryption. Although images are just array of bytes. Once it is converted to bytes, this encryption can be applied.
- Highly secure because of the use of random number generator.

8 CONCLUSION

Encryption is always useful for securing text, image of small size if trying to confuse so tediously so that crypt analysis becomes next to impossible. This algorithm generates a better encryption decryption where automatic keys are used. With randomness, keys become very difficult to be identified. And to get all this in hand we have this simple and compact code which also results in less processing delay and fast response. At the same time gives highly secure data that cannot be easily intruded while being transferred. Also the available character set is increased to support Unicode characters allowing image encryption with the same concept. In future the algorithm can be enhanced further to more refine encryption with high quality images and suitability for hardware implementation can also be rethink.

REFERENCES

Abusukhon Ahmad et al., 2019. A hybrid network security algorithm based on Diffie Hellman and Text-to-Image Encryption algorithm. *Journal of Discrete Mathematical Sciences and Cryptography*. 22(1):65–81.

N. B. F. Silva et.al., 2016. Case studies of performance evaluation of cryptographic algorithms for an embedded system and a general purpose computer. *Journal of Network and Computer Applications*. 60(1):130–143.

Patil, P et.al., 2015. A Comprehensive Evaluation of Cryptographic Algorithms:DES,3DES,AES,RSA and Blowfish,Science Direct, Elsevier. *International Conference on Information Security & Privacy (IGISP)*. :617–624.

Prasad, D et.al., 2014. ENCIPHER: A Text Encryption and Decryption Technique Using Substitution-Transposition and Basic Arithmetic and Logic Operation. *International Journal of Computer Science and Information Technologies*.5(2):2334–233.

Prasad, G.A. et al., 2013. A cipher design with automatic key generation using the combination of substitution and transposition techniques and basic arithmetic and logic operations. *The SIJ Transactions on Computer Science Engineering & its Applications (CSEA)*. 1(1):233–237.

SCHNEIER, B.,2015. Applied Cryptography Protocols, Algorithms and Source Coding, John Wiley & Sons, Inc.

IBM. 1994.The Data Encryption Standard (DES) and its strength against attacks. *IBM Journal of research and Development*, 38(1):243–250.

Jonathan Katz, Yehuda., 2007. Introduction to Modern Cryptography, Lindell Chapman & Hall.

Rick Burgess,2011. *URL: http://www.techspot.com/52011-one-minute-on-the-internet-640tb-data-transferred-100k-tweets-204-million-e-mails-sent.html*.

R. Venkateswaram et.al., 2010. Information Security: Text Encryption and D cryption with Poly Substitution method and combining features of cryptography.

S. G. Srikantaswamy, et.al., 2011. A Cipher Design using the Combined Effect of Arithmetic and Logic Operations with Substitutions and Transposition Techniques. *International Journal of Computer Application*. 29(8):0975–8887.

V. U. K. Sastry et.al., 2010. A block cipher having a key on one side of plaintext Matrix and its Inverse on the other side. *International Journal of Computer Theory and Engineering*. 2(5):1793–8201.

Stallings W.,2004. Cryptography and Network Security: Principles and Practice, Pearson Education:2–80.

Recent Trends in Communication and Electronics – Sharma et al. (Eds)
© 2021 Taylor & Francis Group, LLC, ISBN 978-1-032-04572-6

Vision based bridge safety monitoring system using WSN-IoT

Himanshu Sharma, Ananya Gupta, Anjali Sharma & Yashika Gupta
KIET Group of Institutions, Delhi-NCR, Ghaziabad, U.P., India

ABSTRACT: In metro cities, the bridges are the essential part which connect one place to other such as railway line crossing, river water crossing & to cross the busy roads. In smart cities, the bridge safety monitoring is very critical concern. Heavy traffic passes daily from the over-bridges built over complex terrain such as silica based soil, river water over bridge, railway line over bridges etc. There are chances that the bridge may get collapsed due to poor construction material, earthquakes, heavy loads, vibrations and rain water conditions etc. Therefore, in this paper a remote vision based bridge safety monitoring system is developed using internet of things (IoT) for smart cities

1 INTRODUCTION

Bridges are the important part of a city which connects one side of a city to the other side e.g. railway line crossing, a river water over bridge, a busy road etc. In metro cities like Delhi, Mumbai, Kolkata & Madras the bridges are made to reduce the traffic congestion. In smart cities, most of the tasks are online such as structural health monitoring systems, waste management system, Traffic light system, Smart agriculture monitoring & Bridge safety monitoring system etc (Lee & Yauan 2017). An embedded system connected to the internet is called Internet of Thing (IoT) device, for example, a computer, laptop, smart phone or camera etc (Lazo & Gallardo 2015).

In this paper, an autonomous bridge safety monitoring system is developed using IoT for smart cities as shown in Figure 1. It consists of a wireless sensor network (WSN) (Sharma et al. 2019). The WSN for bridge monitoring consists of Vibration sensors, CCTV Cameras, humidity sensors, Raspberry Pi microcontroller, a router, internet server, and control stations (Sharma et al. 2018). As shown in the flow chart, when crack is detected by the CCTV camera or a deep shock is sensed by vibration sensors then an alert message is sent to the IoT cloud server (Sharma et al. 2018). From IoT cloud server the alert message is sent to the remote control station (Mahmud & Bates 2018). From the control stations the necessary action can be taken such as sending a quick response team at the damage bridge location (Zhang & Zhou 2019).

2 BLOCK DIAGRAM & OPERATION

The various components of a smart car parking system are as follows:

- Camera Module, Vibration Sensors, Water Level Sensors, Moisture Sensors
- Processing Unit: Raspberry Pi Microcontroller
- Wi-Fi Module: ESP 8266
- Android Mobile Application

DOI: 10.1201/9781003193838-69

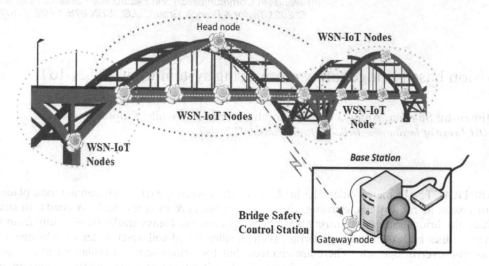

Figure 1.   Bridge safety monitoring system using IoT.

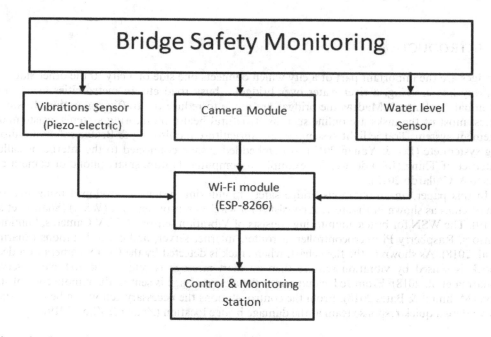

Figure 2.   Framework of bridge safety monitoring system using IoT.

The vibration sensor senses the vibrations of bridge in case of earthquake, heavy rain, heavy vehicle loads (Mahmud & Abdelgawad 2017). If the vibration level reaches a certain threshold level, then an alert messages sent to the remote control system.

The safety parameters of bridge and its pillars are stress, strain, acceleration, deformation, pits, moisture and any physical damage (Lamonaca & Sciammarella 2018).

3 MATHEMATICAL MODELLING

Let us consider the bridge safety system is denoted by S. Now, the bridge safety system (S) depends upon the following factors as:

$$S = \{A, S, D, P, M, V\} \tag{1}$$

Where, A= Acceleration, S = Stress, D = Deformation, P = pits, M = Moisture, V = Vibrations level.

4 FLOWCHART

The flowchart for Bridge safety monitoring is shown in Figure 3. Here, the bridge safety parameters are being monitored continuously (Pritam Paul & Nixon Dutta 2018). When any damage occurs, the corrective actions are taken immediately in real tie conditions.

5 RESULTS AND CONCLUSION

The results obtained by the vibration sensors over a period of time, when a heavy truck passes over it is shown in Figure 4.

Here, the bridge displacement occurs with a peak strength of 200 mm, which is set as the threshold value. But the vibrations reach till -300 mm at time 13:22 minutes. Therefore, an alert message is sent to the control station & a quick response team reaches at the bridge

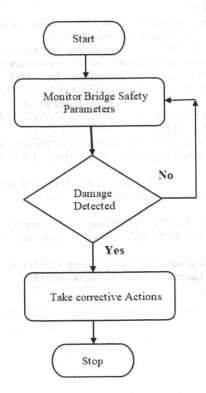

Figure 3.   Flowchart for bridge safety.

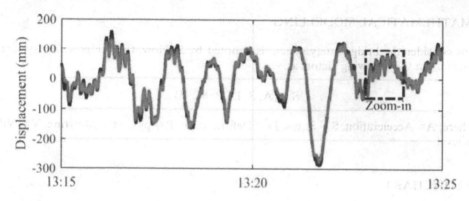

Figure 4. Bridge vibration sensor output monitoring for heavy truck loads.

location for repair work. This system is very efficient and can be used by central & state government departments like Public Works Department (PWD) and National Highways Authority of India (NHAI) and commercial civil construction & maintenance companies. In future, the AI & ML based algorithms will be used with advanced microcontrollers for prediction of possible damages in the bridge safety.

REFERENCES

Lamonaca F., Sciammarella P.F., 2018. "Synchronization of IoT Layers for Structural Health Monitoring", *IEEE Workshop on Metrology for Industry 4.0 and IoT*, IEEE.

Lazo C., Paulo G., 2015, "A Bridge Structural Health Monitoring System Supported by the Internet of Things", *IEEE COLCOM*, IEEE.

Lee J.L., Yauan Y., 2017. "Development of an IoT-based Bridge Safety Monitoring System", *IEEE International Conference on Applied System Innovation*, IEEE.

Mahmud M. A., Abdelgawad A., 2017. "Signal processing techniques for IoT-based structural health monitoring", *29th IEEE International Conference on Microelectronics (ICM)*, IEEE.

Mahmud M. A., Bates K., 2018. "A complete Internet of Things (IoT) platform for Structural Health Monitoring (SHM)", *IEEE 4th World Forum on Internet of Things (WF-IoT)*, IEEE.

Paul P., Dutta N., 2018. "An Internet of Things (IoT) Based System to Analyze Real-time Collapsing Probability of Structures", IEEE *9th Annual Information Technology, Electronics and Mobile Communication Conference (IEMCON)*, IEEE.

Sharma H., Haque A., Jaffery Z. A. 2019. "Maximization of Wireless Sensor Networks Lifetime using Solar Energy Harvesting for Smart Agriculture Monitoring", *Adhoc Networks Journal, Elsevier*, Vol. 94, Netherlands, Europe.

Sharma H., Haque A., Jaffery Z. A., 2018 "Solar energy harvesting wireless sensor network nodes: A survey", Journal of Renewable and Sustainable Energy, *American Institute of Physics (AIP), USA*, vol.10, no.2, pp.1–33.

Sharma H., Haque A., Jaffery Z. A., 2018. "Modelling and Optimization of a Solar Energy Harvesting System for Wireless Sensor Network Nodes", *Journal of Sensor and Actuator Networks, MDPI, USA*, vol. 7, no. 3, pp.1–19.

Zhang L., Zhou G., 2019. "Application of Internet of Things Technology and Convolutional Neural Network Model in Bridge Crack Detection, *IEEE Access*, vol. 6, IEEE.

Recent Trends in Communication and Electronics – Sharma et al. (Eds)
© 2021 Taylor & Francis Group, LLC, ISBN 978-1-032-04572-6

A framework based on security issues in cloud computing

Abu Zafar & Vipul Goyal
Amity School of Engineering and Technology, Amity University, Noida, India

Achyut Shankar
Department of Computer Science & Engineering, ASET, Amity University, Noida, India

ABSTRACT: This paper is focused on the topic framework based on security issues in cloud computing. This paper present's a good description of cloud computing. It tells about different types of cloud SaaS, PaaS, and IaaS and its characteristics. It also describes disadvantages of it which may lead to security issues. We did a good review of previous works which represented many ideas for a framework. This paper discusses threats like malicious insider, data breaches, etc. This paper also describes attacks and ways to protect them. A framework was designed through which we can very easily protect sensitive data from unauthorized use.

1 INTRODUCTION

Cloud computing is basically storing data in a remote area and using it. "Cloud Computing provides shared resources and services via the Internet" . Cloud computing needs a hosting service provider. A hosting service provides storage of data, networking facilities, database access, and many other facilities. It creates a server that helps in maintaining the flow of data sequentially. "The cloud computing is a web-based model which is connected with more than one system".The word "cloud" itself means that several networking devices are interconnected with each other in a particular network design. In some manner, this particular design looks like a cloud. This service allows users to access data kept in a remote area anytime anywhere and it eradicates the need for physical hardware for data storage. Cloud computing is of three types-platform as-a-Service (PaaS), software as a service (SaaS), and infrastructure as a service(Iaas).choosing a correct type of cloud computing helps in deciding what necessary things are needed for developing our model.

1.1 *Software-as-a-Service (SaaS)*

In this service providers provide software to the clients where they host different applications over the internet for their clients. SaaS offers many business software. All the things from programs, codes, algorithms, and the hardware too are shared in SaaS.

Disadvantages-

- It contains some drawbacks which can lead to security issues and other problems-
- Customers need to rely totally on the service provider for security and maintenance, they can't manage the security themselves.
- Customers need to organize their data in such a way it matches the SaaS system because data management and its governance are maintained by it.

DOI. 10.1201/9781003193838-70

1.2 *Platform-as-a-Service (PaaS)*

It provides the platform to the client to enhance their business activities where they can create, execute, and manage their applications. It has to do with the development model and deployment of applications. In PaaS, we can test, deploy, manage, and update at the same location. It also has DBMS|(database management systems), operating systems, development tools. In short, PaaS develops a platform for software development.

Disadvantages-

- The work of the client will be governed by the service provider.
- There are security risks because PaaS is used by many end-users which hare having access to the same resources.
- To make outdated systems work in PaaS solution numerous customization work and configuration changes are necessary.

Some examples of PaaS are-Heroku, Red hat, IBM, Google, Amazon web services.

1.3 *Infrastructure-as-a-Service (IaaS)*

Infrastructure as a service is a type of cloud computing that provides resources to the clients. The IaaS provider provides services to manage infrastructure. These are monitoring, log access, doing load balancing, doing a backup of data, recovery of data. IaaS clients have the responsibility of managing things such as data, applications, and runtime.

Disadvantages-
It has a few limitations –

- It has multitenant security
- To work on IaaS internal training is required and many resources too.

2 LITERATURE REVIEW

In "An analysis of security issues in cloud computing" the authors described cloud computing importance and its growing market in science and industry. This paper described that cloud computing is a computational paradigm and they went on to describe that what is its objective. They described what technologies are combined to make cloud computing successfully work. It portrays that though cloud computing is great field, security is a barrier.

In "Security issues in cloud computing" authors give light on the basic definition of cloud computing. It talked that security is the important issue of cloud computing. The attacks like a breach in knowledge, hijacking, and dos attacks possess a great threat to the cloud. They also described security issues such as server & Application access, data Transmission, the security of the virtual machine, security of network, data privacy, data integrity, physical security.

In "Security Problems in Cloud Computing Environments: A Deep Analysis and a Secure Framework. In Cyber Security and Threats: Concepts, Methodologies, Tools, and Applications" the authors did good research in cloud computing and talked about how cloud computing is providing resources, software, and hardware to the clients. They talked about the security issues of cloud computing. They are trying to give away to solve the security models by a quantitative risk assessment model.

"Security framework for cloud computing environment: A review. " is also good research done by the authors in which they told that cloud computing is a model in which information is processed stored and delivered to the clients. The authors went on to tell about security issues in cloud computing such as risk profiling, data loss. The author's motive was to create a solution for security issues so they went on to describe a framework.

In "A cloud computing security framework based on cloud security trusted authority" cloud computing has been described in a good way. The authors main motive was to design a framework which can resolve security issues so they proposed the idea of using a 3rd party for the protection of data rather than being dependent on the cloud service provider, they use a concept of cloud security trusted authority.

3 THREATS ON CLOUD COMPUTING

Data breaches-The main goal of an attack is to steal important and critical information such as bank details, travel details, social security numbers to obtain information about personal identity and other related information by an unauthorized person.

Lack of security in interfaces and APIs-Cloud computing service has many user interfaces and application programming interfaces. Any unauthorized use may try to access these and reuse these.

System vulnerability-Security in the system may occur due to bugs present in programs that remain in the system. This allows an unauthorized user to try to infiltrate the system and steal important information.

Service hijacking-Service hijacking redirects a user to a website that has illegitimate and obscene content.

Risk profiling-Cloud service providers take care of the maintenance of software and hardware, big companies do not have access to these services so they are unknown that how secure their data is. There are chances that their organization may face great risk.

A malicious insider-A malicious insider can be a person who works for an organization and try to access the confidential data of an organization and may use it against it.

4 ATTACK'S ON CLOUD COMPUTING

Zombie attack-In this attack, the attacker keeps sending requests till the server goes down. The attacker uses hosts that are someone else's, these are known as zombies. The request of the virtual machine is easily reachable by the user via the world wide web. The attacker uses this technique to flood the server such that it may go down and cannot be used by the clients because the cloud gets overloaded because of the number of requests. This type of situation is called denial of service(dos) or distributed denial of service (DDOS). Solution-Using a strong firewall and DDOS protection algorithms that identify the IP address via which multiple requests are coming and block it.

Abuse of cloud services-Hackers and crackers try to attack the cloud by using clouds that provide services like dos and DDOS attacks. Solution- Clouds that harm other clouds should be banned.

SIDE channel attacks-In this a virtual machine is placed in close proximity to the cloud server which needs to be attacked. Solution-"To prevent the cloud from side-channel attack we can use a combination of virtual firewall appliance".[8]

Authentication attacks-This type of attack happens because of the lack of security provided by the clouds. A simple username and password can be guessed by the attacker and it can gain access to the cloud and do the malicious activity. There are many authentication attacks-

Brute force attack-encrypted passwords are tried to break by the attacker by trying all possible combinations of passwords.

A phishing attack-In this the attacker transfers the user to another website to steal the information of the client. A phishing attack is a fraud attack done to steal information which is very delicate.

Link manipulation –In this method link is manipulated in such a way that it looks original link by changing a spelling or any other way possible.

Man in the middle attack-In this attack the attacker tries to change the content of the message which is being transferred from one person to another person by intercepting it in a public key and retransmitting it.In this way, the attacker makes it almost impossible. Solution-To prevents authentication attacks we can use biometrics verification.

5 FRAMEWORK FOR SECURE CLOUD COMPUTING

Cloud security completely depends upon cloud service provider so we are Introducing this new way to defend the cloud using a technique which we call as locker. For secure cloud computing we provide system requirements, then defending the system from any threats and attack and then we try to avoid any possible risks and if any attack do occurs then try to mitigate the loss.

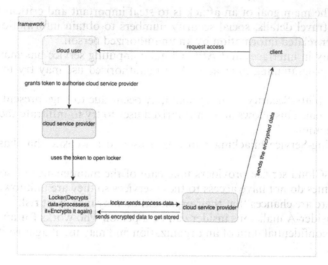

Figure 1. A framework for secure cloud computing and to prevent sensitive data.

5.1 *Methodology*

Client: Anyone can become a user and can access this client-side using the world wide web through the web browser. The client is a data user.

Cloud user-cloud user is the data owner who has taken the service of a cloud provider

Cloud service provider-these These are the ones who provide hardware, software, and accessories.

Locker-here this a concept which we think will provide huge security to the sensitive or ordinary data of the data owner. The locker is something that will be maintained by the cloud service provider. This can only be accessed when both the data owner and cloud service provider authorize it. until cloud user gives the token to authorize cloud security provider to allow it to use the locker to get the data stored out. Locker decrypts the data, processes and sends it to the front side of the cloud service provider, and then it forwards it to the client.This framework proposes an idea that the cloud user will be keeping their sensitive data in a special place called a locker which will be managed by the cloud service provider.

The steps of how this framework should work are

- Client sends a request to access the data, this request directly goes to the cloud user.
- Cloud user grants the token to the cloud service provider and authorizes it to open the locker and use it.
- The locker section decrypts the data and processes it and sends it to the front end of the cloud service provider
- Cloud service provider forwarded it to the client
- After it is used then agalin cloud service provider sends it back to the locker where this data is encrypted and stored.

6 RESULT AND CONCLUSION

Cloud computing is a growing field in computer science area. It has not just provided a boost to businesses but also gave employment to so many people. Cloud computing is a great technology which has brought a huge development in the world. People are very tensed about one thing that whether their data is secured or not. It is expected that once cloud computing overcome this problem of data security then it will bring a revolution which will lead a bright future. The above framework will be able to create a platform where a user can protect its sensitive data easily. In order to defend an attack from happening one must know how attack is done, then one must provide a security measure for it. Thus, cloud computing is found to be a interesting and ever green field where businesses could grow and enhance easily providing jobs to many people.

REFERENCES

Chouhan, P., & Singh, R. (2016). Security attacks on cloud computing with possible solution. *International Journal of Advanced Research in Computer Science and Software Engineering*, 6(1).

Dawoud, M. M., Ebrahim, G. A., & Youssef, S. A. (2016, May). A cloud computing security framework based on cloud security trusted authority. In *Proceedings of the 10th International Conference on Informatics and Systems* (pp. 133–138).

Dutta, S. (2014). Security issues in cloud computing.

Hashizume, K., Rosado, D. G., Fernández-Medina, E., & Fernandez, E. B. (2013). An analysis of security issues for cloud computing. *Journal of internet services and applications*, 4(1), 5.

Malgey, S., & Chauhan, P. (2016). A review on security issues and their impact on cloud computing environment. *Int J Adv Res Comput Commun Eng*, 5, 249–253.

Malik, A., & Nazir, M. M. (2012). Security framework for cloud computing environment: A review. *Journal of Emerging Trends in Computing and Information Sciences*, 3(3), 390–394.

Jouini, M., & Rabai, L. B. A. (2018). Security Problems in Cloud Computing Environments: A Deep Analysis and a Secure Framework. In *Cyber Security and Threats: Concepts, Methodologies, Tools, and Applications* (pp. 926–952). IGI Global.

Wadhwa, A. V., & Gupta, S. (2015). Study of security issues in cloud computing. *International Journal of Computer Science and Mobile Computing IJCSMC*, 4(6), 230–234.

Recent Trends in Communication and Electronics – Sharma et al. (Eds)
© 2021 Taylor & Francis Group, LLC, ISBN 978-1-032-04572-6

Medical imaging and diagnosis using machine learning and deep learning

Ananya Rakhra, Raghav Gupta & Akansha Singh
Department of CSE, ASET, Amity University Uttar Pradesh, Noida

ABSTRACT: This paper reviews various approaches and algorithm for medical imaging and diagnosis with the help of modern technologies such as machine learning and deep learning. There are various models presented in this paper each different in its own way and can help the reader to get a brief idea about the various algorithms that are used or have previously been used for medical imaging and diagnosis. Along with the mentioned knowledge, these authors also suggest and put forth the broad scope offered by these technologies for further advancement for greater human well being. This paper begins with understanding and elaborating the technologies of machine learning and deep learning. Further their scope in the healthcare sector particularly for diagnosis and imaging if focused upon. There's a combination of different topics such as medical imaging, diagnosis, deep learning and machine learning. Over the past decade an exponential difference has been observed in machine learning. There are various algorithms as well as machine learning models which are used and explained in the following paper such as but not limited to CNN and MTANN. This paper encapsulates the latest research undertaken in the field and helps lay a foundation for the researchers willing to further work in this sphere.

1 INTRODUCTION

The broad field of artificial intelligence has exhibited growing potential in today's ever growing and complex world to tackle various challenges. These advancements brought about by machine learning and deep learning have evolved the machine-user interface. The machines are now smarter primarily owing to the predictive capabilities brought about using varied models based on previously available information and data. Machine learning can be understood as a machine's ability to make predictions, without any human intervention based on the data it has been trained on. Deep learning may be seen distinctly as having been trained much more extensively and critically and the machine need not necessary be told about the features to take under consideration, as with deep leaning the machine can figure that out on it's own. Further the applications and advantages of these technologies are explored through the sphere of imaging and diagnosis. This is indeed a critical sphere as the remedy to any ailment is based upon it's diagnosis hence it's accuracy is very important.

The meaning of machine learning can quite be interpreted from it's name itself, i.e. 'machine' and 'learning'. It is the process of training a machine to be able to make predictions without any human interventions. The fundamental element that facilitates the prediction process is the usage of various models, specially designed to analyse the available information. The various approaches that are used to train a machine are supervised learning, unsupervised learning and reinforcement learning. Supervised learning is when data regarding the recognisable features of any object is fed to the computer and the computer can then identify and label

DOI: 10.1201/9781003193838-71

the unknown on the basis of it's knowledge about these features. Unsupervised learning is when the computer is able to find relations and patterns among provided data and use these patterns to make any further predictions, the available data in this case is unlabeled data. Where as, reinforcement learning is based on the feedback mechanism. Here the machine learns from the feedback provided to it. For example, suppose a picture of a boat is given and the machine needs to identify it, if it is identified incorrectly, the user provides feedback regarding the inaccuracy of the output and the machine learns on the same basis.

As the data generated across the globe is increasing every second, this abundant tool holds high utilisation potential to help transform technologies for an enhanced and a more personalised experience. Since it is this data that helps a machine learn better, the scope of the technology of machine learning further widens in the times to come. Many existing applications of machine learning are observable in one's day to day life, from suggesting songs as per the users liking to fetching products based on the person's purchase and surf history, the machine's ability is owed to ML. This technology has seeped into the healthcare sector and has proven highly beneficial in ways that are covered through the course of this paper.

2 DEEP LEARNING

Deep learning is classified as a subcategory under machine learning itself. This technology is also referred to as artificial neural network as it functioning holds similarity to that of the human brain. The further advancement that deep learning offers is that, say if a machine is required to differentiate between two given object, a person will not have to specify the features distinctly as it will be able to recognise the features on it's own. However, this facility needs to be backed up with intensive data to train the machine. The large amount of data fed to the machine during training and learning process helps it to achieve accurate and precise prediction model formulation capability. However, examining clinical results where, minor differences may affirm to varying causes, is way more complex than simply identifying and distinguishing between different physical objects. These technologies hold great potential for improvement in diagnosis.

3 DEEP LEARNING FOR MEDICAL DIAGNOSIS

Medical imaging and diagnosis is a sphere that requires a high level of precision and accuracy as the treatment or medication administered is prescribed based upon the results from these diagnoses and any erroneous decision may lead to lethal consequences. Interpreting the test results has met new heights with the incorporation of machine learning and deep learning. The computers here particularly can be trained on the case data and reports available from numerous patient records that have been diagnosed with various syndromes. These cases can also be analysed for any patterns or trends in the development of the disease through out. The response to different medications and treatments can also be better studied to look for the most effective and efficient means to manage and cure a disorder. Hereafter the current applications of these technologies that have been formulated by various researchers over the years are as follows.

This research paper, documented by (Erickson 2017), talks about how machine learning and deep learning are applied into the field of medical imaging. It looks into the wide range of algorithms available for the task of diagnosing the disorder if any. The advantage that deep learning offers over machine learning is that is does not require a per-existing set of features to be able to spot a disease, it rather is capable of identifying the features itself. However this paper also talks about the pitfalls that these techniques may lead to maybe due to lack of accuracy of an algorithm or if a particular algorithm may not be suitable in a particular case or so. Therefore, overall it is seen how these technologies serve their required purpose in the medical imaging. The work presented by another scholar deals with the machine learning algorithms that particularly apply to offer advancement into neural science particularly brain imaging by (Lemm 2011). The amount of data that is procured from imaging a human brain is tremendous and these technologies serve as efficient means to observe patterns throughout it's entire functioning and to observe it during different stages, which otherwise is a tedious task. However there are certain challenges that sustain in particular to application in brain imaging particularly because each brain functions slightly differently than any other.

This paper provides a brief introduction to machine learning focusing on two different models namely MTANN i.e. massive training artificial neural network and CNN i.e. convolutional neural network (Suzuki 2017). A comparison between the two is also presented. Here one gets to know about the history and development in the field of ML and medical imaging. MTANN, it's working, algorithm and framework are discussed at length. Similar approach has been applied to CNN. Both these have been differentiated on the basis of architecture and performance. Focus is also laid upon the advantages or disadvantages of machine learning. One also gets an insight of progress in machine learning over the years. In this paper, analysis of different algorithms for breast cancer detection with the help of machine learning are undertaken by (Agarap 2018). It provides us a dataset which tells us about algorithms which can be used for accurate detection. In this study Google PensorFlow has been used for the algorithm, they have proposed a viable and a unique approach for medical imaging using ML.

This paper is based upon data analysis and includes the model based on GRU-SVM. This algorithm provides almost 96.09% accuracy. The whole code implementation is based upon the work of Daneil. The work is of extreme importance and a great approach towards breast cancer diagnosis. MDD i.e. unipolar major depressive disorder is the main focus of this research paper, more commonly known as depression (Mumtaz 2019). It follows a structured methodology which includes studying the participants experimental data accusation, EEG noise reduction and proposed deep learning scheme. In this paper, the author presents a deep learning architecture, the model is named 1DCNN-LFCM due to it's different layer of 1DCNN and LSCM. Readers are presented with the accuracy of the data of 1DCNN, along with classification accuracy of 93.5%. However, this design is incomplete and requires a GUI interface to make it more user friendly to find application in clinics. It has a very promising approach towards clinical depression.

Diabetic retinopathy leads to loss of vision if not diagnosed at the early phase. Most patients remain ignorant of this condition and do not see the need to take the existing tests such as fluorescein angiography and optical coherence tomography. Here (Gadekallu 2020) has presented a model that is superior to the existing models implementing deep learning for this diagnosis in terms of accuracy and sensitivity. What makes it different from the other approaches is that apart from examining the general patterns which are insufficient, they further apply the firefly algorithm to focus on the features that are more prominent and hold better significance to be able to predict one's risk of developing this condition and also helps to administer suitable cure thus helping prevent severe consequences.

These technologies have also found application in keeping a check over the heart's condition. The algorithm classified by (Potes 2016) gets it's knowledge of a total of 124 features, which facilitate to predict health risks, if any. The analysis is performed on the sound of the patients heart beat by applying the algorithms of AdaBoost and CNN for final results. This method owes it's accuracy to the large number of features that are utilized to perform operations on the dataset. This paper provides for a model to overcome the need of performing FNA (fine needle aspiration) to check if a thyroid nodule is benign or malignant, especially in cases where malignancy is least expected. Here imaging is performed using ultrasonic imaging technique. Ultrasound is the tool to diagnosing many medical conditions however reading the report needs to be performed accurately. Accommodation of machine and deep learning in this sphere can help bring about ease as proposed by (Song 2019). The machine uses Inception-V3 model and is trained upon 1358 images from reports of patients across varied age groups. The accuracy achieved by the results of this model is approximately 95%. Hence this technique reduces the need for unnecessary FNA, considering the risk and cost of the procedure.

Applications of machine learning have also been extended for diagnosing eating disorders such as Binge Eating disorder. Studies show that this results in more theta activity in the person's brains, and using EEG to figure out this parameter facilitates diagnosis. Considering the importance of one's mental health and the scenario where most of these cases remain undiagnosed (Raab 2020) has formulated a machine learning algorithm utilising random forest classification for the diagnosis of this mental condition. The researcher has been able to achieve an accuracy rate of 81.25%. This approach tends to diagnose this disorder and draw the required care and treatment to those suffering from it.

This research paper puts forth a method for diagnosing Covid-19 test results for suspected cases. It resorts to analyzing the chest x-ray report of the person using machine learning algorithms. The accuracy rate acclaimed by this diagnostic approach is 96-98% as achieved by (Elaziz 2020). The methodology here is to first examine the x-ray and extract features with the help of Fractional Multichannel Exponent Moments (FrMEMs), the computation process is enhanced by a multi-core framework. From the thus obtained dataset the technique of Manta-Ray Foraging Optimization based on differential evolution enables the identification of the key features to ensure efficient diagnosis. Especially in the scenario of such a pandemic, these improvements in diagnosis and imaging help to minimize or at least span the devastation being caused.

In this research paper, the authors propose a combination of residual thought and dilated convolution to develop a deep learning model for the diagnosis of pediatric pneumonia (Liang 2020). It involves implication of image classification to differentiate between normal and affected x-ray images and to accurately diagnose the test results. This algorithm implemented by the researchers has delivered an accuracy rate of 92-96% with is more than the existing ones. The proposed CNN of this model's comprises of 49 convolution layers and 2 dense layers for classification.

As we have read across the work presented by various researchers the concerns that comes into focus with regard to the usage of these technologies in the field of medical imaging and diagnosis is to be able to ensure that no faulty treatment is administered because of the usage of an inappropriate algorithm, and to ensure that all sensitive data related to a patient is kept safely and is not misused.

4   CONCLUSION

This paper encompasses knowledge attained from various papers from different authors and how their approach differs from each other, we have taken over ten different papers under the topic of medical imaging, diagnosis, deep learning and machine learning. One also observes the change in technology with change in time and how the technology became more and more advanced along the way. It can therefore be concluded that machine learning and deep learning have played a vital role in the imaging and diagnosis and it is only getting better.

REFERENCES

Agarap, Abien Fred M. "On breast cancer detection: an application of machine learning algorithms on the wisconsin diagnostic dataset." Proceedings of the 2nd International Conference on Machine Learning and Soft Computing. 2018

Elaziz, Mohamed Abd, Khalid M. Hosny, Ahmad Salah, Mohamed M. Darwish, Songfeng Lu, and Ahmed T. Sahlol. "New machine learning method for image-based diagnosis of COVID-19." *Plos one* 15, no. 6 (2020): e0235187.

Erickson Bradley J,et al. "Machine learning for medical imaging." *Radiographics* 37.2(2017):55–15.

Gadekallu, Thippa Reddy, et al. "Early detection of diabetic retinopathy using PCA-firefly based deep learning model." *Electronics* 9.2 (2020): 274.

Lemm, Steven, et al. "Introduction to machine learning for brain imaging." *Neuroimage* 56.2 (2011): 387–399.

Liang, Gaobo, and Lixin Zheng. "A transfer learning method with deep residual network for pediatric pneumonia diagnosis." *Computer methods and programs in biomedicine* 187 (2020): 104964.

Mumtaz, Wajid, and Abdul Qayyum. "A deep learning framework for automatic diagnosis of unipolar depression." International journal of medical informatics 132 (2019): 103983.

Potes, Cristhian, et al. "Ensemble of feature-based and deep learning-based classifiers for detection of abnormal heart sounds." *2016 Computing in Cardiology* Conference *(CinC.*IEEE, 2016.

Raab, Dominik, Hermann Baumgartl, and Ricardo Buettner. "Machine Learning Based Diagnosis of Binge Eating Disorder Using EEG Recordings." *PACIS*. 2020.

Song, Junho, et al. "Ultrasound image analysis using deep learning algorithm for the diagnosis of thyroid nodules." *Medicine* 98.15 (2019).

Suzuki, Kenji. "Overview of deep learning in medical imaging." Radiological physics and technology 10.3 (2017): 257–273.

Recent Trends in Communication and Electronics – Sharma et al. (Eds)
© 2021 Taylor & Francis Group, LLC, ISBN 978-1-032-04572-6

An efficient smart display menu for customizing the performance using AI cameras

K. Bramha Naidu, B. Seetharamulu, B. Deevena Raju & B.N. Kumar Reddy
Faculty of Science and Technology, IcfaiTech, ICFAI Foundation for Higher Education, Hyderabad, India

ABSTRACT: Now a day, Artificial intelligence is the most trending technology used in any platform. As the technology increases, the burden on the humans gets decreased rapidly. The tasks which need to be performed by the humans are being executed efficiently with the help of AI. Due to more dependence on the human load, the completion of the tasks leads to poor performance and less reliability. So, in order to improve the performance, reduce the execution time and delay in some of the crowded places like restaurant, this paper introduced a methodology named Smart Display Menu. This is used for visualizing the menu display clearly for all kinds of age groups. So, based on the AI enabled cameras, the age of the customer will be detected and based on the age group the Menu display change from static menu to dynamic menu or vice versa. The experimental outcomes of this research provide the ease in the visibility of menu for all age groups which provides improvement in the performance and helps customer making fast decisions there by reducing the time span during the shipment.

Keywords: face recognition, face detection, web API, voice assistance, database

1   INTRODUCTION

Artificial Intelligence is otherwise called as the Machine Intelligence, which defines in place of humans, machines executes the tasks. The intelligence that is exhaled either through software or any machines such type are defined as an Artificial Intelligence (Lu, et al. 2018). The tasks that are needed to be per- formed with an intelligent behavior created through computer or computer software comes under this scientific area. The AI (Artificial Intelligence) is improving exponentially from SIRI to recent Alpha Go (Ben Mabrouk, et al. 2018). Although science fiction frequently depicts AI as human-like robots, AI can include anything from predicting e-commerce algorithms to IBMs Watson machines. Today, indeed, artificial intelligence is commonly referred to as weak AI, which is programmed to perform a special task (e.g., facial recognition or just the internet searches or just driving a car) (Ariel Rosenfeld, et al. 2018). Although weak AI can surpass humans in a particular task, such absolving equations, overall AI will exceed humans in almost every intellectual task (Hoy, et al. 2018).The end users can ask wide variety of questions to their assistance to make their daily life easier. Some of the actions are automating the home appliances, by managing althea daily basic needs of a person etc., (Kepuska, et al. 2018). Figure 1 represents the Artificial intelligence overview and the tasks performed by it (Prentice, and Karakonstantis, 2018). Though it resembles as a substitute to the Human brain, it cannot be self-controlled or self-assisted. A large number of mergers and acquisitions, along with the influx of capital, are accelerating the integration of technology with applications, thereby increasing the already rapid transformation of the related economy (Pan, 2016).

In this research paper, Section 2 explains literature survey which contains various methodologies in the enhancement of future AI. Section 3 consists of the existing Display menu in

DOI: 10.1201/9781003193838-72

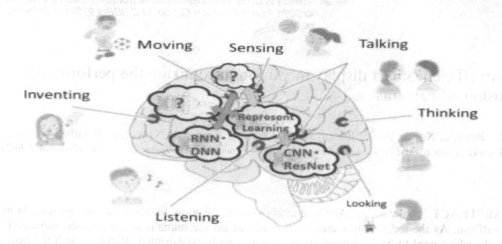

Figure 1. Overview of artificial intelligence.

restaurants, Section 4 explains our proposed Smart Display Menu algorithm, Section 5 contains the results and Section 6 provides the conclusion and future work.

2 RELATED WORK

The literature survey provides the analysis of various technologies in enhancing the Facial recognition in the recent years. Manna Raj et al., (Raj, and Seamans, 2019) has implemented the intervention of the future AI, robotics mainly for economic and the management literature that increases widely in research field. Marc Hanheide et.al.,(Hanheide, et al. 2017.) have introduced how to organize robots to complete the tasks in few situations, having incomplete in-formation and to handle the failure situations in an intelligent way and the results were evaluated on 5 various experiments. Raphael (Leong, et al. 2017). elaborated the overview of the issues regarding the privacy that is composed of few skills like Alexi, the echo framework and third-party hardware components. From the above statement we can justify that the third-party APIs are risky when compared with others.

3 EXISTING METHODOLOGY

These days the customization of the crowded areas or plat- form into a more robust way by making the life of the people simpler is going trend. Let us consider one of the places which contains a greater number of surrounded people i.e. restaurants. As Restaurants plays a vital role in most of the metro or non-metropolitan cities, customizing such places could have a major impact in making the customers life easier (B. Naresh Kumar Reddy et al. 2016.) - (BNK Reddy et al. 2017.). Many of the restaurants are providing Display menu which contains the list of items presenting that respective restaurant. In some places automated robots are used to help the customers in placing their order. Let us have a brief idea of the existing framework represented in Figure 2.From the above diagram, after the customer arrives the restaurant, they either places the order manually by looking at the Display menu or through automated robots. Restaurant is a place where all kind of age groups visit. There is a concern with the aged people that the font or the text available on the display menu or card is not appropriate. To overcome with this issue, proposed an automated Smart display menu in restaurants (NKR Beechu et al. 2017.)-(B. Naresh Kumar Reddy et al. 2019).

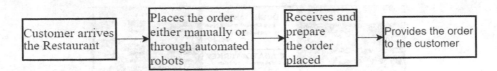

Figure 2.   Existing approach of assistance in restaurants.

4   SMART DISPLAY MENU

To do a brief introduction about this research paper, usually restaurant will have the discrimin-ate display with high graphic content and a bit of low point in size and also restaurant will see customers of different age groups. Considering invaders customer, it will be easy for them to read the menu and understand what they need to order. And it will be smoother experience for them. But the real problem comes for the oldest customers where it will be difficult for them to read and understand because they might have an eyesight problem or any other issues or because of the high graphic low font images getting changed. Considering all these problems, like placing the wrong orders based on assumptions could take away the time and efforts. To overcome this problem AI enabled cameras can be installed into the restaurant digital menu dis-play. Once a customer visits a restaurant and when they start looking at the menu, the Alienable camera will capture the customers face and the AI program will take this as an input and will detect the age of the customer and display the menu accordingly. Let us consider a block dia-gram for this architecture i.e. represented in Figure 3.Based on the age detected by the AI enabled cameras, if it was a young age the display menu will be same the graphical menu. If the age detected was an old age customer, the display menu will automatically convert into static menu. Through this, the customer can see the menu more clearly and can order the correct items without any difficulty. Thus, it provides us most experience for the customer.

5   RESULT ANALYSIS

Based on the Algorithm, the visual sensors sense the presence and capture the images using the Web API. This Web API provides all the information about the customer such as age, gender, emotions, personality factors etc., which is illustrated in Figure 3. This data is stored in the data base in form of tables. This research paper uses the open CV API that uses deep learning concept. This algorithm was developed using the Python, Open CV2 for video captur-ing and age detection. Based on the input that we get from the AI enabled cameras, where the AI enabled cameras give us the information regarding the customers age, gender, emotions and many other factors. This could be similar to the face recognition feature. Once the inputs

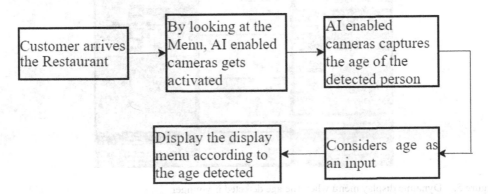

Figure 3.   Improvised approach of smart restaurant assistance.

Figure 4.   Static display menu when the age detected is older.

are received from the AI enabled cameras, the product display menu will change it graphics or text font based on the age of the customer. Let us consider a scenario of two age groups and check how the graphics content or text font gets changed accordingly.

Case 1: Let's say if the customer age that detected was about 55 years. The digital menu display will automatically convert the digital menu into a static menu. The term static menu states that, it is a plain text with bold font and only the price of the item will be visible on the menu. Through this, the customer can see the menu more clearly and can order the correct items without any difficulty. Thus, it provides us most experience for the customer. The above scenario is represented in Figure 4.

From the above figure, the middle area which is of a plain text containing the price of the item order is clearly visible to the persons of older age.

Case 2: Let's take another example of another customer of a different age group standing in front of the menu. Suppose, the customer age group detected by the AI camera states that the person is pretty in 25 to 32 years. Now the static menu gets converted into a graphical menu as illustrated in Figure 5. With this Solution, the acquired results show that awaiting time can be reduced and saved at ordering, billing and at payments which is useful for all kind of people by resolving all complex possible tasks. The two main metrics such as the performance and the cost metrics had a drastic improvement when compared with other automated techniques. This provides more user interaction providing high flexibility for the user to review his selections and interests and increases the productivity.

Figure 5.   Dynamic display menu when the age detected is younger.

6 CONCLUSION AND FUTURE WORK

In this paper we proposed an algorithm for converting the graphical menu to the static menu based on the age of the person. The age of the person is detected through thecae enabled cameras present at the top the display menu. This methodology will have drastic improvement in customizing the restaurant. This helps customer making faster decisions, reduces the time spent during the order or shipment which improvesreliability. In future, this research work could be extended to all the other platforms such asmedical, Grocery shops, cinemas etc. One single solution could be reused to all the platforms that user wishes to integrate.

REFERENCES

Ariel Rosenfeld, et al. 2018. Predicting HumanDecision-Making: From Prediction to Action, Morgan and Claypool.

B. Naresh Kumar Reddy et al. 2015. A Fine Grained Position for Modular Core on NoC IEEE International Conferenceon Computer, Communication and Control, PP. 1–4.

B. Naresh Kumar Reddy et al. 2015. Communication Energy Constrained Spare Core on NoC, 6th International Conference on Computing, Communication and Networking Technologies (ICCCNT), PP. 1–4.

B. Naresh Kumar Reddy et al. 2016. A Gracefully Degrading and Energy-Efficient Fault Tolerant NoC Using Spare core, IEEEComputer Society Annual Symposium on VLSI, pp. 146–151.

B. Naresh Kumar Reddy et al. 2019. Performanceconstrained multi-application network on chip core mapping, International Journalof Speech Technology.

B. Naresh Kumar Reddy, and B. Sireesha 2018. An Efficient Core Mapping Algorithm on Network on Chip, 22 nd International Symposium on VLSI Design andTest (VDAT).

Ben Mabrouk, A et al. 2018. Abnormal behaviour recognition for intelligentvideo surveillance systems: A review, Expert Systems with Applications, 91, 480–491.

BNK Reddy et al. 2017. An energy-efficient fault-aware core mapping in mesh-based network on chip systems, Journal of Network and Computer Applications, Vol 105, pp. 79–87.

BNK Reddy et al. 2018. Energy-Aware and Reliability-Aware Mapping for NoC-Based Architectures, Wireless Personal Communications 100 (2), 213–225.

Hanheide, et al. 2017. Robot taskplaning and explanation in open and uncertain worlds, Artificial Intelligence, 247, 119–150.

Hoy, M. B et al. 2018. An Introduction to Voice Assistants, Medical Reference Services Quarterly, 37(1), 81–88.

Kepuska, V., et al. 2018. Next-generation of virtual personal assistants (Microsoft Cortana, Apple Siri, Amazon Alexa and Google Home), 2018 IEEE 8thAnnual Computing and Communication Workshop and Conference, 99103, CCWC 2018.

Leong, R et al. 2017. Analyzing the Privacy Attack Landscape for Amazon Alexa Devices.

Lu, H., et al. 2018. Brain Intelligence: Go beyondArtificial Intelligence, Mobile Networks and Applications, 23(2), 368–375.

NKR Becchu et al. 2017. System level fault-tolerance core mapping and FPGA-based verification of NoC, Microelectronics Journal, Vol 70, pp. 16–26.

NKR Beechu et al. 2017. High-performance and energy-efficient fault-tolerance core mapping in NoC, Sustainable Computing: Informatics and Systems, Vol 16, pp. 1–10.

NKR Beechu, et al. 2018. Hardware implementation of fault tolerance NoC core mapping Telecommunication Systems 68 (4), 621–630.

Pan, Y, 2016 Heading toward Artificial Intelligence 2.0., Engineering, Elsevier, 2(4), 409–413.

Prentice, C., and Karakonstantis, G, 2018. Smart office system with face detection at the edge, IEEE SmartWorld, Ubiquitous Intelligence and Computing, Advanced and Trusted Computing, Scalable Computing and Communications, Cloudand Big Data Computing, Inter- net of People and Smart City Innovations, Smart-World.

Raj, M., and Seamans, R, 2019. Primer on artificial intelligence and robotics, Journalof Organization Design, 8(1).

Recent Trends in Communication and Electronics – Sharma et al. (Eds)
© 2021 Taylor & Francis Group, LLC, ISBN 978-1-032-04572-6

Analysis of various copy-move forgery detection techniques

Prince Sapra
Delhi Technological University, Delhi, India

Vaishali Kikan
KIET Group of Institutions, Delhi NCR, India

ABSTRACT: In copy-move forgery, a certain area of an image is cut or copied to hide the unrequired parts of an image and then this part is pasted on some other part. Copy-move forgery is among the most popular image tampering methods used these days. The use of this image tempering method is quite common due to its simplicity and the generation of high-quality outcomes. Nowadays, this method is used to serve different purposes with the technological advancement. The various schemes are reviewed in this work for the copy-move forgery detection which are based on the block and pixel matching techniques.

Keywords: forgery detection, block matching, pixel matching

1 INTRODUCTION

At present, humans are capable of accessing remarkable multimedia from the internet and remaking or tampering with it due to the development of technology and convenience of the internet. CMF imaging is a special kind of forgery in which replicated parts of an image are comprised. Afterward, these replicated parts are pasted into similar image. In a Copy-Move forgery, the copying and pasting of a part of the image is done itself into another part of the same image. This is often carried out with intend of making an object disappear from the image for which it is covered with a segment copied from another part of the image. The noise component, color palette and most other important properties become compatible with the rest of the image as the duplicated parts come from the similar image. Therefore, these properties are not detected with the utilization of techniques which seek inaptness in statistical measures within diverse parts of the image. The feathered crop or the retouch tool is useful for masking any traces of the copied-and-moved segments further so that the forgery can be made even complicated for the detection.

A correlation amid the original image segment and the pasted segment is established with the help of any CMF. The deployment of this correlation is acted as a base while detecting this kind of forgery efficiently. The segments is not matched in accurate as well as approximate manner with the possible implementation of the retouch tool or other localized image processing tools and also because of the saving of forgery in the lossy JPEG format. There are four phases comprised in the image processing-based copy-move forgery detection pipeline. The initial phase is of pre-processing that carries out conversion, transformation or decomposition schemes. This phase focuses on arranging and exhibiting the data in such a manner so that the subsequent phase of extracting feature will become more effective. The grayscale conversion is included in the easiest and the significant techniques of pre-processing that assist in transforming the Red, green and blue pixels into a greyscale image. This phase has different

DOI: 10.1201/9781003193838-73

methods of decomposition such as wavelet decomposition or PCA. Subsequently, the feature extraction scheme is carried out. The major purpose of this phase is that a set of short but significant data vectors can be generated in order to represent each part of intended digital image. Numerous schemes are available to extract feature vectors for the digital image including Scale-Invariant Feature transform, Speeded-Up Robust features etc. Once the extraction of feature descriptors is completed by the previous stage, feature matching processes are applied. These processes search for matched patches or parts of the target image carrying same feature descriptors. The stage of matching process contributes significantly to decide the general detection speed of the CMFD framework. Eventually, the post-processing stage filters or processes raw matched detection for enhancing and generating the concluding detection results with maximum quality. There are mainly two categories of CMFD techniques, known as block-based and keypoints-based copy move forgery detection techniques. In Block-based techniques, an image is segmented or cut into overlapping blocks to extract features from those blocks. Next, the similarity between block features is measured to decide the forgery areas. Discrete cosine transform (DCT) is a milestone in the arena of block-based CMFD techniques. Firstly, the image is segmented into overlapping image patches of finite magnitude with a raster scan and then DCT is enforced to every image block. In order to obtain a quantization feature vector, zigzag scanning is carried out on the quantized DCT coefficient matrix. The matrix that composes of feature vectors is available in lexicographic order and similar features are searched using Euclidean distance. Polar-based forgery detection is the most well-known block-based approach. This technique subdivides an image into pixels in overlapping blocks. At first, the transformation of pixels within the block is carried out into log-polar map (LPM) and then one-dimensional descriptor is generated by summing this LPM along the angle axis. Afterward, the Fourier coefficient magnitude is computed followed by the Fourier transformation. The descriptors remain constant to the reflection and rotation. The information entropy is computed as block features using the descriptor of every block. The entropy difference between blocks is computed to detect the similar parts. The key point-based CFMD techniques extract and match image features with the overall image for detecting the forged regions. Scale-invariant feature transform (SIFT) features and speedup robust features (SURF) are the leading and most commonly used feature points. The use of these feature points is quite popular for image retrieval and object recognition as they are highly robust against geometrical transformations such as scaling and rotation. Speeded-Up Robust Features (SURF) provides a much faster and computationally effectual alternate to SIFT. SURF key point detection method searches points of interest in the target digital image in a very effective and speedy way. GLCM (Gray level co-occurrence matrix) is a very popular texture-based feature extraction algorithm. In fact, GLCM used for texture descriptors has represented a statistical matrix of joint conditional probability distribution amongst the Gray levels of image pixel pairs with some distance and direction within a particular space. It is essential for geo-texture observation slot based on GLCM to reflect the original spatial features of the related group. The Grey-Level Co-occurrence Matrix or GLCM as a remarkable statistical approach is used so often to express spatial relations between pixel and its neighboring pixel. The GLCM is a standard and competent approach of texture feature extraction used in image recognition, image segmentation, image retrieval, image classification and texture analysis methods. Homogeneity, energy, correlation and contrast are some generally used statistical values obtained from the co-occurrence matrix.

$$G_{\Delta x, \Delta y}(i,j) = \sum_{x=1}^{n} \sum_{y=1}^{m} \begin{cases} 1, & if I(x,y) = i \text{ and} \\ & I(x+\Delta x, y+\Delta y) = j \\ 0, & otherwise \end{cases} \tag{1}$$

Equation (1) illustrates the mathematical expression of GLCM, where i and j represent the indexes of elements in the output matrix. In addition to this, x and y are referred to represent pixel location of the target digital image. Also, the spatial relationship parameters (so-called "offset") are represented by Δx and Δy. Completely different results may be generated if GLCM of the similar target image are performed with different offset parameters such as G_(1,-1) and G_(1,1). The final result called co-occurrence matrix derived from GLCM, comprises ex- pressive statistical and spatial relationship information that indicates the individuality of the real image. This means that the robust feature vectors may be created using co-occurrence matrix for CMFD (Copy Move Forgery Detection).

2 LITERATURE REVIEW

Gul Muzaffer, et.al (2020) investigated a novel method to detect the copy-move forgery. Quadtree Decomposition had implemented in this method for segmenting the input image into two sub segmented images. The sub segmented images were labeled as smooth image and textured image. The texture form of the smooth labelled segment image was attained on the basis of LBPROT approach. Afterward, the extraction of keypoints was carried out from both textured segment image and texture form of the smooth segment image. These keypoints were matched for attaining forgery clues. The matching determined the image as a forged or normal. The outcomes of experiment validated that the presented technique had robustness for dealing with scaling and rotation attacks.

Ankit Kumar Jaiswal, et.al (2020) suggested a hybrid technique in which DCT and BRISK attributes were included to detect the forgery in copy-move. The keypoints and descriptors of both the techniques were executed to extract the features. The matching was carried out applying Binary Robust Invariant Scalable Keypoints and FLANN matcher. The false matches were eliminated with the utilization of clustering method based on the Euclidean distance. The outcomes of experiment demonstrated that this technique was capable to resist against blurring as well as transformations such as rotation and scaling. Moreover, the suggested technique had potential for detecting multiple forged regions.

Ye Zhu, et.al (2020) intended an E2E NN on the basis of AR-Net. The adaptive attention system fused attributes of position and channel attention for gathering information related to context and enhancing the representation of attributes initially. Afterward, the self-correlation was evaluated among feature maps using deep matching and the scaled correlation maps were fused through atorus spatial pyramid pooling for creating the coarse mask. At last, the residual refinement module in which the object structure boundaries were preserved, had executed for the optimization of coarse mask. CASIAII, COVERAGE, and CoMoFoD were different datasets employed to conduct the experiments. The outcomes revealed that the intended Adaptive Attention and Residual Refinement network had performed better as compared to existing algorithms and it was capable of locating forged and corresponding original regions at the pixel level.

Khushkaran Kaur, et.al (2020) described that a passive technique had implemented for detecting the forgery of copy-move. First of all, the division of image was done into overlapping blocks. After that, KPCA was carried out for extracting the attributes. These attributes were sorted in lexicographic order. The similar block was investigated within an image to find out the replicated image blocks. It was observed in the results of experiment that the designed strategy was adaptable for discovering the duplicated regions even under the attack of translation that crooked the images.

Gul Muzaffer, et.al (2019) discussed that CMF was an easy approach of modifying an image which focused on replicating or eliminating the objects in the image. A novel forgery detection technique based on deep learning was introduced. The feature vectors of image having overlapped sub-blocks were extracted applying AlexNet. The similarity among feature vectors was found to detect and localize the forgery subsequent to attain the features. On the basis of outcomes of testing, it was evaluated that a superior accuracy rate was provided by the introduced technique in comparison with the conventional technique.

Yanfen Gan, et.al (2019) recommended an approach in which PST was integrated with LSH for detecting the copy-move forgery of image. Firstly, the partition of detected image was performed in various overlapping blocks. Later on, feature was extracted from every block deploying Polar Sine Transform. These block features were determined with the utilization of Locality Sensitive Hashing. The features which seemed similar had discovered in similar class as candidate block feature pairs. At last, Euclidean distance was exploited to establish the post-processing operation so that the weak block feature pairs were sorted out. The outcomes of various experiments exhibited that the recommended technique was robust against rotation and JPEG compression and performed much better than the existing techniques.

Gül Muzaffer, et.al (2018) presented a block-based method for detecting the forgery of copy-move. A novel and more effective technique known as LIOP together with Patch Match algorithm was implemented in order to extract the attributes from blocks. Consequently, copy-move forgeries were detected quickly. Additionally, the comparison of presented algorithms was carried out with state-of-art works and their resistance against attacks was also tested. The outcomes obtained from experiments indicated that the presented algorithm had performed efficiently while detecting the forgery of copy-move under various attacks such as noise, JPEGe.

Author	Year	Description	Outcome
Gul Muzaffer	2020	Quadtree Decomposition had implemented in this method for dividing the input image into two sub-segmented images. The sub-segmented images were labeled as smooth and textured.	The outcomes of experiment validated that the presented technique had robustness for dealing with scaling and rotation attacks.
Ankit Kumar Jaiswal	2020	Hybrid technique in which DCT and BRISK attributes were included to detect the forgery in copy-move	The outcomes of experiment demonstrated that this technique was capable to resist against blurring as well as transformations such as rotation and scaling.
Ye Zhu, et.al	2020	The adaptive attention system fused attributes of position and channel attention for gathering information related to context and enhancing the representation of attributes initially.	The outcomes revealed that intended Adaptive Attention and Residual Refinement network had performed better as compared to existing algorithms and it was capable of locating forged and corresponding original regions at the pixel level.
Khushkaran Kaur	2020	The division of image was done in-to overlapping blocks. After that, KPCA was carried out for extracting the attributes.	It was observed in the results of experiment that the designed strategy was adaptable for discovering the duplicated regions even under the condition of translation that crooked the images.
Gul Muzaffer, et.al	2019	A novel forgery detection technique based on deep learning was introduced. The feature vectors of image having overlapped sub-blocks were extracted applying AlexNet.	On the basis of outcomes of testing, it was evaluated that a superior accuracy rate was provided by the introduced technique in comparison with the conventional technique.
Yanfen Gan, et.al	2019	These block features were determined with the utilization of Locality Sensitive Hashing. The similar features were discovered in similar class as candidate block feature pairs.	The outcomes of various experiments exhibited that the recommended technique was robust against rotation and JPEG compression and performed much better than the existing techniques
GülMuzaffer, et.al	2018	A novel and more effective technique known as LIOP together with Patch-Match algorithm was implemented in order to extract the attributes from blocks.	The outcomes obtained from experiments indicated that the presented algorithm had performed efficiently while detecting the forgery of copy move under various attacks such as noise, JPEG etc.

3 CONCLUSION

In this work, it is concluded that copy-move forgery is the major issue of image processing techniques. The various schemes are already proposed for the detection of copy-move forgery detection. The pixel matching schemes and block matching are the popular schemes for the copy-move forgery detection. The GLCM algorithm can be used in future to improve accuracy for the copy-move forgery detection.

REFERENCES

AlSawadi M. et al. 2013, "Copy-Move Image Forgery Detection Using Local Binary Pattern and Neighborhood Clustering", Modelling Symposium (EMS), vol. 5, issue 13, pp. 249–254.

Amerini Irene et al. 2011, "A SIFT-based forensic method for copymove attack detection and transformation recovery", Transactions on Information Forensics and Security vol. 6, issue 3, pp. 109–117.

Babu S B G Tilak & Ch Srinivasa Rao (2020), "Statistical Features based Optimized Technique for Copy Move Forgery Detectiom", 11th International Conference on Computing, Communication and Networking Technologies (ICCCNT), vol. 37, issue 81, pp. 790–798.

Bansal Ankita & Atri Aditya (2020), "Study of Copy-Move Forgery Detection Techniques in Images, 8th International Conference on Reliability, Infocom Technologies and Optimization (Trends and Future Directions) (ICRITO), vol. 6,issue 28, pp. 438–446.

Dua Shilpa et al. 2020 "Passive Copy-Move Forgery Detection using Localized Intensity Feature", 11th International Conference on Computing, Communication and Networking Technologies (ICCCNT), vol. 5, issue 23, pp. 561–569.

Fridrich J. et al. 2003, "Detection of copy move forgery in digital images," Proceedings of the Digital Forensic Research Workshop, vol. 17, issue 3, pp. 58–66.

Gan Yanfen & Yang Jixiang (2019), "An Effective Scheme for Copy-move Forgery Detection Using Polar Sine Transform", 2nd International Conference on Safety Produce Informatization (IICS- PI), vol. 40, issue 16, pp. 733–741.

Ghorbani Mehdi et al. 2011 "DWT-DCT (QCD) based copy-move image forgery detection,", 18th International Conference on Systems, Signals and Image Processing (IWSSIP), vol. 12, issue 4, pp. 141–149.

Hajialilu Somayeh Fatan et al. 2020, "Image copy-move forgery detection using sparse recovery and keypoint matching", IET Image Processing, vol. 4, issue 28, pp. 262–270.

Jaiswal Ankit Kumar et al. 2020 "Detection of Copy-Move Forgery Using Hybrid approach of DCT and BRISK", 7th International Conference on Signal Processing and Integrated Networks (SPIN), vol. 10, issue 37, pp. 699–707

Kaur Khushkaran (2018), "Efficient and Fast Copy Move Image Forgery Detection Technique", Second International Conference on Intelligent Computing and Control Systems (ICICCS), vol. 26, issue 38, pp. 425–433

Koshy Litty & Shyry S. Prayla (2020), "Copy-Move Forgery Detection and Performance Analysis of Feature Detectors", International Conference on Communication and Signal Processing (IC-CSP), vol. 56, issue 3, pp. 675–683.

Muzaffer Gul & Ulutas Guzin (2019),"A new deep learning-based method to detection of copymove forgery in digital images", Scientific Meeting on Electrical-Electronics & Biomedical Engineering and Computer Science (EBBT), vol. 7, issue 5, pp. 1359–1367.

Muzaffer Gül et al. 2018, "A copy-move forgery detection approach based on local intensity order pattern and patchmatch", 2018, 26th Signal Processing and Communications Applications Conference (SIU), vol. 33, issue 8, pp. 837–845.

Muzaffer Gul et al. 2020 "Copy Move Forgery Detection with Quadtree Decomposition Segmentation", 43rd International Conference on Telecommunications and Signal Processing (TSP), vol. 52, issue 111, pp. 653–661.

Popescu A. C. & Farid H. (2004), "Exposing digital forgeries by detecting duplicated image re-gions," Dept. Comput. Sci., Dartmouth College, Tech. Rep. TR2004-515, vol. 5, issue 2, pp. 33–40.

Sharma Parul & Kaur Harpreet (2019), "Copy-Move Forgery Detection with GLCM and Euclidian Distance Technique in Image Processing", International Journal of Recent Technology and Engineering (IJRTE), vol. 8, issue-1C2, pp. 43–47.

Sunitha K & Krishna A.N. (2020), "Efficient Keypoint based Copy Move Forgery Detection Method using Hybrid Feature Extraction", 2nd International Conference on Innovative Mechanisms for Industry Applications (ICIMIA), vol. 50, issue 72, pp. 989–997.

Yao H. et al. 2011, "Detecting Copy Move Forgery Using Non-Negative Matrix Factorization," Third International Conference on Multimedia Information Networking and Security, vol. 8, issue 18, pp. 591–594.

Zhu Ye et al. 2020 "AR-Net: Adaptive Attention and Residual Refinement Network for Copy-Move Forgery Detection", IEEE Trans- actions on Industrial Informatics, vol. 61, issue 12, pp. 308–316.

Recent Trends in Communication and Electronics – Sharma et al. (Eds)
© 2021 Taylor & Francis Group, LLC, ISBN 978-1-032-04572-6

Wild animal species detection using deep convolution neural network

Arshita Verma, Vishu Sangwan & Neha Shukla
KIET Group of Institutions, Delhi-NCR, Ghaziabad

ABSTRACT: Monitoring animals in forest one to one is very tedious job so technology have evolved which include installing cameras in animal living areas and acquire videos and pictures. But the pictures acquired from installed camera are always not good most of the time pictures are blurred and without animals, so it's hard to detect animals, resulting in doubt and error. For solving this obstacle, we proposed a database of installed camera network captured images with multilevel graph of different animal species stored in database in spatiotemporal domain and pictures with animal are captured from forest are compared with the database, identifying whether a graph matches the animal or anything else. We created a model for detecting animals based on oneself training by the method explained in the paper. The technology is used for classification using machine learning (ML) and Artificial intelligent (AI) algorithms, hold up vector machine, k-mean closest neighbor, together with group tree. From the experimental results it is shown that, the proposed system accurately classifies wild animals up to 91% accuracy.

Keywords: machine learning, animal detection, deep learning

1 INTRODUCTION

The animal detection in forest is one of the major challenge for saving wild life animals in this 21st century. Bulk amount of specifics is available now for wildlife animals for any country according to their behaviour, feature and domain. Installed camera techniques including technologies are used in wildlife animal monitoring and image analysis because it is easy to use and low in cost. As large quantity of data is available for wild life animal it become handier to analysis and compare behaviour and growth of specie in different area of the earth and can be study influence with respect to personage interference at wild life creatures as well as variety in various period, region and category. To study and capture wildlife animal's sensor based cameras are installed on trees to create Installed camera grid. The camera starts tracking & taking pictures; every time movement of animal is being detected. [1] Therefore, small recording concerning species be noted along with specifics regarding environment. The installed camera networks are vital for study of wild life animal information with no disturbance. Adept critical requires planning different devices to mechanize preparing of such enormous Installed webcam pictures, for example, creature recognizable proof, division, uprooting, also [2]. Further, we project a technique that recognize natural life creature utilizing convolutional neural networks (CNN) dependent on Installed camera pictures [3].

Figure1 shows the Animal detection image feature extraction for machine learning. Article recognition is broken apart in categories: first of all; find out the areas in image by projected technique which will have all required objects and second; by classification find out whether region have required objects [4]. object location as far as creature identification manages issue of precision and speed because of the profoundly powerful and exceptionally jumbled picture successions

DOI: 10.1201/9781003193838-74

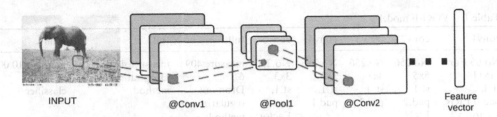

| INPUT | @Conv1 | @Pool1 | @Conv2 | Feature vector |

Figure 1. Animal detection image feature extraction.

acquired through webcam trick [5]. The usable space created outlook creates large quantity of expectant space. Thus, it is essential to examine the error free attributes of Installed camera picture sequence in spatiotemporal area to show a productive locale proposition approach that makes few applicant districts [6]. Thus, we utilized the Installed camera picture groupings that are investigated utilizing Iterative Embedded Graph Cut (IEGC) method of making little gathering through Wild Animal Detection Using Deep Convolutional Neural Network applicant creature locales. Additionally, for better applicant creature locale order, DCNN highlights are utilized with various classifiers, for example, Support Vector Machine (SVM) as well as it's variations and Ensemble classifiers. H.Sharma et al. has developed solar energy harvesting solutions for long life sensors nodes which can be deployed for animal detection in forests [7–9].

2 LITERATURE SURVEY

We see such dependable as well as productive article division including identification need evacuation in regard to suspicion on fixed foundations. With division of powerful pictures, adept various points to acknowledge, i.e., spatial setting, diagram cut at region level, assessment of scenes at pixel level, and closer view foundation distinguishing proof at consecutive level, and so forth. M. M. Ch et al. [10] utilized Installed camera pictures that are gained from exceptionally powerful pictures utilizing a dependable and proficient article cut technique.

Latest work in object identification and detection show the model work of DCNN [11]. To get minimize the in depth study of whole picture animals, with surrounding between is examined coming about quick preparing of DCNN-based article location. A few investigations were done on item identification utilizing item proposition outlook. P.L. StCharles et al. [12] utilized Deep Neural Network (DNN) to foresee items utilizing bouncing box objects district by weaken models.

3 ALGORITHM USED

Solid and strong natural life recognition from profoundly powerful and jumbled picture successions of Installed camera network is a difficult assignment. Subsequently to increase high performance, pictures should be examined at pixel or little locale level. Nonetheless, because of low differentiation and jumbled pictures, it gets hard to distinguish whether a specific district or pixel dependent on nearby data speaks to creature or foundation. Consequently, we have to break down worldwide picture includes moreover. For instance, the district of a creature body might be considered foundation locale. In such case, nearby data supportive of cessing won't be adequate, prompting prerequisite of worldwide preparing (to extricate worldwide picture highlights) to distinguish creature. For instance, perceive whether a creature is available or not, one ought to likewise distinguish body parts like head, legs and so forth as opposed to body as it were.

Table 1. VGG-F model.

conv1	conv2	conv3	conv4	conv5	full6	full7	full8
No.65 For 11x11 st.4, pad 0 LRN, Factor: x2 pool	No.256 5x5 st.1, pad.2 LRN, Factor: x2 pool	No.256 3x3 st. 1 pad 1	No.256 3x3 st.1, pad 1	No.256 3x3 st.1, pad 1 Factor: x2 pool	measure:409 6 Drop-out regularization method	measure:40 96 Dropout regularize method	measure:10 00 Soft-max classifier

3.1 Creature background verification model

For dealing with aforementioned difficulties, we are presenting accompanying plan to creature foundation check design. Our plan possesses three stages: (1) pre-processing, (2) adjusted DCNN highlights, along with (3) characterization from studying calculations.

3.2 Animal detection model

It has demonstrated in existing writings that Deep Convolution Neural Network is a very effective descriptors for the acknowledgment of objects, arrangement and recovery, and so forth. There are different convolutional layers and at any rate one completely associated layer in a DCNN. DCNN has a pooling layer for the interpretation invariant highlights. In our work, we seek after the engineering depicted by Krizhevsky et al. utilizing the VGG-F pre-trained model. The pre-trained model has been scholarly on gigantic assistant ILSVRC 2012 dataset. This pre-trained model has a picture size for contribution of 224 × 224; subsequently, we resize the pictures to 224 × 224, without thinking about its genuine size and proportion.

From the above Table 1, we can see that there are five convolutional layers and three completely associated layers. The five convolutional layers are from conv1-5 and the completely associated layers are full 6–8.

4 EXPERIMENTAL RESULTS

4.1 Datasets

For the experimentation and evaluation, the execution of framework we have utilised the standard installed camera dataset. The dataset contains almost 100 pictures succession for every specie. Whereas there are almost 20 different types of creature from different species. The accessible pictures are available in both daytime arrangement and evening design, bringing about untamed life checking framework for both daytime and evening time. Installed camera networks give complex pictures profoundly jumbled characteristic recordings and furthermore high-goal pictures. The pictures got change in goal from 1920 × 1080 to 2048 × 1536. The quantity of pictures in each grouping differs from 10 to 300 that is very small. The quantity of pictures in a picture arrangement relies upon the time of activity by creature.

4.2 Result and analysis

We used different evaluation standards as
　　True Positive Rate i.e. TPR,
　　Positive Predictive Values i.e. PPV,
　　F1-measure
　　Discovery Rate i.e. FDR
　　Area Under Curve i.e. AUC to study the result of system.

True Positive Ratio (TPR) is given as:

$$TPR = T_P/T_P + T_N \tag{1}$$

Where, T_P= True Positive, F_N = False Negative,

Positive Predictive Values (PPV) is the ratio of number of true positives i.e. number of positive calls as given in eq.2.

$$PPV = T_P/T_P + F_P \tag{2}$$

Where, F_P = False Positive
Discovery rate (FDR) is given as:

$$PPV = F_P/T_P + F_P \tag{3}$$

The overall performance of the classifier is measured as Area Under Curve or AUC. Higher the value of AUC better the classifier.

$$F1 \text{ Measure} = 2X \text{ (Precision*Recall)}/\text{(Precision} \pm \text{Recall)} \tag{4}$$

With SVMs and KNNs and Troupe Classifiers we have achieved the normal exactness of 90.7 percent. We have used various variations of SVM such as quadratic, direct and cubic for the investigation and also medium Gaussian SVM. The most elevated accuscandalous accomplished utilizing nonlinear SVMs, i.e., quadratic and cubic is 91.1% and 91.2% individually. Notwithstanding, weighted KNN and outfit supported tree additionally perform well like SVM with 91.4% and 91.2% precision, separately. We saw that favourable to presented framework acquired huge outcomes with exceptionally jumbled pictures. The general exactness of the framework is acquired in the scope of 89– 91.4% and most noteworthy precision is 91.4% accomplished with weighted KNN classifier. The outcome shows that DCNN highlights furnish great outcome with various AI calculations.

5   CONCLUSION

In this paper, we have proposed an effective, dependable and vigorous technique for wild creature recognition in exceptionally jumbled pictures utilizing DCNN. The jumbled pictures are acquired utilizing Installed camera organizations. The pictures in Installed camera picture groupings likewise give the competitor creature area proposition done by staggered chart cut. We have present a check step in which the proposed area is ordered into creature or foundation classes, Thus, deciding if the proposed district is really creature or not. DCNN for Wild Animal Detection 337 applied DCNN highlights to AI calculation to the better accomplishment and execution. The proposed framework is very effective according to the results from the trial and hearty wild creature location framework for 24hr of a day.

REFERENCES

[1]   H. Sharma, A. Haque, Z. A. Jaffery, "Maximization of Wireless Sensor Networks Lifetime using Solar Energy Harvesting for Smart Agriculture Monitoring", Adhoc Networks Journal, Vol. 94, Elsevier, Netherlands, Europe, November 2019.
[2]   H. Sharma, A. Haque, Z. A. Jaffery, "Solar energy harvesting wireless sensor network nodes: A survey", Journal of Renewable and Sustainable Energy, American Institute of Physics (AIP), USA, vol. 10, no. 2, pp 1–33, March 2018.

[3] H. Sharma, A. Haque, Z. A. Jaffery, "Modelling and Optimization of a Solar Energy Harvesting System for Wireless Sensor Network Nodes", Journal of Sensor and Actuator Networks, MDPI, USA, vol. 7, no. 3, pp.1–19, September 2018.

[4] J. R. Uijlings, K. E. van de Sande, T. Gevers, and A. W. Smeulders, *Selective search for object recognition, in Int. J. Computer Vision, vol. 104, no. 2, pp. 154171, 2013.*

[5] R. Kays et al., *eMammal Citizen science camera trapping as a solution for broad-scale, long- term monitoring of wildlife populations,* in Proc. *North Am. Conservation Biol.,* 2014, pp. 8086.

[6] K. H. Shaoqing Ren and J. S. Ross Girshick, *Faster R-CNN: Towards real-time object detection with region proposal networks, in Adv. Neural Inf. Process. Syst.,* 2015, pp. 9199.

[7] M. M. Cheng, Z. Zhang, W. Y. Lin, and P. Torr, *BING: Binarized normed gradients for objectness estimation at 300fps,* in Proc. *IEEE Conf. Computer Vision. Pattern Recognition. Jun.* 2014, *pp. 32863293V. Mahadevan and N. Vasconcelos, Background subtraction in highly dynamic scenes,* in Proc. *IEEE Conf. Comput. Vis. Pattern Recog., Jun. 2008, pp. 16.*

[8] M. Oquab, L. Bottou, I. Laptev, and J. Sivic, *Learning and transferring mid-level image representations using convolutional neural networks,* in Proc. *IEEE Conf. Computer Vision Pattern Recognition., Jun. 2014, pp. 17171724.*

[9] P.L. St-Charles, G.-A. Bilodeau, and R. Bergevin, *Flexible background subtraction with self- balanced local sensitivity,* in Proc. *IEEE Conf. Computer Vis. Pattern Recognition Workshops, Jun.* 2014, *pp. 414419.*

[10] R. Girshick, J. Donahue, T. Darrell, and J. Malik, *Rich feature hierarchies for accurate object detection and semantic segmentation,* in Proc. *IEEE Conf. Comput Vis. Pattern Recog., Jun. 2014, pp. 580587.*

[11] R. Girshick, *Fast r-CNN,* in Proc. *Int. Conf. Comput. Vis., pp. 14401448, 2015.*

Brain tumor segmentation using CNN

Alok Kumar Tiwari & Neha Shukla
KIET Group of Institutions

ABSTRACT: Brain tumor classification and segmentation with the help of brain Magnetic Resonance Imaging (MRI) is an interesting area. The target is to focus on developing such an automatic detection system that it would aid the radiologists and clinicians. For this detection of brain tumor at the earliest can help in better diagnosis. Brain tumor have different shapes and also differs from patient to patient, MRI scan is a non-invasive technique and could be a contribution in this regard. Less contrasted details is one such drawback. In order to improve this, we present a FastAI based Transfer Learning tumor classification in which pre-trained model with features of segmented tumor area classifies tumor based on its learning. The proposed model first predicts and extracts the tumor from MRI scans. Further classification based on tumor existence. The proposed brain tumor FastAI deep learning is evaluated on BRATS2015 dataset.

Keywords: CNN, Deep Learning, MRI, Segmentation

1 INTRODUCTION

A. *Image Segmentation*

The partitioning of images into different subsections is known as image segmentation. It is considered as one of the toughest task in Digital image processing and Computer vision because generally the images used for study do not have high contrast rate or some time do not proper boundary. Some of the area of image are overexposed while some are under exposed hence preprocessing techniques are used to first prepare the image for study.

1) *Threshold-based methods*

The Various studies conducted have shown that the simplest property shared by the voxels of an image is its intensity. So to perform segmentation a natural way is to set a threshold. This process of setting threshold separates the given object based on the intensity of the voxel. When the object intensity are clearly distinguishable single level threshold technique is used and when the intensity of the object lies between intensities of other objects then multilevel threshold technique is used.

2) *Region-based methods*

These methods search for homogeneous properties which are based on predefined similarity criteria of connected regions of voxel. The two most common algorithm used in this context are the watershed are region growing. The region growing algorithm is a semi-automated algorithm as it works by selecting at least one seed point manually, especially in brain tumor segmentation task. The disadvantage of this method is it's inaccuracy in partial volume effect, which leads to blur border in images. The watershed algorithm works on ridge which divides areas drained by different river systems. The problem with this technique is over segmentation.

B. *Convolutional Neural Network*

Convolutional Neural Network (CNN or ConvNet) is a subtype of deep neural Networks. It is majorly used to visualize the images and because of the shared weights architecture, they

DOI: 10.1201/9781003193838-75

Recent Trends in Communication and Electronics – Sharma et al. (eds)
© 2021 Taylor & Francis Group, LLC, ISBN 978-1-032-04573-3

are also called as Space invariant artificial Neural Networks (SIANN) or shift invariant. They are versions of the multilayer perceptron which means fully connected networks so each neuron in one layer will be connected to the all the neurons in other layers. These are more prone to the overfitting data problem due to being fully connected. CNN use relatively little pre-processing compared to other classification algorithms. CNN take in an input image, assign weights and biases to various objects in the image and make it able to differentiate one image from the other image. The architecture of the ConvNet is similar to the connectivity pattern of Neurons in the Human Brain and was inspired by the organization of the visual cortex. It is successful in capturing the Spatial and Temporal dependencies in an image through various application filters. There is an added advantage to this algorithm the network can be trained to understand the image better. The objective of this algorithm is to extract high level features from the images and this is done only when we have multiple layers.

2 LITERATURE REVIEW

In paper [9] Enhanced convolutional Neural Networks (ECNN) is proposed with the loss function optimization by BAT algorithm for automatic segmentation method. The optimization of the MRIs image segmentation and the results were better as compared to the existing algorithms. There were some categories which were used to compare i.e. precision, recall, accuracy. In order was used by BAT algorithm to perform segmentation of a MRI image in an automatic manner.

In [5] a novel cascaded deep convolutional neural network is used to address the challenging task of the predicting brain tumor with the help of MRI. This contains two feature fusion models: The Gaussian-pyramid multiscale input features-fusion technique (This was used to solve the issues of size variation and weak brainstem gliomas boundaries) and the brainstem-region feature enhancement (This was used to retain higher frequency details for sharper tumor shape). The proposed method achieved good segmentation result with high dice similarity coefficient of 77.03% and also a genotype prediction with average accuracy of 94.85% upon 5 fold cross validation.

By all the study done we have concluded a common point that if the tumor is big in size but is not growing then there is no problem related to it as it will not cause any brain haemorrhage or any brain related problem. Whereas a small tumor will good growth rate is very dangerous, hence the size of tumor does not matter but the growth of tumor matters.

3 METHODOLOGY

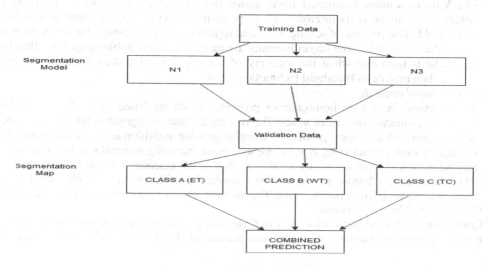

Figure 1.   Generalised Ensembling Technique used for ensemble prediction.

Table 1. Hyper parameter used for training of neural network.

Name	Value
Input Size	128 x 128 x 128
Batch Size	7
Learning Rate	1×10^{-8}
Optimizer	Adams
Epoch	200
Loss Function	Generalized Dice loss + Focal loss

From [7] we know that ensembling technique is used for brain tumor segmentation as it improves performance and results. Segmentation maps in terms of segmented tumors and sub-regions are obtained as the output of these networks. Then these segmented maps are combined to get the final output.

In the pre-processing step the data is first augmented by use of multitude technique like rotation, mirroring or cropping, before it is fed to the neural network for training.

4 RESULTS

The below graph in Figure 2 represents the statistical features that have been obtained on grey-level co-occurrence matrix on high level and low level glioma sub bands of images that have been trained.

The above graph in Figure 3 represents the statistical features that have been obtained on grey-level co-occurrence matrix on high level and low level glioma sub bands of images that have been trained.

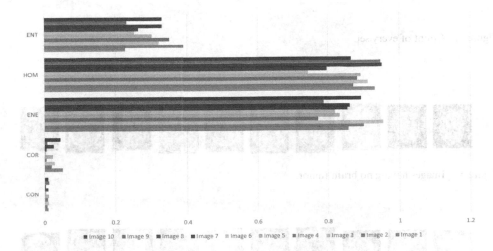

Figure 2. Features obtained from GLCM of LL and HL of training images.

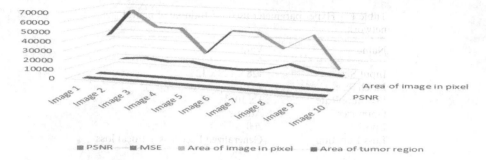

Figure 3.   Features obtained from GLCM of LL and HL of training images.

The Figure 4 represents the distribution of data in various sets i.e. the training set, the validation set and the test set. The orange bar represents a No inclusion label whereas the blue bar represents a yes label.

Figure 5 represents images of size 10 to 4 distributed over training and testing set and having no brain tumor.

Figure 6 represents images of size 10 to 4 distributed over training and testing set and having brain tumor. After this these images are fed into neural network for further training and prediction.

The Figure 7 represents the vgg-16 accuracy that has been obtained while training the network. The orange line is the training line while the blue line is the validation line.

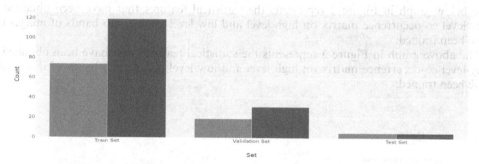

Figure 4.   Count of every set.

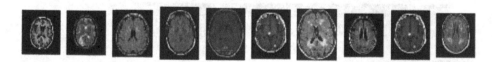

Figure 5.   Images having no brain tumor.

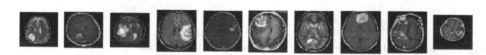

Figure 6.   Images having brain tumor.

414

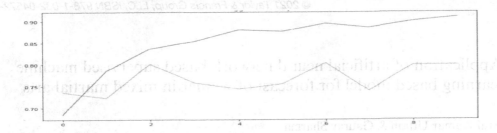

Figure 7.   Vgg-16 accuracy graph.

5   CONCLUSION AND FUTURE WORK

This paper was written with the intention to show case the results that has been obtained with the combination of convolutional neural network model which acts as a classifier to classify weather a person has brain tumor or not and it combines the area of computer vision which helps to automate the system to crop brain images from Magnetic Resonance imaging (MRI) scans. If we further increase the layers of the CNN the accuracy could be above 95% but presently the accuracy is 91.3% (approximately). In future we plan to apply this algorithm with more complex CNN and a real world large dataset.

REFERENCES

[1] Çinar, Ahmet, and Muhammed Yıldırım. "Detection of tumors on brain MRI images using the hybrid convolutional neural network architecture." *Medical Hypotheses* (2020): 109684.

[2] Gordillo, Nelly, Eduard Montseny, and Pilar Sobrevilla. "State of the art survey on MRI brain tumor segmentation." *Magnetic resonance imaging* 31, no. 8 (2013): 1426–1438.

[3] Iqbal, Sajid, M. Usman Ghani, Tanzila Saba, and Amjad Rehman. "Brain tumor segmentation in multi-spectral MRI using convolutional neural networks (CNN)." *Microscopy research and technique* 81, no. 4 (2018): 419–427.

[4] Işın, Ali, Cem Direkoğlu, and Melike Şah. "Review of MRI-based brain tumor image segmentation using deep learning methods." *Procedia Computer Science* 102 (2016): 317–324.

[5] Liu, Jia, Fang Chen, Changcun Pan, Mingyu Zhu, Xinran Zhang, Liwei Zhang, and Hongen Liao. "A cascaded deep convolutional neural network for joint segmentation and genotype prediction of brainstem gliomas." *IEEE Transactions on Biomedical Engineering* 65, no. 9 (2018): 1943–1952.

[6] Liu, Jin, Min Li, Jianxin Wang, Fangxiang Wu, Tianming Liu, and Yi Pan. "A survey of MRI-based brain tumor segmentation methods." *Tsinghua science and technology* 19, no. 6 (2014): 578–595.

[7] McKinley, Richard, Raphael Meier, and Roland Wiest. "Ensembles of densely-connected CNNs with label-uncertainty for brain tumor segmentation." In *IInternational MICCAI Brainlesion Workshop*, pp. 456–465. Springer, Cham, 2018.

[8] Myronenko, Andriy. "3D MRI brain tumor segmentation using autoencoder regularization." In *International MICCAI Brainlesion Workshop*, pp. 311–320. Springer, Cham, 2018.

[9] Thaha, M. Mohammed, K. Pradeep Mohan Kumar, B. S. Murugan, S. Dhanasekeran, P. Vijayakarthick, and A. Senthil Selvi. "Brain tumor segmentation using convolutional neural networks in MRI images." *Journal of medical systems* 43, no. 9 (2019): 294.

[10] Zhang, Nan, Su Ruan, Stéphane Lebonvallet, Qingmin Liao, and Yuemin Zhu. "Multi-kernel SVM based classification for brain tumor segmentation of MRI multi-sequence." In *2009 16th IEEE International Conference on Image Processing (ICIP)*, pp. 3373–3376. IEEE, 2009.

Recent Trends in Communication and Electronics – Sharma et al. (Eds)
© 2021 Taylor & Francis Group, LLC, ISBN 978-1-032-04572-6

Application of artificial neural network based supervised machine learning based model for forecast of winner in mixed martial arts

Atul Kumar Uttam & Gaurav Sharma
Department of Computer Engineering & Applications, GLA University, Mathura, India

ABSTRACT: Mixed Martial Art (MMA) is a one-to-one combat sport in which many techniques are applied, such as punching, wrestling, jiu-jitsu etc. It's a well organized sport that is popular all over the world. It is a one-to-one fighting sport in which a red label identifies one side and a blue label identifies the competitor. The aim of this paper is to study the application and the effects of an artificial neural network based machine learning model to forecast the match winner based on the features of the fighters. In this neural network-based model, the fighters, different characteristics for the prediction of the winner have been used. In this study, a considerable 71 percent accuracy was achieved in this study.

1 INTRODUCTION

This Mixed Martial Arts (MMA) is a one-to-one combat sport in which various techniques such as punching, wrestling, jiu-jitsu of various fighting-based sporting arts are used (Jensen et al. 2013). It is basically an unarmed fight competition where opponent fight with each other using different fighting techniques (Tota et al. 2014). Several rules were added for the safety of fighters since inception of the Ultimate Fighting Championship (UFC). Now it is one of the most popular sports (The Guardian), (Watanabe et al. 2015) around the world (MacDonald et al. 2017) based on pay per view. There are nine weight classes according to which fighters are categorized as shown in Table 1 (Crighton et al. 2015).

In the past several pieces of research has been done for prediction of a match outcome prior to it and it is widely used to improve the winning tactics and sportsmanship of player (Bunker et al. 2017), (Constantinou et al. 2012), (Baboota et al. 2019). This has been the main motivational factor for this research article to study the effect of winner prediction in mixed martial arts-based sports. The rest of the research article is organized as follows: part 2 presents the detailed literature survey of the research article related to this study, part 3 presents the data set and proposed model used in this study, in part 4 result analysis and various evaluation metrics summary has been presented and at last in part 5 conclusion and future work has been discussed.

2 LITERATURE SURVEY

A detailed literature survey has been conducted related to mixed martial art and its related study. In article (Gift 2018) the author has study the favoritism in the judge's decision towards titleholder in detail. In research article (Johnson 2009), the author has study the winner prediction using logistic regression based supervised machine learning model. During cross-validation, concordant 63.4 and discordant of 36.6 has been reported by author. In research article (Collier et al. 2012), the authors probed

DOI: 10.1201/9781003193838-76

Table 1. Weight class of fighters.

Weight Class	Weight(in Kg.)
Fly_weight	56.7
Bantam_weight	61.2
Feather_weight	65.8
Light_weight	70.3
Welter_weight	77.1
Middle_weight	83.9
Light_heavy_weight	93.0
Heavy_weight	120.2
Super_heavy_weight	>120.2

the features that bear a resemblance to the aggression of a fighter and its impact on match conclusion. By using data on individual fights, the authors have studied the probability of winning based on the characteristics of the fighters. In research article (Hitkul et al. 2019), the authors have used different machine learning-based model to predict the outcome of a match. Authors have found in their study that Support Vector Machine based model gives the highest accuracy of 62.8 %. The authors have considered the data from post 2013 onwards. In all these studies (Gift 2018), (Hitkul et al. 2019), (Johnson 2009) and (Collier et al. 2012), the various aspects of mixed martial art based fights have been studied but a detailed study is required to study the artificial neural network based model for such kind of the studies.

In these studies it has been found that data is highly unbalanced (Hitkul et al. 2019), and accuracies of the models used in these studies are not elevated, hence a thorough study is required to predict the outcome of a fight using supervised machine learning based models.

3 METHODOLOGY

3.1 *Data processing and feature analysis*

For the study purpose, we use the Ultimate Fighting Championship (UFC) data from the year 1993 to the year 2019. It consists of 3592 records of different matches. Among these records the blue side has 1211 wins recorded while the red side has 2381 wins. We have not considered the draw matches or no result matches for this study. A total of 117 independent variables and one dependent variable have been chosen for this study. All the null values have been filled with zero and the categorical variable (winner), which is also the dependent variable, is converted to the numeric value as follows: Red is 1 and Blue is 0. A heat map of 10 subset features of the data set is shown in Figure 1.

3.2 *Proposed model*

In this paper, for the prediction of the match outcome based on fighter's available data from past matches, a multilayer neural network-based model has been used as depicted in Figure 2.

A neural network is a computing system that has one input layer of neurons, some hidden layer of neurons and one output layer of neurons. Each layer of a neural network may have a variable no of nodes. In a hidden layer, most of the computation happens with weights (w) and biases (b). The output of one layer is fed forward to the next layer in sequence in this model. In this case, we have taken the number of nodes in the input layer is 117 which is the total number of features. The output layer in our case contains one node as it is a binary class problem as the winner is either red or blue

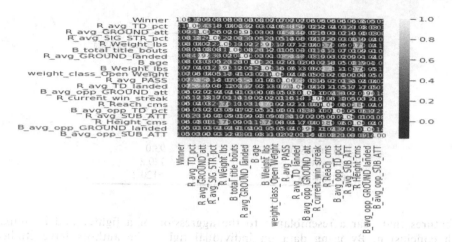

Figure 1.   Heat-map of partial subset of features.

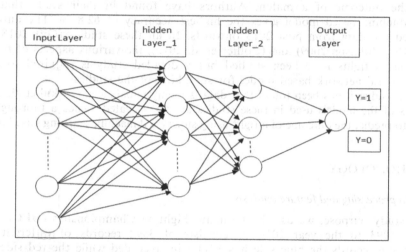

Figure 2.   Sample architecture of multilayer neural network.

which are encoded (red as 1 and blue as 0). We have considered two layers of the hidden layer which contain 50 and 10 nodes respectively.

When the input is received, the weighted sum plus bias is calculated at a neuron (as shown in the circle in Figure 2), then a preset activation function (in our case we used ReLu) is used to check whether it should be fired or activated. The ReLu activation function as shown in Figure 3. It can be defined as equation (1).

$$f(x) = \max\{0, x\} \tag{1}$$

In the next step, the neuron forwards these calculated values as an input to the next layer of neurons. This process is continued until the output value reaches the output layer. The weighted sum is calculated as equation (2).

$$\text{Weighted sum} = \sum W_j.X_j \tag{2}$$

Where, W_j = vector of weights, and X_j = vector of input(features).

418

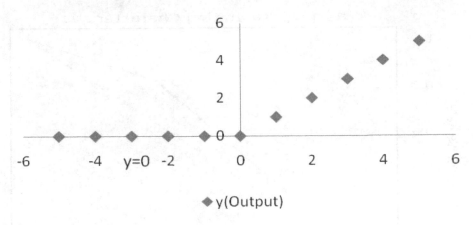

Figure 3.   Activation function: rectified linear unit (ReLu).

$$F(x) = 0, \text{ if } W. X + b < 0$$
$$\qquad\quad 1, \text{ if } W. X + b \geq 0 \qquad\qquad (3)$$

Where b=bias or threshold.

After calculating the weighted sum the activation function is applied. Initially, the weights of the network will get initialized randomly in the range [-1, 1]. After each epoch, the weights will be updated by training the input data. In our experiment there are 117 features thus number of nodes at the input layer will be 117 plus one bias unit. After augmenting the input features with a bias we have 9900 connections with first hidden layer. Similarly at first hidden layer after augmentation, we have 510 connections and at the second hidden layer, we have 22 connections finally. Adding more hidden layers does not improve model performance.

4   RESULT ANALYSIS

In order to evaluate the performance of the model we have used 100000, iterations. The training accuracy of 84 % and test accuracy of 71 % has been achieved by our model. The following parameters have been used in this study: activation function (ReLu for hidden layers and sigmoid for output layer), alpha (0.0001), batch size (auto), initial learning rate (0.01), no of iteration no change (10), solver (adam), validation fraction (0.1), test fraction (0.1), training fraction (0.8). A 10 fold cross validation has been performed to validate the accuracy. A receiver operating characteristic graph is also shown to depict the accuracy of the model used in this study, as shown in Figure 4.

5   CONCLUSION

From this study, it can be concluded that a artificial neural network-based machine learning model can be used for the prediction of the binary class problems with significant results as shown in Table 2. As the input data is highly unbalanced several data balancing techniques like ADSYN could be used in the future to improve the accuracy of the model. Further improvement in the model is required. From this study it has been observed that red player win possibility can be done more wisely than blue player win.

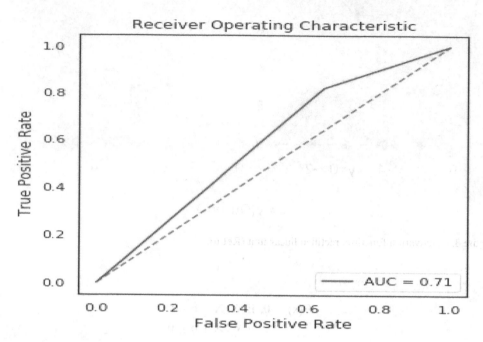

Figure 4.   ROC Curve of the model.

Table 2.   Classification Result.

Metric	Red	Blue
Precision	0.79	0.49
Recall	0.82	0.45
F1-Score	0.81	0.47
Accuracy	0.71	

REFERENCES

Baboota, Rahul & Kaur, Harleen, (2019). "Predictive analysis and modelling football results using machine learning approach for English Premier League," International Journal of Forecasting, Elsevier, vol. 35(2), pages 741–755.

Collier, Trevor & Johnson, Andrew & Ruggiero, John. (2012). Aggression in Mixed Mar-tial Arts: An Analysis of the Likelihood of Winning a Decision. 10.1007/978-1-4419-6630-8_7.

Constantinou, Anthony & Fenton, Norman & Neil, Martin. (2012). pi-football: A Bayesian network model for forecasting Association Football match outcomes. Knowledge-Based Systems, 36, 322-339. Knowledge-Based Systems. 36. 332–339. 10.1016/j.knosys.2012.07.008.

Crighton, Ben & Close, Graeme & Morton, James. (2015). Alarming weight cutting behaviours in mixed martial arts: a cause for concern and a call for action. British journal of sports medicine. 50. 10.1136/bjsports-2015-094732.

Gift, P., (2018), 'Performance Evaluation and Favoritism: Evidence From Mixed Martial Arts'. Journal of Sports Economics, 19(8), 1147–1173.

Hitkul, Aggarwal K., Yadav N., Dwivedy M., (2019): A Comparative Study of Machine Learning Algorithms for Prior Prediction of UFC Fights'. In: Yadav N., Yadav A., Bansal J., Deep K., Kim J. (eds) Harmony Search and Nature Inspired Optimization Algorithms. Advances in Intelligent Systems and Computing, vol 741. Springer, Singapore.

Jensen, Peter & roman, jorge & shaft, barrett & Wrisberg, Craig. (2013). In the Cage: MMA Fighters' Experience of Competition. Sport Psychologist. 27. 1–12. 10.1123/tsp.27.1.1.

Johnson, J. D., (2009), 'Predicting Outcomes in Mixed Martial Arts Fights with Novel Fight Variables'. (Master of Science). University of Georgia.

MacDonald, Katie & Lamont, Matthew & Jenkins, John. (2017). Ultimate Fighting Cham-pionship Fans: Foundations of Subcultural Stratification. Leisure Sciences. 1–19. 10.1080/01490400.2017.1344164.

R. Bunker, F. Thabtah, (2017). A machine learning framework for sport result prediction: J. Appl. Comput. Informatic, 10.1016/j.aci.2017.09.005. Elsevier.

The Guardian home page: https://www.theguardian.com/sport/2016/mar/04/the-fight-game-reloaded-how-mma-conquered-world-ufc.

Tota, Łukasz & Drwal, Tomasz & Maciejczyk, Marcin & Szygula, Zbigniew & Pilch, Wanda & Pałka, Tomasz & Lech, Grzegorz. (2014). Effects of original physical training program on changes in body composition, upper limb peak power and aerobic performance of a mixed martial arts fighter. Medicina Sportiva. 18. 78–83. 10.5604/17342260.1110317.

Watanabe, Nicholas. (2015). Sources of direct demand: The case of the Ultimate Fighting Championship. International Journal of Sport Finance. 10. 26–41.

Recent Trends in Communication and Electronics – Sharma et al. (Eds)
© 2021 Taylor & Francis Group, LLC, ISBN 978-1-032-04572-6

A review on methods for the identification of 3-D geometry for remote sensing

Shivansh Sinha, Shraddha Tripathi, Saurabh Mishra, Ananya Singhal, Abhishek Sharma & Parvin Kumar

CoE- Space Technologies, Department of ECE, KIET Group of Institutions, Delhi- NCR, Ghaziabad, India

ABSTRACT: This manuscript presents a review of various methods for identification of height or altitude, that is generally used for space science & remote sensing applications. The main aim of this review paper is to provide an outline of some basic technological concepts used in astronomy which must be helpful for the researchers involved in low range photogrammetry. This manuscript also includes some common applications of parallax.

1 INTRODUCTION

There are various techniques by which altitude/elevation of an object can be obtained & later on various information can also be extracted that can be used for further analysis. Satellite imagery, aerial pictures, remote sensing, LiDAR mapping and relief displacement are some of the general methods for calculate elevation data on high resolution for an object (Center 2012). One such another method is parallax, which is discussed below to measure height of any given object on the earth surface. The difference or shifting in the position of an object when viewed from a different angle as well as a line of sight is considered as parallax (Hemmings 1949).

Close by objects shows a better parallax than farther objects when viewed from dissimilar orientation, so parallax can be used to establish distances for photogrammetric motives & to map the terrain levels. The closer object continually appears to move in the direction opposite that of our eye, in comparison to the far object, this is the very generic example of parallax.

The parallax is a very phenomenon method used to find heights of objects in stereo pairs of verticals. We experience this anomaly when a moving body shifts its position when compared to a static object considering the camera as an eye. This phenomenon is graphically revealed in Figure 1a. This difference in parallax gives a three-dimensional effect when stereo pairs are viewed stereoscopically.

2 VARIOUS METHODS OF PARALLAX

Parallax is one of the finest means to obtain distances in space. For the scaling of the large distances, between a celestial body or a stellar from globe, astronomers use the concept of parallax (Bailer-Jones 2015). The parallax is the partly-angle of slant between two sight-lines to the stellar, as noticed when Earth is on reverse sides of the Sun in its trajectory. The Trigonometric Parallax method determines the distance to a star or another object by measuring its slight shift in apparent position as seen from opposite

DOI: 10.1201/9781003193838-77

Figure 1(a). Apparent shift in the position of object due to shift in position of inspection.

Figure 1(b). Methods of using parallax in astronomy.

ends of Earth's orbit. Methods of using parallax in astronomy are discussed in Figure 1b.

2.1 Relation between a star's distance & its parallax angle

The concept of parallax can be very useful in calculation of the distance between two objects in space. Apparent shift of position of any close proximity star or any other object present in space in context to background of far proximity objects is known as stellar parallax or trigonometric parallax (Cenadelli et al. 2009). In general, astronomers calculate star's superficial movement. Stellar parallax totally relies on the concepts of geometry. The technique is based upon computing two angles and the involved side of a triangle shaped by the stellar. The Relation between a star's distance and its parallax is presented in Figure 2. For the very far objects in space form the earth, the parallax is a best and simplified method for the distance measurement by the astronomers. Eqn. (1) is revealing a relation between parallax and distance for the very far distance from the earth

$$d = 1/p \qquad (1)$$

Where, d is the distance measured in parsecs, p is parallax angle measured in arc-seconds.

2.2 Calculation for the height measurement

The height of an object can be calculated with the help of the Eqn. (2). This equation is providing a relation between differential, absolute parallax and height of the object from the ground

$$H_\alpha = \frac{H\Delta P}{P + \Delta P} \qquad (2)$$

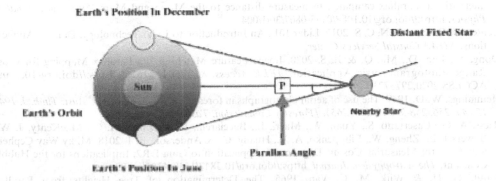

Figure 2. Use of parallax to find distance of far objects from earth using parallax angle.

423

where, "H_a" is Height of the object, "H" is Height of aircraft/satellite above the ground. P is an absolute parallax at the base of the object being calculated. The stereo pair is substituted for absolute stereophonic parallax and ΔP is differential parallax (Schut & Wijk 1965).

3 APPLICATION OF PARALLAX

Parallax has various applications in different domains. Some of them are listed down below-

- Calculation of Cosmic Distance- Parallax is a chief rung in the cosmic distance ladder (Riess et al. 2018). By measuring the distances to a number of nearby stars, astronomers have been able to set up associations between a star's color & its built-in intensity, i.e., the intensity it would come into sight to be viewed from a regular distance.
- Determine the accuracy of Shooting- Parallax takes place when the aim & reticle (i.e. helping aid in locating objects) are on dissimilar planes surrounded by the scope. It is noticeable when you shift your head or eyeball looking through the scope and the reticle appears to be in motion or floating around the point, at which the observer is aiming. Parallax is a visual delusion that must be removed for a perfect shot.
- Parallax in Photogrammetric Applications- Closer object travels quicker than that of the distant distanced object, alike is the case of an above ground camera exposed to overlapping pictures which is caused by the motion of the airplane. Dimensions of this parallax are used to derive the altitude of the buildings, provided that flying tallness and baseline distances are known. This is a chief factor in the process of photogrammetry (Dong et al. 2020).
- 3D Imaging- Parallax can be used for 3D imaging. This process is similar to the approach as human eyeballs do & presents them in such a manner that each eyeball observes only one of the two pictures.

4 CONCLUSIONS

Parallax has been proved very beneficial in many domains especially in the field of astronomy. Parallax has always been a very vital distance calculation technique, preferred by astronomers. It can be used for close proximity stars as it is very perfect. It's also important to be considered while using optical instruments. Parallax will surely play an important role in futuristic exploration of space.

REFERENCES

Bailer-Jones, C. A. L. 2015. Estimating Distances from Parallaxes. *Publications of the Astronomical Society of the Pacific*. https://doi.org/10.1086/683116

Cenadelli, D., Zeni, M., Bernagozzi, A., Calcidese, P., Ferreira, L., Hoang, C. & Rijsdijk, C. 2009. An international parallax campaign to measure distance to the Moon and Mars. *European Journal of Physics*. https://doi.org/10.1088/0143-0807/30/1/004

Center, N. O. & A. A. N. C. S. 2012. Lidar 101 : An Introduction to Lidar Technology, Data, & Applications. *NOAA Coastal Services Center*.

Dong, Y., Fan, D., Ma, Q. & Ji, S. 2020. Image Feature Matching Via Parallax Mapping for Close Range Photogrammetric Application. *IEEE Access*, 8, 32601–32616. https://doi.org/10.1109/ACCESS.2020.2973723

Hemmings, W. D. 1949. The use of aerial photographs in forestry in Tasmania. *1949. Aust. Timb. J. 1949 15 (4), 256, 258-9, 261, 263, 265). [Forestry Commission, Tasmania.]*.

Riess, A. G., Casertano, S., Yuan, W., Macri, L., Bucciarelli, B., Lattanzi, M. G., MacKenty, J. W., Bowers,J. B., Zheng, W., Filippenko, A. V., Huang, C., & Anderson, R. I. 2018. Milky Way Cepheid Standards for Measuring Cosmic Distances & Application to Gaia DR2: Implications for the Hubble Constant. *The Astrophysical Journal*. https://doi.org/10.3847/1538-4357/aac82e

Schut, G. H. & Wijk, M. C. Van. 1965. The Determination of Tree Heights from Parallax Measurements. *The Canadian Surveyor*, 19(5), 415–427. https://doi.org/10.1139/tcs-1965-0091

Preparing application of K-Nearest Neighbor (KNN): A supervised machine learning based model in placement prediction for graduate course students

Gaurav Sharma & Atul Kumar Uttam
Computer Engineering and Applications, GLA University, Mathura, India

ABSTRACT: Placement is one of the important criteria for choosing a university or college for taking admission in any course. Most of the rating agencies in the field of education like NAAC, NBA, and NIRF, etc treat it as a positive sign while awarding accreditation to an educational organization. Every student aims to achieve the best placement in their course period. Predicting placements is very helpful to those organizations that are in the field of education. Forecasting placements help universities to plan accordingly; if they are lacking somewhere then they can sort out the things and improve performance before the time slips from their hand. We have used the previous years of training and placement data, refine and rearrange the data according to the various resources available at that time, and students' grades. To study this problem and for valuable outcomes, we have used KNN which is a supervised machine learning based algorithm. Since applying the KNN based supervised ML algorithm, we have acquired 84 percent of the train data accuracy. To further validate the model's accuracy, tenfold cross-validation was done, which guarantees the 82 percent accuracy of the model.

1 INTRODUCTION

The Information Technology & Business Process Outsourcing (IT & BPM) (ibef)sector in India stood at US$ 177 billion in 2019, witnessing a year-on-year growth of 6.1 percent, and the volume of the business is expected to rise to US$ 350 billion by 2025. India's Information Technology & Internet Technology-Information Powered Services industry breeds in to 2018–19 to US$ 181 billion. Industry exports increased to US$ 137 billion in the financial year 2019 while domestic revenues increased to US$ 44 billion. Engineering colleges needs radical reforms; they have to focus on enforcing quality education and also strengthen the job oriented approach. Many factors define the placement of the student, such as institutional approval status, campus jobs, location, industry relations, field specialization students' performance are some influential features, though these features are not suitable as a qualification for students because these factors cannot be directly controlled by the student and their various attributes like student gender, rural and urban situation, high school performance and college are the qualities that keep it a right prediction of placement. In India, 1.5 million engineers graduate every year according to statistics. Placements are one of the largest lifetime obstacles a student faces. The institution must offer its students full placement potential. The placement department and an institute's teachers should also take appropriate steps to produce a set of students suited to the needs of each company. A predictive placement program may be used to determine a specific student's skill for the given job.

DOI: 10.1201/9781003193838-78

2 RELATED WORK

In paper (Sharma et al. 2014), the supervised machine learning-based logistic regression-based model has been used to predict the placement probability of the students. For this study, the authors have used in-house data of their native college. The data used in this study is of batch 2009 to 2013 batch of a four-year graduation course. In their study, the author has achieved the test accuracy of about 83.33%. Their study uses five dependent variables and one binary independent variable.In paper (Thangavel et al. 2017), the authors have used machine learning based model like logistic regression has used and achieved an accuracy of 71.66%. The authors have used approximately eight dependent variables and one independent variable.In paper (Jeevalatha et al. 2014), the authors have used Decision tree algorithms such as ID3, C4.5 & CHAID for performance analysis of students for placement. The authors have achieved an accuracy of 95.33% using decision tree based ID3 algorithm. The authors have used five dependent variables and one independent variable for this study.In paper (Parekh et al. 2016), the authors have developed a web-based dashboard and data mining tools for placement performance analysis.In paper (Jeganathan et al. 2017), the authors have used a fuzzy logic-based system for student's placement prediction. In paper (Joy et al. 2019), the authors have presented a review of students' placement predictions. In this review, they have studied the logistic regression-based model (Sharma et al. 2014), random forest-based model (Bharambe et al. 2017), the sum of difference based model (Ramanathan et al. 2014), classification and clustering based model (Pruthi et al. 2015).

3 PROBLEM FORMULATION

It becomes more daunting with the task of placing the maximum number of students, especially when there is no availability of concrete tools for the TPO to gain insights into the performance of the student in the placement from the current session. Because of the absence of these tools, the steps taken by the administration to boost students, performance are wasted because they are implemented strangely without any stratum on the whole bunch.

3.1 *System model and dataset*

To solve this problem we use the following system model, first in house data of graduation course of three year duration has been obtained. Then data has been processed and unbalancing of data has been fixed. All the null value record has been eliminated.

3.2 *K-Nearest Neighbors (KNN)*

KNN algorithm is the simplest supervised foundation algorithm for machine learning and can be used to answer classification and regression problems. The KNN algorithm believes identical objects occur next to each other. To put it another way, analogous

Table 1. Features of the system model.

Parameter		Value
Gender	Male	1
	Female	0
10th/12th Board	State Board	0
	CBSE	1
	ICSE	2
	Other	3
	Placed	1
Target	Not Placed	0

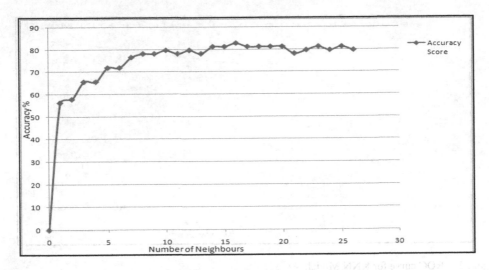

Figure 1.   KNN number of neighbors.

things are close to one another. The KNN algorithm relies on this assumption as being valid enough to be relevant to the algorithm. KNN captures the idea of similarity calculating the distance between points on a graph (sometimes called distance, proximity, or closeness). In KNN suitable number of neighbors is selected and the distance among them is calculated to find the correlation among them. Figure 1 shows the best possible value of k for the placement prediction problem and it has been observed that at value k = 16 the KNN model gives satisfactory training accuracy of approximately 84 %. To validate this result a 10 fold cross-validation has been applied which gives an accuracy of 82%, which validates the training resultobtained by the KNN model. In the KNN model following parameters have been used for this classification problem:

KNeighborsClassifier (algorithm='auto', leaf_size=50, metric='minkowski',
metric_params=None, n_neighbors=27, p=1,weights='uniform')

4   RESULT AND DISCUSSION

While applying the KNN based supervised machine learning model the train data accuracy of 84 % has been obtained, further to validate the accuracy of the model tenfold cross-validation has been performed which ensure the 82% of the accuracy of the model. A receiver characteristic graph has been shown in Figure 2 to show this accuracy.

A confusion matrix of the above model is offered in Figure 3. Furthermore, this model has been compared against other accepted supervised machine learning based model like XGBoost, random forest, Support vector machine, and Naïve Bayes models. All these models' comparative accuracy has been given in Figure 4.

While evaluating these models it has been observed that the KNN model gives the highest accuracy among all the models used here and the Support vector machine-based model gives the least accuracy of about 69 %. The random forest-based model takes the highest time and provides an accuracy of 78%, while XGBoost based model also achieves the same amount of accuracy of 78% but uses less time I comparison to the random forest model as shown in Table 2.

These models give a significant view of the attributes used in this problem study. And their importance score has been shown in Table 3.

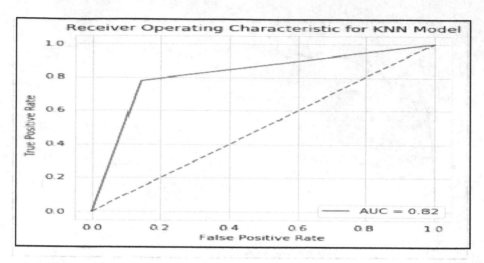

Figure 2.   ROC curve for KNN Model.

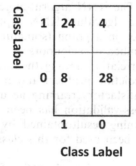

Figure 3.   Confusion matrix of the model.

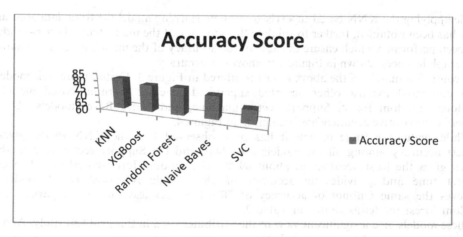

Figure 4.   Accuracy score of the various model.

Table 2. Accuracy score of machine learning models.

Model	Accuracy Score	Runtime Training	Runtime Prediction
KNN	81.75	0.007702	0.006088
XGBoost	78.125	0.037669	0.001722
Random Forest	78.125	0.300677	0.01269
Naive Bayes	75	0.004252	0.000582
SVC	68.75	0.028476	0.004296

Table 3. Importance score of the features.

Feature	Importance Score
CPI	0.865038
III_SPI	0.064186
II_SPI	0.04768
V_SPI	0.023096
Gender, 10th_Board, 10_per, 12th_Board, 12th_per, I_SPI, IV_SPI	0

REFERENCES

A.S. Sharma, S. Prince, S. Kapoor, K. Kumar, "PPS – Placement prediction system using logistic regression", IEEE international conference on MOOC, innovation and technology in education (MITE), pp 337–341, 2014.

Bharambe, Yogesh, "Assessing employability of students using data mining techniques", Advances in Computing, Communications and Informatics (ICACCI), 2017 International Conference on. IEEE, 2017. https://www.ibef.org/industry/indian-iT-and-iTeS-industry-analysis-presentation-report. march.2020/

Jeganathan, Jayanthi. (2017). student prediction system for placement training using fuzzy inference system. ictact Journal on Soft Computing. 7. 10.21917/ijsc.2017.0199.

L. C. Joy and A. Raj, "A Review on Student Placement Chance Prediction," 2019 5th International Conference on Advanced Computing & Communication Systems (ICACCS), Coimbatore, India, 2019, pp. 542–545, doi: 10.1109/ICACCS.2019.8728505.

Parekh, Siddhi & Parekh, Ankit & Nadkarni, Ameya & Mehta, Riya. (2016). Results and Placement Analysis and Prediction using Data Mining and Dashboard. International Journal of Computer Applications. 137. 22–25. 10.5120/ijca2016908985.

Pruthi, P. Bhatia, "Application of Data Mining in Predicting Placement of Students", International Conference on Green Computing and Internet of Things (ICGCIoT), 2015.

Ramanathan.L et al., "Mining Educational Data for Students' Placement Prediction using Sum of Difference Method", International Journal of Computer Applications (0975 – 8887), Volume 99– No.18, August 2014.

T. Jeevalatha, N. Ananthi, D. Saravana Kumar, "Performance analysis of undergraduate students placement selection using Decision Tree Algorithms", International Journal of Computer Applications, vol. 108, pp. 0975–8887, December 2014.

Thangavel, S. Bkaratki, P. Sankar, "Student placement analyzer: A recommendation system using machine learning", Advances in Computing and Communication Systems (ICACCS-2017) International Conference on. IEEE, 2017.

Recent Trends in Communication and Electronics – Sharma et al. (Eds)
© 2021 Taylor & Francis Group, LLC, ISBN 978-1-032-04572-6

A survey on tongue control system

Abhilash Chand, Md. Yasrub Siddique, Prabhat Gautam, Yash Chauhan & Manish Kumar Singh
Department of Electronics & Communication Engineering, KIET Group of Institutions, Ghaziabad, India

ABSTRACT: "Tongue control system" is basically controlled by Assistive Technology (AT) which is developed for people who has severe physical disability or paralyzed who want to control/access their environment. The Tongue control system consists of four hall sensor modules placed on the Mouthpiece and the Hall Effect magnetic field is created by placing a permanent magnet on the tongue of the victim with the help of tissue adhesive or piercing. These fields with the help of wireless transmitter is sent to the prototype receiver which the victim wants to control; in this way a victim can control his/her environment using the tongue that works on AT. Finally, the design was tested, and the result confirms the design approach for a healthy subject.

1 INTRODUCTION

According to the Census report 2001 (India), 21 million people are suffering from on or other kind of disability. This is approximately 2.1% of the total population. [1]. So, there was a lot of need of such type of ATs which can help them carry out their daily needs. In most of the cases disability can be in the full body but the tongue is one such organ which remains intact even in harsh situations. Therefore, tongue is used for this design. Assistive Technology is any item, system or piece of equipment that increases or improves the functional abilities of individuals having disabilities. Disabilities vary widely in type and severity. AT contains a vast range of devices, including the mobility aids, augmentative communication devices (voice synthesizers and communication boards), prosthetic and Orthotic devices and equipment. ATs can be "Low-Tech" (A cup-holder for a wheel-chair tray) or "High-Tech" (Brain interfaces to control prototype) [2]. Assistive Technology improves the standard of life of a disabled person. They contribute economically, socially and increase their independence. Even if having so much potential, ATs sometimes can be problematic for availability of appropriate people with disabilities [3].

Till date, there are very few ATs which had made their way out from the research laboratories to the daily life of the victims. There are lot of factors influencing the acceptance rate of the assistive Technology such as psychological and technical phase. The most important factors would be the convenience in control and ease of usage. Also, working on the fact that the disabled person should not look different from an ordinary person. Hence, working with an aim to bring ATs to the daily life of the needy ones more, than just being a scientific achievement and persisting in the laboratories itself [4].

2 MAIN COMPONENT OF TONGE DRIVEN SYSTEM

2.1 *Hall sensor*

Hall Effect sensor module converts the magnetic field into electrical energy; the sensor does not affect any external environmental condition. The hall sensor has four-bit output. Tongue

DOI: 10.1201/9781003193838-79

is moved to the left sensor which triggers it from high to low. Right sensor output is high, and the command is transmitted to the encoder [5]. It converts parallel data into serial RF receiver, the data from transmitter output is transmitted to the decoder. The output provided to the motor driver IC. Right sensor and front sensor remain at high output. The output Hall Effect sensor connected to the encoder and RF receiver receive data from transmitter output to the decoder. The output is decoded and given to the microcontroller. Microcontroller output given to the motor driver IC and the prototype perform as desired [6].

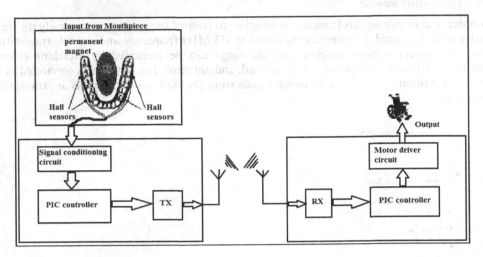

Figure 1. Block diagram of tongue driven system.

2.2 *Button magnet*

A small button type magnet is put on the tongue with the help of biological adhesive or tongue piercing. The magnet is used to trace the tongue movement as when tongue moves, magnetic field around it changes which is being sensed by the Hall sensors on the mouthpiece [7].

2.3 *Arduino uno*

The two sets of Arduino uno is used in the paper, one Arduino uno is used in the mouth piece at the transmitter end and the other one is used at the receiver end. The signals created by the

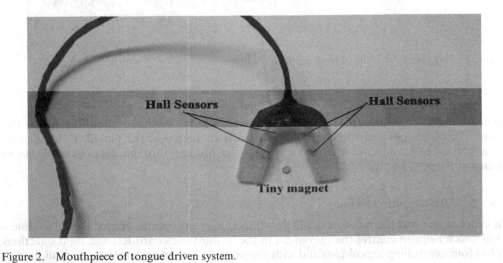

Figure 2. Mouthpiece of tongue driven system.

Hall sensors mounted on the mouthpiece which is a low-level signal are fed into the Arduino Uno. Four pins of it is used as an input pins to sense the signals from the mouthpiece. For the calibration of the signals and for the programming part the Arduino IDE is used which is a open source software used for programming of Arduino. There are four movements possible as for wheelchair prototype left, right, forward and backward. Programming for each part is done separately in Arduino IDE [8].

2.4 Transmitter module

In order to transmit signals from the mouthpiece to control the prototype wirelessly the transmitter module is used. It transmits the signal at 433 MHz frequency and capable transmitting unto 100 meters without antenna and the range can be increased by attaching external antenna. It has four pins power, Data, ground, and antenna. The data pin is connected at the pin 11 of Arduino in order to transmit signals from the Arduino the signals are transmitted serially [9].

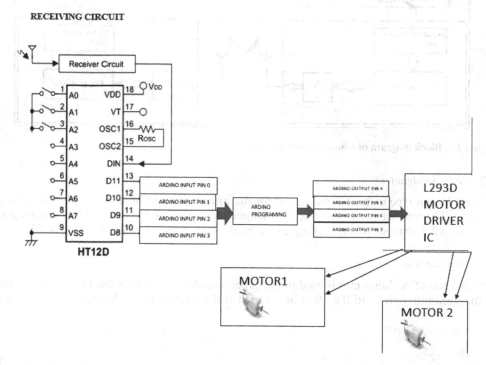

Figure 3.   Transmitter module of tongue driven system.

2.5 Receiver module

In order to receive signals from the mouth piece the receiver module is used. It has four pins power, Data, ground and antenna. The data pin is connected at the pin 12 of Arduino the data received by the receiver module are fed into the Arduino through data pin which is connected at the pin 12 of the Arduino [10].

2.6 L293D motor driver module

In order to control the movement of the wheelchair prototype the motor driver module is used which helps to control the movement in the forward backward left and right directions. It has four controlling logical pins and with the particular pin status combinations all the four

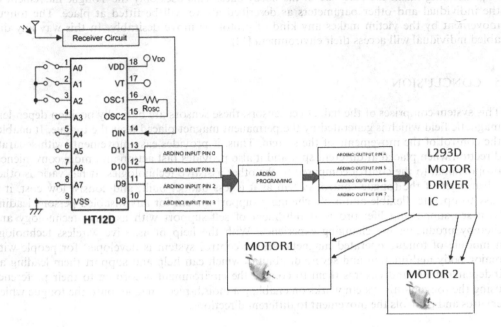

Figure 4.   Receiver module of tongue driven system.

directions can be generated. These pins are then connected to any four pins of Arduino as an output pins and by using the programming all the pins combinations are achieved for all the directions [11].

3   LITERATURE SURVEY

G. Krishnamurthy et. al. stated that tongue-operated assistive technology created for the disabled people to control their environment. The Tongue is a great appendage for the severely disabled individual for operating this system. They used the Hall sensor modules for creating the required magnetic fields. A small neodymium magnet is placed which acts a tracer and tracks the movement of the tongue [12]. Steven F. Barrett described the Arduino's microprocessor ATmega328 and introduction to C programming. It also explores on the Arduino IDE. It also taught the basic syntax and programming on the Arduino IDE. Their approach was to share all the knowledge about micro-controllers so as to enhance more ideas and promoting innovations [13]. M. M. Tahsin et. al. states that the system is a wireless Assistive Technology which works on tongue motion that has minimum invasion & has no obstruction in daily activities. The author also stated that the system can also help those people who have severe spinal cord injuries such as tetraplegia who could not move around with ease and control their surroundings as well. The main advantage is to create more tongue movements which will give combination of outputs. In this approach a wheel chair is controlled using the various tongue movements which moved the wheel chair to different directions [14].

4   OVERVIEW OF TONGE DRIVEN SYSTEM

This paper circumference around the Tongue controlled system operated on Assistive Technology. This gives a cutting edge over other, which are made to create a futuristic world leaving behind the physically challenged or particularly disabled ones. We are focused on implementing thisin the daily life of the needy ones. This uses only the Tongue movement of the individual and other parameters as described above will be fitted at place. The tongue movement by the victim makes any kind of prototype move desirably. In this way any disabled individual will access their environment [14].

5 CONCLUSION

This system comprises of the hall effect sensors, these sensors are basically position dependent magnetic field which is generated by the permanent magnet placed over the tongue. It enables the control of the movements of the system. Thus, it provides easy movements with separate direction when placed at different spot and it also provides fast action and more convenience, proportional to the control compared to the other assistive technologies. It has various other advantages of the tongue control system as it is less budget built, and consist low cost, it is easy to operate, flexible and a life changing support system for any disabled persons leading to a self-supporting life; provided fulfilment of self-support with assistive technology and thereby producing life changing experience. With the help of assistive wireless technology a module of tongue operated magnetic sensor control system is developed for people with major body malfunction and severe disabilities which can help and support them leading an independent life and ensures them to control the environment according to their preference using the tongue, this system works on enabling a mouth-piece magnet onto the tongue which ensures and controls the movement to different directions.

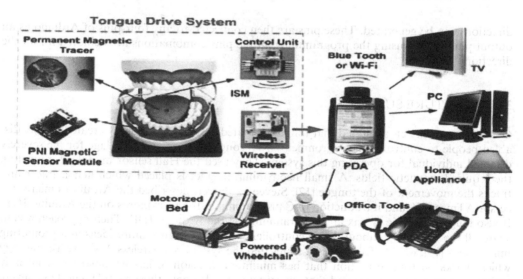

Figure 5. Future technology.

REFERENCES

Badamasi Y. A (2014).The working principle of an Arduino.11th International Conference on Electronics, computer and compution (ICECCO), Abuja, pp. 1–4, doi: 10.1109/ICECCO.2014.6997578.
Monà D. F., Sakomura E. S., Nascimento D. C. (2018). Microstrip-to-Probe Fed Microstrip Antenna Transition. IEEE International Symposium on Antennas and Propagation & USNC/URSI National Radio Science Meeting, Boston, MA, 2018, pp. 1521–1522
Lindberg O. (1952). Hall Effect. in Proceedings of the IRE, vol. 40, no. 11, pp. 1414–1419

Zaman H. U. et al. (2016). A novel design of line following robot with multifarious function ability. International Conference on Mi–croelectronics, Computing and Communications (MicroCom), Durgapur, pp. 1–5

Joo S. et al. (2018). Spin Hall Effect Device for Magnetic Sensor Application. Conference on Precisio Electromagnetic measurement (CPEM 2018), Paris, 2018, pp. 1–2, doi: 10.1109/CPEM.2018.8500932.

Singh M. K. et al. (2018). A Survey of Wireless Sensor Network and its types. 10.1109/ICACCCN.2018.8748710

Sanver U. et al. (2018). Design and implementation of a programmable logic controller using PIC18F4580. IEEE Conference of Ru-ssian Young Researchers in Electrical and Electronic Engineering (EIConRus), Moscow, pp. 231–235

Steven F. Barrett (2020). Arduino II: Systems," in Arduino II: Systems, Morgan & Claypool

Pratap C. B. (2018). Smart Reading Aid using Tongue Drive System for Disabled patients. International Conference on Control, Power, Communication and Computing Technologies (ICCPCCT), Kannur, 2018, pp. 563–566.

Krishnamurthy G., Ghovanloo M. (2006) Tongue drive: a tongue operated magnetic sensor based wireless assistive technology for people with severe disabilities," IEEE International Symposium on Circuits and Systems, Island of Kos, pp. 4.

Tahsin M. M (2016). Assistive technology for physically challenged or paralyzed person using voluntary tongue movement 5th International Conference on Informatics, Electronics and Vision (ICIEV), Dhaka, pp. 293–296.

J. Kim et al.,(2016). Assessment of the Tongue-Drive System Using a Computer, a Smartphone, and a Powered- Wheelchair by People with Tetraplegia," in IEEE Transactions on Neural Systems and Rehabilitation Engineering, vol. 24, no. 1, pp. 68–78.

L. Liao et al., "Control system of powered wheelchairs based on tongue motion detection," 2016 IEEE 15th International conference on Cognitive Informatics & Cognitive Computing (ICCI*CC), Palo Alto, CA, 2016, pp. 411–414

R. Geidarovs and A. Podgornovs (2016). Research of two-dimensional and three-dimensional magnetic fields in an axial Inductor machine IEEE 4th Workshop on Advances in Information, Electronic and Electrical Engineering (AIEEE), Vilnius pp. 1–6, doi: 10.1109/AIEEE.2016.7821824

Recent Trends in Communication and Electronics – Sharma et al. (Eds)
© 2021 Taylor & Francis Group, LLC, ISBN 978-1-032-04572-6

A systematic review on air quality monitoring and qualification systems

Nida Praveen, Lipika Goel & Sonam Gupta
AKGEC, Ghaziabad

ABSTRACT: The exposure of these harmful particles in the atmosphere causes various chronic health diseases like respiratory infections and lung cancer etc. In order to overcome this issue, health experts promote the real-time monitoring of unfavorable situations occurring in the air, hence this is a concerning research topic as various methods are proposed to monitor the quality of air in the real-life scenario. Hence, this paper aims to present a systematic review of some of the Air quality monitoring (AQM) and purification system proposed in the last 3 years, utilizing different microcontroller sensor devices and the Internet of Things (IoT). This study aims to find the most favorable micro-controller device, reliable display technology, the most widely used storage system, and reliable communication technology used in AQM systems and through our proposed review which is found to be Arduino, smartphone apps, cloud servers, and ZigBee and Bluetooth respectively.

1 INTRODUCTION

With the advancement of technologies for better social life, it has been essential to keep track of the purity of the surroundings of society. Air pollution is a leading environmental problem with the establishment of new factories and other factors such as fuels burnt in transports as well as in houses. For example, India has a great population that utilizes biogas, cow dung cakes, and kerosene, etc. in their houses as a means of fuel for cooking food. And while these biomass fuels are not burnt properly due to inappropriate ventilation in houses, these lead to the emission of harmful gases like; carbon monoxide, nitrogen oxides, Particulate matter (PM), and other toxic compounds in the air. The exposure of these harmful particles in the atmosphere causes to various chronic health diseases like respiratory infections and lung cancer etc. this problem has been overcome by various researchers in recent years that promote the real-time monitoring of unfavorable situations occurring in the air, hence this is a concerning research topic as various methods are proposed to monitor AQ in the real-time scenario. As the monitoring of air quality is utilized to apply various purification methods either by the government or by the building owners. These methods work in a way that if the air quality parameter exceeds the set threshold limit, the purification process starts. Thus the researchers focus on the enhancements of the methods to develop Air Quality Monitoring (AQM) systems that can efficiently categorize the air as polluted or non-polluted so that various actions can be performed to purify the surroundings. This study presents various aspects of researches proposed that are: the Microcontrollers used, Air Quality Monitoring (AQM) display interfaces, various storage devices used to store the real-time data, various communication technologies used, and the notification systems used to alert the people about the air quality parameters. Thus the goal of the review is to know about the most favorable micro-controller device, reliable display technology, most widely used storage system, and reliable

DOI: 10.1201/9781003193838-80

communication technology used in AQM systems and through our proposed review which is found to be Arduino, smartphone apps, cloud servers, and ZigBee and Bluetooth respectively.

2 SYSTEMATIC REVIEW PROCESS

We have tried to find the solution to the following research questions based on the research studies reviewed in the proposed literature.

Q1. Which is the most favorable microcontroller sensing device?
Q2. Which is the most reliable Air Quality monitoring (AQM) display technology?
Q3. Which is the most widely used storage system to store the real-time air quality parameters?
Q4. Which is the most reliable communication technology used in AQM systems?

Following steps have been performed in order to execute the proposed review work.

2.1 *Searching process and sources*

We have performed the searching process on Google and Google Scholar platform. In order to find the relevant publications, the related key phrases that have been used are: "Air quality monitoring systems", "Air purification MCU", "Air quality IoT", and "Air purification and monitoring IoT". The relevant publications for AQM and purification systems were taken from 4 well databases: PUBMED, ScienceDirect, IEEE Explore, and Web of Science.

2.2 *Exclusion and Inclusion criteria*

The criteria that are considered to discard the irrelevant studies and select the relevant studies are categorized as exclusion and inclusion criteria respectively. The following points are considered in the exclusion and inclusion criteria in the proposed literature review.

- The papers based on Wireless sensor networks (WSN) instead of IoT have been discarded.
- The papers, providing unambiguous details about the sensor types, have been discarded.
- The papers written only in the English language are taken into consideration.
- Papers published in last 4 years are taken into consideration related to the AQM systems.

3 REVIEW RESULTS

Table 1 shows the studies along with different aspects that have been considered to perform the proposed review process and to find the answer to the proposed research questions.

4 RESULTS ANALYSIS

Table 1 shows the MCU (Micro-controller units) distribution, Display interface, Storage technology, communication technology, and Notification/alert medium that have been utilized for the deployment of AQM system. The results of column 2 show that Raspberry Pi and Arduino were most likely preferred by previous studies presented in Table 1, which are 32% and 34% of the total studies presented. Additionally, the ESP8266 module has also been utilized by many studies that are about 30% of the total studies. All three MCUs are open source platforms that are publicly available to monitor real-life application. Thus, Arduino is mostly used MCU, thereby giving the answer to research question 1. The results of column 3 of Table 1 show the medium to display the real-time statuses of AQM parameters. It can be seen that near about 50% of studies have been used the web platform to display the characteristics of AQM parameters. However,

437

Table 1. Related studies.

Authors	Microcontrollers used	Display interface	Storage technology	Communication technology	Notification medium
Kumar et al. (2019)	ESP8266	Smartphone app (SA)	IoT data storage service	Wifi	SMS
Karami et al. (2018)	Arduino Uno	WebPortal/ server (WP/S)	CS	Zigbee	Email
Rahman et al. (2019)	Arduino Nano, Raspberry Pi	WP/S	IoT datastore service	Wifi	Email
Folea et al. (2020)	Raspberry Pi3	WP/S, SA	CS	Bluetooth	Email
Kodali et al. (2019)	ESP8266	SA	CS	Wifi	Mobile notifications
Pradityo et al. (2019)	Arduino Uno, Raspberry Pi	WP/S, SA	CS	Wifi	Email
Alexandrova et al (2018)	Arduino Uno	SA	Local server	Bluetooth	Mobile notifications
Muladi et al. (2018)	ESP8266, Arduino Mega	WP/S	CS	Wifi	-
Marques et al. (2020)	ESP32	WP/S, SA	CS	Bluetooth	Mobile notifications
Hsu et al. (2020)	Arduino Uno, Raspberry Pi	SA	CS	Wifi	SMS
Chiesa et al. (2019)	Arduino Uno, Raspberry Pi3B+	SA, LCD	CS	Wifi	-
Marques et al. (2019)	ESP8266	WP/S	IoT datastore service	Wifi	Mobile notifications
Mois et al. (2018)	Raspberry Pi, ARM cortex MO	WP/S	CS	Bluetooth	-
Benammar et al. (2018)	Raspberry Pi2	WP/S	IoT datastore service	ZigBee	-

most of these platforms require login-password to check the updates of AQM parameters. Since, smartphone apps allow viewers to be up-to-date concerning the updation of AQM parameters, thereby providing a reliable solution for real-time AQM systems, hence answering the research question 2. Column 4 of the table shows the data storage technologies utilized by studies taken into consideration in this paper. The results show that nearly 60% of studies have preferred cloud servers to store AQM data because they provide easily accessible updates on AQM parameters globally thus giving the answer to research question 3. Nearly 25% of studies used IoT as a platform for data storage for AQM parameters' update and monitoring system. It can be seen from the results that ZigBee, Bluetooth, and wifi are widely preferred communication methods for real-time AQM systems in their increasing order of performance by studies. ZigBee and Bluetooth provide a reliable solution for communication technology requirements in AQM systems due to their low power consumption requirements. Wifi-enabled devices consume more power as compare to the ZigBee and Bluetooth enabled devices, thus answering the research question 4.

5 CONCLUSION

This paper provides a systematic review of AQM and purification systems that have been propounded by various researchers in the last 3 years (2018-2020). IoT based methods have been taken into consideration to select the publications. From the review process, it has been concluded that MCU that is most widely used is Arduino and Raspberry pi. In addition to this wifi has been the most widely used communication technology, however, this is not reliable from the perspective of energy consumption, as this consumes more power than the other two communication technology used that are: Bluetooth and ZigBee. These technologies can be set to appropriate calibrated arrangements in order to boost the performance of AQM systems. From the perspective of further identification of best technology in air quality monitoring and purification systems, the inclusion scope can be extended like only IoT-based papers have been selected in this study, hence publications on WSN based AQM systems can also be selected in addition to the IoT based systems. Furthermore, more databases can be searched to include more studies into consideration. Submission of material to the editor.

REFERENCES

Alexandrova, E.; Ahmadinia, A. Real-Time Intelligent Air Quality Evaluation on a Resource-Constrained Embedded Platform. In Proceedings of the 2018 IEEE 4th International Conference on Big Data Security on Cloud (BigDataSecurity), IEEE International Conference on High Performance and Smart Computing, (HPSC) and IEEE International Conference on Intelligent Data and Security (IDS), Omaha, NE, USA, 3–5 May 2018; pp. 165–170.

Benammar, M.; Abdaoui, A.; Ahmad, S.; Touati, F.; Kadri, A. A Modular IoT Platform for Real-Time Indoor Air Quality Monitoring. Sensors 2018, 18, 581.

Chiesa, G.; Cesari, S.; Garcia, M.; Issa, M.; Li, S. Multisensor IoT platform for optimizing IAQ levels in buildings through a smart ventilation system. Sustainability 2019, 11, 5777.

Folea, S.C.; Mois, G.D. Lessons Learned from the Development of wireless Environmental Sensors. IEEE Trans. Instrum. Meas. 2020, 69, 3470–3480.

Hsu, W.-L.; Chen, W.-T.; Kuo, H.-H.; Shiau, Y.-C.; Chern, T.-Y.; Lai, S.-C.; Fan, W.-H. Establishment of a smart living environment control system. Sens. Mater. 2020, 32, 183.

Rahman, M.; Rahman, A.; Hong, H.-J.; Pan, L.-W.; Sarwar Uddin, M.Y.; Venkatasubramanian, N.; Hsu, C.-H. An adaptive IoT platform on budgeted 3G data plans. J. Syst. Archit. 2019, 97, 65–76.

Karami, M.; McMorrow, G.V.; Wang, L. Continuous monitoring of indoor environmental quality using an Arduino-based data acquisition system. J. Build. Eng. 2018, 19, 412–419.

Kodali, R.K.; Rajanarayanan, S.C. IoT based Indoor Air Quality Monitoring System. In Proceedings of the 2019 International Conference on Wireless Communications Signal Processing and Networking (WiSPNET), Chennai, India, 21–23 March 2019; pp. 1–5.

Kumar Sai, K.B.; Mukherjee, S.; Parveen Sultana, H. Low-Cost IoT based air quality monitoring setup using Arduino and MQ series sensors with dataset analysis. Procedia Comput. Sci. 2019, 165, 322–327.

Marques, G.; Miranda, N.; Kumar Bhoi, A.; Garcia-Zapirain, B.; Hamrioui, S.; de la Torre Díez, I. Internet of Things and enhanced living environments: Measuring and mapping air quality using cyber-physical systems and mobile computing technologies. Sensors 2020, 20, 720.

Marques, G.; Pitarma, R. An Internet of Things-Based environmental quality management system to supervise the indoor laboratory conditions. Appl. Sci. 2019, 9, 438.

Mois, G.D.; Sanislav, T.; Folea, S.C.; Zeadally, S. Performance evaluation of energy-autonomous sensors using power-harvesting beacons for environmental monitoring in the Internet of Things (IoT). Sensors 2018, 1709.

Muladi, M.; Sundari, S.; Widiyaningtyas, T. Real-Time Indoor Air Quality Monitoring Using Internet of Things at University. In Proceedings of the 2018 2nd Borneo International Conference on Applied Mathematics and Engineering (BICAME), Balikpapan, Indonesia, 10–11 December 2018; pp. 169–173.

Pradityo, F.; Surantha, N. Indoor Air Quality Monitoring and Controlling System based on IoT and Fuzzy Logic. In Proceedings of the 2019 7th International Conference on Information and Communication Technology (ICoICT), Kuala Lumpur, Malaysia, 24–26 July 2019; pp. 1–6.

Recent Trends in Communication and Electronics – Sharma et al. (Eds)
© 2021 Taylor & Francis Group, LLC, ISBN 978-1-032-04572-6

Compact circularly polarized asymmetrical shaped DRA for UWB applications

Sachin Kumar Yadav, Amanpreet Kaur & Rajesh Khanna
Electronics and communication engineering, Thapar institute of engineering & technology, Patiala, India

ABSTRACT: A Circularly Polarized (CP) Dielectric Resonator Antenna (DRA) consists of two ceramic blocks, segmented to each other with similar permittivity material ($\varepsilon_d = 9.8$), these blocks are called dielectric resonator (DR), excited by transformer type microstrip feed-line with 50 Ω. The antenna covers the ultra-wideband (UWB) frequency range between 6.3 GHz to 12.5 GHz with impedance bandwidth of 65.95 % is achieved. A maximum gain 5.65 dB is reported at 10.7 GHz frequency. The novelty of this antenna has an axial ratio (AR ≤ 3) for UWB range and simulated AR bandwidth (6.7 to 12.3 GHz) is 58.8%. This antenna is used for military applications, UWB RADAR, wireless short-range.

Keywords: dielectric resonator antenna, circular polarization, UWB

1 INTRODUCTION

In a compact size antenna for ultra-wideband (UWB) communication system (Docket 2002), to achieve convenience attention in industrial and academic communities in telecommunication. The applications of UWB frequency range is defined the 3.1 to 10.6 GHz for the unlicensed band. The designs of UWB antenna can be divided into four types, such as structures, self-complementary structures, travelling wave structures and structure having multiple reflections, Because of remarkable features the DRA have received agreeable consideration for UWB applications (Ryu & Kishk 2010, Majeed et al. 2015, Petosa et al. 1998, Abedian & Rahim 2012). The UWB range 6.6 GHz to 11.3 GHz is defined (Majeed et al. 2015). The DRA is considered to give an efficient solution for achieving the widen impedance bandwidth and size reduction. It has major characteristics like low power losses, light in weight, easy to fabricate, high radiation efficiency, small antenna size and different excitation mechanism. Recently the UWB monopole DRA designs based on linear polarization (LP). UWB antenna have limited distance by low transmitting power. LP have many mismatches or misalignment losses to the receiving side, to control losses by using circularly polarized (CP) antenna in transmitting and receiving side (Pan & Leung 2012, Fakhte et al. 2015, Varshney et al. 2018).

2 ANTENNA GEOMETRY AND CONFIGURATION

The geometry of the proposed circular polarized segmented DRA is shown in Figure 1 and the design of compact size DRA with UWB applications. The antenna was simulated and fabricated on the FR_4 substrate ($25 \times 25 \times 0.8$). As in Figure 1a-b. DR radiator is fed by a transformer type microstrip (MS) feed line with an SMA connector. Since two blocks are considered in superstructure of DR antenna and feedline on the same plane, the bottom side of substrate layer metal part is used as a partial ground plane to achieve UWB range. The

DOI: 10.1201/9781003193838-81

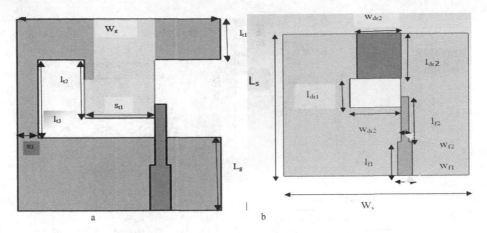

Figure 1. Compact size of DRA (a) front view (b) Back view.

feeding network ensures good impedance matching in a wide range of frequencies. The micro-strip feed-line have two parts $l_{f1} \times w_{f1}$ and $l_{f2} \times w_{f2}$ add to each other with 50 Ω impedance matching. The parameter of the substrate is denoted by L_s, W_s and t_s, ground plane has L_g, W_g, with adding F-shape metallic strip with dimension variables l_{t1}, l_{t2}, l_{t3}, s_{t1}, s_{t2} to increase the gain of the antenna, impedance bandwidth and axial ratio ≤3 shown in Figure 1b. The DR is built from a two rectangular substrate with relative permittivity of 9.8, first block height is 4mm, the second block is 8mm, and side lengths are l_{dr1}, l_{dr2} and side width w_{dr1}, w_{dr2} respectively, shown in Figure 1a.

Transformer type MS feed line with Parameter, side width w_{f1}, w_{f2} and lengths l_{f1}, l_{f2}, it is exciting to DRA beginning side inset corner. A strong E-field distribution within the DR is reported.

The simulated E-Field configuration has been analysed in the proposed super DRA at different frequencies (Varshney et al. 2018), as shown in Figure 2. Resonance verifies

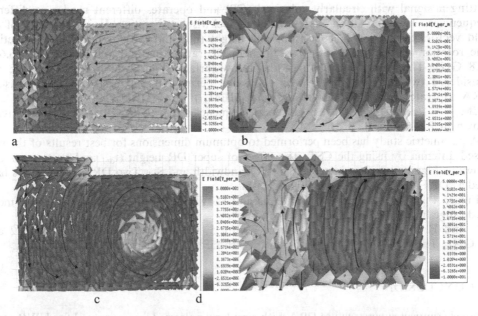

Figure 2. E field distribution inside DR at different frequencies and modes: (a) 6.5 GHz, fundamental mode $HEM_{111+\delta}$ (b) 9 GHz, $HEM_{112+\delta}$ (c) 10 GHz, $HEM_{122+\delta}$ (d) 10.5 GHz, $HEM_{212+\delta}$.

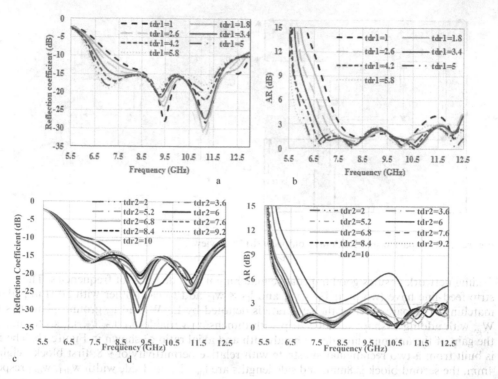

Figure 3. Reflection coefficient and AR for various parameter varies according to antenna dimensions.

from peaks in the Z real impedance. The E – Field distribution depends upon the near field of the antenna geometry. The proposed antenna operates with the hybrid electro-magnetic mode (HEM). The specific structure of the antenna to create the higher-order hybrid modes along with the fundamental mode. So antenna will be efficient for trans-mitting a signal with circularly polarized (CP) and operates different modes at different frequencies. The simulation represents the clockwise and anticlockwise rotation of the field vectors from 0 to 2π concerning the height (t_{dr1} and t_{dr2}) of DRA in Z-direction. The rotation of field vectors is fixed to help CP radiations in UWB range from 6.6 to 10.8 GHz. The field distribution of the DRA surface slightly divided because the DR considered two different height with the same permittivity material. So these reasons, DRA operates HEM mode, modes are $HEM_{111 + \delta}$, $HEM_{112 + \delta}$, $HEM_{112 + \delta}$, $HEM_{122 + \delta}$ and $HEM_{212 + \delta}$ at different frequencies 6.5, 9, 10, 10.5 GHz, respectively, shown in Figure 2 a-d.

A parametric study has been performed to optimum dimensions for best results of the pro-posed antenna by using the CST. The effect of super DR height (t_{dr1} and t_{dr2}) varies that shows in Figure 3 (a-d) to improve the AR bandwidth and S_{11}. Two DR combinations have used as DR1 and DR2 to show in Figure 1a-b.

Figure 4 is shown the fabricated antenna and VSWR screenshot in VNA, covers the meas-ured range from 5.8 to11.8 GHz frequency.

E and H co-polarization is shown in Figure 5, the measured pattern at 7.5, 9.2 and 11 GHz. This antenna radiates efficiently with circular polarization characteristics.

3 CONCLUSION

A novel compact superstructure DRA with circularly polarized was presented for UWB appli-cations. The proposed antenna has been two blocks of DR with different height, which is

Figure 4. Fabricated antenna and VSWR of VNA screenshot.

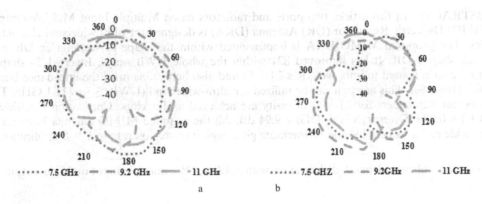

······ 7.5 GHz ── 9.2 GHz ── 11 GHz   ······ 7.5 GHZ ── 9.2GHz ── 11 GHz

a   b

Figure 5. Measured radiation pattern (a) E-Plane (b) H-Plane.

excited by transformer type MS feed-line with 50 Ω and F-shaped metallic strip added to the partial ground plane. By modified the F-shaped metallic strip and DR structure to give the widen bandwidth 65.65% and AR≤3. This antenna peak gain (5.65 dB) is reported at 10.8 GHz, processing range from 6.7 to 12.3 GHz with the circularly polarised band for a satellite communication system with fulfilling C and X band applications.

REFERENCES

Docket, E.T. 2002. FCC, Revision of part 15 of the commission's rules regarding ultra- wideband transmission systems. *Technical Report*; Docket 98–153.

Ryu, K.S. & Kishk, A.A. 2010. Ultra wideband dielectric resonator antenna with broadside patterns mounted on a vertical ground plane edge. *IEEE Transactions on Antennas and Propagation*. 58(4): 1047–1053.

Majeed, A.H. et al. 2015. Balanced dual-segment cylindrical dielectric resonator antennas for ultra-wideband applications. *IET Microwaves, Antennas & Propagation*. 9(13): 1478–1486.

Petosa A.M. et al. 1998. Recent advances in dielectric-resonator antenna technology. *IEEE Antennas and Propagation Magazine*. 40(3): 35–48.

Abedian, M. & Rahim, S.K. 2012. Two-segment compact dielectric resonator antenna for UWB application. *IEEE antennas and wireless propagation letters*. 11:1533–1536.

Pan, Y.M. & Leung, K.W. 2012. Wideband omnidirectional circularly polarized dielectric resonator antenna with parasitic strips. *IEEE Transactions on Antennas and Propagation*. 60(6): 2992–2997.

Fakhte, S. et al. 2015. A new wideband circularly polarized stair-shaped dielectric resonator antenna. *IEEE Transactions on Antennas and Propagation*. 63(4): 1828–1832.

Varshney, G. et al. 2018. Axial ratio bandwidth enhancement of a circularly polarized rectangular dielectric resonator antenna. *International Journal of Microwave and Wireless Technologies*. 10(8): 984–990.

Recent Trends in Communication and Electronics – Sharma et al. (Eds)
© 2021 Taylor & Francis Group, LLC, ISBN 978-1-032-04572-6

High gain MIMO dielectric resonator antenna for UWB applications

Sachin Kumar Yadav, Amanpreet Kaur & Rajesh Khanna
Electronics and Communication Engineering Department, Thapar Institute of Engineering & Technology Patiala, India

ABSTRACT: In this article, two ports and radiators based Multiple Input Multiple-Output (MIMO) Dielectric Resonator (DR) Antenna (DRA) is designed for ultra-wideband characteristics. The proposed MIMO DRA is implemented within the shape of rectangular DR into a rack shaped DR. It has improved S21 within the whole UWB range. Inverted T- shaped parasitic strip is used to improve AR ≤ 24.9 dB and also helps to control the impedance bandwidth (106.2 %). This antenna can be utilized for ultra-wideband (UWB) 3.3 to 11.1 GHz. The important parameters for MIMO diversity are achieved as Envelope Correlation Coefficient (ECC) ≤ 0.006, Diversity Gain (DG) ≥ 9.94 dB. All the acquired MIMO antenna boundaries are inside as far as possible and furthermore gives high information rate to UWB applications.

Keywords: ultra-wideband, dielectric resonator, ECC, DG, multiple input multiple-output

1 INTRODUCTION

After the characterization of the 3.1– 10.6 GHz unlicensed band by the Federal Communications Commission (FCC), the ultra-wideband (UWB) range was restored. UWB marks its place in enormous fields of wireless sensor systems because it demands low power spectral density, the equipment required are of the fundamental design, information transmission rates are quite good, imparts high accuracy extending ability with low power utilization (ET-Docket 2002). Dielectric resonator antenna (DRA) has made its path in recent years for UWB applications. DRA possesses several striking features, including low conductor losses, high radiation efficiency, inherent design flexibility, and easy to excitation. As the dimensions of the dielectric resonator (DR) is inversely correlated with the permittivity, the small size of DR is preferred for high permittivity (Luk & Leung 2003).

Unlike the obstacles in traditional wireless devices, multiple-input multiple-output (MIMO) antenna has been developed to obtain the high data rate, less channel capacity loss, high diversity gain (Abedian et al. 2017). To give high isolation between the spaces, multiple antennas are employed at the transmitting and receiving sites to implement the MIMO (Wu et al. 2018). In recent years, the MIMO antenna concept has been developed for UWB applications (Yadav et al. 2018, Zhang et al. 2009, Li et al. 2013).

In the present study, a new MIMO rectangular two-element dielectric resonators are presented with keeping the distance of $\lambda/2$ to achieve the good isolation. T- shaped MPS is fixed on the top of substrate layer to achieve the widen impedance bandwidth and AR ≤ 24.9 dB. Antenna size is 39.6×29 mm^2 ($\varepsilon_r = 4.4$) with 0.8 mm height. DR structure along with feedline is etched on the upper side of the substrate layer.

DOI: 10.1201/9781003193838-82

2 ANTENNA CONFIGURATION

Figure 1 presents the view of the proposed MIMO DRA. DRA consists of two radiator elements having permittivity (ε_{DR} = 10.2) Rogers RT 6010. The dimensions of DR is 14 × 9 ×5, excited by quarter-wave transformer type microstrip fed (4 × 1.9; 11 × 0.8). As Figure 1 (a & c), Inverted T-shaped parasitic strip (PS) is designed and etched on the top of the substrate, placed between DRs. Ground (37.6 × 9) is etched on the bottom side of the substrate (37.6 × 29 × 0.8 mm³). A stub (9.2 × 3.5) is added in ground to improve the impedance matching and S21. The AR of antenna is controlled by inverted T-shaped PS with P_1, P_2, P_3, and P_4 (16.8, 7.9, 2 and 2.6 mm) parameters.

2.1 Antenna iterations

The iterations of DRA are shown in Figure 2 a-e. The simulated antenna follow five steps to achieve the final design DRA 5. Step1 have rectangular DR with modified ground plane, which provided peak AR 40dB and constant gain. In step2, a matching stub is added in the ground to achieve the better impedance matching and transmission coefficient (S21). Step 3 have modified the shape of DR to achieve the better transmission coefficient with slightly improved the impedance bandwidth. In step4, the electrical path of antenna is increased by using of inverted T-shaped PS element to achieve the UWB range. In step 5, proposed antenna is designed with a rectangular DGS slot cut in the ground, to achieve the high peak gain (9.4 dBi) and UWB range. Figure 3a-d are shown the steps performance results in Table 1. Proposed MIMO DRA5 have best results within UWB range and improved AR bandwidth ≤ 24.9 dB.

2.2 In the metallic part of antenna surface current distribution

In Figure 4, is shown the current distribution in the antenna. It can be observed from Figure 4 a-b with DGS slot all current flow in a single direction and supporting the fundamental mode.

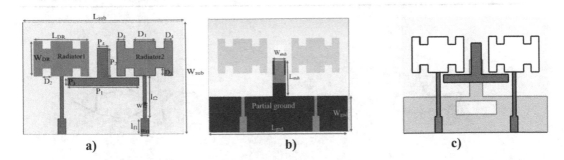

a)　　　　　　　　　　　b)　　　　　　　　　　　c)

Figure 1.　a-c MIMO antenna different views.

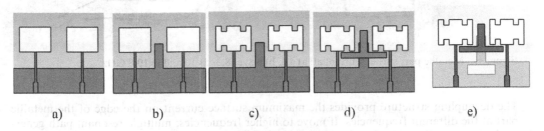

a)　　　　b)　　　　c)　　　　d)　　　　e)

Figure 2.　Steps performance view of antenna a) DRA 1 b) DRA 2 c) DRA 3 d) DRA 4 e) DRA 5.

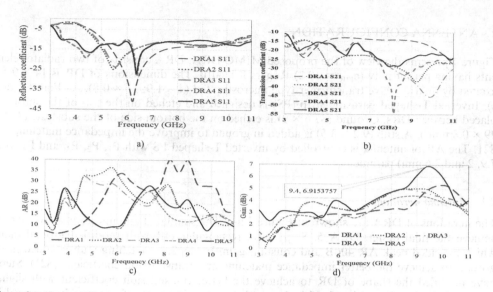

Figure 3. Frequency response concerning a) S_{11}, S_{22} parameters b) S_{21} parameter c) AR bandwidth d) Gain vs. frequency.

Table 1. Step performance of all antennas.

Step antenna	S_{11}, S_{22} 10 dB	Impedance bandwidth %	S_{12}, S_{21} (dB) Isolation	Gain range (dBi)	Axial ratio (dB)
DRA1	5.09 to 10.1 GHz	66	12 to 26	0.5 to 5.4	6 to 41
DRA2	4.9 to 10.12 GHz	69.34	15 to 36	0.6 to 5.1	7.2 to 36.5
DRA3	4.8 to 10.28 GHz	72.67	16 to 46.1	0.8 to 5	7.1 to 33
DRA4	3.54 to 10.89 GHz	101.87	15.5 to 54.2	1.5 to 4.5	8.1 to 25.4
DRA5	3.32 to 11.1 GHz	106.2	16.8 to 29	1.8 to 9.4	8.1 to 27.3

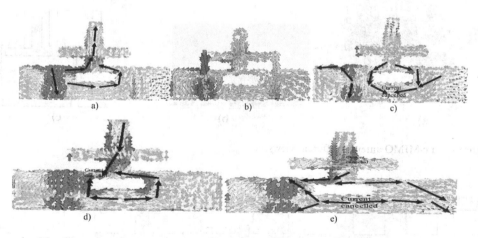

Figure 4. Metallic part surface current at: a) 4.4, b) 6.3, c) 8.3, d) 10 and e) 10.8 GHz.

The decoupling structure provides the maximum surface current on the edge of the metallic part at the different frequencies. If move to higher frequencies, multiple resonant path generated to support the widen bandwidth and higher order modes as Figure 4c-e.

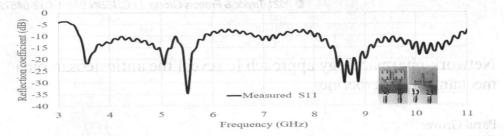

Figure5. S11 measured.

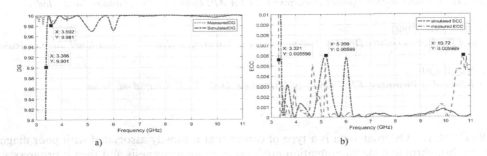

Figur 6. a. DG b. ECC.

Figure 5, shows the measured S11 with cover the frequency from 3.4 to 10.8 GHz. The ECC is shown in Figure 6b with ≤ 0.0059 existing UWB range (Zhang et al. 2009, Li et al. 2013. The other important MIMO parameter DG is shown in Figure 6a to give the informatics signal through the high capability of the diversity, calculated through S-parameters.

3 CONCLUSION

Proposed high gain UWB MIMO DRA excited by a filter type microstrip fed is designed for UWB applications. It covers frequencies from 3.32 GHz to 11.1 GHz (106.2 %), supporting partial S, C, and X bands for high data rate applications. An inverted T-shaped parasitic strip is used to improve AR ≤ 24.9 dB in UWB range. Further, a rectangular DGS slot cut in ground to improve the impedance bandwidth and high gain within UWB range. The performance of MIMO antenna is reported low ECC ≤ 0.0059, DG ≥ 9.94dB, and transmission coefficient ≤ -16.8 dB, it is used for high data rate applications.

REFERENCES

ET-Docket 2002. Revision of part 15 of the commission's rules regarding ultra-wideband transmission systems. Federal Communication Commission. 98–153, FCC 02–48.

Luk, K.M. & Leung, K.W. 2003. Dielectric Resonator Antenna. U. K. Research Studies Press. The UK.

Abedian, M. et al. 2017. Compact ultrawideband MIMO dielectric resonator antennas with WLAN band rejection. IET Microwaves, Antennas & Propagation. 11: 1524–1529.

Wu, Y. et al. 2018. Design of a compact UWB MIMO antenna without decoupling structure. International Journal of Antennas and Propagation Magazine.

Yadav,Dinesh et al. 2018. Two element band-notched UWB MIMO antenna with high and uniform isolation. Prog. Electromagn. Res. 63: 119–129.

Zhang, Shuai et al. 2009. Ultrawideband MIMO/diversity antennas with a tree-like structure to enhance wideband isolation. IEEE Antennas and Wireless Propagation Letters. 8: 1279–1282.

Li, Yingsong et al. 2013. A multi-band/UWB MIMO/diversity antenna with an enhance isolation using radial stub loaded resonator. Applied Computational Electromagnetics Society Journal. 28: 8–20.

Recent Trends in Communication and Electronics – Sharma et al. (Eds)
© 2021 Taylor & Francis Group, LLC, ISBN 978-1-032-04572-6

Network pharmacology approach to reveal the antiosteosarcoma mechanism of berberine

Parul Grover
KIET School of Pharmacy, KIET Group of Institutions, Delhi-NCR, Ghaziabad, India

Lovekesh Mehta
Analytical Research & Development Department, TEVA API India Pvt. Ltd., Greater Noida, India

Monika Bhardwaj
Natural Product Chemistry Division, Indian Institute of Integrative Medicine, Canal Road, Jammu, India

K. Nagarajan
KIET School of Pharmacy, KIET Group of Institutions, Delhi-NCR, Ghaziabad, India

ABSTRACT: Osteosarcoma is a type of cancer that is mostly associated with poor diagnosis. In this, chronic bone inflammation predisposes to tumorigenesis and then it progresses to osteosarcoma. Chemotherapy is mostly used for its treatment but it results in number of toxicities. So, in search of finding safe and effective treatment, plant derived products are used as alternative. Latest research reveals use of berberine in treatment of osteosarcoma but its mechanism is still unknown. So, current study was conducted to identify the potential target of berberine in human body and to identify its mechanism. From current study, it has been concluded that berberine can efficiently suppress the osteosarcoma proliferation and control the TP53, CHEK2, VEGF, EGFR, HER-3, HER-4, JUN MAPK1, ATM and H2AFX expression which may potentially be used for treatment of osteosarcoma.

1 INTRODUCTION

Berberine, a natural origin isoquinoline alkaloid shows various pharmacological effects. It is a principal component that is isolated from roots and stems of plants of Berberis species including *B. darwinii* (Habtermariam 2013), *B. aristata* (Potdar et al., 2012), *B. vulgaris* (Shou et al., 1998) and *B. petiolaris* (Singh et al., 2015). Recent reports suggests that berberine can play a role in treatment of inflammation (Habtermariam 2016), diabetes (Liang et al., 2019), obesity (Tabeshpour et al., 2017), dementia (Shinjyo et al., 2020) and observed to have significant anticancer effects on a spectrum (broad) of carcinomas encompassing osteosarcoma (Luo et al., 2019). Osteosarcoma has a predilection for developing in rapidly growing bone. It is the most common primary bone malignancy and is lethal in both adults and children. Most frequently it occurs in early adulthood and its incidence again peaks in older populations with greater than 65 years of age. Most of the patients with osteosarcoma have micrometastatic or metastatic disease at diagnosis and its treatment is very complicated. Some of the clinical methods can reach to remission and chemotherapy is getting important for remission. But there are number of toxicities and side effects that are associated with chemotherapy viz., cardiotoxicity, renal toxicity, pulmonary toxicity, cutaneous toxicity, immunosuppression, alopecia, etc. So, plant derived products can be used as alternative therapy that may provide chemo protective potential against cancer. Some plant products show promising anticancer effects

DOI: 10.1201/9781003193838-83

like latest research reveals use of berberine in treatment of osteosarcoma. But still the mechanisms underlying the effect of osteosarcoma are poorly understood till date and its explanation is vital for development of improved therapies. To explain the mechanism of berberine in osteosarcoma, network pharmacology approach may be used in which we understand the complex interaction between our molecule(s) and their targets (cellular) that contribute to efficacy of molecule and its side effects. This approach is useful in prediction of pathways and potent targets of a molecule/compound in various diseases.

In the current study, network pharmacology is used to explain different pathway and targets belongs to osteosarcoma. Hence, purpose of current study is (i) potential therapeutic agent identification and (ii) depiction of mechanism of berberine for osteosarcoma treatment.

2 MATERIALS AND METHODS

2.1 Berberine proteins target

SEA database (http://sea.bkslab.org/) was used to obtain target proteins of berberine. Compound information was obtained by entering the simplified molecular input line entry specification (SMILES). Human target proteins, amongst all target proteins were selected and downloaded.

2.2 Potential target proteins in osteosarcoma

The next step was to identify target proteins associated with disease osteosarcoma. Two databases viz. DisGeNET (http://www.disgenet.org) and STRING database (http://string-db.org) were used for this. Protein targets obtained from both the databases were collected and combined and then 112 proteins were considered as potential targets in osteosarcoma regulation.

2.3 Interaction of protein-protein data

Information about experimental and predicted interactions of proteins was gathered from PPI (protein-protein interaction) data with the help of STRING database. Furthermore, confidence range (confidence score (low) <0.4; medium 0.4-0.7; high: >0.7) for PPI data was also established by these databases. PPI data scores which have higher confidence range (>0.7) were chosen for additional research study.

2.4 Construction of network

Berberine-targets-osteosarcoma network, PPI network and network of berberine-target were constructed. SEA database was used to construct the berberine-target network through linkage of berberine and its target. DisGeNET and STRING database were used for establishment of other two network by linking the proteins that relate with another protein. On the basis of these two networks, the merged network of berberine-targets-osteosarcoma network was established. Cytoscape (http://cytoscape.org) was used for visualization of all the three networks.

2.5 Enrichment analysis

For carrying out the experiments of analysis of enrichment, KEEG (Kyoto Encyclopedia of Genes and Genomes) pathway analysis enrichment was accomplished. This is done for analyzing gene expression data that depends on the differentially expressed genes functionality annotation.

3 RESULTS AND DISCUSSION

Thirteen human target proteins were identified for berberine. Based on DisGeNET and String database, for treatment of osteosarcoma, 112 proteins were selected as the infection based targets.

In the osteosarcoma pathological process, thirteen proteins, that includes TP53, VEGFA, RB1, RUNX2, MMP2, MDM2, DHFR, MET, RFC1, JUN and TNFRSF11A, CHEK2, EGFR were reported to play a crucial role among total of 112 proteins. The other remaining proteins (99) can directly control the thirteen proteins and indirectly companion with osteosarcoma. To further define the berberine mechanism against the infection/disease, the proteins which are considered as targets generally combines/associated were linked with the berberine target proteins. The PPI integrated network is shown in Figure 1. After the processing through software, some proteins found to have very crucial targets in berberine plant constituents for treating the osteosarcoma. Among them, most crucial proteins are Checkpoint kinase 2 (CHEK2), E3 ubiquitin-protein ligase Mdm2 (MDM2), ATM serine/threonine kinase (ATM), Serine/threonine-protein kinase (ATR), TP53 (Cellular tumor antigen), CDC25C (M-phase inducer phosphatase 3), JUN (Jun proto-oncogene), BGLAP (Bone gamma carboxyglutamate protein), H2AFX (H2A histone family member X), EGF (epidermal growth factor), CDC25A (M-phase inducer phosphatase 1). The KEGG enrichment analysis evaluated the module of this study.

The result shows the targets in berberine treating osteosarcoma. TP53, is a main active protein in self-investigation of DNA damage, that it can internment the G1 phase cell if there is fault in DNA completeness and continuity. TP53 mutations were linked with 2-year overall survival in osteosarcoma patients. TP-53 carries directions for bring together the p53 protein, that acting a part in repair of DNA and also in death of cells. p53 Inactivation benefits tumour cells to persist therapy of radiations.

Checkpoint kinase 2 (CHEK2) is a suppressor of tumour gene which encrypts the CHK2 protein. It is taken part in repair of DNA and also in apoptosis in case of DNA damage. CHEK2 gene mutations are thought to confer a predisposition to sarcomas, especially the primary bone tumor osteosarcoma. Immune check points can be considered as potential targets for the treatment of osteosarcoma.

VEGF (Vascular endothelial growth factor) expression and osteosarcoma patient's survival rates have been correlated in angiogenesis studies, in which VEGF expression above 30%, the type of surgery in limb sparing reduced to 14 patients as compared to 29 patients (<=30%) VEGF expression) and the amputation in patients dropped to zero as compared to 7 patients (with <=30%) VEGF expression) which clearly indicate that higher the VEGFA expression (>30%), may have improved outcome of results in osteosarcoma patients (Baptista et al., 2014).

The study made by Wang et al., have described that the concept of EGFR/HER-3, HER-3/HER-4m and EGFR/HER-4 hetero dimerization in combinations with HER-4, EGFR and HER-3 expression proved to be clinically give beneficial outcome for the patients suffered from osteosarcoma which revealed that EGFR is very important for treating the patients suffering from osteosarcoma (Wang et al., 2018).

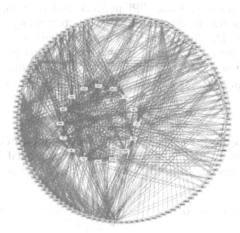

Figure 1. PPI network of osteosarcoma targets (yellow squares represent targets related to osteosarcoma; blue squares represent other human proteins which are directly interacting with the osteosarcoma targets).

One the most crucial mechanism in enhancement of osteosarcoma is DSB (Double stranded DNA break) which is mainly a cause of escaping of self-surveillance which may lead to various kind of errors in gathering of DNA and also cause the gene deviation (oncogenic) results augmentation in the osteosarcoma proliferation. For DSB (Double stranded DNA break) downstream regulation pathway of signalling of C-JUN. The major components of this pathway comprised of JUN and MAPK1 (Luo et al., 2019).

Establishment and phosphorylation of the H2AFX on Ser139 was done by ATM after the occurrence of DSB. This phosphorylation of H2AFX was very helpful in stabilizing the binding of DSB along with TP53 binding protein (P53BP1).

4 CONCLUSION

Till date, mechanisms underlying the of berberine effect as antiproliferative in treatment of osteosarcoma remain poorly understood, therefore in the current study, with the help of network pharmacology, we have predicted the mechanism of berberine for treating osteosarcoma Hence, all the mechanistic approaches suggests clearly that the targets TP53, CHEK2, VEGFA, EGFR, H2AFX and JUN are most promising for osteosarcoma from the 13 targets obtained through network pharmacology with regard to the drug candidate berberine. Further clinical studies are necessary to investigate the same with regard to above promising targets for necessary validation in near future. The present work can be very useful in understanding and will give a new insight to researchers for further demonstrating the mechanism of berberine treating osteosarcoma.

REFERENCES

Baptista, A.M., Camargo, A.F.D.F., Filippi, R.Z., Oliveira, C.R.G.C.M.D., Azevedo Neto, R.S.D. & Camargo, O.P.D. 2014. Correlation between the expression of vegf and survival in osteosarcoma. *Acta ortopedica brasileira*. 22(5): 250–255.

Habtemariam, S. 2013. The hidden treasure in Europe's garden plants: Case examples; Berberis darwinni and Bergenia cordifolia. *Medicinal & Aromatic Plants*. 2(4): 132–135.

Habtemariam, S. 2016. Berberine and inflammatory bowel disease: A concise review. *Pharmacological research*. 113: 592–599.

Liang, Y., Xu, X., Yin, M., Zhang, Y., Huang, L., Chen, R. & Ni, J. 2019. Effects of berberine on blood glucose in patients with type 2 diabetes mellitus: A systematic literature review and a meta-analysis. *Endocrine journal*. 66(1): 51–63.

Luo, Q., Shi, X., Ding, J., Ma, Z., Chen, X., Leng, Y., Zhang, X. & Liu, Y. 2019. Network pharmacology integrated molecular docking reveals the antiosteosarcoma mechanism of biochanin A. *Evidence-Based Complementary and Alternative Medicine*. https://doi.org/10.1155/2019/1410495.

Potdar, D., Hirwani, R.R. & Dhulap, S. 2012. Phyto-chemical and pharmacological applications of Berberis Aristate. *Fitoterapia*. 83(5): 817–830.

Shinjyo, N., Parkinson, J., Bell, J., Katsuno, T. & Bligh, A. 2020. Berberine for prevention of dementia associated with diabetes and its comorbidities: A systematic review. *Journal of Integrative Medicine*. 18(2): 125–151.

Singh, A., Bajpai, V., Srivastava, M., Arya, K.R. and Kumar, B. 2015. Rapid screening and distribution of bio active compounds in differerent parts of Berberis petiolaris using direct analysis in real time mass spectrometry. *Journal of pharmaceutical analysis*. 5(5): 332–335.

Suau, R., Rico, R., López-Romero, J.M., Nájera, F., & Cuevas, A. 1998. Isoquinoline alkaloids from Berberis Vulgaris subsp. australis. *Phytochemistry*. 49(8): 2545–2549.

Tabeshpour, J., Imenshahidi, M. & Hosseinzadeh, H. 2017. A review of the effects of Berberis vulgaris and its major component, berberine, in metabolic syndrome. *Iranian journal of basic medical sciences*. 20(5): 557–568.

Wang, S.L., Zhong, G.X., Wang, X.W., Yu, F.Q., Weng, D.F., Wang, X.X. & Lin, J.H. 2018. Prognostic significance of the expression of HER family members in primary osteosarcoma. *Oncology letters*. 16(2): 2185–2194.

Recent Trends in Communication and Electronics – Sharma et al. (Eds)
© 2021 Taylor & Francis Group, LLC, ISBN 978-1-032-04572-6

Internet of Medical Things (IoMT) based healthcare monitoring system for patients in smart hospitals

Himanshu Sharma, Deepanshi Verma, Shivani Verma & Shekhar
ECE Department, KIET Group of Institutions, Delhi-NCR, Ghaziabad, U.P., India

ABSTRACT: In this modern world, the remote healthcare system using Internet of Medical Things (IoMT) is emerged as the boon for the patients. The elderly patients, rural patients and disabled people who can't reach to the Hospital in emergency situations can take benefit from the IoMT technology for their treatment. The wireless sensors attached to the human body area networks senses the body temperature, heartbeats, brain neurological signals and send to the remote doctor using the internet with the help of smart mobile phone. In this paper, we propose a remote healthcare system using IoMT technology which can be helpful for regular health monitoring and to save the patient's life in emergency conditions.

1 INTRODUCTION

In smart cities, the hospitals are smart such that patient healthcare can be monitored remotely by using Internet of Medical Things (IoMT). An embedded medical equipment connected to the internet is called Internet of Medical Thing (IoMT) device. The IoMT example are X-Ray Machine, Electrocardiogram (ECG) machine, Heartbeat pacemaker computer, laptop, smart phone or camera all connected to the internet (Cao et al. 2020).

As shown in Figure 1(a) & 1(b) there are various sensors connected to the human body wirelessly. This is called wireless body area network (Limaye & Adegbija 2018). The sensors sense various physical healthcare parameters like, glucose level, blood pressure, heartbeat monitoring, body temperature etc. The sensed data is sent to the e-health gateway (Sun et al. 2020). From the gateway router the patient healthcare data is sent to the internet. From internet the patient health care data can be accessed globally for medical services (Habibzadeh et al. 2020). The patient healthcare data can also be stored in large cloud data centers like Amazon AWS, Google Cloud storage etc. (Pasluosta et al. 2015). This is called internet of medical things (IoMT) technology (Liu et al. 2018). The IoMT can also be powered by energy harvesting techniques (Sharma et al. 2019, Sharma et al. 2018) has shown solar energy harvesting for IoT nodes which can be used in IoMT to extend the network lifetime.

The paper is organized as follows: section 1 provides introduction, section 2 presents block diagram & operation, Section 3 gives results & conclusion.

2 BLOCK DIAGRAM AND OPERATION

Figure 2 (a) & 2 (b) shows the basic block diagram of Arduino based patient health care monitoring. Here, the three physical health parameters of the patients are measured using sensors i.e. heartbeat sensor, Temperature sensor, and humidity sensor (Jinhong Guoet al. 2018). The measured health data is sent to the Arduino microcontroller. The Arduino sends

DOI: 10.1201/9781003193838-84

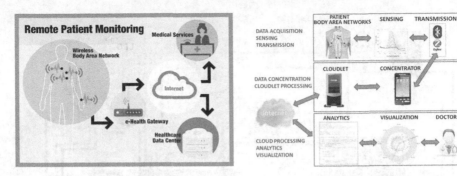

1(a)Remote Patient Monitoring   1(b) Medical Data Cloud Processing

Figure 1.　Remote healthcare system using Internet of Medical Thing (IoMT).

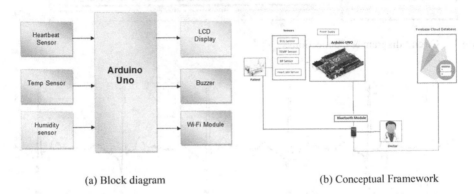

(a) Block diagram   (b) Conceptual Framework

Figure 2.　Block diagram of arduino based patient health care monitoring.

the healthcare data to the LCD display, Buzzer and Wi-Fi module. From the Wi-Fi module data is sent over the internet to the doctor for monitoring & control actions.

The Figure 3 shows the circuit diagram & connections of Arduino board, Wi-Fi module (ESP 8266), and LCD display. The components are Arduino UNO Board, WSP8266 Wi-Fi Module, Pulse Sensor, 16*2 LCD Display, Resistor 1k,2k, LED, Breadboard and Wires. The Arduino UNO board is connected to the system and the android applications. The program code is set on the Arduino studio which measure the pulse rate of the patient using pulse sensor (Zhang 2018). The key challenges in IoMT implementation are security and privacy of patient data.

3 RESULTS AND CONCLUSION

The smart health care monitoring system is designed to work on an android Smartphone or windows computer. The results obtained at the smart phone of a doctor is shown in Figure 4. Here, the remote patient ECG and Blood Pressure (BP) is shown in Figure 4.

If the ECG & BP reaches the normal value, then the doctor immediately alerts the hospital staff by phone call & the patient life can be saved. Furthermore, using the previous recorded data doctor can recommend the solution to the patients before getting into worst situation.

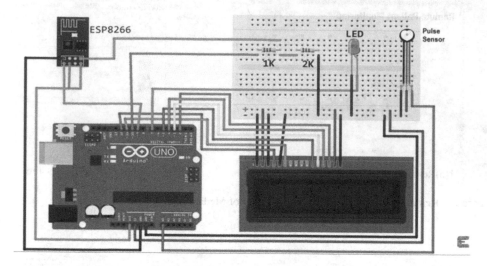

Figure 3.   Circuit diagram & Connections.

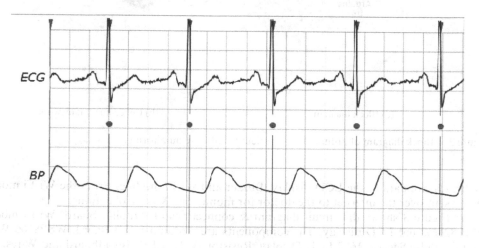

Figure 4.   Patient ECG & Blood pressure monitoring on doctors smart phone.

REFERENCES

Cao R., Tang Z., Liu C., Veeravalli B.,2020. "A Scalable Multicloud Storage Architecture for Cloud-Supported Medical Internet of Things (IoMT), *IEEE Internet of Things Journal*, Volume: 7, Issue 3.

Habibzadeh H., Dinesh K., Shishvan O. R., 2020. "A Survey of Healthcare Internet of Things (HIoT): A Clinical Perspective", *IEEE Internet of Things Journal*, Volume: 15, Issue: 5.

Jinhong Guo, 2018. "Smartphone-Powered Electrochemical Biosensing Dongle for Emerging Medical IoTs Application", *IEEE Transactions on Industrial Informatics*, Volume: 14, Issue: 6.

Limaye A., Adegbija T., 2018. "HERMIT: A Benchmark Suite for the Internet of Medical Things", *IEEE Internet of Things Journal*, Volume: 5, Issue 5.

Liu C., Chen F., Zhao C, 2018. "IPv6-Based Architecture of Community Medical Internet of Things" *IEEE Access*, Volume: 6, USA.

Pasluosta C. F., Gassner H., Winkler J. 2015. "An Emerging Era in the Management of Parkinson's Disease: Wearable Technologies and the Internet of Things", IEEE Journal of Biomedical and Health Informatics, Volume 19, Issue: 6.

Sharma H., Haque A., Jaffery Z. A. 2019. "Maximization of Wireless Sensor Networks Lifetime using Solar Energy Harvesting for Smart Agriculture Monitoring", *Adhoc Networks Journal, Elsevier*, Vol. 94, Netherlands, Europe.

Sharma H., Haque A., Jaffery Z. A., 2018 "Solar energy harvesting wireless sensor network nodes: A survey", Journal of Renewable and Sustainable Energy, *American Institute of Physics (AIP), USA*, vol. 10, no.2, pp.1–33.

Sharma H., Haque A., Jaffery Z. A., 2018. "Modelling and Optimization of a Solar Energy Harvesting System for Wireless Sensor Network Nodes", *Journal of Sensor and Actuator Networks, MDPI, USA*, vol. 7, no. 3, pp.1–19

Sun L., Jiang X., Ren H., Guo Y., 2020. "Edge-Cloud Computing and Artificial Intelligence in Internet of Medical Things: Architecture, Technology and Application", *IEEE Access*, Volume: 8.

Zhang H., Li J., Wen B. 2018, "Connecting Intelligent Things in Smart Hospitals Using NB-IoT", *IEEE Internet of Things Journal*, Volume: 5, Issue: 3.

Recent Trends in Communication and Electronics – Sharma et al. (Eds)
© 2021 Taylor & Francis Group, LLC, ISBN 978-1-032-04572-6

Telemedicine: An application of cloud computing

Ayati Maitra, Swati Tripathi & Achyut Shankar
Computer Science and Engineering, Amity University, Noida, U.P., India

ABSTRACT: Cloud Computing is the part of the ongoing digital revolution and has found its application in many fields. One such domain is Telemedicine in Healthcare. Providing instant and quality medical services at any time irrespective of patient's location to everyone is the aim of any healthcare industry. To realize this goal telemedicine is playing a crucial role. But there are many challenges that continue to impede the growth of Telehealth services. Moreover, Telemedicine is not a new concept but it has only recently seen to gain popularity with improvement in the IT industry and adoption of cloud computing. It is not wrong to say cloud computing is the key for Telemedicine to effectively get utilized. This article briefly presents what is Telemedicine and how Cloud Computing has impacted it to boost its popularity. The importance, benefits, challenges and statistics of people adopting the services of Telemedicine is also discussed.

Keywords: telemedicine, cloud computing, healthcare, telehealth

1 INTRODUCTION

With advancement in the technology sector it is now possible for the healthcare industry to offer many of its services even if doctors are not physically present at the location of their need. By utilizing many technological technique's doctors can remotely come in touch with their patients and examine them. In general, Telemedicine refers to delivery of healthcare services from a distance. A broader term of telemedicine is telehealth. But there is not any significant difference between these two terminologies. In this paper we will use telehealth and telemedicine synonymously. Some examples of Telemedicine include remote health consultations, examination of ECG reports by doctors separated by patients' miles away, treatment discussion carried via video conferencing of family doctors with more skilled doctors from big hospitals and transmission of health reports and medical records for case study. Telemedicine industry has seen improvements in recent years with improving technologies including Cloud Computing.

Cloud computing is a service provider in the field of information technology with the primary functionality being delivering IT resources like computing services. It provides on-demand access to its user, which may be individuals or organizations, to large shared IT resource services which are scalable and remotely manageable.

2 LITERATURE REVIEW

Over the years several research and innovative work has been done in the field of telemedicine. Hao et al. which works upon a biometric encryption key generation structure using iris-scanners which not only helps implement the cryptographic model of 128-bit AES but also provides with an IrisCode which facilitates a 140-bit biometric key [2].

DOI: 10.1201/9781003193838-85

Monrose et al. presented a cloud-based telemedicine structural architecture which works on voice-based recognition techniques for key generation which ensures authentication of the users [6]. Wei-Yen Hsu proposes a clustering-based compression technique to be attached with cloud databases to improve the data transmission rate and storage capacity [3]. Jui-chien Hsieh and Meng-Wei Hsu propose to utilize the power of cloud computing to make the transmission of 12-lead Electrocardiography (ECG) reports to hospitals and doctors for analysis simply using mobile phones from ambulances, small hospitals, etc [8]. Lin et al. proposed a set-top-box system to assist family health care which uses hybrid cloud architecture to manage the system. The study proved that hybrid cloud is the best choice for the proposed model and gives advantages over centralized architecture for storing medical information of patients [4]. Doukas et al. proposed a solution for early skin cancer symptom detection using a smartphone integrating the power of cloud computing. To make early diagnosis easy, low-cost and faster mobile teledermoscopy is proposed. Using the cloud's computational power image classification and assessment can be easily done [1]. Lindsay Ross et al. in his study reports how outpatient neurology follow-up visits can be effectively done using personal device videoconferencing platform [7].

3 BENEFITS OF USIG CLOUD COMPUTING FOR TELEMEDECINE

In this section benefits of cloud computing used for telemedicine services over traditional client/server architecture used on premise will be discussed in detail as shown in Figure 1.

3.1 Cost effective

Cloud Computing offers low cost service as it is based upon a pay-as-you-go model where you only need to pay for services that you really consume.

3.2 Remote delivery of services

The main aim of telemedicine in the healthcare sector is to make good healthcare systems available remotely to all who require it, especially including those patients who would otherwise get no access to good healthcare services due to their lack of means of reaching the service centers physically.

3.3 Ease of access

The simplicity of the cloud framework and the transparency of the services thus provided by it makes the telemedicine easily accessible and usable by the patients and the doctors who make use of this technology to perform the tasks that they were otherwise unable to do due to complications in other physical or geographical factors.

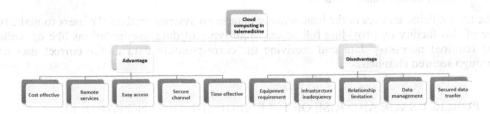

Figure 1. Taxonomic structure of cloud computing in telemedicine.

3.4 Safe and secure communication

In telemedicine framework architecture, a lot of emphasis has been given to providing a secure network between the patients and the doctors to ensure absolute confidentiality in their transactions, as medical information is very private to all users.

3.5 Better time management provider

The large-scale usage of telemedicine as the mode for doctor-patient interaction has proved to be highly efficient on time management for both the users as a lot of tasks which would otherwise increase the total demand of time from the doctors or the patients, are no longer required in this healthcare management model.

4 CHALLENGES IN TELEMEDICINE INDUSTRY

Some of the major problems or challenges the telemedicine technology faces today as shown in Figure 1 are mentioned below.

4.1 Technological equipment's difficulties

Since the cloud computing functionalities are largely dependent on technological systems, telemedicine also primarily works on these technological systems, thus opening the possibilities of difficulties due to the improper functioning of the same.

4.2 Infrastructural inadequacy

Along with the issues related to technological system devices, many issues of the technological infrastructure also hinder the proper deliverance of efficient and effective healthcare services over the cloud framework to the users.

4.3 Limitation to patient-doctor relationship

One of the most important services provided by the healthcare service provider to its users is comfort. This refers to the relationship of trust and comfort between the patient and the doctor or the healthcare worker.

4.4 Data management and exchange issues

In telemedicine healthcare services, the basic interaction between the doctors and the patients occurs by data transferring and receiving, which is primarily managed and synchronized by the cloud service provider.

4.5 Security ensured data portability

The telemedicine services in the healthcare management systems enables the users to make full use of this facility by providing full access to all types of data manipulations like uploading the required necessary data and receiving the corresponding data at the correct user end through secured channels.

5 PUBLIC USAGE AND RISE OF TELEMEDICINE DURING PANDEMIC

Telemedicine is not a new concept but still, people hesitate to adopt it. Acceptance of Telemedicine is still a challenge due to many factors including awareness, lack of technical skills, infrastructure, etc. In India telehealth is expected to grow to $5.4 billion dollars by 2025. There are many start-ups in India including CallHealth, DocPrime, Navia Lifecare, Lybrate, Meddo and mFine which are in the race to boost the usage of telehealth in Indian medical industry. The biggest obstacle they face is establishing the trust of patients as compared to doctors they can visit in-person. During the time of the pandemic, the number of patients increased exponentially. Many hospitals and doctors having no choice left had started adopting it for teleconsultations. In lieu of a growing number of cases, the Indian healthcare ministry (MoHFW) brought the guidelines regarding telemedicine which was long-awaited. Practo which is one of the oldest telemedicine companies in India said that with the pandemic their platform witnessed a 50% increase in doctors registering with them and on a week to week basis there was a rise of 100% on average [5]. Telemedicine makes patients only one click away from getting treatment and consultation as and when needed. Hence, if people embrace this change in the healthcare industry it's will bring a paradigm shift in society.

6 CONCLUSION AND FUTURE SCOPE

In this study we have seen what is Telemedicine, its importance and how Cloud Computing plays a major role in this domain. We also discussed about the benefits and disadvantages of Telemedicine along with how Covid-19 has led to growth in adoption of Telemedicine. However, making Telemedicine a natural choice for patients is still a challenge to overcome which is because of many limitations including lack of clear medical guidelines, lack of infrastructure, etc. But it is clear that the medical industry needs telemedicine to grow and become an alternative choice for in-person visits and consultation because as witnessed in Covid-19 Telehealth is the need of the hour. The future of Medical Industry can overcome many current challenges only when Telemedicine becomes an easy option that is accessible to all. The new wave in Telehealth is also seen to be brought up by the Internet of medical things (IoMT) deployments integrated with cloud. Therefore, there is an urgent need for existing health care mobility management to handle large load of connected devices. Future of Telemedicine is very promising and every year positive changes and improvements are happening due to several innovations taking place. Cloud computing and new technologies will bring limitless possibilities to improve and maximize the capabilities of Telemedicine.

REFERENCES

Doukas, I. Maglogiannis and P. Stagkopoulos, "Skin Lesions Image Analysis Utilizing Smartphones and Cloud Platforms," *Methods in molecular biology*, 2015.

Hao, R. Anderson and J. Daugman, "Combining crypto with biometrics effectively," in *IEEE transactions on computers*, 2018.

Hsu, "Clustering-based compression connected to cloud databases in telemedicine and long-term care applications," *Telematics and Informatics, vol. 34, no. 1*, February 2017.

Lin, P. Hsiao, P. Cheng, I. Lee and G. Jan, "Design and Implementation of a Set-Top Box-Based Homecare System Using Hybrid Cloud," *Telemed J E Health*, 2015.

Mishra, "Factors affecting the adoption of telemedicine during COVID-19," *Indian Journal of Public Health, vol. 64, no. 6, pp.* 234-236, 2020.

Monrose, M. K. Reiter, Q. Li and S. Wetzel, "Cryptographic key generation from voice," in *IEEE Symposium on Security and Privacy*, 2019.

Ross, J. Bena, R. Bermel, L. McCarter, Z. Ahmed, H. Goforth, N. Cherian, J. Kriegler, E. Estemalik, M. Stanton, P. Rasmussen, H. H. Fernandez, I. Najm and M. McGinley, "Implementation and Patient Experience of Outpatient Teleneurology," *Telemed J E Health*, 2020.

Wei and J.-c. Hsieh, "A cloud computing based 12-lead ECG telemedicine service," *BMC medical informatics and decision making*, vol. 12, 2012.

Recent Trends in Communication and Electronics – Sharma et al. (Eds)
© 2021 Taylor & Francis Group, LLC, ISBN 978-1-032-04572-6

Whistle-stop Low-power MCML technique to design Toggle Flip-Flop at nanoscale regime

Pragya Srivastava, Ramsha Suhail & Richa Yadav
ECE department, IGDTUW New Delhi, INDIA

Richa Srivastava
ECE department, KIET group of Institutions, Ghaziabad, INDIA

ABSTRACT: Pronounced digital circuits such as counters, sequential state machines, frequency dividers make an extensive use of Toggle Flip Flop (T FF). Accordingly, a MOS based conventional T FF is studied and investigated for design metrics such as delay (t_p), power (*pwr*), Power Delay Product (*PDP*) and Energy Delay Product (*EDP*). Additionally, MCML (MOS Current Mode Logic) based implementation are examined for T FF. The research article proposes a modish design of MOS based MCML TFF. The proposed design surfaces as proficient candidate for numerous T FF based digital designs as it yields superior results compared to conventional counterparts. Precisely, it offers high speed operation (2×), improvement in power dissipation (2.2×), improvement in PDP (4.53×) and improvement in EDP (9.06×). Thus, the proposed implementation edges out as high-speed low-power MOS based MCML T FF ideal for latest digital design applications.

1 INTRODUCTION

Very large-scale integration is an extremely vast domain that encompass an immense share in modern electronics. Gigantic research is been carried out to achieve low-power digital circuits that can be used in counters and multiplexers [1]. With the paramount increase in device density, integration of Analog and Digital circuitry will become preferable in decreasing overall circuit area [2]. Today, researchers in this filed are evolving new technology that could simulate and design digital logic circuits using advanced implementation techniques for improved performance. With that in mind, a logic style that seems to be encouraging in dispensing the analog affable environment and helps in minimizing the overall power consumption of the portable device is called MOS current mode logic (MCML) [3]. Due to high speed of operation and markable power efficiency, MCML is becoming an emerging trend for current technology stipulation [4]. In addition to this, MCML circuits have very small switching noise and minimized sensitivity to process variations [5]. Moreover, these aspects of MCML makes it profitable and are utilized in wide range of application including digital communications and mixed signal IC's [6] [7].

In the contemporaneous VLSI Design, cyclic logic circuits are employed extensively for high production and reduced size applications. The output of these circuits depends both on the present as well as on previous input values. FFs in electronics is a memory element that stores these values as two stables states and is often called as a bi-stable multi-vibrator. One of the primary and coherent constituents for designing a circuit with sequential logic are T FF, abbreviated as "Toggle FF". Toggling refers to switching the next output state by complement of present output state. The circuit symbol of this edge triggered circuit is shown in Figure 1. It is a single input device which Toggle and clock as inputs and Q and Q bar as output where

DOI: 10.1201/9781003193838-86

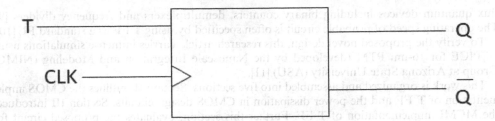

Figure 1.   Circuit symbol of T FF.

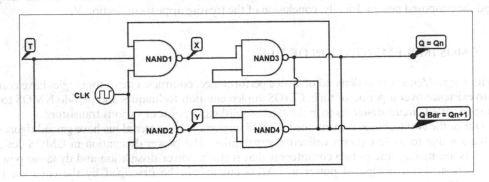

Figure 2.   T FF using NAND Gate.

Table 1.   Truth table of T FF.

T	Q_{n+1}	State
0	Q_n	No Change
1	Q_n	Toggle

Table 2.   Characteristic table of T FF.

Q_n	T	Q_{n+1}
0	0	0
0	1	1
1	0	1
1	1	0

Q bar is complement of Q. The internal circuitry of T FF derived using NAND gates is shown below in Figure 2. Truth Table and characteristic tables of T FF is given in Table 1 and Table 2 respectively.

The characteristic equation of T FF is given by

$$Q_{n+1} = T \otimes Q - (T * \overline{Q}) + (\overline{T} * Q) \tag{1}$$

When we apply clock input to T FF and on giving high input to T input, the corresponding T FF toggles. On the other hand, on giving the low value to T input, the FF holds onto the previous input value [8]. It is one of fundamental element that is employed in designing single

461

flux quantum devices including binary counters, demultiplexers and frequency dividers [9]. The operating speed of probable circuit is often specified by using T FF as a standard FF [10].

To verify the proposed novel design, this research article carries immense simulations using HSPICE for 16-nm PTM (developed by the Nanoscale Integration and Modeling (NIMO) Group at Arizona State University (ASU) [11].

This work is organized and assembled into five sections. Section II outlines the CMOS implementation of T FF and the power dissipation in CMOS design circuits. Section III introduces the MCML implementation of T FF. Further this section, evaluates the proposed circuit for various design metrics, such as delay (t_p), power (*pwr*), PDP and EDP. These design metrics are also compared with its counterpart, convention T FF (discussed in Section II). Section IV concisely concludes the work and depicts the comparative analysis between the conventional design and the proposed design. Finally, conclusion of the treatise appears in Section V.

2 CMOS IMPLEMENTATION OF T FF

With the sudden surge in demand of device performance, countless circuit topologies have come into existence over a period of time. CMOS implementation techniques and pseudo NMOS topology assisted circuit designers to make sure established Q point of various transistors.

Out of the two techniques mentioned, CMOS implementation technique have gained rigorous attention due to its low power delivering capabilities. The power dissipation in CMOS design circuits are mainly due to two constituents that is static power dissipation and dynamic power dissipation. Total dissipated power in CMOS circuit can be described by the equation (2) below [13]

$$P_{Total} = P_{Static} + P_{Dynamic} \tag{2}$$

Today most of the designs have primary trade-off between power and delay. These are, thus two design objectives that needs to be balanced in order to achieve improved speed and relative power and thus power delay product (PDP) comes into picture. PDP is stated in the equation (3) as

$$PDP = P_{avg} \times t_p \tag{3}$$

To avoid this discrepancy, superior low power design parameter known as Energy delay product (EDP) is operative. EDP is stated in the equation (4) as

$$EDP = PDP \times t_p \tag{4}$$

The author has taken this opportunity to substantiate the above-mentioned circuit (Figure 3) and examined the circuit for various design metrics. The inferences drawn are in the Table 3. The CMOS based T FF circuit is investigated and examined on four parameters including propagation delay, power dissipation, PDP and EDP at supply voltage of 0.7V.

3 MCML IMPLEMENTATION OF T FF

Efficient implementation of CMOS based T FF discussed in the last section gives impetus to realize even better configuration for the circuit. A circuit style that delivers high speed with low power can thwart the counterparts when it comes to fabrication requirements of the design industry. MCML (MOS Current Based Logic) based implementations are known for their superior performance, such as, differential operation. This section proposes MCML based T FF.

MCML based T FF as shown in Figure 4. Figure 4 is implemented using four NAND gates. Each NAND gate in turn is implemented using MCML topology.

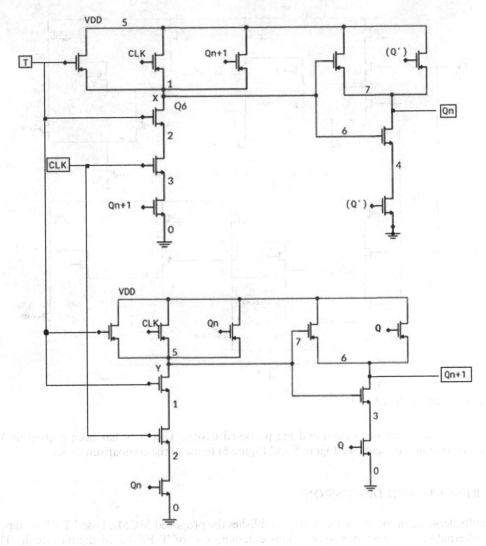

Figure 3. CMOS based T FF [12].

Table 3. Design Metric of MOS based T FF.

Device parameter supply = 700 mV	MOS based TFF
t_p (ns)	0.38496
PWR (μW)	5.7581
PDP (fJ)	2.2166
EDP (fJ-ns)	0.85329

Further this section, evaluates the proposed circuit for various design metrics, such as delay (t_p), power (*pwr*), PDP and EDP. These design metrics are also compared with its counterpart, convention T FF (discussed in Section II).

Both variants of T FF are simulated for 16nm Technology node at 0.7 V supply (V_{DD}). HSPICE based simulated results are tabulated in Table 4. Table 4. brings comparative analysis

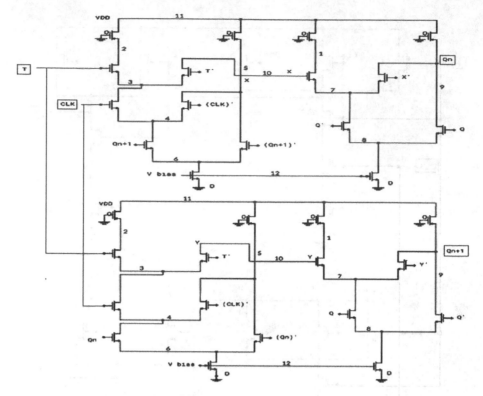

Figure 4. MCML based T FF.

between the conventional design and the proposed design. The same has been plotted as 3D stacked column-bar chart (see Figure 5 and Figure 6) to make the comparison easier.

4 RESULTS AND DISCUSSION

Results depicted in the previous section establishes the proposed MCML based T FF as superior alternative for numerous applications emerging out of T FF based digital circuit. The MCML based T FF is faster (2x), it offers improvement in power dissipation (2.2x), improvement in PDP (4.53x), improvement in EDP (9.06x) compared to conventional counterpart.

Reduced delay of MCML based proposed circuit is achieved by monitoring the voltage swing between two crucial levels of minimum threshold and noise. In addition to delay, the proposed circuit also exhibits less power when operated for submicron dimension (typically 16nm in this work) at nominal V_{DD} - 0.7V (see Figure 5). PDP and EDP are significant design metrics, mathematically derived from delay and power. Figure 6 shows the comparative analysis chart for PDP and EDP.

Table 4. Design Metric of MCML based T FF.

Device parameter supply = 700 mV	MOS based TFF	MOS based MCML TFF
t_p (ns)	0.38496	0.19248 (2)
PWR (μW)	5.7581	2.5431 (2.2)
PDP (fJ)	2.2166	0.48949 (4.53)
EDP (fJ-ns)	0.85329	0.094215 (9.06)

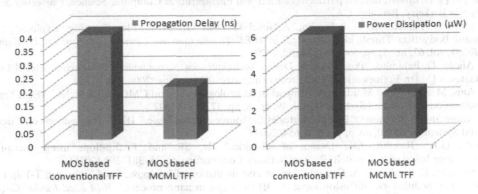

Figure 5. Delay and Power comparison between Conventional and MCML based T FF.

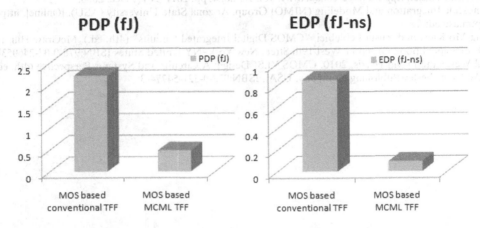

Figure 6. PDP and EDP comparison between Conventional and MCML based T FF.

5 CONCLUSION

This research work targets design engineering-based evaluation of T FF so as to propel its use in latest applications of digital circuit designing. The popular design metrics such as delay (t_p), power (*pwr*), Power Delay Product (*PDP*) and Energy Delay Product (*EDP*) has been considered as touchstone to evaluate the proposed MOS based MCML T FF. Extensive HSPICE based simulation and results at 0.7V supply establishes the proposed circuit as competent candidate to replace the conventional MOS based T FF.

REFERENCES

A. Sahu, M. E. Çelik, D. E. Kirichenko, T. V. Filippov and D. Gupta, "Low-Power Digital Readout Circuit for Superconductor Nanowire Single-Photon Detectors," in IEEE Transactions on Applied Superconductivity, vol. 29, no. 5, pp. 1–6, Aug. 2019

A. K. Dwivedi, P. Srivastava, F. Tarannum, S. Suman and A. Islam, "Performance evaluation of MCML-based XOR/XNOR circuit at 16-nm Technology node," *2014 IEEE International Conference on Advanced Communications, Control and Computing Technologies*, Ramanathapuram, 2014, pp. 512–516.

Musicer, J. (2002). An analysis of MOS current mode logic for low power and high performance digital logic, Ph.D. dissertation, Department of Electrical Engineering & Computer Science, University of alifornia Berkeley, Berkeley, CA.

G. Scotti, A. Trifiletti and G. Palumbo, "A Novel 0.5 V MCML D-Flip-Flop Topology Exploiting Forward Body Bias Threshold Lowering," in *IEEE Transactions on Circuits and Systems II: Express Briefs*, vol. 67, no. 3, pp. 560–564, March 2020.

M. Alioto, G. Palumbo, "Power Aware Design Techniques for Nanometer MOS Current Model Logic Gates: a Design Framework, IEEE Circuits and Systems Magazine, 2006, pp. 41–59.

M. Anis, M. Allam, and M. Elmasry, "Impact of technology scaling on CMOS logic styles," IEEE Transactions on Circuits and Systems II, vol. 49, no. 8, pp. 577–589, 2002.

M. Alioto and G. Palumbo, "Design strategies for source coupled logic," IEEE Transactions on Circuits and Systems I, vol.50, no. 5, pp. 640–654, 2003

Bharti, G.K., Rakshit, J.K. Design of all-optical JK, SR and T flip-flops using micro-ring resonator-based optical switch. Photon Network Communication 35, 381–391 (2018).

E. Abiri, M. R. Salehi and A. Darabi, "Design and simulation of low-power and highspeed T-Flip Flap with the modified gate diffusion input (GDI) technique in nano process," *2014 22nd Iranian Conference on Electrical Engineering (ICEE)*, Tehran, 2014, pp. 82–87.

K. Tsubone *et al.*, "Operation of HTS Toggle-Flip-Flop Circuit with Improved Layout Design," in *IEEE Transactions on Applied Superconductivity*, vol. 16, no. 4, pp. 2011–2017, Dec. 2006.

Nanoscale Integration and Modeling (NIMO) Group, Arizona State University (ASU). [Online]. https://ptm.asu.edu/.

Sung-Mo Kang and Yusuf Leblebici, "CMOS Digital Integrated Circuits" (4th. ed.). McGraw-Hill, Inc. Professional Book Group 11 West 19th Street New York, NY United States ISBN:978-0-07-246053-7

Neil Weste and David Harris. 2010. CMOS VLSI Design: A Circuits and Systems Perspective (4th. ed.). Addison-Wesley Publishing Company, USA., ISBN:978-0-321-54774-3

Recent trends in electroencephalography based biometric systems

T.A. Khan & S. Jabin
Department of Computer Science, Faculty of Natural Sciences, Jamia Millia Islamia, New Delhi, India

ABSTRACT: In recent years, EEG based biometric systems have received great interest from researchers. With ever increasing cases of various types of spoof attacks, and due to vulnerabilities present in the existing systems different biometric modalities have been explored and EEG is one of them. EEG based biometric systems are resilient against cases of forgery and theft as these systems do not work under stress and coerced conditions of users. Due to these favorable characteristics, EEG has been studied as a potential "biomarker" for behavioral analysis of a user. Despite all these favorable features, EEG systems face challenges in the development of on-the-go biometric systems which are needed to be resolved. This paper aims to present recent trends of EEG based biometric systems and discusses major issues faced in the development of these systems.

Keywords: Electroencephalography (EEG), biometrics, machine learning, deep learning

1 INTRODUCTION

Electroencephalography (EEG) is a non-invasive brain imaging technique that measures difference in electrical voltage(micro-voltage), that occur as a part of neural activity in the human brain. EEG devices have been active areas for research in Human-Computer Interaction (HCI). EEG signal data from humans have been used, for Medical diagnosis, Mental task classification, and intelligent tutoring systems, etc. EEG signal data acquired from subjects are unique for every individual and offers stability in signal patterns that are needed in designing biometric systems and thus have been explored as a potential biometric trait.

Biometrics plays a prominent role in design and implementation of today's security systems. These security systems have evolved over time to counter the various types of threats and attacks. A biometric system can perform either identification (Moctezuma et al. 2020) or authentication task (Kumar et al. 2018). In identification, biometric system establishes the identity of the user i.e. it has a one-to-many (1: N) relationship while in authentication a biometric system verifies the claimed identity of the user (1:1) by matching the given sample with the stored template in the database (Kumar et al. 2018).

Biometric is divided into two classes physiological and behavioral. Physiological biometrics deals with the bodily aspects of the user that includes voice, fingerprints, face, iris, retinal, etc. Behavioral biometrics explores the patterns in the user behavior for example dynamic signatures verification, hand movements, keyboard typing recognition, and Gait recognition etc. Physical biometrics, such as face and fingerprints are most commonly deployed in real-world case scenarios. However, these biometrics share their own weaknesses. For example, face recognition system suffers from presentation attacks where similar faces can be regenerated with high accuracy and can be used for authentication purpose. Fingerprints can be easily forged using simple spoofing techniques. A more robust and secure biometric must satisfy two criteria: it should be cancelable, and difficult to be spoofed. EEG has a potential to meet both of

DOI: 10.1201/9781003193838-87

these criteria as compared to other biometrics. EEG signals are hard to steal or spoofed since EEG signals cannot be recorded against users will as EEG doesn't work under stress conditions, thus making them robust against coercion attacks.

This paper presents a review of methods and techniques used in EEG based biometric systems during period 2019-2020. In addition, it discusses major issues and challenges faced in the development of on-the-go EEG authentication systems.

2   EEG DATA FOR BIOMETRIC SYSTEMS

EEG data from the human brain is recorded by placing electrodes on the scalp following the standard international 10-20 system (La Rocca et al. 2013). EEG signals represent the rhythmic activity of brain waves in the form of power spectrum. The frequency and amplitude of EEG signals change from one brain state to another. EEG signal is classified into five different frequency bands these are: delta (δ), theta (θ), alpha (α), beta (β), and gamma (γ). These frequency band or bands are experimentally chosen according to the brain state being explored. EEG based biometric system satisfies the requirements of uniqueness, universality, stability, and performance measures to be accepted as a authentication or identification method (Gui et al. 2019). These EEG based biometric system predominantly have four components: EEG signal acquisition phase, signal pre-processing step, feature extraction process, and classification phase. Complexity of task involved during EEG data acquisition phase has an effect on the performance of the biometric system, resting state is a simple protocol to record EEG data from the subject, where as protocol involving performing cognitive tasks such as mental/motor imagery task introduces mental fatigue in the subject and requires more training time. In general task complexity, setup time and processing time should be kept to minimum for ensuring user-friendliness and effectiveness of the EEG biometric system. The performance of EEG based biometric systems is evaluated in the terms of standard metrics which includes genuine acceptance rate or correct recognition rate for EEG based user identification systems. For EEG based authentication systems the accuracy, false acceptance rate, false rejection rate, half total error rate, and equal error rate are used as performance measures.

2.1   *Sources of EEG datasets*

Various studies in EEG biometrics have used publically available EEG datasets for designing a biometric system. DEAP dataset has EEG recording from 32 subjects for human emotion analysis. EEGMMIDB is a motor imagery dataset having EEG recording from 109 subjects when they performed some tasks. BCI Competition datasets for testing and performance evaluation of brain signal-processing methods and classification techniques for brain computer interface devices. BNCI Horizon2020 is a dataset repository providing access to various BCI datasets. Other publically datasets available are ATR dataset, Alcoholism dataset, and Keirn and Aunon dataset, etc. each having EEG recordings from different number of subjects under various conditions.

2.2   *Feature extraction*

In feature extraction process most relevant and Important features are extracted from EEG signal data, the quality of extracted features has an effect on the performance of recognition system. Extracted features have minimal intra-personal and maximal inter-personal differences. In EEG biometric systems features are extracted under various domains such as frequency (Kumar et al. 2019), time (La Rocca et al. 2013), and time-frequency (Moctezuma et al. 2020) domains.

3 METHODS USED FOR EEG BIOMETRICS

In this section, we describe the most commonly used methods used by various researchers for designing EEG based biometric systems.

Support vector machines (SVM) is a classification technique based on supervised learning. SVM has good generalization capabilities and has been considered as a good choice for identification and authentication problems (Moctezuma et al. 2020; Kumar et al. 2018) in EEG based biometric studies. Hidden Markov Model processes non-stationary EEG signals, the goal of HMM is to make use of observable information, so as to gain insight of various hidden states. For a model λ and an observation sequence $O = (O_0, O_1, \ldots, O_t)$ posterior probability $P(\lambda|O)$ is calculated (Kumar et al. 2018). Linear discriminant analysis (LDA) is another widely used method for classification of subjects (Seha et al.2020) in EEG based biometric systems. Neural network has ability to generalize well, but they are sensitive to noise and may take a lot of time for training purpose (Zeynali et al. 2019). In deep learning methods features are learnt from input data by applying a set of convolutional filters and classification is performed. These methods have achieved good performance in EEG based recognition systems (Wang et al. 2019; Wilaiprasitporn et al. 2019). The other aspect of these methods is that they may require large amount of training data which is a point of concern in EEG biometrics.

4 REVIEW OF RECENT STUDIES IN EEG BIOMETRICS

In the literature various studies have been conducted to design robust EEG based biometric systems, where various data-acquisition protocols, data pre-processing methods, feature extraction methods, and classification techniques have been explored. This section reviews most recent studies during 2019-2020 based on their novel approaches and reported accuracy in EEG Biometrics.

(Zeynali et al. 2019) explored the potential of using the single channel with optimized electrode placement in EEG based authentication systems, for their proposed study they used publically available Keirn and Aunon's dataset. Signals were recorded when subject performed five different mental tasks. Band-pass filter was used to remove noise-artifacts and various features including PSD, wavelet, autoregressive, and entropy were extracted and combined to form a feature vector. Classification was performed using SVM, neural network, and Bayesian network. Neural network achieved highest mean classification accuracy of 97.54% when mental tasks were considered for performance evaluation. Furthermore, O2 was selected as the optimal channel with an accuracy of 95% independent of the type of mental tasks.

Multi-layer security protocols are needed for designing highly secured systems as described in (Wang et al. 2019), author proposed a novel EEG biometric identification model. Two datasets were used, publically available motor movement/imagery dataset EEGMMIDB dataset and other self-collected dataset from 59 subjects with attention specific task, and image description tasks. EEG signals were constructed as a graph based on intra-frequency and inter-frequency connectivity measures. Graph CNN (GCNN) model was used for automated learning of features from the EEG signal graphs and also for classification purposes in subject identification. Training and testing on diverse human states experiment protocol with PLV +GCNN model achieved a CRR of 98.96%. Proposed study addressed the stability issue in EEG biometrics by studying the effectiveness of each method to diverse human states.

(Kumar et al. 2019) proposed a multi-modal biometric authentication framework for person authentication. Person signatures and EEG signals from 58 subjects were acquired simultaneously for the authentication process. EEG data was recorded using EMOTIV EPOC+ headset when subject was performing his signatures. Discrete Fourier Transform (DFT) was used to extract features in frequency domain and BLSTM-NN classifier for performing identification and verification tasks in uni-modal and multi-modal paradigm. Both systems (Signature and EEG) were built and trained individually before being merged. Features from both signature and EEG signal systems were combined using decision fusion approaches and the Borda count classifier combination technique is used to combine both systems features. Their model reported accuracy of 98.78% for subject identification and False Acceptance Rate of 3.75 and an Equal Error Rate (EER) of 4.01% during the verification process.

(Wilaiprasitporn et al. 2019) proposed a deep learning based person identification system. In their study they used benchmark DEAP EEG dataset. This dataset consist of EEG recording from 32 subjects, when subject watched affective music videos and have some score values (valence and arousal) for the watched videos. During pre-processing step independent component analysis (ICA) was used to remove eye-blinking signals. Filters were further applied to extract band related frequency values. CNN-GRU and CNN-LSTM with different layers were used for extracting features and performing classification tasks. Proposed study reported a mean CRR of 99.90% with 32 electrodes and a CRR of 99.17% when five electrodes were used. CNN-GRU took less training time and achieved higher mean CRR than CNN-LSTM.

(Seha et al. 2020) employed auditory evoked potential (AEP) protocol to record EEG data from 40 subjects using a medical device to perform human recognition task. Intra-session and cross-session, Gaussian filter and PSD features were extracted after pre-processing of signal data. LDA was used as a classifier for identification and verification tasks. Proposed study attained a CRR of 96.46% and an EER of 2% under cross-session criteria.

(Moctezuma et al. 2020) devised a multi-objective EEG based biometric system using minimum number of EEG channels. Benchmark EEGMMIDB dataset having EEG recording from 109 subjects was used for the proposed study. Common average reference method was employed to remove noise from the EEG signal data. DWT and empirical mode decomposition methods were used for decomposing signals into sub-frequency bands. For these decomposed sub-bands Higuchi & Petrosian fractal dimension, instantaneous & Teager energy, features were extracted. Further local outlier factor (LOF) algorithm was used for creating a classification model for each subject. Using Non-dominated sorting genetic algorithm (NSGA-III), three channels were selected from Pareto-front region and were further used in subject identification. Using DWT-based features from three channels, a TAR of 0.997±0.02 and the TRR 0.950±0.05 was reported under resting-state with the eyes-closed condition. This study demonstrated that EEG based biometric systems can be optimized by using a smaller number of EEG channels. Table 1 provide the summary of reviewed studies.

5 OPEN ISSUES IN EEG BIOMETRICS

Here we describe most pertinent challenges faced by EEG biometrics.

5.1 *Complexity of tasks and number of subjects*

Studies (Gui et al. 2019) have reported that the complexity of task involved during EEG acquisition protocol has an effect on the system performance, increasing task complexity increases the training time, other protocols involving simple tasks are needed to be explored. In existing literature, researchers have considered less number of subjects for training and have shown the model with better accuracies. But that is not the case with authentication systems where large numbers of subjects are used and has a risk of security failure.

5.2 *Need for spoofing studies in EEG biometrics*

In the EEG biometrics literature, a few studies have explored the spoofing of EEG based recognition systems. A recent study by (Shukla et al. 2020) proposed a novel attack on multimodal EEG authentication systems, based on the idea that hand movements and brain signals from a subject are highly correlated, they claimed to have spoofed brain signals from the subjects' hand movement using a correlation matrix. More similar studies exploring the vulnerabilities present in the uni-modal and multi-modal EEG biometrics are needed.

Table 1. Summary of reviewed EEG based biometric studies.

Author	Protocol	Features	Classification	Performance
Zeynali et al. 2019	mental tasks	PSD, AR, WT, Entropy	SVM, NN Bayesian Network	Accuracy of 97.54%
Wang et al. 2019	resting state, imagery tasks, attention and image description task	Graph CNN extracted features	Graph-CNN	CRR: 98.96%
Kumar et al. 2019	performing signatures	Discrete Fourier Transform features	BLSTM-NN	EER:4.01% FAR: 3.75% Accuracy: 98.78%
Wilaiprasitporn et al. 2019	watching music video	CNN-GRU /LSTM	CNN-GRU /LSTM	CRR: 99.90%
Seha et al. 2020	Auditory evoked potential	Gaussian & PSD features	LDA	CRR: 96.46% EER: 2%
Moctezuma et al. 2020	resting state	DFT	Local Outlier Factor (LOF)	TAR: 0.997±0.02 TRR: 0.950±0.05

5.3 Data sharing

Benchmark EEG data recorded solely for biometrics purpose are needed and data sharing has to be made a common procedure for reproducibility of the research studies conducted, thereby increasing the confidence of results obtained in the EEG biometrics literature.

6 CONCLUSION

In the EEG biometrics literature robustness of biometric systems lies' in the selection of appropriate data acquisition protocols, pre-processing and feature extraction methods, and classification techniques. Multi-modal biometrics have provided good results and are needed to be explored further for permanence, stability and collectability issues. For designing a robust biometric system, a good EER as low as zero is more desirable than achieving 99% accuracy. So, there is scope of reducing EER. Furthermore, these biometric systems can be optimized by selecting the minimum number of EEG channels for identification or authentication tasks. Studies exploring the optimization of various tasks involved in EEG biometrics are needed. Moreover, vulnerabilities present in existing EEG based biometric systems have to be explored both in uni-modal and multi-modal EEG biometrics as future work.

REFERENCES

Gui, Q., Ruiz-Blondet, M.V., Laszlo, S. & Jin, Z., 2019. A survey on brain biometrics. ACM Computing Surveys (CSUR), 51(6), pp.1–38.

Kumar, P., Singhal, A., Saini, R., Roy, P.P. & Dogra, D.P., 2018. A pervasive electroencephalog-raphy-based person authentication system for cloud environment. Displays, 55, pp.64–70.

Kumar, P., Saini, R., Kaur, B., Roy, P.P. & Scheme, E., 2019. Fusion of neuro-signals and dynamic signatures for person authentication. Sensors, 19(21), p.4641.

La Rocca, D., Campisi, P. & Scarano, G., 2013, February. Stable EEG features for biometric recognition in resting state conditions. In International Joint Conference on Biomedical Engineering Systems and Technologies (pp. 313–330). Springer, Berlin, Heidelberg.

Moctezuma, L.A. & Molinas, M., 2020. Towards a minimal EEG channel array for a biometric system using resting-state and a genetic algorithm for channel selection. Scientific RepoRtS, 10(1), pp.1–14,

Seha, S.N.A. & Hatzinakos, D., 2020. EEG-Based Human Recognition Using Steady-State AEPs and Subject-Unique Spatial Filters. IEEE Transactions on Information Forensics and Security, 15, pp.3901–3910.

Shukla, D., Kundu, P.P., Malapati, R., Poudel, S., Jin, Z. & Phoha, V.V., 2020. Thinking Unveiled: An Inference and Correlation Model to Attack EEG Biometrics. Digital Threats: Research and Practice, 1 (2), pp.1–29.

Wang, M., El-Fiqi, H., Hu, J. & Abbass, H.A., 2019. Convolutional neural networks using dynamic functional connectivity for EEG-based person identification in diverse human states. IEEE Transactions on Information Forensics and Security, 14(12), pp.3259–3272.

Wilaiprasitporn, T., Ditthapron, A., Matchaparn, K., Tongbuasirilai, T., Banluesombatkul, N. & Chuangsuwanich, E., 2019. Affective EEG-based person identification using the deep learning approach. IEEE Transactions on Cognitive and Developmental Systems.

Zeynali, M. & Seyedarabi, H., 2019. EEG-based single-channel authentication systems with optimum electrode placement for different mental activities. biomedical journal, 42(4), pp.261–267.

Recent Trends in Communication and Electronics – Sharma et al. (Eds)
© 2021 Taylor & Francis Group, LLC, ISBN 978-1-032-04572-6

Gain enhancement of a microstrip patch antenna using superstrate radiator

Namrata Medhi
Department of Electronics and Communication Engineering, Gauhati University, Guwahati, Assam, India

Sivaranjan Goswami
Department of Electronics and Communication Engineering, Gauhati University, Guwahati, Assam, India
Department of Electronics and Communication Technology, Gauhati University, Guwahati, Assam, India

Kandarpa Kumar Sarma
Department of Electronics and Communication Engineering, Gauhati University, Guwahati, Assam, India

Kumaresh Sarmah
Department of Electronics and Communication Technology, Gauhati University, Guwahati, Assam, India

ABSTRACT: In this paper, a low power microstrip antenna is designed which is then mounted by a superstrate radiator to enhance the effective radiation gain of the antenna without increasing the feeding power feed to the antenna. Here, in such an arrangement, the secondary radiator is a periodic metallic structure over a dielectric substrate that can converse the radiation pattern in the desired direction without consuming extra radiation power. First, a modified curved shaped microstrip patch antenna resonating at a frequency of 2.4 GHz is designed. Later on, a superstrate is loaded over the patch antenna. The superstrate is covered by a thin copper plate, on which copper strip lines are etched to increase the radiation power. The S_{11} parameters and gain of the microstrip antenna loaded with and without superstrate structure are being compared. The results show an enhancement of gain of the antenna from 2.65 dB to 3.75 dB at the first resonating frequency of 2.2 GHz. Relative to the radiation pattern of the original patch antenna, the radiation of the superstrate loaded microstrip antenna seems to be strongly directed towards the positive z-direction.

1 INTRODUCTION

Microstrip antennas are generally simple in construction and can be made using conventional microstrip fabrication techniques. On the other hand, electromagnetic metamaterials are some artificial materials which can be created by arranging metallic structures in a homogeneous manner over a superstrate. These metamaterials can be used over the simple patch antennas to increase their gain and to enhance the antenna frequency and bandwidth.

In a work, a slotted H-Shaped microstrip patch antenna operating at 2.4 GHz WLAN application is shown which is designed on a FR4 (Flame Retardant-4) substrate and a dumbbell-shaped cut is made on the ground termed as defected ground structure (DGS) (Ayyapan et al., 2017). Another paper (Kumar et al., 2017) described the design of a microstrip antenna with a metamaterial containing swastika shaped structures designed on it. Cross-polarization of a single-feed leaky wave antenna is reduced using a M-shaped anisotropic wide-angle impedance matching (WAIM) layer (Farrokhzad et al., 2020). Characteristic mode analysis (CMA) is used to

DOI: 10.1201/9781003193838-88

predict the modal behaviors of metasurface based antennas which can operate at different resonating frequencies (Li et al., 2018). A slotted rectangular patch antenna at 2.4 GHz is designed for high data rate applications (Sarma et al., 2015). In order to focus a propagating wave at a point with high efficiency, phase gradient metasurface (PGMS) is used to transform a spherical wave to a near plane wave (Li et al., 2015). A compact wideband flexible implantable slot antenna is designed for industrial and biomedical applications (Das et al., 2018). Efficiency of a near-field wireless power transfer (WPT) system is improved using metasurface (Shaw et al., 2019). A metasurface based frequency reconfigurable antenna is designed were, with the increase in the rotation angle of the metasurface, the relative permittivity decreases resulting in frequency shifting of the antenna (Pavan et al., 2015). A miniaturized slot antenna is designed using loading wires penetrated into the substrate (Ghosh et al., 2013). A patch antenna with metasurface is designed containing complementary split ring resonators (CSRR) and the ground plane loaded with non-uniform metasurface (NUMs) (Painam et al., 2019). A metamaterial superstrate is used for suppressing mutual coupling in densely packed patch antenna arrays using SC –CSRR (slot combined CSSR) (Jafargholi et al., 2019). Miniaturization of antenna is done using defected ground structures and shorting pins (J et al., 2017).). Metasurface is used to solve the decoupling problem of antenna array in both H-plane and E-plane (Guo et al., 2019). In another work, a magneto dielectric superstrate is used to suppress the surface wave propagation (Jafargholi et al., 2019). In this paper, a rectangular microstrip antenna is modified with a curved shaped patch resonating at 2.4 GHz operating frequency. To enhance the radiation performance such as gain of the antenna, a square shaped superstrate is placed over the antenna. The design details and results are discussed in the following sections.

2   CURVED ANTENNAE WITH METAMATERIAL SUPERSTARTE

Structure and design of the proposed modified rectangular patch antenna is shown in Figure 1(a). The antenna is feed by a microstrip line designed at 2.4 GHz operating frequency. The patch of the antenna is cut at both sides in a curve-shaped design. For a better impedance matching, the antenna is loaded with inset feed at its resonating frequency with a dimension of the feeding geometry of (19×2) mm^2. The curved microstrip antenna is designed over a *FR-4* epoxy substrate with a dielectric constant, $\varepsilon_r = 4.4$. The slotted curved shaped microstrip patch antenna is designed according to the dimensions mentioned in the Table 1 and the antenna simulation is done using Ansys HFSS software.

Strip lines are etched on a copper plate (superstrate cover) is designed and loaded over a high dielectric substrate material Rogers 3006 as shown in Figure 1(b). The substrate used to design the superstrate has dielectric constant (ε_r) of 6.15. The construction of superstrate is summarized in Table 2. The structure of the proposed metamaterial inspired superstrate as shown in Figure 1(b) is loaded over the proposed curved microstrip antenna and is represented in Figure 1(c).

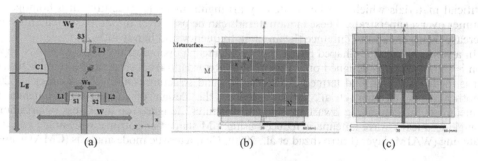

(a)   (b)   (c)

Figure 1.   (a) Schematic diagram of curved microstrip patch antenna (b) Proposed periodic superstrate structure over the substrate (c) Antenna loaded with metamaterial inspired superstrate.

Table 1. Constructional parameters of the proposed microstrip antenna.

Dimension Parameter	Values mm
Lg &Wg (length and width of the substrate and ground plane)	60
L (Length of the patch)	29.5
W (Width of the patch)	38
Ls (Length of the Feedline)	19
Ws (Width of the Feedline)	2
Radius of both the spheres cut as curves C_1 and C_2	26
L_1 (Inset1 length) & L_2 (Inset2 length)	6
S_1 (Inset1 width) & S_2 (Inset2 width)	5
L_3 (Inset3 length)	4
S_3 (Inset3 width)	3

Table 2. Constructional parameters of the proposed metasurface.

Dimension Parameter	Values mm
M (Length of the superstrate and top conducting plane)	52
N (Width of the superstrate and top conducting plane)	65
X (Side of the square elements of the proposed SCR)	8
Y (Gap between two square elements of the proposed SCR)	0.7
d (Air gap between the patch and the superstrate)	0.5

3   RESULTS AND DISCUSSION

The S_{11} plot vs frequency of the microstrip antenna with and without superstrate are compared and shown in Figure 2. The resonant frequency is reduced from 2.3 GHz to 2.2 GHz when the microstrip antenna is loaded with the superstrate. The return losses of the microstrip antenna without superstrate are found to be -20.5 dB at 2.3 GHz and -19 dB at 4.1 GHz. With superstrate it is found to be -24 dB at 2.2 GHz and -8 dB at 3.8 GHz respectively. The separation gap between the proposed patch antenna and the proposed metasurface is tuned at a distance of d=0.5 mm. Without superstrate, the antenna is showing gains of 2.65 dB, 1.86 dB and 1.9 dB at frequencies of 2.3 GHz, 4.1 GHz and 2.4 GHz respectively. With superstrate, it is showing gains of 3.75 dB, 3.32 dB and 1.94 dB at frequencies 2.2 GHz, 4.1 GHz and 2.4 GHz respectively. The average gain has been enhanced from 2.65 dB to 3.75 dB at the first resonating frequency of 2.2 GHz. The enhanced radiation behavior of the proposed antenna

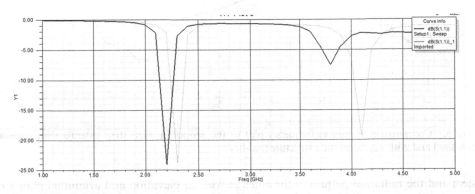

Figure 2.   Comparison of return loss (S_{11}) vs Frequency of the proposed microstrip antenna without superstrate (red-line) and with superstrate structure (black-line).

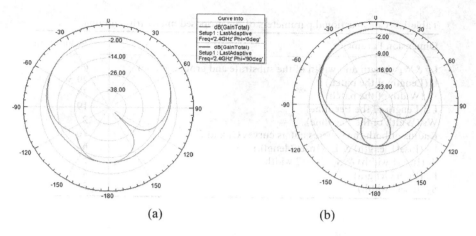

Figure 3. Radiation pattern of the proposed antenna (a) patch antenna without superstrate and (b) patch antenna with superstrate.

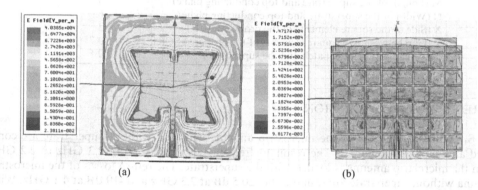

Figure 4. Electric field distribution of the antenna (a) patch without superstrate and (b) patch with superstrate.

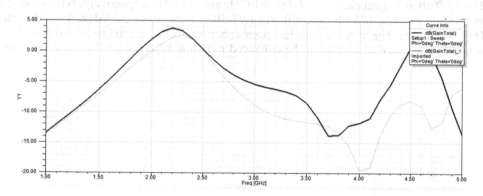

Figure 5. Comparison of gain vs frequency plot of the proposed microstrip antenna with superstrate (black-line) and without superstrate structure (red-line).

system and the radiation pattern of the antenna with its elevation and azimuthal planes both with and without superstrate are being studied and are represented in Figure 3. The magnitude of the electric field distribution of the antenna without superstrate as well as after loading of

4 CONCLUSION

A simple rectangular microstrip antenna working at 2.4 GHz is designed and copper strip
lines are etched as metamaterial inspired by adding a thin copper plate as a superstrate cover
over the substrate. The metamaterial superstrate is then loaded over the modified microstrip
patched antenna. The separation gap between the patch antenna and the superstrate is tuned
at a distance of d = 0.50 mm. The resonant frequency of the microstrip antenna is reduced
from 2.3 GHz to 2.2 GHz and the average directional gain has been enhanced from 2.65 dB to
3.75 dB at the first resonating frequency of 2.2 GHz. The radiation of the superstrate loaded
antenna is highly directional towards the positive z-direction of radiation.

REFERENCES

Das, S. & Mitra, D. May 2018. A Compact Wideband Flexible Implantable Slot Antenna Design with
    Enhanced Gain. *IEEE Transaction in Antenna and Propagation*. 66(8): 4309–4314.
Aravindraj, E. & Ayyappan, K. 5-7 Jan 2017. Design of Slotted H-Shaped Patch Antenna for 2.4 GHz
    WLAN Applications. *International Conference on Computer Communication and Informatics (ICCCI)*:
    1–5.
Fartookzadeh, M. et al. March 2020. Efficiency improvement and cross-polarization reduction of
    single-fed frequency-scan leaky wave microstrip antennas by using an M-shape metasurface as the
    WAIM layer. *Int. J. Electron. Communication*. (AEU). 116: 153057.
Ghosh, B. et al. April 2013. Miniaturization of Slot Antennas Using Wire Loading. *IEEE Antenna and
    Wireless Propagation Letters*. 12: 488–491.
Guo, J. et al. July 2019. Metasurface Antenna Array Decoupling Designs for Two Linear Polarized
    Antennas Coupled in H-Plane and E-Plane. *IEEE Access*. 7: 100442–10045.
Jafargholi, A. et al. Jan 2019. Mutual Coupling Reduction in an Array of Patch Antennas Using
    CLL Metamaterial Superstrate for MIMO Applications. *IEEE Transaction in Antenna and Propagation*.
    67(1): 179–189.
Anjali, K. J. & Suriyakala, C. D. 20-21 April 2017. A Highly Miniaturized Patch Antenna. *International
    Conference on circuits Power and Computing Technologies (ICCPCT)*.:1–6.
Kumar, P., Sreelakshmi, K., B, S.& Narayan, S., 6-8 April 2017. Metasurface based Low Profile Recon-
    figurable Antenna. *International Conference on Communication and Signal Processing*.
Li, H., Wang, G., Xu, H., Cai, T. & Liang, J. Nov 2015. X-Band Phase-Gradient Metasurface for
    High-Gain Lens Antenna Application. *IEEE Transactions on Antennas and Propagation*. 63(11):
    5144–5149.
Li, T. & Chen, Z. N. 5-7 March 2018. A Dual-Band Metasurface Antenna Using Characteristic Mode
    Analysis. *International Workshop on Antenna Technology*, IWAT:1-6.
Painam, S. & Bhuma, C. Feb 2019. Miniaturizing a Microstrip Antenna Using Metamaterials and
    Metasurface. *Antenna Applications Corner, IEEE Antennas Propagation Magazine*. 61(1): 91–13.
Pavan, M. N. & Chattoraj, N. March 2015. Design and Analysis of a Frequency Reconfigurable Antenna
    Using Metasurface For Wireless Applications. *IEEE Sponsored 2nd International Conference on Innov-
    ations in Information, Embedded and Communication systems (ICIIECS)*.
Qamar, Z. et al. Feb 2016. Mutual Coupling Reduction for High-Performance Densely Packed Patch
    Antenna Arrays on Finite Substrate. *IEEE Transaction in Antenna and Propagation*. 64(5): 1653–1660.
Sarma, A., Sarmah, K. & Sarma, K. K. 19-20 Feb 2015. Low Return Loss Slotted Rectangular Micro-
    strip Patch Antenna at 2.4 GHz. *International Conference on Signal Processing and Integrated Net-
    works (SPIN)*.
Shaw, T. & Mitra, D. June 2019. Metasurface-based radiative near-field wireless power transfer system
    for implantable medical devices. *IET Microwaves, Antennas Propagation*. 13(12): 1974–1982.

Recent Trends in Communication and Electronics – Sharma et al. (Eds)
© 2021 Taylor & Francis Group, LLC, ISBN 978-1-032-04572-6

Scorpio shaped multiband microstrip patch antenna for S-band and X-band applications

Piyush Kumar Mishra, J.A. Ansari, Devesh Tiwari & Abhishek Saroj
JK Institute of Applied Physics and Technology, University of Allahabad, Prayagraj, UP, India

ABSTRACT: This paper proposed a planer microstrip patch Antenna for multiband wireless applications. The imitative results exhibit that the designed antenna works at multiband resonance frequencies that cover various wireless applications. The presented antenna contains a quasi-improved quadrate radiating patch with bisque created out at upper and bottom part of the patch. Three stepped cuts are made in radiating patch to achieve multiband features. The presented antenna is simulated and optimized using CST Studio Suit. The presented antenna topology consists of 30×30×1.6mm³. The stimulated results display that presented antenna has S_{11} framework with -10 dB return loss that comply with the need of S-band and X-band application.

Keywords: bisque, computer simulation Technology (CST) studio suit, radiating patch, return loss (S11)

1 INTRODUCTION

Wireless Technology taken on a prominent position in indoor and outdoor environment recently (Boutejdar, 2020) With the fast growth of wireless technology the key issue is to make compact design antenna that gives multiband characteristics across the whole operating band particularly for indoor application 2.4 GHz (S- band applications) and outdoor application 9.2 GHz and 12.2GHz (both for X-band applications) respectively (Kumar, 2018). In reality, the growing requirement for wireless technology requires an antenna with multiple operating frequencies. Thus, it is good to choose only antenna with more than one operating frequency than use a different antenna for every frequency band (Dahele, 1984). Thus, the need of designing multiband antenna has enhanced because such type of antennas are very important for designing integrated communication standards into one compact system (Tiwari, 2020). For this requirement of planar antennas is tempting choice in researcher's community. Microstrip antennas exhibit very engaging physical features, like simple structure, compact size, low-cost manufacturing, simple installation light weight, and easy integration with various feeding techniques. In presented work, a line-fed planar microstrip antenna with a Scorpio - shaped patch and three slots is cut in patch has been presented for multiband wireless applications. It offer multiband features in compact size (15×15×1.6). The presented antenna draft with FR4 substrate (Das, 2014). The detailed examination of antenna fundamentals (Voltage Standing Wave Ratio, radiation characteristics, gain, reflection coefficient etc.) are weigh up under the following parts. The advantages of the presented design are simple structure, small size, increased frequency ratio, better frequency rejection and less interference by dissimilar band of frequency for wireless utilization (*Balanis, 2016*).

DOI: 10.1201/9781003193838-89

1.1 Antenna structuring

The physical framework of presented design is depicted in Figure 1. with Scorpio shaped radiating patch. The design comprises modified rectangular radiating patch in Scorpio shape with bisque coming out at upper and lower side (Boutejdar, 2017). The presented quadrate antenna is connected with 50Ω strip line which is 2 mm wide and 12 mm in length. The presented antenna is designed with FR4 base of thickness 1.6mm, loss tangent $tan\delta = 0.02$, *permittivity* = 4.4 and the overall dimension of the presented structure is 30×30×1.6. Other modifications include one big cut slot of dimension $12 \times 1\ mm^2$ and two comparatively small fine cuts of dimension $4 \times 1\ mm^2$ which provide multiband operation in compact structure (Pei, 2011). Every dimension of presented work is choosing with simulation analysis through multiple stages of design process using CST studio suit platform. The structure with optimum dimensions of the presented antenna is depicted in Figure 1 and Figure 2.

1.2 Working and analysis of presented antenna

The outcomes of retouched geometry on the implementations of the antenna are inspected through CST studio suit platform. The geometry of the presented antenna is completed in various stages of modifications. Firstly the retouched quadrate patch consist three stepped cut slot that improve bandwidth, gain, and multiband features (Wu, 2016). Again quadrate patch loaded with bisque come out from the patch. Another bisque in bottom side in rectangular shape also work as resonator (Garg, 2014). It is observed from the S_{11} characteristics that the resonance behavior of presented antenna varies a lot because of the impact of bisque resonators and retouch patch design. The details of S_{11} characteristics of the presented antenna with two pairs of open loop bisque with fundamental antenna design are depicted in Figure 3. The fundamental antenna design depicts resonant frequencies with S_{11} parameter ≤-10 dB. But due to bisque resonators which has an essential contribution in deciding sensibility of impedance matching of the presented antenna. Due to these changes it can impact the EM coupling amidst the patch and ground plane, so enhance the impedance bandwidth. The observation gives that the presented antenna excited all over 3 resonant modes i.e. 2.4 GHz, 9.2 GHz, 12.2 GHz, to get multiband applications for S11 ≤-10dB (Yoon, 2006).

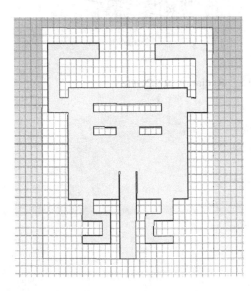

Figure 1.   Proposed antenna physical layout.

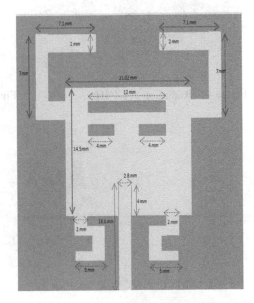

Figure 2.   Proposed antenna dimension.

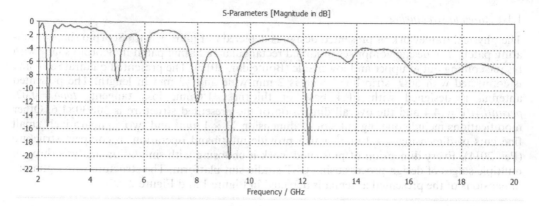

Figure 3. S_{11} parameters of proposed antenna.

1.3 Obtained results and discussion

The imitated S_{11} parameter of the presented antenna is depicted in Figure 3. It shows presented antenna gets multiband characteristics with $S_{11} \leq -10$ dB and gives desirable performance for S-band and X-band applications (Liu, 2011). The VSWR (voltage standing wave ratio) of the presented antenna is depicts in Figure 7. The calculation of mismatch loss (ML) can be obtained by the value of VSWR with the formula, *mismatched Loss (dB)=10 log* $(1-(VSWR-1))/((VSWR+1))^2$. The radiation characteristics obtained from presented antenna are depicted in Figure 4, Figure 5 and Figure 6. The presented design exhibit almost

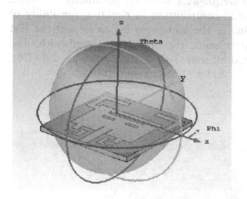 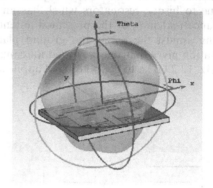

Figure 4. Radiation at 2.4GHz.   Figure 5. Radiation at 9.2GHz.

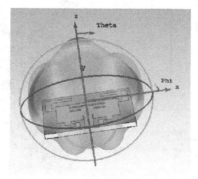

Figure 6. Radiation at 12.2 GHz.

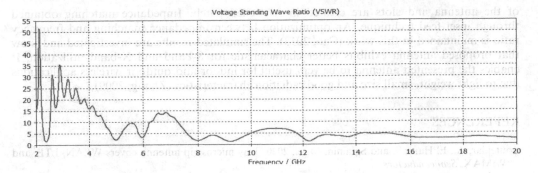

Figure 7.   VSWR.

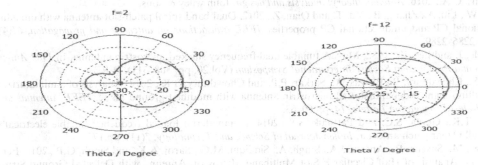

Figure 8.   Gain 2 GHz.

Figure 9.   Gain 9.2 GH.

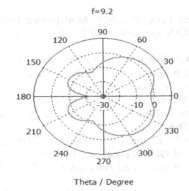

Figure 10.   Gain 12.2 GHz.

steady radiation characteristics with acceptable 3dB beam widths around the multiple frequency bands at the far field region. Obtained gains of the presented antenna are depicted in
Figure 8, Figure 9 and Figure 10 with which we get desirable performance.

1.4   Conclusions

The Presented antenna shows multiband feature for narrowband applications. To enhance
radiation characteristics and impedance bandwidth bisque are created at upper and lower part

of the antenna and slots are created in retouched patch. Impedance matching obtained through inset feed technique. An optimization between impedance matching and frequency ratio is maintained in this work (Cao, 2013). The simulated results are investigated and show that proposed antenna exhibits better characteristic for S-band and X-band utilization. It enables the presented antenna to be very useful for indoor and outdoor wireless applications due to light weight, small size and good radiation characteristics (Zhang, 2009).

REFERENCES

Boutejdar, A., El Hani, S. and Salamin, M.A., 2020. Tiny microstrip antenna covers WLAN, LTE, and WiMAX. *Semiconductors*.

Boutejdar, A., Challal, M., Bennani, S.D., Mouhouche, F. and Djafri, K., 2017. Design and fabrication of a novel quadruple-Band monopole antenna using a U-DGS and open-Loop-Ring resonators. *Advanced Electromagnetics*, 6(3), pp.59–63.

Balanis, C.A., 2016. *Antenna theory: analysis and design*. John wiley & sons.

Cao, W., Liu, A., Zhang, B., Yu, T. and Qian, Z., 2012. Dual-band spiral patch-slot antenna with omnidirectional CP and unidirectional CP properties. *IEEE transactions on antennas and propagation*, 61(4), pp.2286–2289.

Dahele, J. and Lee, K., 1982, May. A tunable dual-frequency stacked microstrip antenna. In *1982 Antennas and Propagation Society International Symposium* (Vol.20, pp. 308–311). IEEE.

Das, S., Chowdhury, P., Biswas, A., Sarkar, P.P. and Chowdhury, S.K., 2014. Analysis of a miniaturized multiresonant wideband slotted microstrip antenna with modified ground plane. *IEEE Antennas and Wireless Propagation Letters*, 14, pp.60–63.

Garg, T.K., Gupta, S.C. and Pattnaik, S.S., 2014. Metamaterial loaded frequency tunable electrically small planar patch antenna. *Indian Journal of Science and Technology*, 7(11), pp.1738–43.

Kumar, M., Saxena, R., Ansari, J.A., Singh, A., Siddiqui, M.G., Saroj, A.K. and Singh, G.P., 2018, February. Analysis of Half Circular F Slot Multiband Microstrip Antenna with Defected Ground Structure. In *2018 Recent Advances on Engineering, Technology and Computational Sciences (RAETCS)* (pp. 1–5). IEEE.

Liu, W.C., Wu, C.M. and Dai, Y., 2011. Design of triple-frequency microstrip-fed monopole antenna using defected ground structure. *IEEE transactions on antennas and propagation*, 59(7), pp.2457–2463.

Pei, J., Wang, A.G., Gao, S. and Leng, W., 2011. Miniaturized triple-band antenna with a defected ground plane for WLAN/WiMAX applications. *IEEE Antennas and Wireless Propagation Letters*, 10, pp.298–301.

Tiwari, D., Ansari, J.A., Saroj, A.K. and Kumar, M., 2020. Analysis of a Miniaturized Hexagonal Sierpinski Gasket fractal microstrip antenna for modern wireless communications. *AEU-International Journal of Electronics and Communications*, 123, p.153288.

Wang, K., Kornprobst, J. and Eibert, T.F., 2016, June. Microstrip fed broadband mm-wave patch antenna for mobile applications. In *2016 IEEE International Symposium on Antennas and Propagation (APSURSI)* (pp. 1637–1638). IEEE.

Wu, R., Tang, H., Wang, K., Yu, C., Zhang, J. and Wang, X., 2016, July. E-shaped array antenna with high gain and low profile for 60 GHz applications. In *2016 IEEE MTT-S International Microwave Workshop Series on Advanced Materials and Processes for RF and THz Applications (IMWS-AMP)* (pp. 1–3). IEEE.

Yoon, J., 2006. Fabrication and measurement of modified spiral-patch antenna for use as a triple-band (2.4 GHz/5GHz) antenna. *Microwave and optical technology letters*, 48(7), pp.1275–1279.

Zhang, Q.Y. and Chu, Q.X., 2009. Triple-band dual rectangular ring printed monopole antenna for WLAN/WiMAX applications. *Microwave and Optical Technology Letters*, 51(12), pp.2845–2848.

Recent Trends in Communication and Electronics – Sharma et al. (Eds)
© 2021 Taylor & Francis Group, LLC, ISBN 978-1-032-04572-6

Smart car parking system using openCV

Vidushi, Manika Goel, Aditya Singh, Divyanshi Goyal, Akash Rajak & Ajay Kumar
Shrivastava
KIET Group of Institutions, Ghaziabad, India

ABSTRACT: Nowadays, system of Smart Parking becomes the pivotal exploration for comfort of people. This article objective is basically to spread as well as scrutinize a well-maintained parking system called smart parking where ongoing already available parking system is in contrast to the freshly recommended new system. However, still many places exist that have not applied the newly introduced parking structure because this needs much expenditure. As we go in any parking site, we found many employees working over there. Couple of people work for the Open and Closing of gate and other used to keep to record. So, we make the System using OpenCV that is used to extract the Vehicle number and Store in the database for records and verification of vehicle owner and for the opening/closing of barriers. Permit Plate Recognition was a computer system that acknowledges any digital image automatically on the car number plate. This system includes many actions such as taking snaps, saving the snap of number plate, deleting characters and OCR from symbols. The primary propose of this mechanism is to depict and establish productive image processing approaches and algorithms to save the permit plate in the snapped image, to divide each character from that number plate and to recognize each character of the segment by using the Open Computer Vision Library.

Keywords: OpenCV, K-NN algorithm, number plate recognition, OCR Optical Character Recognition

1 INTRODUCTION

Traffic generated from automobiles' search of vacant parking space is important in crowded urban areas. Real-time parking space data gives important information for parking space governance system, which is obtained by site-specific sensors. Fitting and maintenance of on-site sensors is quite costly. For resolving the parking organization issue, various techniques are being deployed and researches were coordinated to come up with structured parking mechanism. Although a small number of parking systems are stationed as isolated technologies, many approaches will be merged to accomplish the assigned task. These automated systems comprise of Digital Image processing systems, embedded systems, Internet of Things. The studies interest in solitude covers Access control [3]–[5], Cloud security. [6]–[9] These studies have gained lot of attraction from the society. The potential to transfer data from a system to other without the need of any human intervention. [1]. In this paper, a robust algorithm is developed anticipated the connected capabilities of Gaussian blurring, short end thresh holding, sharp edge detection, edge. Detection combined with the potentiality of IoT to give prospects structured results and entrance to information. [2] Image classification in a broad sense is an intensely contrasting area which obviously cannot be purposeful by a single, ace method. By the usage of Gaussian Blur the image is ransacked of higher constancy and direct contrast for

DOI: 10.1201/9781003193838-90

even analysis. It increases the appearance identification of the snapshot while withdrawing sounds that are attached at the time accession. [3] The side of the snapshot is the utmost at which the gray value of the nearest pixel at the abrupt point of the signal changes exceptionally. Edge detection is an important part of computer vision and image study method of outline detection in image processing. It belongs foundational and still open figure processing work due to the difficulty in study of various types of figures with a huge number of classes of objects taken into consideration. The parking observing system proposed is low consumption, easy implementation, and cost effective. [4] The additional space registration, and exploration ability gives prospect the adaptability and choice to amalgamate the algorithm with other self-sufficient devices.

1.1 Smart parking system

We have two types of parking techniques: Automated, and Traditional.

In the long run, the traditional parking garages will be costly than automated parking systems. Automated multiple levels parking mechanisms are cost efficient per parking space, as they need little building space and flooring space than a traditional establishment with the same dimensions [1]. Using an intelligent car parking mechanism with image processing techniques and OpenCV. The project includes system with infrared transmitter and receiver at entry/exit gates and server motor barriers that is pushed by hydraulic system by a micro controller for car entry/exit. The proposed mechanism is doing all the necessary work like opening and closing of the barriers without any interference of any person. If there is no empty parking slot, the barriers will not open therefore vehicles will not able to enter in the garage. Figure 1 represents proposed figure. In this paper we have developed a system in which:

- We click the 20 images from infrared camera
- Upload them into the mechanism & use OpenCV library for scanning the images. [4]

Table 1. Shows literature review briefly.

2 METHODOLOGY AND PROCESSES

- To meet this objective we use various Hardware and software component. Hardware like ARDUINO, servo motor, 2 cameras (opening/closing). And in software part we used the OpenCV library in Python and SQL for database, PHP,HTML, CSS for the web portal of booking of parking slot. Making this project reduces time in getting parking slot and provide assurity over the availability of the parking slot and other side it get the aim to reduce the no of employee over the parking site and ensures the Security of the vehicle.

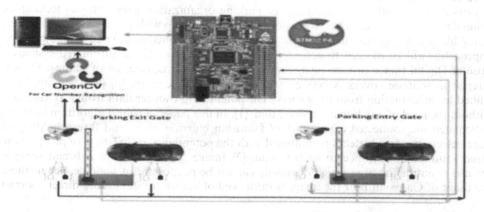

Figure 1. Proposed system.

Table 1. Literature review.

Author	Year	Area of focus	Conclusion	References
S. Mahmud et al	2013	Parking agent is added in the software.	Parking agent allots the parking space.	[8]
J. K. Suhr et al	2014	Detection of empty parking space is based on sensors.	Sensor detects the vacant parking slots.	[11],[12]
T. Rajabioun et al	2015	Suggested the use of multi layered auto regressive model.	On and off the road parking slot availability prediction.	[19]
C.-C. Huang et al	2017	Empty parking slot based on Machine learning algorithms.	Predictive algorithms don't support the variable nature of vehicle network.	[13],[14]
C. Roman et al	2018	Detection of on the road parking spaces in mega cities.	Predictive algorithms don't support to the variable nature of vehicle network.	[15],[16]
C. Tang et al	2018	Analysis of extensive IoT data from IoT sensors.	Provide parking data to the user.	[9]
W. Shao et al	2018	Managing car parking violations.	Sensors send the data to the authorities.	[10]
M. Lee et al	2019	Algorithm suggested the usage of Around-View Monitoring (AVM).	Finds the vacant parking slots.	[19]
J. Lin et al	2019	Algorithm proposed forecasts the probable parking congestion.	Prospects are given directions to various parking spaces.	[17],[18]

- We make the system using OpenCV (Computer Vision) that is used to extract the vehicle no and store in the database for records and verification of vehicle owner. And for the opening/closing of gate (barriers). We used ARDUINO system that instructs the servo motor to the automatic opening and closing of gate. And for the booking of the parking slot over the site we made the Web portal using PHP, HTML and CSS. This project majorly works on the Python i.e. OpenCV library is on python and all transfer of instruction is done on Python. And for the security purpose we used the Android that send the OTP to the user for verification of owner of vehicle.

3   ALGORITHIMS USED FOR DETECTING NUMBER PLATE

1. Begin
2. Giving Input: Primary Captured Image: In this system, we give the input of image which is captured using web cam. Out coming: Alpha Numeric Characters: After the process has completed alpha numeric characters are displayed as the output.
3. Technique: KNN: The technique used to achieve the output is K-Nearest Neighbors throughout the scanning process. LP: We are scanning the License Number Plate or the vehicle to achieve our goal. Transformation of RGB Image to Grayscale Image: In this the original image that was captured by web cam is converted in black and white image. Filtrate Morphological Modification: In this we read the characters from the image as one character.
4. Convert Grayscale image to binary image: In this step, the black and white picture is converted into binary image in form of 0 & 1. Filtrate Gaussian for Blurs image: In this the bell shape structured number plate will be detected.
5. Searching all contours in the captured image: In this step all the characters will be cut and displayed one by one. Find&perceive all feasible characters in image: We find all the possible character of the image by matching the character saved in database.

6. Cut thefragment of image with maximum candidate LP regions: Cutting the license plate into segments of different characters. Cut the LP from theprimary image: In this the LP is cut from primary image captured in the starting.
7. Implement the procedure from 6 to 11 again on cutted image: We have to apply the steps from 6-11 again to get another character of the number plate. Display the Text/Alpha Numeric Characters in LP: We have to display the output on the screen.

4 RESULT

4.1 *Web portal and number plate detection*

A web portal is a custom-made website that brings information from different sources, like emails, online conventions and browsers, together in an unvarying method. Usually, each source of information gets its committed area on the page for presenting information. It gets the entire data from the user for the parking slot.

It closes all the process, proceeding by acquiring the picture, which is attended by the position of the car number plate until it is segmented. The detection of the car number plate is made from the pictures of those characters that are achieved at the end of the separation method. The model that will be used for car number plate detection is able to read it carefully.

4.2 *OTP generator and entry/exit gate working*

It is the app which sends the OTP on the user phone no and also sends the parking slot. This app gets the instruction through the parking_otp_test.py.

Parking_otp_test.py: this python script instructs the app to send the certain 4 digit no which is OTP and a parking lot no.

After the OTP is verified then we used ARDUINO system that instructs the servo motor to the automatic opening and closing of gate.

5 CHALLENGES OF TEXT DETECTION

Recognizing text in can be achieved with the help of heuristic-based approaches, such as using leaning information or the verity that text is unexpectedly categorized into paragraphs and characters that seem in a queue. Figure 2 shows text detection challenges.

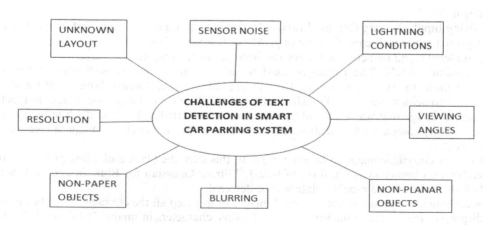

Figure 2. Text detection challenges.

486

6 CONCLUSION AND FUTURE WORK

Parking space determination with traditional algorithms gives wrong results in region where the space lines are misrepresented and when automobiles coincide with spaces itself. The algorithm proposed resolves the issues of excess sounds from random automobile and people lingering gives wrong results. The algorithm binds the image processing abilities of OpenCV with the adaptability of Raspberry Pi for practical along with structured identification algorithms. Using various modules and cases made the algorithm sturdy and enables it to handle real life problems. It allows the data flow from variable points and to be added in the algorithm. The results are exhibited on a web portal where the prospect has a choice to reserve space or to set navigation path to the allotted empty spaces. In suggested method the prospect can select the parking spot of his choice and give the information to any self-sufficient mobile device via web portal. The automatic permit plate recognition system proposed in this research has several restraints. Most major being that the state data position is assumed to be at top part of permit plate. Though most of the plates consists of state data at the upper part of permit plate, the suggested system won't be able to determine the state data if position of state information is changed.

REFERENCES

[1] Qi Luo, Student Member, IEEE, Romesh Saigal, Robert Hampshire and Xinyi Wu, "A Statistical Method for Parking Spaces Occupancy Detection via Automotive Radars", Vehicular Technology Conference (VTC Spring), 2017 IEEE 85th.

[2] Nastaran Reza Nazar Zadeh, Jennifer C. Dela, "Smart Urban Parking Detection System", 2016 6th IEEE International Conference on Control System, Computing and Engineering, 25-27 November 2016, Penang, Malaysia.

[3] K.Malarvizhi, A.Kayathiri, K.Gowrisubadra" Survey paper on vehicle parking slot detection using internet of things", 2017 International Conference on Computation of Power, Energy, Information and Communication (ICCPEIC).

[4] Harshal Sardeshmukh, Dayanand Ambawade," Internet of Things: Existing Protocols and Technological Challenges in Security", 2017 International Conference on Intelligent Computing and Control (I2C2).

[5] S. Xu, Y. Li, R. H. Deng, Y. Zhang, X. Luo, and X. Liu, "Lightweight and expressive fine-grained access control for healthcare internet-of-things," *IEEE Transactions on Cloud Computing*, 2019.

[6] C. Zhang, L. Zhu, C. Xu, K. Sharif, C. Zhang, and X. Liu, "PGAS: privacy-preserving graph encryption for accurate constrained shortest distance queries," *Inf. Sci.*, vol. 506, pp. 325–345, 2020.

[7] S. Mahmud, G. Khan, M. Rahman, H. Zafar *et al.* [7] S. Xu, G. Yang, and Y. Mu, "Revocable attribute-based encryption with decryption key exposure resistance and ciphertext delegation," *Inf. Sci.*, vol. 479, pp. 116–134, 2019. 11, no. 5, pp. 714– 726, 2013.

[8] C. Tang, X. Wei, C. Zhu, W. Chen, and J. J. Rodrigues, "Towards smart parking based on fog computing," *IEEE Access*, vol. 6, pp. 70 172–70 185, 2018.

[9] W. Shao, F. D. Salim, T. Gu, N.-T. Dinh, and J. Chan, "Traveling officer problem: Managing car parking violations efficiently using sensor data," *IEEE Internet of Things Journal*, vol. 5, no. 2, pp. 802–810, 2018.

[10] J. K. Suhr and H. G. Jung, "Sensor fusion-based vacant parking slot detection and tracking," *IEEE Transactions on Intelligent Transportation Systems*, vol. 15, no. 1, pp. 21–36, Feb 2014.

[11] X. Hou, Y. Li, M. Chen, D. Wu, D. Jin, and S. Chen, "Vehicular fog computing: A viewpoint of vehicles as the infrastructures," *IEEE Transactions on Vehicular Tech nology*, vol. 65, no. 6, pp. 3860–3873, 2016.

[12] C.-C. Huang and H. T. Vu, "Vacant parking space detection based on a multilayer inference framework," *IEEE Transactions on Circuits and Systems for Video Technology*, vol. 27, no. 9, pp. 2041–2054, 2017.

[13] R. Garra, S. Mart´ınez, and F. Sebe, "A privacy-preserving ´ pay-by-phone parking system," *IEEE Transactions on Vehicular Technology*, vol. 66, no. 7, pp. 5697–5706, 2017.

[14] S. Faddel, A. T. Elsayed, and O. A. Mohammed, "Bilayer multi-objective optimal allocation and sizing of electric vehicle parking garage," *IEEE Transactions on Industry Applications*, vol. 54, no. 3, pp. 1992–2001, 2018.

[15] C. Roman, R. Liao, P. Ball, S. Ou, and M. de Heaver, "Detecting on-street parking spaces in smart cities: per formance evaluation of fixed and mobile sensing sys tems," *IEEE Transactions on Intelligent Transportation Systems*, vol. 19, no. 7, pp. 2234–2245, 2018.

[16] J. Lin, S.-Y. Chen, C.-Y. Chang, and G. Chen, "Spa: Smart parking algorithm based on driver behavior and parking traffic predictions," *IEEE Access*, vol. 7, pp. 34 275–34 288, 2019.

[17] N. Fulman and I. Benenson, "Establishing heterogeneous parking prices for uniform parking availability for au tonomous and human-driven vehicles," *IEEE Intelligent Transportation Systems Magazine*, vol. 11, no. 1, pp. 15–28, 2019.

[18] M. Lee, S. Kim, W. Lim, and M. Sunwoo, "Probabilistic occupancy filter for parking slot marker detection in an autonomous parking system using avm," *IEEE Trans actions on Intelligent Transportation Systems*, vol. 20, no. 6, pp. 2389–2394, June 2019.

[19] T. Rajabioun and P. A. Ioannou, "On-street and off street parking availability prediction using multivariate spatiotemporal models," *IEEE Transactions on Intelligent Transportation Systems*, vol. 16, no. 5, pp. 2913–2924, Oct 2015.

Recent Trends in Communication and Electronics – Sharma et al. (Eds)
© 2021 Taylor & Francis Group, LLC, ISBN 978-1-032-04572-6

Neural-network's way to refine keystroke-information for high quality classification purposes

Anurag Tewari
PranveerSingh Institute of Technology, Kanpur, India

Bipin Kumar Tripathi
Harcourt Butler Technological University, Kanpur, India

ABSTRACT: A user's typing manner defines identity of a user because typing rhythm of every user is always unique due to its link with behavior, style and mental attitude of user. On account of this, keystroke-feature creates a behavioral biometric recognition system. But timing based typing information can be deceptive also if information is shared by user half-heartedly. In this work neural-network-based classifier systematically investigates, presence of less significance in given dataset. Entries of few users are identified and eliminated from the data set, having relatively large misclassification errors. These entries are result of negative participation of some users during data collection process and during authentication process these entries may deteriorate performance-accuracy. This paper consists of different sets of password-entries based timings (of users) to be observed while classification procedure. Afterwards z–score normalization is applied to compare and analyze results of classification on original dataset and on reduced dataset.

Keywords: keystroke, biometric, neural network, normalization, classification

1 INTRODUCTION

In today's digital world the most precious entity after humanity is information. 'Login-ID, password' protection is the most popular and ancient method. It becomes unsafe if kept as simple and hard to remember if made difficult due to safety. Any physical device as: e-card or e-key is vulnerable towards loss, steal or duplicate-creation. It directed us towards search or development of other means of security having qualities like uniqueness, stability, easy to use, accurately measurable, universally available and highly robust against all type of attacks [Jain et.al, Saeed]. Biometrics is the only panacea, invented and developed with above mentioned qualities. With the help of biometrics all the problems as remembering a difficult password or safely carrying a card or key are being resolved. Biometrics is a medium to identify and verify a person scientifically. It can recognize and authenticate user by measuring physiological or behavioral qualities or mixture of both. Face, retina, DNA, fingerprint, hand geometry, shape and sizes are counted under physiological characteristics. While individualities like signature, voice, keystroke-timings and gait are within behavioral features as per books by Jain et.al and Saeed. Accurate measurement of quantities in biometrics only depends on the efficiency of sensors used to record the response.

On comparing both biometrics: behavioral and physiological, it is obvious that maybe physiological are looking more accurate but they are obtrusive and revealed way of authenti-

DOI: 10.1201/9781003193838-91

cation. In recent past various presentation attacks (duplicate fingerprint, retina lens etc.) have made the physiological biometrics less secure. On the contrary behavioral biometrics is far from these negativities. Keystroke biometrics is a behavioral biometrics, used for authenticating and authorizing a legitimate user present among various imposters. It is a type of pattern recognition based task where a database has various typing styles of different users enrolled in the form of timing features. Then a newly entered user's timing characteristics are matched with these existing values, genuineness is decided as per the match result. This method can work as an additional but urgent way to raise security along with password based verification technique.

In this work we observed chances of betterment in capability of available keystroke dataset to recognize the user. To accomplish our goal we explored one benchmark dataset and divide and conquer method is used by us to keenly search out the odd users with neural network based classifier.

The remaining part of the paper is well thought-out as: In next section an orderly review of previous and correlated work is given. Section 3 mentions about neural network used as classifier. Section 4 consists of material used and methodology applied in experiments. Section 5 has summarized outcome and discussion. Last but not the least Section 6 concludes the work along with future impact of the research.

2 REVIEW OF PREVIOUS RELATED WORK

Keystroke-timing's utilization to authenticate an individual's identity has been started way back in 1980 by Gain et al. After that Joyce and Gupta claimed that delay timing during keyboard utilization can prevent fake user, even using all necessary information as login ID and password. An exhaustive survey presented by Teh et al., has work of last 3 decades having kept all together. According to another review presented by Natasha et al.: enough improvement is urgent in keystroke dynamics dataset, so that its performance as an authenticator could be more accurate and timely. Dwell time, flight time, word graphs, digraph, trigraph, and n-graph are the recent features used in current scenario to represent keystroke dynamics in Literature.

As per the survey done by Shawkat et.al many popular classification methods are mentioned as: linear classifier, SVM, D-trees, boosted trees, KNN, random forest, neural network, gaussian mixture model, and distance based classifier etc. Trendy metrics for performance evaluation in relation with keystroke authentication system are: false acceptance rate (FAR), false rejection rate (FRR), equal error rate (EER), and accuracy etc. About availability of benchmark datasets Giot et al. have presented a well organized comparative study of various public dataset for keystroke dynamics as: CMU, GREYC, PRESURE, WEBGREYC, DSN2009, GREYC-NISLAB, KEYSTROKE100 datasets are mentioned in this study.

As per the reviews and surveys it is obvious that neural network is one of the most popular classification method used with keystroke biometrics. Research work produced by Ozbek has identified, eliminated and classified faulty users using various popular machine learning techniques as: KNN, SVM and D-tree. Kołakowska systematically examined for identifying not only a person but emotional or psychological states are also recognized of that person. Another quality work was shown by Harun et al., which displayed supremacy of ANN based classifier and in this work equalization-transformation was used to preprocess the keystroke dataset. Neural network was used for keystroke identification by Uzun et al and Baynath et al. in their researches.

3 NEURAL NETWORK

Neural network or more fittingly artificial neural network (ANN) was developed to solve problems in which machines were far behind than human. These problems were generally related to

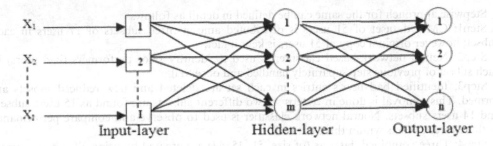

Figure 1.   Feed-Forward neural Network (made up of Multilayered-perceptrons).

human's sensory organs. Human takes decision based on previous learning and error occurred (difficulty felt) with similar or novice type of problems. ANNs also tries to learn using Back propagation method, and result of learning (error or no error) is accumulated then positively modified and again stored in the form of synaptic weights within interconnection of a massively parallel layered network of neurons. This learning is performed based on pair of input-output, presented to network. After learning our trained network becomes capable of predicting correct outputs for unknown inputs in future. A layered supply ahead ANN is generally made up of three layers of neurons as: Input layer, Hidden layer and Output layer as in Figure 1. In this Procedure within feed-forward neural network, few functions have been utilized known as: Activation function, transfer function and error function etc.

4   MATERIAL USED AND METHODOLOGY APPLIED

In this section dataset, model, classifier and evaluation procedure utilized to accomplish the goal of research, are suitably presented.

4.1   *Dataset narrative*

We have used a benchmark data set defined and known as CMU keystroke data (Killourhy, Maxion). It consists of information about using a common keyboard to type a difficult password ".tie5Roanl". This information is actually in the form of timing vector generated while pressing different keys to enter password. 51 users have contributed by entering same password 50 times in 8 different time-slots. Hence per user 400 entries are enrolled. Time slots were created with at least one day gap. The dataset has total 31 timing features which were extracted and systematically placed in array format. For every key features are: delay between 2 keys pressed down- 1 after another, 1 key is made free while other is pressed and 1 key's press and release time-duration. Overall size of ready dataset has ideally 20400 samples with 31 features for each sample. Hence, 20400 samples with 31 features for each sample. Figure 2 is representing entire methodology of the research work.

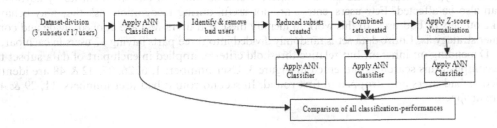

Figure 2.   Methodology.

Stepwise approach for the same can be defined in detail as follows:

Step1: CMU dataset of 51 users is distributed among three subsets of 17 users in each subset; however original copy of 51 users is kept intact.

Step2: Neural network-based classifier is used to identify poor performers (bad users) in each subset of previous step separately handled and observed.

Step3: Identified bad user's entries in each set are deleted and new reduced subsets are formed. This removal is done in 2 steps so two different subsets are found as 15 users subsets and 14 users subsets. Neural network classifier is used to observe and compare performance of these three subsets within three datasets.

Step4: Three combined datasets (of size: 51, 45, 42) are created by using all subsets correspondingly. These combined datasets are evaluated by neural network classifier to observe and compare performance accuracy.

Step5: Z-score normalization is applied on all three combined data sets.

Step6: Combined datasets are investigated by neural network classifier and results are observed and comparatively analyzed in both states normalized and un-normalized.

4.2   *Classifier and software used*

In this research work all the classification oriented experiments are carried upon neural network based classifier. A function applied here to train the network-model is back-propagation based Scaled Conjugate Gradient method which is an efficient algorithm for classification and pattern recognition tasks. Setting of quantity of neurons present in hidden layer is also random initially and then fixed finally. Our experiments thru simulation were executed using MATLAB R2019a software on a personal computer with CPU Intel Core i3-4th generation 1.70GHz, HDD 500GB, RAM 8 GB specification.

4.3   *Evaluation metrics*

For every dataset combined, distributed or reduced; neural network based classifier is used. Data samples are divided and used as 70%, 15%, and 15% for training, testing and validation respectively. Performance of keystroke dynamics is observed with the help of confusion matrix, ROC plot and percentage error data (indicate the fraction of samples which are badly-classified). We used all the timing features available with CMU dataset known as Killourhy and Maxion's database. Neural network used has 3-layered architecture: 1st layer for input + 2nd as a hidden layer + 3rd layer for outputs. All the data sets are trained validated and tested within the neural network exhaustively (every time average is computed by performing training procedure multiple times to get true results of observation).

5   RESULTS AND DISCUSSION

We executed our experiments on a benchmark dataset having 51 users along with their typing statistics captured within 31 features. Here every user is giving data by entering 400 times the same password ".tie5Roanl" in different sessions. Due to human psychology any person can have different type of attitude and mood at different time so filling of password entries by keyboard can also be affected. Few users' entries are not useful during recognition task they may behave like outliers or bad uses. In this work to identify these bad users we have applied divide and conquer strategy as: Entire data set is randomly divided into three parts having 17 users in each part.

Depending on inherent but relative threshold criteria, applied in each part of data subset to accomplish this search as presented in Figure 3. User number: 1, 6, 26, 27, 42 & 48 are identified as bad users and dropped in first round. In second round bad user numbers: 11, 29 & 45 are dropped.

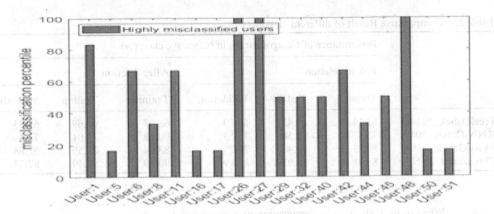

Figure 3. Identified non-significant users having high relatively large misclassification error.

To verify effect of this elimination, accuracy of classification is being observed as in Figure 4 within divided subsets and various other combined (complete and reduced) results are grasped in following tables (Table 1, 2 and 3) in percentage error of classification.

As per mentioned Figures and Table 1, it is obvious that a plan to remove non-significant or faulty user's data from the dataset is neatly beneficial. Distribution or no distribution is also effective since performance depends on size.

According to Table 2, it is logically justified that our neural network based classifier is providing better results on same type of datasets than tree, KNN, SVM.

Now next experiment is done by applying Z-score normalization method on both types of datasets (original and reduced). Finally to see a comparative picture of classification-errors

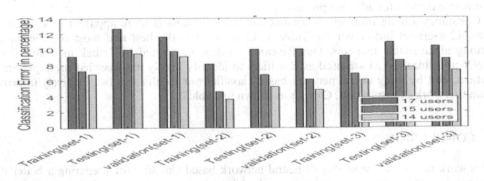

Figure 4.   Classification-Error accuracy in distributed user-sets (17, 15, and 14).

Table 1. Classification-Performance in combined sets.

No. of Users	Operation-type	Classification-Accuracy (%)
51 Users	Training	88.51
	Testing	84.92
	Validation	85.19
45 Users	Training	90.69
	Testing	87.49
	Validation	87.33
42 Users	Training	90.68
	Testing	87.55
	Validation	88.39

Table 2. Comparative Result of different classifiers.

	Performance of Categorization in respective class (%)					
	Before deletion			After deletion		
	Training	Testing	Validation	Training	Testing	Validation
Tree[Ozbek, 2019]	63.19	60.42	60.69	68.97	65.69	65.33
KNN[Ozbek, 2019]	78.42	73.56	73.53	80.12	73.01	75.4
SVM[Ozbek, 2019]	80.83	79.48	79.9	83.04	81.59	81.88
This work (ANN)	88.51	84.92	85.19	90.69	87.49	87.33

Table 3. Effect of normalization in classification-performance (%).

	51 users			45 users			42 users		
Norma lization	Training	Testing	validation	Training	Testing	Validation	Training	Testing	validation
Before	88.51	84.92	85.19	90.69	87.49	87.33	90.68	87.55	88.39
After	88.41	85.28	85.23	91	87.5	87.97	91.7	88.87	88.63

with normalized versus un-normalized datasets having 51, 45, and 42 users respectively, is presented in a tabular manner in Table 3 for view and feel the result obtained. Even the best normalization method applied on all features of this dataset is unable to generate a big difference in classification performance. So by deleting non-significant entries we have almost acquired a superior dataset for advance processing.

Constancy among training testing and validation process is more apparent in 45 users-set than 42 users-set but as per the Table 3, 42 users-set is the best and most reliable dataset among all available user-sets. Our Research work is motivated by Ozbek in which D-tree KNN and linear-SVM are used as classifiers to identify highly misclassified users present in dataset. But by using neural network based classifier our results are comparatively improved than this work of inspiration, Ozbek's as shown in Table 2.

6 CONCLUSION

This work has shown capability of neural network based classifier for preparing a better biometrics dataset to proceed in further research. All the features of benchmark dataset are used to identify relatively insignificant users. Results of classification based authentication depends on correctness, accuracy or purity of dataset used, neural network has played vital role to achieve this. The keystroke-timing Information is a very promising biometrics related to behavior for identification and verification of legitimate user. During identification of poor performers we have used a benchmark dataset and shown that even deletion will reduce dataset but if a dataset used during classification is not improved by normalization then a way to raise its quality is to cut it short by dropping some erroneous or non-significant data existing within it. We have also hypothesized (as a future aspect) that if by removal of some data our dataset will be shortened for training purpose then either of the following two steps could be taken:

A. Carry on with reduced dataset but raise complexity in algorithm used.
B. Fill that gap with some similar type of classifiable data for free text based identification.

REFERENCES

Baynath, P., Soyjaudah, K.M.S., & Khan, M. Heenaye-Mamode. 2017. Keystroke Recognition using Neural Network. IEEE, 5th International Symposium on Computational and Business Intelligence, UAE.

Gaines R. S., Lisowski W., Press S. J., Shapiro N. 1980. Authentication by keystroke timing: some prelimnary results, Rand Corporation, Tech. Rep.

Giot, R., Dorizzi, B., Rosenberger, C. 2015. A review on the public benchmark databases for static keystroke dynamics, Computers & Security 55, 46–61

Harun, N., Woo, W.L, & Dlay, S.S. 2010. Performance of Keystroke Biometrics Authentication System using Artificial Neural Network (ANN) and Distance Classifier Method, International Conference on Computer and Communication Engineering (ICCCE), (pp. 1–6).

Jain, A. K., Ross, A. A. 2015. Biometrics, Overview, in: Li S.Z., Jain A.K. (eds.), Encyclopedia of Biometrics (2nd ed.). New York: Springer.

Joyce R., Gupta G. 1990.User authorization based on keystroke latencies. Communications of ACM, 33(2), 168–176.

Killourhy, K.S., Maxion, R.A. 2009. Comparing anomaly detectors for keystroke dynamics, 39th annual international conference on dependable systems and networks (pp.125–134), Portugal.

Kołakowska, A. 2013. A review of emotion recognition methods based on keystroke dynamics and mouse movements, 6th International Conference on Human System Interactions (HIS), Poland.

ÖZBEK, M. E. 2019. Classification Performance Improvement of Keystroke Data. Innovations in Intelligent Systems and Applications Conference (ASYU), Izmir, Turkey.

Raul, N., Shankarmani, R., Joshi, P. 2019. A Comprehensive Review of Keystroke Dynamics-Based Authentication Mechanism, International Conference on Innovative Computing and Communications, (pp.149–162).

Saeed K. 2017. New Directions in Behavioral Biometrics. Boca Raton: CRC Press, Taylor & Francis.

Shawkat, S. A., Ismail, R. N. 2019. Biometric Technologies in Recognition Systems: A Survey. Tikrit Jour nal of Pure Science, Vol.24 (6).

Teh P. S., Teoh A. B. J., Yue S. 2013. A survey of keystroke dynamics biometrics, The Scientific World Journal, vol. 2013.

Uzun, Y., & Bicakci, K. 2012. A second look at the performance of neural networks for keystroke dynamics using a publicly available dataset. Computers & Security, 31, 717–726.

Recent Trends in Communication and Electronics – Sharma et al. (Eds)
© 2021 Taylor & Francis Group, LLC, ISBN 978-1-032-04572-6

Investigation of forgery in speech recordings using similarities of pitch chroma and spectral flux

Kasiprasad Mannepalli

Department of ECE, K L Deemed to be University, Vaddeswaram, Guntur (district), Vijayawada, India

Swetha Danthala

Department of ME, Koneru Lakshmaiah Education Foundation, Vaddeswaram, Guntur (district), Vijayawada, India

Panyam Narahari Sastry

Department of ECE, CBIT, Hyderabad

Durgaprasad Mannepalli & Bathula Murali Krishna

K L Deemed to be University, Vaddeswaram, India

ABSTRACT: Forgery made by copy and moving of very short segments of speech, followed by placed up-processing operations to remove signs of the forgery, gives the forensic identification incredible challenge.in this paper, an approach with features spectral flux, Pitch chroma and tonal power ratio is proposed to identify the speech forgery. A database is prepared for the speech recordings with a positive sentence and corresponding negative sentence. Then the positive sentence is made negative sentence by forging. These recordings are used as database for the feature extraction. Dynamic time warping is done to measure that function set's similarities. By contrasting the similarity level using threshold limit in speech recording, one can identify the forgeries of replicapasses. The proposed approach can be very useful in detecting and finding replicamoving forgeries, even as fast as one voiced speech section on a solid segment of speech. The recommended technique is stout in resisting to numerous varieties of generally used post-processing practices and highlights the hopeful accuracy of the presented method to locate copy&move forgery in speech.

Keywords: Speech forgery, Spectral features, Pitch chroma, DTW

1 INTRODUCTION

Speech videos are becoming widely used in court as electronic testimony. Unfortunately increasing technology is leading to adapt the fraudulent notion in the society. A common person, with no advanced audio processing experience can make a forged copy of the speech signal. Copy-mix maneuver is one of the most commonly used methods of speech manipulation. The hacker can simply clone a few voice recording fragments and insert these parts into different locations within the same file of audio to swap the sentence's linguistic information. For example, the term "I am innocent" can be modified to "I am guilty" by copying and adding the word "guilty" in the same speech clip using the same Speaker. Usually, this kind of forgery is imperceptible because of the fact that the forged sections were extracted from the

DOI: 10.1201/9781003193838-92

transcription of equivalent expression. Researchers used the method of first extracting syllable level pitch sequences followed by computing their correspondences sequences to sense the replicated speech (Yan et al., 2015).

The biomedical signals analysis can be better performed when good features extracted from the signals (Gattim, 2019) As speech is not as stationary signal, it needs extract the features by windowing the signal. In much speech processing researches temporal features (Mannepalli et al., 2017). Auditory features, like MFCC features (Mannepalli et al., 2015), pitch chroma (Mannepalli et al., 2016), spectral flux (Mannepalli et al., 2018) and tonal power ratio (Mannepalli et al., 2016) were the feature sets. To classify the pattern of signals for a specific claim, classification methods are to be employed. There are many different such techniques are reported for different applications (Srinivasa Reddy, S., & Suman, M., 2018; Reddy, S. S et al., 2018; Bojja, P., & Sanam, N., 2017; Vallabhaneni & Rajesh, 2018). Based on the data acquisition method a signal de-noising may also be incorporated (Gattim, N. K et al., 2017; Bhavana & Rajesh, 2016; Bhavana et al., 2016; Bhavana et al., 2019). Transform techniques are useful to obtain the enhanced image in the applications such as MRI and CT scan and various assessable analyses are reported on the MR Imaging and CT image (Revathi et al., 2019; Bennilo Fernandes, J et. Al., 2018; Bennilo Fernandes, J et. Al., 2019). Basing on the pitch and LPC based formants, the deviations in the speech segments can be traced (Yan, Q., Yang, R., & Huang, J., 2019) The rules package of dynamic time warping(DTW) is implemented for the measure of differences concerning each pitch series and process of formation.

2 METHODOLOGY FOR COPY AND MOVE FORGERY DETECTION

In this segment, the methodology of duplicating and move forgery in advanced speech segments is presented

2.1 Copy-move forgery

Duplicate stream activity is a normally utilized fraud strategy in advanced speech recording. An aggressor, without much technical knowledge can change the speech recording by introducing a word into the sentence or by deleting a word from the sentence into different positions inside the indistinguishable speech recording. These recording by using filtering and resampling the forged speech samples can be made as good as original speech recordings. This may likewise make it more difficult to identify duplicate move falsification by method for simply looking the copied speech sections in waveform.

In this work, it is intended to detect whether the speech signal is forged or not. For speech forgery detection, a database is needed so a database is prepared by selecting few sentences. The positive sentences (for ex: I am coming) were recorded and negative sentences(for ex: I am not coming) for the positive sentence were also recorded. The methodology is shown in Figure 1 as a flow chart. From the sentences in the database we extract features as described in later sections.

3 FEATURES USED IN THE PRESENT WORK

In this section the extracted speech features such pitch Chroma, spectral flux, and power ratio are presented.

3.1 Pitch chroma

The tonal representation of the speech signal can be utilized as a basis for deriving audio features. Application of the derivative of the energy results the features. Here, chroma refers to the 12 classes of traditional tones of the same temperament scale. The chromatic

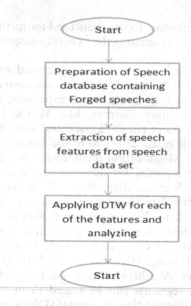

Figure 1.   Methodology involved in the speech forgery detection.

characteristics are Audio functions based on tone and chrominance much more often than the tone characteristics and show a great amount of sturdiness variations of timbre and instrumentation.

3.2   *Tonal power ratio*

Tonal Power ratio: it is used to calculate speech signal tonality. The tonal power ratio is determined by the continuum components tonal power and the overall power (Mannepalli, K., Sastry, P. N., & Suman, M., 2017). Let the input speech signal be snapped and speaks the input voice output spectrum. The value is between 0 and 1 where the weak where the low worth speaks to the commotion design and tonal range is characterized by the high worth. At that point, tonal power proportion of the information speech signal is communicated in condition and the equation 1 represents the tonal power ratio.

$$T = \frac{G(c)}{\sum_{i=0}^{\frac{n}{2}-1} |F(k,c)|^2} \tag{1}$$

Here G(c) represents is the tonal power and is calculated by adding all bins which are local maximum (Mannepalli, K., Sastry, P. N., & Suman, M., 2017).

3.3   *Spectral flux*

This feature can determine the spectral components of the speech signal. These spectral components are the important aspect because if the signal's spectral quality changes over time, the performance identification will be diminished.

You will evaluate the incoming speech signal using the squared differences of normalized magnitudes of successive spectral distributions corresponding to the consecutive signal frames. This is given in the equation 2 and the N denotes the length of the vector.

Table 1. DTW distances of the experiments.

	Pc0	Pc1	Pc2	Pc3	Pc4	Pc5	Pc6	Pc7	Pc8	flux	TPR
DTW(o,o)	0.004405	8.367348	0.112368	1.42979	0.018486	0.1246	3.257171	0.413482	0.955399	3.72E-05	0.000037
DTW(o,f)	0.139941	36.40463	4.509276	1.97297	0.694751	0.76795	4.035274	2.032375	2.931871	3.89E-05	2.415431
DTW(f,f)	0.10939	8.678546	3.573492	2.8492	0.640015	0.91906	5.055357	2.570932	0.886151	0.000053	0.000536

$$P_r = \sum_{k-1}^{N/2} (|F[(k)]| - |F_{r-1}[k]|)^2 \qquad (2)$$

4 RESULTS AND DISCUSSIONS

To detect whether the speech signal is forged, the features such as pitch Chroma, spectral flux, and power ratio are selected for analysis. With dynamic time warping (DTW), the extracted features from the audio data set were compared. The main use of power ratio technique is its simplicity to observe rise and fall ness in the tone. With the help of this power ratio technique, the similarities between Original signals, original to forged are computed and observed. Apart from pitch chroma, the spectral flux. spectral flux will be used in extracting the spectral components. The obtained features are arranged systematically and DTW distance is computed. The distances between the original and original signals and distances between the original and forged signals are computed. The results of the experiment are shown in the Table 1. In the Table 1, 'O' represents the negative sentences (without forging), the positive sentence (which is intentionally made negative sentence by forging a negative word into it) is represented as 'f'.

The Table 1 shows the features extracted in the experiment such as pitch chroma represented as Pc, the spectral flux represented as flux and the tonal power ratio represented as TPR. From these results, it is understood that the distance between original speech to original speech (it is represented as 'o,o') is less when compare to original and forged(it is represented as 'o,f'). The features obtained from the pitch Chroma spectral flux and tonal power ratio arranged in a sequential order and calculated the distance between each and every sample by using dynamic time warping and stated them in table. The dynamic time warping helps us to get normalized distance between two audio signals the audio signals may be original or forged .based on the distance one can identify whether the signal is forged or original. Table 1 represents the comparison of DTW distances for the three cases. This method is compared with DFT method (Liu, Z., & Lu, W., 2017) in the Table 2. The databases used in the compared works are different. In the present work the database collected from the native Telugu speakers and pitch shift operations performed on the audio. The result shows the proposed method is helpful in identifying the forgery using a pitch shift post processing operations by the attackers.

5 CONCLUSION

In this work, a database was created with a positive sentence and the same sentence with negation. The positive sentence is negated by copying the negation from another sentence of same speaker which was recorded at another context. The original negative sentence and synthesized negative sentence were taken for forgery identification. The features such as pitch chroma, spectral flux and tonal power ration have been extracted from the above made

Table 2. Comparison of proposed method with DFT method.

Method	Database used	Nativity of the speakers	Precision for rising pitch	Recall for rising pitch	Precision for falling pitch	Recall for falling pitch
Proposed method	Created using mobile phone	TELUGU	74.16	72.5	75	71.66
DFT method (Liu, Z., & Lu, W., 2017)	TIMIT database	English	69.23	67.42	70.14	70.83

database. As speech is continuously varying, the time series analysis was taken to identify the forged sentence.

REFERENCES

Bennilo Fernandes, J., Mannepalli, K.P., Saravanan, R.A. & Kumar, K.T.P.S. 2019, "Fuzzy utilization in speech recognition and its different application", *International Journal of Engineering and Advanced Technology*, vol. 8, no. 5 Special Issue 3, pp. 261–266.

Bennilo Fernandes, J., Sivakannan, S., Prabakaran, N. & Thirugnanam, G. 2018, "Reversible image watermarking technique using LCWT and DGT", *International Journal of Engineering and Technology(UAE)*, vol. 7, no. 1, pp. 42–47.

Bhavana, D., Rajesh, V. & Kumar, K.K. 2016, "Implementation of plateau histogram equalization technique on thermal images", Indian Journal of Science and Technology, vol. 9, no. 32, pp. 1–4.

Bhavana, D. & Rajesh, V. 2016, "A new pixel level image fusion method based on genetic algorithm", *Indian Journal of Science and Technology*, vol. 9, no. 45, pp. 1–8.

Bhavana, D., Anish Kumar, G., Abhilash, B., Rohit Prathyush, K., Ram Kumar, I. & Kishore Kumar, K. 2019, "Automated end of line product validation for soft drink bottles", *International Journal of Scientific and Technology Research*, vol. 8, no. 12, pp. 953–956.

Bojja, P. & Sanam, N. 2017, "Design and development of artificial intelligence system for weather forecasting using soft computing techniques", *ARPN Journal of Engineering and Applied Sciences*, vol. 12, no. 3, pp. 685–689.

Gattim, N.K., Rajesh, V., Partheepan, R., Karunakaran, S. & Reddy, K.N. 2017, "Multimodal image fusion using curvelet and genetic algorithm", *Journal of Scientific and Industrial Research*, vol. 76, no. 11, pp. 694–696.

Gattim, N.K., Pallerla, S.R., Bojja, P., Reddy, T.P.K., Chowdary, V.N., Dhiraj, V. & Ahammad, S.H. 2019, "Plant leaf disease detection using SVM technique", *International Journal of Emerging Trends in Engineering Research*, vol. 7, no. 11, pp. 634–637.

Liu, Z. & Lu, W. 2017, "Fast Copy-Move Detection of Digital Audio", *Proceedings - 2017 IEEE 2nd International Conference on Data Science in Cyberspace, DSC 2017*, pp. 625.

Mannepalli, K., Sastry, P.N. & Rajesh, V. 2015, "Accent detection of Telugu speech using prosodic and formant features", *International Conference on Signal Processing and Communication Engineering Systems - Proceedings of SPACES 2015, in Association with IEEE*, pp. 318.

Mannepalli, K., Sastry, P.N. & Suman, M. 2016, "MFCC-GMM based accent recognition system for Telugu speech signals", *International Journal of Speech Technology*, vol. 19, no. 1, pp. 87–93.

Mannepalli, K., Sastry, P.N. & Suman, M. 2016, "FDBN: Design and development of Fractional Deep Belief Networks for speaker emotion recognition", *International Journal of Speech Technology*, vol. 19, no. 4, pp. 779–790.

Mannepalli, K., Sastry, P.N. & Suman, M. 2017, *Accent recognition system using deep belief networks for telugu speech signals*.

Mannepalli, K., Sastry, P.N. & Suman, M. 2018, "Emotion recognition in speech signals using optimization based multi-SVNN classifier", *Journal of King Saud University - Computer and Information Sciences*.

Revathi, B., Naveen Kishore, G. & Dheeraj, V. 2019, "A survey on OCR for Telugu language", *International Journal of Scientific and Technology Research*, vol. 8, no. 12, pp. 559–562.

Srinivasa Reddy, S. & Suman, M. 2018, "Microaneurysm extraction with contrast enhancement using deep neural network", *Journal of Advanced Research in Dynamical and Control Systems*, vol. 10, no. 11, pp. 313–320.

Reddy, S.S., Suman, M. & Prakash, K.N. 2018, "Micro aneurysms detection using artificial neural networks", *International Journal of Engineering and Technology(UAE)*, vol. 7, no. 4, pp. 3026–3029.

Vallabhaneni, R.B. & Rajesh, V. 2018, "Brain tumour detection using mean shift clustering and GLCM features with edge adaptive total variation denoising technique", *Alexandria Engineering Journal*, vol. 57, no. 4, pp. 2387–2392.

Yan, Q., Yang, R. & Huang, J. 2015, "Copy-move detection of audio recording with pitch similarity", *ICASSP, IEEE International Conference on Acoustics, Speech and Signal Processing - Proceedings*, pp. 1782.

Yan, Q., Yang, R. & Huang, J. 2019, "Robust copy-move detection of speech recording using similarities of pitch and formant", *IEEE Transactions on Information Forensics and Security*, vol. 14, no. 9, pp. 2331–2341.

Recent Trends in Communication and Electronics – Sharma et al. (Eds)
© 2021 Taylor & Francis Group, LLC, ISBN 978-1-032-04572-6

Road accident analysis using random forest algorithm

Mrudul Dixit, Sai Deshmukh, Mannase Dongaonkar & Snehal Jadhav
Cummins College of Engineering for Women, Pune, India

ABSTRACT: Road accidents are one of the major causes of deadly injuries and deaths. It is possible to predict the possibility of accidents by studying past data. The occurrence of road accidents is associated with multiple factors such as speed, traffic condition, day, time, weather conditions, road construction, etc. Machine learning Algorithms are used to achieve the goal. The key steps involved are data pre-processing, Training the model with supervised learning concepts and creation of the interactive Dashboard. The supervised learning algorithms like Random Forest and Logistic regression are tried and assessed on the basis of accuracy and performance for the training of the model, along with the DBSCAN algorithm for the clustering of the data. The outcome of this project will benefit the public in providing a visualization tool that will evaluate the probability of an accident. In addition, it will help the traffic department in implementing strategies to reduce road accidents.

1   INTRODUCTION

Traffic incidents are extremely common. Traffic collisions are a significant cause of death worldwide, cutting short millions of lives each year because of their frequency. So a system that can predict traffic accidents or accident-prone areas may potentially save lives. Transport departments worldwide are trying to implement strategies and methods to minimize road accidents. Despite their endless efforts, Road Accidents have not significantly reduced due to the difficulty in the prediction of when and where the Accidents will happen.

To begin with, this project the most important step is to obtain sufficient data and process it according to the requirements. The processed data is then required to be used for the training and testing of the unsupervised learning model. Here we will be using 60% of the data for the training and 40% for testing purposes.

2   LITERATURE SURVEY

Many researchers have focussed on Road Accidents Analysis and have predicted the reasons for accidents in their own different way. Xin Zou and Wen Long Yue presented the use of Bayesian network theory which is primarily based on the probability to find the cause of accidents. The K2 Algorithm is used to implement the Bayesian model. In Netica parameter learning is done using EM and for posterior probability reasoning and analysis. They have used sensitivity analysis to determine the accuracy of learning and prediction. Chao Wang, Mohammed A. Quddus, Stephen G. Ison has used a multivariate model for the frequency and severity of accidents. They have used the two-stage mixed multivariate model that connects the frequency of accidents with the severity models. The accident frequency is found using the Bayesian spatial model and the severity is found using the logit model. Hence the final result is produced by combining the two results obtained from the

DOI: 10.1201/9781003193838-93

two models. The paper published by Kassu Jilcha, Ajith Abraham, Daniel Kitaw discusses the application of neural networks in analysis of road traffic accidents. Their goal was to develop a machine learning based intelligent model that could recognize the patterns and neural network performance analysis for severity of injury. Salvatore Cafiso, Grazia La Cava, Giuseppina Pappalardo have employed the logistic regression model to estimate the significance of the independents and to understand covariate control variables. According to the results, bike riders over 41 have a greater risk than the riders between the age of 18 and 21 years. High speed affects the most to the probability of fatal consequences. The accuracy of the logistic model is 90%. Miao Chong, Ajith Abraham analyzed the GES automobile accident data from 1995 to 2000 and investigated the performance of neural network, decision tree, support vector machines and a hybrid decision tree.The classification accuracy obtained in their experiments reveals that, for the non-incapacitating injury, the incapacitating injury, and the fatal injury classes, the hybrid approach performed better than neural network, decision trees and support vector machines.

3 METHODOLOGY

3.1 Data collection

A massive dataset was released by the UK government which was published on Kaggle. This Data consists of 1.6 million Accident records with accurate location coordinates, Date, Time, Year, Road conditions, weather conditions, etc.

3.2 Data pre-processing

Data Clustering: The Dataset consisted of millions of samples of which some are rare case accidents that we can minimize. For this purpose, we used the Density-Based Spatial Clustering of Applications with Noise (DBSCAN) algorithm to find clusters, here known as Accident Hotspots. This process is done for reducing the noise in the dataset. The dataset consists of some extra data which fall under rare cases. For this purpose, we are using the DBSCAN algorithm. The Algorithm finds clusters based on the location. Basically, it will see, in which area the number of accidents occurred is very high and neglect the low accident rate regions.

Now we apply the DBSCAN algorithm to our dataset of locations in order to systematically identify the accident hotspots, we grouped the accident locations into clusters. Each cluster is considered to be a hotspot. This algorithm takes two parameters: Radius (epsilon) and Minimum samples (min).

There are three conditions for the point to be classified as given below:

– Core points: If we draw a circle of a given radius from the core point, we should have a "min" number of points inside that circle.
– Boundary Points: It is a point that is not a core point but lies inside the radius of a core point circle.
– Noise Points: It is a point that does not lie inside the radius of a core point circle.

After the Classification of all the points in the dataset, we will connect all the circles of the core points and also include the boundary points. This Connection all together will be called a cluster. Now each cluster will be an Accident Hotspots.

Negative Sampling: In the Dataset we found all the accident points. After clustering, we were left with accident hotspots, but the accident hotspot is not active for 24 hours. Now we had to find when the accident occurs the most on these accident hotspots. To study with this approach, we required the non-accident data as well. So, for this purpose, we randomly generated 3 non-accidental points for every accident point in the clustered data.

Our primary accident dataset contains features that were captured only when an accident occurs, i.e. all labels now are 1. But the model cannot be trained only with positive samples. So, we were required to generate negative samples as well. In order to generate negative samples, we randomly varied the hour and day of the year of the accident data point. Then the negative generated data was mapped with the positive data to prevent from positive point attributes getting repeated for negative samples. If found similar a new negative data will be generated. In this manner, we created three negative samples for every accident record within the clusters.

3.3 *Supervised learning*

It maps an input to an output based upon input-output combinations of examples. It suggests a feature consisting of a collection of training samples from the labeled data of the training set.

Random Forest Classifier: In data mining Machine learning algorithms are used for data analysis and for prediction generation on the basis of those results. Random Forest, ensemble algorithm. It creates and applies multiple decision trees as baseline classifiers voting by a majority to combine the base tree results.

Features of Random Forest Algorithm:

- Effectively it operates on large databases.
- Without overwriting, it is possible to accommodate upto large number of input variables.
- The algorithm has an efficient data-missing estimation technique and retains precision where a significant part of the data is unavailable.
- This has functions to manage mistakes in unbalanced class population data sets.
- The forests created may be saved for future usage on other data.
- Prototypes are determined and provide information on the relationships.

The random forest is an algorithm developed with the help of several decision trees. Forest here refers to averaging the prediction of trees. This algorithm connects humongous amounts of decision trees.then it trains every decision tree with different sets of observations, after that the nodes are split into each tree acknowledging a bounded number of the features. By taking the average result of the predictions made by each tree the ultimate predictions of the random forest are devised. The main parameter used in this model is n-estimators. The n estimators are the count of the trees of which average has to be taken.

The random forest learning algorithm applied to tree learners the general bootstrap strategy of aggregating, or bagging. Instead of the training set X = x1, ..., xn with responses Y = y1, ..., yn, bagging (B times) consistently chooses a random sample to substitute the training set and compares the samples with the trees, For b = 1, ..., B, Training a tree fb for regression or classification on Xb, Yb. We can make predictions of unknown examples x' after training by combining the predictions made on all the singular regression trees on x'.

This method explains the primary algorithm for bagging a tree. The Random forests vary only in one form from this usual design: they utilize a changed tree learning algorithm that chooses a random subset of characteristics for every candidate break in the process of learning. Sometimes this cycle is named the "bagging part". The purpose of doing this is the correlation of trees in a common bootstrap sample: it will check if any of the features are very good predictors of the desired output, all of these features would be chosen in all the B trees, resulting in their correlation

Figure 1 gives the score of features, so we can select the important features. Also, by selecting important features, the accuracy increases which also helps in increasing F1 score.

Step by step Random Forest Algorithm:

1. Select arbitrary samples from the dataset.
2. For each sample, create a decision tree and obtain a prediction from every decision tree.
3. Conduct a vote for every predicted outcome.
4. The final prediction is determined from the outcome with the highest number of votes.

Performance parameters for the algorithm evaluation:

To understand the Confusion matrix, certain terms and definitions are important to understand.

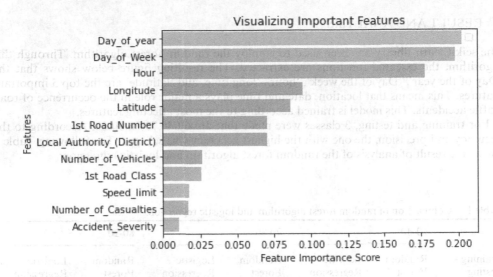

Figure 1. Important features of accident occurrence.

- True Positive:TP: Result is positive and predicted value is also positive.
- False Negative:FN: Result is positive but the predicted value is negative.
- True Negative:TN: Result is negative and the predicted value is also negative.
- False Positive:FP: Result is negative but the predicted value is positive.
- Accuracy is the measure of the number of examples of data correctly identified over the total number of instances of data.

$$ACCURACY = \frac{TN + TP}{TN + FP + FN + TP} \qquad (1)$$

- Precision assesses how effective a model is in predicting optimistic marks. Ideally, the precision for an efficient classifier must be 1 (high).

$$Precision = \frac{TP}{TP + FP} \qquad (2)$$

- Recall is ratio of true positive to sum of true positive and false negative. For an efficient classifier ideally the value of recall should be one.

$$Recall = \frac{TP}{TP + FN} \qquad (3)$$

- F1 score is an overall measure of the accuracy of a model that combines precision with recall. F1 score 1 indicates an ideal model whereas 0 indicates an inefficient model.

$$F1 = 2 \times \frac{Precision \times Recall}{Precision + Recall} \qquad (4)$$

505

4 RESULT AND ANALYSIS

The scikit learn library has been used to employ the random forest algorithm. Through this algorithm, the essential functions are extracted. The resulting figure below shows that the 'Day of the year', 'Day of the week', 'hour', 'longitude, and 'latitude' are the top 5 important features. This means that location, date and time plays a major role in the occurrence of road traffic accidents. This model is trained according to the most affecting features.

For training and testing, 3 classes were made that are 60-40,70-30,80-20. According to the accuracy and precision, the one with the highest accuracy was selected. The following Table 1 shows the result of analysis of the random forest algorithm and logistic regression:

Table 1. Comparison of random forest algorithm and logistic regression.

Training - Testing	60-40		70-30		80-20	
	Random Forest	Logistic Regression	Random Forest	Logistic Regression	Random Forest	Logistic Regression
Accuracy	0.871	0.755	0.893	0.763	0.877	0.752
Precision	0.946	0.8	0.919	1.0	0.978	1.0
Recall	0.514	0.023	0.625	0.031	0.523	0.034
F1 Score	0.666	0.045	0.744	0.061	0.681	0.065

4.1 Conclusions

As road accidents are unexpected, complex, random and Lately, incident inquiries have to be carried out the mechanism and the exact causes are correctly described.

The goal of the study was to show the use of the Random Forest algorithm in the analysis of road accidents. This paper focuses mainly on 12 parameters from 35 parameters available in the original dataset. The accuracy of prediction of the occurrence of accidents is 89.3% with the Random Forest algorithm as compared to 76.3 % in Logistic Regression with training and testing data of 70 - 30 % respectively also the precision, recall and F1 score values are better as compared to others.

REFERENCES

Ajith Abraham, and Marcin Paprzycki, 2005, "Traffic Accident Analysis Using Machine Learning Paradigms", Traffic Accident Analysis Using Machine Learning Paradigms. Published in Informatica.

Chao Wang, Mohammed A. Quddus, Stephen G. Ison, 2011, "Predicting accident frequency at their severity levels and its application in site ranking using a two-stage mixed multivariate model", Transport Studies Group, Department of Civil and Building Engineering, Loughborough University, Loughborough, Leicestershire, LE11 3TU, United Kingdom,2011. Published by Elsevier Ltd.

Kassu Jilcha, Ajith Abraham Machine, Daniel Kitaw, 2013, "Road accident analysis using neural network", Industrial Engineering Chair School of Mechanical and Industrial Engineering Addis Ababa Institute of Technology, Addis Ababa University, Addis Ababa, Ethiopia, Miao Chong.

Salvatore Cafiso,Grazia La Cava, Giuseppina Pappalardo, 2012, "A logistic model for Powered Two-Wheelers crash in Italy", a Department of Civil and Environmental Engineering, School of Engineering, University of Catania, Viale Andrea Doria, 95125 Catania, Italy, 2012. Published by Elsevier Ltd.

Xin Zou, Wen Long Yue "A Bayesian Network Approach to Causation Analysis of Road Accidents Using Netica", 2017, School of Natural and Built Environments, University of South Australia, Adelaide, SA 5095, Australia, Hindawi journals of advanced transportation volume 2017, Article ID 2525481.

Recent Trends in Communication and Electronics – Sharma et al. (Eds)
© 2021 Taylor & Francis Group, LLC, ISBN 978-1-032-04572-6

Analysis of blind and non-blind de-convoludtion methods for de-blurring of ancient manuscript

Vaishali Kikan & Diksha Singh
KIET Group of Institutions, Delhi NCR, India

ABSTRACT: Image restoration is the technique of enhancement of an original image which was degraded due to various factors like noise or other physical parameters in the cnvironment. The process of image restoration involves the use of de-convolution of the degraded or blurred image. The de-convolution of the blurred image is taken with the PSF (point spread function) i.e the degradation function. There are two methods of de-blurring i.e. non-blind and blind de-convolution. In the restoration process of any image, if blur function is random, then the estimation of the PSF is done and this is how the ideal image is generated by using the original image as the input. In case of non blind method iterative method is used, which gives better results as number of iterations can be increased. Blind de-convolution is proposed to increase the quality of the deblurred image, to estimate the proper or true size of the PSF and also to remove the ringing effects in the output a specific weighing function is also proposed. In non-blind deconvolution technique, Lucy Richardson (LR) technique is presented in the paper.

Keywords: de-convolution, PSF, image restoration, de-blurring, ringing effect, lucy richardson method, sobel filter

1 INTRODUCTION

Image restoration is the phenomenon of restoring a degraded image and also it is the concept for improving the quality of the image. The factors like motion blur, de-focus blur, various noise models and some physical phenomenon degrade the quality of the image. Out of the various types of blur models present in the nature, when there is the presence of the relative motion between the camera and the object (motion blur), then the amount of degradation is very significant.

There are two classes for the image de-blurring process one is non blind approach and another one is blind approach. In non-blind technique, the degradation function is known and in case of blind approach the degradation function is unknown. Precise determination of this function can provide us with a properly restored image. We can use the degradation model to restore the image, which is determined by the convolution of the input image with the degradation function i.e. point spread function. In this paper a blind method is approached to precisely determine or to estimate the size of the PSF. Additionally to improve the results a weight function with sobel filter is defined which is applied to the edges of the given image to remove the ringing effect. Limitation with blind deconvolution method over non blind method is that this method does not have any prior knowledge of the degradation function PSF but the advantage of this method is that it improves the quality and resolution of the image in comparison to the non blind one. For our paper we are using Lucy Richardson non blind method.

DOI: 10.1201/9781003193838-94

2 LITERATURE REVIEW

Degradation model is given by the convolution of the given image with the degradation function and then that degraded image with the noise is restored using de-blurring methods.

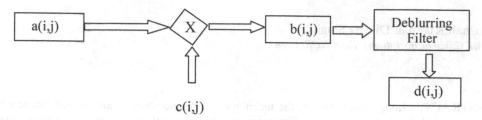

Figure 1. Degradation model c(i,j).

Here a(i,j) is the original input image which is given to the degradation function c(i,j) then result is given by b(i,j).

$$b(i,j) = a(i,j) * c(i,j) \tag{1}$$

where * is defined as convolution. If K(i,j) is the noise present then final model is given as

$$b(i,j) = a(i,j) * c(i,j) + K(i,j) \tag{2}$$

Then this b(i,j) is passed through de-blurring process or filter to get de-blurred or restored image is d(i,j). Convolution gives us the degraded or blurred image so, we need to use the de-convolution process to get the restored image.

2.1 *Gaussian noise*

Gaussian noise is a type of noise which is having normal distribution type density function and its curve is bell shaped as shown in the Figure 2.

$$p(r) = \frac{1}{\sigma\sqrt{2\pi}} e^{-\frac{1}{2}\left(\frac{x-\mu}{\sigma}\right)^2} \tag{3}$$

Where, p(r) is the density function, μ is the centre of the curve and the σ is the standard deviation and x is the random variable. Gaussian filter is the type of filter which is having its response as that of Gaussian function.

2.2 *Non-blind technique*

In Non blind method there is an information available about the degradation function i.e. the size of the PSF. This method is iterative type and non linear type To restore the picture of good quality, the quantity of cycles is to be resolved for the picture according to the PSF measure. The Richardson-Lucy method is an repetitive method to recover the picture which was distorted by recognized PSF

$$Z(i) = \Sigma n(i,j)t(j) \tag{4}$$

Where n(i,j) is point spread capacity, t(j) is pixel esteem at areand Z(i) the watched an incentive at pixel area I. The fundamental thought is to figure doubtlessly t(j) given watched Z(i) and known n(i,j). It is an iterative approach of image restoration.

$$o^{(k+1)}(x,y) = o^k(x,y)[\frac{i(x,y)}{h(x,y) * o^{(k)}(x,y)} * h^*(x,y)] \tag{5}$$

2.2.1 Searching distorted regions

This approach is based on the search of motion vector for image fragments using the previous frame as a reference. I_1 is the current frame and I_0 is the previous reference frame, then the divide the I_1 in the non overlapping blocks of 8×8 pixel size. Next, for each block $I_1(x,y)$ of the current frame the most similar block $I_0(k,l)$ in the reference image I_0 is searched. To define this block the following deviation function D is minimized.

$$\sum\sum|I_0(8k+n, 8l+m) - I_1(8x+n, 8y+m).| \tag{6}$$

To select the blocks corresponding to moved fragments of the reference frame, we perform a threshold separation, taking only those blocks $I_1(x,y)$ which has a matching block in the reference frame $I_0(k,l)$ with deviation D<T, where the threshold value T is set empirically. Otherwise the block $I_1(x,y)$ is not considered as a moved one. After the thresholding some morphological processing is performed to exclude isolated moved blocks which arose accidentally or as a result of insignificant movement. First, an auxiliary binary image B is formed, where we fix each value of pixel B(x,y)=1 if the corresponding 8×8-block $I_1(x,y)$ of the current frame is considered as a moved one; otherwise B(x,y)=0. Then image B is processed by two subsequent erosions followed by two dilations, each morphological procedure having the same bit mask representing a cross of 3×3 pixel size. After this morphological procedure the image B is considered as a bit mask defining those frame blocks which correspond to the regions of moving objects.

2.2.2 Finding the parameters of blurring

To get better quality from Lucy-Richardson image restoration the IR defining distortion model should be given as an input parameter of the restoration algorithm. A good way to determine the parameters of blurring is using image cepstrum. The cepstrum of picture is characterized as

$$V(c,d)=F^{-1}\{log|U(u,v)|\} \tag{7}$$

where F^{-1} denotes Fourier transform inverse.

2.2.3 Estimating the number of iteration

To determine the optimal number of iterations in Lucy- Richardson method an empirical algorithm was presented. It is based on the observation that when the restoration process has converged to the solution, the norm of the difference, $P_k = ||\hat{f}_k - \hat{f}_{k1}||$ between images, which is obtained at the current and last iteration, stabilizes.

i. Find normalization factor (rms) for an area Ω comprising of some central pixels of the image.

$$\sigma = \left[\frac{1}{\Omega(\sum g(i,j) - A)^2}\right]^{0.5} \quad \text{Where, } A = \frac{1}{\Omega\sum g(i,j)} \tag{8}$$

ii. Perform 5 iteration method of Lucy Richardson while assuming k=6
iii. Go to the next (k-th) iteration and find the image estimate $\hat{f}_k$.
iv. Find the norm of the difference $P_k = ||\hat{f}_k - \hat{f}_{k1}||$
v. To flatten changes of P_k do smoothing: $P_k \leftarrow \frac{1}{4}P_k{-}2 + \frac{1}{2}P_k{-}1 + \frac{1}{4}P_k$
vi. If $\max(P_k{-}4, P_k{-}3, P_k{-}2, P_k{-}1, P_k)<\sigma\Omega C*10^{-2}$

If the value of blurring r < 15,
We stop after k + 26 iterations.
Else

We stop after k + 1 iterations.
Else
    Increase k by 1 and go to step 2.

2.3 *Blind technique*

we have seen that if we know how to estimate the PSF and if we can finally estimate this PSF h(x,y,) which causes the degradation or blurring then image can be deblurred by deconvolution.But it no that easy to know the exact size of the PSF.

2.3.1 *Methods to approach*

It has 2 kinds of methods to approach. From true image, recognize the PSF independently and then so as to use it after with popular classical image methods for restoring the image. The methodology prompts basic calculations. Joining the recognizable proof methodology with the restoration algorithm. This union includes that the PSF and the true image need to be evaluated, which prompts the improvement in progressively the algorithms.

2.3.2 *Proposed method*

This algorithm can perform well when no information about PSF is known. Here the Blind process is applied on the Lucy Richardson algorithm (Non blind algorithm) for restoring the blurred image. Image is restored by estimating the near true value of the PSF.

2.3.3 *Algorithm*

i. Select the image then simulate the blur by convolving the image with the gaussian filter.
ii. Three restorations will be done to estimate the near true value of the psf.
iii. Deconvolution is performed for underpsf, overpsf and initial psf.
iv. To remove the ringing effect in the restored image which occurs along the sharp intensity contrast and the borders weight function is defined.
v. Using and passing the structuring element pixel desired and undesired will be be given pixel value 1 and 0 simultaneously.
vi. Then the deconvolution is performed for restoring the image with the help of structured psf.

3 EXPERIMENTAL RESULTS

3.1 *Lucy-Richardson method*

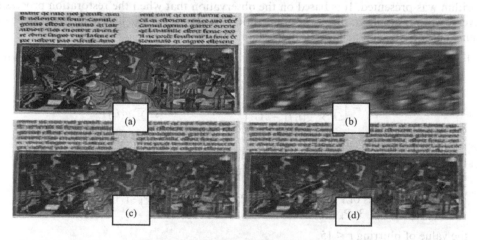

Figure 2. (a) Original manuscript, (b) Blurred manuscript, (c) Deblurring after 100 iterations, (d) Deblurring after 1000 iterations.

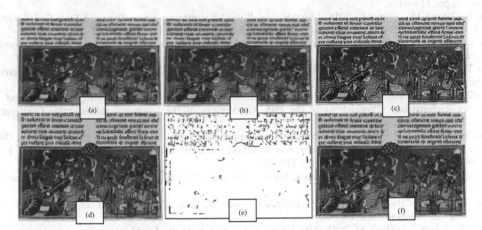

Figure 3.   (a) Blurred with blind deconvolution, (b) Deblurring with UNDERPSF, (c) Deblurring with OVERPSF, (d) Deblurring with INTPSF, (e) Weight Array of manuscript, (f) Final restored manuscript.

Table 1.   Comparison of PSNR and MSE values of blurred and deblurred images with the help of deconvolution methods.

	Deblurring		Observation	
Technique	Yes/No	iteration	PSNR	MSE
Non Blind	Yes	100	30.2330168	56.12
Non Blind	Yes	1000	30.6453829	51.23
Non Blind	Yes	10000	31.0098756	48.11
Blind	No	NA	40.3728929	7.45
Blind	Yes	NA	41.9684930	4.37

3.3   *Conclusion*

(LR) Lucy-Richardson algo is nonlinear and therefore it has very slow convergence and but it is a non blind process where knowledge of PSF is present and also it gives better results than linear methods.Blind method does not have prior knowledge of the PSF but it gives better results than non blind in terms of resolution as its PSNR value is better than the earlier one and MSE is less than earlier one but it is showing ringing effects in the outcome so we should try to make this process more efficient by using efficient filters and also we can work to decrease the processing time and the limitations regarding the accuracy and precision of estimation of the parameters.

REFERENCES

Biemond J., Lagendijk R.L., Mersereau R.M., May 1990 Iterative Methods for Image Debrruing in proceedings of the IEEE", vol 78, №5, pp. 856–883.

Gonzalez R.C., Woods R.E., 2008, Digital Image Processing, 3rd ed., Prentic Hall, 954.

Hanif M. and Seghouane A., Dec. 2012 "Blurred Image Deconvolution using Gaussian Scale Mixtures Model in Wavelet Domain", *2012 International Conference on Digital Image Computing Techniques and Applications (DICTA)*, 3–5.

Jansson P.A., 1997, Deconvolution of images and spectra//2nd ed., Academic Press, CA, 514.

Jones D I., 2000, "Aerial inspection of overhead power lines using video: estimation of image blurring due to vehicle and camera motion[J]", *IEE Proceedings - Vision Image and Signal Processing*, vol. 147, no. 2, pp. 157–166.

Liu S., Wang H., Wang J., 2016, "Blur-Kernel Bound Estimation From Pyramid Statistics[J]", *IEEE Transactions on Circuits & Systems for Video Technology*, vol. 26, no. 5, pp. 1012–1016.

Lucy L. B., 1974, An iterative technique for the rectification of observed distributions. The Astronomical journal, vol. 79 (No. 6).

Ma T H, Huang T Z, Zhao X L., 2017, "Image de-blurring with an inaccurate blur kernel using a group-based low-rank image prior ☆ [J]", *Information Sciences*, vol. 408, pp. 213–233.

Panfilova K., 2015, Compensation of linear blurring digital images using the method of Lucy-Richardson, "Proceedings of the 25th Anniversary of the International Scientific Conference. Graphicon 2015", pp. 163–167.

Richardson I.E., 2010, The H.264 advaced vidoe compression standard. 2nd ed, John Wiey & Sons, 348.

Ohkoshi K., 2014, "Blind Image Restoration Based on Total Variation Regularization and Shock Filter for Blurred Images", Proc. ICCE, pp. 219–220.

Richardson W. H., 1972, Bayesian-Based Iterative Method of Image Restoration. Journal of the optical society of America, vol. 62 (No. 6).

Qin Shen, 2011, "Research and improvement of digital image clarity evaluation function", *Image Processing and Multimedia Technology*.

Zhao B., Chen X., Zhao X., 2018, "Real-Time UAV Autonomous Localization Based on Smartphone Sensors[J]", *Sensors*, vol. 18, no. 12.

Recent Trends in Communication and Electronics – Sharma et al. (Eds)
© 2021 Taylor & Francis Group, LLC, ISBN 978-1-032-04572-6

A review on the novel approach of path planning and exploration of an area for a multi mobile robotic systems

Shubham Shukla
KIET Group of Institutions, Ghaziabad

Sanjeev Sharma
Indian Institute of Information and Technology, Pune

Shashank Kumar & Vibhav Kumar Sachan
KIET Group of Institutions, Ghaziabad

Kamal Kishore Upadhaya & N.K. Shukla
J.K Institute of Applied Physics, Central University of Allahabad

ABSTRACT: In this paper the entire technique of area exploration and path planning has been elaborately explained. Several concepts related to area exploration problem and path planning has been discussed . So far two goals has been considered, one is path planning and the other is area exploration of multi robot, therefore in a separate segment, we present the technique for these activities. Throughout the first segment the idea of directional motion central to both route planning and exploring is discussed.

Keywords: path planning, area exploration, CBDF, ZIG ZAG effect

1 INTRODUCTION

In this paper the entire technique of area exploration and path planning has been elaborately explained. Several concepts related to area exploration problem and path planning has been discussed.So farthere are two goals, one is path planning and the other is area exploration of multi robot Therefore, in a separate segment, we present the technique for these activities. First segment discuss about the directional motion central to both route planning and exploring. We have already explored application related to robotics with the help of Nature's inspired algorithms.The very next study describes the flow chart and robot route planning algorithm. Flow graph provides an outline of the multi robot goal search and tracking route planning process.

BasicApproach

1.1 *Directional movement*

Sanjeev Sharma and Chiranjib Sur presented a paper been published in the proceedings of the ICCVR 2014 which discusses about the Cluster Based Distribution Factor (CBDF) technique. (Sharma, S. Sur, C. Shukla, A. Tiwari,R. et al, 2014) this technique is been developed by two research scholars. It is the all new methodology mainly brought into use in the case of scattering agents in a random manner to traverse the search space and explore the target randomly towards

DOI: 10.1201/9781003193838-95

different parts of the search area for effective and opportunistic exploration of the target rather than relying on any non-repetitive searching pattern of manual implementation which is not feasible. Local scanning is another aspect of the exploring process that plays a critical role, and it must be used synonymously such that the robot agent can still utilize its senses and the range of senses throughout travel. It has been observed that in a known environment where only local search- based decision-making is needed when the goal and path is defined and the only prerequisite is to identify the aim and determine the way from source to destination. Yet there is no concept of the destination of undisclosed target agents and so the hunt must be both random anduncoverable. If NIA had been used alone in the experiment, the agents' spontaneous dispersing effect would have been interrupted by randomly forward and back movement-based scans, and thus trapped in the local region. Hence with an addition of the Clustering Based Distribution Factor (CBDF), the agents are ready to reach out of the local field, and discover new area to search for the target. In the CBDF methodology, the entire place is split into a series of areas say x and we'll have x path from every destination in the work area. While considering the clustering head as the last point and the agent's position as the initial point, then by applying this methodology the robots receives the dimension of directionality and a way out of the local environment. CBDF consists of 2 types

Scheme 1 – Scattering Impact which is Directional in nature

This is used to direct the robot to enter a particular area being explored and where a region has to move, emerge from a trap or enclosure etc. The path element is derived from the head of the cluster and is usually randomly chosen to discover new routes and outings. Like the quest effect in a zig-zag fashion, for quite some time the Spatial Scattering Effect depends on position and relies on volume of workspace. Figure 1 displays the spatial dispersing-effect model. (Wei-Chang, Y. et al, 2013) (S.-T. Lo, R.-M. Chen, D.-F. Shiau, and C.-L. Wu, et al, 2008) (Vahdati, G. Yaghoobi, M. Akbarzadeh M.R. et al, 2010)

Scheme 2 – Search Impact in a ZIG, ZAG mode

Search Impact in zig-zag mode is similar in all respects to the Spatial Scattering Effect but mostly used in the local search. The course here changes more often, and hence continues to flow in the workplace. Here too, the path is automatically and randomly chosen based on cluster heads. The second scheme does not rely solely on one path, but for local searches one of the given location of the clusters is collected and completed with the collective strength of the Nature's-inspired algorithm. (Vahdati, G. Yaghoobi, M. Akbarzadeh M.R. et al, 2010) (Saelim, A. Rasmequan, S. Kulkasem, P. Chinnasarn, K. and Rodtook, A. et al,2013) (Jati, G. K. Manurung, H. M. Suyanto, S. et al 2012)

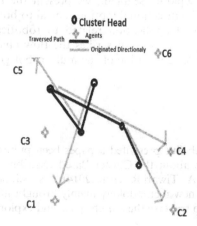

Figure 1.   Search Impact in a ZIG,ZAG mode.

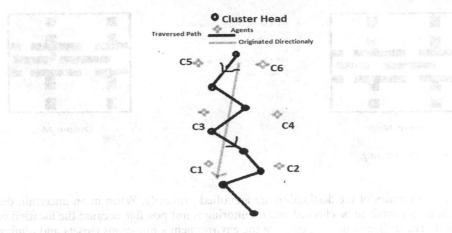

Figure 2. Scattering Impact which is Directional innature.

1.1.1 *Known and Unknown TargetMovement*

Algorithm: Movement for the directional nature

1: Present position P*pr*
    if *Target==Known* **then**
3: location of the known target *Pt* is been detected
4: in the case of known target calculate movement in unit vector   (4.1)
5: end if
6: if Target==Unknown then
7:      Direction of the cluster Pd isselected
8:      for each dimension the unit vector iscalculated
9: **end** *if*   (4.2)
10: for each dimension calculation of fresh variation
      $Vr = \delta r * D$
here $\delta r \in \delta$ and $D \in Rd$ (Range ofDetection)
Here $V \in (V1, V2, ..., Vr)$ and v is thevariation.

Robotic Path PlanningApproach

1.1.2 *Description of the problem*

The question of path planning is known as the most fundamental and primary step in auto-mated systems comprising of robotic agents until it can achieve a specified goal for the ful-fillment of the multiple tasks for which it is intended. By stochastic processes the agents are able to produce their own path and path of local search. For this purpose we used Nature's Inspired Algorithm. Planning for this form of travel is crucial to avoid wasting so much time and energy on movement, overcoming all sorts of barriers, and eventually maintaining and directing other robotic agents to the location desired. The searching based movement is evaluated on two types of situation which includes both, first is when the target path is known to the robot, and second is when the path is not identified and the robot needs to travel to find the location before it get close to the target with its tracking range to locate the destination.

    Known target is one form of target management issue that includes effectively hitting the target and maximizing resource usage.(Sharma, S. Shukla, A. Tiwari, R. et al, 2014) Task management primarily includes overcoming challenges in an insightful and exploratory manner so that potential possibilities are resurrected and remembered. The position or

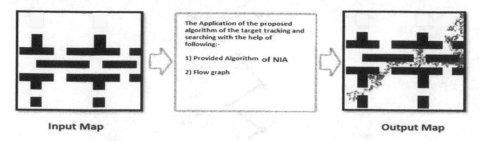

Input Map   Output Map

Figure 3.   Path planning.

location coordinates of the destination are identified primarily. When in an uncertain destination the target needs to be checked and monitoring is not possible because the location is not identified. The difficulty lies in exploring the environment's numerous closets and aiming for the target. There are certain dynamic settings in the enclosed spaces where only the problems associated with the known target are addressed, in the problem associated with the unknown target due to complex wallingand openings there is no enclosed space. In both known and unknown goal systems, multiple outcomes are addressed for specific conditions, and the outcome is compared and analyzed.

Figure 3 explains the mechanism of planning a path. Here feedback is given to a map or environment. Once the technique has been applied, we get explored map. The blue segment indicates the region that the robot traverses to meet the target at the output.

Algorithm: Robotic Movement withNIA

1. Initialization of an environment matrixin side the GraphG= (S,T). (x 1 and x 2 are the obstaclecells)
2. Initialization of N robot-agents and data structure associated to it. {Memory Path, Reach,Nodes, Time, Coverage Energy, Fitness and Memory of a traversedmap.}
3. Initialization of functions to calculate the number of Moves, Consumed Time, Total area coverage and dissipatedEnergy
4. Initialization of the particle from thesource
5. For $^{j=1}$ to n_1 (maxiteration)

    (a)   For k= 1 to n* (no of agents=n*) If condition = =true
          Initiate the agent with NIA
          Else
          Follow Directional Movement
       End of *k*
    End of *j*

6. For *k*= 1 to n* (no ofagents=n*))
       If condition = = true
          Fitness of the particle is determined.
          Iteration and global best is updated
    End of *If*
    End of *k*

7. The loop will continue until the optimum value isreached.
    End

1.2   *Flow graph for pathplanning*

Figure 4 displaying a path planning flow graph for multi-robots. Here you configure first map and agent and other variables. Our task of monitoring or looking for aim is accomplished after the test for accessibility for some robots, all of the robots reach up to mark. Next the

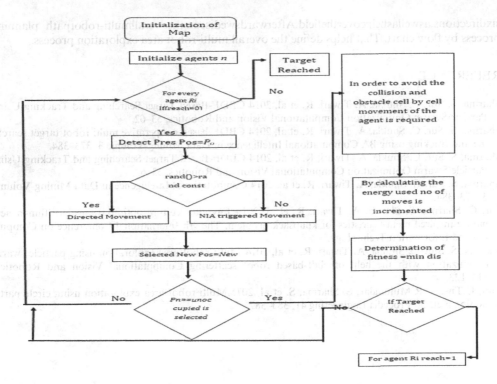

Figure 4. Path planning flow graph.

suggested technique for driving the robots is implemented. As the robots'motiona dvances parameters like time, Energy, coverage and move, is to be measured. Fitness value is determined, to repeat the cycle before the target hits the robot.

Area ExplorationTechnique

1.2.1 *Description of the problem*
Steps of area exploration:

1) Initialize the robots at the initial position first. The robot would then travel to itsallocated location.
2) Assuming areas of similar scale, there are some boundary overlaps and someunexplored regions would bepresent.
3) Fromthepossibilityoftheallocatedrobot,unexploredregionsseemimpossibletoachieve. Target is achieved until the exploration is completed, consuming iterationtime
4) The operation of each robot depends on testing the importance of each cell as ittravels through bio-inspired computational guidance andjudgment.
5) The requirement for exploring and extraction shall be decided by a random generator and it shall decide either to use a spatial movement of NIA drivenmovement.
6) The gradual section saretested step by step,and the pseudo-bestis determined by temporary fitness.

2 CONCLUSION

The conceptual framework for millirobot planning of path and exploration of are using NIA is given in this paper. This involves path planning approach, and multi-robot area exploration. Next, it addresses the clustering-based distributed component. It will lead the robots going in sixdirections,aswellashelpcoverthefield.Afterwardswediscusstheoverallmulti-robotpath planning process by flow chart. That helps define the overall multi-robot area exploration process.

REFERENCE

Sharma, S. Sur, C. Shukla, A. Tiwari, R. et al, 2014 CBDF-Based Target Searching and Tracking Using Particle Swarm Optimization Computational Vision and Robotics, 53–62.

Sharma, S. Sur, C. Shukla, A. Tiwari, R. et al, 2014 CBDF based cooperative multi robot target searching and tracking using BA Computational Intelligence in Data Mining-Volume 3, 373–384.

Sharma, S. Sur, C. Shukla, A. Tiwari, R. et al, 2014 CBDF-Based Target Searching and Tracking Using Particle Swarm Optimization Computational Vision and Robotics, 53–62.

Sharma, S. Sur, C. Shukla, A. Tiwari, R. et al, 2014 Computational Intelligence in Data Mining-Volume 3, 373–384.

Sur, C. Sharma, S. Shukla, A. Tiwari, R. et al, 2014 Egyptian vulture optimization algorithm–a new nature inspired meta-heuristics for knapsack problem, The 9th International Conference on Computing and Information Technology.

Sharma, S. Sur, C. Shukla, A. Tiwari, R. et al, 2014 Multi-robot area exploration using particle swarm optimization with the help of cbdf-based robot scattering Computational Vision and Robotics, 113–123.

Jain, U.Tiwari, R.Mujumdar, S. Sharma, S. et al, 2015,Multi robot area exploration using circle partitioning method Procedia Engineering 41, 383–387.

Multi patch truncated ground microstrip patch antenna for Wi-MAX, X and C band applications

Ritik Jain, Sanskriti Tyagi & Ruchita Gautam
KIET Group of Institutions, Delhi-NCR, India

ABSTRACT: The design presented here is best suited for low loss applications and depicts a quad band microstrip patch antenna. The design consists of multiple U and L shaped patches that are used to improve impedance bandwidth. By adding inverted rectangular patches of U and L shape multiband characteristics of antenna are obtained, by cutting a square slot in the ground increase in bandwidth is observed .The volume of proposed antenna is 20×20×0.8 which as compared to previous designs is compact with frequency bands of (3.2963-3.5093 GHz)/3.23GHz, (3.9359 – 4.4630 GHz)/4.199 GHz, (6.5370 – 6.8889 GHz)/6.712 GHz,(8.2593-8.7407 GHz)/8.49 GHz which covers operating bands of WiMAX, C and X band applications according to IEEE standards of 802.11a/b/g/n 6.39%, 12.55%, 4.70% and 5.67% of impedance bandwidth respectively.

Keywords: C Band, Microstrip patch antenna, HFSS, Radiation pattern, S-parameters, Return Loss, Wi-MAX

1 INTRODUCTION

Microstrip patch antennas are the most popular type of feeding technique used in antennas that are printed directly on printed circuit boards which reduces the space requirements, also making them economical.

The microstrip patch antenna holds a special place in the field of electronics like in mobile phones, satellites etc. wherever wireless communication is required due to their small size which consists of patches of metal foil of various shapes. As microstrip antennas due to their small size are broadly used in WiMAX and have application in military, colleges, and schools. Therefore, WiMAX is a very important antenna application. According to IEEE standards 802.11 a/b/g/n operating bands for WLAN is 2.4 GHz (2.40–2.484 GHz), 5.2 GHz (5.15–5.35 GHz), 5.8 GHz (3.585–5.825 GHz) and opera"ting bands for the WiMAX are 2.4 GHz (2.5–2.8 GHz),3.5 GHz (3.2–3.8 GHz) and 5.5 GHz (5.2–5.8 GHz). So for meeting the above requirements an antenna is needed at which multiple frequencies can operate. The most desirable feature for the present communication system is compact size and low cost due to which there is a need for microstrip antennas for multiple applications. In this paper keeping in mind the above requirements quad band antenna has been proposed.

The past studies reveal multiple different designs applicable for WiMAX, C and X band applications using meandering split ring slot, antenna consisting of L shaped slots, Pi slit, K Shaped slots, etc. By adding rectangular and U shaped patches we can have multiple frequency bands. A simple and effective design in compact antenna having four bands with good radiation pattern and gain have been presented in this paper.

DOI: 10.1201/9781003193838-96

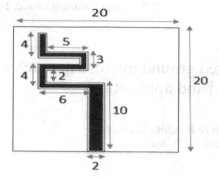

Figure 1.   Antenna stage I.

2   GEOMETRICAL MODEL

The antenna design presented in this paper is as shown in Figure 1. The dimensions of trun-cated ground are W_g x L_g (20 X 5 mm^2) and then patches of different dimensions are made over the substrate of dimensions W_{sub} x L_{sub} (20 X 20mm^2). The dimensions of patches are calculated with the help of the following formulas:

$$W_p = c/2f_r * (2/\epsilon_r + 1)^{1/2} \quad \varepsilon_{reff}=(\epsilon_r+1)/2+(\epsilon_r-1)/2 * [1+12h/W_p]^{1/2}$$
$$L_{eff} = c/ (2f_r (\epsilon_{reff})^{1/2} \quad \Delta L=0.412h[(\epsilon_{reff}+0.3)(W_p/h+0.264)]/[(\epsilon_{reff}-0.258)(W_p/h +0.8)]$$
$$L_p =L_{eff} - 2\Delta L \quad \text{where c (= 3} \times 10^8 \text{ m/s) denotes the speed of light.}$$

The substrate used for fabricating the antenna is FR4 epoxy with a thickness 0.8 mm and having dielectric constant of 4.4.FR4 epoxy is preferred due to its easy availability and low cost. It is fed by a 50 ohm microstrip line of dimensions W_fx L_f (2 x 10 mm^2). The dimensions are marked in the design.

3   SIMULATED ANTENNA PARAMETERS

The antenna design presented in the article is simulated using HFSS R2013 software, based on S_{11} parameters, Current Distribution and Radiation Pattern, its characteristics are studied and analyzed. The S parameters here are used to represent the input output relationship between ports of the antenna.

3.1   *Return loss*

In antennas, the most commonly used parameter is S_{11} which reflects how much power is reflected back. This is also known as return loss curve or reflection coefficient. According to studies S_{11} is equal to -10 dB i.e. in that 3 dB power is delivered to the antenna and -7 dB is reflected back. In ideal conditions if S_{11} is considered as 0 dB all the power is reflected from the antenna and nothing is radiated. Discussion of results at different stages of design with their return loss curves are described in this section. The first stage of the design is shown in Figure 1 and the return loss curve for stage 1 has been depicted in Figure 2 but this consists of two bands so as to get multiple bands.

Now one more rectangular patch i.e. inverted L shape is introduced to obtain third bands which is depicted in Figure 3 return loss curve S_{11} for second stage is depicted in Figure 4.

Further to obtain a quad band a rectangular patch of I shape is added to right hand side which gives the proposed design of antenna with (L_g = 5 mm) length of ground and is depicted in Figure 5(a) and Figure 5(b), its return loss curve is given below in Figure 6, the proposed antenna operates for four bands with four resonant frequencies. The frequency

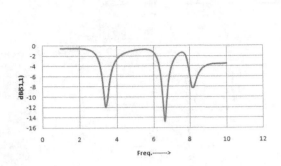

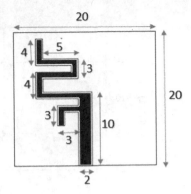

Figure 2. S_{11} curve for stage I.   Figure 3.   Antenna stage II.

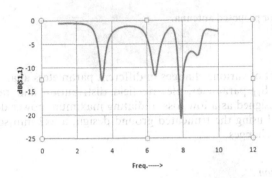

Figure 4. S_{11} Curve for stage II.

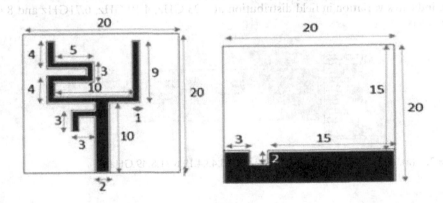

Figure 5.   (a),(b) Proposed design, antenna's bottom view.

bands are obtained at (3.2963-3.5093 GHz)/3.43 GHz, (3.9359 – 4.4630 GHz)/4.199 GHz, (6.5370 – 6.8889 GHz)/6.712 GHz, (8.2593-8.7407 GHz)/8.49 GHz which covers operating bands for WiMAX, C and X band applications according to IEEE 802.11 a/b/g/n standards.

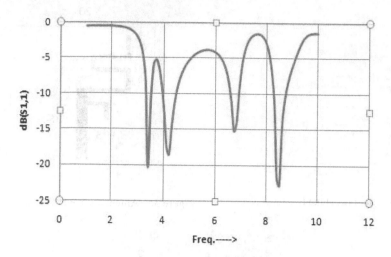

Figure 6.   Return loss for proposed antenna.

To observe the effect of various changes in different parameters analysis is done on various parameters including S_{11} parameter, electric field distribution and radiation pattern. The antenna is typically designed as a low loss, radiating maximum power device. The return loss parameter is improved using the truncated ground design, a 2x2 mm square slot is cut from the ground to observe changes.

3.2   *Current distribution*

Figure 7(a), Figure 7(b), Figure 7(c) and Figure 7(d) represents a surface current that is the electric current induced by applying electromagnetic field and the distribution pattern on patch indicates variation in field distribution at 3.23 GHz, 4.19 GHz, 6.71GHz and 8.49 GHz respectively.

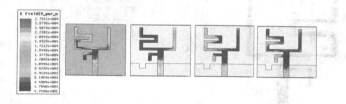

Figure 7.   (a), (b), (c), (d): Current distribution at 3.43,4.19,6.71,8.49 GHz.

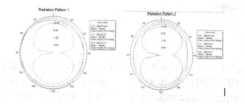 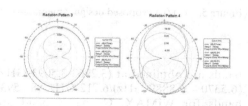

Figure 8(i).   Radiation pattern at 3.43Ghz.   Figure 8(ii).   Radiation pattern at 4.19Ghz.

522

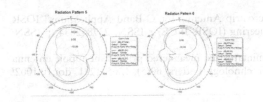 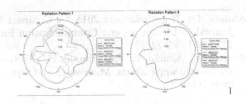

Figure 8(iii). Radiation pattern at 6.71Ghz.   Figure 8(iv). Radiation pattern at 8.49Ghz.

3.3 Radiation pattern

The radiation pattern of an antenna basically refers to the directional dependence of radio waves. Since they have low power and therefore they have broad radiation patterns. To observe radiations, 2-D radiation patterns of antennas are observed at frequencies 3.43GHz, 4.19GHz, 6.71 GHz and 8.49 GHz respectively.

4 CONCLUSION

A patched microstrip antenna with truncated ground for WiMAX, C and X Band applications has been designed and simulated successfully. The frequencies at which the proposed antenna operates are (3.2963-3.5093 GHz)/3.23GHz, (3.9359 – 4.4630 GHz)/4.199 GHz, (6.5370 – 6.8889 GHz)/6.712 GHz, (8.2593-8.7407 GHz)/8.49 GHz frequency bands. The proposed design is best suited for low loss antennas for Wi-MAX, C and X Band applications.

REFERENCES

Brar, R. S., Saurav, K., Sarkar, D., & Srivastava, K. V. (2018). A quad-band dual-polarized monopole antenna for GNSS/UMTS/WLAN/WiMAX applications. Microwave and Optical Technology Letters, 60(3), 538–545. doi:10.1002/mop.31008

Chandan, Srivastava Toolika and Rai B.S., 2017 "L-Slotted Microstrip Fed Monopole Antenna for Triple Band WLAN and WiMAX Applications," Springer in Proceedings Theory and Applications, Vol. 516, pp. 351–359,

Chandan, Bharti, G., Bharti, P. K., & Rai, B. S. 2018. Miniaturized Pi (Π) - Slit monopole antenna for 2.4/5.2/5.8 applications. doi:10.1063/1.5031997

Chandan, Bharti G.D., Srivastava Toolika and Rai B.S., 2018 "Dual Band MonopoleAntenna for WLAN 2.4/5.2/5.8 with Truncated Ground," AIP Conference Proceedings, American Institute of Physics pp. 200361–200366, 2018. doi:10.1063/1.5031998

Chandan, Bharti G.D., Srivastava Toolika and Rai B.S., 2018 "Miniaturized Printed K shaped Monopole Antenna with Truncated Ground Plane for 2.4/5.2/5.5/5.8WLAN Applications," AIP Conference Proceedings, American Institute of Physics pp. 200371–200377, 2018. doi: 10.1063/1.5031999

Ch. Priyanka, Kumar K. Ranjith, B. Sonia, Nayak Deepak Kumar, 2019 "A Substantial Rectangular Shaped X-band Slot antenna for Satellite and Terrestrial Applications" International Journal of Innovative Technology and Exploring Engineering (IJITEE) ISSN: 2278–3075, Volume-8 Issue-7 May,.

Hamza Ahmad,Wajid Zaman,ShahidBashir, Rahman, M. 2019. Compact triband slotted printed monopole antenna for WLAN and WiMAX applications. International Journal of RF and Microwave Computer-Aided Engineering. doi:10.1002/mmce.21986

Liu Pingan, Zou Yanlin, Xie Baorong, Liu Xianglong, & Sun Baohua. (2012). Compact CPW-Fed Tri-Band Printed Antenna With Meandering Split-Ring Slot for WLAN/WiMAX Applications. IEEE Antennas and Wireless Propagation Letters, 11, 1242–1244. doi:10.1109/lawp.2012.2225402

Anusha, Dr. B Lethakumary 2015 "A Compact Microstrip Antenna for C Band Applications" IOSR Journal of Electronics and Communication Engineering (IOSR-JECE) e-ISSN: 2278–2834, p-ISSN: 2278–8735 PP 12–15

Zaman, W., Ahmad, H., &Mehmood, H. (2018). A miniaturized meandered printed monopole antenna for triband applications. Microwave and Optical Technology Letters, 60(5), 1265–1271. doi:10.1002/mop.31149

Recent Trends in Communication and Electronics – Sharma et al. (Eds)
© 2021 Taylor & Francis Group, LLC, ISBN 978-1-032-04572-6

Miniaturized dual cavity MIMO antenna with slotted ground structure and vias

Ragini Sharma
KIET group of Institutions, Delhi-NCR, India

Vandana Niranjan
IGDTUW, New Delhi, India

Vibhav Kumar Sachan
KIET group of Institutions, Delhi-NCR, India

ABSTRACT: A dual Cavity SIW MIMO antennawith reduced dimensions has been developed for 4.84GHz frequency. RT Duroid (5880 Lossy)substrate whose dielectric constant value is 2.2 and height is 1.6mm has been used for designing of MIMO antenna. Size of a single cavity is 15×15mm and size of overall structure is 51.2×34.2mm.Gain of proposed geometry is 6.290dB. S_{11}, S_{22}, S_{21} parameters are -33.783dB, -25.429dB, -10.86dB respectively. Both ports of MIMO antenna are perfectly matched with input impedance. ECC is 0.005 which is less than 0.5 and considered under allowable limit of ECC. Simulation of MIMO design has been carried by CSTMW studiosoftware.

Keywords: RT duroid material, MIMO antenna, Vias, Slot

1   INTRODUCTION

Antennas play a crucial role in our communication systems. Thus, increasingly more number of professionals is attempting to do advancements and improvement in the field antennas (Sharma et. al., 2020). However, with speedy development of the communication business, the need of antennas is going to be achieved with top quality. These days, all sorts of antennas are available in the market like dipole, patch antenna, loop antenna, meander-line antenna, SIW, MIMO etc (Aghwariya et. al., 2019). So as to satisfy the increasing demand of current systems the analysis of antennas target on some important aspects, such that cutting down the size of antennas and at the same time maintaining higher radiation efficiency (Sharma et. al., 2018). Multiple-input-multiple-output antenna consist rare features such as high date rate and channel capacity(Sippal et. al., 2019). MIMO antenna comprises many antennas at transmitter side as well as at receiver (Deshlandes et. al., 2001).

SIW also known as Substrate integrated waveguide design in such a manner that it combines features of Microstrip structure and waveguide (Niuet. al. 2019). MIMO antennas suffer with high mutual coupling, SIW can be a great factor to reduce mutual coupling (Sharma et. al.2020).

DOI: 10.1201/9781003193838-97

1.2 *Antenna configuration*

The Layout of the presented MIMO has been shown in Figure 1. This layout has been simulated for RT duroid substrate of dimensions 51.2 mm × 34.2 mm and height 1.6 mm. Dielectric constant of RT duroid is 2.2. Designed two rectangular SIW cavities of size 15×15mm. One slot of size 1 mm × 17 mm has been cut on ground plane to reduce coupling between both cavities.

Dimensions of presented design are as follows:

Lgn(length of ground structre)= 51.2mm

Wgn(Width of ground structure)= 34.2 mm

diameter of vias= 0.4mm

Distance between vias=1.00mm

Width of patch= 15mm

Distance between cavity= 2 mm

Length of patch= 16mm

2 RESULT

CST MW software has been used for simulation work. Two identical rectangular SIW cavities and two input output port have been used for this MIMO prototype. Simulation results of presented antenna are in good match with standard characteristics of MIMO antenna.

Figure 2 has shown return loss at port 1 which is -33.783dB.

Figure 3 has shown return loss at port 2 which is -25.429dB.

Figure 4 shows S_{21} parameter at 4.84GHz which is -10.86 assured high isolation between both cavities.

Figure 5 shows Smith chart calculation for presented antenna. By analysing Smith chart, MIMO antenna is matched with standard 50 ohm impedance. Presented MIMO is terminated at 50 ohm and is perfectly matched.

Figure 6 shows gain of presented antenna at port 1 and achieved value of 6.290dB value.

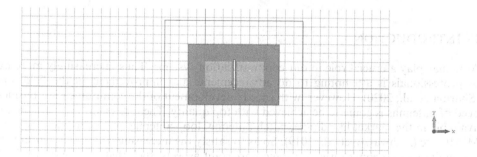

Figure 1.   Layout of proposed design.

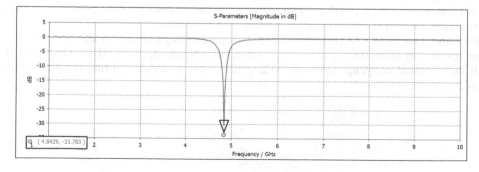

Figure 2.   Return loss at port 1 S_{11} in dB.

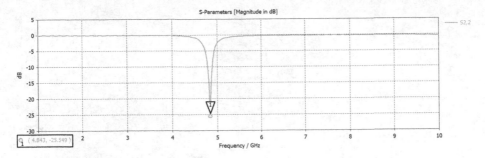

Figure 3. Return loss at port 2 S_{22} in dB.

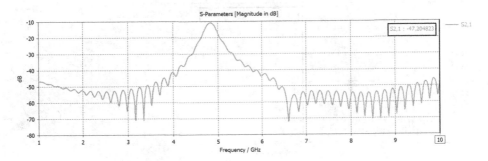

Figure 4. S_{21} parameter for isolation between two ports.

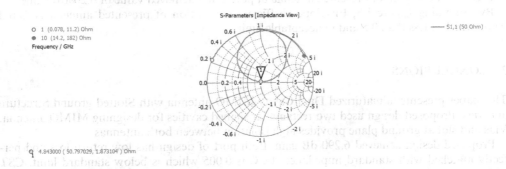

Figure 5. Smith chart.

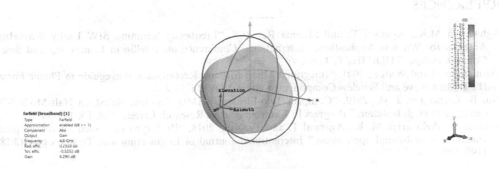

Figure 6. Gain of propsed antenna geometry in dB for port 1.

527

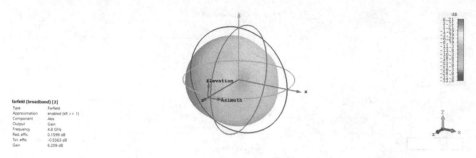

Figure 7. Gain of propsed antenna geomtry in dB for port 2.

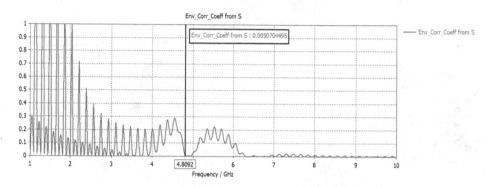

Figure 8. Envelope correlation coefficient of proposed geometry.

Figure 7 shows gain of presented antenna at port 2 and achieved value of 6.209dB value.

By examining Figure 1.8, Envelope coefficient correlation of presented antenna is found 0.005 which is less than 0.5 and in acceptable limit.

3 CONCLUSIONS

This paper presents miniaturized Dual Cavity MIMO antenna with Slotted ground Structure and vias. Proposed design used two rectangular shaped cavities for designing MIMO antenna. Vias and slot at ground plane provides high isolation between both antennas.

Proposed design achieved 6.290 dB gain. Both port of design has low return loss and perfectly matched with standard impedance. ECC is 0.005 which is below standard limit. CST MW software is used for Simulation of presented prototype.

REFERENCES

Aghwariya M.K., Agarwal T. and Sharma R., 2019, "Frequency Scanning SIW Leaky WaveHorn Antenna for Wireless Application," International Conference on Intelligent Computing and Smart Communication, THDC IHET, Tehri.

Deshlandes D.and WuKe., 2001, "Integrated Microstrip and Rectangular Waveguide in Planar Form" IEEE Microwave and Wireless Components Letters, VOL. 11, NO. 2.

Niu B. J. and Tan J. H., 2019, "Compact Two-Element MIMO Antenna Based on Half-Mode SIW Cavity with High Isolation," Progress In Electromagnetics Research Letters, Vol. 85, 145–149.

Sharma R., Aghwariya M. K., Agarwal T. and Tyagi S., 2018, "Prototype of slotted microstrip patch antenna for multiband application," International Journal of Engineering and Technology, 7(3.18), 1199–1201.

Sharma R., Sachan V. K., Niranjan V. and Baral R. N, 2020, "INVERTED L SHAPED SLOTTED MIMO antenna with HWSIW," Telecommunications and Radio Engineering, 79 (2):143–148.

Sharma R., Sachan V. K., Niranjan V. and Baral R. N., Sept. 2020, "Dual cavity SIW MIMO antenna," Integrated Intelligence Enables Networks and Computing, Uttarakhand.

Sippal D, Abegaonkar M and Kaul S. K, 2019, "Highly Isolated Compact Planar Dual-Band Antenna with Pattern Diversity Characteristic for MIMO Terminals, IEEE Antennas and Wireless Propagation Letters.

Recent Trends in Communication and Electronics – Sharma et al. (Eds)
© 2021 Taylor & Francis Group, LLC, ISBN 978-1-032-04572-6

Design and analysis of circular microstrip patch antenna wdith SRR on modified and defected ground structures for UWB applications

D. Subitha

Department of Electronics and Communication Engineering, Vel Tech Rangarajan Dr. Sagunthala R&D Institute of Science and Technology, Vel Nagar, Chennai, India

R. Vani

Department of Electronics and Communication Engineering, SRM Institute of Science and Technology, Ramapuram, Chennai, India

M. Vishal Kumar

Department of Electronics and Communication Engineering Saveetha School of Engineering, SIMATS Chennai

ABSTRACT: Modified Slit-Ring Resonators (SRR) are used develop a high gain Ultra Wide Band (UWB) antenna for microwave imaging applications. This paper analyzes patch radiator with modified SRR with the rectangular slotted ground. Two SRRs are placed on the circular radiator patch element, each in the shape of alphabet letters 'S' and 'V'. The results of the with DGS demonstrate that the reflection coefficient is well below -10 dB in the frequency range from 1.9 GHZ to 11.9 GHz with the fractional bandwidth of 110%. Further modification in the resonant modes of UWB is obtained by adding single rectangular slot on the ground plane. This helps in eliminating the lower band from 1.9 GHz to 6 GHz. This band rejection helps in alleviating the interferences caused by other unlicensed wireless applications lying in sub 6GHZ band

Keywords: Circular patch antenna, SRR, DGS, RGS, UWB, microwave imaging

1 INTRODUCTION

Ever since, the government has declared an unlicensed spectrum of 3.1 GHz to 10.6 GHz authorized by Federal Communication Commission (FCC) is ready for use, its reach in health sector also extended. Hence, the recent developments in imaging techniques make use of the available UWB. The major area of research of UWB technology is going on in the area of breast cancer detection based on radar-based microwave imaging technique (Fear et al. 2002, Hossain et al. 2007 & Kanj et al. 2008). It utilizes the high frequency RF wave which penetrates through the tissue towards the underlying area and getting reflected depends on the degree of variation between the abnormal tissue (tumor) and the surrounding area. No reflection occurs if there is no abnormality present and if the tissue is homogeneous. Hence, in order to focus the RF wave towards the tissue and to receive the reflected wave, antenna is required.

Hence, many works are involved in designing antennas for UWB microwave imaging techniques. Many UWB antennas are designed in different shapes such as elliptical, rectangular,

DOI: 10.1201/9781003193838-98

pentagon shaped, trapezoidal shaped, lamp post shaped and square (Gharakhili et al. 2007, Kwaha et al. 2008, Madhan et al. 2011, Majid et al. 2019 & Pingale et al. 2009). But, the limitation of all the above designs is their radiation patterns are almost omni-directional radiation patterns which is not recommended for microwave imaging antennas applications which requires highly directional narrow beams. In works detailed in (Sukanya et al. 2020) various ground structures and modifications on ground structures using DGS is recommended for improving bandwidth and directivity. But, the designs discussed in the above literatures complexity and sometimes size is the serious problem encountered as to compromise on bandwidth and directive gain. The work on (Mazhar et al. 2013) has designed the miniaturized antenna with meta-material design of SRR and capacitance-loaded strip (CLS).

In UWB based imaging techniques, the entire bandwidth is divided into two bands (i) sub 6 GHz (ii) above 6 GHz. The lower sub 6 GHz band or higher 6 GHz band requires lesser impedance bandwidth of approximately 50 % for full coverage whereas for covering the entire UWB range of 3.1 GHz- 10.6 GHz over 100% bandwidth is needed. In this paper, a monopole antenna with metamaterial based patch antenna radiator (Huang et al. 2010, Islam et al. 2015, Odabasi et al. 2013, Wang et al. 2009 & Anusudha & Karmugil 2016) is designed. The two slots etched are the initial of the authors -'SV' that makes SRR structures as well. The ground is also defected (DGS) so as to obtain wide bandwidth Vishal kumar & Subitha (2020). This model provides excellent results in the entire UWB range (3.1-12 GHz) in terms of bandwidth, reflection coefficient and gain.

2 DESIGN OF PROPOSED ANTENNA

Antenna design proposed in this work is shown in Figure 1 (a, b) is explained in this section. This antenna is printed on FR4 substrate of thickness 1.6 mm with relative permittivity of 4.4 and loss tangent of 0.02.

As shown in Figure 1(a), the disc antenna with radius of 12 mm is printed on the substrate mentioned above, with the dimensions given in Table 1. The optimized feed line dimensions are also mentioned in the Table 1. This proposed model mainly focuses on upper frequency band of above 6 GHz. This is mainly obtained by simply inserting a rectangular slot at the ground structure (RSGS). This creates a band reject filter from 2 GHz to 6 GHz. This helps in reducing the interference of other wireless applications that makes use of WLAN,

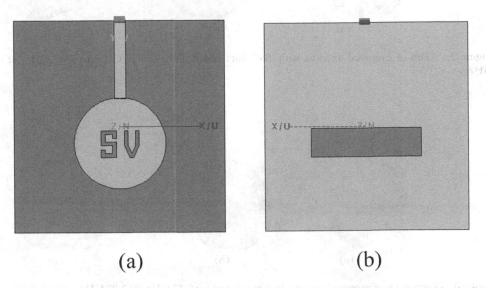

(a)   (b)

Figure 1.   Monopole disc antenna with 'SV' slots and RSGS (a). Model front view (b). Model back view.

Table 1. Proposed antenna module dimensions.

Length of the substrate	56mm
Breadth of the substrate	55mm
Height of the substrate	1.6mm
Length of the line feed	3mm
Width of the line feed	20.59mm
Defected ground length	20mm
Radius of the circular patch	12mm

BLUETOOTH, WIMAX, LTE and helps in projecting the function of unlicensed UWB (3.1-12 GHz) RF wave operations in the microwave imaging applications mostly in above 6 GHZ (6-12 GHZ). This proposed antenna provides a maximum gain of 8 dBi and VSWR of less than 2 in the desired band of interest.

3 SIMULATION RESULTS OF PROPOSED ANTENNA

3.1 *Gain, directivity*

The polar plots at the designed resonant frequencies are shown below in Figure 2 (a-c). The obtained the gain 8.0 dB, 2.5 dB and 5.0 dB at the frequencies 8.79 GHz, 7.07 GHz and 9.47 GHz. The directivity of the proposed antenna module at 9.47 GHz, 8.79 GHz, 7.08 GHz are 7.5 dB, 8 dB, 5.0 dB respectively as shown n Figure 3 (a-c).

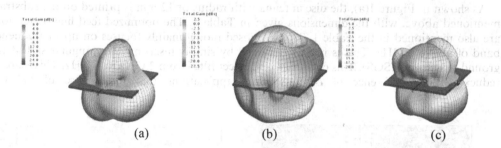

(a)   (b)   (c)

Figure 2. Gain of proposed antenna with 'SV' slots and RSGS (a) 8.79 GHz (b) 7.07 GHz (c) 9.47 GHz.

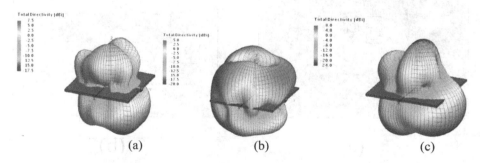

(a)   (b)   (c)

Figure 3. Directivity of proposed monopole disc antenna (a) 9.47 GHz (b) 7.07 GHz (c) 8.79 GHz.

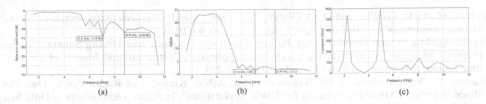

| (a) | (b) | (c) |

Figure 4. (a) Reflection co-efficient, (b) VSWR (c) Impedance bandwidth.

3.2 *Return loss, VSWR and Impedance bandwidth*

The following graph in the Figures 4 (a-c) illustrates the reflection coefficient, VSWR and impedance bandwidth respectively. The results show that VSWR is less than 2 dB in the 7.11 GHz to 11.9 G Hz frequency range. Reflection coefficients also less than -10 dB in the above specified frequency range.

4 CONCLUSION

The work proposed here explicit the performance of circular patch antenna design with SRR based patch resonator, circularly polarized in FEKO tool. The design shows very high gain of 8 dB at the resonant frequency of 7.1 GHz. This exhibits multiple resonant modes at 7.08 GHz and 10.52 GHz with less than -10 dB return loss and VSWR of less than 2. The operating frequency is 1.6 to 11.6 GHz with BSF in the frequency spectrum of 1.59-6 GHz. The simplicity of antenna and good results in terms of gain, directivity, VSWR and return loss proves that the antenna is suitable for applications in sub 6 GHz range.

REFERENCES

Anusudha, K., and M. Karmugil. "Design of circular microstip patch antenna for ultra wide band applications". In 2016 International Conference on Control, Instrumentation, Communication and Computational Technologies (ICCICCT), pp. 304–308. IEEE, 2016.

Fear, E.C.; Li, X.; Hagness, S.C.; Stuchly, M.A. Confocal microwave imaging for breast cancer detection: Localization of tumors in three dimensions. IEEE Trans. Biomed. Eng. 2002, 49, 812–822.

Gharakhili, Fatemeh Geran, Masum Fardis, Gholamreza R. Dadashzadeh, Akram Kalateh Ahmad Ahmadi, and Nasrin Hojjat. "Circular slot with a novel circular microstrip open ended microstrip feed for UWB applications". Progress In Electromagnetics Research 68 (2007): 161–167.

Hossain, Iftekhar, Sima Noghanian, and Stephen Pistorius. "A diamond shaped small planar ultra wideband (UWB) antenna for microwave imaging purpose". In 2007 IEEE Antennas and Propagation Society International Symposium, pp. 5713–5716. IEEE, 2007.

Huang, C., Z. Zhao, Q. Feng, J. Cui, and X. Luo. "Metamaterial composed of wire pairs exhibiting dual band negative refraction". Applied Physics B 98, no. 2-3 (2010): 365–370.

Islam, Md, Mohammad Tariqul Islam, Md Samsuzzaman, Mohammad Rashed Iqbal Faruque, Norbahiah Misran, and Mohd Fais Mansor. "A miniaturized antenna with negative index metamaterial based on modified SRR and CLS unit cell for UWB microwave imaging applications". Materials 8, no. 2 (2015): 392–407.

Kanj, Houssam, and Milica Popovic. "A novel ultra-compact broadband antenna for microwave breast tumor detection". Progress In Electromagnetics Research 86 (2008): 169–198.

Kwaha, P. Amalu, and O. N. Inyang The Circular Microstrip Patch Antenna Design and Implementation", IJRRAS, July 2011.

Madhan M.D, Subitha D, Chandra I "Research on millimeter-wave microstrip patch antenna design for 5G", International Journal of Recent Technology and Engineering, 8, 11, (2019): 2841–2847.

Majid, Huda Abdul, Mohamad Kamal Abd Rahim, and Thelaha Masri. "Microstrip antenna's gain enhancement using left-handed metamaterial structure". Progress in Electromagnetics Research 8 (2009): 235–247.

Mazhar, W., M. A. Tarar, F. A. Tahir, Shan Ullah, and F. A. Bhatti. "Compact microstrip patch antenna for ultra-wideband applications". PIERS Proceedings, Stockholm, Sweden (2013).

Odabasi, H., F. L. Teixeira, and D. O. Guney. "Electrically small, complementary electric-field-coupled resonator antennas". Journal of Applied physics 113, no. 8 (2013): 084903.

Pingale, Ajinkya, Aniket Shende, Ashay Jadhav, Chaitanya Ghanote, and Umang Sonare. "Design of elliptical microstrip antenna for ultra wide band applications". International Journal of Engineering and Technical Research (IJETR) 3, no. 3 (2015): 276–279.

Sukanya, Y., Viyapu Umadevi, PA Nageswara Rao, Ashish Kumar, and Rudra Pratap Das. "DGS-Based Wideband Microstrip Antenna for UWB Applications". In Smart Technologies in Data Science and Communication, Springer, Singapore, 2020; 181–195.

Vishal kumar M, Subitha D. "Design of Circular Microstrip Patch Antenna with Defected Ground Structure for UWB Applications." Journal of critical reviews 7, no. 5 (2020): 1905–1917.

Wang, Jiafu, Shaobo Qu, Jieqiu Zhang, Hua Ma, Yiming Yang, Chao Gu, Xiang Wu, and Zhuo Xu. "A tunable left-handed metamaterial based on modified broadside-coupled split-ring resonators". Progress in Electromagnetics Research 6 (2009): 35–45.

Recent Trends in Communication and Electronics – Sharma et al. (Eds)
© 2021 Taylor & Francis Group, LLC, ISBN 978-1-032-04572-6

Segmentation of digital images with various edge detection techniques using OpenCV

Saumitra Shukla, Vaishali Kikan, Ritvi Sachdeva, Yash Chaturvedi & Ronak Jain
KIET Group of Institutions, Delhi-NCR, India

ABSTRACT: In the domain of computer vision or image segmentation, Edge detection is the most widely used technique to determine and outline the presence of lines or objects in an image appropriately. The main aim is to segment the image data so that the complexity of the image is reduced also the amount of data to be processed in further steps of advanced algorithm is reduced. In simple language, we can say that edges are the sudden and significant changes in the intensity of the image which occur at the boundary of an object in the image. In this paper, the Sobel operator technique, the Laplacian technique, and the Canny technique are used and compared with one another for edge detection in an image. This paper represents the detailed analysis of the performance of these edge detection tools by using the input images with different types of backgrounds for the computer vision or image segmentation or any other advanced algorithm. Further, we have done the comparison among these techniques in presence of different types of noise by using the OpenCV library and PythonSoftware.

Keywords: Image segmentation, edge detection, canny operator, sobel operator, laplacian operator

1 INTRODUCTION

Edge detection means the extraction of information from the image or an in-depth analysis of any image, i.e., size and shape of the object present in the given input picture and detection of various objects present in the foreground of the given input image from different types of backgrounds. An image is simply a combination of different pixels (picture elements) of different intensities, which are denoted using different numbers assigned to them and in addition to that, noise (e.g., salt and pepper noise, Rayleigh noise, Speckle noise, Gaussian noise) is degrading the quality of the image by tempering these pixel intensities. An Edge is defined as the local change in the intensity of an input image. Edges divides the image into different segments or regions as edges are the boundary between segments or regions.

2 LITERATURE REVIEW

The Image segmentation using edge detection will be giving a significant reduction in the amount of the data to be processed in advanced applications of image processing and computer vision. The amount of information to be processed is reduced while all important information from the images is preserved as mostly information is contained in the edges of the image, which is used in the advanced algorithms.

DOI: 10.1201/9781003193838-99

2.1 Laplacian technique

The Laplacian of an image is found out using the second derivative image processing kernel after the image is smoothened out in the pre-processing steps. Only a single kernel is used, which is applied to the full image. The derivative is calculated to find out the regions of rapid changes in the intensity of the image and there is the usage of the zero-crossing concept, due to which, this technique is also termed as zero-crossing technique. The Laplacian is applied to an image after getting it smoothened with the help of various image processing filters.

2.2 Sobel technique

This tool involves the processing of all pixels present in the image with the help of a 3*3 matrix kernel to find out the gradient-based upon different weights (center weight is highest as the center pixel contains more information) as shown in the matrix in Figure 1. This tool is widely used in various advanced algorithms where the convolution with the kernel is done first in the X direction (Sobel-X) and then in the Y direction (Sobel-Y) and then both of these results are combined to get the final edge.

2.3 Canny technique

This is the most efficient but typical edge detection technique which is mainly used to enhance the edge detection process. It includes some steps to complete the whole process as:

2.3.1 Converting RGB image to grayscale mode for reducingcomplexity

2.3.2 Smoothing the image to remove the unwanted noise by applying Gaussianblur

2.3.3 Use of Sobeldetectorforintensitygradientcalculationtolocateallrealandfalse-positive edges

2.3.4 Non-Maximum Suppression reduces the false positives

2.3.5 Double Thresholding and

2.3.6 EdgeTracking

Non-Maximum Suppression is the conversion of pixels that cant be considered as edges to zero, This step is an optimization of what we obtained from the Sobel detector as the results obtained after the Sobel technique are not perfect and the edges obtained are thick, also some false positives are present. So with this process, edges are precise by converting thick edges into thin edges and removing undesired edges or false positives.

Although one of the oldest and sophisticated edge detection technique is the Canny edge detection technique, which is used as a part of a lot of advanced computer vision algorithms, it cannot remove the Gaussian noise from highly tempered images. For those images, an improved algorithm suggested by Tai Kuang in 2011 can be used. involves the processing of all pixels present in the image with the help of a 3*3 matrix kernel to find out the gradient-based upon different weights (center weight is highest as the center pixel contains more information

3 RESULT

OpenCV software has been used for simulation work. Noise type and amount must be considered while applying any kind of edge detection technique in the image because according to the amount of noise the parameters like kernel size and depth, minimum and maximum threshold should be adjusted to obtain the best possible results.

Figure 1 has shown that the Laplacian technique is performing better while the Canny edge detection will find better usage in object detection applications.

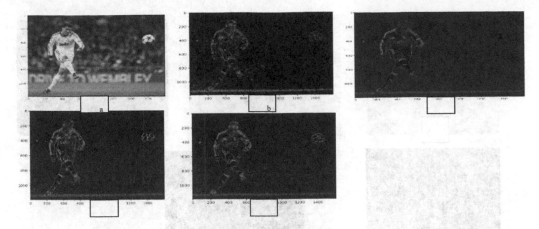

Figure 1. (A) Original Input, (B) Output With Canny Technique, (C) Output With Laplacian Technique, (D) Output With Sobel-X, (E) Output With sobel-Y.

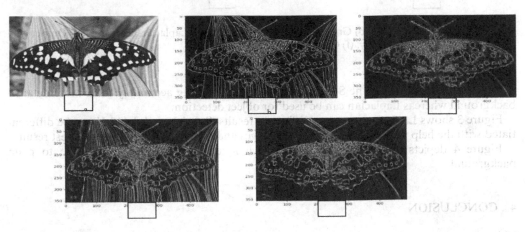

Figure 2. (a) Original Input, (b) Output with canny technique, (c) Output with Laplacian technique, (d) Output with Sobel-x, (e) Output with Sobel-y.

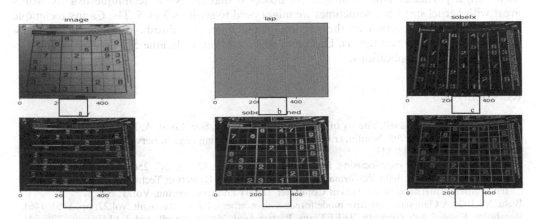

Figure 3. (a) Original Input, (b) Output with Laplacian technique, (c)Output with Sobel-x,(d) output with Sobel-y, (e) Output with Sobel-combined, (f) Output with Canny technique.

537

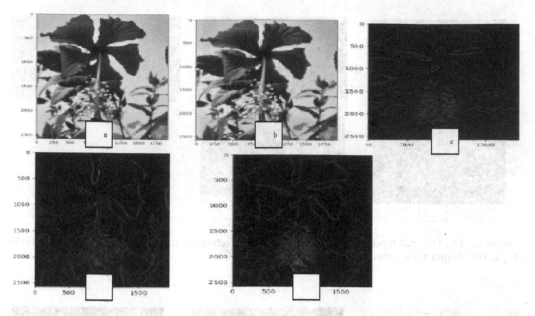

Figure 4. (a) Original Input, (b) Greyscale Image, (c) Output with Laplacian technique, (d) Output with Sobel-x, (e) Output with Sobel-y, (f) Output with Canny technique.

Figure 2 has shown that the Sobel technique is not working accurately for a more detailed background whereas Laplacian can be used for object detection.

Figure 3 shows Laplacian gave unsatisfactory results, Sobel-X and Sobel-Y can be differentiated with the help of horizontal and vertical edges and canny technique gave the best results.

Figure 4 depicts that all edge detection techniques are performing alike due to plain background.

4 CONCLUSION

This paper presents the analysis of the canny technique, laplacian technique, Sobel (Sobel-x & Sobel- y) technique with a variety of different types of images and varying backgrounds as well.

We proposed various advantages and disadvantages of these edge detection methods when used with a particular type of image. We observed that the Sobel technique usually works great with kernel size 1 but sometimes we might need to set it as 3 or 5. The Canny technique depends upon the selection of the maximum and minimum threshold values in a precise manner to obtain the best results. Laplacian edge detection technique can be used in 70-80 % of object detection applications.

REFERENCES

Marr, D & Hildreth, E. 1980. Theory of edge detection, Proc. R. Soc. Lond. A, Math. Phys. Sci.
Aurich, V & Weule, J. 1995. Nonlinear Gaussian filters performing edge preserving diffusion in Proc.17th DAGM Symp: pp. 538–545
Goshtasby, A. 1994. On edge focusing. Image Vis. Comput., vol. 12: pp. 247–256.
Juneja, M & Sandhu, P. 2009. Performance Evaluation of Edge Detection Techniques for Images in Spatial Domain. International Journal of Computer Theory and Engineering, Vol. 1, No. 5.
Basu, M. 1994.A Gaussian derivative modelforedgeenhancement. PatternRecognit.,vol.27: pp. 1451–1461.
Bergholm, F. 1987. Edgefocusing,"IEEETrans. PatternAnal. MachineIntell., vol. PAMI-9: pp. 726–741.

Clark, JJ. 1989. Authenticating edges produced by zero-crossing algorithms, IEEE Trans. Pattern Anal. Machine Intell., vol. 11: pp, 43–57.

Schunck, B.G.1987. Edge detection with Gaussian filters at multiple scales, in Proc. IEEE Comp. Soc Work.CompVis: pp. 208–210.

Tagare, H.D. 1990. On the Localization Performance Measure and Optimal Edge Detection, IEEE Trans. on Pattern Analysis and Machine Intell., v.12 n.12: p.1186–1190.

Shah, M, Sood, A & Jain, R. 1986. Pulse and staircase edge models, Comput. Vis. Graph. Image Process., vol. 34: pp. 321–343.

Deng. G & Cahill, L.W. 1994. An adaptive Gaussian filter for noise reduction and edge detection, in Proc. IEEE Nucl. Sci. Symp. Med. I'm. Conf., pp. 1615–1619.

Berzins, V. 1984. Accuracy of laplacian edge-detectors, Comput. Vis. Graph.Image Process., vol. 27: pp. 195–210.

Williams, D.J & Shah, M. 199. Edge contours using multiple scales, Comput. Vis. Graph Image Process., vol. 51: pp. 256–274.

Gonzalez, R.C & Woods, R.E. 2010. Digital Image Processing. 3rd ed.

Recent Trends in Communication and Electronics – Sharma et al. (Eds)
© 2021 Taylor & Francis Group, LLC, ISBN 978-1-032-04572-6

A Study on low noise amplifier topologies in 5.2 GHz band applications

Diksha Singh & Vaishali Kikan
KIET Group of Institutions, Delhi-NCR

ABSTRACT: RF transceiver plays a crucial role in signal reconstruction at the receiver end. LNA is placed at the front end of the receive block and LNA noise performance dominates the noise performance of the complete receiver block. In this paper we present a detailed analysis of the topologies prevalent in the 5.2 GHz frequency band and the parameters optimised by every topology. LNA performances are compared on the basis of certain design parameters and the best topology found suitable for being used in 5.2 GHz band applications is listed. The topology is found to be the best to be used in 5.2 GHz band applications and provide a much superior performance than the others reported so far.

1 INTRODUCTION

Low Noise Amplifier (LNA) is widely used in many applications which includes RF transceiver systems, Bluetooth low energy (BLE), wireless communication systems, biomedical applications and many more. As the domains are explored further performance constraints on a LNA have also increased. LNA amplifies the signals with a very little noise being added to the signals i.e. noise contribution of the LNA is very less. LNA's find use in the applications where the Signal to Noise (SNR) is very sensitive factor in determining the system performance. LNA design is determined by the application for which it is to be designed. Applications in biomedical domain focus more on the gain and the noise performance of the LNA. Applications in IoT domain focus more on reducing the power consumption of the circuit.

2 TOPOLOGIES

WLAN applications require the operating frequency to be equal to 2.4 GHz or 5.2 GHz. The frequency at which LNA is to be operated is a major step in design process of LNA. However, LNA's which are capable of providing optimum performance at 2.4 GHz as well as 5.2 GHz have also been reported. Operating an LNA at two different frequencies involves the use of suitable circuitry which is capable of switching the operating frequencies of the amplifier and providing finest performance at both the operating frequencies. Dual band LNA's increase the utility of the device as it can be made to operate in two different bands utilised in WLAN applications.

Designing an LNA is a challenging task as it needs to be operated at high frequencies and power consumption needs to be kept low to make it suitable for use in Bluetooth and IoT applications. High performance requirements of LNA make the design process complex. However choosing a suitable topology for LNA circuit can be helpful in keeping many parameters optimized. Choice of suitable topology for LNA requires consideration of many factors that include Noise Figure (NF), gain, reverse isolation etc. Most widely used topology for implementing LNA in wireless applications is the cascode topology.

DOI: 10.1201/9781003193838-100

S.A.Z. Murad et. al 2019 reported an amplifier that was designed to operate at 5.2 GHz frequency. This LNA was also capable of providing optimum performance at 2.4 GHz which is also used in WLAN applications i.e. dual-band LNA. Dual band LNA's are more versatile as compared to single band LNA's as they are capable of being used at two different frequencies without making any alteration to the existing circuit.

The LNA used CMOS 130 nm technology and was designed in cascode configuration to achieve required performance in terms of gain, NF and other circuit parameters. Cascode topology was chosen as it provided a higher gain and a better reverse isolation measured in terms of S_{11}.

Dual band LNA requires matching for two bands of frequencies and the LNA used LC networks at the input as well as output to attain the required matching at both the frequencies of interest i.e. 2.4 GHz and 5.2 GHz. Reported LNA achieved a gain of 14.2 dB at the frequency of interest i.e. 5.2 GHz. However the circuit provided an impressive gain of 21.8 dB at the other frequency of operation i.e. 2.4 GHz.

Matching at two different frequencies in thus LNA was achieved with the help of concurrent technique rather than using a relevant switching technique. Concurrent technique is preferred over switching technique to achieve matching at two different frequencies as it uses a smaller chip area and also consumes less power than the other technique.

The input return loss at 5.2 GHz was found to be -16 dB and the input return loss at 2.4 GHz was found to be -18 dB. The parameter values signify that the concurrent technique used provided good matching at both the frequencies of interest.

Noise Figure of the LNA was reported to be 3.5 dB at a frequency of 5.2 GHz which exceeds the specified limit for LNA's. Noise figure of the circuit is compromised in achieving matching at two frequencies as there is a trade-off between the parameters of a circuit.

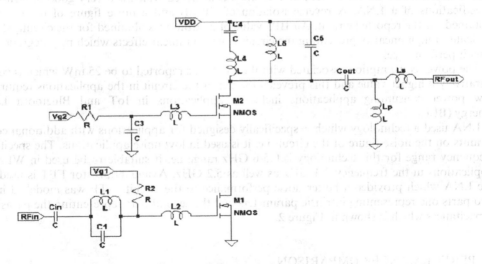

Figure 1.   Dual band LNA (reproduced from Murad et.al 2019).

Mou Shouxian et. al 2003 also reported a LNA that was dual band LNA and could be operated at two different frequency bands that were lying around 2.4 GHz and 5.2 GHz frequencies which are used in WLAN, Bluetooth applications.

The LNA incorporated used a matching network at the input which provided impedance matching at both the frequencies of interest. Matching circuit used were designed to provide a matching of 50 Ω at the input. LNA reported provided a gain of 27 dB in the frequency band around 2.4 GHz and a gain of 21 dB in the frequency band around 5.2 GHz. LN used 250 nm technology and the noise figure of the LNA was a bit higher than the LNA designs that are reported in new CMOS technologies.

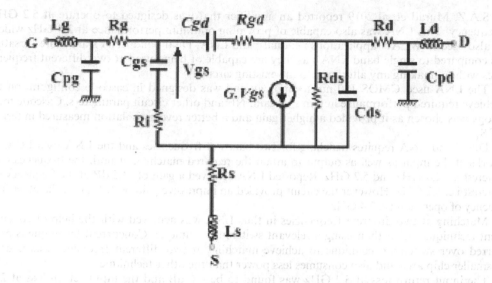

Figure 2. ATF transistor model (Reproduced from Moustapha et. al 2019).

Moustapha El Bakkali et. al 2019 reported a LNA that was designed to operate at a frequency of 5.2 GHz. The LNA reported was fabricated on a substrate (FR-4) and occupied a considerable die area. Gain of 11.3 dB was achieved which is a very good in terms of specifications of a LNA. A reverse isolation of -10 dB and a noise figure of 0.7 dB was obtained for the reported circuit. An IIP3 value of 27 dBm was obtained for the circuit which indicates a high linearity preventing the circuit from non-linear effects which may degrade the circuit performance.

The power consumption associated with the LNA was reported to be 25 mW which is comparatively a higher value and this prevents the use of the circuit in the applications requiring low power consuming applications including applications in IoT and Bluetooth Low Energy (BLE).

LNA used a technology which is specifically designed for applications with additional constraints on the noise figure of the circuit i.e. it is used in low noise applications. The specified frequency range for this technology is 0.5-6 GHz range i.e. it suitable to be used in WLAN applications in the frequency 2.4 GHz as well as 5.2 GHz. Avago Transistor FET is used in the LNA which provided a better noise performance to the circuit. ATF was modelled into two parts one representing intrinsic parameters and the second part representing the parasitic capacitances which is shown in Figure 2.

3 PERFORMANCE COMPARISON

Circuit performance is compared in terms of some common parameters which in case of an amplifier can be collated in terms of gain (S_{21}), Noise Figure (NF), Input Return Loss (S_{11}), power consumption etc.

In IoT applications the parameter which becomes demanding is the power consumed by the circuit and it is crucial in determining the feasibility of using the amplifier in IoT applications. The topologies discussed have a power consumption which is higher and hence it creates a hurdle in using these in low energy applications.

Table 1. Comparison table.

Parameter	Moustahpha et. al 2019	Kao et. al 2008	Murad et. al 2019	Shouxian et.al 2003	Hsiao et. al 2016
Gain(dB)	11.3	10.06	14.2	22	15
Noise Figure(dB)	0.7	3.73	5.0	2.95	2
IIP3(dBm)	27.8	-0.4	10	-	-4
Reverse isolation(S_{11})	-17	-13.56	-16	-24 - -11	-7
Technology	ATF21xx	CMOS	CMOS	CMOS	HEMT
Power(mW)	50	13.5	32.9	21	37.8
Frequency(GHz)	5.2	5.2	5.2	5.0-6.0	5

4 CONCLUSION

Comparison of the various topologies reported so far give the best performance in terms of all the design parameters except for a exception of having a high power consumption. Analyzing the circuit reported can help us to overcome this drawback of higher power consumption. Higher power consumption is a hurdle in utilizing this circuit in IoT applications. We also conclude that dual band LNA's can also be designed to increase the utility of the LNA in more applications involving different frequency bands. However dual band LNA's are seen to acquire a larger chip area and may also have additional noise in the circuit due to increased complexity of the circuit.

REFERENCES

Bakkalia Moustapha El, Touhami Naima Amar, Elftouh Hanae and Marroun Abdelhafid, 2019, Design of 5.2 GHz Low Noise Amplifier for Wireless LAN in proceedings of International conference Inter disciplinarity in Engineering.

Bansal Malti, Singh Diksha, 2019, Low Noise Amplifier In Bluetooth And Bluetooth Low Energy (Ble) Applications In Proceedings Of National Conference On Emerging Trends in Electronics and Communication, pp. 110–113.

Hsiao Y.-C., Meng C., and Yang C., 2016, Design optimization of single-/dual-band FET LNAs using noise transformation matrix, IEEE Trans. Microw. Theory Techn., vol. 64, no. 2, pp. 519–532.

Kim Sinyoung, Kim Taejong, and Kwon Kuduck, 2017, An Ultra-Low-Power 2.4 GHz Receiver RF Front-End Employing a RF Quadrature Gm-Stage for Bluetooth Low Energy Applications", in proceedings of International SoC Design Conference.

Kao C. Y., Chiang Y. T., Yang J. R., 2008, A Concurrent Multi-Band Low-Noise Amplifier for WLAN/WiMAX Applications. IEEE.

Murad S.A.Z., Hasan A. F., Azizan A., Harun A., Karim J., 2019, A concurrent Dual-band CMOS low noise amplifier at 2.4/5.2 GHz for WLAN applications, International Journal of Electrical Engineering and Computer Science Vol. 14, No. 2, pp. 555~563.

Razavi B., "Design of analog CMOS integrated circuits", Mc Graw-Hill 2001.

Roy Niladri, Najmabadi Mani, Raut Rabin, Devabhakutani Vijay, 2006, A systematic approach towards the implementation of a Low-Noise Amplifier in Sub-Micron CMOS technology", in proceedings of IEEE CCGEI.

Shouxian Mou, Jianguo Ma, Seng Yco Kiat, and Anh Do Manh, 2003, An Integrated Si-Ge Dual-band Low Noise Amplifier for Bluetooth, HiperLAN and Wireless LAN Applications", in proceedings of European Microwave Conference.

Recent Trends in Communication and Electronics – Sharma et al. (Eds)
© 2021 Taylor & Francis Group, LLC, ISBN 978-1-032-04572-6

Maximizing the efficiency of portable air conditioning units using heat from exhaust gasses and refrigerant gasses

Kartikeya Dwivedi, Vaishali Kikan, Shambhavi Kaushik, Arjun Singh Jadon & Shruti Talyan
KIET Group of Institutions, Delhi NCR, India

ABSTRACT: The demand of useful energy is continuously increasing in the world while the portable air conditioning unit is one of the most power-consuming devices in our day-to-day lives, which has completely transformed our world. As we know according to the law of conservation of energy, energy only transform from one form to another, the detailed analysis of electrical power converted in the form of heat energy i.e. waste energy is used to determine the efficiency of the given air-conditioning system. The air conditioner is the device, which transfers heat from indoor to outdoor using pumps. One of the most important factor that reduces the efficiency of the portable air conditioning system is this exhausted heat and also it is critical to analyse the composition of gasses emitted. Therefore its measurement becomes extremely critical as it is the low grade energy. Maximizing the efficiency of the portable ac by re-using the exhaust gas and the heat generated outdoor due to it, in various general applications in order to convert it into high grade energy has become important today.

Keywords: efficiency, exhaust gasses, condenser, refrigerant, heat

1 INTRODUCTION

The traditional air conditioning system uses the process of vapor compression to provide the required cooling indoor with the help of refrigerant gas. The electric power provides the required energy to transform the refrigerant from vapor to liquid phase and vice versa in order to change the temperature of the refrigerant gas in the chiller tube. There is a cooling and a heat rejection device connected in opposite direction in the portable ac(air conditioner), which creates the required pressure condition for the warm air indoor to be sucked inside the chamber containing chiller tubes to create the desired cooling effect and the excess moisture particles are also removed at this stage only, which comes out as an exhaust liquid from the exhaust pipe. This hot air transfer heat to the refrigerant, which is vaporized into gaseous form from the previous liquid form and in the result, cool air is formed to transmit to the room. Compression of the gaseous refrigerant is done in order to convert it into liquid phase again so as to make it ready for the next cycle. And in this process, the heat energy is exhausted, which is the low grade energy. Thus, there is a need for the conversion of this low grade energy into high grade energy. So concrete efforts must be taken for the recovery of this waste heat.

1.1 *Study of refrigerant gasses*

i. Chloroflouro carbons (CFC + R12): This gas is a very critical gas for the green house effect and ozone layer depletion. Thus its production stopped in 1994.

DOI: 10.1201/9781003193838-101

ii. Hydrochlorofloro carbons (HCFC +R22): This gas is also contributing towards the damage of the ozone and due to this, with the clean air act of 2010, EPA has mandated the ban on its usage completely by the end of 2020.

iii. Hydrofluoro carbons (HFC + R410A + R134): As there is no chlorine used, this gas is comparatively safer for the environment. Usage of R410A gives better comfort, reliability with the additional benefit of providing improved air quality but it is not 100% safer gas and due to this EPA has allowed its usage with some additional laws for its proper handling and environment safety.

The advantages of these refrigerant gasses is that they are non-flammable and non-toxic. They provide good insulation, are used for aerosol for spray cans and further these are utilized in the making of plastic foams for furniture. Freons (CF_2Cl_2) is photoed in the upper atmosphere by UV light. In this process, chlorine is formed from the Freons and it reacts with O_3 to form ClO radical and oxygen. This ClO is a very reactive radical, which reacts with O to form O_2 and another Cl atom to repeat the cycle again and in this manner ozone layer is depleted.

1.2 *Methods to increase efficiency of portable AC*

One of the efficient way to increase system efficiency is by reusing excess heat or cooling air that would normally waste as the passive energy. The portable ac can be made more efficient by using the heat recovery system, where heat emitted out by the portable ac is converted/recycled to provide hot water or for other heating operations. Heat is recovered by two methods i.e. by passive heat recovery and by heat pumps. Heat pumps are expensive in comparison to the passive heat recovery system and this is the reason this technology is used only by the larger consumers like Supermarkets in developed countries like US, Europe whereas its usage in the domestic system may further increase the efficiency of the air conditioning system.

The end users may take various measures to improve the efficiency of their portable AC. The most commonly used techniques are by preventative maintenance and smart technology, by combining sensible commissioning, contractors and refrigeration OEM s. Further the hot exhausted air can be used for various domestic heating purposes like to heat up utensils in the kitchen for cooking, heat up the water for bathing purpose. The exhaust air can be used to rotate small efficient turbines that can be used to produce electricity and power the air conditioning unit thus reducing the overall operating costs of the unit. It may be used for disinfection purposes and drying up the floors after mopping. In laundries, the hot air from exhaust may be used to dry up clothes after washing and at last, it can be used as an air blanket.

1.3 *Domestic applications analysis*

a. Integrated hot water system: The exhaust heat rejected by Portable AC has been taken to make an integrated hot water system. The water is cyclically rotated in this system and temperature is increased after each circulation. The water keeps on rotating until desired temperature is transferred to it from the pump. An insulated tank can be used to store this hot water. There is a significant reduction in the use of electricity or the demand of LPG in this application. Further the heat rejected by the pumps can be used to make the boilers which finds applications in industry as well as in domestic usage.

b. The water emitted out from the portable ac can be used in various domestic and industrial applications. This water is generated from the collection and condensation of the moisture from the surrounding and this is the reason the purity level is very pure. In addition to that, this water can be used in various industrial applications because of its distilled nature and its demineralized properties results in its usage in industrial cleaning purposes as it will significantly reduce the amount of contamination.

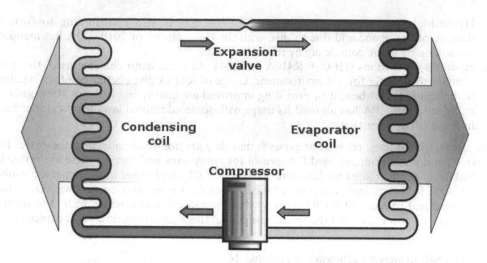

Figure 1.   Working of portable AC: Movement of refrigerant gas.

1.4   *Disadvantages of portable ac on health*

Gases released from AC: Mainly oxides of Carbon and Surface cause get pollute the air.
   Examples for these gases are Carbon Monoxide, Sulfur Dioxide and Carbon dioxide etc
   Carbon dioxide(CO_2): cause green house effect.
   Sulfur dioxide(SO_2): Acid Rain.
   Carbon monoxide(CO): Lungs diseases and cancer
   Air conditioners do not produce harmful gases but the refrigerant gasses(freon) have the
ability to create an ozone hole if they got leaked out due to poor maintenance work. There is
a lubricant inside the air conditioner, which seal the freon inside it but when ac is not used for
a long time, freons can slowly leak into the atmosphere. Freons are used as the refrigerant gas
because of its property of absorbing heat. On compression of freons, it is converted into gaseous state from liquid state as its temperature rises, which is passed through a loop of coils to
decrease its temperature and to convert it back into the liquid phase as shown in Figure 1.
Due to high temperature and pressure, sometimes these freons get leaked and this is the
reason proper maintenance and cleaning of the coils must be regularly done.

2   BASIC CHANGES REQUIRED

Some basic changes are required for prevention from disadvantages of air conditioner on
health of any individual. First of all, the manufacturer's instructions for cleaning/changing the
air filter must be properly followed. In order to flush out pollutants and for the circulation of
fresh air, windows must be opened. AC should be replaced after every 10 years and during the
working period, HVAC technician must perform annual maintenance checks. In addition to
that, the fan only mode will enhance the performance of AC unit.

3   CONCLUSION

We have observed the various effective ways to increase the efficiency of the portable AC by
using the electrical energy lost as heat in various day-to-day applications. The gasses released
from the exhaust of the portable AC are studied to analyse the composition of various gasses
inside it and according to it, various applications have been presented in the given article to

reduce the heat losses and to maximize the utilization of the electrical energy used to run this portable AC unit. In addition to that, the water exhausted from the moisture of the room can also be used for various general day-to-day applications. The work can be extended in future in the manufacturing of the portable AC design with effective application of exhaust gasses and exhaust liquid and to also minimize the various disadvantages and limitations.

REFERENCES

Benstead R, 1988, Industrial heat pumps, IEEE Power Engineering Journal., vol. 2, pp. 173–178

Hepbasli A. and Kalinci Y., 2009, A review of heat pump water heating systems, Renewable and Sustainable Energy Reviews, vol. 13, pp. 1211–1229.

Jarnagin R. E., 2006, Heat Recovery from Air Conditioning Units in Fact Sheet EES-26 Florida Cooperative Extension Service, University of Florida.

Langley B.C., 2001, Heat Pump Technology in, New York: Prentice Hall

Mc Quiston M.C, Parker J.D. And Spitler J.D., 2005, Heating Ventilating and Air Conditioning Analysis and Design in,New York: John Wiley&Sons, Inc.

Reardon C, S Woodcock and Downton P, 2008, Technical Manual of Design for Lifestyle and the Future: Australia's guide to environmentally sustainable homes, Commonwealth of Australia.

Somsuk N., Wessapan T. and Teekasap S., 2008, Conversion of conventional commercialized window type air conditioning unit into a portable air conditioning-heat pump unit, Sustainable Energy Technol ogies 2008 ICSET 2008 IEEE International Conference, pp. 728–732.

Vaivudh S., Rackwichian W., and Chindaruksa S., 2008, Heat transfer of high thermal energy storage with heat exchanger for solar trough power plant, Energy Conversion and Management, vol. 49, pp. 3311–3317.

Novelty approach for hospitals in COVID-19 pandemic

Ayush Bhardwaj, Dhanesh Shukla, Parvin Kumar, Abhishek Sharma, Abhishek Gupta, Apoorva Chand & Anushka Sharma

KIET Group of Institutions, Delhi-NCR, Ghaziabad, India

ABSTRACT: As we all are aware of the fact that the entire world is under the influence of the Corona Virus, due to which several precautionary measures are been rigorously taken to degrade the effect of Corona Virus. Amidst all the safety measures and precautions of ongoing pandemic of covid-19, we have come up with a novelty approach based on the field of Robotics and its associated technology, we will be designing a multitasking Robot for hospitals and houses. The main function of our robot will be to sanitize the entire space within Hospitals and houses with the help of sprinkler and it will also behaves like a serving agent, where it can deliver the essential goods to the staff and patients inside the hospitals as well as it can deliver necessary food items from one room to another room without human interactions.

1 INTRODUCTION

Covid illness 2019 (COVID-19) is an irresistible infection brought about by serious intense Respiratory Syndrome Coronavirus-2 (SARS-CoV-2). It was first distinguished in December 2019 in Wuhan, China and has since spread universally, bringing about a continuous pandemic (Huang W.H. et. al. 2020). In any case, the main case might be followed back to 17 November, 2019 in China. Basic manifestations incorporate fever, hack, weakness, windedness and loss of smell and taste. While most of cases it brings about mellow indications, some advancement to intense respiratory pain disorder (ARDS) likely hastened by a cytokine storm multi-organ disappointment, septic stun, and blood clumps (Lee P.I. et. al. 2020). The time from introduction to beginning of manifestations is commonly around five days yet may go from two to fourteen days. The complexities due to Coronavirus is Pneumonia, viral stepsis, intense respiratory trouble disorder, kidney disappointment, cytokine discharge condition (Lu X. et. al. 2020).

2 PROCESS METHODOLOGY

As explained earlier, COVID-19 pandemic is a worldwide pandemic and its index case was found on date 30 january, 2020 at Thrissur, Kerala. At that time this disease was considered as a normal SARS disease and we all were unaware about its fatality. As this paper is a study about its spread, impact and strategies against pandemic in India, severe effects were seen from march. Indian Prime Minister took an appreciated initiative and declare total lockdown in country to inhibit this pandemic in its early phase. On 4 may, country has to face 3rd phase for 14 days with country divided in three zones(i.e., Green, Orange, Red) and facilities to avail was depending zone and meanwhile government tried to provide each and every hand to all people of country but due to positive slope this phase ended with 53636 confirmed cases, 1774 deaths and 23503 recovered in this phase . Early decision of lockdown has impacted the

DOI: 10.1201/9781003193838-102

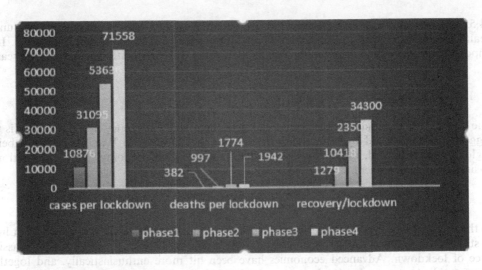

Figure 1. Total Cases, Deaths, Recovery per lockdown.

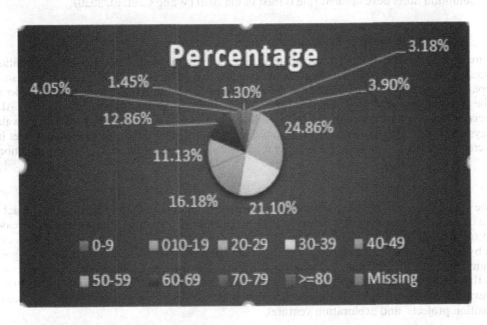

Figure 2. Age wise impact of corona disease.

spread of virus and WHO appreciated India for its strategy against pandemic which has grounded superpower of world. Phase 4 of lockdown started by 18 may 2020 for 14 days with same strategy of zone division but ended with 71558 confirmed cases, 19484 deaths and 34300 recovered cases in this phase till 26 may 2020. By this time country has reached 1.5 lakh of confirmed cases with more than 50,000 recovered cases.

This pandemic has affected every continent in various ways.Staying at homes and maintain social distancing are only ways to stop its spread. By helping each other and with patience we can overcome this disease. Till now around 57,16,621 confirmed cases, 3,52,964 deaths and

24,55,170 recovered cases has been recorded(Perlman S. et. al., 2020). Many of research units, healthcare units are working continuously in frontline for defence of human civilization. It is important to understand the situation and support them who are fighting against such disease.

3 IMPACT AND DISCUSSIONS

The COVID-19 19 pandemic has been declared as the global emergency by WHO and it is the biggest challenge faced by humankind faced since the 2nd World War. Apart from being a global health hazard, this has also extended its impacts on economy, GDP, international business and environment.

3.1 *Effect on global economy*

In the midst of the COVID-19 pandemic, legislatures of a few countries have figured out how to straighten the ceaselessly rising chart of disease spread by Corona infection, with the assistance of lockdown. Advanced economies have been hit more enthusiastically, and together they are relied upon to develop by - 6 percent in 2020. Developing business sectors and creating economies are relied upon to decrease by - 1 percent. In the event that China is prohibited from this pool of nations, the development rate for 2020 is relied upon to be - 2.2 percent. The consumption hotel development rate is least in the 2020 (Wang C. et. al. 2020).

3.2 *Effect on environment*

Coronavirus 19 has affected the climate in an aberrant manner. The primary examinations assessed a positive roundabout effect on the climate. This result is basically because of the social separating arrangements embraced by the legislatures following the presence of the pandemic. The air contamination level has diminished much after the lockdown has been forced. Also, social separating measures received by individuals have caused the cleaning of sea shores, waterways and a few other water sources. Diminishing being used of public and private vehicles has decreased the air contamination levels, yet it has likewise decreased the clamour contamination.

3.3 *Effect on employment*

The greater concern, be that as it may, at the forefront of everyone's thoughts is the impact of the illness on the business rate. Because of COVID 19 lockdown, joblessness rate has increased up to 23.4 %. CMIE report says India's metropolitan joblessness rate takes off to 30.9% even as by and large rate increases to 23.4%, showing Coronavirus' effect on the economy. Ongoing alumni in India are dreading withdrawal of propositions for employment from corporates due to the current circumstance. There is likewise an incredible dread of downturn. Prompt measures are needed to relieve the impacts of the pandemic on bids for employment, temporary position projects, and exploration ventures.

3.4 *Effect on labourers*

There is a serious level of revenue in the numerous different effects of COVID-19, including the numerous monetary and work market impacts, which have been quick and critical, and liable to proceed soon or Past, conceivably. Lockdowns have legitimately impacted a tremendous number of workers across an immense number of nations on account of the job market. Via teleworking or far off working game plans, some may continue with their work. Many others have seen a reduction or total loss of their jobs.Many others have seen a reduction or total loss of their jobs. Others will also undergo an alternative kind of advancement, such as workers in welfare or public protection, in particular a gigantic expansion in working weight even with the emergency.

4 APPROACH

Due to this ongoing pandemic of covid-19, there were many approachesthat have been taken by many countries to stop its transmission, which lead to development of new tech-innovations in fighting against this pandemic. We have also come up with a novelty approach to contribute some help to hospitals in this pandemic.After analysis the data provided in methodology, we came with our novelty approach towards this pandemic is mainly for hospitals so that hospital staff should easily know about the people coming to hospital about whether they are infected or not. It basically includes two flir cameras on the main entry of hospital which will help in determining thermal temperature of the people and these two cameras are in that position that they can easily capture the person on main entry and the data would be send to person's monitor sitting on reception which would be inside the main entry separating by glass window and the person sitting on the reception would have control to open the main entry gate. The main entry gate has two parts: one part is the isolation ward for infected person and other part is the main hospital entrance. If the person coming to hospital has been tested positive after thermal screening from cameras, then the person sitting on reception would open the isolation ward gate for that person or if that person would not test covid-positive that hospital main entry door for general hospital would be open

The isolation ward should have two rooms, one room for patients and one room has disinfecting chamber. The robots would be there for helping covid-19 patients for delivering foods and many things. The robots would also be used for disinfecting clothes and utensils of infected patients in disinfecting chamber. This novelty approach would be really helpful for hospital staffs and helpful in prevention them from being infected from this deadly virus.

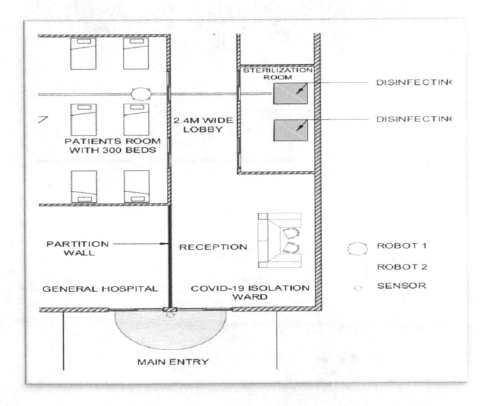

Figure 3.   Represents the novelty approach for hospitals in covid-19.

5 CONCLUSION

Covid-19 has been the deadliest virus which shattered the complete world through its outbreak and affected millions of people. The preventive measures should be taken so that it would not affect us and our approach would be very helpful for the hospitals in this pandemic so that many people would be treated carefully without being in contact with them in hospitals and should reduce the workload of hospital staff during this pandemic with the help of robotics and cameras so that people can be recovered easily and fast without being in contact with others to stop its further transmission in other people.

REFERENCES

Hou F., Yu T., Du R., Fan G., Liu Y., Liu Z (2020). Clinical course and risk factors for mortality of adult inpatients with COVID-19 in Wuhan, China: a retrospective cohort study. Lancet. 2020 Mar 11 doi: 10.1016/S0140-6736(20)30566-3. pii: S0140–6736(20)30566–3.

Huang W.H., Teng L.C., Yeh T.K., Chen Y.J., Lo W.J., Wu M.J. (2020). Novel coronavirus disease (COVID-19).in Taiwan: reports of two cases from Wuhan, China. J Microbiol Immunol Infect. 2020; 53:481–484. doi: 10.1016/j.jmii.2020.02.009.

Lee P.I., Hu Y.L., Chen P.Y., Huang Y.C., Hsueh P.R. (2020). Are children less susceptible to COVID-19? J Microbiol.Immunol Infect. 2020 Feb 25. doi: 10.1016/j.jmii.2020.02.011. pii: S1684–1182(20)30039–6.

Lu X, Zhang L, Du H, Zhang J, Li YY, Qu J. (2020). SARS-CoV-2 Infection in Children. N Engl J Med. 2020 pmid:32187458.

Perlman S. (2020). Another Decade, Another Coronavirus. N Engl J Med. 2020. pmid:31978944.

Wang C., Horby P.W., Hayden F.G., Gao G.F.(2020). A novel coronavirus outbreak of global health Concern. Lancet .2020. pmid:31986257.

Recent Trends in Communication and Electronics – Sharma et al. (Eds)
© 2021 Taylor & Francis Group, LLC, ISBN 978-1-032-04572-6

Advanced encoding scheme for security

Keshav Bhardwaj & Aaisha Makkar
Chandigarh University, Chandigarh

ABSTRACT: Encryption is a process of creating cypher text from plain text. In this process, plain text is the input and cypher text are the output. Decryption is complete opposite of encryption, it's the process of converting cypher text again to plain text using a secret encryption key. There are various types of encryption and decryption processes classified based on their respective methods. Here, we present an encoding algorithm based on ASCII values of text entered. The secret key in this method is designed in such way that it becomes more difficult for any intruder to crack the encryption process. And as result we will get an encrypted text which will be impossible to understand until you have secret key. At the process of decryption, just have reversed the steps and by using secret key, plain text can be returned.

Keywords: Encryption, decryption, ASCII values, plain text, cypher text, encryption key

1 INTRODUCTION

Cryptography is a method to communicate using codes so that only those for whom the information is intended can read and process it. Encryption is basically a process to hide some information. Encryption is a process which is subset of cryptography. In encryption process we have a secret key also known as encryption key. This secret key is used to encrypt the data. Here at result, we got cypher text which will not at all match with plain text we entered earlier. In the process of communication, we transfer this encrypted cypher text just to keep our data safe from an intruder or hacker.

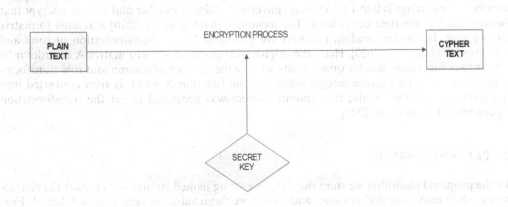

Figure 1. Encoding process.

DOI: 10.1201/9781003193838-103

Decryption is complete opposite process of previous one. It converts cypher text into the plain text using the same secret key. In this process of taking encoded or encrypted data and converting back to plain text which is understandable easily. This plain text which we achieve after the process of decryption will resemble to the text entered at start of cryptography.

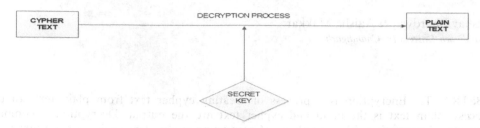

Figure 2.   Decoding algorithm.

2   LITERATURE REVIEW

There are many existing encryption algorithms which work based on ASCII values. In this algorithm [Mathur et. al. 2012] first limitation is user must enter a key to encrypt the data equal to the length of data which sometimes becomes impossible if we take data set of large size. Then we must encrypt that the key which is the key used for cryptography. And the other limitation of this algorithm is that the execution time of this algorithm is much more than others. In another [Satyajeet et. al 2014] encryption and decryption algorithm instead of asking user to enter the encryption key, the user generates a key randomly equal to the length of the message then perform other functions and converted to another which is further used to encrypt and decode the message, it takes less execution time as compared to existing algorithms but not the optimal one. Another limitation is as conversion process of message in ASCII value is initiated and then a random key is generated, makes the process a little complex and complicated to understand. According to the authors [Chandravathi et. al. 2010], the characters in message gets encoded by its ASCII code whose ASCII value is encoded to a point which is encrypted to two cipher text points but in proposed method the private key technique is used where each character is encoded by its ASCII value, but each character get encoded by one pixel only whereas there are three integer values i.e., R for Red, G for Green, B for Blue. Another technique [Singh et. al. 2011] is to encrypt the text to a float point number whose range is 0 to 1 which gets converted to binary number and then to encrypt that binary number one-time key is used. The technique [Kumar et. al. 2010] is related to matrix disordering. In this the random numbers are generated for the transformation of rows and columns [Kumar et. al. 2020]. Here, the original text gets ordered into matrix A of order m*n where two directional circular queue exists and for the number of column and row transformation, it generates a positive integer using random function X which is then converted into a binary number then again, the random number was generated to get the transformation operation [Makkar et. al. 2020].

3   PROPOSED WORK

In the proposed algorithm we store the data in a string named str and will convert the characters to their respective ASCII values and then store those values to vector named ASCII. Further after storing the ASCII values in a vector we generate a random number between 1-5 and will store it in variable n. Then a loop generates n numbers dynamically between 1-100 and will store in array named q and of size n and then a sequence of operation has been operated on ASCII which will be further explained and then resulting ASCII values will be converted in characters and stored in new string named newstr.

[1] Convert all characters of input into its ASCII value and store it in vector named ascii.
Str[]- i n d i a
Ascii[]- 105 110 100 105 97
[2] Now generate a random number between 1-5 and store it in variable n.
Now just for explanation let us assume n=2.
[3] Now generate n numbers between 1-100 and store it in array named q.
let us assume the values of array just to explain the complete process.
q[2]=[80,50]
[4] Now a loop will be followed which will create cypher text and consists of secret key and add it in vector named cypher.

```
int i=0;
while(i!=sizeof(ascii))
{
    for(int j=0;j<n;j++)
    {
        code=3*q[j]+ascii[i]-j;
        i++;
        cypher.push_back(code);
    }
}
```

[5] Now if ASCII value are exceeding the values of 255 then we must make then again in range by subtracting 255 from it.
cypher[]- 345 259 340 299 337
cypher[]- 904 854482(after subtraction)
[6] Convert ASCII values in cypher into characters and store them in new string.
cypher[]- 904 854482
newstr[]- Z ? U , R

By use of this encoding algorithm, we encoded 'india' to 'Z?U,R'. this encoding algorithm is beneficial because even if we have same letter twice in a word it will not be encoded same, which make it more difficult to for intruders to hack it. now to decode this data we will use n and q[] as secret keys and follow the process in reverse and we can get the plain text.

4   CONCLUSION

This algorithm converts data to ASCII values and then store it in vector datatype which will make program data efficient and then randomly generates few numbers on which basis forward operations were done. Here, we generate a secret key based on randomly generated numbers which will make this algorithm more tough for intruders to hack and which also means that there will be different key for each time we execute this program over any message. The beauty of this algorithm is it does not generate any pattern which means even for same letter in the data there will be different cypher text each time.

REFERENCES

A. Mathur, "A research paper: An ASCII value-based data encryption algorithm and its comparison with other symmetric data encryption algorithms",International Journal on computer science and engineer ing (IJCSE), vol. 4, pp. 1650–1657, sep 2012 ISSN: 0975-3397.

Bh., P., Chandravathi, D., Roja, P. (2010) Encoding and decoding of a message in the implementation of Elliptic Curve cryptography using Koblitz's method.International Journal of Computer Science and Engineering, 2(5).

Kiran Kumar, M., Mukthyar Azam, S., and Rasool, S. (2010) Efficient digital encryption algorithm based on matrix scrambling technique.International Journal of Network Security and its Applications (IJN SA), 2(4).

Kumar, N., & Makkar, A. (2020).Machine Learning in Cognitive IoT. CRC Press.

Makkar, A., Garg, S., Kumar, N., Hossain, M. S., Ghoneim, A., & Alrashoud, M. (2020). An Efficient Spam Detection Technique for IoT Devices using Machine Learning. IEEE Transactions on Industrial Informatics.

Makkar, A., Kumar, N., Zomaya, A. Y., & Dhiman, S. (2020). SPAMI: A cognitive spam protector for advertisement malicious images.Information Sciences, 540, 17–37.

Satyajeet R. Shinge et al, /(IJCSIT) International Journal of Computer science and information technolo gies, Vol. 5(6), 2014,7232-7234 ISSN: 0975-9646.

Singh, A., Gilhorta, R. (2011) Data security using private key encryption system based on arithmetic coding.International Journal of Network Security and its Applications (IJNSA), 3(3).

Analysis of big data paradigms: Constraints, solutions and challenges

Pooja Arora
Research Scholar, Chandigarh University, Gharuan, India

Aaisha Makkar
Chandigarh University, Gharuan, India

ABSTRACT: With the rapid advancement in the digital world and the extensive use of internet technologies, the amount of data generated and stored is increasing exponentially. The term "Big Data" refers to the large amount of data which cannot be managed by conventional tools available for data handling. The field of "Big Data" deals with the analysis, storage and management of complex data sets. This includes all the work done starting from gathering data, storing data, analyzing data, querying data, updating the data as and when required. In this paper, we are going to concentrate on the 4 V's of Big Data and the various security issues that are currently being faced in this regard. We will also put some light on the probable solutions for the same.

Keywords: big data, hadoop, mapreduce, data lake

1 INTRODUCTION

In the most general words, "Big Data" refers to the massive collection of complex data which is difficult to store and process with the available means today. There is not any specific amount which can be called as "Big Data". It is generally described in terms of Zettabytes (1,024 exabytes) or Brontobytes (1,024 yottabytes) which is generated from diverse sources [6]. This data can be available in any form, viz., structured,semi-structured or unstructured. Structured data is a repository of data which is organized in a certain format such as a database. A database stores the data in the form of tables with rows and columns, for example: relational data. Unstructured data is unorganized data or the data which does not have a predefined data model, for example: Word, Text, PDF,Media logs. Semi-structured data does not follow any formal data model, but does contain some markers or tags that can separate the elements into various hierarchies. An example of such data is JSON (thestructure that DataAccess uses by default), .csv files, XML, tab delimited files etc.

Various studies and research in the field of Big Data have shown that if we can find a way to manage and process the data in an effective way then Big Data has the capacity to save time and money,boost efficiency and improve decision making in the fields of fraud control, weather forecasting, health and medicines, national security, business areas, education and traffic control.

2 THE 4 V'S OF BIG DATA

The whole theory of Big Data revolves around the 3 V's, namely, volume, variety and velocity. A fourth V has also been now introduced which expands to veracity.

Below is the detailed description of these V's which form the building block for Big Data:

DOI: 10.1201/9781003193838-104

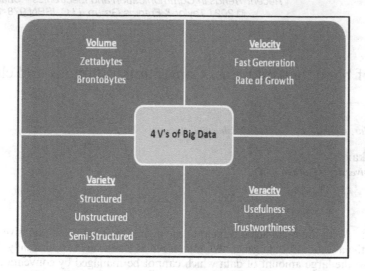

Figure 1.   The four V's of big data.

2.1   Volume

It is the huge amount of data that exists today which is the reason for the discovery of the term "Big Data". Therefore, volume of the data is the core feature of "Big Data". Today, data generation is increasing exponentially which can be quantified not in terms of terabytes but zettabytes and brontobytes. Today, the data generated every minute is the same amount which was generated between the oldest date and 2008. Hence, the available conventional means are of no use in the management of such a massive amount of data. Although, today we have low cost storage mediums and storage solutions like Hadoop which have been very helpful in the processing of this data, however this is not enough if we keep in mind the ever-growing quantity of data every second in the digital world.

2.2   Velocity

It denotes the speed at which the data is generated that needs to be analyzed. With the improvement in technology today, we can analyze the data as it is generated, we don't need to put it in a database first.This technique is often called in-memory analysis. Every minute,data is created in the form of watching movies on YouTube, sending emails, viewing photos, tweets, Facebookaccess,google queries etc. The speed of data generation is infinitely large which poses a challenge for the researchers to cope up with the enormous speed of data creation.

2.3   Variety

Variety applies to the form of the data - structured, unstructured or semi-structured. These have already been described above. We have tools for analyzing structured data. Our focus is on unstructured data which alludes to messages, conversations on social networks, sensor data, photos, and voice orvideo recordings. We need tools which can take this unstructured data and make sense of it.

2.4   Veracity

Veracity is related with the usefulness, noise and irregularity in data. The data being explored should be related to the problem being analyzed. It is therefore necessary to clean the data before using it so that useless data can be kept away.So, the accuracy of big data depends not only on the quality of the data itself but also the trustworthiness of the source from where it comes.

3 STATISTICS OF BIG DATA - WHAT THE FIGURES HAVE TO SAY?

If we take a look at the actual facts and figures related to big data, the following statistics has been recorded in terms of "internet users and population" and it is mindblowing:

- In March 2019, 4,383,810,342 internet users were recorded. This means that 56.8% of the world's population is online[7].
- In January 2019, there was 8.842 billion mobile subscriptions around the world [8].
- In January 2019, 45% of the world's total population which is around 3.484 billion were active social media users [4] and 42% of the population were mobile social media users[4].

Considering the figures in terms of "data",we have the following numbers:

- Total Data - By 2020, there will be around 40 trillion gigabytes of data (40 zettabytes) [4].
- Emails - The "Radicati Group" explored the data and found that in 2019, more than 50% of the world population uses email and by the end of 2023, this number is expected to grow to over 4.3 billion. As far as the business and consumer emails (both sent and received) is concerned, it was 293 billion in 2019, and is expected to grow to over 347 billion by the end of 2023 [3].
- Facebook Statistics –As per the latest update about Facebook statistics, it is found that Worldwide, there are over 2.38 billion MAUs for December 2019[5].

4 CHALLENGES OF BIG DATA

In recent years, there have been many applications of big data in the fields of health care, bio-chemistry, retail, and other interdisciplinary scientific researches. Not only this, big data has contributed a lot in web-based applications such as social computing, internet text and documents, and internet search indexing. Keeping these advantages of big data in mind, we can say that knowledge processing can take many benefits from big data. However, there are also many challenges which follow these opportunities. The various challenges that are being faced in the research of big data includes:

1. *Storing and analyzing big data*
   First and foremost, challenge in managing big data is to find out the means to store and analyze big data. In recent times, bulk of data is created and stored every minute. To store this data, we need to have proper storage medium with high input/output speed. Data must be accessible easily from these mediums so that knowledge discovery becomes easy for further analysis.The conventional computing methods in data mining fail to meet the expectations when it comes to the processing and analysis of huge and fast changing data in big data. Traditional computing methods like machine learning works fine with only small quantity of data.

2. *Computational Complexities*
   Not only the quantity but the quality of data also poses a challenge when we talk about analyzing the data in big data. The data types, structures and patterns in big data are so complex that the representation, understanding and computation become very difficult. This results in an increase in computational complexity. The traditional and conventional tools available for knowledge discovery and representation work well only when the amount of data is small or average but when we talk about big data, then these tools fail to

live up to our needs. The main area of concern here is to have a clear vision on the complexity of big data. The further findings on its complexity theory will help figure out vital features and organization of complicated patterns in big data, clear its representation, get better discovery of knowledge, and govern the design of computing models and algorithms on big data.

3. *Security*

The main features of security include confidentiality, authenticity, integrity and availability. First and foremost issue related to the security of big data is the intentional diminishing of the quality of data by cybercriminals. Useless data can be fabricated by cybercriminals and poured into our data lake which may harm our data analysis. Another issue exists if we have untrusted mappers. After the data has been collected,it is processed in parallel using the MapReduce paradigm in which the data is split into various batches and the mapper allocates each batch to a particular storage option.If the mapper is an untrusted one or if an outsider gets access to our mapper's code,they can change the setting by producing inadequate list of key/value pairs.

One security issue in big data occurs due to our carelessness in encrypting the big data.Sensitive data is stored on the cloud without any encryption algorithm since repeatedly encrypting and decrypting the data slows down the speed.As the main advantage of big data is high speed access, so we have to chose this advantage over the disadvantage which occurs due to storing the sensitive data as it is.

Another data security issue exists in the information about the source or origin of data often referred to as metadata.Big data itself is very large in volume and when we say that every data it contains has a detailed information about its history also,we can very well imagine how vast the size of big data will go.This metadata is also vulnerable to hackers as any unauthorized changes in metadata may lead us to wrong information and we may not be able to find out the actual cause of security breaches.

4. *Dealing with Unstructured Data*

As cited above,unstructured is the free-form information that is extracted from sources like email, notes made by a call center agent, social media posts, or Twitter conversations with customers[1]. Most of the enterprises have very little insightabout their unstructured data and its management.

First issue that exists in unstructured data is its relevance. A google search on any topic may not be required any time in future so that search is insignificant to any future user but the system analyzing big data wouldn't know that.In this way,useless data goes on accumulating which results in increase in unstructured data.As per the studiesreport,the growth rate of the volume of unstructured data is 62% annually.

Another problem is that a large volume of unstructured data is "unverified".Themostappropriate is Facebook profiles. There are many cases about Facebook users in which their Facebook updates are more fantasy than reality. People not even posts false updates but also saved their important details such as marital status and hometown wrongly. This unverified data greatly affects enterprises and businesses which tend to create their customers on the basis of their profiles such as determining insurance rates on the basis of a user's social media posts.

5. *Recruiting and retaining big data talent*

With the exponential increase in big data,there is an inherent need to have resource persons who can work on solving the various problems that are being faced in big data analytics. Today,there is an acute shortage of talents who can work on the studies related to big data and generate algorithms for the extraction and processing of information in big data.The need of current hour is to have the managers and HR professionals arrange seminars and workshops for the employees so that they can be educated in this field.They themselves need to learn how big data will be a strategic driver for competitive advantage in their organizations.

5 CONCLUSION

Big data has become the most trending topic in today's world.A lot has been explored in this regard but there is a scope for much more discovery and innovation. In this paper, we have observed the 4 V's which are the core of big data.I have also penned down some statistics in terms of users of data and amount of data that exists today.We have also seen the various challenges that are posed by big data and the general solutions in each regard.To end up, I would like to state that much study and research is required in this field to take the maximum advantage from big data analysis.

REFERENCES

Blog on "Big Data and the Challenge of Unstructured Data" at ciklum.com.

Damanik, D., Wachyuni, S., Wiweka, K., & Setiawan, A. (2019). The Influence of Social Media on the Domestic Tourist's Travel Motivation Case Study: Kota Tua Jakarta, Indonesia. Current Journal of Applied Science and Technology, 36(6), 1–14.

Ghani S. (10/06/2019)– Success Factors of Email Marketing – Topical Research.

Jason C. Young, Renee Lynch, Stanley Boakye-Achampong, Chris Jowaisas, Joel Sam, Bree Norlander (08 April, 2020) - Volunteer geographic information in the Global South: barriers to local implementation of mapping projects across Africa – GeoJournal.

McManaman, R. (2019). Strategic Audit of Facebook Through the Lens of International Reputation. Undergraduate Honors Thesis. University of Nebraska-Lincoln.

Raphael R., Raj Kumar T. (3, March 2016) —Big Data, RDBMS and HADOOP - A Comparative Study —, Volume 5 Issue.

Sara Daniela Soares Rodrigues Silva - ATÉ QUE PONTO OS COMPORTAMENTOS ETICAMENTE QUESTIONÁVEIS DOS INFLUENCIADORES DIGITAIS SÃO PERCEBIDOS PELOS SEGUIDORES?

Wen-Jang Kenny Jih & Su-Fang Lee (2004) - An Exploratory Analysis of Relationships between Cellular Phone Uses' Shopping Motivators and Lifestyle Indicators, Journal of Computer Information Systems.

A survey on state-of-the-art of cloud computing: Its challenges and solutions

Mayur Rahul
Department of Computer Application, UIET, CSJM University, Kanpur, India

Rati Shukla
GIS Cell, MNNIT Prayagraj, Allahabad, India

Shailendra Singh
Lucknow Polytechnic, Lucknow, India

Vikash Yadav & Anurag Mishra
ABES Engineering College, Ghaziabad, U.P., India

ABSTRACT: Cloud computing provides a good services to internet users on low cost basis. Going towards the concept of online services, resources are shifting towards the distributive systems. Security is the main issue of cloud computing. This paper surveys the idea of cloud computing and challenges of security issues in cloud computing and its solutions which effects the performance. We also investigate the security threats and its solutions. We survey three current methods for threat detection and evaluate its performances.

Keywords: Cloud computing, threats, security, infrastructure

1   INTRODUCTION

In the beginning of computer's era, computers require large rooms and consumes large amount of electric, slow processing and outputs. The larger computers are replaced by small and high speed computers. These small computers and its infrastructures are becomes the basis for distributed systems (Rahul et al., 2018). In today's era, an information accessing and online user increase, effective cost is also increases as well as difficult to manage. This increase in rapid Growth of online users, we are bending towards the use of new area called cloud computing (Rahul et al., 2017).

The cloud computing is based on the concept of "Pay As You Go"(PAYG) model means you have to pay services you have used. The most benefitted advantage of this model is that we able to decrease our budget by using limited resources. The people can chose according to their requirement from hard disk, memory, access control, operating systems. It provides lots of benefits to users from home and industries and also attracts researchers (Dimitrios et al, 2012, Fotiou et al., 2015).

Cloud computing (CC) is based on virtualization techniques to give efficiently resources to customers. The properties of cloud comprise of scalability, availability and manageability, expedient, on-demand services, elasticity, ubiquitous and stability. It provides only 3 service delivery models: Platform as a Service (PaaS), Infrastructure as a Service (IaaS) and Software as a Service (SaaS) (Hashizume et al., 2013). The four-development model defined by NIST: Private, Public, Community and Hybrid. CC is based on client server where client acts as front end and server as back end

DOI: 10.1201/9781003193838-105

Figure 1.   Services of Cloud Computing.

and service lies on the middle given as in Figure 1. The top level has applications which provide services to the users and users don't need to spend money on software but they have to pay according to their usage (Ku et al., 2013).

NIST provides necessary standards and guidelines to the security of CC. The architecture of cloud computing is describe as follows: (1) Basics of cloud computing (2) Cloud based models (3) Cloud delivery service models (4) Cloud security fundamentals. The cloud architecture is explained in details:

1.1   *Basics of cloud computing*

In the given section, we explain some basic concepts based on cloud computing. They are having very wide range of services used in internet. Some important concepts discussed are as follows:

- Multi-tenancy: Multi-tenant are those peoples who do not share their information but share their resources according to requirements in cloud environment. This will results in the utilization of resources like data storage and hardware (Tan et al., 2011).
- Cloud storage: It is used to remotely managed, backup and maintained, it can available where user access data (Laura et al., 2011).

1.2   *Cloud based models*

Cloud Computing is based on resources used by various servers. It is used to get maximum consistency with the help of resource sharing. It is also tell us about nature and purpose of the cloud. It minimizes the servers load, expenditure and cost (Laura et al., 2011). NIST defines two type of model:

- Private cloud: Managing and operating of information centre with in the same organisation is known as private cloud. Many clients of same infrastructure have been given the exclusive permission to use same organisation.
- Public cloud: It is applicable where customer and its vendor have agreement to maintain the trust. It is basically the good representation of cloud computing. Both public and organisation have strong permission in public cloud. Government organisation, businesses and academics have open access public cloud. Difficult to detect who is the owner and location of the resources, increasing the problem of identifying threats (Modi et al., 2013).

1.3   *Cloud delivery service models*

The development of big data and internet technology raise a new concept of technology. This new technology favours the growing online activities and interconnects them. The recent research shows that Internet of Things (IoT) favours the development and capabilities of cloud (Modi et al., 2013). The three delivery models are known so far: PaaS, IaaS, SaaS. More models are available according to their capabilities and services. Anything-as-a-Service (AaaS), is another model, was created to increase the services and functionality.

- Infrastructure as a Service (IaaS): It deals with hardware like memory, storage, and processor as a service. It is basically a new concept of investment in infrastructure. It is also used to provide provisions and scalability of infrastructure without the help of big time and funds. It focuses on security like intrusion detection, prevention, firewall and monitoring of virtual machine (Flavio et al., 2011).

- Platform as a Service: PaaS acts as a middle layer of service framework. Its services are to deliver services like framework, programs, architecture and development tools. The Google search engine is the perfect example of PaaS. It gives an environment which supports many programming languages such that Go programming, Java, Python etc. PaaS is more flexible than SaaS.
- Software as a Service: SaaS is a service provider of computing working remotely. It position is in top of model. It works remotely to give permission to third party clients. It provides the permission to clients to work on cloud infrastructure using internet. GoogleApp is the SaaS providers which work on the concept of remote computing services.
- Anything as a Service: AaaS is concept of combining various things as service or anything as service. This services are interchangeable among each other in cloud computing. The cloud which are easy to support systems such as Security as a Service, Data as a Service and Communication as a Service.

4.1 *Cloud security fundamentals*

Protecting from threat is the most challenging task in the distributed systems nowadays. It is very difficult task in client-server also. It is important because lots of data transfer takes place in cloud computing environment. The Zhaolong et al. focuses on the unauthorised access of app in any organisation. It gives theory about the types of data used in these apps, its risks and resolves the problem arise in these apps. In this section, we discuss various security problems in cloud computing environment.

- Infrastructure security: Basic challenge is to develop physical and virtual infrastructure which protects it from the various malicious attacks. The trust on third party may not good for proper functioning of business. It is very important for organisation to show its secureness of the underlying infrastructure.
- Software security: The idea of security in software is to protect system from malicious activities comes from software development process. The most critical problem in building the efficient cloud computing environment is software security. It detects the various problems like bugs, overflow and error handling.

Our study is different from other existing methods. We concentrate on the broad description of the existing methods. The aim of this survey is to provide benchmarks for the new comers and researchers of this field. The deep analysis of various existing papers is explained along with challenges and solutions. This paper also address some future aspects of the cloud computing.

The remaining paper is organised as follows: Section 2 summarizes the issues and challenges in cloud security. Section 3 explains the basic idea and analysis of attacks on cloud system. Suggestion and discussions of security in cloud computing is explained in section 4. Finally, concluded in section 5.

2 ISSUES AND CHALLENGES IN CLOUD SECURITY

The data used for communication have some problem due to cryptographic techniques. Problem also occurs in client side due to CRM. Clustering also affects the performance of the cloud computing environment. Some problems in virtual and physical structure are also responsible for the threat attacks in cloud network.

2.1 *Issues in embedded systems*

Security is the main problem in CC nowadays. So many challenges are present which affect the potential of the embedded systems. The improvements of these systems are done using efficient tools with the embedded systems. The debugging in these systems is done by connecting it with local network. Virtualization is the main issue which affects the security of embedded systems. Different problems in embedded systems are explained as follows:

- Programmability: Routers uses the additional functionality for programmable packets in each port in CC environments. The main challenge of developing software in CC is to develop software with packet monitoring. The software gives high performance rate by applying low abstraction.
- Isolation in Virtual Machine: The isolation is primary advantage of virtualization. It should be employed efficiently to protect from the threats. The sharing of workload among virtual machine is one of the major issues in cloud system. It can create the problem of cross-VM attacks and data leakage. Isolation should be incorporated carefully in adding virtual machine in cloud environment.

2.2 *Issues in software*

Security in software is very critical part in cloud computing. The software has platform, front end, back end, frameworks. Each has its own vulnerability. The software in general consists of millions line of codes. Different programmers have different logic, languages with some vulnerability. In this section, different security issues have been explained as follows:

- Front end: The security risk in front end is very high due to its layered structure. The unauthorised access to the software is at very high probability. The software developer must know the security features of all programming languages like CSS/PHP/HTML. The isolation can be easily breakup by the hacker to get the information. If a person hack the database, then there must be a front end to represent problem (Modi et al., 2011).
- Back end: The back end of the system mainly focus on the three things: hardening, development, testing. These three have so many security problems needs to be monitor.

2.3 *Issues in web*

The security in web application is also a major concern in cloud computing. It will produce some flaws in website. Attacker can also defects the security system and perform some abnormal activities. Web based application security system is also similar to the internet based application system and creates some problem like spoofing, port scanning and some malicious activities.

- Web technology: Integrity is the biggest issue in web application. The attackers perform eavesdropping and injecting in the web server for unauthorised access to the web server. Increase in the people on the internet, hacker tries to break the integrity of the web application.
- Servers: DoS is the attack which attacks the slave computers on the internet. Proxy attacks are depends on the server. Proxy server attacks the server and perform unauthorised access to access the sites which are banned from the server.

3 SECURITY IN CLOUD COMPUTING

Researchers have proposed the various techniques to evaluate the effects of cloud computing security. The Fernandes et al. gives survey on different topics of security and its effects and solutions. They addressed several keys topics related to security like vulnerability, threats and taxonomy. The Ali et al. discussed the comprehensive review in the field of cloud computing and also highlighted some issues like architectural, communication, legal and contractual aspects. Saripalli et al. discussed the concepts of trust and privacy in the cloud computing. They also able to analyse threats based on trust and privacy.

There are so many security risks are generated from the cloud computing paradigm. To overcome the security threats, we have to identify the threats with respect to security attributes. It basically means which threats effects which attributes and what are the suggestions of researchers to challenges of attacks. CSA reports that nine common threats are found due to sharing of common resources in 2013. In next two years, threats were increases tremendously like DDoS, hijacking, data loss etc.

4 SUGGESTIONS AND DISCUSSION

In this section, we put forward 5-tier framework in Figure 2, where safety is independent to each other. The introduced security level comprise of 5 levels: backend level, infrastructure level, platform level, software level, frontend level. IT businesses which are keen to adopt security need to analyse and change software. The organisations should focus on important points which are used to enhance the security and its capabilities.

Figure 2. Proposed 5-tier architecture of security.

The organisation used cloud computing environment to save money, increase security and efficiency, so proper security should be adopted to achieve higher level of security.

In public cloud system, organisation gives and gets the information with integrity and confidentiality. The risk of the threats is classified according to what they affect. Confidentiality and integrity threats like eavesdropping, attack on information etc. makes the organisation less efficient.

- Backend level: Storage security focuses mainly on the cryptography, data leakage, snooping and malware. All the data of end user has been kept in cloud and no control over the storage of data and its location. It is always responsible for the standard of the service. So, security at this level protects the data from outside world.
- Infrastructure level: The most fundamental challenges is to develop the physical and virtual infrastructure which protects it from the various malicious attacks. The trust on third party may not good for proper functioning of business. This level of security prevents the physical and virtual structure of the organisation.

Multi-tenancy is used to maximize the use of the resources and is very essential part. The challenges so far in multi-tenant is to maintain privacy, trust in cloud computing environment. So, some solution should be developing to maintain the said issues.

5 CONCLUSION

The progress in cloud computing is going to benefit many private and government organisation. Despite of having many advantages, some challenges like vulnerability still makes this field difficult. Due to this reason, security is the major challenge to apply cloud computing in the organisation. This paper gives the state-of-the-art survey on issues and challenges in cloud computing. We also surveyed vulnerabilities from characteristics of CC like resource pooling and virtualisation.

REFERENCES

Dimitrios, Z. & Dimitrios, L. 2012. Addressing cloud computing security issues. *Future Gener. Comput. Syst.* 28(3), 583–592.

Flavio, L. & Pietro, R. 2011. Secure virtualization for cloud computing. *J. Netw. Comput. Appl.* 34(4), 1113–1122.

Fotiou, N. Machas, A. Polyzos, G. C. & Xylomenos, G. 2015. Access control as a service for the Cloud. *J. Internet Serv. Appl.*

Hashizume, K. Rosado, D.G. Fernández, M. Fernandez, E.B. 2013. Analysis of security issues for cloud computing. *J. Internet Serv. Appl.* 4(1), 1–13.

Ku, C.Y. & Chiu, Y.S. 2013. A Novel Infrastructure for Data Sanitization in Cloud Computing. In: Diversity, Technology, and Innovation for Operational Competitiveness: *Proceedings of the 2013 International Conference on Technology Innovation and Industrial Management*. 25–28.

Laura, S. 2011. Cloud computing: deployment models, delivery models, risks and research challenges. *In: Proceedings of International Conference on Computer and Management (CAMAN)*.

Modi, C. Patel, D. Borisaniya, B. Patel & Avi, R. 2013. A survey on security issues and solutions at different layers of cloud computing. *J. Super comput*. 63(2), 561–592.

Rahul, M. Kohli, N. & Agarwal, N. 2017. Facial Expression Recognition using Multi-Stage Hidden Markov Model. *Journal of Theoretical and Applied Information Technology, 95 (23)*.

Rahul, M. Kohli, N. & Agarwal, N. 2018. Partition Based Technique for Facial Expression Recognition using Multi-Stage Hidden Markov Model. *Journal of Engineering and Applied Sciences*. 13(9).

Tan, X. & Ai, B. 2011. The issues of cloud computing security in high-speed railway. *In: Proceedings of IEEE International Conference on Electronic and Mechanical Engineering and Information Technology (EMEIT)*. 8. 4358–4363.

Recent Trends in Communication and Electronics – Sharma et al. (Eds)
© 2021 Taylor & Francis Group, LLC, ISBN 978-1-032-04572-6

Simulation and analysis of various structures of organic TFTs in 2D and 3D geometrical plane

S.K. Tyagi

Department of Electronics and Communication Engineering, KIET Group of Institutions, Ghaziabad, India

Poornima Mittal

Electronics and Communication Engineering Department, DTU, New Delhi

ABSTRACT: In this paper, various 2D and 3D geometrical structures have been examined with a variety of comparison between various designs of Organic TFTs like Single Gate (SG) design, Dual Gate (DG) design, Vertical Gate (VG) design and Cylindrical Gate (CG) design. The Dual Gate (DG) design of Organic TFTs performs better than Single Gate (SG) design mainly in terms of mobility (μ), I_{on}/I_{off} current ratio and sub-threshold slope. The simulation has been carried out using silvaco software for simulation and analysis of various structures of organic TFTs in 2D and 3D geometrical plane. In this paper, simulation and analysis of the Organic TFTs in terms of cylindrical geometry has also carried out to overcome the short channel effects in 2D Organic TFTs designs. We have achieved the 3.8 μA drain current at $V_{GS} = V_{DS} = 3V$ and channel length of 50nm for Cylindrical Gate (CG) design.

1 INTRODUCTION

"Wearable devices" has become the latest application of electronics over the last few years and this application signifies the first step of the next electronic devices phase. "Wearable" may refer to devices and systems that can be small and compact. Because of their small size, it can be mounted on a textile substrate. Organic transistors are physically highly rugged and versatile. The organic electronics have very significant benefits over its rival silicon, first is that it is possible to chemically customise several semiconductors so that new applications can be found and explored more and more. Second, it can be manufactured and processed at low temperatures, make it possible to find the applications for the organic LED display and sensors in designing the radio frequency identification (RFID) tags.

Organic TFTs has an active layer consists of organic semiconductor. OTFTs has three electrodes terminals called source (S), drain (D) and gate (G) and dielectric material. Here Gate terminal acts as channel conductivity controller. The Organic TFTs are identical to MOSFET structure wise, but in terms of device physics and channel forming process, both are distinct. The proposed devices operate in an accumulation mode. The general 2D models considered to be planar OTFTs will be adapted to the modern 3D models having cylindrical geometry.

2 DIFFERENT OTFT STRUCTURES IN DIFFERENT GEOMETRICAL PLANES

Generally, OTFTs are distinguished on the basis of their geometrical planes. First is 2D OTFTs and another one is 3D OTFT. The 2D Organic TFTs are again categorized into Single Gate (SG) design, Dual Gate (DG) design, Vertical Gate (VG) design. Similarly the 3D OTFTs are designed with cylindrical geometry illustrated as in Figure 1 below.

DOI: 10.1201/9781003193838-106

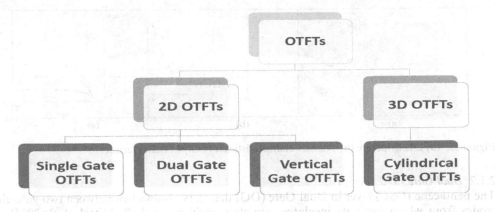

Figure 1.   Different OTFT structures.

2.1   *Two - dimensional organic TFTs*

2.1.1   *Single Gate (SG) design*

A Single Gate (SG) design consists of a thin OSC film, typically made with only one Gate (G) terminal at the bottom as the inverted structure. Depending upon the location of Gate (G) terminal, the design of Single Gate (SG) is of two types as Top Gate (TG) design and Bottom Gate (BG) design. It is also possible to categorise Single Gate (SG) design as Top Contact (TC) design and Bottom Contact (BC) design as per the location of the source (S) and drain (D) terminals in reference to the OSC film layer. So there are total four categories of Single Gate Organic Thin Film Transistor named as Top Contact Top Gate (TCTG) design, Top Contact Bottom Gate (TCBG) design, Bottom Contact Top Gate (BCTG) design, Bottom Contact Bottom Gate (BCBG) design. The Organic TFTs designs with Top Contact (TC) gives best results as compared to Bottom Contact (BC) design in terms of current [B. Kumar et al. 2012].

Here, the study has been conducted to examine the p-type device characteristics and the output parameters of BGTC Structure. For simulating the device, the key parameter values used in the study are 100μm, 800μm, 100 nm, 10μm, 10μm and 150 nm, respectively, for device channel width (W), device channel length (L), dielectric thickness (tox), source terminal and drain terminal length, and pentacene (OSC) thickness. Gate electrodes are known to be made of 50 nm thick gold. The parameter values of pentacene (OSC) used for simulation are 2.2eV E_d, 2.8eV E_a, ESD (Valence Band) of 2.0 x 10^{21} per cm^3, ESD (Conduction Band) of 1.7 x 10^{21} per cm^3 and ε (permittivity) of 4.0.

The threshold voltage extracted is -8.06186 V. Further Transconductance and Ion/Ioff are extracted as $2.23761e^{-07}$ A/V and $6.68458e^{+06}$ respectively.

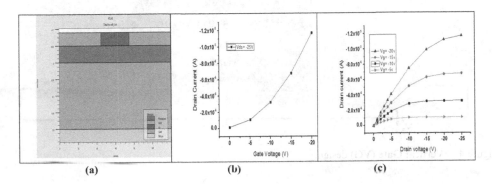

Figure 2.   (a) BGTC Structure (b) Transfer characteristics (c) Output characteristics.

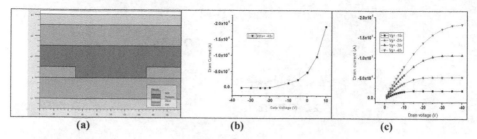

| (a) | (b) | (c) |

Figure 3.   (a) DG Structure (b) Transfer characteristics (c) Output characteristics.

2.1.2 *Dual Gate (DG) design*

The pentacene (OSC) layer in Dual Gate (DG) design is sandwiched between two gate electrodes from which two gate insulators are electrically isolated [P. Mittal et al. 2015]. By taking same parameters mentioned above for simulation, the threshold voltage 1 and threshold voltage 2 extracted are -16.0279V and -2.27348V respectively. Further Transconductance and I_{on}/I_{off} are extracted as $7.598e^{-07}$ a/v and $2.23179e^{+13}$ respectively.

2.1.3 *Vertical Gate (VG) design*

An Organic TFTs Vertical Gate (VG) design was manufactured to overcome the limitations of Top Contact (TC) design. The efficiency of the Single Gate (SG) design and Dual Gate (DG) design is limited by Thin Film morphological disorder, modified device channel length and low mobility charge carriers. A Vertical Gate (VG) design OTFT has five separate layers consisting of S, D, and G metallic layers, coinciding with two organic active semiconducting layers.

2.2   *Three dimensional organic TFTs*

The advancement in modern electronics has led to an innovative design of devices in 3D geometrical plane for various applications. Cylindrical 3D geometry has been used for e-textile applications with long yarn like designs. Cylindrical geometry is also intended for reducing short channel effects with high pressing thickness [S. Locci et al 2007]. In view of the geometry wise assumptions, the radius of gate electrode is (r_{GATE}), thickness and radius of dielectric layer ($t_{DIELECTRIC}$) and radius ($r_{DIELECTRIC}$) respectively. The thickness and radius of OSC layer is t_{OSC} and r_{OSC} respectively. Two external rings have designed for the source and drain electrodes and the gap between them is denoted by $L_{CHANNEL}$. After simulating the device, the threshold voltage and I_{on}/I_{off} extracted are - 0.031V and 2.23 x 10 ^ 6 respectively.

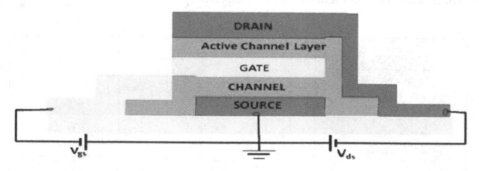

Figure 4.   Vertical Gate (VG) design.

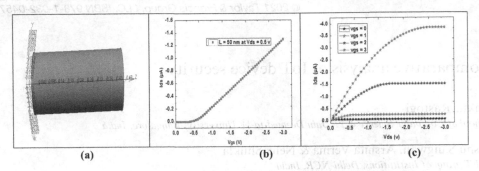

Figure 5. (a) Cylindrical 3D Structure (b) Transfer characteristics (c) Output characteristics.

3 CONCLUSION

In this paper, different 2D and 3D geometrical structures have examined with a variety of comparison between various designs of Organic TFTs. Among 2D geometrical structures, the Organic TFTs designs with Top Contact (TC) gives best results as compared to Bottom Contact (BC) design in terms of current. The threshold voltage extracted is -8.06186 V. Further Transconductance and I_{on}/I_{off} are extracted as $2.23761e^{-07}$ A/V and $6.68458e^{+06}$ respectively. For Cylindrical 3D geometry, the threshold voltage and I_{on}/I_{off} extracted are - 0.031V and 2.23 x 10 ^ 6 respectively. Discrepancies among the 2D and 3D geometrical designs are very crucial since the gate radius is diminished. The dielectric thickness may also have a major function in calculating the width and threshold voltage of the depletion region, which may vary the output characteristic curves to some extent.

REFERENCES

Brijesh Kumar et al., 2012, "Channel length variation effect on performance parameters of organic field effect transistors", ELSEVIER Microelectronic Engineering others.

Cheng-Jung Lee et al., 2019,"Biodegradable Materials for Organic Field-effect Transistors on a Paper Substrate", IEEE Electron Device Letters.

Hong Wang et al., 2015, « »Contact Length Scaling in Staggered Organic Thin-Film Transistors", IEEE Electron Device Letters.

Jun-Young Park et al., 2019,"Curing of Hot-Carrier Induced Damage by Gate-Induced Drain Leakage Current in Gate-All-Around FETs", IEEE Electron Device Letters.

Poornima Mittal et al., 2015," An analytical approach for parameter extraction in linear and saturation regions of top and bottom contact organic transistors", Springer Science & Business Media New York.

Poornima Mittal et al., 2015," Mapping of performance limiting issues to analyze top and bottom contact organic thin film transistors", Springer Science & Business Media New York.

Poornima Mittal et al., 2016," A depth analysis for different structures of organic thin film transistors: Modelling of performance limiting issues", ELSEVIER Microelectronic Engineering others.

Rajesh Agarwal et al., 2018," Floating Drain-Based Measurement of ON-State Voltage of an OTFT for Sensing Applications", IEEE Transactions on Electronic Devices.

Simone Locci et al., 2007," An Analytical Model for Cylindrical Thin-Film Transistors", IEEE Transactions on Electronic Devices.

Subhash Singh et al., 2020," Flexible PMOS Inverter and NOR Gate Using Inkjet-Printed Dual-Gate Organic Thin Film Transistors", IEEE Electron Device Letters.

Recent Trends in Communication and Electronics – Sharma et al. (Eds)
© 2021 Taylor & Francis Group, LLC, ISBN 978-1-032-04572-6

Comparative analysis of IoT device security

Ankur Rastogi
School of Computer Science & IT, Jain Deemed-to-be University, Bangalore, India

Vishu Sangwan, Arshita Verma & Neha Shukla
KIET group of Institutions, Delhi-NCR, India

ABSTRACT: Internet of things is one of the latest technology in today's era, IoT devices are increasing day by day and become a part of our life. Sometimes we even didn't know that we are using and surrounded by such technolgy. Smart devices like smart phone, tablets and any wearables are the example of IoT technology. These all are IoT devices and there are many more. As we all are in 2020 and facing the pandemic we all are become more connected with IoT devices to be safe. In this way the number of IoT devices per person will increase so much rapidly and several new innovations either of hardware or software take place. But the Internet of Things (IoT) devices are becoming more popular, the security measures are not upto the mark, and several issues will come in knowledge at different time on different devices. So we have to think about the security and privacy of user and should make it correct. We should learn from the past issues and try to fix them accordingly. To overcome the issues of IoT devices we can have to ways, one is to use ECC which will enchance the security in IoT device and another possible solution can be Blockchain. In this paper we will discuss about ECC and blockchain technology both and try to find a comparitive study.

Keywords: IoT, Blockchain, ECC, Security

1 INTRODUCTION

The Internet of things is changing nearly everything thing in our environmental factors. From the manner in which we shop, the manner in which we get power for our homes, or the manner in which we speak with one another. It has gotten one of the fundamental pieces of our every day life. Little chips and sensor are implanted in physical gadgets which communicate important data. This data gives us a superior comprehension of how these gadgets work and how they are getting basic for our everyday life. IoT gadgets gather the information from people as well as include information offering to outsiders (for example voice acknowledgment, unique mark) while getting to various gadgets or playing computer games. Such sort of information can make a protection issue as the client is ignorant of the presence of any purposeful and malignant assault/assailant on the grounds that the spillage or robbery of the data doesn't have any immediate impact over the current application where the client is locked in. Then again, the client is additionally uninformed of the way that where by whom, and in what way that data can be used. The essential security challenge of IoT is that it expands the quantity of gadgets behind an organization firewall. The current cryptographic security calculations, actualizing them on a low fueled machine and dissects the general effect it makes on the gadget. It additionally shows the utilization of cryptographic calculations and calculations which utilize mysterious key understanding conventions that permits two gatherings, each

DOI: 10.1201/9781003193838-107

having an elliptic bend public–private key pair, to build up a common mystery over a shaky channel. Blockchains decentralized cryptographic model permits clients to confide in one another and make distributed exchanges, dispensing with the need of go-betweens. This innovation isn't just influencing the manner in which we use web, yet the worldwide economy is additionally being upset.

2 MAJOR CHALLENGES IN IOT SECURITY

All Data can be secured by applying cryptographic algorithms. These algorithms can be classified into two parts:

1) Symmetric Key Algorithms
2) Asymmetric Key Algorithms

Both the types will contain several algorithms and we will consider DES, AES, 3DES, RSA and creating a comparitive analysis of these algorithms on the basis of key length, block size, no of rounds, power consumption, throughput and security risk in Figure 1.

After seeing this comparative analysis we come to know that these algorithms are good in some perspectives but they are not working well in IoT devices, so need some other more efficient and lightweight algorithms. The examination of the IoT gadgets and their IoT put together security with respect to the framework level is acted in 3 angles that are sensors, actuators, and streamlining. A significant number of the IoT gadgets whether wired or remote are a lot of open to assaults and there makers can't give the related security patches on normal premise. The ECC (elliptic bend cryptography) is the new essence of encryption in the realm of information security. ECC creates the encoded code based on an elliptic bend condition that produces codes based on bend, that too by utilizing a 164 digit framework however the aggressor or the breacher will require 1024 cycle framework to break. ECC at each layer so the security increments at each layer the standard conventions each time related with another encryption unscrambling code will no uncertainty make it hard for each aggressor to jump into the framework or the gadget. There are works with figure calculations where they present an open structure of lightweight square codes on a huge number of implanted stages. individuals are going for crossover philosophy, they have tended to some lightweight codes, contrasted them and came up and another calculation. All the gadgets must convey and coordinate with one another in a protected manner. As the investigation of data for web of things is significant. It is critical to secure the

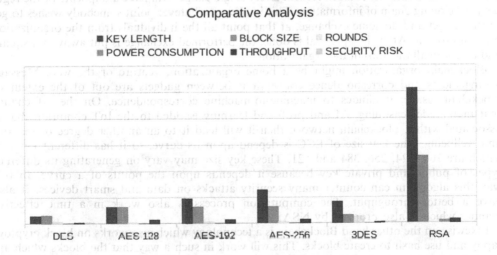

Figure 1. The comparison chart.

information through its whole lifecycle. Square chain utilizes the decentralized technique to make sure about the information instead of having a devoted framework. It is an information base which keeps up the persistent records of informational index which are continually developing. Blockchain doesn't have any ace framework which holds the whole chain, it resembles a circulated record. Each time another information record is included as it is continually developing and every hub that is partaking in the chain has a duplicate of the chain. The decentralized methodology of blockchain kills single point disappointment and making a more adaptable condition for the gadgets to run. Blockchain utilizes cryptographic calculations which make the information safer.

3 PROPOSED WORK

ECC the recently developing method for encoding the information dependent on some awe-inspiring calculations is the procedure which can engage IoT security whenever utilized proficiently. Securing the remote gadgets on this IoT network is an alternate errand overall, since these gadgets which get associated with the organization assumes a significant function to make an organization. These gadgets are the most widely recognized individuals over an organization and are a lot of inclined to dangers and weaknesses since it is the simplest objective for the dark caps. ECC can improve the security of such gadgets and shield the information on or over the organization. ECC is emergin gas extraordinary compared to other security rehearses for encryption over an organization anyway any sort of interference between the exchange while utilizing ECC is a serious troublesome undertaking to be finished. The encryption technique for ECC is significantly more secure and solid to discovery and adding this to the entire IoT organization will grasp the degree of IoT security over all the gadgets, etc remote gadgets since they are the most weak dangers to the organization and are generally inclined to dangers. In this manner, the general investigated result says that the proposed thought will expand the degree of IoT security over the entire organization while on a similar note will grasp the utilization of remote gadget in or on IoT making them more dependable and less defenseless.

Of course blockchain centers are the people from the net-work, sharing successfully in the trade cycle. They endorse the trade through mining; they can be PCs, undertaking laborers, or also cloud-based center points. Clients are the IoT contraptions; they don't store the spread record. Blockchain client's square chain center points speak with each other through APIs. IoT devices make trades and these trades are given off to blockchain center points for taking care of the data into the coursed record. Every single gadget will have a duplicate of the regularly developing chain of information. This implies at whatever point somebody wishes to get to the gadget and do some exchange, at that point all the individuals from the organization must approve it. After the approval is done, the performed exchange is put away in a square and is sent to all the hubs of the organization.

Blockchain organization might be a home organization, venture or the web. Message arrangements and correspondence conventions between gadgets are out of the extent of blockchain usage: it alludes to machine-to-machine correspondence. On the off chance that man-made reasoning, AI and profound learning isadded to the IoT condition that is associated with a blockchain network then it will lead it to an another degree of security and wellbeing. The key size of ECC is depend upon its curve, so it has different key size which are 160, 224, 256, 384 and 521. These key size may vary on generating its different types of public and private key because it depends upon the points of a curve. In this way this algorithm can counter many security attacks on data and smart devices. It also have a better throughput. The computation process is also work in a time effective manner which is also prooved by NSA.

Likewise on the other hand Blockchain is a technology which also works on block cryptography and use hash to create blocks. This will work in such a way that the blocks which are encrypted using the hash are irreversible and uneditable, while a block is created and send in a network then it will be validated by number of nodes called miners which make it more safe and secure. So if any attacker get any block in any network then it is not useful for attacker and miner will validate the blocks then it will add to the chain.

4 CONCLUSION

In this paper we try to find an analysis regarding the ECC and Blockchain to get which work better with IoT but get that ECC works on a mathematical concept of curve, in which we get number of points and on those point we get number of keys, which make it more secure under several attacks. On the other hand Blockchain is a technology which works in decentralized and distributed public environment, where number of transactions will take place. In blockchain miners will authenticates the blocks and then it will add to the chain. So in this way both cryptographic algorithms plays a vital role in the security of IoT. So it depends on different devices and scenario that which process will work on it.

5 FUTURE SCOPE

As we know that today's time is all about new technology, this new technology will lead us to some area like AI, machine learning and deep learning. These areas will affect our daily life concepts and technology so how IoT will remain untouch from it. So the future of IoT will also lead us to some generic and more secure real life algorithms to prevent our data from risk and attacks. Several new algorithms are going to design on the basis of AI, Machine learning and deep learning to fullfill the current time requirements and to achieve a better security for our data.

REFERENCES

Ankur Rastogi, Saurabh Singhal and Amit Garg, "Analysis and Evaluation of Current Security Trends in IoT" IPEM Journal of Computer Application and Research, 2018, India, pp. 38–47, ISSN No. 2581 – 5571.

Angin P., Mert M.B., Mete O., Ramazanli A., Sarica K., Gungoren B. (2018) A Blockchain-Based Decen tralized Security Architecture for IoT. In: Georgakopoulos D., Zhang LJ. (eds) Internet of Things – ICIOT 2018. ICIOT 2018. Lecture Notes in Computer Science, vol 10972. Springer.

A. Shantha, J. Renita and E. N. Edna, "Analysis and Implementation of ECC Algorithm in Lightweight Device," 2019 International Conference on Communication and Signal Processing (ICCSP), Chennai, India, 2019, pp. 0305–0309.

E. Gyamfi, J. A. Ansere and L. Xu, "ECC Based Lightweight Cybersecurity Solution For IoT Networks Utilising Multi-Access Mobile Edge Computing," 2019 Fourth International Conference on Fog and Mobile Edge Computing (FMEC), Rome, 2019, pp. 149–154.

E. H. Teguig and Y. Touati, "Security in Wireless Sensor Network and IoT: An Elliptic Curves Crypto system based Approach," 2018 9th IEEE Annual Ubiquitous Computing, Electronics & Mobile Com munication Conference (UEMCON), New York City, NY, USA, 2018, pp. 526–530.

F. Meneghello, M. Calore, D. Zucchetto, M. Polese and A. Zanella, "IoT: Internet of Threats? A Survey of Practical Security Vulnerabilities in Real IoT Devices," in IEEE Internet of Things Journal, 2019, vol. 6, no. 5, pp. 8182–8201.

Fan S., Song L., Sang C. (2019) Research on Privacy Protection in IoT System Based on Blockchain. In: Qiu M. (eds) Smart Blockchain. SmartBlock 2019. Lecture Notes in Computer Science, vol 11911. Springer.

J. M. Carracedo, "Cryptography for Security in IoT," 2018 Fifth International Conference on Internet of Things: Systems, Management and Security, 2018, pp. 23–30.

Khalid, U., Asim, M., Baker, T. *et al*. A decentralized lightweight blockchain-based authentication me chanism for IoT systems. *Cluster Comput* **23**, 2067–2087 (2020).

Kudithi, T., Sakthivel, R. High-performance ECC processor architecture design for IoT security applica tions. *J Supercomput* **75**, 447–474 (2019).

M. Singh, A. Singh and S. Kim, "Blockchain: A game changer for securing IoT data," 2018 IEEE 4th World Forum on Internet of Things (WF-IoT), Singapore, 2018, pp. 51–55.

R. Gauniyal and S. Jain, "IoT Security in Wireless Devices," 2019 3rd International conference on Electronics, Communication and Aerospace Technology (ICECA), Coimbatore, 2019, pp. 98–102.

S. Garg, K. Kaur, G. Kaddoum, S. H. Ahmed, F. Gagnon and M. Guizani, "ECC-based Secure and Lightweight Authentication Protocol for Mobile Environment," IEEE INFOCOM 2019 - IEEE Conference on Computer Communications Workshops (INFOCOM WKSHPS), Paris, France, 2019, pp. 1–6.

V. Hassija, V. Chamola, V. Saxena, D. Jain, P. Goyal and B. Sikdar, "A Survey on IoT Security: Application Areas, Security Threats, and Solution Architectures," in IEEE Access, 2019 vol. 7, pp. 82721–82743.

Recent Trends in Communication and Electronics – Sharma et al. (Eds)
© 2021 Taylor & Francis Group, LLC, ISBN 978-1-032-04572-6

MOSFET based comparator with static reduction logic for SAR ADC in biomedical applications and its comparison with conventional comparator

Mohit Tyagi
KIET Group of Institutions, Delhi-NCR, India

Poornima Mittal
Delhi Technological University, New Delhi

ABSTRACT: In this paper we have designed and compared MOSFET based conventional comparator (without static power reduction logic) and controlled comparators (with static reduction logic) for Low Power SAR ADC at 45nm technology. The peculiar advantage obtained after simulating controlled comparator with conventional comparator is power consumption gets reduced by forty four times with almost same propagation delay as of conventional comparator. Both comparators are simulated at 0.6V in 45nm CADENCE Virtuoso technology node. Dynamic power dissipation has come out to be 1.6u and 20u for conventional and controlled comparator respectively. Simulated controlled comparator is suitable for comparison of two signals of the order of 50 MHz frequency range. Minimum voltage difference between two signals to be compared is 0.3 volt for conventional comparator and its value is 0.2 volt for controlled comparator.

Keywords: SAR ADC, conventional comparator, controlled comparator, SNR

1 INTRODUCTION

ADC needs to be optimized for ultra-low power applications. A SAR ADC comprises of an assortment of vital blocks; such as, comparator, ring counter, DAC and controlling unit. Though, the power optimization of each individual module is imperative, nevertheless, for now conventional MOSFET based comparator and comparator with static power reduction using controlled comparators are designed and compared to achieve an efficient one for ultra-low power SAR ADC.

(Goll B. et al. 2000) has worked on a comparator with reduced delay time in 65nm CMOS for supply voltage reduced to 0.65V. Further, (Cheong Jia Hao et al. 2011) designed a SAR ADC based on trilevel switching scheme and time domain comparison for biomedical application. Further, (Harpe Pieter et al. 2013) worked on segmented charge redistribution DAC for sensor applications. (Ahmadi et al. 2015) has developed an accurate and simple model to accomplish the noise analysis of comparator so as to model the noise effect which may deteriorate the performance of SAR ADC. Further, (Wang et al. 2018) has developed a dynamic proximity comparator for bypass switching SAR ADC.

DOI: 10.1201/9781003193838-108

2 ARCHITECTURAL VARIETY FOR COMPARATORS

2.1 *Conventional comparator*

In architecture design of conventional comparator, a cross coupled CMOS is used to create pre-charge nodes; outn and outp, as depicted in Figure 1. The transistors; M3 and M4 switches to 'ON' state on a low clock signal and thus outn and outp attains the levels equal to V_{DD}. The pulsating clock signal establishes a timing reference in the circuit.The inputs to be compared is applied to the transistors M7 and M8, whereas; M1, M2, M5 and M6 form a cross coupled CMOS structure. The transistor M9 is utilized to create the path between given inputs and the ground. On the contrary, M3 and M4 become 'OFF' while clock switches to a high logic.

Now, if INA>INB, then the outn node will discharge at a faster rate as compared to outp. Owing to this, the outn node turns M2 in 'ON' condition and consequently, M1 switches in 'OFF' condition. Thus, the outn node attains a logic low state while outp node charges to VDD level.

2.2 *Controlled comparator*

Controlled comparator is intended with latch regeneration capabilities with the help of two cross coupled control transistors MP1 & MP2 placed in parallel to M5 & M6, as depicted in Figure 3. It also employs a cross coupled CMOS latch formed by M7, M8, M9 and M10 with outn and outp as output nodes. Further compared to conventional comparator, this design utilizes two tail transistors; Mtail1 and Mtail2 to facilitate the static power reduction capabilities. Inclusion of M3 and M4 transistors provide a significant improvement in the speed than that of conventional one. In this design, the clk is applied to Mtail1, M5 and M6 transistors, while Mtail2 operates according to the switching condition of nclk voltage.

On the clock at low logic, nodes F_p and F_n are charged to V_{DD} which in turn switches the MT1 and MT2 to ON state and resultantly nodes; outn and outp remains in logic-low under this situation. Further, switching clk to logic-high, leads to the transistors Mtail1, Mtail2, M8 and M10 in ON state owing to the pre-charging of outn and outp nodes to V_{DD}.

3 SIMULATION RESULTS

3.1 *Static and dynamic power of conventional comparator*

The output static power consumption of conventional comparator for full operating voltage range of 0 to 1V at 45nm node is depicted in Figure 2 (a). The maximum static power consumption is

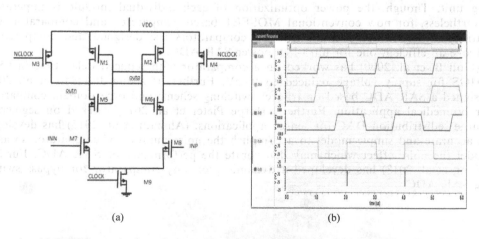

(a)   (b)

Figure 1.   (a) Cell Schematic view of conventional comparator (b) transient output for INN>INP.

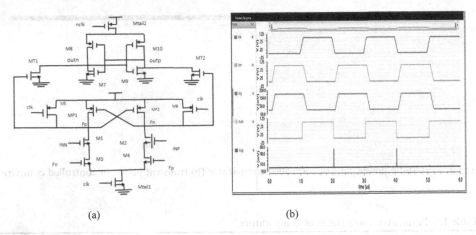

(a)   (b)

Figure 2.   (a) Structural design of controlled comparator (b) transient output for INN>INP.

obtained as 1.2nW. As this comparator does not avoid the direct path connection between V_{DD} and ground, the static power dissipation of the order of 1.2nW is sustained in this design under the specified operating conditions.

Dynamic power, Pd= C* V_{dd}^2 * f, C is the capacitance and f is the switching frequency Static power, Ps= V_{dd} *I. The transient power curve of conventional comparator is shown in Figure 2 (b). In conventional comparator controlling transistors as well as tail transistors are not incorporated within design due to which there is not any control on the switching of nodes in the circuit which leads to the simulated transient power of conventional comparator of the order of 1.6 µW.

3.2   Static and dynamic power of controlled comparator

In this comparator controlling transistors reduces the switching frequency of nodes which reduces the transient power consumption of this comparator. In this comparator design, the maximum static power consumption of the order of 28 pW is achieved for the complete operating voltage range of 0 to 1 V (Figure 4), which is lower by a factor of 44 than that of conventional comparator. The tail transistors Mtail1 and Mtail2 exclude the connection between V_{DD} and ground under the low clk signal which rules out the leakage current of transistors. In addition to this, M5, M6 tail transistors avoid the standby current drawn continuously from V_{DD} to ground and thus reduce the standby or static power consumption of the comparator. In this comparator design, the maximum static power consumption of the order of 28 pW is achieved for the complete operating voltage range of 0 to 1 V (Figure 4), which is lower by a factor of 44 than that of conventional comparator.

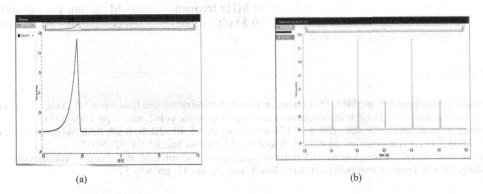

(a)   (b)

Figure 3.   (a) Static power plot (b) transient power of conventional comparator.

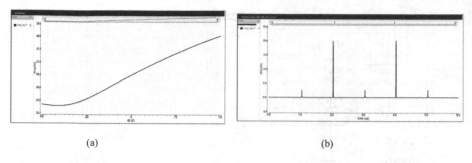

(a)   (b)

Figure 4.   (a) Static power plots of controlled comparator (b) transient power of controlled comparator.

Table 1.   Parametric comparison of comparators.

S.No.	Parameter	Conventional	Controlled comparator
1	Supply (V)	0.6	0.6
2	Delay (ns)	800n	1000n
3	Static Power Dissipation	1.2nW	27pW
4	Dynamic Power Dissipation	1.6u	20u
5	Total Offset Voltage(mV)	13.5	14.8

Table 1. Gives a comparative analysis of conventional comparator (without static reduction logic) and controlled comparator with static reduction logic, both comparators are simulated at 0.6 V in 45nm technology node. Though the obtained propagation delay of both comparators is of same order i.e. 0.8μ for conventional comparator and 1.0μ for controlled comparator but as we know that at lower operating voltage range leakage or static power reduction holds a great concern and in this regard controlled comparator dissipates approximately 44 times lesser power compared to conventional comparator.

4 CONCLUSION

Simulation of controlled comparator with conventional comparator is carried out at 45nm technology node of CADENCE Virtuoso. Simulation results depicts that static power consumption of controlled comparator is approximately forty four times less as compared to conventional comparator with around same propagation delay of the order of 0.9. Dynamic power dissipation has come out to be 1.6u and 20u for conventional and controlled comparator respectively. Simulated controlled comparator is suitable for comparison of two signals with maximum frequency of the order of 50 MHz frequency range. Minimum voltage difference between two signals to be compared is 0.3 volt for conventional comparator and its value is 0.2 V for controlled comparator.

REFRENCES

Ahmadi Muhammad et al. 2015, "Comparator power minimization analysis for SAR ADC using multiple comparators", IEEE transactions on circuits and systems, vol62, no.10, pp 1549–8328.
Cheong J. et al. 2011, "A 400 –nW 19.5 FJ/conversion step 8 ENOB 80-KS/S SAR ADC in 0.18- μm CMOS," IEEE Transactions on circuits & systems-11 express briefs VOL-58, No. 7.
Goll B. et al. 2009, "A comparator with reduced delay time in 65-nm CMOS for supply voltages down to 0.65," IEEE Trans. Circuits Syst. II, Exp. Briefs, vol. 56, no. 11, pp. 810–814.

Harpe P. et al. 2013, "A 10b/12b 40 KS/s SAR ADC with Data-Driven noise reduction achieving up to 10.1b ENOB at 2.2 fJ/conversion –step," IEEE Journal of Solid state circuits, VOL 48,No. 12.

Liu T. et al. 2020, "A 12 –bit 120 MS/s SAR ADC with improved split capacitive DAC and low noise dynamic comparator," Springer, 102, 403–413.

Mao W. et al. 2018, "A Low power 12 bit 1-KS/s SAR ADC for Biomedical signal processing SAR-assisted Time interleaved SAR (SATI-SAR) ADC," IEEE Transactions on circuits & systems-1, Regular papers, 1549–8328.

Mandrumaka K. et al. 2019, "A low power 10 bit SAR ADC with variable threshold technique for bio-medical applications," Springer Nature, s-42452-019-0940–03.

Yun Tzu et al. 2018,"A bypass switching SAR-ADC with a dynamic proximity comparator for biomedical applications", IEEE journal of solid-state circuits, pp 0018–9200.

Recent Trends in Communication and Electronics – Sharma et al. (Eds)
© 2021 Taylor & Francis Group, LLC, ISBN 978-1-032-04572-6

Novel method to design 4-bit BCD adder using and-type inhibitor

Ayush Bhardwaj, Sachin Tyagi, Dhanesh Shukla & Apoorva Chand
KIET Group of Institutions, Delhi-NCR, India

ABSTRACT: A collection of programmable virtual tools has been designed and evaluated in the Combinational Logic Circuits (VIs) using the Lab-VIEW environment. The aim of this paper is to improve the programmability of the circuits of Combinational Logic (Blume A. P. et al. 2007). This paper has the designed modules (VIs) for studying four-bit BCD Adder (Binary-coded decimal) in interactive graphical user interface (GUI)(Drew SM. et al.1996). Though we know that we can realized logic gates and adders or subtractors with the help of NAND (N-G) and NOR(NR-G) but now we are realizing the four-bit BCD adder with the use of full adder by creating their sub VI's designed with the help of another universal gate (Univ-G) called AND-type inhibitor implemented through Lab-VIEW (LAB-V) software.

1 INTRODUCTION

Logic Gates are the basic electronic circuit building blocks. These are the most basic circuits that can be constructed from coupled diodes, transistors and resistors in such a way that the output of the circuit is the result of a simple logical operation carried out on the inputs(Moriarty P.J. et al. 2003). It is simply a system with a single output and two or more inputs. The styles of logical gates are simple gates (OR, AND, NOT), Universal Gates (NAND, NOR) and others (EXOR, EXNOR). Adders are combinational circuits that are used in circuitry in the form of sum and carry to add the bits and produce output.

A four-bit Binary-Coded-Decimal (BC-D) adder is used to add the four-bit number that have Binary-coded-decimal (BC-D) format that results in the BCD-format four-bit output number which represents the decimal sum of the number getting after the addition, and also a carry is generated when the sum becomes greater than the decimal value of nine.

Lab-VIEW (Laboratory Virtual Instrument Engineering Workbench) is a G-language, graphical programming environment. Lab-VIEW (LAB-V) operates on a data flow model in which information flows from data sources to data sinks linked by wires within a Lab-VIEW application, called a virtual instrument (VI) (Lauterburg et. al. 2001). This paper includes the designed front panel and block diagram of four-bit BCD adder realized with the help of and-type inhibitor through Lab-VIEW software.

2 AND-TYPEINHIBITOR

If one of the inputs of an AND Gate is inverted then it is called and-type inhibitor. It is universally complete function. It makes it complete because it can realize any logic gate and also it can be very useful in reducing propagation delay in adders which makes it more efficient in digital circuits. The table below represent its truth table(T-T).

DOI: 10.1201/9781003193838-109

Table1.   Represents truth table of AND type Inhibitor.

X	Y	Result
0	0	0
0	1	1
1	0	0
1	1	0

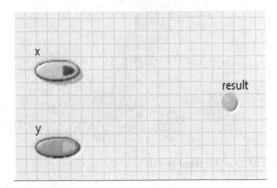

Figure 1.   Represent and-type inhibitor front panel.

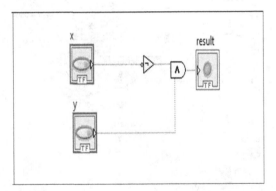

Figure 2.   Represent and-type inhibitor block diagram.

In above figures, the and-type inhibitor structure is represented with the help of block diagram and one of its value in truth table (Table-1) has also been verified in front panel in which it can been seen clearly when x=0 and y=1 then output is 1.

3 IMPLEMENTATION OF 4-BIT BCD ADDER USING AND-TYPE INHIBITOR

The above figures represent the novel approach to design the four-bit BCD (Binary-Coded Decimal) using and-type inhibitor in which Figure 3 is showing the output of the BCD adder on front panel and Figure 4 is showing the graphical programming of BCD adder on block

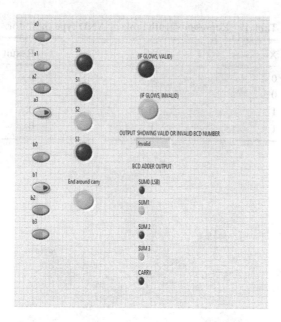

Figure 3.   Represent 4-bit BCD adder front panel.

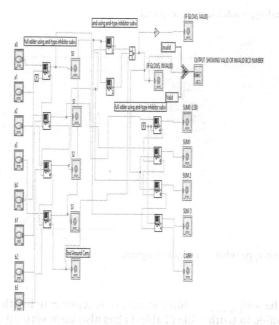

Figure 4.   Represent 4-bit BCD adder block diagram.

diagram in Lab VIEW. In this programming, full adder has been designed using and-type inhibitor sub-VI and then full adder sub-Vi's has been called. This program checks whether the given binary sequence is valid or invalid after adding two binary numbers and then give the valid BCD sequence as output by checking the conditions of valid BCD sequence that is if the number getting after adding two 4-bit binary numbers is less than equal to decimal value of nine then the same number will be represented in output as valid BCD sequence. But if the number getting after addition is greater than decimal value of 9 or there is generation of end around carry then the output will represent invalid sequence and convert that number to valid BCD sequence by adding six(0110) to that number and then show that output in form of Boolean led lights.

4 CONCLUSION

Logic gates and adders have been designed easily using and-type Inhibitor and then realised four-bit BCD adder with the help of the full adder and gate designed by the inhibitor by creating their SUBVIs through LabVIEW. The implementation works quite well however there is no room for improvement as user can easily see the output and checks whether the number is four-bit valid or invalid BCD number and also the conversion of invalid number to valid number in LabVIEW. This software helps to visualize the designing of different gates and different bits adder to help the user to visualize it clearly. LabVIEW does offer a wide range of features in comparison with MATLAB tool and due to its G programming language and Dataflow programming, it is easier indeed to make the designing more efficient and less complex.

REFERENCES

Belletti A., Borromei R., Ingletto G., A. Borromei, R. Ingletto, G., 2006, Teaching physical chemistry experiments with a computer simulation by LabVIEW, Journal of Chemical Education. ACS. 83 (9): 1353–1355.

Blume A.P., 2007, The LabVIEW Style Book, Prentice Hal, Part of the National Instruments Virtual Instrumentation Series. ISBN 0-13-145835-3.

Drew SM, Steven M., 1996, Integration of National Instruments' LabVIEW software into the chemistry curriculum, Journal of Chemical Education. ACS. 73 (12): 1107–1111.

Jerome J., 2010, Virtual Instrumentation Using LabVIEW.

Lauterburg, Urs, 2001, LabVIEW in Physics Education (PDF). A white paper about using LabVIEW in physics demonstration and laboratory experiments and simulations.

Moriarty P.J., Gallagher B.L., Mellor C.J., Baines R.R., P. J., 2003, Graphical computing in the under graduate laboratory: Teaching and interfacing with LabVIEW". American Journal of Physics. AAPT. 71 (10): 1062–1074.

Muyskens MA, Glass SV, Wietsma TW, Gray TM, Mark A.; Glass, Samuel V.; Wietsma, Thomas W. Gray, Terry M., 1996, Data acquisition in the chemistry laboratory using LabVIEW software, Journal of Chemical Education. ACS. 73 (12): 1112–1114.

Ogren PJ, Jones TP, Paul J.; Jones, Thomas P., 1996, Laboratory interfacing using the LabVIEW software package". Journal of Chemical Education. ACS. 73 (12): 1115–1116.

Travis J., Kring J., 2006, LabVIEW for Everyone: Graphical Programming Made Easy and Fun, 3rd Edition, Prentice Hall". Part of the National Instruments Virtual Instrumentation Series. ISBN 0-13-185672-3.

Design of scientific calculator based on event structure by employing labview software

Dhanesh Shukla, Mohit Tyagi, Ayush Bhardwaj & Apoorva Chand
KIET Group of Institutions, Delhi-NCR, Ghaziabad (U.P.), India

ABSTRACT: A Scientific Calculator is a calculator which is designed to help us in calculating science, engineering, and various mathematical problems but we cannot perform complex functions implementation in that device. This paper has been focused on the designed and implementation of a simple scientific calculator that should perform complex operations along with arithmetic operations, reciprocal and square functions with features of memory clear, memory recall, memory store and add memory. This has been designed to solve the strategy of many sophisticated problems along with mathematical knowledge to focus and advanced the simulated engine field in mathematic (Moriarty P.J.et al. 2003). This work will help in easy calculations of tedious problems of mathematics and errors can be easily revealed from that calculations and help to solve many complex calculations through LabVIEW software.

1 INTRODUCTION

A Scientific Calculator is a device intended to assist us in measuring problems in science, engineering and mathematics. Now, standard scientific calculators cannot perform complex functions used in engineering such as calculation with matrices, fast Fourier transforms (FFT) matrices, CA calculation with transfer functions along with Bode plot and simulation of DC motor control. This can be achieved by using engineering software LabVIEW.

LabVIEW (Laboratory Virtual Instrument Engineering Workbench) is a programming environment based on the G language of graphical programming. It basically operates on a data flow model in which information flows from data sources to data sinks linked by wires within a LabVIEW application, called a virtual instrument (VI) (Travis J. et al. 2006). This paper includes the designed block diagram of every performed function of scientific calculator based on event structure and will be part of our use for science, engineering, and mathematical problems.

2 SCIENTIFIC CALCULATOR

A scientific calculator is a type of electronic calculator designed to measure issues in science, engineering and mathematics, usually but not always handheld. Scientific calculators have been superseded in some contexts, such as higher education, by graphing calculators, which provide a superset of scientific calculator features along with the ability to graph input data and write and store device programs (Ogren P.J.et al.1996). The below Figure 1 and Figure 2 are representing the design of scientific calculator with the help of event structure through LabVIEW programming and also the output of the scientific calculator can be easily visualizing in LabVIEW.

DOI: 10.1201/9781003193838-110

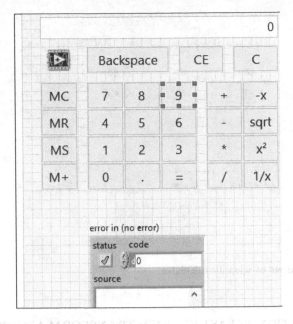

Figure 1. Represent scientific calculator front panel.

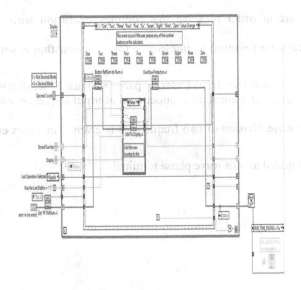

Figure 2. Represent scientific calculator Block Diagram.

In this scientific calculator, there are various events created for simple arithmetic operations along with square, negate and reciprocal functions to perform these operations and also it has error checking function in case of any error occurred while running this scientific calculator. This will help in clear visualization of various operations in scientific calculator one by one through LabVIEW.

The above Figure 3 is representing one of the arithmetic operations. The operation performed in above figure is square function. For example, if the number is 12 then its square will be 144.

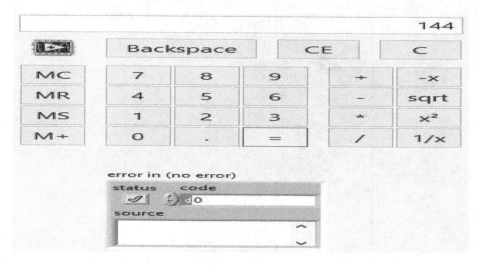

Figure 3.   Showing output of scientific calculator.

3   IMPLEMENTATIONOF FAST FOURIER TRANSFORM ALGORITHM

In this scientific calculator, there is also implementation of Fast Fourier transform with the help of

LabVIEW software in order to solve this complex operation with other operations in calculator.

The following steps are followed to implement this function that is shown in the Figure 4 and 5:

i. The Function of FFT is chosen from Function pallet≪ Signal Processing≪Transforms≪FFT.

ii. At the FFT size terminal a constant is added and at input sequence, output of sine wave is given.

iii. At input of sine wave division of two frequencies is given. For index entry, cosine wave is needed

so, phase of 90 is added at sine wave phase terminal.

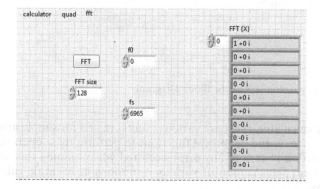

Figure 4.   Represent FFT Front Panel.

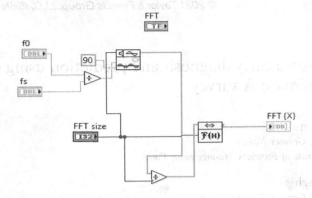

Figure 5.   Represent FFT Block Diagram.

4   CONCLUSION

We designed a scientific calculator based on event structure through LabVIEW that can act as a conventional scientific calculator with features addition, subtraction, quick access to pie function, negate, reciprocal, square, and other conventional functions. In addition, we are able to implement a fast Fourier transform algorithm. The implementation works quite well however there is a room for improvement. The FFT algorithm can be implemented using graphs too. Although we have used TAB Control to switch over the different controls and functionalities. So, basically, we have created the nodes in which the user can operate the calculator. LabVIEW does offer a wide range of features in comparison with MATLAB tool and due to its G programming language and Dataflow programming, it is easier indeed to make the designing more efficient andless complex.

REFERENCES

Belletti A., Borromei R., Ingletto G., A. Borromei, R.Ingletto, G., 2006, Teaching physical chemistry experiments with a computer simulation by LabVIEW, Journal of Chemical Education. ACS. 83 (9): 1353–1355.

Drew SM, Steven M., 1996, Integration of National Instruments' LabVIEW software into the chemistry curriculum, Journal of Chemical Education. ACS. 73 (12): 1107–1111.

Jerome J., 2010, Virtual Instrumentation Using LabVIEW.

Lauterburg, Urs, 2001, LabVIEW in Physics Education (PDF). A white paper about using LabVIEW in physics demonstration and laboratory experiments and simulations.

Moriarty P.J., Gallagher B.L., Mellor C.J., Baines R.R., P. J., 2003, Graphical computing in the under graduate laboratory: Teaching and interfacing with LabVIEW". American Journal of Physics. AAPT. 71 (10): 1062–1074.

Ogren PJ, Jones TP, Paul J.; Jones, Thomas P., 1996, Laboratory interfacing using the LabVIEW software package". Journal of Chemical Education. ACS. 73 (12): 1115–1116.

Travis J., Kring J., 2006, LabVIEW for Everyone: Graphical Programming Made Easy and Fun, 3rd Edition, Prentice Hall". Part of the National Instruments Virtual Instrumentation Series. ISBN 0-13-185672-3.

Recent Trends in Communication and Electronics – Sharma et al. (Eds)
© 2021 Taylor & Francis Group, LLC, ISBN 978-1-032-04572-6

Alzheimer disease early diagnosis and prediction using deep learning techniques: A survey

Nilanjana Pradhan
Galgotias University, Greater Noida
Analytics, Pune institute of Business Management, Pune

Ajay Shankar Singh
Galgotias University, Greater Noida

Akansha Singh
Department of CSE, ASET, Amity University Uttar Pradesh, Noida, India

ABSTRACT: Alzheimer's Disease (AD) is the most well-known type of dementia which in the long run prompts neurological disorder which causes progressive declination of cognitive abilities. The advancement of Alzheimer's illness (AD) is related with critical shortages in patient's body functioning necessary for long term care. The particular manifestations of Alzheimer's disease are: decrease in the cognitive abilities resulting in memory deficiencies, trouble in talking or perceiving objects, hindered movement control, and behavioral issues. MRI (Magnetic resonance imaging) is used to scan brain imagery and hence we can get the aid of Artificial intelligence techniques for identification as well as prediction. The Deep Learning (DL) algorithms have been extremely helpful in recent years for the diagnosis of Alzheimer's Disease as DL algorithms perform well with huge datasets as a predictor of this disease and categorize AD patients whether or not they will be affected by this condition in the future or not. In this paper, we have explained various deep learning techniques used for prediction and detection of AD.

Keywords: Alzheimer's Disease (AD), MRI, cognitive abitilites, Deep Learning (DL)

1 INTRODUCTION

Several research have been conducted to analyze the unusual structures of brain which eventually leads to identify Alzheimer disease using medical images. The early detection and diagnosis of Alzheimer's Disease is possible to detect and provide relevant treatment to the patients. Magnetic resonance images having good quality are used for biomedical image processing. The challenges faced in this context are overcome with AI technologies. Super resolution is one such technology in which images with high resolution is obtained from low resolution images. Hence this method provides convenient way of diagnosis of diseases. Deep Learning techniques are also used to obtain high quality reconstructed images (Altinkya et.al 2020). Deep Learning Techniques are used for retrieving features and patterns from data. Early diagnosis of Alzheimer disease is important as its prevents the advancement of the disease so that it will be treated as early as possible. Deep neural networks (DNN) cocnsists of input layer,output layer and hidden layers respectively (Altinkya et.al 2020). Unsupervised approach is used for feature extraction and clustering.

DOI: 10.1201/9781003193838-111

2 CONVOLUTIONAL NEURAL NETWORK (CNN)

Convolutionary Neural Networks (CNN) is a common approach of deep learning in which the data from the image is taken from the input and transmitted in one direction to the output and achieves the output according to the desired category.

CNN consists of input, output and several hidden layers, (Altinkya et.al 2020) Deep Automatic Encoder (DA) has the same number of input and output nodes and is trained to reconstruct the vector. DA input is a type of neural network that performs unattended learning. The DA model consists of three layers: input (encoder), output (decoder) and hidden layer. The number of hidden layer nodes is less thant the same number of input and output node layers. Deep Boltzman Machine (DBM) is an unsupervised model which does not have a complete relationship between its layers. (Altinkya et.al 2020). Computational complexity is very high in the process of retrieving information from the medical images. However Deep neural network easily detects essential features from images and is one of the growing medical diagnosis application (Raj et. al 2020). The following figure represents a deep neural network:

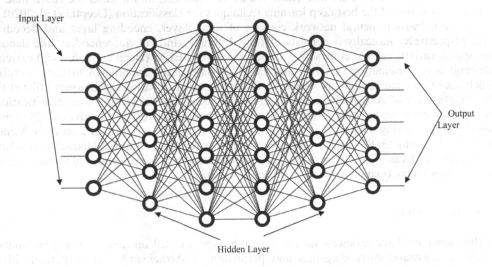

Figure 1. A deep neural network.

In case the data set is huge it can be easily recognized. Deep Learning methods helps in avoiding regular issues that takes lot of time. Image classification is a vital domain and confronts great challenge. This can overcome by deep learning which is a subset of machine learning and which makes use of many computations that attempt to show the aberrant state in the current data by using a profound course of action that has less number of layers (Raj et. al 2020).

3 DEEP LEARNING TECHNIQUES FOR EARLY DIAGNOSIS AND PREDICTION OF ALZHEIMER'S DISEASE

As per Salehi, Ahmad Waleed, et al. MRI data collection and consecutive preprocessing of the collected data followed by training and testing of the data and subsequently the model development is done. Data is collected validated and is divided into three classes ie. mild, normal control and Alzheimer disease. The use of two data sources (Salehi et. al 2020) is used for increasing image number and using data from various sources to improve model performance. Implementation of the model is made from 8 GB RAM and Intel HD6000 1536 MB using anaconda for python and tensor Flow. As for CNN is basically a feedforward neural network and it is implemented in image processing and pattern recognition as well as classification problems. (Salehi et. al 2020) CNN have different layers, such as convolution, activation, pooling and fully-connected layer. The first layer is convolutional which receives the image input and decides whether or not it is that of an Alzheimer's patient (Salehi et. al 2020).

Deepthi et al. 2020 have proposed a method in which falling is a major issue for older people suffering from Alzheimer disease (AD). The falling rates causes other damages like fractures and injuries.Hence the proposed a Convolutional neural network (CNN) which is one of the most popular deep learning models which is used for predicting complicated outcome of any issues. There is convolutional layer together with a pooling layer and fly connected layer (Suzuki et. al 2020) which are responsible extraction of important features and minimize the dimensionality of features. CNN in their proposed method is used to predict the time of falling (TF) based on various factors like age, dementia etc. They have used 42 actual data sets which includes Multiple Complicating factors. Hence CNN togetherwith MCF can predict the TF amoing patients suffering from Alzheimer's Disease (AD) (Suzuki et. al 2020). In this proposed method Electroencephalography (EEG) signals are used rather than MRI which is very cost effective. The features from the pre-processed signals are retrieved usin fast fourier transform. The features retrieved are then fed to the CNN which classifies whether the AD is mild or severe. CNN is one of the best deep learning techniques for classification (Deepthi et al. 2020).

This deep learning neural network consists of input layer, encoding layer and decoding layer respectively. Basically it consists of four different kinds of autoencoders like denoising, sparse variational and contractive autoencoder().DC-ELM (Deep convolutional extreme learning) is a combination of CNN and extremely fast paced training of extreme learning which uses Gaussian probability. Salehi et al have compared various techniques utilized for the classification of AD (Alzheimer's Disease) and came into conclusion that classifications are accurate when we use deep CNN (Convolutional neural network). Harshit et al. 2020 proposed a deep learning method that uses the 3D CNN network to predict the onset of Alzheimer's disease using fMRI datasets. As a result, both spatial and temporal characteristics have been recovered from 4D volume and all the complicated forms of conventional feature extraction processes have been removed (Harshit et. al 2020).

4 CONCLUSION

In this paper we have reviewed various research papers based on deep learning techniques which demonstrates early diagnosis and prediction of Alzheimer's disease patient. Most papers are based on the implementation of convolutional neural network. Hence we come into conclusion that deep learning which is the subfield of machine learning can be used to solve many problems in medical data analysis. Alzheimer's Disease early diagnosis and prediction system can be built using deep learning model. Hence it will bring a revolutionary change in our society providing immense support to the aged population suffering from dementia. Hence we hope to develop more and more research proposal to prevent AD progression and getting converted into severe issues.

REFERENCES

Altinkaya, Emre, Kemal Polat, and Burhan Barakli, 2020, Detection of Alzheimer's Disease and Dementia States Based on Deep Learning from MRI Images: A Comprehensive Review, *Journal of the Institute of Electronics and Computer* 1: 39–53.
Raj, R. Joshua Samuel, 2020, Optimal Feature Selection-Based Medical Image Classification Using Deep Learning Model in Internet of Medical Things, *IEEE Access* 8 (2020): 58006–58017.
Salehi, Ahmad Waleed, 2020, A CNN Model: Earlier Diagnosis and Classification of Alzheimer Disease using MRI, *International Conference on Smart Electronics and Communication (ICOSEC)*. IEEE.
Suzuki, Makoto, et al., 2020, Deep learning prediction of falls among nursing home residents with Alzheimer's disease, *Geriatrics & Gerontology International*.
Deepthi, L. Dharshana, D. Shanthi, and M. Buvana, 2020, An Intelligent Alzheimer's Disease Prediction Using Convolutional Neural Network (CNN)." *International Journal of Advanced Research in Engineering and Technology (IJARET)* 11.4: 12–22.

Salehi, Ahmad Waleed, Preety Baglat, and Gaurav Gupta, Alzheimer's Disease Diagnosis using Deep Learning Techniques.

Parmar, Harshit S., et al. 2020, Deep learning of volumetric 3D CNN for fMRI in Alzheimer's disease classification, *Medical Imaging 2020: Biomedical Applications in Molecular, Structural, and Functional Imaging*. Vol. 11317. International Society for Optics and Photonic.

Recent Trends in Communication and Electronics – Sharma et al. (Eds)
© 2021 Taylor & Francis Group, LLC, ISBN 978-1-032-04572-6

An arduino based voltage controlled oscillator for biomedical applications

Suraj Prasad Barnwal & Toushik Maiti
Department of Electrical Engineering, National Institute of Technology Durgapur (NITDgp), Durgapur, India

Sunil Choudhary
Department of Electrical Engineering, B. C. Roy Engineering College (BCREC), Durgapur, India

Tushar Kanti Bera
Department of Electrical Engineering, National Institute of Technology Durgapur (NITDgp), Durgapur, India
Center for Biomedical Engineering and Assistive Technology, National Institute of Technology Durgapur, Durgapur, India

ABSTRACT: A voltage-controlled oscillator has been developed for electrical impedance spectroscopy (EIS) system. In EIS, VCO generates an alternating voltage signal which is utilized for measuring the impedance of the sample under test (SUT). In this paper, an Arduino UNO based multifrequency-VCO (Mf-VCO) is developed for EIS and the VCO performance is studied, tested and evaluated. The VCO is developed with a MAX038 IC and interfaced with the Arduino UNO to generated sinusoidal voltage signals at different frequencies. The digital control signals generated by the Arduino board are fed to an analog multiplexer circuit and the signal frequencies are controlled by connecting the different passive components to the VCO. The digital signal generation and multiplexer operation are assessed by an LED-based digital signal evaluator (LED-DSE) to assure the connectivity of the passive components and frequency control. Results show that the Mf-VCO is not only suitable for EIS studies of biological samples within a wide bandwidth but also for other multifrequency instrumentation.

1 INTRODUCTION

A signal generator is a key component of the electrical impedance spectroscopy (EIS) (Macdonald, 1992; Bera et al. 2016; Bera 2014; Basak eta al, 2020; Soares et al., 2020) system. EIS is a powerful non-invasive technique for impedance analysis in our modern era of biomedical (Bera and Nagaraju, 2011a; K'Owino and Sadik, 2005; Bera et al., 2017; Santos et al., 2013; Keshtkar et al., 2006; Tidy et al., 2013) and non-biomedical (Bera and Nagaraju, 2011a; Bera et al., 2014; Aguedo et al. 2020; Christensen et al., 1994; Rout et al., 2009; Bredar et al., 2020, DuToitet al., 2021) research fields. An EIS system measures the electrical impedance of a material at different frequency points within a particular bandwidth to study its frequency response of the material (Figure 1a). The impedance measurement could be made by applying a voltage (or current) signal to a sample under test (SUT) and measuring the developed current (or voltage) signal across the SUT (Bera et al.., 2014; Almuhammadi et al. 2017; Bera et al. 2019; Bera and Nagaraju, 2011a). The voltage-current measurement is conducted using either two probe method (Bera et al. 2019) or four probe method (Bera et al. 2019). The signal generator of an EIS instrumentation could be developed as a voltage-controlled oscillator (VCO)

DOI: 10.1201/9781003193838-112

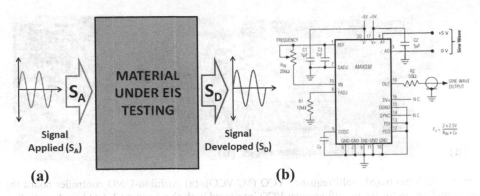

(a)   **(b)**

Figure 1.   (a) Applied signal (sine wave in red colour) and developed signal (sine wave in blue colour) in Electrical Impedance Spectroscopy (EIS) system (b) Operational block diagram of the Mf-VCO.

which is used as an alternating voltage signal generator circuit (Liu, 1999, Bera and Nagaraju, 2009a). In EIS measurement, the voltage signal generated by the VCO is either applied directly or indirectly (after converting into a current signal) to the SUT. Therefore, VCO is found as one of the key parts of impedance measuring instrumentation (Bera and Nagaraju, 2009a; Bera and Nagaraju, 2010; Bera and Nagaraju, 2011b, Hartov 2000; Bera and Nagaraju, 2014) to generate an alternating voltage (or current) signal to estimate the electrical impedance of the SUT by measuring the current (or voltage) signal developed. On the other hand, the voltage signal is also, sometimes, directly injected to the SUT to measure its electrical impedance from the current signal developed due to the voltage applied. An automatic EIS instrumentation essentially needs an automatic sinusoidal voltage generation at different frequencies which is generally obtained by a VCO controlled by a microcontroller or any other electronic controlling unit such as PC based Data Acquisition systems (Payne & Menz, 1995, Bera et al., 2015; Bera et al., 2013). In this paper an Arduino UNO (Arduino, 2015) based multifrequency-VCO (Mf-VCO) is developed for EIS studies. The VCO is developed with function generator circuit which generates the sinusoidal voltage signal at different frequencies by changing the value of few passive components connected to the IC as the frequency controlling components. The frequency controlling components are connected to the VCO through a multiplexer circuits operated by the digital signals to generate the signals with required frequency values. The digital signals are generated in Arduino board through its digital I/O pins and fed to the multiplexer to control the frequency generation in VCO. The digital signal generation and multiplexer operation are assessed by LED based digital signal evaluator (LED-DSE) to assure the connectivity of the passive components and frequency control. Results show that the Mf-VCO is suitable for EIS studies of biological samples within a wide bandwidth.

2   MATERIALS AND METHOD

2.1   *Voltage Controlled Oscillator (VCO)*

A voltage-controlled oscillator (VCO) is an electronic circuit which generates sinusoidal or other periodic waveforms by oscillation frequency is controlled by a voltage input. The applied input voltage determines the instantaneous oscillation frequency.

2.2   *MAX038 based VCO*

A voltage controlled oscillator circuit has been developed with MAX038 IC (MAXIM Inc., USA) (MAX038 IC data sheet). MAX038 is a high-frequency, precision waveform generator IC which produces accurate, high-frequency sinusoidal, triangle, square, waveforms with a few resistors (R_f) and capacitors (C_f) (Bera and Nagaraju, 2010) connected

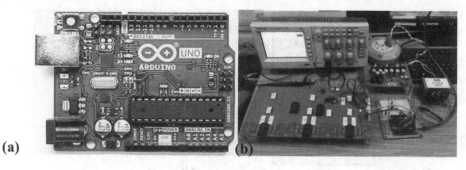

(a) **(b)**

Figure 2. Arduino based multifrequency VCO (Mf-VCO): (a) Arduino-UNO controller board (b) the experimental set up for the multifrequency-VCO interfaced with the Arduino-UNO based controller and connected with power supply, and CRO.

as the external components (Figure 1b). The amplitude of the generated is 2V (peak to peak) signal for all waveforms. Controlling the value of the R_f and C_f a particular frequency of an AC signal is obtained. The VCO is connected with a resistor-capacitor bank which is controlled by a multiplexer (MUX) circuit. An Arduino based controller board is used to generate the digital signals to control the MUX to connect a particular resistor and capacitor.

2.3 Arduino UNO based VCO controller

In the present work, different values of resistors and capacitors are connected and a particular set of the R_f and C_f is connected to obtain a particular frequency. The resistors and the capacitors are connected to the MAX038 IC through a multiplexer circuit developed with CD4067BE (Texas Instruments Inc., USA) (CD4067BE IC data sheet). The digital signals are generated in the Arduino Board and the digital signals are fed to a 32:1 multiplexer circuit. A 32:1 multiplexer circuit is developed with two16:1 multiplexers ICs (CD4067BE, Texas Instrument Inc., USA). The Arduino-UNO controller (Figure 2a) board (Arduino, 2015; Kumar et al., 2016; Bawa et al., 2013; Fransiska et al., 2013; Bereziuk et al., 2018; Hidayanti et al., 2020; Sihombing et al., 2019) uses an ATMEGA328 Microcontroller (ATMEGA328 Microcontroller IC data sheet) which acts as a brain in the entire circuit. The digital bit generator code has been written in the program to generate 5-bit parallel digital data ($D_4D_3D_2D_1D_0$) which are required to feed the selector/controller pins/channels of the 32:1 multiplexers to operate it for R-C switching. The picture of circuit board of the VCO is shown in the Figure 2b.

3 RESULTS AND DISCUSSIONS

The VCO response is studies for different waveforms coming out of the impedance analyzer and we did some sort of analysis over it, such as SNR evaluation, FFT evaluation etc. with the help of CRO. The signals from the VCO output are obtained with sinusoidal, triangular and square waveforms. Figure 3 shows the screen shots of the output waveforms obtained from CRO. The VCO is also controlled by the controller and the sinusoidal signals with different frequency are generated and tested. It is found that the multiplexer is switched and different frequency signals are generated from 200 Hz to 2 MHz as shown in the Figure 3 below. The calculated values of the R_{IN} and C_F are listed in the table below for which different frequencies are. The practical VCO is found capable of producing sine wave from 100 Hz to 2 MHz at 32 different frequency points. The resistance (R_{IN}) and capacitance (C_{IN}) valued required for generating the alternating signals have been shown for few of the frequency points in.

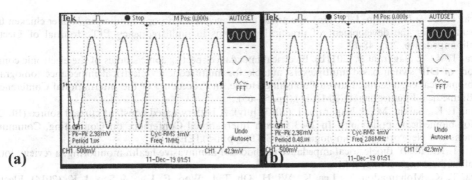

(a) **(b)**

Figure 3. VCO output signals with amplitude of 1 V (RMS) value for different frequencies (a) sinusoidal wave at 1MHz, (b) sinusoidal wave at 2 MHz.

4 CONCLUSIONS

Multifrequency constant VCO is developed with MAX038 IC interfaced with PC through Arduino-UNO board. The VCO board is connected with a resistor-capacitor board which is developed with high precision resistors and capacitors to connect a required set of resistor-capacitor with the VCO to generate an alternating signal with a specific frequency. The resistors and capacitors are selected and switched by an analog multiplexer board developed with CD4067BE ICs. Arduino board sequentially generates the set of 32 different digital bits which connects 32 different sets of resistors and capacitors with the MAX038 IC based VCO. The alternating signals are generated and studied. The VCO efficiently generates constant Voltage with minimal amount of noise at variable frequency levels up to 2 MHz. Results demonstrate that the Arduino based multifrequency voltage signal generator is suitable for impedance measurement applications such as EIS and othermultifrequency instrumentation applications.

REFERENCES

Aguedo, J., Lorencova, L., Barath, M., Farkas, P., & Tkac, J. (2020). Electrochemical Impedance Spectroscopy on 2D Nanomaterial MXene Modified Interfaces: Application as a Characterization and Transducing Tool. Chemosensors, 8(4), 127.

Almuhammadi, K., Bera, T. K., & Lubineau, G. (2017). Electrical impedance spectroscopy for measuring the impedance response of carbon-fiber-reinforced polymer composite laminates. Composite Structures, 168, 510–521.

Arduino, S. A. (2015). Arduino. Arduino LLC.

ATMEGA328 Microcontroller IC data sheet, Atmel Corporation 1600 Technology Drive, San Jose, CA 95110 USA.

Basak, R., Wahid, K., & Dinh, A. (2020). Determination of leaf nitrogen concentrations using electrical impedance spectroscopy in multiple crops. Remote Sensing, 12(3), 566.

Bawa, D., & Patil, C. Y. (2013). Fuzzy control based solar tracker using Arduino Uno. International Journal of Engineering and Innovative Technology, 2(12), 179–187.

Bera, T. K., & Nagaraju, J. (2009a, May). A study of practical biological phantoms with simple instrumentation for electrical impedance tomography (EIT). In 2009 IEEE Instrumentation and Measurement Technology Conference (pp. 511-516). IEEE.

Bera, T. K., & Nagaraju, J. (2009b, August). A Simple instrumentation calibration technique for Electrical Impedance Tomography (EIT) using a 16-electrode phantom.In 2009 IEEE International Conference on Automation Science and Engineering (pp. 347-352). IEEE.

Bera, T. K., & Jampana, N. (2010, December). A multifrequency constant current source suitable for Electrical Impedance Tomography (EIT). In 2010 International Conference on Systems in Medicine and Biology (pp. 278-283). IEEE

Bera, T. K., & Nagaraju, J. (2011a). Electrical impedance spectroscopic studies on broiler chicken tissue suitable for the development of practical phantoms in multifrequency EIT. Journal of Electrical Bioimpedance, 2(1), 48–63.

Bera, T. K., & Nagaraju, J. (2011b). Electrical impedance spectroscopic studies of the electronic connectors of DIP switch based multiplexers suitable for multifrequency electrical impedance tomography, biomedical engineering, Narosa Publishing House. In Proceeding of the International Conference on Biomedical Engineering (ICBME'11) (pp. 58–65).

Bera, T. K., Saikia, M., & Nagaraju, J. (2013, July). A battery-based constant current source (Bb-CCS) for biomedical applications. In 2013 Fourth International Conference on Computing, Communications and Networking Technologies (ICCCNT) (pp. 1-5).IEEE.

Bera, T. K. (2014). Bioelectrical impedance methods for noninvasive health monitoring: a review. Journal of medical engineering, 2014.

Bera, T. K., Mohamadou, Y., Lee, K., Wi, H., Oh, T. I., Woo, E. J., ...& Seo, J. K. (2014). Electrical impedance spectroscopy for electro-mechanical characterization of conductive fabrics. Sensors, 14(6), 9738–9754.

Bera, T. K., & Nagaraju, J. (2014). A Low Cost Electrical Impedance Tomography (EIT) Instrumentation for Impedance Imaging of Practical Phantoms: A Laboratory Study. In Proceedings of the Third International Conference on Soft Computing for Problem Solving (pp. 689-701). Springer, New Delhi.

Bera, T. K., Chowdhury, A., Mandai, H., Kar, K., Haider, A., & Nagaraju, J. (2015, February). Thin domain wide electrode (TDWE) phantoms for Electrical Impedance Tomography (EIT). In Proceedings of the 2015 Third International Conference on Computer, Communication, Control and Information Technology (C3IT) (pp. 1-5). IEEE.

Bera, T. K., Nagaraju, J., & Lubineau, G. (2016). Electrical impedance spectroscopy (EIS)-based evaluation of biological tissue phantoms to study multifrequency electrical impedance tomography (Mf-EIT) systems. Journal of Visualization, 19(4),691–713.

Bera, T. K., Bera, S., Chowdhury, A., Ghoshal, D., & Chakraborty, B. (2017, March). Electrical impedance spectroscopy (EIS) based fruit characterization: A technical review. In Proc. Comput., Commun. Elect. Technol. (pp. 279-288).

Bera, T. K., Jampana, N., & Lubineau, G. (2019). A LabVIEW-based electrical bioimpedance spectroscopic data interpreter (LEBISDI) for biological tissue impedance analysis and equivalent circuit modelling. Journal of Electrical Bioimpedance, 7(1),35–54.

Bereziuk, V., Lemeshev, M. S., Bogachuk, V. V., & Duk, M. (2018, October). Means for measuring relative humidity of municipal solid wastes based on the microcontroller Arduino UNO R3. In Photonics Applications in Astronomy, Communications, Industry, and High-Energy Physics Experiments 2018 (Vol. 10808, p. 108083G).International Society for Optics and Photonics.

Bredar, A. R., Chown, A. L., Burton, A. R., & Farnum, B. H. (2020). Electrochemical Impedance Spectroscopy of Metal Oxide Electrodes for Energy Applications.ACS Appl. Energy Mater., 3 (1),66–98.

CD4067BE IC data sheet, Texas Instruments Inc., USA

Christensen, B. J., Coverdale, T., Olson, R. A., Ford, S. J., Garboczi, E. J., Jennings, H. M., & Mason, T. O. (1994). Impedance spectroscopy of hydrating cement-based materials: measurement, interpretation, and application. Journal of the American Ceramic Society, 77(11),2789–2804.

Dos Santos, M. B., Agusil, J. P., Prieto-Simón, B., Sporer, C., Teixeira, V., & Samitier, J. (2013). Highly sensitive detection of pathogen Escherichia coli O157: H7 by electrochemical impedance spectroscopy. Biosensors and Bioelectronics, 45, 174–180.

DuToit, M., Ngaboyamahina, E., & Wiesner, M. (2021). Pairing electrochemical impedance spectroscopy with conducting membranes for the in situ characterization of membrane fouling. Journal of Membrane Science, 618, 118680.

Fransiska, R. W., Septia, E. M. P., Vessabhu, W. K., Frans, W., & Abednego, W. (2013, November). Electrical power measurement using Arduino Uno microcontroller and LabVIEW. In 2013 3rd International Conference on Instrumentation, Communications, Information Technology and Biomedical Engineering (ICICI-BME) (pp. 226-229).IEEE.

Hartov, A., Mazzarese, R. A., Reiss, F. R., Kerner, T. E., Osterman, K. S., Williams, D. B., & Paulsen, K. D. (2000). A multichannel continuously selectable multifrequency electrical impedance spectroscopy measurement system.IEEE transactions on biomedical engineering, 47(1),49–58.

Hidayanti, F., Rahmah, F., & Wiryawan, A. (2020). Design of Motorcycle Security System with Fingerprint Sensor using Arduino Uno Microcontroller. International Journal of Advanced Science and Technology, 29(05), 4374–4391.

Keshtkar, A., Keshtkar, A., & Smallwood, R. H. (2006). Electrical impedance spectroscopy and the diagnosis of bladder pathology. Physiological Measurement, 27(7), 585.

K'Owino, I. O., & Sadik, O. A. (2005). Impedance spectroscopy: a powerful tool for rapid biomolecular screening and cell culture monitoring. Electroanalysis: An International Journal Devoted to Fundamental and Practical Aspects of Electroanalysis, 17(23), 2101–2113.

Kumar, N. S., Vuayalakshmi, B., Prarthana, R. J., & Shankar, A. (2016, November). IOT based smart garbage alert system using Arduino UNO. In 2016 IEEE Region 10 Conference (TENCON) (pp. 1028-1034). IEEE.

Liu, T. P. (1999, February). A 6.5 GHz monolithic CMOS voltage-controlled oscillator. In 1999 IEEE International Solid-State Circuits Conference. Digest of Technical Papers. ISSCC. First Edition (Cat. No. 99CH36278) (pp. 404-405). IEEE.

Macdonald, J. R. (1992). Impedance spectroscopy. Annals of biomedical engineering, 20(3), 289–305.

MAX038 IC data sheet, MAXIM Inc., USA

Payne, J. R., & Menz, B. A. (1995, October).High speed PC-based data acquisition systems. In IAS'95. Conference Record of the 1995 IEEE Industry Applications Conference Thirtieth IAS Annual Meeting (Vol. 3, pp. 2140-2145). IEEE.

Rout, S. K., Hussian, A., Lee, J. S., Kim, I. W., & Woo, S. I. (2009). Impedance spectroscopy and morphology of SrBi4Ti4O15 ceramics prepared by soft chemical method. Journal of Alloys and Compounds, 477(1-2), 706–711.

Sihombing, P., Tommy, F., Sembiring, S., & Silitonga, N. (2019, June). The citrus fruit sorting device automatically based on color method by using tcs320 color sensor and arduinouno microcontroller. In Journal of Physics: Conference Series (Vol. 1235, No. 1, p. 012064). IOP Publishing.

Soares, C., Machado, J. T., Lopes, A. M., Vieira, E., & Delerue-Matos, C. (2020). Electrochemical impedance spectroscopy characterization of beverages. Food chemistry, 302, 125345.

Tidy, J. A., Brown, B. H., Healey, T. J., Daayana, S., Martin, M., Prendiville, W., & Kitchener, H. C. (2013). Accuracy of detection of high-grade cervical intraepithelial neoplasia using electrical impedance spectroscopy with colposcopy. BJOG: An International Journal of Obstetrics & Gynaecology, 120(4), 400–411.

Recent Trends in Communication and Electronics – Sharma et al. (Eds)
© 2021 Taylor & Francis Group, LLC, ISBN 978-1-032-04572-6

Resource management and allocation in Fog computing

Sunakshi Mehta & Akansha Singh
Department of CSE, ASET, Amity University

Krishna Kant Singh
KIET Group of Institutions, Delhi-NCR, Ghaziabad, India

ABSTRACT: In this article, what we shall highlight upon is the resource allocation along with edge computing in Internet-of-Things (IoT) networks - that too through machine learning approaches. To be specific, every single end device is categorized as an agent, helping in making decision whether the computation task should be offloaded to edge devices or not. In order to reduce long-term weighted sum cost, such as task execution latency and spiraling levels of power consumption, we take the channel conditions between the gateway and the end devices into consideration.

1 INTRODUCTION

Owing to the right data management and advanced analytics systems in place, today, people want to make cities smarter by harnessing the power of big data. Backed with the power of collecting, storing and processing big data, it can uncover hidden patterns, hidden insights and correlations.

In a traditional setup, big data is used to receive real-time healthcare information. The backend cloud helps with performing functions, such as monitoring, foreseeing and identifying disease and giving treatment solutions. It is with deploying conventional getaways for Internet connection that one can access services of restricted networks.

In order to fully harness the potential and extract data from it, today, cloud architectures face the problem of controlled scalability. This, as a result, leads to lack of computing environments fulfillment, which is focused on the centralized Internet of Things (IoT). The main problem arises from the fact that most of the budding IoT technologies, right from machine learning at scale and other learning technologies, are computationally intensive. Owing to the issues of heat and volume considerations, mobile devices and sensor nodes that have a huge role in collecting big datafor the smart cities tend to exhibit shortened processing and battery power, along with storage space. This makes it hard to meet the requirements of exuberant computing tasks. As a result, the data moving in copious amount will become extremely large and processing of Big Data will gingerly appear.

Its solution lies in the way smart cities would establish efficient channels of communication with people, along with higher levels of latency that will minimize the user experience. But with applications that require real-time involvement and requirements, it's foreseeable that the issue of high latency will become intolerable. In fact, in some extreme cases it can also endanger the safety of the user. Especially in cases, such as automatic navigation boats, automated driving, and intuitive healthcare.

It's since 2015 that globally enterprise firms, such as Cisco, Microsoft, Dell and Intel pioneered the concept of Edge Computing. In the scheme of IoT network, Edge Computing has

DOI: 10.1201/9781003193838-113

revolutionized the way computing capabilities are being provided to the users. But, it's also imperative to keep in mind how a large number of users require enough spectrum resource in IoT for the transmission of their computation tasks to an edge server. Even though the IoT devices are developed to possess powerful computation abilities, allowing users to execute some tasks locally because of a powerful computation. Just like in the traditional cloud computing, the data is often dispersed instead of being in a concentrated format in Edge Computing.

The objective goal of this study is to bring forth a resource allocation algorithm to distribute IoT application modules to the whole network in an Edge Computing environment. As latest and budding distributed cloud architecture, what brings forth a novel solution for the smart cities is Edge Computing. It is simulations that assist us in validating the feasibility of our projected algorithm. It helps in achieving an improved trade-off between the task execution latency and power consumption as compared to the edge computing along with local computing modes. It is to be noted that often edge servers are equipped with limited computing resources, owing to the lack of scalability of deployment and economic benefits. This is mainly because the edge servers are distributed in large quantities that too, in close proximity to the use, hence marking the importance of economic benefits of deployment. This means that when compared to a cloud center, a single edge server is incapable of having as many resources, making the edge server unable to process all the tasks fully. If these tasks are transmitted to the edge server as it is, without any difference, that will lower the level of processing efficiency and exaggerate the processing time for thegiven task. Hence, studying the process of offloading tasks based on the needs and demands remain a query worth studying.

Offloading is nothing but understanding the process of migrating a computing task from one single mobile device to an appropriate edge server or cloud center. Developing an offloading strategy is a classic and pivotal issue when it comes to decoding the field of edge computing. When using edge serve to offloading tasks it will inevitably result in higher latency when compared to local execution, and also considerably reducing power consumption.

It is thereby important that mobile devices select the most appropriate offloading strategies in order to match the needs, like reducing power consumption, lowering task execution time, managing delays, and power consumption that the algorithm must resolve the issue of cloud computing latency.

2   RELATED WORK

The launch of the concept of edge computing in IoT networks has brought up a range of research challenges. Right from dynamic computation offloading scheme designing, resource (for example, computing resource, spectrum resource) allocation, transmitting power control, it has brought forth a range of issues. The trick lies in how these problems cannot be resolved in an independent manner. When drafting a viable computation-offloading scheme, the issues of limited resources of the gateway and transmission power of users must be considered.

Lately, a score of literature have invested in studying resource allocation strategy in varied edge computing scenarios.

Wanget al. had put forward an integrated framework for computation offloading as well as interference management in a wireless cellular network with Mobile Edge Computing. By making use of graph coloring techniques, they performed resource allocation techniques. But alas, they did not pay much heed to competence between different applications. Munoz et al had put much focus on wireless application offloading for radio and computational resources, along with optimization of trade-off between energy consumption and latency in femtocloud. This type of allocation framework could help in saving upon the terminal battery, while lowering latency in the executing application.

Game theory is one powerful concept, which has been taken up by varied works to solve the issues in data offloading. Zhang et al had put forward a coalitional game, which was based upon the pricing scheme of offloading big data to the Mobile Edge Computing server. Lian et al have designed a game-theoretic framework, which performs under the constraint of cost of communication in order to enable some economical communications. This all helps in optimizing the performance of Mobile Edge Computing Server, all based upon the location of edge servers.

Even though these areas of the work have been assigned to separate Edge Computing aspects, majority of these are revolving around power control or resource allocation for the servers or the base stations. In the sector of healthcare taking into consideration the requirements of high QoS and real-time response of a number ofemergency delay-sensitive applications, it becomes imperative that we establish a methodical and well distributed framework to guarantee lower levels of interference, while being based on Edge Computing.

3 FOG BASED ARCHITECTURE

3.1 *Requirements for Fog computing with IoT*

Fog computing in IoT applications requires distributed coordination, latency, awareness of location and mobility support. The latency of the network can be considered lowered if we have the capability of analyzing and processing data that is nearer to its originating source. This leads to speeding up the data computation, which has to be executed. In sectors such as energy, travelling, health and banking, where lower levels of latency is required when it comes to responding to an action, IoT applications play a significant role. IoT applications tend to play a decisive role in effective computing with as low latency as possible.

3.2 *System model*

An appropriate architecture is mainly a cluster of four cities in which two clouds are doing the job of maintaining levels of connectivity. Below each cloud, lies a set of defined Fog-computing system network. Each of these controls a set of smart homes in its specific city. Here Fog-To-Fog communication is established by utilizing a cloud-link controller, in which a cloud system is responsible for the management of shared knowledge, data, settings, and every environmental parameter. Environmentally, the main focus of this research is on the transfer of data models from one Fog node to another, and simultaneously maintaining the characteristics of the target at an abstract level.

In an IoT domain data communication amongthe nodes can take place between edge-edge, edge-fog, fog-fog, or fog-Cloud nodes. Nevertheless, if a Fog node in area A needs to exchange data with a Fog node from area B, the data has to be transmitted through a basic Internet connection, and it is indeed a time-involving and costly process.

An IoT Edge Computing network has a three-layer hierarchical architecture, consisting of a cloud platform, multiple different gateways, and a number of IoT users, over several independent IoT networks. Providing quality services to a large number of IoT users over different IoT devices by each IoT network providers, this is further accomplished by the gateway or, the edge devices that collect data from IoT users in its coverage area and then process them through its edge server. It is to be noted that an edge server has a finite capacity of computation and spectrum resource so that the gateway can allow only a streamlined number of devices that can be accessed at any given point of time. Every IoT user through different sizes creates computation tasks in a continuous manner - something that has limited computation capacity as well as battery power. This means that the part of the offloading computation of tasks to the gateway might amplify the whole computation experience when it comes to energy consumption latency of task execution.

3.3 *Computation offloading scheme through deep reinforcement learning*

Post the clustering of centralized user IoT users are categorically classified into various clusters by their gateway as per the distinctive characteristics and priorities of users. In each cluster, every single IoT user is categorized by the similar user priorities. In this, a cluster with the highest priority is assigned as the edge computing while a cluster with the lowest priority is meant for local computing.

4 CONCLUSION

What we have studied in this paper is the creation of a smart computation task offloading policy for an end device in the IoT. We can conclude after reviewing multiple references that the most efficient way to achieve resource allocation with task offloading for edge computing in the IoT Networks is through reinforcement learning. When end devices have pending tasks, then the current task is offloaded or is executed locally to reduce latency and optimize performance.

REFERENCES

Aazam, Mohammad, Sherali Zeadally, and Khaled A. Harras. "Offloading in fog computing for IoT: Review enabling technologies, and research opportunities." Future Generation Computer Systems 87 (2018): 278–289.

Bittencourt, Luiz, et al. "The internet of things, fog and cloud continuum: Integration and challenges." Internet of Things 3 (2018): 134–155.

Chen, Yung-Chiao, et al. "Cloud-fog computing for information-centric Internet-of-Things applications." 2017 International Conference on Applied System Innovation (ICASI). IEEE, 2017.

Clohessy, Trevor, Thomas Acton, and Lorraine Morgan. "Smart City as a Service (SCaaS): a future roadmap for e-government smart city cloud computing initiatives." 2014 IEEE/ACM 7th International Conference on Utility and Cloud Computing. IEEE, 2014.

George, Anita, et al. "Internet of Things in health care using fog computing." 2018 IEEE Long Island Systems, Applications and Technology Conference (LISAT). IEEE, 2018.

Guo, Shaoyong, et al. "Mobile edge computing resource allocation: A joint Stackelberg game and matching strategy." International Journal of Distributed Sensor Networks 15.7 (2019): 1550147719861556.

He, Xiangyu, et al. "Energy-efficient mobile-edge computation offloading for applications with shared data." 2018 IEEE Global Communications Conference (GLOBECOM). IEEE, 2018.

Hong, Cheol-Ho, and Blesson Varghese. "Resource management in fog/edge computing: A survey." arXiv preprint arXiv:1810.00305 (2018).

Lian, Feier, Aranya Chakrabortty, and Alexandra Duel-Hallen. "Game-theoretic multi-agent control and network cost allocation under communication constraints." IEEE journal on selected areas in communications 35.2 (2017): 330–340.

Liu, Xiaolan, Zhijin Qin, and Yue Gao. "Resource allocation for edge computing in iot networks via reinforcement learning." ICC 2019-2019 IEEE International Conference on Communications (ICC). IEEE, 2019.

Mach, Pavel, and Zdenek Becvar. "Mobile edge computing: A survey on architecture and computation offloading." IEEE Communications Surveys & Tutorials 19.3 (2017): 1628–1656.

Muñoz, Olga, Antonio Pascual-Iserte, and Josep Vidal. "Joint allocation of radio and computational resources in wireless application offloading." 2013 Future Network & Mobile Summit. IEEE, 2013.

Nowicka, Katarzyna. "Smart city logistics on cloud computing model." Procedia-Social and Behavioral Sciences 151.Supplement C (2014): 266–281.

Wang, Pan, et al. "A fog-based architecture and programming model for iot applications in the smart grid." arXiv preprint arXiv:1804.01239 (2018).

Wang, Xiaojie, Zhaolong Ning, and Lei Wang. "Offloading in Internet of vehicles: A fog-enabled real-time traffic management system." IEEE Transactions on Industrial Informatics 14.10 (2018): 4568–4578.

Yousefpour, Ashkan, et al. "All one needs to know about fog computing and related edge computing paradigms." (2018).

Zhang, Tian, Wei Chen, and Feng Yang. "Data offloading in mobile edge computing: A coalitional game- based pricing approach." arXiv preprint arXiv:1709.04148 (2017).

Brief enlightenment of applications for deep learning

Sandeep Singh Bindra, Shivani Gaba & Aaisha Makkar
Chandigarh University, India

ABSTRACT: As deep learning is finest, regulated, time and cost effective AI approach. Deep Learning is definitely not a limited learning approach, yet it withstands different techniques and geographies which can be applied to number of issues. The procedure learns the illustrative and differential highlights in a proper manner. Deep learning strategies have made a huge discovery with evident execution in a wide selection of uses with valuable security tools. Deep learning has made critical improvements in various applications, the generally utilized areas of deep learning includes Government agencies, businesses and medical line that is further expanded in health, cyber security, agriculture, natural language processing, speech recognition, computer vision, face recognition, object detection, cancer detection share market ups and down analysis, handwriting verification and lot more. The primary goal of this paper is to briefly enlighten the ideas of deep learning and its applications based on ideas of different researchers.

Keywords: Artificial Intelligence, Deep Learning, Cyber Security, Natural Language Processing

1 INTRODUCTION

Deep Learning (DL) has seen tremendous movement now a days and is currently set to affect almost every community in which individuals are locked in. Artificial intelligence being in a territory of software engineering that manages enabling machines to seem like a human brilliance. Systems which depends upon AI, can be also known as cognitive systems, are helping us in automation of almost all the jobs and can solve many complex tasks that a human can dream, which is outstanding (B.Geluvaraj, P.M. Satwik 2019) (Cao and Nevatia 2016). Deep Learning (DL) coined first time in 2006 as another field of exploration inside AI for researchers. It was first known for hierarchical learning (Mosavi 2017), and it typically included many fields for research. Deep learning adapting primarily thinks about two key components: nonlinear processing and supervised/unsupervised learning. Order is set up among layers to sort out the significance of the information to be considered as helpful or not. Supervised/unsupervised learning accessibility implies an administered framework, its appearance shows supervised while its nonappearance implies an unsupervised framework. Deep learning is form a class of system, where numerous layers of data preparing stages in hierarchical models are misused for pattern arrangement and for highlighting representation learning. It is in almost all the fields of neural network, pattern recognition & signal processing (Diba et al. 2017). Three significant explanations behind the numerous adapting today are definitely expanded chip processing capacities, the fundamentally less expense hardware, and ongoing advances in AI data handling research. The analysts have exhibited the applications of deep learning in different areas of computer vision, agriculture, natural language processing, image processing, handwriting recognition, pattern recognition, speech recognition, object recognition. (Deng 2014)

DOI: 10.1201/9781003193838-114

2   METHODOLOGY AND ANALYSIS

Two primary advances are based on this review: Searching the latest research with the maximum findings of authors after 2014. The assortment of the connected work was performed from 2014 to 2019 predominantly depending. The search was open for almost 24 papers. The examination of the connected work expects particularly in picture preparing and vision-related assignments and has along these lines resuscitated analysts' interest in various fields such as i) Deep Learning; ii) Deep Learning in Cyber Security; iii) Deep Learning in Health and Medicine; iv) Deep Learning in Natural Language Processing; v) Deep Learning in Agriculture and audits headway in the applications of deep learning, mainly dependent on the examples of the preparing units.

3   DEEP LEARNING APPLICATIONS

3.1   *Deep learning in agriculture*

Main objective of agricultural history is to make it economically efficient for humans. However the objective is not achievable due to the difficulty in balancing the cost and quality. Emerging technology like Internet of things and remotely monitor and manage the crops and also can make it cost effective (Ayaz et al. 2019). However, continuous observing agricultural activities aren't sufficient to make it smart. Regular observations are required with the help of sensors for the agriculture which includes soil quality, water level, livestock, atmosphere and climatic conditions. Human needs to be expert in knowledge of agriculture, data analysis and to be smart enough to achieve goals of sustainable agriculture (Unal 2020), but in some geographical areas majority of farmers consist of family farms, for them high level expertise is non-realistic thing. (Varghese and Sharma 2019).

3.1.1   *Disease monitoring*

Plant disease plays horrifying role in the loss of production. It is very difficult for farmer to monitor the farm conditions time to time for plant disease and moreover, diseases vary from plant to plant.

Table 1.   Analysis of deep learning application.

S. No	Applications Used	Analysis	Reference Used
1.	**Agriculture**	Disease Monitoring, Plant Categorization, Land Cover, Smart Irrigation, Weed Detection	(Ayaz et al. 2019), (Unal 2020), (Varghese and Sharma 2019)
2.	**Cyber Security**	Malware, Domain Generation Algorithm, Download Attacks, Border Gateway Protocol & Anomaly Detection	(B.Geluvaraj, P.M. Satwik 2019), (Berman et al. 2019)
3.	**Health Informatics & Medical Imaging**	Fibroid and irregularities, Customized Medical Services, Monitoring health with sensors	(Fakoor and Ladhak 2018), (Ravi et al. 2017), (Johnson et al. 2016)
4.	**Natural Language Processing**	Part-of-speech, Named Entity Recognition, Neural Machine Translation, Image Description	(Ng et al. 2013), (Huang, Xu, and Yu 2015), (Lample et al. 2016), (Karpathy and Fei-Fei 2017), (Vinyals et al. 2017), (Chen et al. 2013), (Xu et al. 2015)

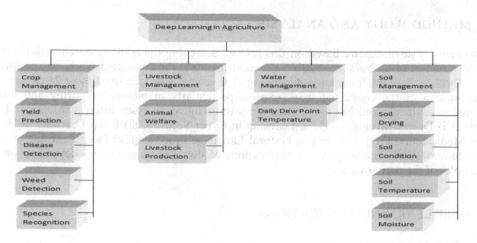

Figure 1. Areas of deep learning in agriculture.

3.1.2 *Plant categorization*
When we are talking about fruit production, the most time consuming part is harvesting of fruit, so there is a need of new development towards automated robots for harvesting.

3.1.3 *Land cover monitoring*
Multi-source satellite pictures are regularly used to catch explicit plant development stages. A few investigations utilized deep learning for land profitability evaluation and land cover monitoring.

3.1.4 *Livestock*
Farming dealing with livestock farming is another challenging task for agriculture and need new techniques which will monitor health, behavior detection, comfort of animal and their production indicators.

3.1.5 *Smart irrigation*
As the decreasing water resources that are available in the world, there is a need of a system that can be used for efficient use of water and also to check the status of water in plants.

3.1.6 *Weed detection*
Undesirable plants that grow rapidly in the crops and can cause major losses as weeds can take over the resources of crops. Using advanced techniques we can work upon detected weeds.

3.2 *Deep learning in cyber security*

Cyber security is the rules and regulations, some ethical policies that are used to secure the system integrity and confidentiality from any attacks. There is end number of tools or mechanisms like firewalls, antivirus and intrusion detection system that work to secure the systems from security breach. However some vulnerability exists in the systems that need to be taken care by the advanced technologies to secure our systems from threats. As numbers of systems that are connected to the internet are increasing day by day so our systems are leading to be attacked. Influenced by neural network based brain's ability to learn, deep learning is advanced technique which is known for its greater perfection and timely action on any work, so most of the problems can be solved by applying deep learning to cyber security (B.Gelu-varaj, P.M. Satwik 2019) (Berman et al. 2019)

3.2.1 *Malware*
Malware attacks are constantly expanding, making it more hard to guard against them utilizing standard strategies. Deep learning can be used to construct generalized models to identify

malware automatically. This can give protection against known malware and huge scope of utilizing new kinds of malware to attack any associations or people.

3.2.2 *Domain generation algorithms*
It is very common tool that is used to generate domain names that are very difficult to monitor the communication link within servers. This is very commonly used in cyber-attacks to grab the personal information.

3.2.3 *Download attacks*
There must be some flaws in plugins which attackers might be looking for and can easily take the user from known websites to unknown websites where user might be forced to download the malware and execute it.

3.2.4 *Border gateway protocol and anomaly detection*
The Border Gateway Protocol (BGP) is used to exchange the routing information of autonomous systems. By exposing the information of BGP, its vulnerable to DDoS attacks, so timely identification of anomalous BGP to weaken any damages.

3.3 *Deep learning in health informatics and medical imaging*

In areas, for example, health informatics, the feature of automation without human interference has an edge.

Some features can be generated in medical imaging that is more enlightened. By understanding the features that can determine fibroid and irregularities can be characterized in tissues such as tumors (Fakoor and Ladhak 2018) (Doulamis 2018). The accomplishment of DNNs is generally because of their capacity to learn novel highlights/designs and comprehend information portrayal in both a solo and regulated various leveled habits. In this way, it is obvious that deep learning

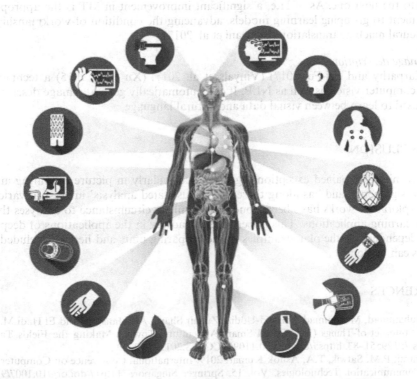

Figure 2. Health monitoring data applications captured using sensors (Ravi et al. 2017).

has immediately been embraced in clinical informatics research. Deep learning has made ready for customized medical services by offering a remarkable force and effectiveness in mining huge multimodal unstructured data put away in hospitals, cloud suppliers and exploration association. Noisy and inadequate datasets bring about Sensors, like wearable device and implantable sensors (Ravi et al. 2017) (Doulamis and Voulodimos 2016) allow the monitoring of health. From Figure 2, we can get the data like food intake and energy loss per day which is used to handle the problem of obesity. For critical care patients, nonstop monitoring or status of blood pressure, sugar level, heartbeat, temperature can be closely monitored. (Johnson et al. 2016).

3.4 Deep learning in natural language processing

NLP uses the programs of computer to process language data which can be used in speech recognition, understanding natural language and generating natural language. (Yang et al. 2019) (Lin et al. 2016)

3.4.1 Part-of-Speech (POS) tagging
Speech Recognition target marking each word with a remarkable label that demonstrates its syntactic job, for example plurals and adverbs. Speech recognition typically used as normal information for different NLP undertakings, like data recovery, machine interpretation, sentence structure checking (Ng et al. 2013), etc.

3.4.2 Named Entity Recognition (NER)
Named entity recognition (NER) is used to locate named entity such as name, place, organization, date etc. LSTM and CNN are used to establish neural network models. (Huang, Xu, and Yu 2015) (Lample et al. 2016).

3.4.3 Neural machine translation
The goal of machine translation (MT) is to decipher text or speech starting with one language then onto the next one. As of late, a significant improvement in MT is the appropriation of arrangement to grouping learning models, advancing the condition of-workmanship strategy called neural machine translation (Vaswani et al. 2017).

3.4.4 Image description
It is (Karpathy and Fei-Fei 2017) (Vinyals et al. 2017) (Xu et al. 2015) a technique which require computer vision as well as NLP. It can automatically generate image description. This can be used to learn between visual data and natural language.

4 CONCLUSION

Deep Learning has gained exceptional ground, particularly in picture preparing and vision-related assignments, and has along these lines resuscitated analysts' interest in various fields. A few exploration works have been done in this unique circumstance to analyses the fortune of deep learning applications. This paper audits headway in the applications of deep learning, mainly dependent on the plan examples of the preparing units and hence concluded in which area this can be done.

REFERENCES

Ayaz, Muhammad, Mohammad Ammad-Uddin, Zubair Sharif, Ali Mansour, and El Hadi M. Aggoune. 2019. "Internet-of-Things (IoT)-Based Smart Agriculture: Toward Making the Fields Talk." IEEE Access 7: 129551–83. https://doi.org/10.1109/ACCESS.2019.2932609..

B. Geluvaraj, P.M. Satwik, T.A. Ashok Kumar. 2019. International Conference on Computer Networks and Communication Technologies. Vol. 15. Springer Singapore. https://doi.org/10.1007/978-981-10-8681-6.

Berman, Daniel S, Anna L Buczak, Jeffrey S Chavis, and Cherita L Corbett. 2019. "A Survey of Deep Learning Methods for Cyber Security." https://doi.org/10.3390/info10040122.

Cao, Song, and Ram Nevatia. 2016. "Exploring Deep Learning Based Solutions in Fine Grained Activity Recognition in the Wild." Proceedings - International Conference on Pattern Recognition 0: 384–89. https://doi.org/10.1109/ICPR.2016.7899664.

Chen, Dong, Xudong Cao, Fang Wen, and Jian Sun. 2013. "Blessing of Dimensionality: High-Dimensional Feature and Its Efficient Compression for Face Verification." Proceedings of the IEEE Computer Society Conference on Computer Vision and Pattern Recognition, 3025–32. https://doi.org/10.1109/CVPR.2013.389.

Deng, Li. 2014. "A Tutorial Survey of Architectures, Algorithms, and Applications for Deep Learning." APSIPA Transactions on Signal and Information Processing 3 (January 2014). https://doi.org/10.1017/atsip.2013.9.

Diba, Ali, Vivek Sharma, Ali Pazandeh, Hamed Pirsiavash, and Luc Van Gool. 2017. "Weakly Supervised Cascaded Convolutional Networks." Proceedings - 30th IEEE Conference on Computer Vision and Pattern Recognition, CVPR 2017 2017-Janua: 5131–39. https://doi.org/10.1109/CVPR.2017.545.

Doulamis, Nikolaos. 2018. "Adaptable Deep Learning Structures for Object Labeling/Tracking under Dynamic Visual Environments." Multimedia Tools and Applications 77 (8): 9651–89. https://doi.org/10.1007/s11042-017-5349-7.

Doulamis, Nikolaos, and Athanasios Voulodimos. 2016. "FAST-MDL: Fast Adaptive Supervised Training of Multi-Layered Deep Learning Models for Consistent Object Tracking and Classification." IST 2016-2016 IEEE International Conference on Imaging Systems and Techniques, Proceedings, 318–23. https://doi.org/10.1109/IST.2016.7738244.

Fakoor, Rasool, and Faisal Ladhak. 2018. "Using Deep Learning to Enhance Head and Neck Cancer Diagnosis and Classification." 2018 IEEE International Conference on System, Computation, Automation and Networking, ICSCA 2018, no. June 2013. https://doi.org/10.1109/ICSCAN.2018.8541142.

Huang, Zhiheng, Wei Xu, and Kai Yu. 2015. "Bidirectional LSTM-CRF Models for Sequence Tagging." http://arxiv.org/abs/1508.01991.

Johnson, Alistair E.W., Mohammad M. Ghassemi, Shamim Nemati, Katherine E. Niehaus, David Clifton, and Gari D. Clifford. 2016. "Machine Learning and Decision Support in Critical Care." Proceedings of the IEEE 104 (2): 444–66. https://doi.org/10.1109/JPROC.2015.2501978.

Karpathy, Andrej, and Li Fei-Fei. 2017. "Deep Visual-Semantic Alignments for Generating Image Descriptions." IEEE Transactions on Pattern Analysis and Machine Intelligence 39 (4): 664–76. https://doi.org/10.1109/TPAMI.2016.2598339.

Lample, Guillaume, Miguel Ballesteros, Sandeep Subramanian, Kazuya Kawakami, and Chris Dyer. 2016. "Neural Architectures for Named Entity Recognition." 2016 Conference of the North American Chapter of the Association for Computational Linguistics: Human Language Technologies, NAACL HLT 2016 - Proceedings of the Conference, 260–70. https://doi.org/10.18653/v1/n16-1030.

Lin, Liang, Keze Wang, Wangmeng Zuo, Meng Wang, Jiebo Luo, and Lei Zhang. 2016. "A Deep Structured Model with Radius–Margin Bound for 3D Human Activity Recognition." International Journal of Computer Vision 118 (2): 256–73. https://doi.org/10.1007/s11263-015-0876-z.

Mosavi, Amir. 2017. "Deep Learning : A Review Deep Learning : A Review." Queensland University of Technology, no. July.

Ng, Hwee Tou, Siew Mei Wu, Yuanbin Wu, Christian Hadiwinoto, and Joel Tetreault. 2013. "The CoNLL-2013 Shared Task on Grammatical Error Correction." CoNLL 2013-17th Conference on Computational Natural Language Learning, Proceedings of the Shared Task, no. July: 1–14.

Ravi, Daniele, Charence Wong, Fani Deligianni, Melissa Berthelot, Javier Andreu-Perez, Benny Lo, and Guang Zhong Yang. 2017. "Deep Learning for Health Informatics." IEEE Journal of Biomedical and Health Informatics 21 (1): 4–21. 10.1109/JBHI.2016.2636665.

Unal, Zeynep. 2020. "Smart Farming Becomes Even Smarter with Deep Learning - A Bibliographical Analysis." IEEE Access 8: 105587–609. https://doi.org/10.1109/ACCESS.2020.3000175.

Varghese, Reuben, and Smarita Sharma. 2019. "Affordable Smart Farming Using IoT and Machine Learning." Proceedings of the 2nd International Conference on Intelligent Computing and Control Systems, ICICCS 2018, no. Iciccs: 645–50. https://doi.org/10.1109/ICCONS.2018.8663044.

Vaswani, Ashish, Noam Shazeer, Niki Parmar, Jakob Uszkoreit, Llion Jones, Aidan N. Gomez, Łukasz Kaiser, and Illia Polosukhin. 2017. "Attention Is All You Need." Advances in Neural Information Processing Systems 2017-Decem (Nips): 5999–6009.

Vinyals, Oriol, Alexander Toshev, Samy Bengio, and Dumitru Erhan. 2017. "Show and Tell: Lessons Learned from the 2015 MSCOCO Image Captioning Challenge." IEEE Transactions on Pattern Analysis and Machine Intelligence 39 (4): 652–63. https://doi.org/10.1109/TPAMI.2016.2587640.

Xu, Kelvin, Jimmy Lei Ba, Ryan Kiros, Kyunghyun Cho, Aaron Courville, Ruslan Salakhutdinov, Richard S. Zemel, and Yoshua Bengio. 2015. "Show, Attend and Tell: Neural Image Caption Generation with Visual Attention." 32nd International Conference on Machine Learning, ICML 2015 3: 2048–57.

Yang, Haiqin, Linkai Luo, Lap Pong Chueng, David Ling, and Francis Chin. 2019. "Deep Learning and Its Applications to Natural Language Processing," 89–109. https://doi.org/10.1007/978-3-030-06073-2_4.

Recent Trends in Communication and Electronics – Sharma et al. (Eds)
© 2021 Taylor & Francis Group, LLC, ISBN 978-1-032-04572-6

Author Index